Hofmann
Rubber Technology Handbook

Werner Hofmann

Rubber Technology Handbook

Hanser Publishers, Munich Vienna New York

Distributed in the United States of America by
Oxford University Press, New York
and in Canada by
Oxford University Press, Canada

The author:
Dr. Werner Hofmann
Kappelerstr. 5, D-4000 Düsseldorf 13, West Germany

Revised version of "Kautschuk-Technologie", translated by **Dr. Rudolf Bauer,** Ontario, Canada, and **Prof. Dr. E. A. Meinecke,** Akron, Ohio

Distributed in U. S. A. by
Oxford University Press
200 Madison Avenue, New York, N. Y. 10016

Distributed in Canada by
Oxford University Press, Canada
70 Wynford Drive, Don Mills, Ontario, M3C IJ9

Distributed in all other countries by
Carl Hanser Verlag
Kolbergerstr. 22, D-8000 München 80, West Germany

Library of Congress Cataloging-in-Publication Data

Hofmann, Werner, 1924-
[Kautschuk-Technologie. English]
Rubber technology handbook / Werner Hofmann; translated by Rudolf Bauer and E. A. Meinecke.
p. cm.
Revised translation of: Kautschuk-Technologie.
Includes bibliographies.
ISBN 0-19-520757-2 (U.S.)
1. Rubber industry and trade, 2. Rubber, Artificial.
3. Elastomers, I. Title.
TS 1890, H69313 1988
678--do19

Deutsche Bibliothek Cataloging-in-Publication Data

Hofmann, Werner:
Rubber technology handbook / Werner Hofmann. Transl. by Rudolf Bauer and E. A. Meinecke. - Munich ; Vienna ; New York : Hanser ; New York : Oxford Univ. Press, 1989
Einheitssacht.: Kautschuk-Technologie ⟨engl.⟩
ISBN 3-446-14895-7 (Hanser) Pp.
ISBN 0-19-520757-2 (Oxford Univ. Press) Pp.

ISBN 3-446-14895-7 Carl Hanser Verlag, Munich, Vienna, New York
ISBN 0-19-520757-2 Oxford University Press
Library of Congress Catalog Card Number 88-16251

Printed in the Federal Republic of Germany

Preface

The rubber industry, though ranking among the "old established" industries, which are generally held to be conservative, has shown a markedly progressive development over the last decades. Developments in the field of synthetic elastomers have progressed so rapidly that the whole concept of rubber technology has changed. In other fields, rubber technology has expanded greatly.

Although numerous monographs have been written about partial aspects of rubber technology, I know of no modern textbook that treats all fields equally. 25 or more years have passed since the publication of earlier textbooks. A variety of new rubber types, rubber chemicals, processing technologies, analytical techniques and test methods have been developed or improved on since then. Many old concepts have had to be modified in the face of new discoveries, or even completely abandoned and replaced with new ones. The previous exact definitions of rubber and elastomer terminology have become progressively more ambiguous since the boundaries between elastomers and thermoplastics and between elastomers and thermoset materials have broken down. New structural modifications have enabled rubbers to penetrate into previously inconceivable border areas.

The time therefore seems to have come to collect the experience in the field of rubber technology. Two procedures are in principle possible for this, each with its advantages and disadvantages. One is for the author to act essentially as editor, coordinating the sections written individually by experts. This method has the advantages of depth of information and topicality, but the disadvantages of uneven structuring, overlap, repetition, etc. The other possibility is for a single author to conceive the entire work himself. This has the advantage that the structure is a unified whole, all individual aspects are interlinked by cross-references, and a uniform bibliography can be drawn up; the disadvantage of this method is that no author can have specialist knowledge in all fields equal to that of the sum of the experts. Despite this disadvantage, I have chosen the second method. To overcome the stated disadvantage of a non-uniform depth of information across the different areas, I have laid special importance on comprehensive bibliographies.

In addition to complete descriptions, these bibliographies, extending to over 2500 literature references, provide, monographs, current publications, lecture topics and important patents. As far as is necessary and feasible, each major bibliographic section has been subdivided into literature that provides a general survey and the special literature that is quoted in the text. The main importance of the bibliographies for the reader should be the possibility of quick reference to more advanced literature for practically any topic of rubber technology. I have therefore paid particular attention to preparing them. During the two years of compiling this book, I have included the most important new literature sources in the already indexed bibliography, so that the literature is largely up to date as far as 1987/88.

This method allowed me, and this was my first priority, to describe all the partial aspects of rubber technology concisely and with a uniform subdivision. This made it possible to deal with and present all areas in one handy volume. The publication of this English-language international edition was encouraged by a positive review of an earlier German-language edition in Rubber Chem. Technol. **55** (1982), whose last sentence read "An English-language version would be of great value".

The book is subdivided into the following sections: Definition (9 pages); Natural rubber (25 pages), which because of its special means of manufacture and importance is treated separately; the large chapter "Synthetic Rubber" (179 pages), which is subdivided into over 20 sections describing the individual rubber classes, and includes the section on thermoplastic elastomers, which in turn includes numerous sections for all important classes of materials; the large chapter "Rubber Chemicals and Additives" (136 pages), also with numerous subsections for all important classes of materials; the chapter on processing technology (113 pages) and on testing and analysis (37 pages), which each describe all the essential problems and techniques in a brief space. To increase the value of the new book as a reference work I have compiled a trade marks index based on the documents accessible to me.

This work is founded on my postdoctoral thesis on "Rubber Technology" at the Technical University of Rheinland-Westphalia in Aachen, which was intended as duplicated lecture notes to help my students in approaching this field. My thanks to Bayer AG for the opportunity of preparing these manuscripts. Main emphasis in these manuscripts was placed on an understanding of the materials and a knowledge of the processing technology. Through my innumerable publications (books, book sections, papers in encyclopedias, and technical papers) on the most varied aspects of rubber technology over the 35 years of my professional life, and my cooperation with numerous companies in the supply industry and in rubber manufacturing, I have had to read so much technical literature and have collected so much information on the most varied specific fields, that it was time to prepare an all-embracing unified work form the previous manuscripts. In doing this I have attempted strict neutrality as regards companies.

A book of this kind cannot and does not try to be a substitute for personal experience; rather it should, on the one hand, be a textbook to help the emerging generation of rubber technologists obtain a rapid grasp of this material. On the other hand it should aid the experienced user as a reference book for various topics, and permit rapid access to the original literature. If it has succeeded in this, then the attempt to create a comprehensive description of rubber technology has been worthwhile.

I have received much help in conceiving this manuscript, for which I would like to use this opportunity to give my heartfelt thanks. The following persons helped me by conceiving, updating or correcting the individual sections: *Mr. Stagraczinski* (Weber und Schär, Hamburg), and *Dr. Klinkhammer* (D.O.G. Hamburg), in correcting and updating the chapter "Natural Rubber", *Mr. Hübsch* and *Mr. Heiling* (Du Pont, Bad Homburg), as well as other employees of this company for checking and supplementing various sections on elastomers that concerned Du Pont; *Mr. Grigat* (Cabot, Hanau) for checking and updating the chapter "Fillers"; *Dr. Graf* (Klöckner-Ferromatik-Desma, Achim) for updating and supplementing the chapter "Injection Molding", and the supply of graphic materials; *Mr. Cappelle,* (Berstorff, Hannover), *Mr. Koch* (Werner & Pfleiderer, Stuttgart), *Dr. Küttner* (Speyer) and *Dr. Targiel* (Troester, Hannover) for correcting parts of the manuscript on "Process Technology" and providing graphic material; *Dr. Seeberger* (Akzo Chemie, Düren), and *Mrs. Wülfken* (D.O.G., Hamburg) for correcting and supplementing the "Trade Marks Index". I thank *Dr. Bauer* (Dunlop, Mississauga, Canada) for his expert translation of Chapters 1-3 and *Prof. Dr. Meinecke* (University of Akron, USA) for his expert translation of Chapters 4-6; *Mrs. Gertrud Wülfken* (D.O.G. Hamburg) for her help with the proofreading, Clouth, Cologne, for the opportunity to use their library, D.O.G., Hamburg, for the realization of this work. Finally I thank my daughter *Regine* for her help in preparing the Trade Names Index, and, last but not least, my wife *Ruth* for her tireless patience with me during the drafting of this work.

Düsseldorf, Spring 1989 *Dr. Werner Hofmann*

Contents

1 Rubber Technology – Introduction, Definitions, Historical Background

1.1 Introduction

[1.1–1.28]

The starting material for the production of elastomers is caoutchouc.

Elastomers or rubbers are classes of materials, like metals, fibres, concrete, wood, plastics, or glass, without which modern technology would be unthinkable. At present, the annual consumption of rubber amounts to more than 13 million tons, and added to this is an equal amount of compounding additives, with the annual consumption rising by about 4%. About one third of the total global rubber usage is natural rubber, produced in plantations or by smallholders in Malaysia, Indonesia, or other Southeast Asian countries, as well as in West Africa, and South or Central America. The remaining two thirds of the required rubber is produced synthetically by a great number of industrial countries, well distributed throughout the world. Today's raw material for producing synthetic rubber is still mostly oil.

More than half of the global production of natural and synthetic rubber is used in tires, and the remainder for a great variety of industrial and consumer products, ranging from motor mounts and fuel hoses over window profiles and heavy conveyor belts to membranes for artificial kidneys.

The predominant property of elastomers is the elastic behaviour after deformation in compression or tension. It is, for instance, possible to stretch an elastomer ten times its original length, and after removal of the tension, it will return under ideal circumstances to its original shape and length. In addition to this, elastomers are characterized by a great toughness under static or dynamic stresses, an abrasion resistance which is higher than that of steel, by an impermeability to air and water, and in many cases, by a high resistance to swelling in solvents, and to attack by chemicals. These properties are exhibited at room temperatures and above, and are, under certain conditions, retained under most climatic conditions and in ozone-rich atmospheres.

Rubbers are also capable of adhering to textile fibres and to metals. In combination with fibres, such as rayon, polyamide, polyester, glass, or steelcord, and depending on the properties of the reinforcing member, the tensile strength is increased considerably with an attending reduction in extendibility. This use in composites increases the range of applications of rubbers considerably. By joining elastomers with metals, one obtains, for instance, components which combine the elasticity of elastomers with the rigidity of metals. This can be of great importance to designers.

The property profile which can be obtained with elastomers depends mainly on the choice of the particular rubber, the compound composition, the production process, and the shape and design of the product. Properties which do justice to elastomers can only be obtained by proper compounding with chemicals and other additives, of which there are about 20,000 different ones, and subsequent vulcanization. Depending on the type and amount of rubber chemicals and additives in a compound, and depending on the degree of vulcanization, a given rubber can yield vulcanizates with considerably different properties with respect to hardness, elasticity, or strength. And yet, the typical properties of the specific rubber, namely oil, gasoline, aging resistance remain unaltered in the different vulcanizates.

1.2 The Basic Concepts of Caoutchouc

[1.7, 1.25, 1.27, 1.28 a–1.50]

The first material known as caoutchouc (derived from the Indian word "caa-o-chu", or "weeping tree"), is polyisoprene recovered from the sap of Hevea Brasiliensis. Today, this material is referred to as natural rubber (NR), in contradistinction to the synthetically produced rubbers. NR can be reacted with sulphur at high temperatures to form crosslinks (vulcanization). Thus, it is transformed from a sticky and largely plastic state into a highly elastic one, or, it is transformed from caoutchouc to an elastomer or rubber.

In the course of developing synthetic analogues to NR, similar compounds were found, which could also be crosslinked with sulphur.

However, only macromolecular compounds which have unsaturation can be crosslinked with sulphur. This unsaturation in the polymer chain will have been derived in part or totally from diene monomers, and examples of such polymers are polyisoprene, polybutadiene, polychlorobutadiene, styrene-butadiene, or acrylonitrile-butadiene copolymers, to name a few.

This first group of synthetic rubbers (SR) was soon extended, after the discovery of other macromolecular compounds, which cannot be cured with sulfur, but which can be converted into elastomers in a similar fashion using other crosslinking agents.

Therefore, the term "caoutchouc" now covers a great variety of raw macromolecular compounds, which can all be crosslinked, albeit with different systems, to form network structures.

In addition to their ability to form crosslinked network structures, caoutchoucs have to satisfy the following requirements:

Preferably they should be *long chain molecules* forming *coils*, which can be extended when subjected to stresses, however small.

The individual *chain segments* should be *flexible* to undergo micro Brownian motion at normal temperatures. Thus, the molecules assume some statistically ordered conformation (hypothetical conformation) when tensile stresses are applied to their ends, and on removal of stress, they return again to their ideal statically random conformation (state of maximum entropy).

The deformation process can be described thermodynamically by assuming that under ideal conditions, there is no change in the internal energy of the system. Starting with the first law of thermodynamics

$$dF = dU - Tds$$

where F is the free energy, U the internal energy, T the absolute temperature, and S the entropy, and for the ideal conditions of rubber elasticity

$$dU = 0$$

one obtains

$$dF = -Tds$$

This means that, in the ideal case, the deformation process of an ideal elastomeric material is associated solely with changes in configurational entropy of the polymer chain, i.e. it is an *entropic elasticity*. After removal of the stress causing deformation, the original random configurations are spontaneously assumed again. NR and some other SR's exhibit this ideal behaviour in the region of very small deformations.

The change in entropy on rapid deformation is an exothermic process, and on removal of the deforming stress, the heat is completely used up again, so that there is a zero net energy balance. If the deformed rubber molecules are cooled down to remove the entropic heat, the oriented comformations are frozen, but this can be reversed again on heating.

Caoutchouc or raw rubbers should be largely *amorphous* at ambient temperatures, so that the chain flexibility is not inhibited by crystallization, which would interfere with the elasticity of the material and make it rigid or stiff instead. This leads to the following additional requirements:

The *freezing temperature* (glass transition temperature T_g) of a rubber has to be below ambient temperatures, and if possible, below $-50\,^{\circ}C$. There should also be interaction between chain molecules, so that individual chains cannot move entirely freely and independently. With raw rubber, this interaction is of a physical nature due to the coiling of chain molecules. In vulcanized rubbers, there is an additional interaction due to the formation of intermolecular bridges (chemical bonding), which reduces the mobility of chains (macro Brownian motion). As a result of this, the tensile strength and the elasticity of the material are improved, and the plastic deformability is reduced. There is also a lesser dependence on temperature of the elastic properties. With thermoplastic elastomers, the interaction between chains is in main cases of a physical nature, and due, for instance, to crystallinity.

A raw rubber should also have as broad a molecular weight distribution as possible, so that it can be processed by means of conventional rubber machinery.

At higher temperatures, and after long-term or very large deformations, raw rubbers behave more like plastics.

Accordingly, the properties of a raw rubber or coautchouc can be defined as follows: a macromolecular material, which is amorphous at room temperature and which has a glass transition temperature considerably below ambient temperatures. It can be crosslinked to form network structures and thus it becomes elastomeric, whereby it is immaterial whether or not the crosslinks had originated from a chemical reaction with sulphur or another crosslinking agent, or from physical interactions. Because of the high extendibility of their coiled macromolecules, caoutchoucs exhibit a considerable amount of elasticity at room temperature, while at higher temperatures, they tend to creep and display thermoplastic properties.

For example, although polyethylene can be crosslinked to a polymeric network, it cannot be regarded as rubber, since, at room temperature, it is crystalline, and since it assumes only elastomeric properties in the melting range at higher temperatures. Only by interfering with the crystallization of polyethylene, through chlorination or sulfochlorination, or through copolymerization of ethylene with propylene or other comonomers, it is possible to produce crosslinkable polymers, whose melting range is below ambient temperatures. Being amorphous at room temperature, these materials represent rubbers, and indeed, they occupy a very important position among rubbers.

Most types of SR have a different chemical structure than NR. They, therefore, have a unique property spectrum, depending on their chemical makeup.

In addition to the chemical nature of polymers, given by homo-, co-, and terpolymerizations or by polycondensations and additions, the molecular structure has a determining influence on properties (see page 48).

1.3 The Basic Concepts of Rubber and Elastomer

[1.7, 1.25, 1.28a-1.50], compare page 477ff

A distinction is made between raw rubbers or caoutchoucs, and crosslinked rubbers, that means elastomers. As discussed in the preceding chapter, the former are completely deformable in a plastic-like manner, particularly at higher temperatures, because they do not have a rigid network structure. By contrast, the crosslinked rubbers do not have a plastic transition zone, since they are transformed into three dimensional networks, which restrain the movement of macro-molecular chain molecules. Thus, these materials can only be deformed in a plastic-like manner after undergoing some chemical or physical structural changes, caused by chemorheological effects due to aging, decomposition, or rearrangements, or they can be deformed after cleavage of the chemical or physical crosslinks.

The elastomers have a much more open network structure, and they occupy an intermediate position between the non-crosslinked caoutchouc (plastomer) and the tightly crosslinked ebonite (thermoelast) or duromer (see Figure 1.1).

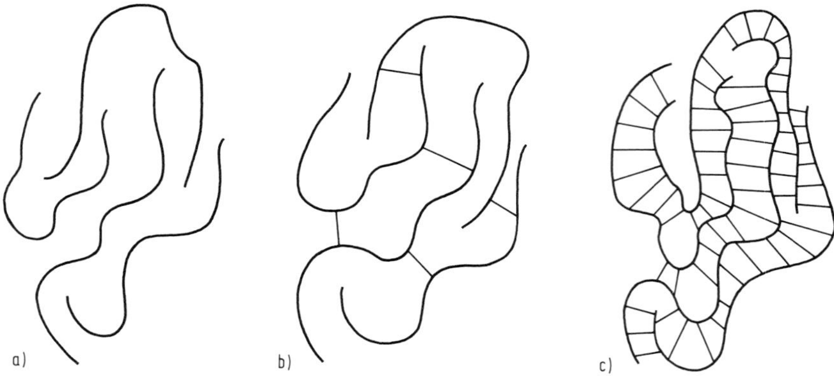

Figure 1.1 Crosslinking of caoutchouc
a) Plastomer (thermoplastic); no crosslinks
b) Elastomer (soft elastomer); loosely crosslinked network structure
c) Duromer (hard rubber, thermosetting plastic); tightly crosslinked network structure

Thus the terms "plastomer", "thermoplast" and "caoutchouc", "elastomer" or "duromer" describe specific states of macromolecular materials [1.48].

Regarding the temperature dependence of the elastic moduli of plastomers (caoutchouc), elastomers, and duromers, the following can be said by reference to Figure 1.2:

Ideally, the elastic modulus of a duromer does not change in a step-wise fashion over a wide range of temperatures. A plastomer, on the other hand, exhibits thermoplastic flow in the melting region and above the melting region, the elastic modulus of the plastomer drops to nearly zero. In contrast to this, an elastomer goes through a rubbery phase after softening, and it retains a relatively high elastic modulus until it reaches the decomposition temperature (see also Figure 1.3). With thermoplastic elastomers, which have been known for some time now, the physical crosslinks dissolve at higher temperatures, but reform again on cooling. As shown in Figure 1.2,

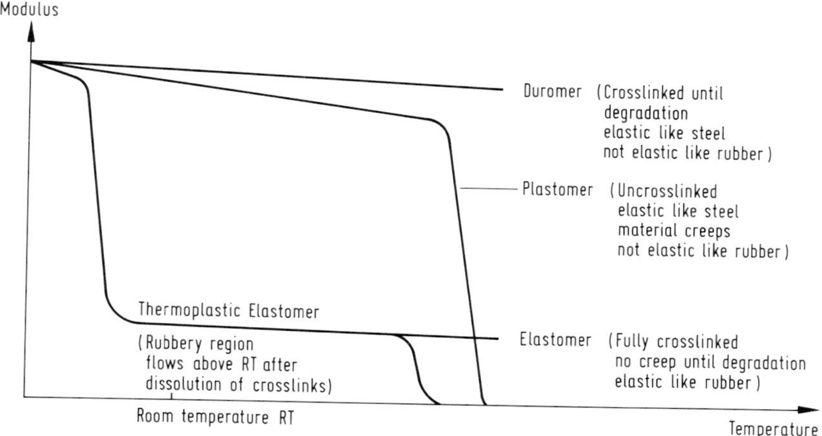

Figure 1.2 The temperature dependence of the elastic modulus of duromers, plastomers, elastomers, and thermoplastic elastomers.

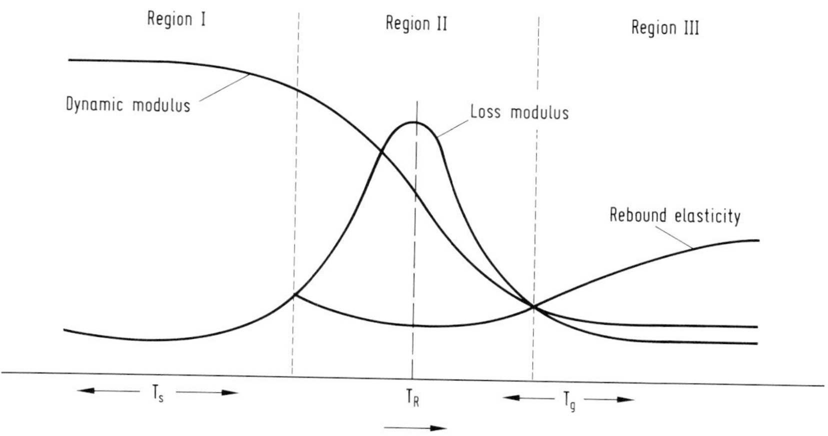

Figure 1.3 The temperature dependence of the elastic modulus and damping properties. Dynamic and loss moduli as a function of temperature (schematic)
Region I: · Frozen state (glassy)
Region II: Transition region from glassy to rubbery state
Region III: Rubbery state

this has the effect, that, above a certain temperature, the elastomeric phase changes into thermoplastic flow.

Analogous to caoutchouc, an elastomer can be characterized as follows:

Under the influence of an already relatively small force, the material should be capable of deformation of at least 100%, if not several hundred percent, of its original length without breaking. (Crystalline materials can be deformed in most cases only by about 1%).

After removal of the external force, the material has to return spontaneously to its original length.

The elastic modulus of most elastomers ranges from about 10^6 to 10^8 dynes/cm^2, whereas that of crystalline polymers is about $10^{10}$to 10^{13} dynes/cm^2 or more.

The deformation of ideal rubber-like materials is an iso-volumetric process.

In the rubbery region and at constant levels of strain, the retractive force of an ideal rubber-like material is proportional to the absolute temperature. In thermodynamic terms, this can be expressed as

$$\gamma = \left(\frac{\delta\ U}{\delta\ L}\right)_{T,V} - T\left(\frac{\delta\ S}{\delta\ L}\right)_{T,V}$$

and

$$\left(\frac{\delta\ U}{\delta\ L}\right)_{T,V} = 0$$

where γ is the retractive force, L, the length, and U, the internal energy.

Thus

$$\gamma = -T\left(\frac{\delta\ S}{\delta\ L}\right)_{T,V}$$

which means that the force or tension of a stretched ideal rubber is proportional to the entropy S.

Consequently, the elastomers are polymeric networks with generally thermally stable crosslinks. When subjected to a small external force, they can be extended to at least double their original length, and after removal of the external force, they retract immediately to their original length.

With thermoplastic elastomers [1.26], one is dealing with macromolecular substances which consist of "amorphous" molecular segments, whose softening range is below ambient temperatures, and of crystalline segments, whose melting range is above ambient temperatures. The amorphous segments behave like elastic springs, and the crystalline segments from different macromolecules form the crosslinking sites, which melt at higher temperatures, but crystallize again on cooling (see page 144).

1.4 The Basic Concepts of Vulcanization

[1.7, 1.12, 1.13]

The vulcanization is a process which transforms the predominantly thermoplastic caoutchouc or raw rubber into an elastic rubbery or hard ebonite-like state [1.12, 1.13] (see Figure 1.1, page 4). This process, which involves the association of macromolecules through their reactive sites, is also called crosslinking or curing.

Thus, the term "vulcanization" embraces not only the crosslinking reaction itself, but also the process which is used to achieve this goal.

As long as only NR and synthetic SR's, which are similar to NR, and which are crosslinkable with sulphur, were known, the historically-based term "vulcanization" was only used to describe the crosslinking reaction with sulphur. With the availability of high molecular weight compounds, which do not react with sulphur, but have to be crosslinked by other means instead, it was recognized that the sulphur-vulcanization is only a special case of crosslinking mechanisms, of which there are indeed many.

The vulcanization reaction is determined in large measure by the type of vulcanizing agents (or curatives), the type of process, temperature, and time. The number of crosslinks formed, also referred to as degree of vulcanization or state of cure, has an influence on the elastic and other properties of the vulcanizate. Therefore, the type of vulcanization process is the important connecting link between the raw material and the finished product (see page 221).

1.5 The History of Caoutchouc

[1.1, 1.2, 1.6, 1.15, 1.19, 1.25, 1.51–1.66]

The historical developments of NR and SR have been discussed in great detail by many authors. Therefore, in the place of a comprehensive discussion, only literature references are listed here to provide the interested reader with a starting point for the study of this subject.

1.6 References on Rubber, Definitions, History

1.6.1 General References on Rubber

[1.1] *Allen, P. W.:* Natural Rubber and the Synthetics. Crosby Lockwood, London, 1972.
[1.2] *Blow, G. M., Hepburn, C. (Eds.):* Rubber Technology and Manufacture, 2nd Ed. Butterworth, London, 1982.
[1.3] *Boström, S. (Ed.):* Kautschuk-Handbuch, Vol. 1–5. Berliner Union, Stuttgart, 1959–1962.
[1.4] *Brydson, J.:* Rubber Chemistry. Elsevier Applied Science Publ., Barking, 1978.
[1.5] *Butt, L. T., Wright, D. C.:* Uses of Polymers in Chemical Plant Constructions. Elsevier Applied Science Publ., Barking, 1981.
[1.6] *Craig, A. S.:* Rubber Technology. Oliver & Boyd, Edinburgh, 1963.
[1.7] *Elias, H. G.:* Makromoleküle, Struktur, Eigenschaften, Synthesen, Stoffe, 4th Ed. Hüthig & Wepf, Heidelberg, 1985.
[1.8] *Evans, C. W.:* Developments in Rubber and Rubber Composites. Elsevier Applied Science Publ., Barking, 1980.
[1.9] *Evans, C. W.:* Practical Rubber Compounding and Processing. Elsevier Applied Science Publ., Barking, 1981.
[1.10] *Heinisch, K. F.:* Dictionary of Rubber. Elsevier Applied Science Publ., Barking, 1974.
[1.11] *Heinisch, K. F.:* Kautschuk-Lexikon, 2nd Ed. A. W. Gentner-Verlag, Stuttgart, 1977.
[1.12] *Hofmann, W.:* Vulkanisation und Vulkanisationshilfsmittel. Berliner Union, Stuttgart, 1965.
[1.13] *Hofmann, W.:* Vulcanizing and Vulcanizing Chemicals. MacLaren, London–Palmerton, New York, 1967.
[1.14] *Hofmann, W., Koch, S. et al.:* Kautschuk-Handbuch (Ed.: Bayer AG). Berliner Union–Kohlhammer, Stuttgart, 1971.
[1.15] *Hofmann, W.:* Kautschuk. In: Ullmanns Encyklopädie der Technischen Chemie, 4th Ed. Vol. 13. Verlag Chemie, Weinheim, 1977, pp. 581–594.
[1.16] *Hofmann, W.:* Kautschuktechnologie, 1st Ed. Gentner Verlag, Stuttgart, 1980.
[1.17] *Kleemann, W.:* Einführung in die Rezepturentwicklung der Gummiindustrie, 3rd Ed. Deutscher Verlag für Grundstoff-Industrie, Leipzig, 1982.
[1.18] *Kluckow, P., Zeplichal, F.:* Chemie und Technologie der Elastomere, 3rd Ed. Berliner Union, Stuttgart, 1970.
[1.19] *Kolb, H., Peter, J.:* Natürliche und synthetische Elastomere. In: *Winnacker/Küchler (Eds.):* Chemische Technologie, 3rd Ed., Vol. 5. Carl Hanser Verlag, München, Wien, 1972, pp. 142–251.

[1.20] *Mark, F., Gaylord, N.G., Bikales, N.M. (Ed.):* Encyclopedia of Polymer Science and Technology. Interscience Publ., New York, 1970.
[1.21] *Morton, M.:* Rubber Technology, 2nd Ed. Van Nostrand-Reinold, New York, 1973.
[1.22] *Schnetger, J.:* Lexikon der Kautschuk-Technik. Hüthig Verlag, Heidelberg, 1981.
[1.23] *Sheldon, R.P.:* Composite Polymeric Materials. Elsevier Applied Science Publ., Barking, 1982.
[1.24] *Stahl, G.A. (Ed.):* Polymer Science Overview: A Tribute to Herman F. Mark. American Chemical Society, Washington D.C., 1981.
[1.25] *Staudinger, M. (Ed.):* Das wissenschaftliche Werk von Hermann Staudinger. Hüthig & Wepf, Basel, 1969.
[1.26] *Walker, B.M.:* Handbook of Thermoplastic Elastomers. Van Nostrand-Reinold, New York, 1979.
[1.27] *Ward, I.M.:* Mechanical Properties of Solid Polymers, 2nd Ed. J. Wiley & Sons, New York, 1983.
[1.28] *Whelan, A., Lee, K.S.:* Developments in Rubber Technology, 1st Ed. Elsevier Applied Science Publ., Barking, 1979; 2nd Ed. 1981; 3rd Ed. 1982.

1.6.2 References on Definition and Physical Properties of Caoutchouc

[1.28a] *Billmeyer, F.W.:* Textbook of Polymer Science, 3rd Ed. J. Wiley & Sons, New York, 1984.
[1.29] *Bueche, F.:* Physical Properties of Polymers. Interscience Publ., New York, London, 1962.
[1.30] *Ferry, J.D.:* Viscoelastic Properties of Polymers. J. Wiley & Sons, New York, London, 1962.
[1.31] *Flory, P.J.:* Theory of Rubber Elasticity. Gordon Res. Conf. of Elastomers, July 18-22, 1983, New London, NH.
[1.32] *Furukawa, J., Onouchi, Y., Yamada, E., Inagaki, S.:* Unsolved Problems in Rubber Elasticity. IRC '85, Oct. 15-18, 1985, Kyoto.
[1.33] *Holzmüller, W., Altenberg, K.:* Physik der Kunststoffe. Akademie-Verlag, Berlin, 1961.
[1.34] *Mark, J.E.:* The Constance 2 C_1 and 2 C_2 in Phenomenological Elasticity Theory and their Dependence on Experimental Variables. RCT *48* (1975).
[1.35] *Mark, J.E.:* Model Elastomeric Network. RCT *54* (1981), p. 809.
[1.36] *Mark, J.E.:* Rubber Elasticity. RCT *55* (1982), p. 1123.
[1.37] *Mark, J.E.:* Recent Studies of Rubber-like Elasticity. Paper 10, 125th Meeting ACS Rubber Div., May 8-11, 1984, Indianapolis, IN.
[1.38] *Mark, J.W.:* Novel Methods for Reinforcing Networks. IRC '85, Oct. 15-18, 1985, Kyoto, Proc. p. 383.
[1.39] *Menges, G., Haack, W.:* Vorausberechnung der mechanischen Eigenschaften von Natur- und Chloroprenkautschuken sowie Ermittlung des Verformungsverhaltens von Elastomerbauteilen. KGK *36* (1983), p. 973.
[1.40] *Nitschke, R., Wolf, K.A.:* Chemie, Physik und Technologie der Kunststoffe, Vol. 1: Struktur und physikalisches Verhalten der Kunststoffe. Springer Verlag, Berlin, Göttingen, Heidelberg, 1962.
[1.41] *Oberst, H. (Ed.):* Elastische und viskose Eigenschaften von Werkstoffen. Deutscher Verband für Materialprüfung. Beuth Verlag, Berlin, Köln, Frankfurt, 1963.
[1.42] *Questel, J.P.:* A Molecular Model of Rubber Elasticity with Constraints on Junctions and Trapped Entanglements. RCT *57* (1984), p. 145.
[1.43] *Stuart, H. (Ed.):* Die Physik der Hochpolymeren, Vol. 4: Theorie und molekulare Deutung technologischer Eigenschaften. Springer-Verlag, Berlin, Göttingen, Heidelberg, 1956.
[1.44] *Timm, Th.:* Bemerkungen zur Definition des Elastomer-Begriffes nach ASTM D 1566 im Rahmen allgemeiner Betrachtungen über die Terminologie hochpolymerer Werkstoffe. KGK *16* (1963), p. 253.
[1.44a] *Timm, Th.:* Was ist Gummi? KGK *34* (1981), p. 927.

[1.45] *Tobolsky, A.:* Properties and Structure of Polymers. J. Wiley & Sons, New York, London, 1960; revised and translated version from *Hoffmann, M.:* Mechanische Eigenschaften und Struktur von Polymeren, Berliner Union, Stuttgart, 1967.
[1.46] *Treloar, L. C. G.:* The Elasticity and Related Properties of Rubber. RCT *47* (1974), Review.
[1.47] *Zachmann, H. G.:* Untersuchung von temporären Vernetzungsstellen und des Einflusses dieser Vernetzungsstellen auf das mechanische Verhalten von Elastomeren. KGK *34* (1981), p. 99.
[1.48] DIN 7724 u. Beiblatt: Gruppierung hochpolymerer Werkstoffe aufgrund der Temperaturabhängigkeit ihres mechanischen Verhaltens. Grundlagen, Gruppierung, Begriffe, vereinfachte Zusammenfassung. Beuth-Verlag, Berlin, Febr. 1972.
[1.49] ASTM D 1566-75a: Rubber and Rubber-like Materials. ASTM, 1975, Annual Book of ASTM Standards, Part 37, Philadelphia, PA, 1975, pp. 376-382.
[1.50] ISO 1382: Rubber Vocabulary. ISO, Geneva, Suisse, 1972.

1.6.3 References on the History of Rubber

[1.51] *Boström, S.:* Naturkautschuk. In: *Boström, S. (Ed.):* Kautschuk-Handbuch, Vol. 1. Berliner Union, Stuttgart, 1959, p. 40 ff.
[1.52] *Corvey jr., B. S.:* History and Summary of Rubber Technology. Edited by *Morton, M.,* Van Nostrand Reinold Co., New York, 1935.
[1.53] *Ditmar, R.:* Synthese des Kautschuks. Th. Steinkopf, Dresden, Leipzig, 1912.
[1.54] *Dunbrock, R. F.:* Historical Review in Synthetic Rubber. Edited by American Chemical Society, Rubber Div., New York, 1954, pp. 32-55.
[1.55] *Eck, L.:* Chronologischer Überblick über die frühe Geschichte des Kautschuks. Gummi-Ztg. *53* (1939), pp. 1015, 1032; *54* (1940), p. 385.
[1.56] *Hauser, E. A.:* A Contribution to the Early History of India Rubber. India Rubber J. *94* (1937) 18a, p. 7.
[1.57] *Hofmann, W.:* Synthetischer Kautschuk - Klassifizierung, Geschichte und wirtschaftliche Bedeutung. In: *Boström, S. (Ed.):* Kautschuk-Handbuch, Vol. 1. Berliner Union, Stuttgart, 1959, pp. 227-231.
[1.58] *Hofmann, W.:* Kautschuk. In: Bayer AG (Ed.): Beiträge zur hundertjährigen Firmengeschichte. 1963, pp. 208-227.
[1.59] *Hofmann, W.:* Kautschuk und Elastomere. Kunststoffe *75* (1985) 7, pp. V-XV.
[1.60] *Hofmann, W.:* Kautschuk und Elastomere - Natur und synthetisch. In: *Glenz, W. (Ed.):* Kunststoffe - ein Werkstoff macht Karriere. Carl Hanser Verlag, München, Wien, 1985, pp. 111-130.
[1.61] *Howard, F. A.:* Buna Rubber, The Birth of an Industry. Van Nostrand Reinold, New York, 1947.
[1.62] *Morton, M.:* Retrospect on Synthetic Rubber during and after World War II. Gordon Res. Conf. on Elastomers, July 14-18, 1980, New London, NH.
[1.63] *Schidrowitz, P., Dawson, T. R.:* History of the Rubber Industry. Edited by Institute of Rubber Industries. Cambridge, Heffer & Sons, 1952.
[1.64] *Stern, H. J.:* History. In: *Blow, G. M. (Ed.):* Rubber Technology and Manufacture, 1st Ed. Butterworths, London, 1971, pp. 1-19.
[1.65] *Törnquist, E. G. M.:* The Historical Background of Synthetic Elastomers with Particular Emphasis on the Early Period. In: *Kennedy, J. P., Törnquist, E. G. M. (Eds.):* Polymer Chemistry of Synthetic Elastomers, Vol. 1. Interscience Publ., New York, London, Sydney, 1968, pp. 21-94.
[1.66] Synthetic Rubber - The Story of an Industry. Edited by International Institute of Synthetic Rubber Producers, New York, Brussels, 1973, pp. 10-20.

2 Natural Rubber (NR)

[1.1, 1.2, 1.15, 1.17–1.19, 1.51, 2.1–2.29 a]

2.1 Sources and the Plantation Economy

Many plants produce a milky sap, also referred to as latex, which is a colloidal caoutchouc dispersion in an aqueous medium. The biosynthesis of caoutchouc is largely understood. It proceeds via mevalonic acid which is converted in a series of reactions into isopentenyl pyrophosphate [1.15, 2.16 c, 2.30, 2.30 a].

The latex-producing plants, of which there are many hundreds known, belong to different botanical families, and they are predominantly found in tropical climates. The main species are euphorbias, moras, apiums, asclepias, and composites as discussed in the following references: [1.15, 2.9, 2.30–2.32].

Of course, not all caoutchouc-producing plants are harvested for industrial purposes, because the yield is either too small, the caoutchouc content in the latex too low, or the caoutchouc contains too many resinous impurities.

Early plantation economies used Ficus elastica, Funtumia, de Castilloa, and Manihot plants, but they were soon displaced by the Hevea brasiliensis, because the latter gives a much greater yield of a superior caoutchouc.

Therefore, in the majority of modern plantations the Hevea brasiliensis (also known as Hevea with 20 subspecies), is cultivated. It is a tree, up to 20 meters high, with a deep taproot, and the first harvests can only be expected from trees which are at least six years old. More recently, Guayule has also been cultivated for caoutchouc production.

In modern plantations, the preferred method of propagation is by various vegetative methods (as opposed to seeds). A group of plants which has resulted from the vegetative propagation of an individual is referred to as "clone". In modern grafting methods, plants are created by bringing together a wood bud scion and a rootstock to combine in a tree a strong and disease-resistant root system, a stem with a tap-resistant bark, and a canopy that is strong and well developed. By propagation through seed, new hybrid strains were developed in modern plantations to increase the yield from about 500 kg/ha to a present 2000 kg/ha with potential yields of as high as 3000 kg/ha [2.11, 2.13 a]. By contrast, the average yields of smallholders is about 1,135 kg/ha, and increases of up to 2000 kg/ha can be expected [2.11]. Another, albeit limited method of improving yield is the stimulation of latex flow by means of various chemicals, such as 2-amino-3.5.6-trichloropicolinic acid, or 2-chloroethyl phosphoric acid. After application, these chemicals penetrate the bark and produce ethylene within the plant [2.31].

2.2 Producers

By far, the largest producer of NR remains the Federation of Malaysia, with 1.41 million tons in 1985 [2.13 a, 2.29 a]. In 1985, Malaysia had a 34.9% share of the global NR production – a reduction from the 41% share in 1970. In 1979, about 40 large plantations provided about 55% of the total Malaysian NR production, but their share has decreased to about 41% in 1983, since many have switched to the more profitable palmoil and cacao production [2.13 a]. It is expected that the area under cultivation for NR production will continue to decrease, although the total Malaysian production will increase because of higher yielding clones [2.11, 2.13 a].

At 2.5 million ha, Indonesia now has a larger area under cultivation for NR production than Malaysia with 2 million ha. But because of poorer yields, Indonesia is in second place, with 1 million tons NR, or a 25% global production share. This is followed by Thailand, with 1,5 million ha and about 0.6 million tons or 15% production share. In addition, there are India and China, both with a 1983 production of 0.17 million tons (4.3% share), (in China, mainly for local consumption), Sri Lanka with 0.14 million tons (3.4%), and numerous other Asiatic, African, and American countries located in latitudes of about 10^0 on either side of the equator. The total NR production in 1985 amounted to 4.044 million tons [2.13 a, 2.29 a].

2.3 Tapping of NR Latex

[2.9]

The latex is contained in a system of capillary vessels, which can be found throughout the living plant, and it exudes through the septum. In the Hevea, these capillary vessels are longitudinally continuous cells arranged as sheaths concentric with the outer bark. The majority of these vessels is found near the cambium (living cell structure) in a 2 to 3 mm thick zone, and the diameter of individual vessels is about 20 to 50 µm. If the vessels are severed by a cut in the bark of the tree, latex flows along the cut very slowly and it coagulates after 2 to 5 hours, due to evaporation, thus plugging the vessels to prevent further flow. However, the vessels fill up with fresh latex, which can then be tapped after a few days by making a fresh incision.

The most frequently used system of tapping is the "half spiral" method, which allows tapping every other day (Method S2, d2). Using gouges and, more recently, specially designed tapping knives, about 0.5 mm to 1–2 mm of cambium is removed. Expert tappers are also capable of making full spiral cuts around the complete tree trunk. This very difficult tapping method is called S1/d4, denoting a full spiral cut every fourth day.

2.4 Recovery of NR from Latex

[2.9]

The by far largest proportion of the tapped latex is further processed to recover the solid NR, but it is not necessary to concentrate the latex for this purpose. The solid NR can either be recovered from the latex by evaporating the water (recovery of wild caoutchouc and spray drying), or, in the by far most important method, by coagulation, drying, and further processing of the coagulate. Since the recovery of wild caoutchouc and spray drying are only of minor importance, no further reference is made her [1.15].

2.4.1 Coagulation, Processing of the Coagulate, Sheets, and Crepe

At plantations, the preferred method of recovering the caoutchouc is by acid coagulation. The coagulating agent is mostly formic acid or acetic acid. By collecting the latex in large tanks, a good measure of crossblending of latex from trees of different ages and from different locations is achieved.

Normally, the latex is diluted with water to a solids content of up to 12 to 18%. The more diluted the latex, the greater amount of acid is required for coagulation. The iso-electric point is reached at a pH of 5.1 to 4.8, under which conditions the latex coagulates.

The coagulum has to be processed immediately, since it changes properties in air under the influence of bacteria. If the coagulum is kept submersed in water or in its own serum, impurities in the caoutchouc are decomposed by bacteria into gases, such as carbon dioxide, methane, and nitrogen-containing compounds.

The coagulation method for the various NR grades differs but little. There are two standard ways of processing the coagulum: drying by exposure to hot wood smoke to produce a so-called smoked sheet, and air drying to produce a pale crepe.

2.4.1.1 Smoked Sheets

[2.9, 2.33]

These are smoked slabs of dried coagulum, which still contain a considerable amount of impurities.

In large rectangular tanks, about 1 part of 0.5% formic acid is added slowly under constant stirring to about 10 to 12 parts of the diluted latex. After removing excess foam, aluminium partitions are inserted in the tank about 4 cm apart. On the following day, soft coagulum slabs which have formed between the aluminum partitions, are carefully removed and processed in a sheeting battery consisting of 5 pairs of rollers arranged in "line-ahead" manner. The coagulum sheets of about 2 to 3 mm final thickness are then draped over wooden partitions and dried in a smokehouse by exposure to hot smoke from fresh burning wood and nut shells. The creosote compounds in the smoke protect the caoutchouc against oxidation and mildew. After an initial smoke period, the temperature is raised to 60 °C, so that the smoked sheets become completely dry within 2 to 3 days.

In addition to the standard ribbed smoked sheets, whose quality is graded from 1 to 5, depending on appearance and purity, there are the non-smoked grades, which are traded under the name of "air-dried" sheets. The high quality, light-brown coloured grades, which are practically odour-free, are characterized by a high degree of purity and their vulcanizates have properties of a high level.

To produce these grades, the coagulum is cut into small pieces while still wet, extruded, cut up again, and then sheeted out on rollers, while being thoroughly washed. Subsequently, it is dried in air for several days, and pressed into sheets of several millimeter thickness.

2.4.1.2 Pale Crepe

[2.9, 2.33]

In pale crepe, the impurities of the caoutchouc have been largely removed by thorough washing in a creping battery consisting of a series of rolls.

The concentration of the coagulant is chosen such that within a few hours, the latex with 15 to 20% solids gives a soft, coherent coagulum, separating out from a clear serum. To remove the acid and a major portion of the non-caoutchouc substituents, the coagulum is thoroughly washed while being worked and torn on profiled friction rolls. By passing the coagulum through a series of these rolls with evenly decreasing nip settings, thin crepe-like sheets are obtained, which are then air-dried for 10 to 12 days.

By using vacuum driers at 70 °C, completely dry sheets can already be obtained within two hours, and after removal from the drier, the sheet has to be cooled rapidly, in order to avoid surface oxidation.

The pale crepe obtained in this process has a brownish colour. To obtain extra white crepes, 0.5 to 0.75 weight % of sodium bisulfite, calculated on the caoutchouc content, is added to the latex prior to coagulation.

The various crepe grades are classified according to differences in colour and strength of the sheets.

2.4.1.3 Special Grades

[2.9, 2.33–2.52]

The special grades have to satisfy specific requirements.

Initial Concentration Rubber (ICR) is produced from undiluted latex [2.34].

Superior Processing Rubber (SP/PA Grades). These are caoutchouc grades from a blend of normal and partially crosslinked latex. They have an improved processing behaviour, particularly regarding extrudability, calenderability, and injection mouldability. Several different grades are marketed. Those containing up to 50% of a crosslinked phase are labelled SP caoutchouc, e.g. SP20, SP40, or SP50, with a 20%, 40%, or 50% crosslinked blend component. Grades with more than 50% of a crosslinked phase are called PA caoutchouc, as, for instance, PA80, with 80% [2.36–2.40]. A special PA80 grade, which is extended with 40 phr of a non-staining oil, is available under the designation of PA57 [2.27–2.29, 2.40].

Oil Extended Natural Rubber (OE-NR), which contains 5 to 40 phr of either, naphthenic or aromatic oil, is assuming greater importance [2.41].

Deproteinated Natural Rubber (DP-NR), also referred to as low-nitrogen natural rubber (LN-NR), is gaining markets in those applications, where the presence of NR impurities is detrimental. Through enzymatic protolysis (e.g. by treatment with papain) or alkaline hydrolysis, the caoutchouc is deproteinated, and the alphaglobulins which are insoluble in the caoutchouc, are then removed by solvation with glycerol. Vulcanizates of the purified caoutchouc have very good mechanical properties, good heat aging, low heat build-up, and a low water swell [2.42]. It is possible to obtain a low-nitrogen NR by papain coagulation of the latex [2.43, 2.43 a].

Heveaplus MG Grades. These are graft copolymers of NR and polymethylmethacrylate and there is an increasing interest in applying these materials not only as selfreinforcing, hard vulcanizates for hard, impact-resistant moulded goods, but also for solvent-based adhesives, primers, etc. [2.44]. These grades are supplied with codes 30, 40, or 49, indicating the nominal polymethylmethacrylate content as percentage. By grafting antioxidants, NR with bound antioxidants can also be obtained [2.45]. These occupy a similar position as NBRs with bound antioxidants (see page 72). These are experimental grades and of no commercial importance as of yet.

Epoxidized Natural Rubber (ENR) is a new class of NR, and it is available with a degree of epoxidation of 10 to 50%. A new process of epoxidation prevents the ring opening reaction [2.46 a], so that it is possible to produce ENR grades with a high degree of uniformity, and whose vulcanizates have a swelling resistance like medium-nitrile NBR. Furthermore, ENR vulcanizates compounded with silica fillers have similar properties as carbon black filled NR vulcanizates [2.46–2.50]. The ENR silica vulcanizates give tire treads which combine a low rolling resistance with an improved wet traction [2.46]. Modern ENR types compete with tire rubbers or with NBR according to the degree of epoxidizing [2.49 a].

Thermoplastic NR is a recently developed blend of NR and polypropylene, which has been crosslinked with peroxides (see page 156) [2.16 c, 2.51–2.52].

Depolymerized NR is a liquid NR of syrup-like consistency. It was specially developed for liquid rubber processing technology, but it has no commercial importance as of yet.

Powdered or Particulate NR is obtained by grinding NR, but it is of no particular importance.

Peptized NR has proven itself in the rubber industry, because it is particularly pure, it can be processed very economically, and it gives good vulcanizate properties. Special attributes are rapid mastication rates which allow short mixing cycles, low Mooney viscosities, good flow properties, and a particularly good building tack.

2.4.2 Classification of Hevea Rubber (TSR)

Although the specifications for the coagulation and processing are, as a rule, strictly obeyed, the standard smoked sheet and pale crepe from different plantations will vary with respect to processibility and cure rate. This has necessitated regular quality control measures by users of NR. Therefore, classifications for caoutchouc were proposed – the first in 1949 by the French Caoutchouc Institute in Indochina. Caoutchouc, which is classified according to these guidelines, is referred to as Technically Classified Rubber (TC Rubber). However, this classification scheme has never been fully accepted, and interest in it seems to vanish. A more refined scheme has been developed by Malaysia, which is known as the SMR (Standard Malaysian Rubber) scheme. This is now also embraced by other NR-producing countries. Indeed, today all rubber which has been classified according to this scheme is generally referred to as Technically Specified Rubber (TSR).
Modern TSR is supplied as compact, pressed bales, contrary to crepe or sheets. Towards this end, the coagulum has to be granulated prior to pressing, which can be done by different methods, namely by the comminuted or the Hevea crumb processes. In the first process, the coagulum is comminuted in hammer mills, while in the second, small amounts of an incompatible oil, such as 0.7% castor oil are added to the still wet coagulum. As a result of this, the coagulum crumbles on passing through the conventional crepe rolls. Because of their large surface area, the crumbs can be readily washed to remove impurities. Subsequently, the material is dried in a pit drier in circulating hot air of 80 to 100 °C for 3 to 4 hours [2.35]. While the comminution process has no influence on the quality of the caoutchouc, the material will contain about 0.4% castor oil when treated by the Hevea crumb process. However, this has no detrimental effect on tack, processibility and the physical properties of the vulcanizates.

2.4.2.1 Technically Classified NR (TC)

[2.53, 2.54]

TC caoutchouc is a visually classified NR, for which also the cure properties are evaluated and graded. The cure properties can be distinguished by reference to a colour code – blue for fast, yellow for average, and red for slow rates of cure and the circle indicates a medium plasticity. Thus, bales of TC caoutchouc show either a blue, yellow, or red circle. The other originally proposed specifications were not practical [2.54]. After the introduction of SMR (see Section 2.4.2.2), and other Technically Specified Rubbers, (see Section 2.4.2.3), no significant markets could be developed for TC-NR.

2.4.2.2 Standardized Malaysian Rubber (SMR)

[2.19, 2.22, 2.35, 2.53–2.66]

SMR was the first, and today still is the most important NR grade belonging to the group of Technically Specified Rubbers (TSR). As already mentioned, TSR's, including SMR, are marketed as compressed bales. The first SMR scheme was introduced in 1965 [2.56], and revised in 1979 [2.58]. As shown in Table 2.1, it comprises latex grades SMR L and SMR WF, sheet material SMR 5, field grade material SMR 10, 20, and 50 [2.59], the viscosity-stabilized latex grades CV and LV [2.60], and the viscosity-stabilized general-purpose grade SMR GP [2.61], which is a blend of 60 parts of factory processed latex and 40 parts of field coagulum. Specifications

Table 2.1: Standard Malaysian Rubber Specifications Mandatory from 1 January 1979

Parameter[a]	SMR CV	SMR LV[b]	SMR L	SMR WF
	Latex			
	Viscosity stabilized			
Dirt retained on 44μ aperture (max., % wt)	0.03	0.03	0.03	0.03
Ash content (max., % wt)	0.50	0.50	0.50	0.50
Nitrogen content (max., % wt)	0.60	0.60	0.60	0.60
Volatile matter (max., % wt)	0.80	0.80	0.80	0.80
Wallace Rapid Plasticity-minimum initial value (P_0)	–	–	30	30
Plasticity Retention Index, PRI (min., %)	60	60	60	60
Colour limit (Lovibond Scale, max.)	–	–	6.0	–
Mooney viscosity ML+4, 100 °C	–[c]	–[d]	–	–
Cure	R[f]	R[f]	R[f]	R[f]
Colour coding marker[g]	Black	Black	Light green	Light green
Plastic wrap colour	Transparent	Transparent	Transparent	Transparent
Plastic strip colour	Orange	Magenta	Transparent	Opaque white

[a] Testing for compliance shall follow ISO test methods.
[b] Contains 4 p.h.r. light, non-staining mineral oil. Additional producer control parameter: acetone extract 6%–8% by weight.
[c] Three subgrades, viz. SMR CV50, CV60 and CV70 with producer viscosity limits at 45–55, 55–65 and 65–75, units respectively.
[d] One grade designated SMR LV50 with producer viscosity limits at 45–55 units.
[e] Producer viscosity limits are imposed at 58–72 units.
[f] Cure information is provided in the form of a rheograph (R).
[g] The colour of printing on the bale identification strip.

and test procedures are summarized in reference [2.62], and the main producers are listed in [2.57].

SMR's are graded for dirt content (indices in grade designations indicate dirt as multiples of 0.1%), ash content, and nitrogen content, in addition to volatile matter, and the so-called plasticity retention index (PRI) [2.63–2.65]. To determine this index, the ratio in plasticity of rubber before and after it has been heated for 30 minutes at 140 °C is multiplied by a factor of 100. The higher the index, the less the rubber degrades, and the better the uniformity of the compound viscosity during mixing processes [2.66]. The index also gives some indication as to the aging resistance that can be expected of the rubber. Depending on their purity, the different SMR grades have PRI values of 30 to 60.

Table 2.1 (continued)

Parameter[a]	SMR 5	SMR GP	SMR 10	SMR 20	SMR 50
	Sheet material	Blend	Field Grade Material		
		Viscosity stabilized			
Dirt retained on 44μ aperture (max., % wt)	0.05	0.10	0.10	0.20	0.50
Ash content (max., % wt)	0.60	0.75	0.75	1.00	1.50
Nitrogen content (max., % wt)	0.60	0.60	0.60	0.60	0.60
Volatile matte (max., % wt)	0.80	0.80	0.80	0.80	0.80
Wallace Rapid Plasticity-minimum initial value (P_0)	30	–	30	30	30
Plasticity Retention Index, PRI (min., %)	60	50	50	40	30
Colour limit (Lovibond Scale, max.)	–	–	–	–	–
Mooney viscosity ML+4, 100 °C	–	–[e]	–	–	–
Cure	–	R[f]	–	–	–
Colour coding marker[g]	Light green	Blue	Brown	Red	Yellow
Plastic wrap colour	Transparent	Transparent	Transparent	Transparent	Transparent
Plastic strip colour	Opaque white	Opaque white	Opaque white	Opaque white	Opaque white

By treating NR latices with, for instance, hydroxyl amine salts [2.64], it is possible to obtain NR grades which are, in large measure, stabilized. These are the SMR CV, LV, and GP grades [2.60, 2.61], which do not have the tendency to harden, and for which the user is guaranteed those Mooney viscosities, which are mentioned in Table 2.1. The viscosity-stabilized grades facilitate the production of particularly uniform compounds, which is, for example, important for suspensions with uniform spring constants. It has, however, been reported that the dynamic properties of NR suffer somewhat through the treatment with hydroxyl amines. The standard viscosity-stabilized SMR CV (constant viscosity) grades are supplied at three different levels of viscosity. The SMR LV (low viscosity) grade contains 4 phr of a non-staining oil, and the SMR GP (General Purpose) grade, being a blend, is of lesser purity and it has, therefore, a lower PRI index.

Starting in 1970, the cure properties were, for a short time, graded by adding MOD together with an index. Since the grading was done through single point measurements, which involves some degree of uncertainty, this grading was abandoned again.

The classification of NR which was first introduced in 1965, has become so popular, that in 1985, about 42% of the global NR production was marketed as Technically Specified Rubber (TSR) [2.11], and an even larger market share is anticipated.

2.4.2.3 Other Technically Specified Rubbers (TSR)

Encouraged by the strong marketing successes of SMR grades, other rubber-producing countries have joined this classification scheme. These specified NR grades are, for instance, SIR from Indonesia, SLR from Sri Lanka, TTR from Thailand, and CSR from China. African countries now also supply specified grades. The specified grades from Malaysia, Indonesia, Sri Lanka, and Thailand can now be considered equal in quality. It should, however, also be remembered, that in spite of elaborate specification schemes, even graded NR's can have a wider range of properties than synthetic rubbers, since they are, after all, natural products. Thus, SMR 20 behaves not necessarily like SIR 20. SIR now sells in larger quantities than SMR.

2.4.3 Other Types of Caoutchouc

2.4.3.1 Guayule

[2.67-2.78]

The shrub guayule (Parthenium argentatum), which is found in Mexico, contains a resin-like caoutchouc in the lignified part of the plant. This caoutchouc has been harvested on plantations since 1900, and consists on average of 70% rubber hydrocarbon, 20% resin, and 10% benzene-insoluble components, like cellulose and lignin.

The caoutchouc component of guayule is cis-1.4-polyisoprene, i.e. NR, [2.68, 2.69], but since it also contains a high proportion of resins [2.68], compounds from it are very tacky and plastic-like, they degrade easily [2.70, 2.71], and cure only slowly to give a lower state of cure than hevea-NR-compounds [2.72]. Since it has a higher plasticity, there have been attempts to use guayule as a thermoplastic NR [2.73, 2.74]. The rubber contaminants, and also the small amount of polymer gel, have a favourable effect on the extrusion properties [2.75]. However, in order to produce a caoutchouc which compares with hevea NR, it is necessary to remove the resin from guayule in an expensive process [2.76], which then makes the production of guayule very involved [2.77, 2.78]. At present, it is only of local importance, and it depends on the development of adequate and economical production methods, whether or not guayule will penetrate world markets.

2.4.3.2 Kok-Saghyz

Of other latex-producing plants, the cultivation of kok-saghyz had assumed some importance for a short time. This plant is of the same species as the dandelion. It is harvested after a 1 to 2 year growing period and the roots are then processed to recover the latex. In the U.S.S.R., about 2000 tons of latex were produced in 1940 from this source, and an estimated threefold amount in 1950. However, the latex from this source can only be considered an emergency supply.

2.4.3.3 Ficus Elastica

Ficus elastica is a latex producing tree which occurs in Burma and Assam, and which, in the early days, was cultivated for a short time in plantations of West Java. However, after 1886, plantations of Ficus elastica ceased to exist.

2.4.3.4 Guttapercha and Balata

Guttapercha and balata are obtained from trees of the Sapotaceae family, and they are very similar to NR, since they are the trans isomer of polyisoprene. Guttapercha is harvested in Malaysia and Indonesia from the sap of the palaquium, isonandra, and payana trees. Balata is produced in South America by drying the sap of the wild native mimusop or ecclinusa balata trees. Guttapercha and balata also differ from NR by their high resin content. They also do not have the typical rubberelastic properties and the typical cure behaviour of NR, and they change in the temperature region of 70 to 100 °C from a hard, horn-like material into a plastic-like one (see Table 3.1, page 49).

Previously, guttapercha was used for cable insulations and balata to some extent for the production of transmission belts. Today, both raw materials are of no technical importance any longer, since they have been replaced by plastics.

2.5 Structure, Composition, and Properties of Natural Rubber (NR)

[2.16 c]

2.5.1 Structure, Composition, and Chemical Properties of NR

[2.79–2.87]

Raw NR, as supplied by plantations, always contains, in addition to rubber hydrocarbons, a certain amount of impurities, which also precipitate out in the coagulum of the latex. The amount of these impurities depends somewhat on the processing conditions.

For instance, the following average composition was measured for # 1 latex crepe: 89.3 to 92.35% rubber hydrocarbons, 2.5 to 3.2% acetone extractables, 2.5 to 3.5% protein, 2.5 to 3.5% moisture, and 0.15 to 0.5% ash.

The rubber hydrocarbon component of NR consists of over 99.99% of linear cis-1.4 polyisoprene. As already mentioned, the trans-1.4 polyisoprene (guttapercha and balata) has entirely different properties than NR. From research into synthetic rubbers, it is also known that a small amount of trans configuration of a few percent only, results in greatly different properties of cis polyisoprene (see page 85). The average molecular weight of the polyisoprene in NR ranges from 200,000 to 400,000 with a relatively broad molecular weight distribution. This corresponds to about 3000 to 5000 isoprene units per polymer chain. As a result of its broad molecular weight distribution, NR has an excellent processing behaviour.

The chain structure of the polyisoprene depends to some extent on the biogenesis of NR [2.30a, 2.79]. It was established by gel permeation chromatography, that there can be a certain amount of long chain branching, depending on growing conditions of NR [2.80]. The amount of gel in NR is very small.

If NR is heated above 300 °C, a destructive distillation occurs. If the rate of heating is high, over 95% of the distillate is dipentene and isoprene. At temperatures of 675 to 800 °C and in a vacuum of 15 mbar, over 58% of the product is isoprene.

In the polyisoprene of NR there is one double bond for each isoprene unit. These double bonds and the α-methylene groups are reactive groups for vulcanization reactions with sulphur. Thus, the double bonds are a prerequisite for the sulphur vulcanization reaction.

The double bonds can, however, also enter into additional reactions with oxygen or ozone, to degrade (age) the rubber. Other reactions can take place with hydrogen (hydrogenated rubber), chlorine (chlorinated rubber), or hydrogen chloride (hydrochlorinated rubber). Cyclization reactions are also possible (cyclized rubber) [1.15].

In high shear mixing processes and under the influence of small amounts of oxygen, a scission of polymer chains takes place. In this so-called mastication process, the molecular weight of the very tough rubber is broken down to facilitate the processing of the material (see page 217). This mastication of NR is required for making production compounds, which are mixtures of rubber, fillers, and specific chemicals. With only 1% bound oxygen, the rubber is already degraded and useless. Therefore, a degradation process, which can already be initiated by small amounts of oxygen and which proceeds autocatalytically, has to be inhibited or prevented by special chemicals, the so-called stabilizers or antioxidants.

Since NR is an unsaturated hydrocarbon, it can also react readily with other oxidizing agents, like peroxides, peroxy acids, potassium permanganate, ozone, chlorine, etc ...

Since NR also contains an innate amount of antioxidants, it is stable for longer storage periods in air at room temperature. However, on storage at higher temperatures, and on exposure to light, NR oxidizes by forming hydroperoxides through oxygen radicals. Under the influence of small amounts of copper or manganese salts, the hydroperoxides reform into reactive radicals in an autocatalytic process [2.81, 2.82]. NR in solution reacts readily with ozone, and on complete ozonolysis, about 90% of the reaction products are derivatives of levulinic acid [2.83].

More recently, the reactivity of NR has been exploited by plantations to produce chemically modified caoutchoucs, where other monomers or polymers have been grafted onto the polyisoprene chains (see also Section 2.4.1.3). These modifications also yield block copolymers. The grafting reactions alter the molecular structure of NR to such an extent, that its main properties are completely changed [2.45, 2.84].

2.5.2 Physical and Technological Properties of NR (Raw Rubber and Vulcanizates)

2.5.2.1 Physical Properties

The specific gravity of raw NR at 20 °C is 0.934, and it increases somewhat, if the rubber is either frozen or stretched. The heat of combustion at constant volume is 44.16 kJ/g, and the specific heat at 20 °C is 0.502. NR, which is purified after extraction with acetone, has a refractive index at 20 °C of 1.5215 to 1.5238. Thin films of NR start to absorb light with a wavelength of shorter than 3100 Å, and below 2250 Å, there is complete absorption.

The electrical properties of NR are determined by its water soluble impurities. For instance, the specific resistivity of sheets is $1 \cdot 10^{15}$ and that of crepe is $2 \cdot 10^{15}$ ohms-cm.

2.5.2.2 Processibility

As supplied by plantations, most NR grades are too hard to be processed directly. Instead, they have to be masticated first. After mastication, the raw NR, or viscosity stabilized types, processes very well, as judged by a good and fast banding on mills, by a good strength of the unvulcanized compounds (also referred to as "green strength"), and by a good building tack. The last two properties are of particular importance for the manufacture of composite components, where several layers of different compounds have to be joined. The extrudability and calenderability of NR is also excellent. Another beneficial property of NR is its high rate of cure.

NR can be deformed in a reversible manner, which is a significant characteristic of its elastic behaviour. However, this is also superposed by some plastic deformations, which only disappear after vulcanization, whereupon the caoutchouc is transformed into a true elastomeric state (see Section 2.8 and 4.2.1.2, page 223).

When caoutchouc is deformed, it exhibits some unique thermal behaviour, which is referred to as the Gough-Joule effect, and which is a result of the entropic elasticity.

On rapid deformation, caoutchouc heats up contrary to what is observed with other materials. Similarly, a caoutchouc sample, which fixed at one end and extended from the other by hanging a weight from it, will contract, when it is heated.

If NR is cooled very slowly from 10 to −35 °C, it becomes opaque and it loses its elasticity. This is due to a partial crystallization. When NR is stretched more than 80% of its original length, crystallization also occurs due to the orientation of polymer chains (strain crystallization). This is also revealed in x-ray diffraction patterns. The crystallization enhances the intermolecular attractive forces, and this reinforces the strength of the polymer. Thus, a higher tensile strength is observed along the axis of deformation, and a lower tensile strength perpendicular to it (anisotropic mechanical properties). The consequences are a good green strength and good building tack [2.85, 2.86, 2.86a, 2.87].

With crosslinked NR, x-ray diffraction patterns can be observed only at higher strain levels than for non-crosslinked NR. This strain orientation of the macromolecules gives also a reinforcement to the vulcanizates of NR (self reinforcement). Therefore, even without the aid of reinforcing fillers, NR vulcanizates have higher levels of tensile strength than vulcanizates from rubbers which do not show strain crystallization, i.e. most synthetic rubbers.

Uncrosslinked NR can be extended to 800 to 1000% of its original length without breaking. As the temperatures become lower, the forces required to extend the rubber become larger. At high rates of strain, the deformations are completely reversible, while at low strain rates, some deformation remains on removal of stress. After a considerable length of time, or on heating, this deformation will, however, recover, and the original length of the sample is restored. Because of this incomplete recovery, the stress-strain curves for extension and retraction do not coincide. This hysteresis, which is a conversion of mechanical into heat energy, leads to the so-called heat build-up, and it is most pronounced in the first strain cycle, and less so in subsequent cycles. The hysteresis is also very pronounced in vulcanizates.

2.5.2.3 Behaviour in Solvents

Another characteristic property of caoutchouc is its behaviour when in contact with solvents. Organic liquids, like benzene, gasoline, vegetable oils, mineral oils, carbon tetrachloride, swell raw rubber to a considerable extent to form a highly viscous solution or a gel. The physical bonds between polymer chains are largely broken in this process. The intrinsic viscosity of the solution is higher, if the caoutchouc has not been masticated (or degraded).

By contrast, a crosslinked rubber swells considerably less, depending on the nature of the solvent. Here, the chemical crosslinks inhibit the swelling, and one consequence of vulcanization is therefore the improvement of the solvent swell resistance of rubbers.

2.6 Compounding of NR

[2.16c, 2.19, 2.24, 2.66, 2.88–2.121]

The choice of compounding chemicals and fillers is governed by the guidelines which will be discussed in the pertinent chapters of this book. Therefore, only some general comments are made here.

Many references, including [2.19] and [2.24], list numerous compound compositions for different applications, and some specific examples are also given for extruded [2.88], microwave-cured [2.89] and injection-moulded [2.90, 2.93] products. Other compound recipes are given for damping devices [2.94–2.97], tires [2.98], conveyor belts [2.99], hose [2.100], ebonite [2.101], and for adhering rubber to metal [2.102] or textiles [2.103].

2.6.1 Blends with NR

Being non-polar, NR can be readily blended with a great number of other non-polar rubbers. Blends with SBR and BR, and also to some extent with NBR, are technically exploited. In these instances, some of the properties of the SR are transferred to NR (regarding abrasion resistance, heat resistance), or from NR to the SR (regarding processibility, building tack, dynamic properties, heat build-up, price etc.) [2.16c, 2.86, 2.104]. With blends it is important that the NR has been masticated such that it has the same viscosity as the blend partner, or viscosity stable types have to be used. Otherwise, it is not possible to achieve an intimate blending of the polymers. The vulcanization system has to be chosen such that it meets the requirements of each blend partner. This can be a problem in blends of NR with SRs which have quite different cure properties, like IIR, or EPDM [2.104].

2.6.2 Vulcanization Chemicals

[1.12, 1.13, 2.3, 2.16c, 2.105–2.113]

Sulphur and Accelerators: Although NR can also be crosslinked with peroxides, or high energy radiation, in practice, sulphur and accelerators are predominantly used. Compared to SBR and NBR, NR requires generally higher sulphur concentrations (2–3 phr), and lesser amounts of accelerator (0.2–1.0 phr). With appropriately high sulphur dosages of 30 to 50 phr, an ebonite is obtained. In addition to sulphur, one uses sulphur donors also. Sulphur dichloride is used for room temperature vulcanization. Guidelines for selecting accelerators are discussed in Chapter 4 of this book. In general, all bases have an accelerating influence, and acids, a retarding one on the cure reaction.

For lower sulphur concentrations, larger amounts of accelerator are required, in order to keep up the level of crosslink density. This will, at the same time, reduce the sulphur rank of the crosslinks, which results in an improvement in the heat resistance and reversion resistance [2.105], and in compression set, but also worsens the notched tear and dynamic properties of the vulcanizate. Therefore, low sulphur vulcanization systems play an increasingly important role in addition to the conventional cures with about 1.5 to 2.0 phr sulphur. For optimum levels of mechanical and dynamic properties of vulcanizates with a good heat resistance, low sulphur (0.5–1.5 phr) or so-called "semi-efficient" (Semi-EV) vulcanizations are used. "Efficient" vulcanizations (EV) with very low sulphur and correspondingly high accelerator concentrations are employed in vulcanizates for which an extremely high heat and reversion resistance is required. Efficient vulcanization systems are also those without elemental sulphur, but with sulphur donors instead, together with a correspondingly high concentration of accelerators. To optimize EV cures, some new systems have been proposed [2.106] namely, soluble ones to reduce creep in suspension elements [2.106, 2.107], OTOS systems [2.108, 2.109], as well as urethane crosslinkers [2.106–2.113] (see page 227ff, 246ff, 261).

Metal Oxides. Metal oxides are required in a compound to develop the full potential of accelerators. The main metal oxide is zinc oxide, but other oxides are used at times to achieve specific results, namely magnesium oxide in the presence of acidic compounding ingredients, such as factice, based on sulphur monochloride, or lead oxide, to obtain an especially low water absorption of vulcanizates.

Activators. Many accelerator systems require additional activators, like fatty acids, or salts of fatty acids, namely stearic acid, zinc soaps, or amine stearates. Glycoles or triethanol amine also serve as activators, the latter primarily in compounds with reinforcing silica fillers. Fatty acids have a lower activity in low Nitrogen NR Grades [2.43a].

Vulcanization Inhibitors. These are used to prevent premature vulcanization or scorching. In many instances, a sufficient scorchlife is obtained by using either an appropriate combination of accelerators, or acidic compounding ingredients. When these measures still do not give sufficient scorch protection, special inhibitors are used, primarily those based on phthalimide sulfenamides. They not only delay the onset of cure, but also the time to completion of cure.

2.6.3 Protective Agents

[2.114, 2.115]

Because it is highly unsaturated, NR has to be compounded with protective agents to achieve a sufficient aging resistance. The level of protection is determined by the chemical nature of the protective agent [2.45]. Most effective are aromatic amines, such as p-phenylene diamine derivatives, which not only protect the vulcanizate against oxidative degradation, but also against dynamic fatigue and degradation from ozone and heat. It is remarkable that some agents such as PAN or PBN impart a good fatigue resistance to NR, but none or very little to SBR vulcanizates. Since the most effective protective agents more or less discolour the vulcanizate, less effective ones, like bisphenols, phenols, polymeric hindered phenols, or MBI have to be used in light-coloured vulcanizates. MBI also has a beneficial synergistic effect when used in combination with other agents. The chapter of this book on protective agents (see page 273) gives detailed guidelines for choosing appropriate ones. For ozone protection, one uses, as a rule, waxes in combination with p-phenylene diamine in dark-coloured vulcanizates, or with enol ethers in light-coloured ones [2.115].

2.6.4 Fillers

[2.66, 2.116, 2.117]

Contrary to most types of SR, NR does not require the use of fillers to obtain high tensile strengths. However, the use of fillers is necessary, in order to achieve the level and range of properties that are required for technical reasons. The active fillers do not give quite the same amount of reinforcement as in most SRs, but the efficiency of the fillers ranks in the same order in both cases. Reinforcing fillers enhance the already high tensile properties of gum NR, and they improve, in particular, the abrasion and tear resistance. Less reinforcing fillers, such as N 770 (SRF) or N 990 (MT), and light inactive fillers, such as kaolin, calcium carbonate, barium sulphate, zinc oxide, or magnesium carbonate are used for a number of reasons. These include an improvement of processibility, and the attainment of particular specifications, such as density, colour, or even price. Depending on their activity, the fillers determine, more or less, the hardness, and they also reduce the rebound elasticity of NR vulcanizates. With inactive fillers, and especially with zinc oxide and N 990 black, one can formulate filled vulcanizates, which have almost the same elasticity as unfilled gum vulcanizates. NR compounds require and therefore contain considerably less filler than SR compounds. For instance, with highly active fillers, one generally uses up to about 50 phr, and somewhat higher concentrations with non-active fillers. The choice of fillers is discussed in greater detail in the chapter on fillers (see page 282).

2.6.5 Softeners and Resins

[2.118, 2.119]

Softeners. A great number of different materials serve as softeners, the most important ones being mineral oils. These oils include a wide range of products, from paraffinic to aromatic. Animal and vegetable oils are also important softeners or process aids, to include wool grease, fish oil, pine tar, and soya oil. NR requires lesser amounts of softener than most SR's. Synthetic softeners, which are commonly used with CR or NBR, play only a minor role in NR compounding. When selecting softeners for NR, the potential of blooming or migration has to be considered [2.118].

Resins. While it is important to add resins to compounds from most SRs, so that they can be fabricated, this is not generally necessary with NR compounds. If an exceptionally good building tack is required, such as compounds for frictioning of textiles, it is advantageous to add rosin, tar, pitch, or other tackifiers to NR compounds. Those tackifiers which were specially developed for SRs are of little relevance in compounding of NR.

Factices. These are, however, very important in NR compounds, since they facilitate the processing in extrusion and calendering operations. They also prevent deformation of compounds during cures at ambient pressures, and they improve the appearance of vulcanizates. Factices also impart a desirable softness to NR vulcanizates, and larger concentrations are used in compounds designed for rubber erasers [2.119].

2.6.6 Process Aids

[2.120, 2.120a]

Stearic acid, zinc and calcium soaps, and residues of fatty alcohols are some process aids which are used in NR compounds in addition to softeners and resins. Zinc soaps of unsaturated fatty acids are usually the best process aids. These materials are important in facilitating the dispersion of fillers in the rubber compounds, and they ensure smooth processing [2.120, 2.120a]. In pure white or pastel coloured articles, calcium soaps of saturated fatty acids are preferred.

2.7 Processing of NR

[2.121-2.123]

The rubber processing technology discussed in Chapter 5 also applies to NR. For non-Technically Specified grades it is advisable to crossblend different NR batches, in order to level out differences in properties.

2.7.1 Mastication

Unless NR has been modified by the producer to a specific processing viscosity, it is very tough, and therefore, requires mastication prior to compounding. During mastication, the NR molecules are mechanically broken down by means of high shear forces. The mastication can be carried out on mills at low temperatures, or at elevated temperatures in the presence of peptizing agents (see Section on peptizing agents, page 219, 359) [2.121, 2.122].

High degrees of mastication are only required, if the NR is to be used in very soft compounds, or if the compound is to be dissolved in solvents (sponge, frictioning, rubber solutions). Low degrees of mastication make it already possible to incorporate chemicals and fillers readily in NR. Many applications also require a hard and nervy NR, which then necessitates very little mastication only. This results in extrudates which have a high green strength and do not deform easily. On the other hand, it also results in reduced extrusion rates. If NR has been masticated only to a small degree, it can accept higher filler loadings, and the vulcanizates also give better mechanical, elastic, and dynamic properties than those from NR which has been severely masticated. For these, and also for economic reasons, it is desirable to keep mastication and mixing cycles as short as possible.

2.7.2 Mixing

Mixing on Mills. The mixing of compounds on open mills proceeds as follows: The rubber is first worked on the mill until a coherent band is formed on the mill rolls. Subsequently, those chemicals are added which are difficult to mix, and which are used in small quantities only, namely protective agents and accelerators, so that they will be well dispersed during the mixing cycle. Next, part of the filler is added, together with stearic acid, if required. On adding softeners, the band will generally split, and it has to heal before additional fillers are added to the compound. Finally, the sulphur is mixed in, if no ultrafast accelerators are present. Otherwise, the sulphur is added on the warmup mill before processing. During the mixing process, the band must not be cut, and only after all ingredients have been incorporated in the compound, the band is cut and folded, that is, the compound is homogenized. When the mixing cycle is completed, the compound is cut from the mill as slabs, and cooled in a waterbath and stored, or it is cut into slabs on batch-off equipment (see Section 5.1.2.4, page 366). Because of the fact, that mixing on mills is very time extensive, mixing on internal mixers is mostly prefered.

Mixing in Internal Mixers. When mixing is carried out in internal mixers a relatively hard and nervy rubber is required for good and efficient dispersion of compounding ingredients. The usual mixing temperatures are about 140 to 150 °C, and through intensive cooling, this temperature can be reduced to 120 to 130 °C for heat-sensitive compounds. If no means of cooling is available, the mixing temperature can go as high as 180 to 190 °C. Mixing at such high temperatures, also referred to as "hot mixing", has been popular for some time now, since it allows very short mixing cycles. However, the properties of the compounds and of the vulcanizates are considerably changed in this mixing process [2.123]. Particularly, the elasticity of the compound is increased and the scorchtime is considerably shortened. The rebound elasticity also becomes higher, while the aging resistance of compounds deteriorates. It goes without saying that vulcanizing agents and other chemicals, which decompose at higher temperatures, must not be present in the compound in "hot mixing" processes. More recent trends in mixing technology are towards lower temperatures, to prevent deterioration of the rubber. When mixing NR compounds in internal mixers, or Banburies, the rubber is first added, followed by fillers and softeners. With low mixing temperatures, the accelerators are added early, together with the fillers, while, with high mixing temperatures, it is necessary to add accelerators later on in a separate mixing pass. Sulphur, and possibly accelerator, are either added after the compound has cooled down, or separately on a mill, after the compound has been warmed up again (see Section 5.1.2.3, page 365).

Cooling Mills. After mixing, the compound is dumped from the internal mixer onto a cooling mill, which could also be equipped with blending bars. It is then cut into slabs or festooned in batch-off equipment for it to cool down. In some large-scale operations, the mixed compound is also pelletized, cooled, and then stored in silos. Subsequently, the pellets are automatically conveyed [2.123 a] measured out, and weighed (see Section 5.1.2.4, page 366).

2.7.3 Further Processing

Prewarming. Further processing occurs only after the NR compound is finished by adding the still missing vulcanizing agents. This is generally done on warmup mills.

Processing in Extruders. Previously, mainly hot-feed extrusion processes were used. This requires the prewarming of compounds which are then fed as strips into extruders of a generally compact design with a 5 to 6 D screw length. More recently, cold feed extruders have become increasingly popular, where, as the name implies, the compound is fed cold into the extruder. Longer screws are required here (up to 20 D), since the first zone of the extrusion barrel serves to warm up the compound (plastication). In cold feed extrusions, the compound is more intensely worked than in hot feed extrusions, and this can have some consequences on the mechanical properties of the vulcanizate. If particularly high demands are made on the elastic properties and the heat build-up of vulcanizates, hot feed extrusions are still preferred today, since this is less demanding on the compound. It should, however, be remembered that, for highly elastic compounds, the die swell is correspondingly high, but this can be controlled by appropriate degrees of mastication (breaking down the elastic component), through the use of semi- or non-reinforcing fillers, or by using softeners, and especially factices. For optimized extrusion temperatures, a very good surface of the extrudate is obtained, if the compounds contain semi- or non-reinforcing fillers, softeners, factices, and other process aids. The extrusion properties are also greatly improved, if SP rubbers are added to the compound (see Section 5.3.1, page 387) [2.4., 2.35–2.41].

Processing on Calenders. Properly compounded NR can be readily calendered. If the compounds are very elastic, one can expect a considerable amount of anisotropic shrinkage of the calendered sheet (which can be compared with die swell). By taking the same procedures used for reducing die swell, the calendering shrinkage can be kept to a minimum, namely adequate mastication, or compounding with semi- or non-reinforcing fillers, softeners, factices, and SP rubber. On the calender, NR compounds can be processed in a variety of operations, such as sheeting, doubling, skimming, or frictioning.

2.7.4 Vulcanization

NR compounds can be vulcanized in all commonly used processes – hot air, with or without pressure, steam, hot water, press, transfer moulding, injection moulding, rotocure, molten salt bath, hot air tunnel, high frequency radiation, lead cure, etc. Because of the non-polar character of NR, there may be problems with preheating light-coloured NR compounds by high frequency radiation. Therefore, one should either add carbon black, or polar compounds, like triethanol amine, polar factices etc., to the compound. Since it has a relatively poor heat resistance, NR can comparatively easily revert during cure. Therefore, curing is generally done at relatively low temperatures, and the length of cure cycles have to be strictly controlled. The higher the cure temperatures, the poorer are the mechanical properties of the vulcanizates, and the shorter is the plateau. By using appropriate vulcanizing systems and good protective agents, the tendency towards reversion can be reduced. Such vulcanization systems are mainly semi-EV and EV systems (see page 246ff), using TMTD, OTOS, triazines, or urethane crosslinkers, and also peroxides. For hot air cure processes, there is a limitation on cure temperatures, since NR tends to oxidize (see page 426).

2.8 Properties of NR Vulcanizates

[2.16 c]

NR vulcanizates combine a range and level of properties which are of great interest from a technological point of view. Regarding individual properties of NR, they can be surpassed by those of SR's. However, the combination of high tensile strength with a high rebound elasticity, very good low temperature flexibility, excellent dynamic properties and very low heat buildup make NR and its synthetic analog IR indispensable in several applications, despite the availability of a great number of SR's. It should also be remembered that the optimum level cannot always be obtained for the properties discussed in the following sections. This depends very much on the particular compound composition.

2.8.1 Mechanical Properties

Hardness. NR vulcanizates can be produced in a wide range of hardnesses, from very soft (Shore A 30 to 50) up to ebonite hardness. This can be done by changing the amount of fillers and softeners in the compound on the one hand, or through the sulphur concentration on the other. Leathery compositions, obtained by using 10 to 20 phr of sulphur, have, however, poor strength and aging properties. Therefore, this hardness range is of little interest technologically, except for a few applications, like floor tiles and roll covers.

Tensile Strength. Due to the previously discussed strain crystallization of NR, which also occurs in vulcanizates, NR has, contrary to most types of SR, a high tensile strength of 20 MPa and more, even in gum vulcanizates. This property is exploited in the design of soft, thin-walled, and very strong products, such as surgical gloves, prophylaxis, or balloons. By adding reinforcing fillers to compounds, the tensile strength can rise up to 30 MPa. Even at higher temperatures, NR vulcanizates have a good tensile strength [2.124].

Elongation at Break. The ultimate elongation depends, naturally, very much on the nature and amount of fillers in the compound, and on the degree of vulcanization. It generally is about 500 to 1000%, or even greater.

Tear Resistance. The tear resistance is also influenced by the strain crystallization [2.85, 2.85 a], and it is therefore very good, and much better than that of most SR vulcanizates. Only isocyanate-crosslinked polyurethanes (AU-I) have a better resistance than NR vulcanizates. Highly reinforcing fillers in NR compounds give, of course, a much better tear resistance than non-reinforcing fillers.

Elastic Rebound. NR vulcanizates have a high rebound, and in this respect, they are only surpassed by vulcanizates from BR. With small or conventional amounts of zinc oxide in the compound, rebound values of 70% or more are achievable with NR vulcanizates, but this value reduces somewhat, if NR is compounded with reinforcing fillers.

2.8.2 Damping Properties, Dynamic Fatigue Resistance

[2.94, 2.125–2.128]

The favourable elastic properties manifest themselves in very low damping (low hysteresis), and a low heat build-up in dynamic deformations. This behaviour, combined with very short relaxation times, qualify NR especially for products which function in dynamic applications. These include vibration and suspension elements and tires. The dynamic fatigue resistance of NR vulcanizates is also excellent [2.127, 2.128].

2.8.3 Heat and Aging Resistance

[2.105–2.114, 2.129, 2.130]

Heat Resistance. The heat resistance of NR vulcanizates is insufficient for many technical applications. According to VDE (Association of German Engineers) tests, the heat resistance, as measured by the temperature at which a vulcanizate has a % maximum elongation still in excess of 100% after 20,000 hours (2 years), is about 60 to 70 °C. This temperature is exceeded with most SR vulcanizates. The heat resistance of NR vulcanizates is primarily determined by the choice of vulcanizing agents, vulcanization conditions, fillers, and secondarily, by the choice of protective agents. EV cure systems, peroxides, or urethane crosslinkers [2.52], low vulcanization temperatures, some under-cure, the use of silica fillers, and a combination of ODPA, SDPA, and MBI as protective agents, give the best level of heat resistance (about 100 °C at 1000 hours exposure time).

Aging resistance. In order to obtain a good aging resistance of NR vulcanizates, it is necessary to use protective agents in the compound, and to use thiazol accelerators in short cure cycles with not too high temperatures. However, even under optimum conditions, the aging resistance of NR vulcanizates does not reach that of most SR vulcanizates.

Weather and Ozone Resistance. [2.115] Even after vulcanization, the NR has double bonds in the polymer chain. Therefore, it has an insufficient weather and ozone resistance, particularly in light-coloured vulcanizates. This can be improved, if carbon black is added to the compound, or especially, if paraffine, microcrystalline waxes, or certain enol ethers are added. Nevertheless, the ozone and weather resistance of NR vulcanizates cannot compete with the excellent performance of vulcanizates of saturated rubbers, like ACM, CO, CSM, EAM, ECO, EPM, EVM, FKM, or Q.

2.8.4 Low Temperature Flexibility

[2.132, 2.133]

Regarding low temperature flexibility, NR vulcanizates deserve full marks. Even without the aid of special softeners, the low temperature flexibility of NR vulcanizates is better than that of most SR vulcanizates, and it is only surpassed by BR and Q vulcanizates.

2.8.5 Compression Set

[2.134]

At ambient and slightly elevated temperatures, the compression set of NR vulcanizates is relatively low. At lower temperatures the compression set appears to be poor due to the tendency of the rubber to crystallize, while at higher temperatures, the poor heat resistance of the NR vulcanizate has a detrimental effect on the compression set.

2.8.6 Swelling Resistance

[2.135]

Since NR is non-polar, its vulcanizates have little resistance to swelling in non-polar solvents. When in contact with mineral oils, benzene, and gasoline, the volume of NR vulcanizates increases several hundred percent. In alcohols, ketones, and esters, the vulcanizates swell less, however.

2.8.7 Electrical Properties

When properly compounded, the electrical properties of NR vulcanizates are also unique. Specific resistivities of 10^{16} Ohm-cm can be obtained, and thus, NR qualifies readily as an electrical insulator,

2.9 Uses of NR

[2.136–2.140]

On account of its unique chemical and physical properties, NR is a very versatile raw material. It is mainly used as solid rubber, and to a lesser extent, as latex. A minor proportion is directly used for making adhesive tapes, rubber solutions, or art gum. By far, the largest proportion is vulcanized, and sold as elastic rubber, and also to some extent, as hard rubber (ebonite). In the past NR was used in the production of virtually all rubber products, because it has such a well-balanced spectrum of physical properties. Therefore, it was considered as general purpose rubber. However, due to the increasing specialization and improvements of SR grades, NR has since been replaced in many applications, especially for most technical parts with requirements to heat and swelling resistance.

Except for tires, there are relatively few products today, for which NR is preferably used on account of its properties, and even here, it has to compete with other materials. Since rubber is a poor conductor of heat, the heat build-up is a major concern in large tires. For rubbers with a low elasticity or too high a hysteresis, there can be a considerable accumulation of heat within products, and this can lead to an internal combustion. NR vulcanizates have, however, a relatively low heat build-up, and therefore, they have always been of great importance for producing truck tires [2.140]. With the introduction of passenger radial tires, NR (IR) has also been increasingly used in this application. Its excellent flexibility makes it particularly suitable for sidewall compounds, and its low heat build-up results in cooler running tires.

An important application of NR is in the production of thin-walled, soft products with a high strength, such as balloons, surgical gloves, or sanitary rubber products. Due to its strain crystallization and thus, self-reinforcing properties, NR still dominates in these applications.

Its high elasticity, combined with a low hysteresis, also makes it an important material for producing suspension elements and bumpers [2.94–2.97, 2.125–2.127, 2.135–2.138].

2.10 Derivatives of NR

[1.15]

Although the chemical behaviour of NR resembles that of a hydrocarbon with a low reactivity, it is possible to produce a number of derivatives from it. Besides additions to the double bonds, substitutions, isomerization, and other reactions play a role. Of the great number of possible derivatives, only the chlorination, hydrochlorination, and cyclization products have achieved technical importance. Since, compared with the raw material, they have a higher hardness, and a better resistance to corrosive agents, they are frequently used for paints, for packaging of food, for coatings of paper, and for the production of adhesives. These derivatives can also be obtained from synthetic rubbers, as discussed in Reference [1.7].

2.11 Competitive Products of NR

NR competes with IR, SBR, BR, IIR, EPDM, PNR, and GPO.

2.12 References on NR

2.12.1 General References on NR

[2.1] *Allen, P. W., Mullins, L.:* Natural Rubber Achievements and Prospects. Rubber J. *149* (1967) 5, p. 104.

[2.1a] *Allen, P. W.:* Future Prospects of Natural Rubber - Production, Processing, End Uses. KGK *35* (1982), p. 189.

[2.2] *Baker, H. C.:* Natural Rubber Faces the Future. IRI Trans. *39* (1963), p. 8.

[2.3] *Bateman, L. (Ed.):* The Chemistry and Physics of Rubber-like Substances. MacLaren & Sons, London; J. Wiley & Sons, New York, 1963.

[2.4] *Bateman, L.:* Science Sociology and Change in the Plantation Industry. Rajiv Printers, Kuala Lumpur, Malaysia. J. IRI *5* (1971), 131.

[2.5] *Boström, S.:* Allgemeine Technologie des Naturkautschuks. In: *Boström, S. (Ed.):* Kautschuk-Handbuch, Vol. 2. Berliner Union, Stuttgart, 1960, pp. 125-135.

[2.6] *Brantley, H. J.:* The Outlook for NR vs. SR in Tire Compounding. 118th ACS-Meeting, Rubber Div., Oct. 7-10, 1980, Detroit, MI.

[2.7] *Brydson, J. A. (Ed.):* Developments with Natural Rubber. MacLaren & Sons, London, 1967.

[2.8] *Davies, C. C., Blake, J. T.:* Chemistry and Technology of Rubber. Reinold Publ., New York, 1937.

[2.9] *Drake, G. W.:* Planting and Production of Natural Rubber and Latex. Progr. of Rubber Technol. *34* (1970), pp. 25-32.

[2.10] *Elliot, D. J., Tidd, B. K.:* Developments in Curing Systems for Natural Rubber. Progr. of Rubber Technol. *37* (1973/74), pp. 83-126.

[2.11] *Farouk, R., Ishak, S. M.:* Recent Developments and Future Trends in Natural Rubber. IRC '85, Oct. 15-18, 1985, Kyoto.

[2.12] *Grosch, K. A., et al.:* Oil Extended Natural Rubber, its Compounding and Service Testing. Rubber J. *148* (1966), 9, p. 76.

[2.13] *Harries, C. D.:* Untersuchungen über die natürlichen und künstlichen Kautschukarten. Springer Verlag, Berlin, 1919.

[2.13a] *Hofmann, W.:* Der Natur- und Synthesekautschukmarkt der Welt. Kunstst. *76* (1986), pp. 1150-1157.

[2.14] *Hurley, P. E.:* Natural Rubber Trends. Paper 30. 120th ACS-Meeting, Rubber-Div., Oct. 13-16, 1981, Cleveland, OH.

[2.15] *Memmler, K.:* Handbuch der Kautschukwissenschaften. S. Hirzel, Leipzig, 1930.
[2.16] *Mullins, L.:* Prospects for Natural Rubber in the 1990th. 124th ACS-Meeting, Rubber Div., Oct. 25-28, 1983, Houston, TX.
[2.16a] *Nadarajah, N. et al.:* Control of Volatile Matter Content to Produce Consistant Quality Natural Rubber. Rubbercon '87, June 1-5, 1987, Harrogate GB, Proceed. A3
[2.16b] *Nair, S.:* Characterization of Natural Rubber for Greater Consistancy. Rubbercon '87, June 1-5, 1987, Harrogate, GB., Proceed. A2.
[2.16c] *Roberts, A. D. (Ed.):* Natural Rubber, Science and Technology. Oxford University Press, Oxford, New York, Kuala Lumpur, 1988.
[2.17] *Vennels, W. G.:* Recent Advances in Natural Rubber Technology. IRI Trans. *42* (1966), p. 227.
[2.18] *Zhou, G.-Y.:* Development of Natural Rubber and Synthetic Rubber in China. IRC '85, Oct. 15-18, 1985, Kyoto.
[2.19] The Natural Rubber Formulary and Property Index. Publ.: MRPRA, 1984.
[2.20] Natural Rubber Technology. Publ.: MRPRA.
[2.21] NRPRA Technical Bulletins, No. 1-11. Publ.: The Natural Rubber Bureau.
[2.22] SMR Bulletins. Publ.: MRPRA.
[2.23] Natural Rubber Progress and Development. Publ.: MRPRA.
[2.24] Natural Rubber Technical Information Sheets, No. 9-120. Publ.: MRPRA.
[2.25] MRPRA Special Brochures. Publ.: MRPRA.
[2.26] Statistical Bulletin. Publ.: Rubber Res. Inst. Malaysia, Kuala Lumpur.
[2.27] MRPRA Publications 1938-1974, An Author Bibliography; Supplement: Publications 1975-1984. Publ.: MRPRA.
[2.28] Internationale Normvorschriften über Qualität und Verpackung von Naturkautschuksorten (Grünes Buch). Publ.: Wirtschaftsverband der Deutschen Kautschukindustrie. See also Ref. [2.24], No. 41, 1979.
[2.29] *Mark, F., Gaylord, N. G., Bikales, N. M. (Eds.):* Encyclopedia of Polymer Science and Technology, Vol. 12: Rubber Natural. Interscience Publ., New York, 1970, pp. 179-256.
[2.29a] GAK *39* (1986), p. 93.

2.12.2 References on NR Sources, Production and Grades

Sources and Production

[2.30] *Lynen, F., Eggerer, H., Henning, U., Kessel, J.:* Angew. Chem. *70* (1958), p. 738; *Lynen, F., Henning, U.:* Angew. Chem. *72* (1960), pp. 820, 826.
[2.30a] *Tanaka, Y.:* The Structure and Biosynthesis Mechanism of Naturally Occuring Polyisoprenes. Rubbercon '87, June 1-5, 1987, Harrogate, GB, Proceed. A4.
[2.31] *Les Bras, J.:* KGK *15* (1962), p. 407.
[2.32] *Ulmann, M.:* Wertvolle Kautschukpflanzen des gemäßigten Klimas. Akademie Verlag, Berlin, 1951.
[2.33] *Leveque, J.:* Rev. Gén. Caoutch. Plast. *43* (1966), p. 1304.
[2.34] *Sekhar, B. C., et al.:* Natural Rubber. Progr. Developm. *18* (1965), p. 78.

Superior Processing Grades

[2.35] Ref. [2.21], No. 2, 1965.
[2.36] *Baker, H. C., Foden, R. M.:* Recent Developments in Superior Processing Natural Rubber. RCT *33* (1960), p. 810.
[2.37] *Baker, H. C., Stokes, S. C.:* Manufacturing Advances Obtained by the Use of SP Rubber. Nat. Rubb. Res. Conf., Kuala Lumpur, Malaysia, 1960, Proceed., p. 587.
[2.38] *Sekhar, B. C., Chin, P. S.:* KGK *19* (1966), p. 80.
[2.39] *Segeman, S. T.:* Rubber World *154* (1966) 1, p. 75.
[2.40] Ref. [2.24], No. 47, 1979.

Oil Extended Grades (see also Ref. [2.12])

[2.41] Ref. [2.24], No. 33, 1979.

Deproteinated Grades

[2.42] *Gupta, S.K.:* A New Process for Deproteinization of Hevea Latex. 13th IRMRA-Conference, March 22–23, 1985, Bombay.

[2.43] *Fernando, W.S.E., Yapa, P.A.J., Jayasinghe, P.P.:* Preparation and Properties of Low Nitrogen Natural Rubber. KGK *38* (1985), p.1010.

[2.43a] *Fernando, W.S.E.:* The Influence of Stearic Acid Activator on the Performance of Low Nitrogen Natural Rubber Vulcanizates. Rubbercon '87, June 1–5, 1987, Harrogate, GB, Proceed. A6

Graft Grades (Hevea Plus)

[2.44] Ref. [2.24], No. 101, 1982.

[2.45] *Barnard, D.:* Chemical Modification of NR – The Future. KGK *35* (1982), p.747.

Epoxidized Grades

[2.46] *Baker, C.S.L.:* Epoxidierter Naturkautschuk. GAK *38* (1985), p.410.

[2.46a] *Gelling, I.R.:* Modification of Natural Rubber Latex with Peracetic Acid. RCT *58* (1985), p.86.

[2.47] *Baker, C.S.L., Gelling, I.R., Newell, R.:* Epoxidized Natural Rubber. RCT *58* (1985), p.67.

[2.48] *Gelling, I.R., Morrison, N.J.:* Sulfur Vulcanization and Oxidative Aging of Epoxidized Natural Rubber. RCT *58* (1985), p.243.

[2.49] *Gupta, S.K., Narayanan, E.:* Epoxidizing of Hydroxy-terminated Natural Rubber. 13th IRMRA Conference, March 22–23, 1985, Bombay.

[2.49a] *Smith, M.G., Baker, C.S.L., Gelling, I.R.:* Compounding and Application of ENR. Rubbercon '87, June 1–5, 1987, Harrogate, GB, Proceed. B30.

[2.50] Epoxidized Natural Rubber, 1984. Publ.: Rubber Res. Inst. Malaysia, Kuala Lumpur.

Thermoplastic Grades

[2.51] *Mullins, L.:* Thermoplastic Natural Rubber. SRC-Conference 1979, Copenhagen; Thermoplastischer Naturkautschuk. KGK *32* (1979), p.567.

[2.51a] *Elliott, D.J.:* Natural Rubber Systems. In: *Whelans, A., Lee, K.S. (Eds.):* Developments in Rubber Technology, Vol. 3. Elsevier Appl. Sci. Publ., Barking, 1982, p.203.

[2.52] Ref. [2.24], No. 94, 1982.

Technical Classified Grades (SMR)

[2.53] *Heinisch, K.F.:* GAK *4* (1951), p.40.

[2.54] Ref. [2.21], No. 9–11, 1979.

[2.55] Ref. [2.24], No. 100, 1982.

[2.56] *Sekhar, B.C.:* Malaysian Natural Rubber, New Presentation Process. Publ.: Rubber Res. Inst. Malaysia, Kuala Lumpur, 1970.

[2.57] Ref. [2.24], No. 48, 1979.

[2.58] Ref. [2.24], No. 60, 1979.

[2.59] Ref. [2.24], No. 62, 63, 65, 1979.

[2.60] Ref. [2.24], No. 61, 1979.

[2.61] Ref. [2.24], No. 64, 1979.

[2.62] Ref. [2.24], No. 66–69, 1979.

[2.63] *Bateman, L., Sekhar, B.C.:* RCT *39* (1966), p.1608.

[2.64] *Smith, M.G.:* Rubber J. *149* (1967), p.28.

[2.65] *Sambhi, M.S.:* Degradative Studies Related to the Plasticity Retention Index of the Standard Malaysian Rubber Scheme. I. Kinetics of Degradation. RCT *55* (1982), p.181. II. Kinetics of Chain Scission. IRC '85, Oct. 15–18, 1985, Kyoto.

[2.66] Ref. [2.24], No. 70, 1979.

Guayule Rubber

[2.67] *Eagle, F.A.:* Guayule. RCT *54* (1981), p.662. (Review).

[2.68] *Black, L.T., Hammerstrand, G.E., Kwolek, W.F.:* Analysis of Rubber, Resin and Moisture Content of Guayule by Near Infrared Reflectance Spectroscopy. RCT *58* (1985), p.304.

[2.69] *Banigan, T.F., Verbiscar, A.J., Ode, T.A.:* An Infrared Spectrometric Analysis for Natural Rubber in Guayule Shrubs. RCT *55* (1982), p.407.

[2.70] *Roduner, L.D., Stephens, H.L.:* The Influence of Resin Content on the Mastication of Blends of Guayule Rubber with Natural and Synthetic Rubbers. Paper 48, 120th ACS-Meeting, Rubber Div., Oct. 13-16, 1981, Cleveland, OH.

[2.71] *Keller, R.W., Stephens, H.L.:* Degradative Effect of Guayule Resin on Natural Rubber. RCT *55* (1982), p.161.

[2.72] *Ramos da Valle, L.F.:* Vulcanization of Guayule Rubber. RCT *54* (1981), p.24.

[2.73] *Ramos da Valle, L.F., Ramirez, R.R.:* Thermoplastic Guayule Blends, Composition and Mechanical Properties. RCT *55* (1982), p.1328.

[2.74] *Ramos da Valle, L.F.:* Thermoplastic Guayule Blends, Rheological Properties. RCT *55* (1982), p.1341.

[2.75] *Montes, S.A., Ponce-Vélez, M.A.:* Effect of Gel and Non-Rubber Constituents on the Extrusion Properties of Guayule Rubber. RCT *56* (1983), p.1.

[2.76] *Budiman, S., McIntyre, D.:* The Deresination of Guayule, Physical Model. I. Effect of Water and Temperature on the Diffusion Coefficient. RCT *57* (1984), p.352. II. A Two-component Analog for Deresination. RCT *57* (1984), p.370.

[2.76a] *Heinisch, K.F.:* Guayule - Kautschuk - eine nutzvolle Ergänzung. KGK *34* (1981), p.1013.

[2.76b] *Mersch, F.:* Augenblicklicher Stand der Entwicklung und Anwendung von Guayulekautschuk. KGK *35* (1982), p.567.

[2.77] *Davila-Castañada, A., Angulo-Sanchez, J.L.:* Study of Rubber Production in Guayule Plants. IRC '85, Oct. 15-18, 1985, Kyoto, Proceed. p.177.

[2.78] *Hamerstrand, G.E., Montgomery, R.R.:* Pilot-Scale Guayule Processing Using Countercurrent Solvent Extraction Equipment. RCT *57* (1984), p.344.

2.12.3 Structure and Properties of NR

[2.79] *Tanaka, H., Sato, H., Kageyu, A.:* Structure and Biosynthesis Mechanisms of Natural-cis-Polyisoprene. RCT *56* (1983), p.299.

[2.80] *Angulo-Sanchez, J.L., Caballero-Mata, P.:* Long-Chain Branching in Natural Hevea Rubber, Determination of Gel Permeation Chromatography. RCT *54* (1981), p.34.

[2.81] *Wibaut, G.P.:* Discuss. Faraday Soc. *10* (1951), p.332.

[2.82] *Hofmann, W.:* Das Oxidationsverhalten von Polymeren und Copolymeren des Isoprens und Butadiens. GAK *20* (1967), pp.602, 714.

[2.83] *Pummerer, R., et al.:* Kautschuk *10* (1934), p.149.

[2.84] *Porter, M.:* Comb Graft Copolymers of Natural Rubber. Ann. Congr. Royal Soc. Chem., University of Aston, March 30-April 2, 1982, Birmingham.

[2.84a] *Payne, A.R., Scott, J.R.:* Engeneering Design with Rubber. Mac Laren Sons, London; Intersciener Publ., New York, 1980.

[2.85] *Hamed, G.R.:* Tack and Green Strength of NR, SBR and NR/SBR-Blends. RCT *54* (1981), p.403.

[2.85a] *Mc Gill, W.J., Joosk, D.A.:* A Theory of Green Strength in Natural Rubber. Rubbercon '87, June 1-5, 1987, Harrogate, GB, Proceed. A5.

[2.86] *Campbell, D.S., Fuller, K.N.G.:* Factors Influencing the Mechanical Behavior of Raw Unfilled Natural Rubber. RCT *57* (1984), p.104.

[2.87] *Eisele, U.:* Spezifische Merkmale von Gummi im Vergleich zu anderen Werkstoffen. DKG-Conference, Sept. 24-26, 1986, Celle, West Germany.

2.12.4 Compounding of NR (see also Ref.[2.66])

[2.88] Ref. [2.24], No. 36–38, 1979; No. 114, 115, 1982.

[2.89] Ref. [2.24], No. 76–79, 1979.

[2.90] Ref. [2.24], No. 20–27, 53–56, 87–89, 1979; 90, 1982.

[2.91] *Whelan, M.A.*: Rezepturaufbau von Naturkautschukmischungen für Spritzgießverarbeitung bei hohen Durchsatzraten. GAK *38* (1985), p. 518; PRI Rubber Conf. 84, March 12–14, 1984, Birmingham.

[2.92] *Woods, W.C., Paris, W.W.*: Accelerator-Systems for Injection Moulding, Paper 4. 124th ACS-Meeting, Rubber Div., Oct. 25–28, 1983, Houston, TX.

[2.93] Ref. [2.24], No. 12, 13, 1979.

[2.94] Ref. [2.24], No. 11, 73–75, 1979; No. 110, 111, 1982.

[2.95] *Derham, C.J.*: Vibration Isolation and Earthquake Protection of Buildings. Progr. Rubber Plast. Technol. *1* (1985) 3, pp. 14–27.

[2.96] *Paris, W.W., Doney, C.*: Improved Service Life of Automotive Suspension Bushings through the Use of Semi-Efficient Vulcanization Systems. RCT *53* (1980), p. 368.

[2.97] *Lemieux, M.A., Killgoar jr., P.C.*: Low Modulus, High Damping, High Fatigue Life Elastomer Compounds for Vibration Isolation. RCT *57* (1984), p. 792 (See also [2.127]).

[2.98] Ref. [2.24], No. 45, 46, 1979; No. 113, 1982.

[2.99] Ref. [2.24], No. 57, 58, 80, 81, 1979; No. 112, 1982.

[2.100] Ref. [2.24], No. 104, 105, 117–120, 1982.

[2.101] Ref. [2.24], No. 18, 19, 1979.

[2.102] Ref. [2.24], No. 16, 44, 85, 86, 1979.

[2.103] Ref. [2.24], No. 17, 81, 1979.

[2.104] *Wingrove, D.E.*: Natural Rubber, Neoprene, Ethylene-Propylene-Diene Blends in White Sidewall Compounds, Paper 7. 122nd ACS-Meeting, Rubber Div., Oct. 5–7, 1982, Chicago, IL.

[2.105] *McSweeney, G.P., Morrison, N.J.*: The Thermal Stability of Monosulfide Crosslinks in Natural Rubber. RCT *56* (1983), p. 337.

[2.106] *Lewis, P.M.*: High Temperature Resistance of Natural Rubber. PRI-Conf. 84, March 12–14, 1984, Birmingham.

[2.107] Ref. [2.24], No. 71, 72, 1979.

[2.108] *Hofmann, W.*: Erweiterte Einsatzgrenzen von Elastomeren unter der Motorhaube. GAK *37* (1984), p. 278; PRI-Conf. 84, March 12–14, 1984, Birmingham.

[2.109] *Hofmann, W.*: Review of Improvements in Sulfur Cured Natural Rubber and Nitrile Rubber Compounds. Plast. Rubb. Proc. Appl. *5* (1985), p. 209.

[2.110] *Baker, C.S.L.*: Latest Developments in the Urethane Crosslinking of Natural Rubber. KGK *36* (1983), p. 677.

[2.111] *Rim, Y.S.*: A New Scorch Safe Urethane Cure System for NR and SBR, Paper 25. 122nd ACS-Meeting, Rubber Div., Oct. 5–7, 1982, Chicago, IL.

[2.112] *Stanford, J.L., Stepto, R.F.T., Still, R.H.*: Formation and Properties of Elastomers based on Polyurethane Networks. IRC '85, Oct. 15–18, 1985, Kyoto, Proceed. p. 359.

[2.113] *Baker, C.S.L.*: Vulcanization with Urethane Crosslinking Agents. Natural Rubber Techn. Bulletin, Publ.: MRPRA.

[2.114] *Fletcher, W.P., Fogg, S.G.*: Compounding Natural Rubber for Heat Resistance. BRPRA Techn. Bulletin, No. 3. 1959, Publ.: The Natural Rubber Development Board.

[2.115] *Braden, M., Fogg, S.G.*: Compounding Natural Rubber for Ozone Resistance. NRPRA Techn. Bulletin, No. 9, Publ.: The Natural Rubber Bureau.

[2.116] *Pal, P.K., Bhowmick, A.K., De, S.K.*: The Effect of Carbon Black-Vulcanization System Interactions on Natural Rubber Network Structures and Properties. RCT *55* (1982), p. 23.

[2.117] *Chen, C.H., Koenig, J.L., Shelton, J.R., Collins, E.A.*: The Influence of Carbon Black on the Reversion Process in Sulfur-Accelerated Vulcanization of Natural Rubber. RCT *55* (1982), p. 103 (See also Ref. [2.130]).

[2.118] *Možišek, M., Bitýška, V.*: Diffusion einiger Weichmacher in Naturkautschuk. KGK *34* (1981), p. 473.

[2.119] Factice. DOG-Kontakt 27, 1982, Publ.: DOG Deutsche Oelfabrik, Hamburg.

[2.120] Dispergum-Zincsoaps. DOG-Kontakt 24, 1979, Publ.: DOG Deutsche Oelfabrik, Hamburg.

[2.120a] *Hofmann, W., Haverland, A.:* New Rating Techniques for Process Aids, II. General Purpose Rubbers. IRC '86, June 3-6, 1986, Gotenburg, Sweden; Aficep-Meeting, Dec. 11, 1986, Paris; DOG-Kontakt 31, 1986, Publ.: DOG Deutsche Oelfabrik, Hamburg.

[2.121] Dispergum 24 - A High Efficient Mastication Agent. DOG-Kontakt 25, 1980, Publ.: DOG Deutsche Oelfabrik, Hamburg.

2.12.5 Processing of NR (see also Ref. [2.121])

[2.122] Ref. [2.24], No. 82, 1979.

[2.123] *Fromandi, G., Reissinger, S.:* Kautschuk u. Gummi *11* (1958), WT p. 3; *13* (1960), WT p. 255; RCT *32* (1959), p. 295.

[2.123a] *Sorgatz, V.:* Experience Gained with Fully-automatic Mixing Rooms. Rubbercon '87, June 1-5, 1987, Harrogate, GB, Proceed. B1

2.12.6 Properties of NR Vulcanizates and Application (see also Ref. [2.95]-[2.97], [2.104])

[2.124] *Bell, G. L., Stinson, D., Thomas, A. G.:* Measurement of Tensile Strength of Natural Rubber Vulcanizates at Elevated Temperatures. RCT *55* (1982), p. 66.

[2.125] *Snowdon, J. C.:* Vibration Isolation - Use and Characterization. RCT *53* (1980), p. 1041.

[2.126] *Gregory, M. J.:* Dynamic Mechanical Properties of Rubber Relevant to Automotive Engineering. PRI-Conf. '84, March 12-14, 1984, Birmingham.

[2.127] *Killgoar jr., P. C., Lemieux, M. A., Tabar, R. J.:* A Low Modulus, High Fatigue Life Elastomer Compound for Suspension Applications. RCT *54* (1981), p. 347 (see also Ref. [2.97]).

[2.128] *Lindley, P. B., Stevenson, A.:* Fatigue Resistance of Natural Rubber in Compression. RCT *55* (1982), p. 337.

[2.129] *Yehia, A. A., Ismail, M. N.:* Heat Resistant Rubbers based on NR and SBR. IRC '85, Oct. 15-18, 1985, Kyoto, Proceed. p. 911.

[2.130] *Chen, C. H., Collins, E. A., Shelton, J. R., Koenig, J. L.:* Compounding Variables Influencing the Reversion Process in Accelerated Curing of Natural Rubber. RCT *55* (1982), p. 1221 (see also Ref. [2.117]).

[2.131] *Datta, N., Das, M. M., Basu, D. K., Chandhuri, A. K.:* Thermal and Age-Resistance Properties of NRGum Vulcanizates in EV-Systems. RCT *57* (1984), p. 879.

[2.132] *Stevenson, A.:* Crystallization Stiffening of Rubber Vulcanizates of Low Environmental Temperatures. KGK *37* (1984), p. 105.

[2.133] *Fogg, S. G., Swift, P. M.:* Compounding Natural Rubber for Service at Low Temperatures. NRPRA Techn. Bulleting No. 4, 1962, Publ.: The Natural Rubber Bureau.

[2.134] *Wood, L. A.:* Representation of Long Time Creep in a Pure Gum Vulcanizate. RCT *54* (1981), p. 331.

[2.135] Ref. [2.24], No. 52, 1979.

[2.136] *Lindley, P. B.:* Design and Use of Natural Rubber Bridge Bearings. NRPRA Techn. Bulletin No. 7, 1962, Publ.: The Natural Rubber Bureau.

[2.137] *Lindley, P. B.:* Engineering Design with Natural Rubber. NRPRA Techn. Bulletin No. 8, 1964, Publ.: The Natural Rubber Bureau.

[2.138] *Freakley, P. K., Payne, A. R.:* Theory and Practice of Engineering with Rubber. Elsevier Appl. Sci. Publ., Barking, 1978.

[2.139] *West, J. P.:* Comparison of the Dynamic Performance of Hydraulically Damped and Conventionally All Rubber Automotive Engine Mounts. PRI-Conf. '84, March 12-14, 1984, Birmingham.

[2.140] *Barnard, D., Baker, C. S. L., Wallace, I. R.:* Natural Rubber Compounds for Truck Tires. RCT *58* (1985), p. 740.

3 Synthetic Rubber (SR)

[1.1–1.7, 1.14–1.21, 1.23–1.28, 1.59, 1.60, 1.66, 2.8, 3.1–3.21]

3.1 Classification of SR

The term synthetic rubber (SR) not only denotes the synthetic analogue of natural rubber (NR), namely cis-1.4 polyisoprene (IR), but also a great variety of other rubbery materials, which are produced by chemical synthesis.

Originally, only those materials were considered SR, which are chemically similar to NR in that they contain olefinic double bonds, and therefore could be vulcanized by means of sulphur [1.12, 1.13, 1.59, 1.60, 2.3]. As a rule, these materials are obtained through homo- or copolymerization of conjugated dienes. In addition, mono-olefins and other monomers were increasingly used for the synthesis of saturated polymers, which can be crosslinked through reactions other than those involving sulphur. These types of SR can be produced by vinyl polymerization, by polycondensation as well as by polyaddition reactions. Rubbers produced from diene monomers are the most important and widely used ones, whereas saturated rubbers constitute, as a rule, specialty products. By now, the number of SR types offered by the chemical industry to the user has grown very large, so that it has become useful to classify the different grades of SR.

Depending on the method used for synthesis, one distinguishes chain addition, polymerization, polycondensation, and polyaddition products.

The polymerization of monomers of the same kind results in homopolymers, while that of different monomers produces copolymers. Those copolymers obtained from three different monomers are referred to as terpolymers. The so-called diene rubbers are produced by the homo- or copolymerization of conjugated diolefins alone or in combination with olefins, so that the polymer chain retains unsaturation, which allows vulcanization with sulphur, analogous to NR.

By polymerizing simple unsaturated monomers (mono-olefins), and by the polycondensation or polyaddition of saturated components, one obtains fully saturated polymers, which cannot be crosslinked with sulphur. By far, the greatest number of these polymers are plastics. However, using special methods, they can also be loosely crosslinked, so that their physical state can be modified, similar to that of diene rubbers through sulphur vulcanization. Thus, these polymers can be used as precursors for producing elastomers.

The different types of rubbers are classified according to ASTM-D 1418-76 or ISO R1629, 1987, as follows, whereby those types identified with "*" constitute the main ones with about 95% share of the total market. Those identified with "**" are as well commercially important materials, being specialty rubbers, while rubbers with "***" are of lesser importance or merely of academic interest. The abbreviations in brackets have not been proposed by ASTM or ISO.

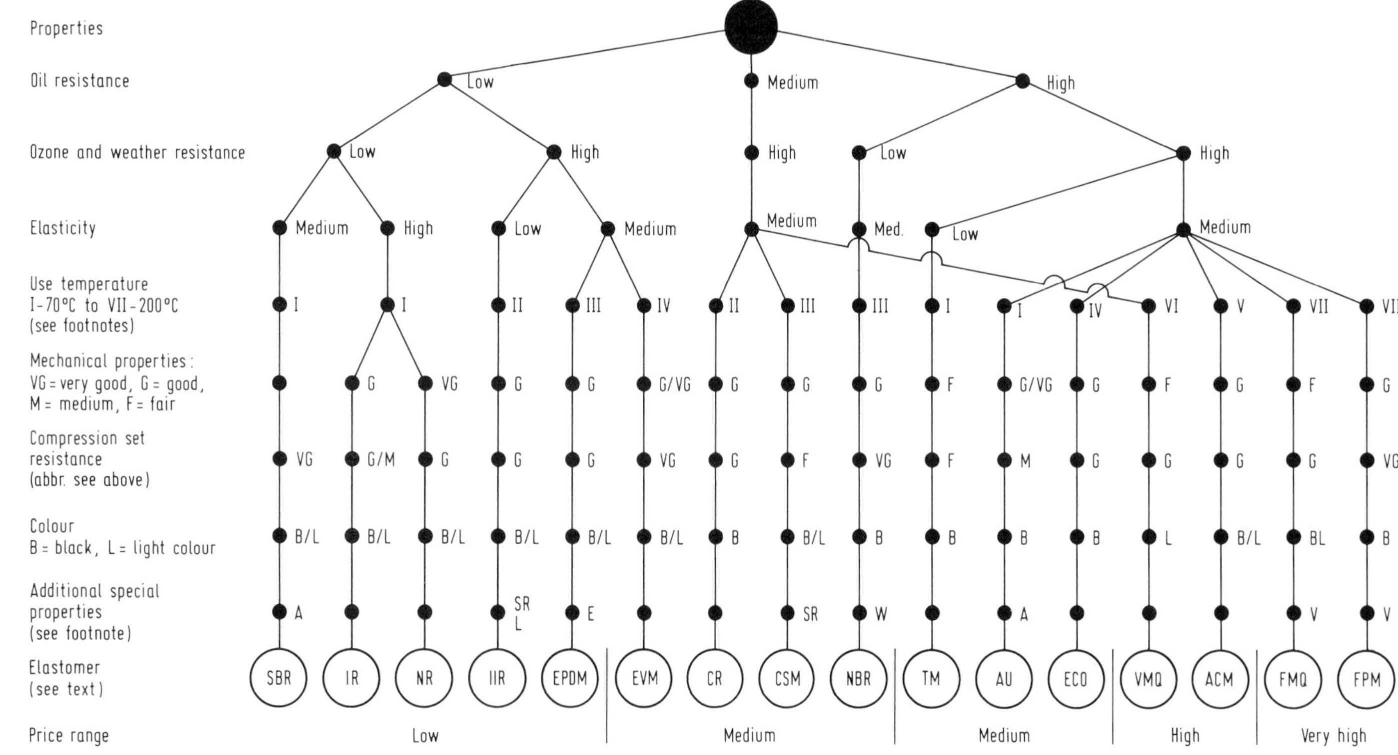

Figure 3.1 Selection scheme for rubbers [3.8]
Max. Service Temperatures: I 70 °C, II 90 °C, II 110 °C, IV 130 °C, V 150 °C, VI 180 °C, VII 200 °C
Special Properties: A – Good abrasion resistance, E – Largely resistant to swell in esters, L – Very low gas permeability, 0 – Medium oil resistance, SR – Resistant to strong acids, V – Difficult to process, W – Resistant to hot water

ABR***	acrylate-butadiene rubber
ACM**	copolymer of ethylacrylate (or other acrylates) and a small amount of a monomer which facilitates vulcanization (acrylic rubber)
AFMU***	terpolymer from tetrafluoroethylene, trifluoronitrosomethane, and nitroso-perfluorobutyric acid (nitroso rubber)
ANM***	ethylacrylate-acrylonitrile copolymer (acrylate rubber)
ASR***	alkylene sulfide rubber
AU**	urethane rubber based on polyester
BR*	butadiene rubber (polybutadiene)
BIIR**	bromobutyl rubber
CIIR**	chlorobutyl rubber
CFM**	polychlorotrifluoroethylene (fluoro rubber)
CM**	chloropolyethylene (previous designation CPE)
CO**	epichlorohydrin homopolymer rubber (polychloromethyloxiran)
CR*	chloroprene rubber (polychloroprene)
CSM**	chlorosulfonylpolyethylene
(EAM)**	ethylene-ethyl acrylate copolymer (e.g. Vamac)
ECO**	copolymer of ethylene oxide (oxiran) and chloromethyloxiran (epichlorohydrin rubber)
(ENM)**	proposed code for hydrogenated NBR, (see H-NBR)
(ENR)**	epoxidized NR
EPDM*	ethylene-propylene-diene terpolymer
EPM**	ethylene-propylene copolymer
EU***	urethane rubber based on polyether
(ETER)**	epichlorohydrin-ethyleneoxide terpolymer
(EVM)**	ethylene-vinylacetate copolymer (previous code: EVA or EVAC)
FKM**	see FPM
FMQ**	methyl silicone rubber with fluoro groups (previous designation FSI)
FPM**	rubber having fluoro and fluoroalkyl or fluoroalkoxy substituent groups on the polymer chain
(FZ)***	see PNF
(GPO)**	copolymer from propylene oxide and allyl glycidyl ether
(H-NBR)**	hydrogenated NBR (proposed code also ENM)
(HSN)	proposed code for H-NBR
IIR*	isobutylene-isoprene rubber (butyl rubber)
IM**	polyisobutene
IR*	isoprene rubber (synthetic)
(LSR)**	liquid silicone rubber
MQ**	methyl silicone rubber (previous designation SI)
NBR*	acrylonitrile-butadiene rubber (nitrile rubber)
NCR***	acrylonitrile-chloroprene rubber
NIR***	acrylonitrile-isoprene rubber
NR*	isoprene rubber (natural rubber)
OT**	polyglycol ether
PBR***	vinylpyridine-butadiene rubber
PE**	polyethylene
PMQ**	methyl silicone rubber with phenyl groups (previous designation PSi)
(PNF)***	polyfluoralkoxyphosphazene (also FZ)
(PNR)**	polynorbornene
PO***	propyleneoxide rubber
PP**	polypropylene

PSBR***	vinylpyridine-styrene-butadiene rubber
(PUR)	generic code for urethane rubbers
PVMQ**	methyl silicone rubber with phenyl and vinyl groups
Q	generic code of silicone rubbers
SBR*	styrene-butadiene rubber
(SBS)*	styrene-butadiene-styrene block copolymer (thermoplastic elastomer)
SCR***	styrene-chloroprene rubber
(SEP)***	polysiloxane treated EPDM
SIR***	styrene-isoprene rubber
(SIS)**	styrene-isoprene-styrene block copolymer (thermoplastic elastomer)
(SR)	generic code for all synthetic rubbers
(ST)***	polythioglycol ether (see ASR)
(TFE)***	tetrafluoroethylene
(TM)**	polysulfide rubbers
(TOR)***	trans-polyoctenamer
(TPA)***	trans-polypentenamer
(TPE)	generic code for thermoplastic elastomers
(TPO)**	thermoplastic polyolefins
(TPU)**	thermoplastic polyurethanes
VMQ**	methyl silicone rubber with vinyl groups
X-LPE	crosslinkable polyethylene

Additional prefixes to designations are:

E-	emulsion polymer
L-	solution polymer
OE-	oil-extended rubber
X-	reactive groups (primarily important for NBR and latices)
Y-	thermoplastic rubber

A guide for the choice of rubbers for specific applications is given in Figure 3.1 [3.8] A list of abbreviations see Section 8.2.

3.2 Some Basics about the Production of SR

3.2.1 Monomers for SR Production

The following compounds or monomers are some of the most important starting materials:

	Structure	Boiling Point °C	°F
ethylene	$CH_2{=}CH_2$	–104	–155
propylene	$CH_2{=}CH{-}CH_3$	– 50	– 58
isobutylene	$CH_2{=}C(CH_3)_2$	– 6	21
1.3-butadiene	$CH_2{=}CH{-}CH{=}CH_2$	– 4.5	24

	Structure	Boiling Point °C	°F
isoprene	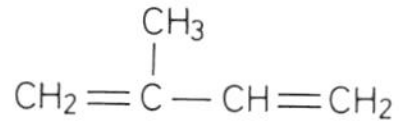 $CH_2{=}C(CH_3){-}CH{=}CH_2$	34	93
chloroprene	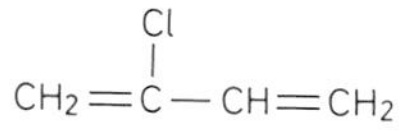 $CH_2{=}C(Cl){-}CH{=}CH_2$	59	138
styrene (vinylbenzene)	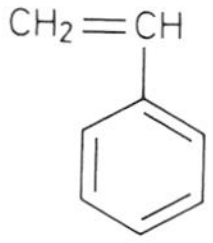 $CH_2{=}CH{-}C_6H_5$	145	293
vinyl acetate	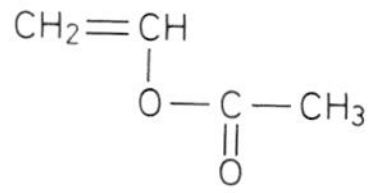 $CH_2{=}CH{-}O{-}C({=}O){-}CH_3$	72	162
methyl methacrylate	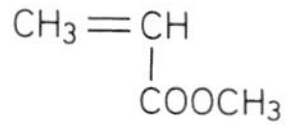 $CH_3{=}CH{-}COOCH_3$	80	176
acrylonitrile	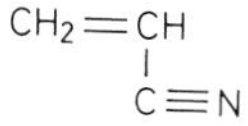 $CH_2{=}CH{-}C{\equiv}N$	77	171

3.2.2 Some Basics about Polymerization

[1.7, 1.59, 1.60, 3.3–3.5, 3.13–3.15, 3.22–3.24]

The basic requirement for the polymerizability of a monomer is the availability of double bonds, where the carbon atoms share two pairs of electrons, namely σ-type and π-type electrons. The π-type bonds, being less stable, bring about the reactivity of the double bond, and they can be broken under the influence of a suitable external agent. The bond may break in two ways: either one electron goes to each atom creating two free radicals (homolysis).

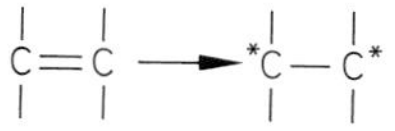

$$>C{=}C< \longrightarrow {}^{*}C{-}C^{*}$$

or the pair of electrons stays with one or the other of the two atoms (heterolysis), leading to ions.

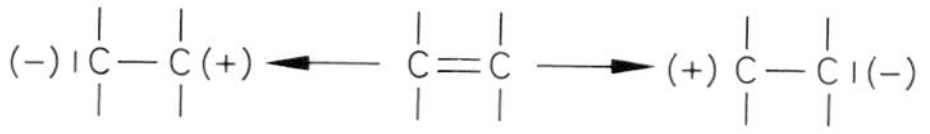

$$(-)\,|C{-}C\,(+) \longleftarrow >C{=}C< \longrightarrow (+)\,C{-}C\,|(-)$$

While the first instance is the starting point of a free radical chain addition polymerization, the second mechanism gives rise to an ionic addition polymerization, where, depending on the nature of catalyst used, the reaction proceeds as cationic, anionic, or coordination polymerization.

There are always three reaction steps in a polymerization: initiation, propagation, and termination, and it can take place in a homogeneous phase (bulk, solution polymerization) or in a heterogeneous phase (almost exclusively emulsion polymerization in SR production). Preferably, the free radical polymerization is conducted in a heterogeneous phase, and the ionic polymerization in a homogeneous one.

3.2.2.1 Free Radical Polymerization

[1.7, 3.3–3.5, 3.13–3.15, 3.22–3.29]

Free Radical Homopolymerization. In free radical polymerizations, the dissociation of the π-electrons in the double bond constitutes the initiation reaction. Heat, but also photochemical or electrochemical stimuli can cause this dissociation. To start with, primary radicals are formed from the decomposition of peroxides, hydroperoxides, or azo compounds, which are initiators, although frequently falsely referred to as polymerization catalysts. The fragments from these initiators then react with a monomer molecule to form a monomer radical. At present, the most frequently used initiators are so-called redox-systems, where the reaction of a reducing agent (e.g. amine, bisulfite, mercaptane, persulfate) with an oxidizing agent (hydroperoxide, peroxide, oxygen) is most effective, particularly at low temperatures. After initiation of the polymerization follows the propagation step, where the polymer chain grows through stepwise addition of monomer molecules, until the reaction is terminated through addition of a transfer molecule.

The free radical chain polymerization can be represented in the following way, where the individual reaction steps succeed each other:

1. Chain initiation

$I \rightarrow 2R^*$	(initiator decomposition forming initiator radicals)
$R^* + M \rightarrow RM^*$	(formation of a primary radical)

2. Chain propagation

$RM^* + M \rightarrow RMM^*$

3. Chain transfer

$RM_n^* + HX \rightarrow RM_nH + X$-	(formation of a chain transfer radical)
$X^* + M \rightarrow XM^*$	(formation of a new primary radical)

4. Chain termination

$RM_m^* + RM_n^* \rightarrow P_{m+n}$	(recombination)
$\rightarrow P_m + P_n$	(disproportionation)

The abbreviations in this representation have the following meaning: I = initiator molecule; R* = initiator fragment; M = monomer molecule; RM* = primary monomer radical; RM_m^* = polymer radical of chain length m; RM_n^* = polymer radical of chain length n; P = polymer molecule; HX = chain transfer molecule; X* = chain transfer radical; XM* = new primary monomer radical.

The time elapsed between the formation of a primary monomer radical and chain termination, that is, the propagation time, is statistically distributed. Hence, the number of monomer molecules varies in different polymer chains. For this reason, and due to the constant interchange of chain initiation and chain termination or transfer, a specific polymer consists of macromolecules with different degrees of polymerization and different distributions of molecular weight.

According to Step 3 in the above reaction scheme, the growing polymer chain can react with other molecules, the so-called chain transfer agents, which are also present in the polymerization charge. In this reaction, the polymer radical becomes deactivated, and a chain transfer radical forms, which, in turn, starts a new monomer radical. Monomers, polymer molecules, initiators, solvent molecules - all can act as chain transfer agents, but primarily the so-called chain modifiers, which are especially added for this purpose to the polymerization charge. Preferred chain modifiers are those compounds, which do not reduce the rate of polymerization, and for which the resulting radicals have virtually the same reactivity as those, which the chain modifier had deactivated. Alcohols, alkyl halides, mercaptanes, and xanthogen disulfide are examples of practical chain modifiers. Depending on the concentration of chain modifiers in a polymerization charge, it is possible to adjust the average molecular weight of a polymer, and to avoid excessive branching and crosslinking (gelation) of chains. Therefore, the use of chain modifiers results in a greater uniformity of the molecular structure and the molecular weight of a polymer.

Reactions between polymer radicals can lead to chain termination, either by recombination or disproportionation.

All the above reaction steps have an influence on the rate of polymerization (rate of conversion of monomer), the average degree of polymerization (average number of monomer units per polymer molecule), and the molecular weight distribution (the relative frequency with which polymer molecules of a specific degree of polymerization occur). Thus, through the appropriate design of a polymerization recipe one can control important technological properties of rubbers, such as processibility.

Free Radical Copolymerization. In principal, the copolymerization of two or more monomers follows the same pattern as the homo polymerization, except that now monomer radicals can be formed by all of the different monomers present. These monomers can react with radicals of their own kind (homo propagation step), or of the other kind (hetero propagation step), the occurance of which depends on the relative concentrations of the different monomers, and the relative reactivity of each of the different kinds of growing chain ends towards each of the different kinds of monomers (copolymerization reactivity ratios).

The following limiting cases can occur, depending on the magnitude of the reactivity ratios of two monomers:

- alternating polymerization - the two types of monomer units alternate in the polymer chain until the monomer in the minor concentration is exhausted;
- azeotropic polymerization - the two monomers enter into a polymer chain in a statistically random manner with their average concentration in the chain corresponding to their charge ratio;
- block polymerization - complete polymerization of one monomer prior to polymerization of the second.

In practice, however, the last limiting case does not occur with free radical polymerizations. With azeotropic polymerizations the reactivity ratios and the concentrations of both monomers in a charge determine the average chemical composition (i.e. the non-uniformity of composition), the average sequence length, and sequence distribution of the polymer. (See also page 68).

3.2.2.2 Emulsion Polymerization

[3.25–3.29]

Emulsion polymerizations are of great importance in free radical chain polymerizations. At least four components are required here: monomers (water insoluble), water, emulsifiers, and initiators (water soluble), in addition to chain modifiers, polymerization stoppers (short stops), and stabilizers.

Through the emulsifier, the monomer is emulsified in the form of small droplets with a diameter of about 10^{-4} cm, or 10^{10} droplets per cm^3. In addition, monomer molecules are solubilized in so-called soap micelles, formed by the aggregation of about 20 to 100 emulsifier molecules. The diameter of these micelles is about $10^{-6} - 10^{-7}$ cm, and there are about 10^{18} of them in one cm^3.

Newly formed initiator radicals react primarily with the monomer that is dissolved in the micelles, so that this becomes the locus for polymerization. As the reaction proceeds, the micelles grow progressively by imbibing new monomer from the surrounding monomer droplets, thus forming latex particles (solution of monomer in the polymer and conversion). After depletion of monomer droplets, the emulsion consists only of polymer-monomer particles, which are stabilized against coagulation by the surface charge of the adsorbed emulsifier molecules.

Termination reactions of polymer radicals are inhibited by the emulsifier shell of the particles, and therefore high conversion rates of monomer and high degrees of polymerization are obtained – in fact, much higher than those obtainable with solution polymerization processes.

After reaching the desired extent of monomer conversion, the polymerization reaction is terminated through the addition of a shortstop. Before recovering the polymer, a stabilizer is added to the latex to protect the polymer against oxidative degradation.

In comparison with polymerization processes that are carried out in a homogeneous phase, the emulsion polymerization offers the following advantages:

- The use of water as reaction medium facilitates the dissipation of the heat of polymerization, and this has a great influence on the consistency of polymer properties.
- There is a uniform distribution of monomer and additives, so that, for instance, at a specific time a shortstop can be added, acting instantaneously to terminate the polymerization reaction.
- Since the viscosity of an emulsion is virtually independent of the molecular weight of the polymer in the latex particles, it is possible to produce, without problem, rubbers with very high molecular weight.
- As a result of the low viscosity of emulsions, it is possible to conduct the emulsion polymerization in relatively simple reaction vessels, and also continuously in a cascade-like arrangement. This results in good space and time yields.
- Emulsion polymerizations produce stable dispersions or latices of synthetic rubbers, which can be used in some applications in place of natural rubber latices. This is important to the rubber industry.

A disadvantage of emulsion polymerizations is the lack of uniformity of polymer molecules with respect to stereoregularity. This is due to the different possibilities with which diene monomer molecules can react, and it applies to all free radical polymerization processes of diene monomers.

While originally for the production of rubbers, the emulsion polymerizations were conducted at 45 °C (113° F), the later use of redox catalysts permitted a drastic lowering of the polymerization temperature to 5 °C (41° F). Lower polymerization temperatures give polymers with greater steric uniformity, and considerably better rubber properties, and thus, a distinction is made between "cold rubbers" and "hot rubbers". Worldwide, the cold rubbers have achieved the greater prominence.

3.2.2.3 Ionic Polymerization

[3.30–3.57]

With free radical initiators, the homolytic cleavage into two free radicals is virtually not governed by electrostatic or solvation effects. This is in contrast to ionic initiators, where the dissociation is greatly influenced by the reaction medium, that is, the solvent. Anions or cations formed from these initiators start the polymerization, which then propagates through the corresponding counterion following the mechanism of a chain reaction. While in a free radical polymerization, the initiator radical transfers the free radical to a monomer molecule, in ionic polymerizations, there is a transfer of charge from the initiator ion to the monomer. Therefore, positively or negatively charged initiator ions result in cationic or anionic polymerizations, respectively. Monomers can exist in different mesomeric forms, and enter into a polymerization reaction in that form, into which they have been polarized under the influence of electrostatic forces.

Example of mesomeric forms of butadiene

$$(-)|CH_2{-}CH{=}CH{-}CH_2(+) \longleftrightarrow CH_2{=}CH{-}CH{=}CH_2 \longleftrightarrow (-)|CH_2{-}\overset{(+)}{C}H{-}CH{=}CH_2$$

Example of mesomeric forms of acrylonitrile

$$(+)CH_2{-}CH|(-)(C{\equiv}N) \longleftrightarrow CH_2{=}CH(C{\equiv}N) \longleftrightarrow (+)CH_2{-}CH{=}C{=}\underline{\bar{N}}(-)$$

A cationic polymerization can be initiated by *Bronsted* or *Lewis* acids (e.g. H_2SO_4, $HClO_4$, $AlCl_3$, BF_3, $C_2H_5AlCl_2$), requiring, as a rule, the presence of small amounts of cocatalyst, such as water, alcohol, or a halogenated hydrocarbon. [3.38–3.41]

$$AlCl_3 + H_2O \longrightarrow H^+[AlCl_3OH]^-$$

For anionic polymerizations, initiators such as alkyl lithium compounds are used. [3.42–3.46]

Since the activation energies for the initiation reactions are often significantly smaller for ionic than for free radical polymerizations, the ionic chain polymerizations can often be carried out at much lower temperatures.

The dissociation of the initiator is governed by the electrostatic effects of the reaction medium. Depending on the magnitude of the ionization constant, the extent of dissociation can increase merely through enhanced polarization, which is conditional on the polarity of the solvent.

Example of the dissociation of a compound

$$R{-}X \rightleftharpoons \overset{(+)}{R}\,{-}{-}{-}\,\overset{(-)}{X} \rightleftharpoons R^{\oplus} + X^{\ominus}$$

Therefore, cationic polymerizations are preferably carried out in polar solvents.

On dissociation, a cation and an anion result, but for thermodynamic reasons, only one of these ions can initiate the polymerization of a monomer.

At the start of the polymerization, the initiator ion transfers the π-electrons of a monomer molecule to one or the other carbon atom, depending on the charge. Subsequently, the initiator ion attaches to the monomer, which then assumes the role of initiator, attaching an additional polarized monomer molecule to form a growing polymer ion.

$$>\overset{(+)}{C}-\overset{(-)}{\underline{C}}< \rightleftharpoons >C=C< \rightleftharpoons >\overset{(-)}{\underline{C}}-\overset{(+)}{C}<$$

Example of the polarization of an olefin

Because of the regular pattern of attaching monomer molecules to polymer ions, ionic polymerizations are capable of producing stereospecific polymers, and in special instances strictly linear polymers as well.

Since in ionic polymerizations the growing polymer ions have the same charge, virtually no chain terminations occur from recombinations and disproportionations. Only impurities and polymerization stoppers can terminate the chain reaction.

Chain transfer reactions can also occur, with cationic polymerizations [3.38-3.41]. However, as the polymerization temperature is lowered, these reactions become less prominent and therefore cationic polymerizations are frequently carried out at very low temperatures.

Under corresponding conditions, chain transfer reactions do not occur with anionic polymerizations - hence very narrow molecular weight distributions are obtained. After all monomer molecules have been used up, the macromolecules still remain reactive, and are so-called "Living Polymers", since they are capable of further growth on addition of fresh monomer. If a different kind of monomer is added, block copolymers can be produced in this way, such as polybutadiene-polystyrene-polybutadiene block copolymers [3.38].

3.2.2.4 Coordination Polymerization or Metal Complex Polymerization [3.47-3.57]

In coordination polymerization reactions, the initiator exists as a complex, attached to the growing end of the polymer chain, and the chain reaction progresses by the insertion of new monomer molecules into this complex - hence the name "Insertion Reaction".

The most commonly used initiators for these polyinsertion reactions are the so-called *Ziegler*-Complex catalysts. These are reaction products of organometallic alkyl- or acyl-compounds of Group I to III elements, and transition metal salts of Group IV to VIII elements of the Periodic Table. Typical examples of Ziegler catalysts are mixtures of aluminum alkyls and halides from titanium, cobalt, nickel, vanadium, tungsten, or similar metals.

The catalytic activity of these transition metal compounds may be due to their ability to form coordinate bonds with unshared pairs of π-electrons from olefins or dienes. These compounds are also capable of forming covalent bonds with σ-electron pairs. This ability allows the catalysts to exchange one type of bond for another, and this exchange is the pertinent feature of the propagation (or insertion) reaction step at the transition metal ion. Since a monomer unit enters the polymer chain in a sterically controlled manner during chain propagation, sterically regular polymers are obtained with coordination catalysts. Depending on the composition of the catalysts, it is possible to synthesize polymers, which have almost pure cis-1.4, trans-1.4, or 1.2 microstructures [3.57].

3.2.2.5 Bulk Polymerization

[3.5]

In this process, the polymerization takes place in the pure, liquid monomer as reaction medium, that is, without solvent being present, or in the gas phase.

The heat of polymerization is dissipated through external cooling, which is satisfactory only in reactors with small charges of about 20 to 50 kg. A continuous polymerization process with a significantly improved heat exchange can be carried out in so-called polymerization screws or reactor towers.

The first synthetic polymers that were produced before and during World War I, and the numbered Buna grades of the IG Farbenindustrie were made in a bulk polymerization process. By now, these bulk polymers have almost completely lost importance, but for ecological reasons, there could be renewed interest in bulk polymerization processes.

3.2.2.6 Solution Polymerization

[3.58, 3.59]

In solution polymerizations, the monomer is dissolved in an organic solvent of high purity. The preparation of catalyst often poses special problems. After initiation of the polymerization reaction, the viscosity of the reaction medium rises in proportion to the degree of polymerization, and this limits the extent of chain growth because of the viscosity of the solution, which is too high. High molecular weight polymers can be obtained by the so-called Mooney Jump (or Molecular Weight Jump) technique [3.59] (See also page 54). After deactivation of the catalyst, the polymer is recovered from the reaction charge.

3.2.2.7 Suspension Polymerization

[3.5]

Suspension polymerizations are carried out using a liquified monomer as the solvent and adding to it other monomers, catalysts, etc. After reaching high degrees of polymerization, the resulting polymer becomes insoluble in the monomer, but remains suspended in the reaction medium. The high solution viscosity limits the molecular weight obtainable in solution polymerizations. However in suspension polymerizations, the viscosity of the reaction medium increases only fractionally. Therefore, polymers with very high molecular weights can be made by this process. For instance, suspension polymerization processes are of importance for the production of ultra high molecular weight EPDMs (see page 94) [3.60], where ethylene, the diene monomer, and polymerization aids are dissolved in the liquified propylene, and then copolymerized.

3.2.3 Some Fundamentals about Polyaddition and Polycondensation Reactions

[1.7]

In contrast to polymerization reactions, polyadditions and polycondensations play a minor role in the production of SR. TM, AU, and Q are examples of polymers obtained by these processes.

In these polymerizations, multifunctional compounds react to form macromolecules. If bifunctional compounds are used exclusively, linear chain molecules result. The presence of at least one trifunctional component per macromolecule yields branched or crosslinked macromolecules, whose degree of branching or crosslink density depends on the concentration of the trifunctional component.

If the reagents are coupled by simple addition, the process is referred to as polyaddition. On the other hand, if the functional groups in the chain form by elimination of water, hydrogen chloride, or other small molecules, one is dealing with a polycondensation reaction.

The synthesis of a polyurethane (AU) serves as an example of a polyaddition:

$$...O-(CH_2)_y-OH + O{=}C{=}N-(CH_2)_x-N{=}C{=}O + HO-(CH_2)_y-O...$$

$$\downarrow$$

$$...-O-(CH_2)_y-O-\overset{\overset{O}{\|}}{C}-\overset{\overset{H}{|}}{N}-(CH_2)_x-\overset{\overset{H}{|}}{N}-\overset{\overset{O}{\|}}{C}-O-(CH_2)_y-O-...$$

Polyaddition of bifunctional esters with bifunctional isocyanates

An example of a polycondensation is the formation of polysulfides (TM):

$$n\ ClRCl + n\ Na_2S_x \longrightarrow (RS_x)_n + 2n\ NaCl$$

Polycondensation of alkylene chlorides with sodium polysulphide (R=alkyl; x=2 and greater).

3.2.4 Structure of Polymers and the Determination of Structure

[1.7, 3.61–3.76]

Repeat Unit Structure of Polymer Chains. Dienes can react in various ways in a polymerization and thus yield polymer molecules with significantly different structures. These can determine to a great extent the general performance capability of polymers, in addition to processibility and cure behaviour. Specific conditions under which polymerizations are carried out can also have a significant influence on the chain structure and therefore on the physical properties [3.64–3.76] to include rheological properties [3.64, 3.66, 3.69, 3.71, 3.72] as well as on autoadhesion, cohesion and tack [3.65, 3.67a, 3.73, 3.75].

In the production of linear, long chain polymers, where the diene monomers react in the 1.4 fashion, and which are the most important reactions in rubber synthesis, cis-trans isomerism plays an important role. Polymers with completely different physical properties are obtained by 1.2 addition in the isotactic or syndiotactic form, and through further aromatization or cyclization reactions, as shown in the following examples (see Table 3.1) [1.7]:

Table 3.1: Influence of the Polybutadiene Structure on the Property Spectrum of the Polymers [1.7]

Addition	Structure	Melting Point	Properties
1.4-cis	$—CH_2—CH{=}CH—CH_2—$	2 °C	Elastomeric
1.4-trans	$—CH_2—CH{=}CH—CH_2—$	140 °C	Thermoplastic
1.2-iso-tactic	$CH_2—CH(CH{=}CH_2)—CH_2—CH(CH{=}CH_2)—CH_2$	126 °C	Thermoplastic
1.2-syndio-tactic	$CH_2—CH(CH{=}CH_2)—CH_2—CH(CH{=}CH_2)—CH_2$	210 °C	Impact Resistant Thermoplastic
1.2-cyclized	$CH_2—CH—CH_2—CH—CH_2$ / $CH_2—CH—CH_2—CH—CH_2$	-	Plastic (Insulator) Insoluble due to Intermolecular Crosslinking
1.2-aroma-tized	$CH{=}C—CH{=}C—CH{=}$ / $CH{=}C—CH{=}C—CH{=}$	-	Plastic (Semiconductor) Insoluble due to Intermolecular Crosslinking

With isoprene, a 3.4 addition is also possible besides 1.2.

$[CH_2{=}C(CH_3)—CH{=}CH_2]_n$ (C atoms numbered 1 2 3 4)

1.4-Polym. → $[—CH_2—C(CH_3){=}CH—CH_2—]_n$

1.2-Polym. → $[—CH_2—C(CH_3)(CH{=}CH_2)—]_n$

3.4-Polym. → $[—CH_2—CH(C(CH_3){=}CH_2)—]_n$

Example of 1.4-, 1.2-, and 3.4-structures of polyisoprene.

Where there is little steric control during polymerization, a variety of configurations can be formed in a single polymer chain.

Ideal elastomeric properties are, for instance, obtained from a pure cis-1.4 configuration of polyisoprene, while a deviation from this structure will influence more or less strongly its performance properties. On the other hand, a small amount of trans structure, or a small amount of 1.2-addition during polymerization, i.e. vinyl structure, can be desirable. Examples are BR or SBR, where a certain vinyl content in the chain structure is important for the development of specific properties (see page 55 and 61).

Cis-Trans-Isomerism. A naturally occuring example of a pure cis-trans-isomerism is natural rubber/gutta percha.

Cis-1.4 Configuration (natural rubber type)

```
     CH3          CH2—CH2          H     CH3          CH2—
        \        /       \        /         \        /
         C=C            C=C                  C=C
        /    \         /    \               /    \
  —CH2        H     CH3      CH2—CH2              H
```

Trans-1.4 Configuration (gutta percha type)

```
     CH3          H     CH3          H     CH3          H
        \        /         \        /         \        /
         C=C                C=C                C=C
        /    \             /    \             /    \
  —CH2        CH2—CH2            CH2—CH2            CH2—
```

The chain configuration of NR is 100% cis-1.4, as can be determined within experimental accuracy. For commercially available IR grades it is only 95% or less.

Tacticity. By contrast, differences in tacticity are of lesser importance in rubber grades.

Molecular Weight and Molecular Weight Distribution. In addition to the arrangement of building blocks in the macromolecules, the distribution of molecular weights determines in great measure the processibility of rubbers. Polymers with a narrow molecular weight distribution soften only in a very limited range of temperatures, which has a negative influence on mixing processes. Because of their easy processing characteristics and better filler distribution, it is desirable to have rubbers with a broad molecular weight distribution, where the low molecular weight portions behave somewhat like plasticizers.

The degree of polymerization or the molecular weight is important for determining the spectrum of properties of SR grades.

If the molecular weight is too high, the processing of the rubber becomes a problem because of the high viscosity. Such products either have to be masticated prior to use, or they have to be extended with appropriate amounts of oil. The latter improves the processibility and it reduces the price of the product.

Today, very high molecular weight rubbers which are extended with oil - the so-called "oil extended" rubbers - play an important technological role. In many instances, the oil extension does not impair the mechanical properties of the vulcanizate. A very high molecular weight based polymer with a high tensile strength is used for this purpose, and the oil-extension only reduces the processing viscosity to a practical level. Without oil-extension the high strength properties of the base polymer cannot be exploited. Since the cure behavior depends, amongst others, on the length of the primary polymer molecule, low molecular weight rubber grades are difficult to vulcanize. For instance, liquid rubbers crosslink only with difficulty, while for very long molecular chains, relatively few crosslinks already suffice to form networks. At very low degrees of polymerization, the property spectrum of a vulcanizate is not fully developed. Therefore, SR grades with very low molecular weights are often not desirable; instead, an intermediate range of molecular weights is favoured as a compromise between good processibility and good mechanical properties, or high molecular weights with appropriate oil extension.

Linearity, Branching. There are additional structural elements of the polymer molecule, such as linearity, short or long chain branching, and crosslinking (gel), which have a considerable influence on the processibility of SR as well as on the property spectrum of the resultant vulcanizate. On the one hand, the gel is unwelcome in most instances, since it reduces the mechanical properties of the vulcanizate. But for some specific applications, such as extruded goods, specially crosslinked SR grades with a well defined gel content are produced. This gel improves the processing behaviour of some rubbers (see, for instance, Section 2.4.1.3 SP Rubbers, see page 14).

```
Linear Chain - Stretched        —C—C—C—C—C—

                                  C-C
                                C'    `C
Coiled Chain (in practice)     / C   /
                              C-C C C      C-C-C-C-C-C
                                / `C-C-C'            `C
                              C'                     /
                             /                      C
                                                   /
                                                  C-C-C-C-

                                    C         C
                                   /         /
Short Chain Branching          —C—C—C—C—C—
                                      \
                                       C

                                            C—C<
                                           /
                                      C—C—C
                                     /
Long Chain Branching           —C—C—C
                                     \
                                      C—C     C—C<
                                       \     /
                                        C—C—C—C
                                             \
                                              C

                               —C—C—C—C—C—C—
                                   |       |
Crosslinked Network            —C—C—C—C—C—C—
                                |     |       |
                               —C—C—C—C—C—C—
```

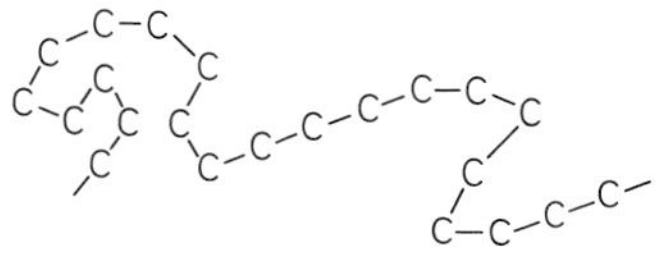

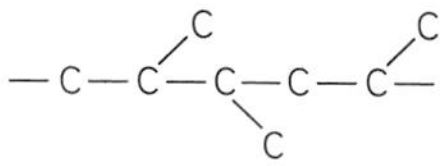

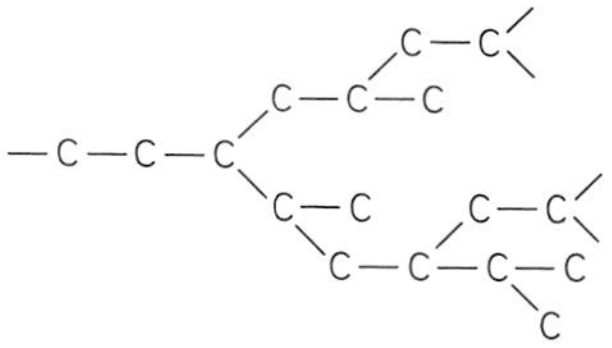

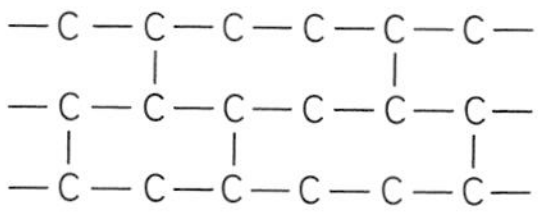

The effect of macrostructure on polymer properties will not be discussed here in greater detail, since this will be done later on for the individual types of rubber.

Monomer Sequence in Copolymers. With copolymerizations using several different monomers, the sequence distribution of monomers in the copolymer can differ significantly, depending on the polymerization process. Monomer sequences can be statistically random (random polymerization), alternating, or in blocks.

Statistically Random Sequence	—A—B—A—A—B—A—B—B—B—A—B—
Alternating Sequence	—A—B—A—B—A—B—A—B—A—B—A—B
Block Sequence	—A—A—A—A—A—A—B—B—B—B—B—

In addition to the pure chain propagation polymerization of monomers, graft polymerizations are also possible, where, in a secondary reaction, a monomer is grafted onto an existing polymer chain.

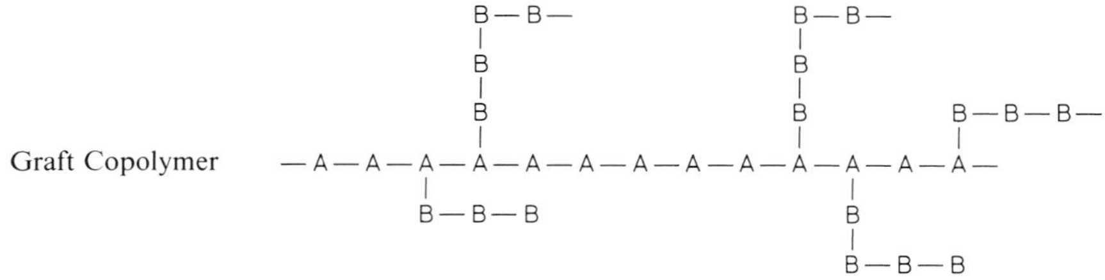

Depending on the particular arrangement of monomer, the resulting polymers have considerably different rubber technological properties.

A completely linear chain structure is practically impossible with free radical polymerizations, such as emulsion polymerization. It is only obtainable with ionic chain polymerization reactions. A high degree of steric uniformity of the polymer molecules, such as a high cis- or trans-content, or alternating or block sequences in copolymers, can be achieved only with ionic polymerizations.

The structure of polymers can be determined by the following means: (see [3.61])

Cis- and trans- structure can be probed by infrared (IR) absorption spectroscopy, since the two structures absorb IR of two distinctly different wavelengths. For polyisoprene the cis-1.4 structure absorbs IR of 724 cm^{-1}, the trans-1.4 of 967 cm^{-1}, and the 1.2 structure of 911 cm^{-1}. The extent of absorption at these wavelengths can be used for the quantitative determination of these structures.

The vinyl content (from 1.2 addition of monomer units) can also be determined by IR spectroscopy, or by nuclear magnetic resonance spectroscopy (NMR).

The molecular weight can be measured, as for other polymers, by means of ultracentrifugation and light scattering (M_W), or osmosis (M_n).

Molecular weight distributions are obtained by fractional precipitation, by gel permeation chromatography, and by ultracentrifugation.

Gel permeation chromatography and comparative viscosity measurements can determine the long chain branching, namely the average number of branches per molecule.

The gel content of a polymer can be determined by means of gel permeation chromatography as well.

3.3 Polymers

3.3.1 Polybutadiene (BR)

[3.77-3.134]

3.3.1.1 General Background about BR

Around the turn of the century, in the early days of SR research, attempts were made to polymerize not only isoprene (methyl butadiene), the structural unit of NR, but also its simplest analogue, butadiene. *(C. D. Harries, F. Hofmann,* 1911). Alkali metals were proposed as catalysts for the bulk polymerization process. Later polymerization of **bu**tadiene using sodium (in German - **na**trium), from which the name Buna (1926) was derived, has led to the development of the numbered Buna grades. These early research efforts, and the first commercial grades that were marketed in Germany and Russia, were only a temporary success, and no substantial markets developed [1.59, 1.60].

Only through the use of coordination catalysts of the *Ziegler–Natta* type, and the use of alkyl lithium catalysts was it possible to produce solution polymers, which are widely applied in tire compounds, particularly in blends with NR or SBR. Today, with an annual consumption in excess of one million tons, BR ranks second behind SBR in usage of the SRs. In 1985, the annual global production capacity of BR was 1.7 million metric tons, which is about 13.7% of the SR production capacity. [2.13a]

BR is composed of butadiene units which can have joined linearly by 1.4 (preferred in cis-1.4, but also in certain measure, trans-1.4 conformation), as well as by 1.2-addition.

```
|                          |                          |              |
+ CH2—CH=CH—CH2 + CH2—CH=CH—CH2 + CH2—CH— |
|                          |                          |      |       |
|                          |                          |      CH      |
|                          |                          |      ||      |
|                          |                          |      CH2     |
```

1.4 Butadiene 1.2 Butadiene Units

3.3.1.2 Production of BR

[3.5, 3.77–3.88]

Today, the by far largest proportion of the globally produced BR is obtained by solution polymerization. The initiators used are primarily coordination catalysts, namely titanium [3.115–3.117], cobalt [3.118], nickel [3.119], and neodyn compounds [3.119a] or alkyl lithium compounds. While with coordination catalysts the butadiene units add linearly in excess of 92% cis-1.4 configuration or even higher (stereospecific polymerization), the alkyl lithium catalysts yield a BR with a intermediate cis-1.4 content, with a correspondingly higher 1.2 structure. From the free radical polymerization of butadiene in emulsion, a less uniform BR is obtained (E-BR), which, on account of some unfavourable rubber technological properties, enjoys only limited market acceptance. By using a $RhCl_2$ catalyst in emulsion polymerization, a predominantly trans-1.4 configuration is obtained.

Ti-BR. Originally, Phillips [3.115] used TiJ_4 and Al(isobutyl)$_3$ as catalyst, but today, because of easier handling, $AlR_3/J_2/TiCl_4$ [3.116] or $AlR_3/TiJ_3OR/TiCl_4$ [3.117] are used in continuous polymerization processes. Preferred solvents are benzene or toluene, which, like the butadiene, have to be highly purified through azeotropic distillization, and virtually free of water.

Co-BR. According to a process developed by Goodrich [3.118], $CoCl_2$ or Co-(acetylacetonate)$_2$ or Co(octenate)$_2$ in combination with diethyl aluminiumchloride, or ethylaluminum sesquichloride are used as catalysts. Benzene is a preferred solvent, and trace amounts of water can enhance the catalytic activity [3.122].

Ni-BR. A Bridgestone process [3.119, 3.123] uses initiators such as Ni(naphthenate)$_2$/BF_3, ethylene oxide/aluminium trialkyl, and aliphatic or cycloaliphatic solvents.

Nd-BR. A new Bayer development [3.119a] leads to an neodyn-BR with high cis-1.4 content and outstanding properties in tread and sidewall compounds for car and truck tires due to improved fatigue resistance, abrasion resistance and lower heat build-up compared to the conventional BR grades.

Li-BR. A process developed by Firestone [3.120] uses alkyl lithium catalysts for the production of BR. The polarity of the solvent has a great influence on the microstructure and the vinyl content. The latter can be adjusted by the use of ethers or tertiary amines [3.77, 3.125].

Alfin-Rubber. A process developed by A. A. Morton [3.126] for the production of BR with so-called alfin catalysts, is a specialty. This catalyst consists of such compounds as Na-allyl/Na-isopropyl/NaCl. The polymerization is carried out in solvents like aliphatic hydrocarbons. Contrary to the other processes, this produces BR with a high trans-1.4 content, and extremely high molecular weights, which are difficult to control through the catalyst concentration.

In the production of BR the following *parameters* are important, and these distinguish the individual grades:

- Initiator type (BR-type);
- Type of stabilizer and its concentration (difference in staining and storage stability);
- Chain modifiers, molecular weight (differences in Mooney viscosity and processibility);
- Type and quantity of extender oil used (oil extended rubber);
- Type and quantity of carbon black used (carbon black masterbatch).

3.3.1.3 Structure of BR and its Influence on Properties

Macrostructure. The macrostructure of BR grades and their molecular weight distribution, which has a particular influence on the processibility, depend in great measure on the polymerization process. While Li-BR usually exhibits, a very narrow molecular weight distribution, accompanied by considerable cold flow, this is less pronounced with Ti-BR. In order to reduce packing, transportation, and handling problems caused by cold flow, one strives for a broadening of the molecular weight distribution, using Mooney Jump techniques [3.127]. With Li-BR, these can be carried out mainly through copolymerization with divinyl benzene [3.127], or through a coupling reaction. This reaction uses, for instance, dimethyl phthalate, silicon tetrachloride, or divinylchloride [3.128], and with Ti-BR through Friedel-Crafts catalysts or inorganic acid chlorides, such as $POCl_3$, $SOCl_2$, S_2CL_2, SCl_2 [3.128–3.132]. These very high molecular weight BRs are of special use for the production of oil extended rubbers [3.133].

A greater amount of long chain branching results in reduced cold flow, longer mixing cycles, better filler distribution, higher green strength, higher extrusion rates, and greater die swell. A broadening of the molecular weight distribution gives better banding on the mill (less bagging), lower compound viscosity, shorter mixing cycles, and lower extrusion temperatures.

The average molecular weight of commercial BR grades is in the range of 250,000 to 300,000, corresponding to Mooney viscosities ML 1+4 at 100 °C of 35 to 55.

Microstructure. As mentioned before, the catalysts also have a considerable influence on the microstructure of BR (see Table 3.2), which determines to a great extent the vulcanizate properties.

The higher the cis-1.4 content of BR, the lower is its glass transition temperature T_g. Pure cis-1.4 BR grades have a T_g temperature of about −100 °C, while commercial grades with about 96% cis-1.4 content have one of below −90 °C. Pure cis-1.4 polymers have a melting point of +1 °C and do not exhibit strain crystallization at room temperature. The glass transition temperature rises linearly as the concentration of 1.2 structure (vinyl content) increases (see Figure 3.2). Such products have attracted greater interest in recent times. [3.92, 3.94–3.97 a, 3.99, 3.101, 3.124, 3.134]

Table 3.2: Microstructure of some BR-Types (in percent)

BR-Type	Ti-BR Phillips	Co-BR Goodrich	Ni-BR Bridgestone	Li-BR Firestone	Alfin-BR	$RhCl_3$ (Emulsion) Shell	Peroxide Emulsion
cis-1.4 Content	93	96	97	35	5	–	15
trans-1.4 Content	3	2	2	55	70	99.5	70
1.2 Content	4	2	1	10	25	0.5	15

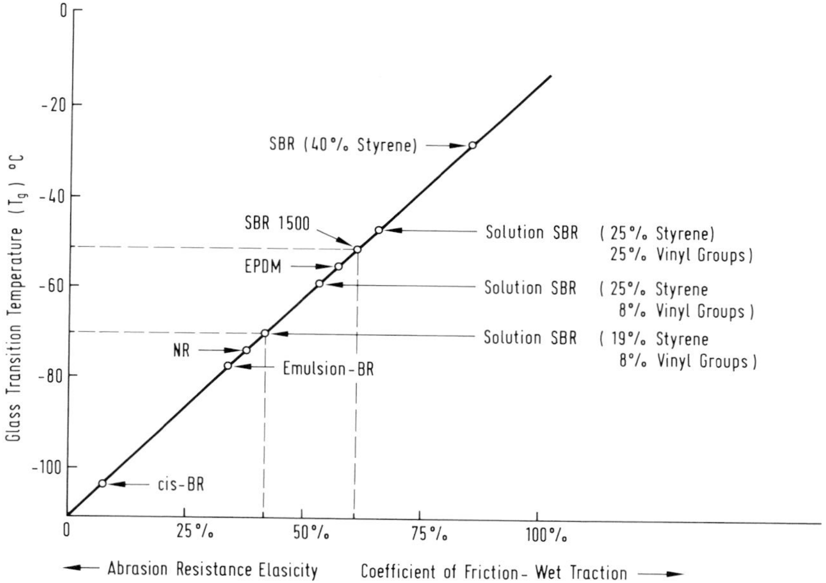

Figure 3.2 The dependence of the glass transition temperature T_g on the 1.2-content of BR, in contrast with the glass transition temperature of some other SR's [3.134]

The tendency towards crystallization is determined by the 1.2 content. Regarding low temperature brittleness, BR grades with about 35% 1.2 content (Li-BR) are comparable to blends of E-SBR/BR, while BR grades with 50 to 60% vinyl content compare with E-SBR. This comparison also extends to the vulcanizate properties. Pure cis-1.4 BRs have the best abrasion resistance, but poor wet traction. As the 1.2 content increases, the abrasion resistance becomes poorer and the wet traction improves, so that a compromise has to be found for specific applications. With the exception of ENR (epoxidized NR) [2.46, 2.49 b], the general rule applies that abrasion resistance is gained at the expense of wet traction. A BR grade with, for instance, 35% vinyl structure has in tread compound formulations the same abrasion resistance and wet traction as 50/50 blends of E-SBR and BR. The tensile strength of these BR vulcanizates is, however, lower than that of the comparable E-SBR blends, but they give less heat build-up in dynamic applications. Appropriate molecular structures can result in a reduced rolling resistance without sacrificing traction [3.87, 3.92].

Pure syndiotactic or isotactic 1.2-polybutadienes are not rubbery. They are, instead, impact resistant thermoplastics (see page 49).

3.3.1.4 Compounding of BR

[1.14, 3.102–3.107]

Blends. Because of its poor behaviour on mills and for vulcanizate property reasons, especially wet traction, BR is mostly used in blends with NR or SBR. These blends have the following advantages:

- capability of higher loadings of carbon black and oil;
- higher extrusion rates;
- higher green strength;
- better mould flow.

BR can be easily blended with all non-polar diene rubbers. The blend ratio depends on the desired result, and ranges from about 30–50% for the BR blend component.

Vulcanizing Agents. BR requires less sulphur for vulcanization than NR, and for blends, it ranges from 1.6 to 1.9 phr [3.105].

Sulphenamides are useful primary accelerators. As secondary accelerators (kickers), one frequently uses thiurams, whereby, for reasons of scorch safety, TMTM is preferred. More recently, OTOS has been used by itself, or in combination with benzothiazyl sulfenamide. These systems offer greater storage stability of mixed compounds, and improved dynamic properties, that is, a reduction in heat build-up, and flow under shear.

Protective Agents. For the choice of protective agents, the same guidelines apply as for SBR. (see page 62).

Fillers and Plasticizers or Softeners. BR compounds only reach their optimum level of properties at high filler and oil loadings, in contrast to NR and some other diene rubbers. Guidelines for choosing fillers and plasticizers for BR compounds are the same as for SBR. Furnace grades are the preferred carbon blacks. BR blends readily accept oils, and aromatic or naphthenic types are favoured choices. Aromatic oils improve the building tack of compounds and they also improve the traction rating of vulcanizates.

Process Aids. Compounding guidelines using fatty acids, resins, and processing aids are the same as for SBR (see page 63).

3.3.1.5 Processing of BR

For the mixing and processing of BR compounds one uses conventional equipment and processes as discussed for NR on page 64.

3.3.1.6 Properties of BR Vulcanizates

[1.14]

Mechanical Properties. The tensile strength of vulcanizates from BR with high cis-1.4 is considerably lower than that of comparative vulcanizates based on NR or SBR. However, in blends with NR or SBR, BR can satisfy technically demanding properties for high quality vulcanizates.

Vulcanizate properties of NR or SBR are improved in several respects through blending with cis-1.4 BR, because of its low glass transition temperature. These blends have a particularly high abrasion resistance, good low temperature flexibility, and high resilience.

Dynamic and Aging Properties. Dynamic properties, such as heat build-up, and resistance to groove cracking, are also improved for NR and SBR vulcanizates, when they are compounded as blends with BR. In addition, the reversion resistance on overcure, and the aging resistance of NR vulcanizates are improved, when they contain BR as blend component. [2.82]

Traction. By increasing the BR content in a blend with NR or SBR, the rolling resistance of tires built from the resulting vulcanizates becomes smaller. This is particularly advantageous for the fuel consumption of cars. At the same time, however, the traction, and in particular, wet traction, becomes poorer, making an optimized balance of properties necessary. On the other hand, a higher BR content in these blends (about 40%), has a favourable influence on ice traction, which is important in formulating tread compounds for winter tires.

3.3.1.7 Uses of BR

[1.14]

Tires. Over 90% of the total BR production is used in tires. In Europe, at least, the use of BR in compounds for specific tire components has changed in recent years due to the advancement of radial tires. While originally BR had been widely used in tread compounds to improve the abrasion resistance, with radial tires, this rubber has found greater use in carcass, sidewall, and bead compounds, because the tread of a radial tire has already an intrinsically higher abrasion resistance than that of bias tires. In winter tire treads, however, BR still plays an important role because of the improved ice traction.

Technical Rubber Goods. In these products, and particularly in shoe soling or conveyor belting compounds, BR is preferably used, if there is a requirement for high abrasion resistance. Compounds containing BR also have improved flow properties, and for this reason, BR is recommended for compounds which are processed by injection moulding. Successfully proven uses of BR are in compounds for shoe soles, bumpers, roll covers, conveyor belts, transmission belts, pads for chain tracks, shock absorber pads, and other products, which require a reversion resistant compound.

Liquid Reactive BR Grades. BR grades with a low molecular weight and reactive end groups (so-called telomers or reactive liquid polymers) can be converted into elastomers through a reaction with other suitable compounds in conventional methods for converting liquid polymers [3.108–3.114a].

Impact Modifiers for Plastics. Li-BR is used in considerable quantities as a modifier to improve the impact resistance of plastics.

3.3.1.8 Competitive Materials of BR

BR competes primarily against NR, IR and SBR.

3.3.2 Styrene-Butadiene Rubbers (SBR)

[1.2, 1.17-1.21, 1.58-1.66, 2.6, 2.82, 2.85, 2.111, 2.129, 3.3-3.5, 3.13-3.17, 3.135-3.176]

3.3.2.1 General Background about SBR

In 1929, E. Tchunkur and A. Bock discovered that mixtures of butadiene and styrene in a 75:25 ratio can be copolymerized in emulsion [3.152]. These E-SBR grades, called Buna S, could be more easily processed than the numbered Buna grades (bulk sodium polymerized BR), and they also gave improved vulcanizate properties. Therefore, E-SBR enjoyed a progressively stronger market acceptance than BR [1.59, 1.60].

Early on, E-SBR was produced at higher temperatures (about 50 °C) (Hot Rubbers). The first E-SBR grades made in Germany were of such high molecular weight, that they first had to be depolymerized under the action of high temperatures, to render them processible. Later use of chain modifiers enabled the producer to control the molecular weight (Buna S_3, GR-S), and the application of redox initiators permitted a lowering of the polymerization temperature to 5 °C. This resulted in more easily processible rubbers [3.153].

Large scale production of Hot E-SBR started in Germany in 1937 at Schkopau and Huels. In 1942, the United States began the construction of government owned plants to produce GR-S (Government Rubber-Styrene), which were privatized after 1954. At the same time, the general generic name of SBR (Styrene Butadiene Rubber) was introduced. Until 1948, only Hot rubbers were produced, and subsequently, also Cold rubbers, which already, by 1953, amounted to about 62% of the total SBR production. Today, Hot rubbers have become relatively unimportant. A considerable proportion of the Cold rubbers are marketed as oil extended rubbers (OE-SBR).

Since catalysts based on organo metallic compounds have become available, stereospecific SBRs have also been produced in solution processes (L-SBR), but today, by far the greatest amount of SBR is still produced in emulsion.

An additional recent development are the block copolymers, made up of butadiene and styrene blocks. At room temperature, the styrene blocks separate into discrete phases, resulting in a physically crosslinked network structure. These block copolymers are thermoplastic elastomers (see also page 150).

SBR can be considered a general purpose rubber, same as NR or IR, since it can be used in many applications and especially in tire compounds. Besides NR, it constitutes the most important rubber and 1985 the total annual global production capacity for SBR was about 6.936 million tons, or about 57.2% of the total SR production capacity [2.13 a].

The styrene content in SBRs ranges usually from about 23 to 40% and the copolymer has the following composition:

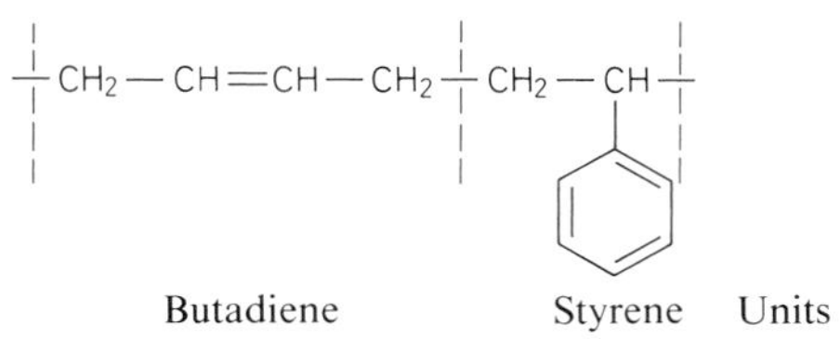

$$-[CH_2-CH=CH-CH_2]-[CH_2-CH(C_6H_5)]-$$

Butadiene Styrene Units

3.3.2.2 Production of SBR

[3.5, 3.135–3.142]

The largest share by far of SBR is produced in emulsion using redox initiators, but solution SBR is steadily gaining in importance. In emulsion polymerizations, latices are first obtained, which are either used as such, or which are further processed to yield solid rubber.

E-SBR (Cold Rubber). The preferred emulsifiers for the dispersion of monomers during polymerization are of an anionic nature, and are, for instance, mixtures of sodium salts from fatty and rosin acids [3.154]. The ratio of butadiene to styrene is mostly 76.5 to 23.5% by weight.

The free radicals required for the initiation step are obtained by the reaction of iron (II) salts with p-menthane hydroperoxide or pinane hydroperoxide.

As redox activators, one uses chelating agents, such as the sodium salt of ethylenediaminetetraacetic acid, together with sodium formaldehyde sulfoxylate [3.155]. The pH of the reaction medium is usually adjusted to about 11 to 12.

Tert.-dodecyl mercaptan is most frequently used as a chain modifier [3.156], and since the polymerization is a chain reaction, it only terminates itself when all the monomer has reacted. At monomer conversion rates above 70%, chain branching and gel formation occurs, despite the presence of chain modifiers. This adversely affects the processibility of rubbers. Therefore, polymerization reactions are usually terminated at monomer conversion rates of about 60% through the addition of sodium dimethyldithiocarbamate, dialkylhydroxy amine [3.157], or the sodium salt of dithionic acid. Prior to coagulation, stabilizers are added to the latex, and depending on the desired qualities of the rubber, the stabilizers are either non-staining (preferably a phenolic type), or staining (amines). After recovery of unreacted monomer, and after coagulation, the rubber crumb is washed and dried.

OE-E-SBR, Carbon Black Masterbatch. With some SBR grades, extender oils (OE-E-SBR) and carbon black (Black Masterbatch) are added already at the end of the polymerization stage. For oil extension, the rubber is usually produced to significantly higher molecular weights [3.158–3.162]. The oil, which is added to the latex, and which is co-precipitated with the rubber from the latex, acts as plasticizer during processing. Similarly, after dispersion in the latex, and after co-precipitation, it is possible to incorporate carbon black finely dispersed in the rubber [3.163].

E-SBR (Hot rubber). The polymerization processes for Hot and Cold rubbers are very similar. For Hot rubbers, one uses soaps from fatty acids, or at times, also alkylaryl sulfonates. The polymerization temperature is normally about 50 °C, or, sometimes even higher. At these temperatures, free radicals are created from potassium persulfate and mercaptan, to initiate the polymerization.

For the production of SBRs, the following *parameters* are important, and they differentiate the individual grades:

- Monomer ratio (mostly 23.5% styrene, and, in some instances, also 40%);
- Polymerization temperature (Cold and Hot rubbers);
- Chain modifiers (differences in Mooney viscosity, processibility);
- Emulsifier (differences in building tack);
- Stabilizer (differences in staining and storage stability);
- Coagulant (differences in electrical properties);
- Oil, type and amount (oil extended rubbers);
- Carbon black, type and amount (carbon black masterbatches).

L-SBR. Analogous to the polymerization of BR, it is also possible to copolymerize butadiene and styrene in aliphatic or aromatic hydrocarbon solvents, using, for instance, alkyl lithium catalysts [3.164]. The properties of the resulting products depend on the specific polymerization process that was used. Due to their different reactivities, the butadiene monomer polymerizes first, followed by the styrene, so that blocks or long segments of these monomers are found in the copolymer [3.165]. In the presence of small amounts of ethers or tertiary amines, the reactivities of the two monomers become more similar, so that a statistically random copolymerization occurs [3.166]. Although the overall composition of block copolymers and random copolymers can be the same, their physical properties will differ significantly [3.167].

A sequential copolymerization of styrene and butadiene in the presence of alkyl lithiums, and in aromatic or aliphatic solvents produces block copolymers [3.168]. At the end of the polymerization process, the "living" polymer has to be deactivated, unless one wishes to produce a three-block polymer by charging additional styrene monomer [3.169]. These triblock polymers of the SBS type have attracted special interest, being thermoplastic elastomers (see page 150).

3.3.2.3 Structure of SBR and its Influence on Properties

[3.5, 3.143, 3.144]

E-SBR (Cold Rubber). With E-SBR broadens the molecular weight distribution to a certain extent as the average molecular weight rises. However, in spite of the widening of the molecular weight distribution, and attending long chain branching, the processibility does not always improve, since E-SBR frequently contains fractions with a very high molecular weight. The latter tends to form gel. On the other hand, this can be alleviated to some extent with a gradual improvement in processibility, if the chain modifier is added step-wise during the polymerization process (oligo increment polymerization process).

E-SBR is commercially available in Mooney viscosities ranging as ML 1+4 (100 °C) from 30 to about 120, which correspond to average molecular weights of about 250,000 to 800,000. Furthermore, E-SBR is supplied as gum, or extended with oil or carbon black, and can usually be processed directly without the need for prior mastication.

Since SBR, like most of the other SRs, cannot be readily masticated, the viscosity of the available material is of special importance regarding its processibility. Lower viscosity grades band more easily on mills, they incorporate fillers and oil more readily, show less heat generation during mixing, are more easily calendered, shrink less, and often give a higher extrusion rate and a superior appearance of the extrudate, than the higher viscosity grades. On the other hand, the higher viscosity SBR's have better green strength, tend to give less porosity in the vulcanizate, and also accept higher filler and oil loadings, which is attractive price-wise. The higher the molecular weight of the SBR, the higher the resilience of its vulcanizate, and there is generally also an improvement in the mechanical properties, particularly tensile strength and compression set. However, the differences in vulcanizate properties of high and low molecular weight SBRs can be largely eliminated through the choice of active fillers in the rubber compound.

E-SBRs produced at low polymerization temperatures, the Cold rubbers, have less chain branching than the Hot rubbers. Consequently, the polymerization temperature has also an influence on the processibility, as does the viscosity. At an equivalent viscosity, the Cold rubbers can be more easily processed than the Hot rubbers, and this applies particularly to a better banding on mills, less shrinkage after calendering, and a superior surface of green tire compounds. The reduced tendency of the Cold rubbers towards cyclization is particularly advantageous when higher mixing temperatures are used in internal mixers. On the other hand, the Hot rubbers give better green strength, because they have more chain branching.

In Cold SBRs, the butadiene component has, on average, about 9% cis-1.4, 54.5% trans-1.4, and 13% of 1.2 structure. At a 23.5% bound styrene level, the glass transition temperature, T_g, of SBR is about $-50\,°C$. As the styrene content in the SBR increases, the glass transition temperature becomes higher, and the resilience becomes less. However, this also brings about an improvement in processibility (extrusion rate, green strength, surface smoothness). At very high styrene contents, that is, with styrene resins, there is nearly a complete loss of elastomeric properties, but these compositions are still very important as processing aids in rubber compounds (see page 293).

Rubbers with very low T_g values, as, for instance, cis-1.4 BR, are characterized by a high resilience and very good abrasion resistance, but have poor wet traction. By contrast, those rubbers with high styrene content and correspondingly high T_g, as, for instance, SBR 1516 or SBR 1721, exhibit a low resilience and poor abrasion resistance with an excellent wet traction [3.143]. Depending on the quantity of styrene in the butadiene chains, it is, however, possible to adjust property levels anywhere between these two extremes [3.174].

The stabilizer that was added at the end of the polymerization process determines the storage stability of the raw rubber, its tendency to cyclize at higher temperatures, and the discolourization after exposure to light. Generally, the more effective stabilizers also tend to cause greater discolourization, and therefore they cannot be used for certain applications.

The emulsifier from the polymerization process which remains in the rubber can also have an influence on the processibility. Emulsifiers derived from rosin acids improve the building tack (see page 302), but their disadvantage is the tendency to discolourization. For this reason, E-SBRs are frequently produced using a blend of emulsifiers from fatty and rosin acids.

L-SBR. By choosing appropriate polymerization systems and conditions, it is possible to custom-make solution SBRs (L-SBRs) with a wide range of specific properties [3.77, 3.94–3.97, 3.124, 3.164–3.173]. In this respect there is a great similarity to solution BR's, as discussed in Section 3.3.1.3, (see page 54).

L-SBR with Higher 1.2 Content. In Section 3.3.1.3, it was already mentioned that the 1.2 butadiene structure, which has the vinyl units in the side group, has largely the same effect on the polymer properties as the styrene [1.143, 3.175, 3.176]. For instance, a blend of E-SBR/BR in the ratio of 60:40 has a T_g temperature of $-70\,°C$. A similar T_g value can also be obtained by the copolymerization of 20% styrene with a butadiene containing 10% vinyl units in the polymer, or from a styrene-free polybutadiene with 35% vinyl units (see Figure 3.2). Thus, it is evident that through the choice of concentrations of styrene and 1.2 butadiene units the property spectrum can be tailor-made to suit the specific requirements of, for instance, tire compounds [3.95–3.97, 3.124]. The width of the molecular weight distribution can be adjusted through the branching reactions [3.62].

Random L-SBR. The L-SBR grades with a random distribution of styrene units are very similar to E-SBR, except that a high proportion of the butadiene units are of a cis-1.4 structure. Furthermore, the random L-SBRs have a narrower molecular weight distribution and less long chain branching than the E-SBRs. Therefore, L-SBR is more difficult to process, but its vulcanizates have a better abrasion resistance and less heat build-up under dynamic conditions than E-SBR [3.164].

Segmented L-SBR. As a result of blockiness of the monomer units in the chain, these polymers exhibit two distinct regions of glass transition, namely at −85 °C for the polybutadiene block and +75 °C for the polystyrene block. Because of the thermoplastic behaviour of the polystyrene blocks, these rubbers process well, and they can replace blends of E-SBR and styrene resins in certain applications. Vulcanizates from segmented L-SBR are hard and of good abrasion resistance [3.167].

Tri-Block L-SBR. Polymers of the SBS-type are already elastomers without vulcanization, because of the phase separation of the terminal styrene blocks. These discrete styrene phases have the same effect as crosslinks (physical crosslinks), and the butadiene blocks constitute the elastomeric connecting elements. On heating, the polystyrene phases soften, thus rendering the rubber processible, and on cooling, the rubber receives its elasticity again. Thus, one is dealing here with a thermoplastic elastomer [3.169]. The range of softening temperatures of the polystyrene blocks is around 75 °C (see page 150).

```
                  B—B—B—B
                 /       \
—S—S—S—S                  S—S—S—S—
—S—S—S—S                  S—S—S—S—
                 \       /
                  B—B—B—B
```

3.3.2.4 Compounding of SBR

[3.145–3.151]

Blends. Being of low polarity, SBR can practically be blended with all non-polar rubbers over the whole range of blend concentrations. Blends with BR or NR are of great importance in tire applications. In this example, the BR improves, for instance, the abrasion resistance and the hysteresis of the vulcanizates. Blends of SBR with polar rubbers, such as NBR, are restricted to those NBR grades with a low acrylonitrile content (see page 71).

Vulcanizing Agents. In contrast to NR, SBR requires less sulphur and a greater amount of accelerator, and the same concentration of ZnO. SBR vulcanizates of high hardness can be produced by using high sulphur concentrations in the compound. Besides ordinary elemental sulphur, insoluble sulphur is also used, and this reduces the tendency towards sulphur bloom of the mixed compound. Sulphur donors play an important role in semi-EV and EV cure systems.

The most widely used accelerators are sulfenamides and MBTS, because they offer good processing safety. These accelerators can be activated in the presence of OTOS, dithiocarbamates, thiurams, or guanidines. Conventional, but also semi-EV and EV systems, are used. More recently, urethane crosslinking agents have also been suggested [2.111].

Prevulcanization inhibitors are used where there is danger of compound scorch. In many instances, it is also possible to retard the cure by using an appropriate combination of accelerators and acidic compounding ingredients. Where this is not possible, special retarders are added to the compound, and examples are n-(cyclohexylthio)-phthalimides. However, in some instances, these retarders not only delay the onset, but also the time to completion of the vulcanization.

Protective Agents. Effective stabilizers are already added to SBR during the production stage, so that the rubber has an excellent storage stability, and the vulcanizate some protection against oxidative degradation. It is, however, necessary to add for many applications additional stabilizers resp. antioxidants to the rubber compound. Stabilizers are especially required for those uses, where the product is exposed to high temperatures or dynamic applications. To improve the oxidative stability, one uses the conventional antioxidants based on p-phenylene diamine or other aromatic amines. The dynamic fatigue resistance of vulcanizates is also improved by p-phenylene diamine, and particularly by IPPD. Special aging resistance at high temperatures is achieved by TMQ or ODPA, and if necessary, also a combination with MBI. MBI by itself gives good resistance to aging in steam. When using amine based protective agents, one has, however, to take into account a strong discolourization of the vulcanizate. For white or light coloured products from SBR, one uses as non-staining antioxidants those compounds which are based on bisphenol, phenol, or MBI. Light coloured SBR vulcanizates are effectively protected against ozone degradation by non-staining enol ethers in combination with microcrystalline waxes (see also page 273).

Fillers. Gum vulcanizates of SBR, or those which contain non-reinforcing fillers only, have a much lower tensile strength and resistance to tear propagation than similar vulcanizates based on NR or CR. However, reinforcing carbon blacks and reinforcing white fillers give the same level of mechanical properties for SBR as for NR or CR vulcanizates. The relative order of activity of fillers is the same in SBR as in NR, but the degree of reinforcement of the fillers is higher in SBR than in NR (see also page 24). Non-reinforcing carbon blacks and inert white fillers are added to SBR compounds, same as with other compounds, to reduce compound cost, and as hardness builders. This generally requires high loadings, and it also improves the processibility of compounds. The use of carbon black masterbatches obviates in many instances the mixing step for incorporating the filler, and this can save plant mixing capacity.

From SBR grades with hydrophilic functional groups and the use of silica, it is possible to produce vulcanizates with a high abrasion resistance and an improved resistance to tear propagation [3.148].

Softeners. Since SBR is not masticated, the softeners assume an important role for adjusting the required compound viscosity. A wide range of compounds can serve as softeners for SBR, the same as with NR. The most important softeners are mineral oils, ranging from paraffinic to aromatic grades. Animal and vegetable oils are also important plasticizers or processing aids. Oil loadings are generally considerably higher in SBR than in NR compounds. Synthetic plasticizers which are used in polar rubbers, such as NBR, CR, etc., are of little use in SBR compounds.

Factices. These materials serve to improve the processibility and green strength of SBR compounds. Factices used in SBR compounds are similar to those used in NR compounds [2.119].

Resins. It is more important to add resins to SBR than to NR compounds in order to achieve good building tack. SBR grades which have been polymerized using rosin acid emulsifiers have already a better tack than those grades from fatty acid emulsifiers, and they require therefore, less resin in the compound. Resins used are xylene formaldehyde compounds, Koresin, rosin, pitch and tar, to name just a few.

Process Aids. These materials are important additives to a compound to achieve a good performance during mixing and further processing, and particularly, to reduce sticking to mill rolls, and to improve the dispersion of fillers. Representative materials are stearic acid, calcium and zinc soaps, residues from fatty alcohols, pentaerythritol tetrastearate etc.

3.3.2.5 Processing of SBR

The preparation and processing of SBR compounds is carried out using conventional rubber machinery and procedures as already discussed for NR on page 26.

3.3.2.6 Properties of SBR Vulcanizates

3.3.2.6.1 Properties of E-SBR Vulcanizates

Mechanical Properties. The tensile properties of E-SBR vulcanizates depend in great measure on the type and amount of filler in the compound. Gum vulcanizates have only poor tensile properties since the rubber lacks self reinforcing qualities, and they are therefore of little technical interest. Instead, reinforcing fillers are usually required in SBR compounds. At optimum loadings with reinforcing carbon blacks, it is possible to reach virtually the same level of excellent tensile properties for SBR as for NR vulcanizates, but the tear resistance of the SBR vulcanizates is still inferior. The elastic properties of E-SBR vulcanizates are also poorer than comparable ones of NR.

The compression set of the vulcanizate, which is important for many applications, depends in great measure on the compound formulation, the cure conditions, and the specific test method. Through proper compounding and at optimum cure, it is, however, possible to obtain very low values of compression set for E-SBR vulcanizates.

Dynamic Properties, Aging Resistance, and Abrasion Resistance. Particularly advantageous are the dynamic fatigue resistance, the aging resistance, and the heat resistance of E-SBR vulcanizates, and in this respect they surpass NR vulcanizates by far. Without antiozonants, the SBR vulcanizates are, however, not resistant to weathering and ozone degradation like NR (see page 28).

In order to achieve an optimum heat and aging resistance, high quality protective agents are used in E-SBR compounds, same as in NR compounds. Sulphur free and low sulphur vulcanization systems (EV and semi-EV systems) play as well an important role, and this applies also to NR. The heat resistance of optimized E-SBR vulcanizates allows service temperatures which are about 20 °C higher than those for NR vulcanizates. E-SBR vulcanizates are reversion resistant.

Regarding abrasion resistance, E-SBR vulcanizates formulated with reinforcing fillers, give a superior performance than comparable ones from NR, and in passenger tire treads this superiority amounts to about 15%.

Since, on account of their good abrasion and aging resistance, the E-SBR vulcanizates are so durable, they have replaced NR in many applications.

The dynamic properties of E-SBR are inferior to those of NR, and, therefore, E-SBR vulcanizates give a higher heat buildup in dynamic applications. However, since they have a better heat resistance, E-SBR vulcanizates can also tolerate higher service temperatures without adverse effects (e.g. higher equilibrium running temperatures in passenger tire treads). Nevertheless, in thick walled rubber products, such as in heavy truck tires, high speed tires or other extreme dynamically stressed rubber parts, heat can accumulate because of the poor thermal conductivity of rubber, and this can lead to temperatures in excess of those tolerable by E-SBR vulcanizates.

For these applications, blends with BR, or a vulcanizate with lower heat build-up based on NR or IR in combination with BR, are better suited.

Electrical Properties. E-SBR grades, being non-polar, and their vulcanizates, are poor conductors of electricity, and in this respect they resemble NR. Aged E-SBR vulcanizates have, however, a lower electrical conductivity than aged ones from NR. Yet, the electrical properties of E-SBR depend to a great extent on the production process, namely residual emulsifier and electrolyte content.

Resistance to Fluids. While E-SBR vulcanizates are resistant to many non-polar solvents, dilute acids and bases, they will swell considerably when in contact with gasoline, oils, or fats. Although the swelling is less than for NR, SBR cannot be used in applications which require a resistance to swelling on exposure to the latter fluids.

3.3.2.6.2 Properties of L-SBR Vulcanizates

Vulcanizates from L-SBR with a Random Distribution of Styrene. These compounds have a lower hysteresis, less heat build-up, and a superior abrasion resistance than the ones from E-SBR. Furthermore, L-SBRs are purer than E-SBRs, and since they do not contain residual emulsifier, have a lower water absorption and particularly good electrical properties. L-SBRs are also odour free and light coloured [3.164].

Vulcanizates from L-SBR with a Blocky Monomer Distribution have, in addition to very low brittleness temperatures, good elastic properties, low water absorption, and particularly low electrical conductivity. They are also abrasion resistant and of higher hardness than normal SBR vulcanizates.

L-SBR with SBS Triblocks. The block copolymers can be processed in injection moulding and extrusion processes same as conventional thermoplastics. At room temperature, they have a relatively high elasticity and tensile strength, and resemble rubber vulcanizates. Since the crosslinks in the SBS polymers are of a physical nature, and due to the separation of styrene into discrete phases, these polymers can only be used in products which require only a low heat resistance. Above temperatures of 65 to 70 °C, the SBS polymers lose their elasticity and strength. Their resistance to organic solvents is also poor [3.169].

3.3.2.6.3 Properties of Vulcanizates from OE-SBR and Black Masterbatches

OE-SBR (Oil Extended SBR). The base polymer of these SBR grades has a very high molecular weight, and through the addition of extender oil (25 to 30%), the processing viscosity is adjusted to that of conventional SBR grades. Thus, there is little difference in the technological properties. OE-SBR grades are produced from E-SBR or L-SBR, and since the oil extension offers economic advantages, OE-SBRs have become an important class of rubbers. In tire applications, OE-SBRs give particularly good traction and low tire noise.

Carbon Black Masterbatches. The resulting vulcanizates resemble those from conventional SBR's with equivalent filler loadings.

3.3.2.7 Uses of SBR

E-SBR. Mostly in combination with BR, E-SBR is predominantly used for the production of car and light truck tires. For heavy truck and high speed tires, E-SBR is practically not used at all, because of the higher dynamic heat build-up, in comparison with NR, IR, or BR. Other applications for E-SBR are belting, moulded rubber goods, shoe soling, cable insulation and jacketing, hose, roll coverings, pharmaceutical, surgical, and sanitary products, food packaging, etc.

Table 3.3 lists some specific applications for the main E-SBR grades.

Table 3.3: Some Fields of Application of E-SBR

E-SBR Grade	Emulsifier	ML 1+4 (100 °C)	Colour	Oil		Carbon Black		Uses
				Grade	PHR	Grade	PHR	
1500	R	50–52	S	–	–	–	–	General purpose for pass. tire treads and technical rubber goods.
1502	F	50–52	NS	–	–	–	–	Light coloured technical rubber goods, for blends requiring good flow properties in injection moulding or calandering.
1507	F	30–35	NS	–	–	–	–	
1509	F	30–35	NS	–	–	–	–	Because of low ash and attending low water absorption used for products in the cable and electrical industry.
1516	F	40	NS	–	–	–	–	Because of high styrene useful for injection and compression mouldings with good surface finish.
1573	R	115	NS	–	–	–	–	Break and transmission pads, belting, adhesives.
1707	R	49–55	NS	NAPH	37,5	–	–	Light coloured and transparent hose, profiles, soling, floor tiles.
1712	F	49–55	S	HAR	37,5	–	–	Pass. tire treads, belting, dark coloured rubber products.
1778	F	49–55	NS	NAPH	37,5	–	–	Same as 1707, also cable insulation.
1609	R	61–68	S	HAR	5,0	N 110	40	Particularly used for tire retreads.
1618	F	70	NS	NAPH	5,0	N 550	50	
1808	F	48–58	S	HAR	47,5	N 330	76	Pass. tire treads, camel back, electrical products.
1843	F	86	NS	NAPH	15,0	N 770	100	Dynamic applications such as V-belts.

R = resin acid
F = fatty acid blend
S = staining
NS = non-staining
NAPH = naphthenic
HAR = highly aromatic oil

Random L-SBR is, for instance, used in blends with E-SBR to improve the extrudability, and in particular, the sharpness of edges and the surface smoothness of extrudates.

Blocky L-SBR. The preferred use is in hard shoe soling, roll coverings, and special technical rubber products.

Thermoplastic SBR is used in instances where products requiring only low heat resistance have to be produced in great quantity, in roofing applications, shoe applications, and for nipples of baby bottles.

3.3.2.8 Competitive Materials of SBR

SBR competes primarily against NR, IR, and BR.

3.3.3 Acrylonitrile Butadiene Rubber, Nitrile Rubber (NBR)

[1.2, 1.3, 1.8, 1.9, 1.14, 1.17, 1.18, 1.21, 1.59, 1.60, 2.82, 2.108, 2.109, 3.5, 3.8, 3.13–3.15, 3.177–3.285]

3.3.3.1 General Background about NBR

The first copolymerization of acrylonitrile and butadiene was carried out in 1930 by *E. Konrad* and *E. Tschunkur* [3.251]. *P. Stoecklin* recognized the special technical advantages of NBR vulcanizates, which, unlike those from NR or SBR, do not swell when in contact with gasoline, oils, or fats. Full scale production of NBR commenced in 1934 in Leverkusen, in 1939 in the U.S.A. (Goodrich), and thereafter in many other countries. The majority of NBR is a Cold rubber, although some Hot NBR is still being produced.

A great number of NBR grades are produced, their acrylonitrile content ranging from 18 to 51% by weight. The structural formula of NBR is as follows:

$$\left[-CH_2-CH{=}CH-CH_2-\right]\left[-CH_2-\underset{\underset{C\equiv N}{|}}{CH}-\right]$$

Butadiene Acrylonitrile Units

In 1985, the global production capacity for NBR was about 0.508 million tons, which represents roughly 4.1% of the total SR capacity [2.13 a].

In the last few years, the use of NBR has been threatened to some extent by a demand for a steadily increased heat resistance, and an attending substitution by other polymers. Therefore, there has been a great deal of development work to improve the heat resistance of NBR, and this is also reflected in the considerable amount of literature on this subject.

3.3.3.2 Production of NBR

[1.19, 3.14, 3.184–3.186]

There is a great similarity of the emulsion polymerization processes of butadiene with acrylonitrile and of butadiene with styrene (see page 44 and 59), and of the corresponding polymerization recipes [3.180, 3.181]. However, in the production of NBR, one has to take into account especially the different reactivity ratios of butadiene and acrylonitrile [3.182]. At 38% acrylonitrile and 62% by weight of butadiene, there exists an azeotropic blend of the two monomers at 25 °C, but the ratio changes somewhat depending on the polymerization temperature. For an azeotropic mixture, the relative concentrations of the components remain the same in the monomer phase of the emulsion and in the polymer throughout the whole polymerization process. If the acrylonitrile concentration in the monomer blend is less than the azeotropic concentration, the polymer will be rich in acrylonitrile at the start of the polymerization reaction, and similarly, if the acrylonitrile concentration in the monomer blend is above the azeotropic concentration, the polymer will have an acrylonitrile content below that of the relative acrylonitrile/butadiene ratio of the monomer charge. The difference between the concentration ratio of the monomers in the polymerization charge and in the polymer becomes greater, the further one deviates from the azeotropic concentration ratio for butadiene and acrylonitrile. In order to counteract this non-uniformity, the acrylonitrile is often added stepwise during the polymerization reaction (oligo increment production), when low nitrile NBR grades are produced (e.g. with 18 to 23% acrylonitrile). Another difference between acrylonitrile and styrene is the greater water solubility of acrylonitrile, which, of course, influences the NBR emulsion polymerization.

The choice of stabilizers is important if NBR vulcanizates with a good heat resistance are desired [3.182, 3.253].

3.3.3.3 Structure of NBR and its Influence on Properties

[3.187–3.189]

As with SBR, many *parameters* come into play in the production of NBR, which results in a great variety of commercially available grades.

The important parameters are:

- Acrylonitrile content (about 18 to 51%, influencing the oil and gasoline resistance as well as the low temperature flexibility);
- Polymerization temperature (Cold and Hot rubbers);
- Chain modifier (differences in Mooney viscosity, processibility);
- Stabilizer (differences in colour and storage stability);
- Gel (processibility);
- Incorporation of reactive groups (ability to crosslink without sulphur and accelerator, preferably used for latices);
- Addition of softener;
- Blends with PVC.

Viscosity. The viscosity of the rubber has an influence on the processibility, and the comments for SBR apply here also (see page 60). There is no difference in the solvent swell resistance and low temperature flexibility properties of vulcanizates from NBR's with different viscosities, as long as the nitrile content of the rubbers is the same.

NBR grades with extremely low viscosities (liquid NBRs) can be used as compatible, non-volatile plasticizers in blends with other NBRs. During vulcanization, these plasticizers are then partially chemically bound to the rubber network, and therefore cannot be easily extracted.

Polymerization Temperature. NBR grades produced at low polymerization temperatures (Cold rubbers) show less chain branching than the Hot NBR grades. Same as the viscosity, the polymerization temperature has therefore an influence on the processibility of NBR's, and the same comments as for SBR apply here (see page 60).

Microstructure. The polymerization temperature not only has an influence on the long chain branching, but also on the monomer sequence distribution, and the cis-1.4, trans-1.4, and 1.2 microstructure of the butadiene. For instance, a NBR polymerized at 28 °C with 36% by weight of acrylonitrile in the polymer, has a butadiene component with a randomly distributed microstructure of 12.4% cis-1.4, 77.6% trans-1.4, and 10% 1.2 addition. Due to the lack of compositional uniformity along polymer chains, it is not possible for NBR, nor for SBR, to form crystallites on extension. This lack of strain crystallization (lack of self reinforcement) results in relatively poor tensile properties of NBR gum vulcanizates.

Precrosslinking. NBR grades, which have been precrosslinked by adding a small amount of divinyl benzene during the polymerization, can be blended with normal NBR grades, to improve their processing behaviour, in addition to lowering the compression set and solvent swell of the final vulcanizates. However, the blending with precrosslinked NBRs also lowers the tensile strength, maximum elongation, and the tear strength of conventional NBR vulcanizates.

Acrylonitrile Content. The glass transition temperatures of polyacrylonitrile at +90 °C and of polybutadiene at −90 °C differ considerably, and therefore with an increasing amount of acrylonitrile in the polymer, the T_g temperature of NBR rises, together with its brittleness temperature (see Figure 3.3). The elastic behaviour of NBR vulcanizates also becomes poorer as the concentration of bound acrylonitrile in the NBR increases, but at the same time the copolymer becomes more thermoplastic, which is advantageous regarding the processibility of compounds.

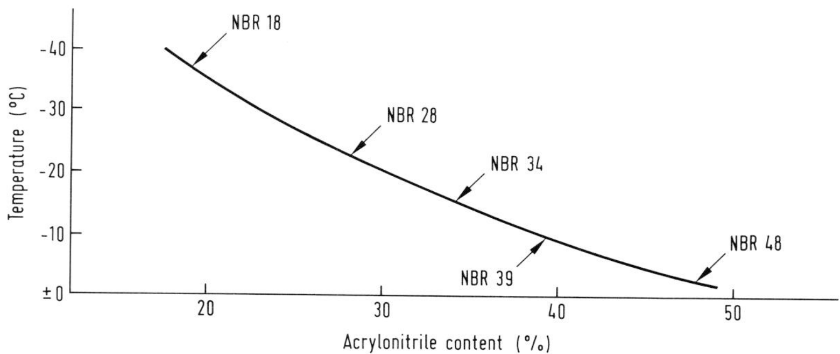

Figure 3.3 The influence of the acrylonitrile content on the glass transition temperature of NBR [3.254]

The polarities of acrylonitrile and butadiene are very different, and the polarity of the copolymer becomes greater with increasing amounts of acrylonitrile. (The solubility parameter of the acrylonitrile unit in the polymer is $\delta = 12.8$, and that of the butadiene unit is 8.4) [3.182, 3.190–3.194, 3.253].

The concentration of the acrylonitrile in the copolymer has therefore a considerable influence on the swell resistance of the vulcanizates in non-polar solvents. The greater the acrylonitrile content, the less the swell in motor fuels, oils, fats, etc. However, the elasticity and low temperature flexibility also become poorer. For the same reasons, the compatibility with polar plasticizers (e.g. those based on esters or ethers) and polar plastics (e.g. PVC or phenolics) improves with increasing acrylonitrile concentrations in the NBR. The solubility of the polymers in non-polar solvents, and the gas permeability of the vulcanizates become lower as the acrylonitrile concentration in the NBR becomes greater.

Acrylic Acid Terpolymers (X-NBR). NBR grades, which are terpolymers of butadiene, acrylonitrile, and a diene monomer with carboxylic acid groups (e.g. one based on acrylic acid), are particularly reactive due to the potential crosslinking site of the carboxylic acid sidegroup [3.180–3.182, 3.229–3.233, 3.256–3.262]. The sidegroup can react with multifunctional reagents in a heteropolar reaction to crosslink the polymer chains. The most important reagents of this type are metal oxides, such as ZnO. Therefore, during the vulcanization, using conventional reagents, such as sulphur, accelerators, and zinc oxide, not only sulphur crosslinks are formed, but also zinc salt (ionomer) bridges [3.229]. For this reason, it is important to choose a suitable grade of zinc oxide for the crosslinking of X-NBR. For instance, it is advantageous to specify a medium BET value of about 5 to 7, and a surface treated grade, and the use of a pre-masterbatched zinc oxide is also recommended. Since the processing safety of X-NBR compounds with zinc oxide and conventional MBTS-based cure systems is poor, one frequently uses zinc peroxide as masterbatch instead of zinc oxide, and this results in greater scorch safety without impairing the other cure properties [3.233]. Since the carboxylic sidegroups also have a hydrophilic character, there is a greater affinity for non-black fillers, and hence these fillers give a better reinforcement in X-NBR than in NBR vulcanizates [3.230]. Thus, soft clays are already effective fillers in X-NBR compounds. Otherwise, some multifunctional reagents, such as di- or oligoamines or isocyanates should not be used, unless these reagents are capped [3.180–3.182].

NBR grades with reactive acrylate groups are also produced in graft polymerizations, and these grades can be particularly easily crosslinked with peroxides.

X-NBR vulcanizates, where the carboxylic groups have also been used as crosslinking sites, show a considerably better solvent swell resistance and abrasion resistance, but they give a higher hardness than comparable non-carboxylated NBR grades. While the dynamic properties of the X-NBR grades are also very good, their resistance to heat and compression set is poorer [3.231].

Stabilizers. The stabilizers added during polymerization give NBR a good storage stability and they protect it against cyclization reactions when subjected to high temperatures during mixing and further processing. Staining amine stabilizers generally give better protection than the non-staining phenolic types. As with SBR, these stabilizers do not normally protect the vulcanizates sufficiently against long-term oxidative hardening, and dynamic or thermal degradation. Besides NBR grades that contain the usual stabilizers, those with chemically bound antioxidants play an increasingly important role [3.182, 3.225–3.228].

Thermoplastic NBR [3.182, 3.234–3.235 a]. With a greater use of thermoplastic elastomers (see page 157), there have been many attempts to develop thermoplastic NBRs by means of different methods. Several investigators have attempted to achieve thermoplasticity through thermal or mechanically reversible crosslink structures [3.182]. This, however, did not lead to commercial products. Another method is the blending of fully compounded NBR with polypropylene or polyamides, polystyrene, [3.234], or with PVC [3.235, 3.235 a]. Thermoplastic NBRs, which recently came on the market, have a better heat resistance than thermoplastic SBRs.

3.3.3.4 Compounding of NBR

[1.14, 3.180–3.182]

There are many similarities in the rules governing the make up of NBR, NR, and SBR compounds, but there are also a number of deviations which are characteristic of NBR compounds and which are due to the higher polarity of NBR than that of NR or SBR. This higher polarity determines the compatibility of NBR with certain compounding ingredients.

Blends. There is only a limited blend compatibility of NBR and non-polar rubbers, such as NR or BR. Yet, sometimes small amounts of NR (to improve building tack) or BR (to lower the brittleness temperature) are used as blend partners. NBR, and in particular those grades with a low acrylonitrile content, can be blended with SBR over the full range of concentrations, without a significant deterioration of mechanical vulcanizate properties. These blends are frequently used for economic reasons, and in applications, where there is only a moderate demand on solvent swell resistance. There is a possibility to improve the ozone resistance and weatherability of NBR vulcanizates through blends with EPDM [3.182], ETER [3.191], or BIIR [3.201]. Blends of NBR and ETER can at the same time be compounded for an improved swell resistance, lower temperature flexibility, and higher heat resistance [3.191]. Particularly well established are blends of NBR and PVC, which are compatible in all blend ratios as long as the NBR grades have a sufficiently high acrylonitrile content (the compatibility limit is about 25% acrylonitrile). These blends have better ozone resistance, improved swell resistance, better tensile and tear strength, but lower elasticity, poorer low temperature flexibility, and higher compression set than conventional NBR vulcanizates [3.180–3.182, 3.191, 3.236–3.240]. Phenolics are also compatible with NBR. In gum and filled NBR compounds and vulcanizates, the phenolics act as hardeners and reinforcing agents, and the higher the acrylonitrile content of the NBR, the higher are these effects [3.180–3.182]. Phenolic resins also impart an extremely high abrasion and swell resistance, high hardness and high tensile strength to NBR vulcanizates, but they reduce the elasticity and compression set resistance. Epoxy resins also reinforce NBR vulcanizates [3.263].

Vulcanizing Agents. [1.12, 1.13, 3.264]. Virtually the same guidelines apply to NBR as to SBR (see page 62). However, by comparison, besides sulphur, which is less soluble in NBR than in SBR, sulphur donors, and particularly semi-EV and EV systems, play a special role in NBR vulcanizates to obtain a high heat resistance and a low compression set. The conventionally used accelerators – MBTS, sulfenamides, thiurams, dithiocarbamates, and guanidines – have given the best results. A particularly good heat resistance is obtained if one uses, for instance, TMTD without or with little (0.25 phr) sulphur or a sulphur donor (0.5 phr). Cadmium oxide, which is, however, suspect because of its toxicity, gives vulcanizates with a high heat resistance [3.213, 3.214].

More recently, the use of OTOS, or that of a new optimized accelerator blend, coded DEOVULC EG 28, has been proposed in place of CdO-based cure systems [2.108, 2.109]. These new cure systems give a particularly low compression set at high temperatures and a high heat resistance.

X-NBR needs multifunctional reagents as curing agents such as ZnO of suitable grades or Zn peroxides. Sulphur and accelerators are also suitable curing agents.

Besides sulphur-based vulcanizations, those using organic peroxides are important, particularly for producing high heat resistant vulcanizates (see page 256 ff). With the additional use of coactivators for peroxides, namely sulphur or monomeric acrylates, such as ethylenediamine dimethacrylate (EDMA), vulcanizates can be obtained which have a high hardness, and which still have a sufficiently low processing viscosity [3.182]. With peroxide crosslinking systems, one has, however, to be satisfied with a lower level of tensile properties, and especially lower tear resistance, a less favourable swell resistance, and poorer dynamic properties. Furthermore, peroxides restrict flexibility in compound developments [3.265].

Protective Agents. The choice of protective agents is governed by the same rules as for SBR. The presence of many protective agents can interfere with the crosslinking reaction by means of peroxides.

Since NBR vulcanizates are often subjected to high temperatures, the volatility of the protective agents plays an important role [3.182]. In such cases, less volatile products should be used, such as DNPD, p-dicumyl-diphenyl amine (e.g. Naugard 445), but also in combination with 2,2'-oxamidobis [ethyl-3-(3.5-ditert.butyl-4-hydroxy-phenyl) propionate] (e.g. Naugard XLI) [3.216], the styrene derivative of diphenyl amine (e.g. Vulkanox DDA), TMQ, and ADPA (e.g. Aminox). These reagents are preferably used in place of the more volatile products such as DPPD, ODPA, PBN, to name a few [3.182]. Network bound antioxidants are, of course, not volatile [3.225–3.228]. Yet, practical experience has shown that the less expensive normal NBR grades in compounds formulated for optimum heat stability have virtually the same performance level as NBR grades with network bound antioxidants [3.182].

Protective agents against ozone and flex degradation [3.266] are, because of their solubility characteristics, less effective in NBR than in SBR vulcanizates. Blends with PVC, EPDM, or other ozone resistant diene rubbers are important for improving the ozone resistance of NBR vulcanizates [3.182].

Fillers. As already mentioned, NBR has to be compounded with reinforcing fillers, in order to obtain vulcanizates with adequate mechanical properties. This is due to the lack of self-reinforcing properties of the rubber. The choice of fillers is governed by the same rules as for SBR.

Plasticizers. The viscosity, tackiness, and processibility of NBR compounds can be adjusted through the use of plasticizers, which, at the same time, also have an influence on the elasticity, low temperature flexibility, and swelling resistance of the vulcanizates. Many plasticizers or softeners, which are important in NR or SBR compounds, namely paraffinic or naphthenic oils, have a limited compatibility with NBR, because of the different polarities. Process aids to improve the calendering and extrusion behaviour, and also the tackiness, are xylene formaldehyde resins, rosin, Koresin, coumarone resins, wool fat, swell resistant and liquid factices, as well as, to a limited extent, aromatic oils. In concentrations of up to 30 phr, the typically used plasticizers based on ethers or esters (for instance, esters from thioglycolic or alkylsulfonic acids, adipates, phthalates, polyglycol ethers, polythio ethers) are very effective in increasing the rebound elasticity and the low temperature flexibility of vulcanizates. However, it should also be remembered, that all plasticizers reduce the mechanical properties and the swell resistance of the vulcanizates, and that they volatilize at high use temperatures. This has an influence also on the dimensional stability of the vulcanizates under practical application conditions.

For high temperature resistant vulcanizates, plasticizers of low volatility should be used only, such as butylcarbitol formal and polyester polythioethers, which, at the same time, improve the low temperature flexibility. Some other examples are polyesters, polyester polyethers, pentaerythritol ester, but also xylene formaldehyde resins, which are also good tackifiers [3.182].

Factices. In order to improve the mixing behaviour of compounds, particularly at higher plasticizer loadings, factices are used. They also improve the extrudability and calenderability of the compounds, in addition to green strength, and the knitting and building properties. Due to the good plasticizer retention, factices are especially used in compounds for low hardness vulcanizates, which require good mechanical properties and no plasticizer bloom. Factices used for NBR compounds are reaction products of castor oil with sulfur which have a good swelling resistance [2.119].

Process Aids. Particularly efficient process aids in NBR compounds are zinc soaps of unsaturated fatty acids, pentaerythritol tetrastearate, which improve the flow behaviour, followed by esters of fatty alcohols [3.267, 3.268]. Although there are often recommended in the literature some of the emulsion plasticizers are of low efficiency. Polar compounds, namely those based on glycol or glycerin, have a better compatibility with NBR than non-polar ones. By choosing appropriate processing aids, it is possible to prepare polymer blends from, for example, NBR and EPDM, where the continuous polymer phase is dispersed in the matrix on a microscopic scale [3.182, 3.269].

3.3.3.5 Processing of NBR

[1.14, 3.180, 3.181, 3.270]

NBR compounds are mixed and processed using conventional rubber machinery, and by following general procedures which have been discussed for NR [3.270] (see page 26). The use of particulate NBR grades, which have been highly promoted at times, has not become very important [3.182].

3.3.3.6 Properties of NBR Vulcanizates

[1.14, 3.180–3.182]

Mechanical Properties. When used in combination with reinforcing fillers, vulcanizates with excellent mechanical properties can be obtained from NBR. The optimum tensile strength of up to about 25 MPa occurs at a hardness of Shore A 70 to 80, but one can formulate NBR compounds such that a wide spectrum of vulcanizate hardness can be obtained, ranging from about Shore A 20 to ebonite.

The mechanical properties depend also to a great extent on the vulcanization temperature [3.271].

With an appropriate compound formulation and optimum vulcanization conditions, it is possible to obtain very low values for compression set [3.218–3.224]. For this purpose, it is important to use a semi reinforcing black (e.g. N550 or N770), and an EV system, such as TMTD, also in combination with OTOS, sulphur donors, and about 1.0 phr of sulphur [3.220]. Peroxide cures give also particularly low compression set values.

The elastic properties of unplasticized vulcanizates depend very much on the NBR grade, but, in general, the elasticity of NBR is distinctly less than that of comparable vulcanizates from NR or SBR. A relatively high elasticity can be obtained from NBR grades with a low acrylonitrile content, which are compounded with ester- or ether-based plasticizers and non- or semi reinforcing blacks (e.g. N770). NBRs with a high nitrile content do not give very elastic vulcanizates.

NBR vulcanizates formulated with reinforcing fillers have an abrasion resistance which is better by about 30% than that of comparable NR vulcanizates and about 15% better than that of comparable ones from SBR [3.272]. X-NBR vulcanizates show extreme high abrasion resistance.

Heat and Aging Resistance [3.212–3.217b]. NBR vulcanizates have a distinctly better heat resistance than those from NR or SBR [2.82]. Therefore, it is possible to produce NBR vulcanizates which are still useful even after aging for 6 weeks at 120 °C [3.198, 3.199]. If oxygen is excluded (e.g. by aging the vulcanizate submersed in oil), this aging resistance becomes even better [3.213]. Naturally, the heat resistance depends very much on the compound formulation. Particularly advantageous are EV systems, based, for instance, on TMTD and sulphur donors, or peroxide cures, and silica fillers in combination with silane coupling agents, or non- or semi reinforcing blacks, such as N770, N990, and N550. Also advantageous is the use of effective protective agents as, for instance, TMQ, SDPA, or ADPA, also in combination with MBI, if necessary. The best heat resistance of NBR vulcanizates is obtained when cadmium oxide and cadmium containing accelerators are used, but these compounds are of dubious value because of their high toxicity. It has already been mentioned that cure systems based on OTOS have been suggested instead [2.108, 2.109, 3.220]. Blends of NBR with heat resistant rubbers, such as ETER [3.191], ACM or FKM [3.182] also improve the heat resistance of vulcanizates. It goes without saying that heat resistant vulcanizates require a plasticizer of low volatility.

The weather and ozone resistance of NBR vulcanizates is comparable with that of NR vulcanizates, but it is more difficult to improve these properties of NBR vulcanizates by means of antiozonants [3.266]. In black compounds, the ozone and weather resistance can be somewhat improved by p-phenylene diamine, and in light compounds, by non-staining enol ethers in combination with micro crystalline waxes, but especially through blending the NBR with ozone resistant rubbers, such as EPDM, ETER, or PVC. Because of the great solubility of polar p-phenylene diamines in NBR they show less migration, and therefore their activity is reduced.

Low Temperature Flexibility. The same comments apply to the low temperature flexibility and the elasticity of NBR vulcanizates. An insufficient low temperature flexibility can be improved by using ester- or ether-based plasticizers, or by blending the NBR with other rubbers, such as BR. However, the latter worsens the solvent swell resistance, unless ETER is used as blend partner.

Swelling Resistance. [3.195–3.211] Vulcanizates from NBR, which is a polar rubber, have a good resistance to swelling when submersed in media, which are non-polar, or weakly polar, such as gasoline, grease, mineral oil, or animal and vegetable fats and oils. The swelling resistance depends, however, very much on the NBR grade, the compound components, filler loading, type and amount of plasticizer, and the degree of vulcanization. NBR vulcanizates also resist swelling in alcohols (methanol, ethanol). Blends of gasoline and alcohols, the so-called gasohols, which are frequently discussed as substitutes to extend the supply of gasoline, swell NBR vulcanizates more significantly than the individual components. Thus, there is a negative synergism [3.206]. The resistance of vulcanizates to gasohol can be improved by choosing a NBR grade with a high acrylonitrile content, and by using a cadmium-based cure system and especially by using NBR/PVC. However, vulcanizates from H-NBR, CO, and FKM have a significantly better gasohol resistance.

In unleaded gasoline, small amounts of hydroperoxide can form under certain conditions, resulting in the so-called "sour gas" [3.199–3.205]. This can alter the structure of NBR vulcanizates. High grade protective agents and UV stabilizers, such as 1,2-di-[(3,5-ditert.butyl-4-hydroxy)hydrocinnamoyl] hydrazine, i.e. Irganox 1024, will deactivate hydroperoxides, and impart to NBR vulcanizates a better resistance to sour gas (see also page 116).

Depending on their amount and chemical nature, additives which are frequently present in oils, can also cause more or less serious surface hardening of NBR vulcanizates [3.273]. In these instances, the limitations of NBR vulcanizates become apparent, and one has to resort to the use of the more expensive ACM, Q, or FKM rubber.

When compounding for NBR vulcanizates, one always has to bear in mind the oils with which the vulcanizate will be in contact. For instance, if NBR components, will be exposed to oils with sulfur and nitrogen containing additives, one has to take into account a potential post vulcanization from the sulphur. Therefore, one would use in this instance a TMTD concentration of less than 2 phr, to limit the formation of ZDMC. Since the optimum sulphur concentration in a compound is at about 0.7 phr (smaller amounts give high compression set, and higher amounts excessive hardening), the TMTD concentration must not be less than 1.5 phr. As an activator, one could use MBTS (but not CBS, because of the amine formation) in concentrations of up to 3 phr. Because they generate amines during the vulcanization reaction, sulphur donors of the DTDM type will also have a detrimental effect on the hardening of NBR vulcanizates with age.

If there is a simultaneous demand for good swelling resistance and acceptable low temperature flexibility, one should use NBR grades with a high acrylonitrile content (for good swelling resistance and impermeability) and specially selected plasticizers for good low temperature flexibility. Criteria for the choice of plasticizers have already been discussed in Section 3.3.3.4.

Polar media swell NBR vulcanizates considerably, and therefore these vulcanizates should not be in contact with esters, ketones, or other polar solvents. Aromatics and chlorinated hydrocarbons swell NBR, considerably, as any other vulcanizate, and examples of these solvents are benzene, toluene, xylene, styrene, methylene chloride, ethylene chloride, trichloro ethylene, carbon tetrachloride, or chlorobenzene. While CO, AU, and TM vulcanizates have a certain swelling resistance in toluene and xylene, for the use in the other solvents listed above, one has to resort to the expensive FKM vulcanizates. NBR vulcanizates can be used in contact with aromatic solvents, as long as the aromatics are blended with aliphatic hydrocarbons, such as in gasoline, and as long as the aromatics' concentration does not exceed 50%.

Permeation. [3.274] NBR vulcanizates are less permeable to gases than the ones from NR or SBR. As the acrylonitrile content increases, the permeation decreases, and with the high acrylonitrile grades, it approaches that of IIR or ECO, but not CO. The diffusivity in NBR vulcanizates is influenced by the amount and type of fillers in the compound and by the degree of crosslinking. Fillers with a laminar structure, such as mistron vapor, reduce the permeability considerably [3.182].

The permeation of gasoline is, however, significant in NBR vulcanizates. To reduce this, one has to use NBR grades with a high acrylonitrile content, compounded with mistron vapor, or NBR blends with PVC. Vulcanizates from CO, from blends of CO/ECO, or from FKM, have a lower permeability of gasoline than NBR and they therefore compete with NBR vulcanizates, particularly in the U.S.A.

Other Properties. Due to their high polarity, NBR vulcanizates have a considerably higher electrical conductivity than the ones from non-polar rubbers. Therefore, NBR is hardly used at all in components which require a low electrical conductivity. The thermal conductivity, and the thermal expansion coefficient are of the same order as for NR and SBR. The dynamic properties of NBR vulcanizates are good [3.275], but NBR vulcanizates are harmed by exposure to hydrogen sulfide [3.207].

3.3.3.7 Hydrogenated NBR (H-NBR or ENM or HSN)

[3.242a–3.250, 3.276–3.284]

In non-aqueous solutions, and by using suitable catalysts, such as pyridine-cobalt complexes [3.276], or complexes from rhodium, ruthenium, iridium, and palladium [3.277–3.281], it is possible to hydrogenate NBR partially or even completely. Particularly effective is, for instance, a catalyst based on transition metals with a trivalent rhodium halide [3.281].

Completely hydrogenated NBRs have a fully saturated polymer chain, which, conceptually, could have been derived from ethylene and acrylonitrile. Therefore, the

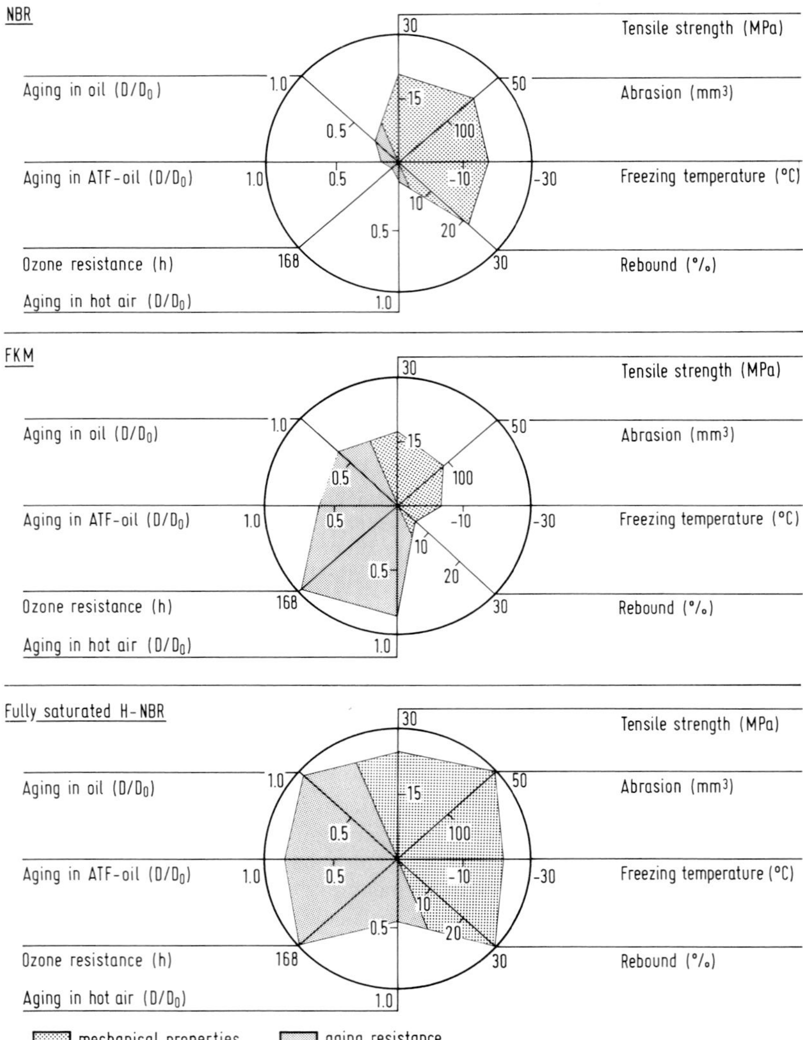

Figure 3.4 Profile of properties of H-NBR versus NBR and FKM [3.282]

code ENM has been proposed for this copolymer [3.248]. Since, however, most of the hydrogenated NBRs still contain double bonds, the designation of H-NBR will be used here.

Completely saturated NBR grades can be crosslinked with peroxides. The vulcanizates give the highest resistance to hot air and hot oils that can be achieved with NBRs, a high resistance to oxidative and ozone degradation, high resistance to sulphur-containing oils, even hydrogen sulfide, sulfur and nitrogen-containing oil additives and a high resistance to industrial chemicals [3.282]. In addition, the fully saturated H-NBRs have an excellent tensile strength, good low temperature flexibility, and very good abrasion resistance. After exposure to high temperatures, these H-NBRs retain their mechanical properties much better than, for instance, FKM vulcanizates, so that, in spite of their lesser heat and swelling resistance, fully saturated H-NBRs can compete with FKM. Because of the excellent mechanical properties also at high application temperatures, the good low temperature flexibility and the good chemical resistance, fully saturated H-NBR is able to replace even FKM in various applications [3.281 a]. Figure 3.4 shows a profile of the properties of fully hydrogenated H-NBR in comparison with NBR and FKM [3.283]. H-NBR can even be used in peroxidic crosslinked blends with FKM.

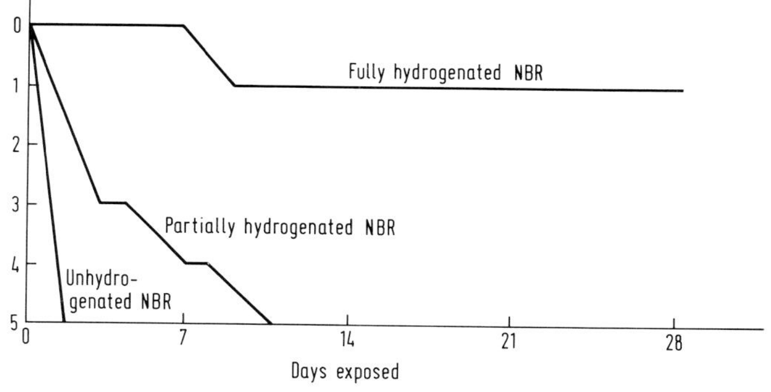

Figure 3.5 Ozone resistance of saturated and unsaturated elastomers (NBR-HNBR) [3.243]

Already relatively small amounts of unsaturation reduce the performance of H-NBRs considerably [3.243, 3.283] (see Figure 3.5). At a level of unsaturation of as low as 3 to 5 mole%, H-NBR can be crosslinked with sulphur. Such sulphur crosslinked H-NBR does not have the good environmental stability of a fully saturated H-NBR any longer [2.283], but it has better dynamic properties, e.g. for use in toothed drive belts working under oil.

Price developments will decide whether H-NBR can satisfy the requirements of the oil and automotive industry for rubber components with a good oil, gasoline, and ozone resistance, and with a high level of property retention after exposure to high temperatures of about 150 °C.

3.3.3.8 Uses of NBR

[1.14, 3.217 a]

Because of its relatively high price, NBR is used in applications where, besides good mechanical properties, there is also the requirement for good resistance to swelling in oils and gasoline, and a good resistance to heat aging and abrasion. Typical uses are in static seals, O-rings and packings for crank shafts and valves, in membranes, bellows for coupling, in hose, including high pressure hose, for hydraulic and pneumatic applications, in roll coverings, conveyor belts, friction coverings, linings, containers, work boots, shoe soling and heels, printing blankets, and stereo type plates. Because of the growing requirements by the automotive industry for higher temperature and oil resistance [3.182, 3.217, 3.217a, 3.217b, 3.252–3.252d, 3.284], NBRs have been increasingly replaced in this market by more heat resistant rubbers. NBRs are also used in considerable quantities in products for the food industry. Liquid NBR grades with reactive end groups on the polymer chain can be reacted using conventional liquid rubber technology to produce elastomers, or they can be used as impact modifiers in plastics [3.108–3.114a, 3.285].

3.3.3.9 Competitive Materials of NBR

NBRcompetesprimarilyagainstACM,AU,CO,CR,CSM,CM,ECO,FKM,andTM

3.3.4 Poly-2-Chlorobutadiene, Chloroprene Rubber (CR)

[2.104, 3.146, 3.286–3.321c]

3.3.4.1 General Background about CR

The polymerizability of 2-chlorobutadiene, or chloroprene, was discovered in 1930 by *W. H. Carothers* and coworkers [3.319]. In 1931/1932, large scale production began in the U.S.A. using a bulk polymerization process (1932 production was 250 tons), and the material was initially marketed in the U.S.A. under the name of Duprene as an oil resistant rubber. The first general purpose CR became known in 1939 as Neoprene GN. With a global production capacity in 1985 of 0.556 million tons, which is 4.6% of the SR capacity [2.13a], CR is besides IIR and NBR the most widely used specialty rubber.

Depending on how it was produced, a distinction is made between sulphur modified and mercaptan modified CRs. In addition, one distinguishes grades of low, medium, and high crystallizability. The latter are especially useful for the production of adhesives.

The structural formula of CR is as follows:

$$-\!\!\!\!\!\!+ CH_2-\overset{\displaystyle Cl}{\overset{|}{C}}=CH-CH_2 +\!\!\!\!\!\!- CH_2-\overset{\displaystyle Cl}{\overset{|}{C}}=CH-CH_2 -\!\!\!\!\!\!+$$

Chloroprene Unit

3.3.4.2 Production of CR

[3.8, 3.288, 3.292–3.295]

Today, CR is exclusively produced in emulsion, and using free radical initiators. Preferred emulsifiers are sodium soaps of rosin acids, and sodium naphthalin sulfonate is frequently added as an additional stabilizer. Redox systems are suitable initiators, as, for instance, sulfinic acid in the presence of oxygen, or persulfate. In order to prevent the formation of peroxides, small amounts of inhibitors are added, such as tert.-butyl pyrocatechol, but these reduce the rate of monomer conversion. With the so-called sulphur modified CR grades (or thiuram grades), the molecular weight is controlled by adding sulphur, which copolymerizes and forms S_x segments in the polymer chain [3.291, 3.292, 3.320, 3.321]. These CR grades are stabilized with thiurams, such as TETD. With the so-called mercaptan CR grades, the molecular weight is controlled with n-dodecyl mercaptan [3.287]. After reaching the desired degree of monomer conversion, of about 70%, the polymerization reaction is terminated by adding phenothiazine or tert.butyl pyrocatechol. After removing unreacted monomer and adding the stabilizer, the polymer is coagulated on a freezing drum, washed, and dried.

In the production of CR a similar number of *parameters* have to be considered as in the case of SBR (see page 59), which results in a wide range of commercially available grades.

The important parameters are:

- Sulphur modification (thiuram grades, ability to masticate the rubber);
- Mercaptan modification (mercaptan grades);
- Polymerization temperature (tendency to crystallize);
- Modifier (different Mooney viscosities, processibility);
- Stabilizer (colour and storage stability);
- Copolymerization with other monomers (tendency to crystallize);
- Precrosslinking (processing behaviour);
- Reactive groups (ability to crosslink without sulphur and accelerator. Only important for latices.).

3.3.4.3 Structure of CR and its Influence on Properties

[1.14, 3.292–3.295]

Viscosity. The molecular weight distribution, the degree of long chain branching, and the average molecular weight have the same effect on CR as on NBR. The Mooney viscosity has an influence on the ability of the rubber to band on mills, the heat build-up during mixing, the acceptance of filler, the extrudability, die swell and calenderability, to name a few.

Microstructure. The microstructure of CR influences the processing behaviour and the elastic properties of the rubber, and it depends very much on the polymerization temperature. With increasing temperatures there will be less uniformity in the chain structure due to increasing proportions of 1.2 and 3.4 moieties and different isomers in the monomer sequences [3.313]. This formation of more irregular chain structures reduces the rate of crystallization of the polymers. On the other hand, those CR grades, which have been polymerized at low temperatures, have a strong tendency to crystallize at high rates, which is an important requirement for adhesives with good immediate tack. These CR grades are, however, less suited for the production of rubber products, because they harden very rapidly with an attending loss of elasticity. Therefore, CR grades which, on account of their low tendency to crystallize, are useful for the production of rubber goods, are usually polymerized at higher temperatures. The copolymerization of chloroprene with small amounts of 2,3-dichloro butadiene, acrylonitrile, or styrene, also result in the desirable structural irregularity of the polymer chains, and this, therefore reduces the tendency to crystallize [3.293]. Since the rate of vulcanization of CR depends to a great extent on the amount of allylic chlorine in the polymer chains, the higher polymerization temperatures have also a beneficial effect in this respect.

As previously mentioned, S_x groups are built into the polymer chains of sulphur modified CR grades. In addition, fragments from the thiuram stabilizers attach to the polymer chains. As a result of these sulphur segments, it is possible to depolymerize these CR grades during processing. As with NR, the molecular weight is reduced during mastication of CR, and thus the initial nerviness of the rubber diminishes with an attending improvement in processing behaviour, such as die swell. Due to this inherent instability, the viscosity of the rubber does not remain constant and the storage is not very good. On the other hand, the presence of sulphur and thiurams obviates in many cases the need for accelerators for the vulcanization.

The considerably more stable mercaptan modified CR grades have found a much greater market acceptance. As with NBR, the compound viscosity of these CRs is determined by the degree of polymerization and the type and amount of plasticizer used in the compound. The curing properties of these CRs are adjusted with accelerators.

Pre-crosslinked CR grades, which consist almost completely of gel, can be blended in concentrations between 10 and 50% with conventional CR grades to improve their processing properties by reducing nerviness and die swell. This is similar to NBRs (see page 69). These blends have, however, a lower tensile strength.

Chlorine Concentration. CR is a polar rubber because it contains one chlorine atom for every four carbon atoms in the chain of homopolymers. Therefore, by comparison with non-polar diene rubbers, CR has a better resistance to swelling in mineral, animal, and vegetable oils and fats. The swelling resistance is, however, less than for NBR. The chlorine atoms also impart to CR a better flame-, weather-, and ozone resistance, than normally encountered with diene rubbers. These properties are particularly enhanced in copolymers of chloroprene and dichlorobutadiene, because of the even higher chlorine content in the copolymer.

3.3.4.4 Compounding of CR

[1.14, 3.287, 3.306]

Blends. The typical property spectrum of CRs can be modified through blends with other rubbers. NR improves, for instance, the elasticity and low temperature flexibility. BR reduces the brittleness temperature considerably. Blends with NBR improve the swelling resistance in industrial oils. With all these blends, it is a problem to adjust the cure system in such a manner, to make it adequate for the individual blend components.

Therefore, frequently used cure systems in these blends are sulphur, and thiourea derivatives, with the addition of thiuram and guanidine accelerators.

Vulcanizing Agents. [3.298–3.304] Contrary to other diene rubbers, the vulcanization of CR compounds is not carried out using sulphur, but metal oxides. In general, the best proven cure systems are combinations of, for instance, 4 phr MgO with 5 phr ZnO, or if very low water absorption is required, lead oxides (PbO or Pb_2O_4 in concentrations of up to 20 phr or even higher).

With mercaptan CR grades, sulphur can be added to enhance the degree of vulcanization, but this also reduces the heat resistance of the vulcanizate.

The choice of accelerators for CR is also governed by different rules than for other diene rubbers. In general, cure systems that provide faster cure rates tend to be more scorchy. Also those, that give a higher degree of vulcanization, i.e. higher tensile strength, lower elongation at break, higher rebound elasticity, and less compression set, tend to have less scorch safety.

As already mentioned, in most instances the thiuram CR grades do not require additional accelerator. The presence of metal oxides is by itself already sufficient in most cases to achieve a sufficient rate of vulcanization. However, by adding accelerators, the cure rate of these compounds can be even further increased, with an attendant reduction in storage stability and time to completion of the vulcanization reaction.

By contrast, the mercaptan CR grades require accelerators besides MgO and ZnO or lead oxides. Most conventionally used accelerators are not very effective in CR, and instead, the most important one is thiourea (ETU). However, in spite of being a good and most effective vulcanizing agent for CR, it is, because of its toxicity, being progressively replaced by DETU, or other recent developments in thioketones and thiadiazines, such as thiadiazole [3.298–3.300, 3.304]. A newly developed thioketone derivative as well as 3-methyl-thiazolidin-thione-2 appear to surpass even ETU with respect to scorch safety and cure rate, and the obtainable degree of vulcanization [3.300, 3.304].

For a good compromise between scorch safety and cure rate on the one hand, and state of cure on the other, combinations of small amounts of ETU or one of the other CR accelerators with thiurams or guanidines should be used. For extremely rapid cure rates (self curing), DPTU in combination with aldehyde amines are employed.

There are no typical **vulcanization inhibitors** for CR compounds, although MBTS or thiurams are capable of reducing the cure rates somewhat. A cure inhibition, however, also reduces the state of cure, which, in turn, can be counteracted by raising the cure temperature, so that cure states obtainable with ETU can be reached.

Peroxide cures are, as a rule, not employed with CR compounds.

Protective Agents. In spite of the inherently excellent oxidative stability of CR, additional agents are added to protect CR vulcanizates against oxygen ond ozone degradation, to meet particularly stringent requirements for some applications. Such agents are especially aromatic amines, such as IPPD, and diphenyl amine derivatives, like ODPA or SDPA, as well as sterically hindered phenols for light coloured compounds. With the use of MBI, one has to consider its accelerating influence on the cure of CR compounds. Therefore, it is seldomly used. To improve the ozone resistance of black vulcanizates, one uses primarily non-symmetric substituted p-phenylene diamines, and for light coloured vulcanizates, enol ethers or benzofuran derivatives [1.14].

Fillers. For gum vulcanizates or those formulated with non-reinforcing fillers, the tensile and tear strengths are higher for CR than for corresponding ones based on SBR or NBR. Yet, vulcanizates from CR do not reach the property level of those from NR, unless reinforcing or semi reinforcing blacks or reinforcing white fillers are used with CR. Particularly noteworthy is the high tear resistance which is obtainable with reinforcing-silica in compounds of CR, to reach a level which cannot be obtained with other rubbers.

Softeners. To reduce compound cost and to improve processibility, CRs are compounded with softeners, mainly mineral oils. Naphthenic oils of relatively low molecular weight (i. e. light process oils) are primarily used, because they are easily incorporated in the rubber compound and give a good level of properties. However, these oils are relatively volatile during heat aging, cause bloom at high loadings, and are expensive. Less costly aromatic oils can also be used even in high concentrations, but they give less favourable vulcanizate properties at low loadings, give poorer low temperature flexibility at high loadings, and they discolour light compounds. On the other hand, aromatic oils reduce the tendency of compounds and vulcanizates to crystallize. Depending on the average molecular weight, paraffinic oils are only compatible with CRs in low concentrations, and they serve generally only as processing aids, that is, to reduce sticking of compounds to mill rolls. For special formulations of CR, synthetic plasticizers must be used instead of oils. This is especially important if improved flame resistance, low temperature flexibility, and elasticity are required of the vulcanizate. Of particular importance are the flame resistant esters of phosphoric acid to replace oils which burn readily. A further improvement in the flame resistance of CR compounds is achieved through the addition of chlorinated hydrocarbons together with hydrated alumina. Elasticity and low temperature flexibility of CR vulcanizates can be improved by using ester- and ether-based plasticizers, same as for NBR.

Various resins, such as xylene formaldehyde resins, Koresin, rosin and coumarone resins, can be used in CR compounds to improve the filler dispersion and the building tack.

Factice. Since CR vulcanizates have a relatively good strength for gum compounds and at low filler loadings, it is possible to formulate very soft vulcanizates with adequate mechanical properties by using factice or plasticizers. Such vulcanizates are required for printing roll covers, seals, and hoses. In addition, as with other rubbers, factices improve the processibility (extrudability, calenderability, green strength, and dimensional stability) of CR compounds, and the compatibility of CRs with oils. Particularly successful are the factices, that were especially developed for CR, to give a high resistance to solvent swell [2.119].

Process Aids. Those particularly suitable for CR compounds are esters of fatty alcohols, to improve flow behaviour [3.267, 3.268]. Due to the interaction with zinc ions, zinc soaps strengthen the CR compounds, but reduce their processibility and Ca soaps tend to bloom. Other process aids recommended in the literature, such as emulsion plasticizers, have only little effect. Therefore esters of fatty alcohols are of great importance. Stearic acid and paraffinic oils are also important for reducing the stickiness of CR compounds and to aid in the filler dispersion.

3.3.4.5 Processing of CR

[1.14, 3.305, 3.306]

The mixing and processing of CR compounds is done using conventional rubber machinery and processes. Because of their tendency to scorch, the heat history of CR compounds has to be kept as short as possible during processing.

3.3.4.6 Properties of CR Vulcanizates

[1.14, 3.287, 3.306, 3.311-3.316]

Mechanical Properties. Gum compounds or those with non-reinforcing fillers have a higher mechanical strength than those from most types of SRs, because, due to their ability to strain-crystallize, CRs have a certain self-reinforcing ability. The greatest influence on the mechanical properties has, of course, the activity of the filler in the compound, same as with other SRs. With highly reinforcing blacks and silicas, one obtains tensile properties which are only slightly below those of comparable NR vulcanizates. The resistance of CR vulcanizates to tear initiation and propagation is excellent, if low structure blacks and, in particular, active silicas are used. Semi-reinforcing blacks together with ester- and ether-based plasticizers give a high elasticity for CR vulcanizates. To obtain a good resistance to compression set at high temperatures, mercaptan modified CRs are used compounded with semi-reinforcing blacks, but with as little plasticizer as possible, and with a higher degree of crosslinking. Crystallization can aggrevate the compression set at low temperatures. Therefore, the same compounding measures are to be implemented for good low temperature compression set, as for depressing the freezing point and the tendency towards crystallization.

Heat and Aging Resistance. To obtain a good heat resistance, mercaptan grades should be used instead of the thiuram grades, and a sulphur-free vulcanization. Light coloured fillers, especially talc, and the use of protective agents based on diphenyl amine, such as ODPA or SDPA, give an optimum in heat resistance, namely of the same order as NBR vulcanizates with EV cures. In addition to very good oxidative stability, unprotected CR vulcanizates have already a significantly better weather and ozone resistance than other diene rubbers. Of fillers, carbon blacks have a positive influence and of plasticizers, aromatic oils and unsaturated fatty acid esters have a negative one on the ozone resistance of vulcanizates. Since the ozone resistance improves as the vulcanizates become less stressed, any measure which reduces the stress in vulcanizates has a positive effect regarding ozone resistance. The presence of appropriate protective agents gives, however, a high level of ozone resistance, which suffices for many applications.

Depending on the specific compound composition, light coloured vulcanizates tend to discolour under sustained influence of light.

Flammability. [3.307-3.310] Because of the chlorine content, CRs have a favourable flame resistance, and in this respect they are superior to other rubbers. Although CR vulcanizates burn when exposed to high temperatures and open flames, they selfextinguish within a short time after removal of the flame. This requires, however, that the vulcanizate has not been compounded with larger quantities of combustible ingredients, such as oils. Vulcanizates with a high degree of flame resistance can be obtained by using non-burning plasticizers, like esters of phosphoric acid, and it can be improved even further by using chlorinated hydrocarbons in combination with antimony trioxide or hydrated alumina.

When chlorinated polymers, including CR, burn, they develop great quantities of toxic and corrosive fumes, which are extremely dangerous and cannot be tolerated in areas where there is a high demand for safety. Therefore, halogen free vulcanizates have been recently specified for these applications, and examples are oxygen containing polymers, like EVM (see page 103) or EAM (see page 112), which are compounded with hydrated alumina. Optimized EVM vulcanizates, which are halogen free and develop little smoke, have oxygen indices of 50% (LOI according to ASTM D 2863), and thus approach those of 55% for the optimum compositions based on CR and chlorinated paraffins [3.310]. Since, in the event of fire, the EVM vulcanizates create less danger regarding toxicity and corrosion, they will be increasingly used in the future, and they will replace CR in one of its major applications - cable jackets.

Low Temperature Properties. The stiffening of CR vulcanizates on cooling is often superposed by crystallization processes. Thus, the freezing temperature of CRs depends on the grade and its tendency to crystallize. Fillers have little effect on the freezing point, but plasticizers determine to a large degree the dynamic freezing point of CR vulcanizates. While aromatic oils have a detrimental effect, ester- and ether-based plasticizers improve the dynamic freezing point to a certain extent.

Tendency towards Crystallization. This undesirable property of CR has already been discussed on several occasions. Crystallization is detrimental in the production of certain technical rubber goods, since it reduces the building tack. Also with cold feed extruders, which are almost exclusively used in modern production processes, a strong crystallization can have a negative effect because of an excessive hardness of compounds. The reversible hardening due to crystallization is most pronounced in the raw rubber, less so in the compounds, and least, of course, in the finished product. And yet, products from CR harden in a reversible manner after extended exposure to low temperatures. This tendency to crystallize depends primarily on the CR type. Fillers have little influence on this phenomenon, and it can be reduced by high degrees of crosslinking of the rubber, and by adding high molecular weight aromatic oils or some of the ester- or ether-based plasticizers.

The crystallizability of CRs results also in a high degree of adhesive strength between CR and other rubbers, such as SBR [3.146].

Contrary to the requirements for rubber goods, the crystallizability of CRs is of great benefit in the production of adhesives. In solution, no crystallization occurs, but after evaporation of the solvent, the adhesive film hardens rapidly due to crystallization, giving good adhesion to substrates and good bonding of two surfaces by simply pressing them together. Therefore, these adhesives are referred to as "contact adhesives".

Resistance to Solvent Swell and Chemicals. As already mentioned, CR vulcanizates are, on account of their polarity, sufficiently oil resistant for many applications. Their solvent swell compares with that of an NBR with about 18% acrylonitrile, or with MQ [3.194].

The oil resistance depends, of course, significantly on the type of oil. CR vulcanizates have a good resistance to paraffinic and naphthenic oils of high molecular weight, but swell extensively in aromatic oils of low molecular weight. CR vulcanizates degrade when in contact with motor fuels, but with increasing levels of filler loading and degrees of crosslinking, the swelling resistance improves. Preferred plasticizers are those which are difficult to extract, such as polyadipates.

Because of the extensive use of CRs in cables, the swelling in water is an important property. Gum vulcanizates of CR absorb relatively large amounts of water, same as all the other emulsion polymers, and they are, in this respect, even inferior to comparable NBR and SBR vulcanizates. To reduce the swelling of CR vulcanizates in water, 10 to 20 phr of PbO_2 or Pb_3O_4 are used instead of MgO and ZnO. When choosing fillers, consideration should be given to high loadings and a low electrolyte content (for instance, N-990 black, calcined kaolin, fine talc, etc.). The usefulness of plasticizers depends on whether they are unsaponifiable. Thus, all compatible plasticizers based on mineral oils, chloro paraffins, ether and unsaponifiable esters are suitable. Finally, a high degree of crosslinking is important for a low water swell of the vulcanizate.

In general, CR vulcanizates are fairly resistant to chemicals. Contrary to aliphatic compounds, esters, ketones, aldehydes, and chlorinated or aromatic hydrocarbons swell and soften CR vulcanizates considerably. However, the vulcanizates are, resistant to alkalis (even in concentrated form) to dilute acids, aqueous salt solutions, and reducing agents. Oxidizing agents and concentrated mineral acids cause a surface hardening or even complete decomposition of CR vulcanizates. To improve the resistance to chemicals, compounds highly loaded with inorganic fillers are recommended.

Permeation. Regarding gas permeability, CR is distinctly superior to NR or SBR, but it does not reach the low permeabilities of CO, ECO, IIR, or NBR.

Electrical Conductivity. Due to its polarity, CR is a better conductor of electricity than the non-polar NR and SBR, but a poorer conductor than NBR. In order to obtain technically acceptable insulations based on CR, the proper choice of fillers, plasticizers, and resins assumes particular importance. At normal concentrations, protective agents have practically no influence on the conductivity of CR compounds. Regarding fillers, special grades of talc, and hard and calcined kaolin, reduce the electrical conductivity. These fillers perform particularly well when the vulcanizates are submerged in water. Adding resins, such as phenol formaldehyde resins, as well has a beneficial effect on the specific resistivity. Depending on the type of carbon black it is possible to design compounds which cover at the same concentration of black the full property spectrum from conductors to insulators. The activity of a black also has a significant influence. Thus, for the same type of black, the conductivity of a compound increases with a growing activity of the black. Regarding plasticizers, mineral oils with a high viscosity are preferred, while ether-based plasticizers reduce the resistivity significantly. With vulcanizing systems it should be remembered that a high degree of crosslinking improves the resistivity, and therefore, a vulcanization using ETU is recommended. Since moisture has a great influence on the electrical properties, compounds for insulations should absorb very little water; therefore they should be activated with lead oxides. Although NR and SBR vulcanizates are better dielectrics, optimized CR compounds are well suited as low voltage insulators for nominal voltages of up to 1 kV.

3.3.4.7 Uses of CR

[1.14, 2.104, 3.287, 3.306, 3.317, 3.318, 3.321 a, 3.321 b]

CR grades with little or a moderate tendency towards crystallization are used in many different technical rubber goods, which are flame resistant, and resistant to oils and fats, weathering and ozone. These rubber goods include mouldings, extrusions, seals, hoses, profiles, rolls, belts, V-belts [3.317, 3.318], bearings, linings, rubberized fabrics, shoe solings, and many applications in the construction industry, such as window and construction profiles, roofing membranes, and also cable jackets. For the latter application, the CR grades with a moderate crystallizability are often preferred. In several automotive applications, the use of CR has shrunk for similar reasons as with NBR. In weather stripping, the use of CR has suffered due to the strong competition of EPDM. SBR/ETER and NBR/EPDM blends have also made inroads on CR usage, because of their more favourable price.

Strongly crystallizable CR grades are preferably used for the production of contact adhesives.

3.3.4.8 Competitive Materials of CR

The following SRs compete with CR: CM, CO, CSM, EVM, ECO, EPDM, ETER, NBR, EPDM/NBR, and ETER/SBR.

3.3.5 Polyisoprene, Isoprene Rubber, Synthetic (IR)

[1.2, 3.5, 3.13, 3.322–3.351]

3.3.5.1 General Background about IR

In the early days of SR research, attempts were made to prepare the synthetic analogue of NR, using isoprene as the starting material (*F. Hofmann*, 1909 [3.337]). By employing the then used polymerization methods (heat polymerization in bulk), this goal was only partially reached. In 1954 Goodrich succeeded in the synthesis of cis-1.4 polyisoprene (IR), the so-called "synthetic natural rubber", using *Ziegler-Natta* catalysts from $TiCl_4$ and Al-trialkyl [3.338]. Shortly thereafter, Firestone discovered a synthetic route for IR using finely dispersed Li [3.339] and alkyl lithium [3.340] as catalysts, which had already been proposed by *C. D. Harries* in 1917 [3.341]. Large scale production of Li-IR was started in 1960 by Shell. A Ti-IR has only been marketed since 1962 by Goodyear.

The global annual production capacity amounted in 1985 to 1.224 million tons, which is about 10.1% of the total SR capacity [2.13 a].

Since NR and IR are chemically and structurally very similar, the latter can be considered a commodity rubber with similar applications as for NR.

By using a coordination catalyst modified with Vanadium, a trans-1.4 polyisoprene can be produced which is similar to the naturally occuring polymers Guttapercha and Balata [3.342]. Since the commercial importance of these polymers has declined with the rise of synthetic plastomers, the synthetic analogue plays only a minor role.

IR exhibits a high degree of steric uniformity, consisting of over 92% of cis-1.4 polyisoprene.

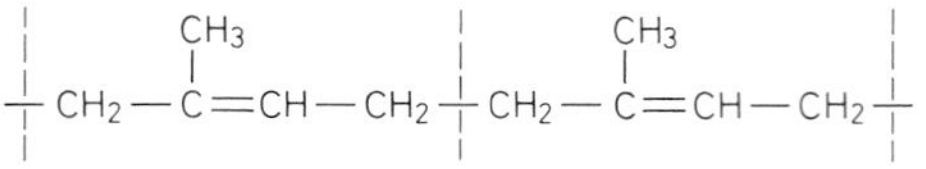

$$\left[-CH_2-\underset{}{\overset{CH_3}{\overset{|}{C}}}=CH-CH_2- \right]\left[-CH_2-\overset{CH_3}{\overset{|}{C}}=CH-CH_2- \right]$$

1.4 Isoprene Unit

3.3.5.2 Production of IR

[3.5, 3.77, 3.81, 3.83, 3.85, 3.323 a, 3.324, 3.326, 3.329]

The polymerization of isoprene is carried out in low boiling aliphatic hydrocarbons pentane or hexane.

Ti-IR. As a catalyst one uses $TiCl_4$ and $Al(alkyl)_3$ in a 1:1 molar ratio. By choosing a suitable organo-aluminium catalyst component [3.343], and by modification with aliphatic [3.344–3.346] or aromatic ethers [3.347], a very high catalyst activity can be achieved. As a result, catalyst concentrations can be kept low, and the formation of objectionably low molecular weight components can be supressed with a concomitant increase in monomer conversion rates and reactor productivity. Processes which are employed for deactivating catalyst residues in the rubber are important, since, already during workup and drying of the rubber, active catalyst residues can reduce the molecular weight and the stability of the polymer. Recommended deactivators are compounds like sodium methoxide [3.348] and certain amines [3.349]. After adding stabilizers, such as BHT, the IR is finished using conventional methods.

Li-IR. For the synthesis of this grade, Li-n-butyl is used as initiator [3.340]. Otherwise, the polymerization process corresponds largely to that used for Li-BR. In order to obtain a molecular weight of at least 1 million, which is required for adequate vulcanizate properties of the rubber, the isoprene monomer and the solvent have to have a very low and constant water content.

3.3.5.3 Structure of IR and its Influence on Properties

[3.5]

Ti-IR. Since this IR has high molecular weights ($1-1.5 \times 10^6$), its Mooney viscosities ML1+4 at 100 °C, range from about 80 to 100. The viscosities depend on the microgel content of between 15 and 25%. The molecular weight distribution is relatively broad. Because of this, and because it can be masticated, Ti-IR processes well, and is very similar to NR in this respect. Ti-IR is also very similar to NR regarding its molecular micro structure, but unlike NR, which is practically all cis-1.4 structure (>99% within experimental error), Ti-IR has about 98% cis-1.4 and 2% 3.4 structure, which interrupts the pure linearity of chains. Although being only a very small component, the 3.4 structure gives significantly different technological properties of Ti-IR compared with NR. With a half life at −25 °C of 5 hours, the crystallization rate of Ti-IR is distinctly shorter than that of NR with 2 hours. Also, the self-reinforcement due to strain crystallization is less pronounced for Ti-IR than NR, which results in a poorer green strength of compounds and lower tensiles for gum and filled vulcanizates.

Li-IR. At about 1.5 to 2×10^6, the molecular weights of Li-IR are considerably higher than those of Ti-IR, but the molecular weight distributions are narrower. Therefore, Li-IR is more difficult to process. Also regarding microstructure, Li-IR differs to a greater extent from NR than Ti-IR, as shown by IR and NMR spectroscopy. The cis-1.4 content is only 90–92%, trans-1.4 is about 2–3%, and the 3.4 structure amounts to about 6–7% [3.350, 3.351]. Due to this greater irregularity in chain structure, Li-IR hardly crystallizes at all, and this can also be seen in the considerably poorer green strength of compounds, and poorer tensile strength of gum and filled vulcanizates in comparison with NR. In this respect, Li-IR compares more closely with BR or SBR, and it is distinctly inferior to Ti-IR.

3.3.5.4 Compounding of IR

Ti-IR. Regarding compounding, Ti-IR can be compared with NR (see page 22). However, since Ti-IR shows less self-reinforcement than NR, the activity of the fillers plays a more important role. Also, since Ti-IR does not contain the natural impurities of NR with its accelerating effects, the sulphur/accelerator ratio has to be more closely adjusted to that required for SBR. Furthermore, to obtain a building tack which compares to that of NR, it is necessary to use somewhat higher concentrations of tackifier resins in Ti-IR compounds. Since Ti-IR has to be stabilized by adding an antioxidant during production, its vulcanizates have a better heat resistance than those from NR. However, once additional protective agents have been added, the difference between NR and Ti-IR vulcanizates vanishes.

Li-IR. Guidelines for compounding Li-IR are largely the same as for SBR (see page 62).

3.3.5.5 Processing of IR

For the mixing and processing of IR, the same rules apply as for NR (see page 26), with the exception that Ti-IR and Li-IR are more difficult to process than NR.

3.3.5.6 Properties of IR Vulcanizates

Ti-IR. The vulcanizate properties of Ti-IR resemble those of NR (see page 28) and the values of most properties reach those of NR. Only the tensile strength and the tear resistance are somewhat inferior, which is due to the reduced tendency of Ti-IR to crystallize.

Li-TR. The vulcanizate properties of this rubber compare more closely with those of BR and SBR (see page 65).

3.3.5.7 Uses of IR

In general, IR, and particularly Ti-IR, is used in the same applications as NR. Li-IR can replace NR only in exceptional cases, and it is preferably used in blends with other rubbers, such as BR or SBR, to improve their processibility.

3.3.5.8 Competitive Materials of IR

These are NR, SBR, and BR.

3.3.6 Isoprene-Isobutylene Copolymers, Butyl Rubber (IIR)

[3.352–3.382]

3.3.6.1 General Background about IIR

[3.352–3.356]

IIR is the oldest specialty rubber beside NBR and CR, but its importance has diminished after EPDM became available. Based on the BASF process of 1931 for making polyisobutylene, Standard Oil developed a method for producing IIR in 1937, and large scale production commenced in 1943 [3.378, 3.379]. In 1985 the global production capacity of IIR amounted to 0.696 million tons, which is 5.7% of the total SR capacity [2.13 a].

IIR consists of 97 to 99.5 mole% isobutylene and 0.5 to 3 mole% isoprene which provides the double bond required for sulphur vulcanization.

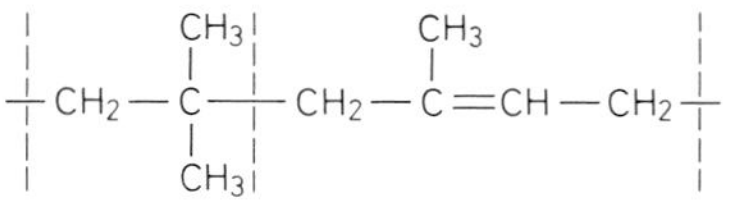

$$\left[CH_2-\underset{CH_3}{\overset{CH_3}{\underset{|}{\overset{|}{C}}}} \right]\left[CH_2-\overset{CH_3}{\overset{|}{C}}=CH-CH_2 \right]$$

Isobutylene Isoprene Units

More recently, chlorinated and brominated butyl rubbers have also been produced (see page 92).

3.3.6.2 Production of IIR

[3.5, 3.352–3.356]

IIR is produced by a cationic copolymerization of isobutylene and isoprene in methylene chloride solvent with $AlCl_3$ as a catalyst. The polymerization temperature is $-100\,°C$, and small amounts of water or HCl act as cocatalysts. The low polymerization temperature is required to obtain suffiently high molecular weights. During the polymerization, which proceeds at very high conversion rates, high molecular weight copolymers precipitate from the solvent in discrete particles (precipitation polymerization), which are then purified in separate reaction vessels from unreacted monomer and solvent, before drying. In order to assure reproducible cationic polymerization processes, a high and constant degree of purity is required of the monomer [3.39].

More recently, it has been suggested that, instead of $AlCl_3$, $Al(alkyl)_2Cl$ should be used as catalyst, hexane as solvent, and higher polymerization temperatures of -40 to $-50\,°C$. This makes possible a real solution polymerization process and it still leads to sufficiently high molecular weights [3.380, 3.381].

3.3.6.3 Structure of IIR and its Influence on Properties

[3.5]

Macrostructure. The average molecular weight of IIR lies in the region of 300,000 to 500,000, which corresponds to Mooney viscosities ML1 + 4 at 100 °C, of about 40 to 70. The Mooney viscosity influences primarily the processibility and the amount of filler and oil, which the rubber can accept. The molecular weight distribution of IIR is relatively broad. This, and the somewhat thermoplastic nature of the polyisobutylene component are the reasons for the relatively good processing characteristics of IIR.

Microstructure. The isobutylene monomer units polymerize predominantly in a head-to-tail arrangement, and the isoprene units are built into the polymer chains in a trans-1.4 configuration.

Unsaturation. The largely saturated chain structure of IIR determines its main properties – good resistance to oxidative and ozone degradation to chemicals, and a low gas permeability. As already mentioned, the double bonds in the chain permit sulphur vulcanization reactions, but at low levels of unsaturation, these vulcanization reactions naturally proceed very slowly. With increasing levels of unsaturation, vulcanization rates increase, but the other beneficial properties mentioned above, become poorer. Therefore, for a good balance of properties and sulphur vulcanization rates, a compromise is sought by using products with moderate levels of unsaturation.

Stabilizers. IIR is sold containing staining or non-staining stabilizers.

3.3.6.4 Compounding of IIR

[3.352, 3.356–3.363]

Blends. Since IIR cures relatively slowly, it does not readily covulcanize with other diene rubbers in blends. For certain applications, blends with CIIR and CSM (up to 25% by weight) are used. CIIR and BIIR have assumed a greater importance as blend partners with other diene rubbers.

Vulcanization Systems. Three different types of vulcanization systems are used for IIR [1.12, 1.13, 3.352, 3.360–3.363].

As previously mentioned, *sulphur vulcanization* is preferably used with IIR grades of high unsaturation, and particularly reactive accelerators are required in this application. These are, for instance, thiurams or dithiocarbamates, also in combination with thiazoles. With these, sufficiently high degrees of crosslinking are obtained. Frequently used systems consist of relatively high quantities of a combination of TMTD and MBT. The use of sulphur donors instead of sulphur is also of interest (see page 231 ff). In addition to sulphur and accelerators, the usual quantities of ZnO and stearic acid are used.

The use of *quinone dioxime* (CDO), together with oxidizing agents like PbO_2 or Pb_3O_4 and/or MBTS gives particularly heat-resistant vulcanizates [3.360, 3.361]. This crosslinking reaction leads to tight cures even with low unsaturation IIR grades. The combination of CDO and lead oxide gives very scorchy cures, and therefore CDO is, at times, replaced by dibenzo-CDO. A combination with MBTS also gives better scorch protection (see page 260).

The third cure system – *resin cure* – has assumed greater importance with IIR. For this purpose, phenol formaldehyde resins with reactive methylene groups and additional activators, such as $SnCl_4$ or $FeCl_3$, or small amounts of chlorinated polymers, such as CR, are used. If halogenated phenolic resins are used, the presence of the reactive activators may not be required [3.363].

In contrast to most of the other rubbers, it is not possible to crosslink IIR with peroxides, since they tend to depolymerize the rubber [1.12, 1.13].

Protective Agents. Although IIR vulcanizates are very resistant to aging, there are applications which require special protection against heat and ozone degradation, and the use of highly efficient protective agents, like aromatic amines, is recommended here.

Fillers. [3.357, 3.358] Gum IIR vulcanizates already have higher tensile strengths than those from SBR or NBR. To obtain vulcanizates which satisfy a sufficiently high level of properties, different types of fillers are used in varying quantities. In general, the same compounding rules apply here as for other rubbers.

Softeners. In IIR compounds, one uses especially paraffinic and naphthenic oils as softeners. The use of highly aromatic and unsaturated oils in IIR compounds leads to a poor aging resistance of vulcanizates. Since IIR is non-polar, it accepts high loadings of oils and paraffins. This has a positive effect because it gives very smooth surfaces of vulcanizates, but higher oil loadings also lead to higher compression or tensile set. The low temperature flexibility of IIR vulcanizates can be improved by using higher molecular weight esters of low volatility.

The addition of coumarone-indene resins of up to 10 phr improves the hardness of vulcanizates considerably, while keeping the level of tensile strength virtually constant. Appropriate quantities of resin also improve the surface smoothness of vulcanizates and make the compound more nervy. The most important function of these resins is to facilitate the processing of compounds at higher temperatures, since the resins melt above 100 °C and thus act as softeners and lubricants. Styrene-butadiene resins also act as processing aids in IIR compounds, and they also improve the hardness of the vulcanizates. However, the compatibility of most styrene-butadiene resins with the non-polar IIR leaves something to be desired, but can be improved if about 10% of the IIR is replaced by the more polar CIIR. The addition of polyethylene in quantities of about 3 to 15 phr improves the extrudability and mould flow, and it also facilitates the mould release of the vulcanizate.

Process Aids. In order to obtain good processing properties and a rapid and uniform incorporation of fillers, stearic acid, zinc soaps, fatty alcohol esters, factices, and other process aids are added to IIR compounds.

3.3.6.5 Processing of IIR

[3.356]

IIR is processed using similar procedures and rubber mixing and processing equipment as for other rubbers (see page 26).

It is peculiar with IIR that, in order to achieve a high state of cure, a prevulcanization can be carried out by mastication under heat in internal mixers at 150–175 °C for 5 minutes using methyl N-dinitrosoaniline or poly-p-dinitroso benzene [1.12, 1.13].

3.3.6.6 Properties of IIR Vulcanizates

[3.356, 3.359, 3.364–3.369, 3.382]

Mechanical Properties. With reinforcing fillers, very high tensile strengths can be obtained, corresponding to those of SBR or NBR vulcanizates.

Rebound at room temperatures is very small, and it increases significantly at higher temperatures. To obtain an adequate elasticity, it is important to use paraffinic oils or dioctyl sebacate.

Normally, IIR vulcanizates can be produced without problems in hardnesses of up to Shore A 85. For harder vulcanizates the use of resins or blending with CSM (up to 25% by weight) is recommended.

As with other rubbers, the compression set of IIR vulcanizates depends, primarily on the degree of vulcanization, and on the type and amount of filler and softener used in the compound formulation. Measures which lead to relatively low values of compression set include those which increase the state of cure and the elasticity, namely prevulcanization during hot mixing with nitroso derivatives, choice of softeners, and the use of semi-reinforcing fillers.

Heat and Aging Resistance. The heat resistance of IIR vulcanizates, and especially those that were cured with CDO and with phenolic or halogenated phenolic resins, is high. It is even considerably better than that of NBR vulcanizates, but not as good as that of ACM, CO, or even EVM vulcanizates. Sulphur vulcanizates of IIR have, however, a considerably poorer heat resistance, and it compares with that of sulphur cured EPDM vulcanizates. The replacement of sulphur donors improves the heat resistance of IIR vulcanizates, same as with other rubbers.

Because they contain only a small amount of unsaturation, IIR vulcanizates have an excellent resistance to weathering and ozone degradation. To optimize these properties, IIR grades with low unsaturation, carbon black and vulcanization systems, which give low stress levels, should be used.

Low Temperature Flexibility. In spite of the low rebound and stiffening of the vulcanizates at lower temperatures, the brittlenesss point is very low at $-75\,^{\circ}C$.

Permeation. Vulcanizates from IIR have a very low gas permeability, which is even lower than that of vulcanizates from high nitrile NBR. However, the gas permeability of IIR does not reach the extremely low values of CO vulcanizates.

3.3.6.7 Uses of IIR

[3.352, 3.356, 3.359, 3.370–3.377]

The main applications of IIR are in cable insulations and jacketing, innertubes of tires, innerliners of tubeless tires, curing bladders, pharmaceutical stoppers (mainly resin cured), and roofing membranes. Tire treads based on IIR have not performed successfully.

3.3.6.8 Competitive Materials of IIR

These are, in particular, BIIR, CIIR, CM, CSM, EVM, and EPDM.

3.3.7 Halogenated Copolymers from Isoprene and Isobutylene (CIIR and BIIR)

[3.355, 3.356, 3.359, 3.383–3.398]

3.3.7.1 General Background about CIIR and BIIR

CIIR and BIIR are obtained by halogenation of IIR.

3.3.7.2 Production of CIIR and BIIR

[3.388, 3.389]

On adding chlorine or bromine to IIR in an inert organic solvent like hexane, a rapid electrophilic substitution reaction occurs, whereby one halogen atom is added per isoprene unit, mainly in the allylic position. The addition of a second halogen atom proceeds at a much slower rate. Thus, only a relatively small number of halogen atoms are built into the polymer chain. An addition of halogen atoms to the double bond occurs hardly at all. With chlorine, one obtains CIIR, and with bromine, BIIR.

3.3.7.3 Structures of CIIR and BIIR, and their Influence on Properties

Compared with IIR, the halogenated IIRs have two advantages: the cure reactivity of the double bond is enhanced by both the halogen atom and allylic halogen structure. As a result, improvements occur in the vulcanization rates, the states of cure, and the reversion resistance, and covulcanization with other diene rubbers is also possible.

3.3.7.4 Compounding of CIIR and BIIR

[3.359, 3.390–3.396]

The same cure systems can be used for CIIR and BIIR, (see page 89), and in addition, crosslinking with ZnO or diamines is possible [3.382, 3.393–3.396]. BIIR gives faster vulcanization rates, and for the same cure systems, it gives a higher state of cure. Since with ZnO alone, the cure state is often inadequate, it is frequently used together with sulphur or sulphur donors. Good states of cure are also obtained with ZnO when it is used with MBTS (which reduces the danger of scorch with BIIR), or TMTD, and particularly with OTOS. If accelerators are used, MgO is also commonly added to improve the scorch safety. For vulcanizates which require good hot water or steam resistance, Pb_3O_4 is preferably used instead of ZnO. Otherwise, CIIR and BIIR compounds are formulated and processed virtually the same as IIR compounds.

3.3.7.5 Processing of CIIR and BIIR

There is practically no difference between the processing of BIIR and CIIR or IIR.

3.3.7.6 Properties of CIIR and BIIR Vulcanizates

[3.359, 3.383, 3.397, 3.398]

The properties of the vulcanizates are more pronounced than those of IIR vulcanizates, that is, BIIR vulcanizates have an even lower gas permeability, better weather and ozone resistance, higher hysteresis, better resistance to chemicals, and better heat resistance than those of IIR. BIIR compounds also cure faster with lower amounts of curatives, and give a better adhesion to other rubbers. The property level of CIIR is between that of BIIR and IIR.

3.3.7.7 Uses of CIIR and BIIR

Because of the properties listed above, and particularly because these properties are, to a certain extent, transferred to other rubbers through blending, CIIR and especially BIIR are used in the following products: innerliners of tubeless tires to improve covulcanization and adhesion, innertubes for demanding applications in buses and trucks, tire sidewalls, linings, belts, hoses, seals, injection moulded components, or pharmaceutical stoppers.

3.3.7.8 Competitive Materials of CIIR and BIIR

These are especially IIR.

3.3.8 Ethylene-Propylene Rubber (EPM and EPDM)

[2.104, 2.108, 2.109, 3.399-3.456]

3.3.8.1 General Background about EPM and EPDM

[3.399-3.402]

At room temperature, polyethylene is a crystalline plastomer, but on heating, it passes through an "elastomeric" phase. By interfering with crystallization of polyethylene, that is, by incorporating in the polymer chain elements, which inhibit crystallization, the melting temperature and therefore the elastomeric phase can be reduced considerably to below room temperature.

Such amorphous and curable materials can be considered rubbers, and they can be obtained by copolymerizing ethylene and propylene with certain catalysts of the *Ziegler-Natta* type. The resulting, so-called EPMs are amorphous and rubbery, but since they do not contain unsaturation, they can only be crosslinked with peroxides. [3.453]

If, during the copolymerization of ethylene and propylene, a third monomer, a diene, is added, the resulting rubber will have unsaturation, and it can then be vulcanizated with sulphur. These rubbers are the so-called EPDMs.

Large scale production of EPM and EPDM started in 1963, and in 1985, with rising growth rates, the global production amounted to 0.540 million tons, which is about 4.4% of the total SR capacity [2.13 a]

The majority of commercial EPM grades contain about 40 to 80% by weight or 45 to 85 mole% of ethylene, and the most important grades, about 50 to 70 mole%.

$$\left[CH_2 - CH_2 \right] \left[CH_2 - \underset{\displaystyle CH_3}{\underset{|}{CH}} \right]$$

Ethylene Propylene Units

The literature quotes many compounds as termonomers, but in commercial rubbers, only three dienes are used. These are non-conjugated, so that the double bond resides in sidegroups of polymer chains.

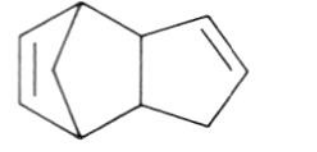

Dicyclopentadiene (DCP)

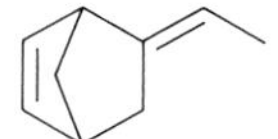

Ethylidene Norbornene (ENB)

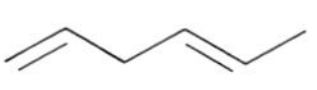

Trans-1.4 Hexadiene (HX)

In the production of EPM and EPDM rubber, an equally large number of parameters apply as with other diene rubbers, and this explains the large number of commercially available grades. The main *parameters* are:

- Concentration ratio of ethylene and propylene (amorphous or segmented grades);
- Co- or terpolymerization (EPM or EPDM);
- Type and amount of termonomer (vulcanization properties, mechanical properties);
- Solution and suspension polymerization (highest obtainable molecular weight);
- Molecular weight (differences in Mooney viscosity and processibility);
- Oil extension (processibility, price).

3.3.8.2 Production of EPM and EPDM

[3.5, 3.399, 3.401, 3.403–3.405]

In large scale copolymerization of ethylene and propylene, Vanadium containing coordination catalysts, such as VCl_4, $VOCl_3$ with $Al_2(ethyl)_3Cl_3$, $Al(ethyl)Cl_2$ or $Al(ethyl)_2Cl$, are almost exclusively used [3.401]. In solution polymerization processes, the solvents are aliphatic hydrocarbons, such as pentane or hexane. In suspension polymerization processes, no solvent is used; instead, an excess of propylene serves as one.

The choice of a suitable termonomer poses several problems [3.399]. Firstly, the two sets of double bonds of the diene should have different reactivities, so that one will copolymerize with the second remaining unreacted in the polymer chain, enabling it to be used in subsequent vulcanization reactions. The other requirement is that of a high reactivity of the second double bond in sulphur vulcanization reactions. Similar copolymerization parameters of the monomers are important as well. Finally, price plays an important role. Therefore, the above requirements have narrowed the choice of termonomers to the three mentioned above.

3.3.8.3 Structures of EPM and EPDM, and their Influence on Properties

[3.5, 3.399, 3.406–3.412]

Termonomer. Dicyclopentadiene (DCP) was the first termonomer used, but since there is relatively little difference in the reactivity of the two sets of double bonds, a certain amount of crosslinking or gel formation occurs during polymerization. In addition, the polymers do not give sufficient cure rates. With DCP, only 3 to 6 double bonds can be introduced per 1000 carbon atoms of the polymer chain. Ethylidene norbornene (ENB) behaves quite differently. The copolymerizability, as well as the cure reactivity of the polymer, is excellent. With ENB it is possible to introduce 4 to 15 double bonds for each 1000 carbon atom chain length, with the polymer remaining free of gel. Hexadiene (HX) occupies an intermediate position between ENB and DCP, and the resulting terpolymers can contain about 4 to 8 double bonds for every 1000 carbon atoms.

Sequence Distribution. Copolymers, which contain between 45 and 60% ethylene, are completely amorphous and are not self-reinforcing. At higher ethylene contents of the order of 70 to 80%, the polymers contain long ethylene sequences, which are partially crystalline. These polymers are referred to as "sequential" grades, and their processing behaviour differs considerably from that of the normal amorphous grades. The partially crystalline domains form thermally reversible physical crosslinks, which, as with the thermoplastic elastomers, give the elastomers an already high mechanical strength without chemical crosslinks [3.454]. At higher temperatures, though, this strength deteriorates.

Viscosity and Processibility. The processibility of EPDM depends on their molecular structure [3.427 a, 3.427b]. The development in time of carbon-black dispersion during mixing of EPDM rubbers in internal mixers depends on the molar mass distribution of the EPDM grade. This has to be suitably chosen, depending on the desired degree of carbon-black dispersion in the application concerned: a narrow distribution for average carbon-black dispersions, and a broad distribution for applications requiring excellent carbon-black dispersion. EPDM grades with broad molar mass distributions based on long chain branching can be masticated in some respect.

The molecular weights of most commercial grades are between 200,000 and 300,000, and the Mooney viscosity ML 1+4 at 100 °C lies between about 25 and 100. The grades with Mooney viscosities in the range of 25 and 50 can, of course, be most easily processed, same as with other SRs, but their acceptance of fillers and plasticizers is limited to low concentrations. If high filler loadings are required, grades with high Mooney viscosities have to be used, but this makes processing more difficult, although it can be facilitated somewhat by using, in addition, high plasticizer loadings.

Ultra high molecular weight EPDMs, which cannot be processed as such, are commercially available as oil extended grades, with either paraffinic or naphthenic oils. These grades process well, and are very economical.

Stabilizers. EPDM grades are almost exclusively sold with non-staining stabilizers, so that problems with staining are largely non-existent.

Building Tack. The building tack of EPM and EPDM leaves much to be desired, and this creates problems when used, for instance, in tire components. Some investigations for improving the self-adhesion (building tack) have been done in the past [3.445 a], but obviously this problem could not yet sufficiently be solved. The addition of tackifiers alone is not sufficient, much more the addition of light-sensitizers and radiation with visible light is necessary. Tetraphenyl porphyrine was reported to be efficient sensitizer also in combination with benzoanthracenes, which are existing in highly aromatic oils [3.445 a]. The building tack is also exceptionally influenced by the applied EPDM grades. The best influence from termonomers has ENB because of its ability to allow singular oxygen reaction in contrast to DCP and HX, which are quite less active. The ethylene propylene relation should be suitable for EPDM grades. Other shortcomings are a moderate level of adhesion to textile and metal, and an insufficient ability to covulcanize with other diene rubbers [3.450].

Degree of Unsaturation. Since in EPDM the unsaturation resides in sidegroups, the polymer chain is completely saturated. This gives the polymer an excellent resistance to degradation from oxygen, ozone, and chemicals, and therefore products made from EPDM vulcanizates are very durable. The vulcanization properties, including rate and state of cure of an EPDM, depend on the amount and type of double bonds from the termonomer. DCP-EPDMs cure considerably more slowly and give lower cure states than ENB-EPDMs. HX-EPDMs occupy an intermediate position. Depending on the ENB concentration in the polymer, one distinguishes with ENB-EPDMs the normal (about 4% ENB), fast (about 6% ENB) and ultra fast curing grades (about 8% and more ENB) [3.450b]. With increasing cure rates, the cure states and along with it, the mechanical properties, e.g. compression set, also improve, but the reversion resistance at high cure temperatures becomes poorer at the same time. The high unsaturation EPDM grades also command higher prices.

3.3.8.4 Compounding of EPM and EPDM

[3.400, 3.402, 3.426, 3.429–3.440]

Blends of EPM or EPDM with other polymers play an important role [3.413–3.425]. It is possible to improve the ozone resistance of other diene rubbers, such as NR, IR, BR, SBR or NBR through blending with EPDM in concentrations of about 30% by weight [3.418, 3.422]. Since covulcanization between the polymer blend components can be a problem, one uses here high-ENB-EPDM grades (ultra fast grades), highly reactive accelerators or peroxides. Blends of EPDM with NBR are of interest as economical chlorine-free substitutes for CR to produce weather and oil resistant components (3.269, 3.416, 3.419, 3.421, 3.425]. EPDM grades with low unsaturation or EPM can also be used in blends with saturated polymers. Examples are blends of EPM or EPDM with polyolefins, sometimes already pre-crosslinked, to produce "thermoplastic elastomers" (TPE), or "elastomer-modified plastomers" (EMP). These blends play an important role [3.413], and the sequential EPDM grades are predominantly used here. The EPM or EPDM blend components also act as elastomeric modifiers to improve the impact strength, and since the modifiers have to be very elastomeric and often pre-crosslinked with peroxides, EPDMs are preferred in this application over EPMs, because of their higher reactivity.

Vulcanizing Agents. [3.400, 3.429–3.440, 3.455]. While EPM can only be crosslinked with *peroxides,* for EPDM one can use peroxides or sulphur plus accelerators. Although unvulcanized sequential EPDM grades already have a reasonably high mechanical strength, they are, as a rule, also crosslinked with peroxides or with sulphur and accelerators.

The methyl groups in polypropylene inhibit a crosslinking reaction with peroxides, and instead, they lead to chain scission. However, copolymers of propylene and ethylene can be crosslinked with peroxides, as long as the ethylene concentration is sufficiently high. The crosslinking reaction with peroxides is further improved as fewer propylene monomer sequences form in the polymer chain during polymerization of EPM or EPDM, since they tend to cause chain scission [3.400].

Therefore, if peroxide cures and high degrees of crosslinking are required, the rubber should have at least 50 mole% of ethylene, and this should be well distributed along the polymer chain. The termonomer also has a influence on the peroxide cure. DCP-EPDMs give higher cure states with peroxide than EPM [3.400]. Therefore, for cable insulations, which should be free of sulphur, one prefers DCP-EPDM over EPM in peroxide cures. Instead of being higher, the reactivity of ENB-EPDM is lower in peroxide cures than that of DCP-EPDM. The choice of peroxides depends on the required vulcanizate properties [3.400], and guidelines for selecting peroxides are discussed in greater detail on page 256ff. For peroxide cures of EPM or EPDM, activators are essential, such as sulphur, sulphur donors, acrylates, maleimides, quinones, [3.400], and criteria governing their choice are discussed on page 259.

For cable insulations, EPDM is frequently crosslinked with *high energy radiation* as well [3.400]. In this process no chemical curing agent is required; however, by using specific coactivators, a given degree of crosslinking can be obtained with a reduced dose rate, such as 150 Mrad without versus 10 to 15 Mrad with coactivator.

With *sulphur cures*, the required amount of sulphur and accelerator depends on the type and amount of termonomer in the polymer. For ENB grades, 0.5 to 2 phr (mainly 1.2 to 1.5 phr) sulphur or equivalent amounts of sulphur donor are generally sufficient, although the amount of sulphur donors should be limited to below 0.8 phr, if bloom is to be avoided.

It has also been commonly experienced that a single accelerator is not sufficient, and instead, a combination of several accelerators is required to obtain the desired fast cure rates and high states of cure. Commercial accelerator blends, which are meant to be used solely, are generally blends of thiazoles, thiurams, and dithiocarbamates, whereby the latter are specifically recommended for DCP and ENB grades. The commonly observed synergistic interaction of accelerators is therefore of great importance in EPDM vulcanization. If, for a specific accelerator, the cure rate does not significantly improve any more as the dosage is increased, it is useful to consider the use of an additional accelerator instead. [2.108, 2.109, 3.301, 3.434, 3.455]. This generally results in the very complex vulcanization recipes for EPDM. A general guideline for a suitable recipe is as follows: a moderate amount of sulphur (about 1.0 to 1.5 phr), sulphur donor (about 0 to 0.5 phr), thiazol accelerator (about 0.5 to 1.5 phr), dithio-carbamyl sulfenamide (1.2 to 1.8 phr), thiuram accelerator (about 0.4 to 0.9 phr), dithiocarbamate or dithiophosphate (about 0.3 to 3.0 phr), and perhaps some additional thiourea derivative. Since a number of accelerators of the above subgroups tend to bloom already at low concentrations, and since there can be a synergistic effect regarding bloom, the design of a suitable vulcanizing system often creates problems. This may be the reason for the wide acceptance of optimized preblended accelerator systems, particularly for no bloom (such as Deovulc EG 28 resp. EG 3, or nitrosoamine-free Deovulc BG 1/87, DOG) [3.434]. For an additional activation of the sulphur cure of EPDM, about 5 phr or more zinc oxide are also required.

Protective Agents. In EPM compounds one seldomly uses protective agents. For EPDM, which contains unsaturation, the use of protective agents is desirable if there is a special demand for good aging resistance of the vulcanizate. If staining can be tolerated, aromatic amines, and especially p-phenylene diamine, can be used. EPM and EPDM vulcanizates do not require protection by antiozonants.

Fillers. Unlike sequential EPM/EPDMs, amorphous and non-self-reinforcing EPM/EPDM grades have to be compounded with reinforcing fillers, if high levels of mechanical properties are desired. EPM/EPDM grades with high Mooney viscosities are particularly capable of accepting high loadings of filler (e.g. 200 to 400 phr), and plasticizer (e.g. 100 to 200 phr), and still give useful vulcanizates. At comparable Mooney viscosities, EPDM can be loaded more than EPM.

To lower cost and improve processibility of light compounds, or to lower cost of black compounds, calcined clay or fine particle size calcium carbonate are used. These fillers are economically priced, they disperse well in compounds, and do not influence the cure reaction. Calcined clays also reduce the water absorption of vulcanizates, which is important in cable applications. Calcium carbonate with very fine particle size can be advantageously used for soft vulcanizates with a good tear resistance at higher temperatures. Although hard clays give a good reinforcement, they retard the vulcanization, and are therefore seldomly used. A good compromise of vulcanization properties, mechanical properties and compound cost is also achieved by using a blend of hard and calcined clays. Soft clays and ordinary calcium carbonate are non-reinforcing fillers in EPM/EPDM, as with other rubber vulcanizates. These fillers can be used only to reduce compound cost at lower levels of mechanical properties, and they are often used in vulcanizates for sponge products.

Softeners. The most widely used softeners are naphthenic oils. For applications which specify high use temperatures, or for peroxide cures, paraffinic oils of low volatility are recommended. However, since paraffinic oils exude at low temperatures from EPDM vulcanizates, or from high-ethylene-EPDMs, they are often blended with naphthenic oils. On the other hand, naphthenic oils interfere with the peroxide cure. Aromatic oils reduce the mechanical properties of vulcanizates, and they also interfere with peroxide cures. Therefore, they are not recommended for EPM/EDPM.

EPM compounds have a very poor building tack, and if tack is required, the use of resins is absolutely necessary. Coumarone resins, Koresin, or xylene formaldehyde resins improve the compound tack to a certain extent, although there is a room for improvement.

Process Aids. To aid in the mixing of EPM/EPDM compounds, and particularly to improve the filler dispersion, stearic acid, zinc soaps, calcium soaps of fatty acids, or fatty alcohol residues are used as process aids [3.267, 3.456]. Ca soaps give the best extrudability and injection flow.

3.3.8.5 Processing of EPM and EPDM

[3.399, 3.427, 3.428]

Only compounds from low Mooney EPM or EPDM grades can be mixed on open mills. Therefore, EPM/EPDM compounds are almost exclusively mixed in internal mixers, and preferably by the so-called "upside down" technique. For further processing, one uses conventional methods. Since the EPM/EPDM is reversion resistant, high cure temperatures are generally used.

3.3.8.6 Properties of EPM and EPDM Vulcanizates

[3.399, 3.401, 3.441–3.452]

Mechanical Properties. The level of mechanical properties depends considerably on the type and amount of fillers in the compound, here the same applies as with SBR. Optimized vulcanizates have lower level of tensile properties as SBR. The mechanical properties of the sequential EPDM grades are particularly high, and they reach the level of those for NR vulcanizates. Tear resistance, particularly at high temperatures, compares also with that achieved by NR vulcanizates.

A wide range of hardnesses can be obtained with EPM or EPDM vulcanizates. Blends of EPDM and polypropylene give already in the uncrosslinked state high hardnesses, together with an acceptably high elasticity (Elastomer Modified Plastics, EMP).

The elastic properties of EPM or EPDM vulcanizates are by far superior to those of many other SR vulcanizates, and particularly IIR, but they do not reach the level obtained with NR vulcanizates.

The resistance to compression set of EPM or EPDM vulcanizates is surprisingly good, and this particularly applies to ENB-EPDM with a high ENB concentration, which is cured with either peroxides or sulphur with a highly active accelerator system [3.450 b]. Compression set of sulfur cured EPDM vulcanizates increases very rapidly with increasing temperatures in contrast to peroxidic crosslinked EPDM, which stay relatively stable [3.450 c].

Dynamic Properties. The dynamic properties and the dynamic fatigue resistance of EPDM vulcanizates are also very good, and comparable to those of SBR vulcanizates. This applies particularly to sulphur cured vulcanizates.

Heat and Aging Resistance. The resistance to heat and aging of optimized EPM and EPDM vulcanizates is better than that of SBR or NBR vulcanizates. It is comparable to that of sulphur cured IIR vulcanizates, but distinctly inferior to that of EVM or Q. Peroxide-cured EPM can, for instance, be exposed for 1000 hours at 150 °C without significant hardening. The heat resistance of sulphur-cured EPDM vulcanizates is somewhat poorer. Therefore, because of the higher heat resistance of peroxide cures, combined with a very low compression set, this cure system is also widely used for EPDM.

The aging resistance of EPM and EPDM vulcanizates is excellent. Under normal use conditions an oxidative degradation of the vulcanizates will not take place.

The ozone resistance is also excellent. Even after an exposure for many months to ozone-rich air of 100 pphm, the vulcanizates will not be seriously harmed. The ozone resistance of peroxide-cured EPM grades is best, and followed by sulphur-cured EPDM. The ozone resistance of EPDM vulcanizates is still superior to that of CR and IIR, although it does not quite reach the level of that for EVM and Q.

The resistance to weathering is also excellent, same as the ozone resistance. On exposure to open air, EPM and EPDM vulcanizates are not harmed by atmospheric conditions.

Low Temperature Flexibility. The low temperature flexibility of EPM and EPDM vulcanizates compares with that of NR vulcanizates.

Resistance to Swelling and Chemicals. EPM and EPDM vulcanizates have an excellent resistance to chemicals. They are not attacked, and if so, only slightly, by dilute acids, alkalis, by acetone, alcohol, and hydraulic fluids. Concentrated inorganic acids harden or destroy the vulcanizates.

This is in contrast to the swelling resistance in aliphatic, aromatic or chlorinated hydrocarbons. Since they are non-polar, EPM and EPDM vulcanizates swell considerably in these media.

Electrical Properties. The electrical insulating and dielectric properties, or the breakdown and corona resistance of EPM and EPDM vulcanizates are all excellent. In these respects, one should, however, differentiate between EPM and EPDM. EPM vulcanizates exhibit the best behaviour, and they compare with NR and Q. Therefore, EPM is used in insulations for rated voltages of over 25 kV. For voltages of up to 25 kV, it is preferable to use peroxide-cured EPDM, because it can be more easily processed and it accepts higher filler loadings. The electrical properties of these vulcanizates are also good at high temperatures and after heat aging of the vulcanizates. Because EPM and EPDM vulcanizates absorb little moisture, their good electrical properties suffer minimally when they are submersed in water.

3.3.8.7 Uses of EPM and EPDM

[2.104, 3.376, 3.441–3.452]

The main uses are in profiles, cable insulations and jacketing, as well as in products which have to be heat, weather, and salt water resistant. Additional uses are in white sidewall compounds of tires. In 1978, of the 150,000 tons of EPM and EPDM that were produced in the United States, 11% were used for tires, 42% for non-tire, automotive products (e.g. radiator hose, window, door and trunk mouldings, and bumper strips), 7% for non-automotive products (e.g. washing machine hose), 6% for electrical cable, 20% for plastic modifiers, and 14% for diverse applications such as window and architectural profiles, seals, or dock fenders.

3.3.8.8 Competitive Materials of EPM and EPDM

These are BIIR, CIIR, CM, CR, CSM, EVM, IIR, NR, and SBR.

3.3.9 Ethylene-Vinylacetate Copolymers (EVM)

[3.310, 3.457–3.464]

3.3.9.1 General Background about EVM

If ethylene is copolymerized with vinyl acetate instead of propylene, another specialty rubber is obtained, namely EVM. The presence of vinyl acetate units in this polymer chain suppresses the crystallization of polyethylene and also its melting point, as with EPM, as discussed on page 93.

Copolymers of ethylene with more than about 30% vinyl acetate give predominantly plastomeric properties. At vinyl acetate concentrations of between 30 and 75% by weight, the copolymers become rubbery. Copolymers ranging in vinyl acetate concentrations from 40 to 50% by weight are technologically the most important ones [3.458].

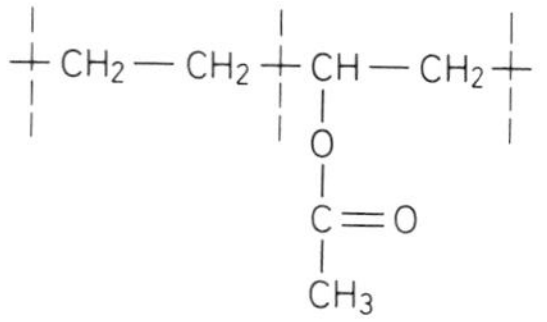

$$\left[CH_2 - CH_2 \right]\left[\underset{\underset{\underset{\underset{\Large CH_3}{|}}{C=O}}{|}}{\underset{O}{\underset{|}{CH}}} - CH_2 \right]$$

Ethylene Vinyl Acetate Units

3.3.9.2 Production of EVM

The high pressure, bulk polyethylene process produces copolymers with low vinyl acetate concentrations, and these products cannot be considered rubbers. The intermediate pressure, solution process, using pressures of 200 to 400 bar, produces high molecular weight copolymers with about 30% vinyl acetate [3.459]. Tert. butanol is used as a solvent, because it has a low chain transfer coefficient. The polymerization is initiated by free radicals, using azo compounds or peroxides as initiators. By this process, translucent, high molecular weight polymers are made, which are practically gel-free, and have little long chain branching.

EVM grades with over 60% by weight of vinyl acetate, which are produced by emulsion polymerization, are mainly used as adhesives and binders for pigments, and they are of little use as rubbers, because of their high gel content and high degree of chain branching.

3.3.9.3 Structure of EVM and its Influence on Properties

Vinyl Content. The crystallizability of polyethylene is already suppressed by small amounts of vinyl acetate in the chain. However, the dynamic freezing point is still very high, so that these polymers represent thermoplastics. Only at higher vinyl acetate concentrations, does the dynamic freezing point fall to below room temperature, and at 40 to 50 weight% of vinyl acetate, it is in the range of -20 to $-25\,°C$. These polymers are amorphous.

Because of the polar nature of vinyl acetate, the swelling resistance of the copolymer in animal, vegetable, and mineral oils improves with increasing vinyl acetate content. At 40 to 50 weight% of vinyl acetate, the swelling resistance is already sufficient for many applications. Copolymers with 70% vinyl acetate have a good oil resistance.

Viscosity. EVM grades produced by the medium-pressure solution process have a intermediate molecular weight of about 200,000 to 400,000, and Mooney viscosities ML 1+4 at 100 °C of 20 to 30. Since it behaves like thermoplastics, EVM processes very well. EVM grades that were produced in bulk polymerization have very low molecular weights of about 50,000, and therefore have little use in the rubber industry.

3.3.9.4 Compounding of EVM

[1.14]

Blends. It is possible to blend EVM with diene polymers like NR or SBR, to improve the weather resistance of the latter. EVM remains uncrosslinked in these blends, and behaves like a polymeric plasticizer. In this application, primarily low viscosity EVM grades are used. It is, of course, also possible to crosslink EVM with peroxides, and in this instance, a co-curing of the blend polymers takes place.

Vulcanizing Agents. Since EVM is a fully saturated polymer, it cannot be crosslinked with sulphur and accelerators. Instead, peroxides have to be used, and their suitability depends on the processing conditions (decomposition temperature, odour). Peroxides alone do not usually give a sufficient state of cure, and therefore, special coactivators have to be used, such as triallyl cyanurate, triallyl phosphate, and so on [3.458, 3.460]. Since these coagents improve considerably the state of cure without raising the peroxide concentration, their use is absolutely necessary. Metal oxides are not required in these cure systems.

Protective Agents. The oxidative stability of the saturated EVM is such that, as a rule, no protective agents are required. These would also interfere with peroxide cures. For optimum heat resistance it is, however, advisable to protect the ester group against hydrolysis by adding a polycarbodiimide.

Fillers. Because EVM is amorphous, and since it lacks self-reinforcing properties, EVM compounds have, therefore, to be formulated with reinforcing fillers, which can be selected following generally applicable rules. The use of light fillers, such as precipitated silicas or silicates, can cause problems, since, in spite of their good reinforcing qualities, they can have a detrimental effect on cure. Neutral kaolin grades have given good results. Since the EVM grades have a relatively low viscosity, they cannot accept as high filler loadings as EPM and EPDM.

Softeners. While the use of softeners is necessary to improve the tack of EVM compounds, their choice is very limited. This is so because softeners can interfere with the peroxide cure, and EVM grades have low Mooney viscosities. Paraffinic oils with very low unsaturation are most frequently used, and very highly saturated paraffinic oils are particularly well suited. To improve low temperature flexibility, ester- and ether-based plasticizers which do not react with peroxides are used, such as dibutylphthalate, dioctyladipate and dioctylsebacate or the condensation products of phenol and alkylsulfonic acids. Resins can react with peroxides and are therefore hardly used at all.

Process Aids. Since stearic acid, pentaerythritol tetrastearate and fatty alcohol residues do not interfere with peroxide cures, they can be used in compounds to improve filler dispersion, and the extrudability or calenderability.

3.3.9.5 Processing of EVM

[1.14]

EVM compounds are mixed and processed using methods which are generally applied in the rubber industry. Special attention should be given to the cleanliness of equipment, since impurities, and especially sulphur, have a detrimental influence on peroxide cures. For the vulcanization process, the same precautions have to be observed as for peroxide cures in general (see page 256).

3.3.9.6 Properties of EVM Vulcanizates

[1.14]

Mechanical Properties. Optimum mechanical properties of EVM vulcanizates are found in the hardness range of Shore A 60 to 85. With reinforcing fillers, high tensile strengths can be obtained, whereas the tear resistance leaves much to be desired. (This is a result of the peroxide cure.) The abrasion resistance also does not compare with that of good diene rubber vulcanizates, and the elasticity of EVM vulcanizates is often obscured by a stiffening effect due to the thermoplastic character of the copolymer.

The compression set is also influenced by the residual thermoplastic behaviour, and it is therefore poor at room temperature. At higher temperatures, however, the compression set becomes extremely low.

Heat and Aging Resistance. The main technical superiority of EVM vulcanizates in the rubber field is their excellent resistance to degradation in hot air. The heat aging resistance of EVM vulcanizates in closed systems, such as in special cables, can be superior even to that of Q.

Other advantageous properties are the extremely good weather and ozone resistance.

Swelling Resistance. Unlike EPM vulcanizates, those from EVM are sufficiently resistant to swelling in aliphatic oils to qualify them for many applications. The swelling resistance improves with increasing vinyl acetate content [3.310].

Electrical Properties. Because of the polar nature of EVM, its electrical resistivity is poorer than that of EPM. However, it can still be sufficient for many applications when EVM grades with about 40 weight% vinyl acetate are used.

Flame Resistance. Since EVM contains oxygen, its oxygen index (LOI according to ASTM D2863) is intrinsically relatively high, and it depends, of course, on the vinyl acetate concentration in the copolymer. EVM grades with 40% vinyl acetate have an oxygen index of 35%, while those with 70% vinyl acetate reach a value of higher than 50% [3.310]. By using hydrated alumina as filler, halogen-free compounds can be formulated for cable insulations, which have a good flame resistance [3.464]. These compounds can compete with those from CR, CM, and CSM (see pages 83 and 103).

Other Properties. As previously mentioned, the ozone resistance of NR and SBR can be improved when they are blended with EVM. Also worth mentioning are the good colour stability of light EVM vulcanizates, and the possibility using them for products whose properties have to comply with physiological requirements.

3.3.9.7 Uses of EVM

[1.14, 3.464]

EVM is frequently used in heat resistant rubber products (seals), and in cables (heating tubes). There are also important uses in profiles and membranes. EVM is also applied in blends to improve the ozone and weather resistance of NR and SBR vulcanizates and it is increasingly used for melt adhesives and for plastic modifiers.

Peroxides can be used to graft different vinyl monomers on to EVM as readily as for curing reactions. EVM copolymers with 40 to 50% vinyl acetate grafted with PVC have gained some importance. Those with a relatively low EVM concentration of 5 to 10% are impact resistant PVCs, while those with a high EVM content are more closely related to soft PVCs [3.461–3.463].

3.3.9.8 Competitive Materials of EVM

These are BIIR, CIIR, CM, CSM, CR, EPDM, EPM, and IIR.

3.3.10 Chlorinated Polyethylene (CM)

[3.465–3.474]

3.3.10.1 General Background about CM

As a result of chlorination, the crystallinity of polyethylene is interfered with, so that the polymers become rubbery, and can be crosslinked by means of peroxides and other chemicals. The CM grades differ mainly in the degree of chlorination (e.g. 25 to 42% by weight of chlorine), Mooney viscosity, and amount of crystallinity.

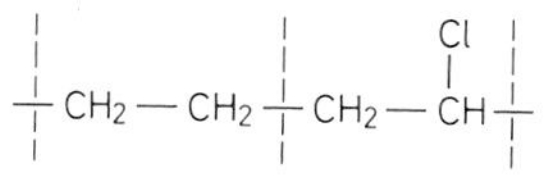

$$\left[CH_2 - CH_2 \right] \left[CH_2 - \underset{}{\overset{Cl}{\overset{|}{C}}}H \right]$$

Ethylene Chloroethylene Units

3.3.10.2 Production of CM

[3.5]

In principle, it is possible to produce CM by chlorination of polyethylene in solution, in emulsion, or in suspension. The commercially available grades are produced by a random chlorination in suspension of high density polyethylene. Since high density polyethylene does not readily dissolve because of its high crystallinity, a uniform chlorination of the polymer requires the use of higher temperatures.

3.3.10.3 Structure of CM and its Influence on Properties

The remaining crystallinity in the polymer depends on the extent of chlorination. CM grades with 25% by weight of chlorine are distinctly crystalline and contain long polyethylene sequences. Therefore, they are relatively hard. At about 35% chlorine, the crystallinity vanishes considerably, and these CM grades have the best low temperature flexibility. The brittleness temperature is about −40 °C. At significantly higher chlorine concentrations, the polymers become even harder and more brittle (above 45% by weight chlorine), because of the stronger interactions of the carbon-chlorine moieties [3.469].

The polymer also becomes more polar due to the chlorine, and this results in a better swelling resistance when in contact with animal, vegetable, and mineral oils. Increasing chlorine concentrations also improve the flame resistance of the polymer.

Because they lack double bonds, and because of the random incorporation of chlorine atoms in the polymer chain, CM can be crosslinked with peroxides to give vulcanizates which have a better heat resistance than those from CR.

Through dehydrochlorination of CM above 160 °C in the presence of ZnO, one can produce unsaturated products, which can be crosslinked with sulphur and accelerators, but they do not play a significant role technologically.

Because of its distinct thermoplastic properties, CM processes extremely well.

3.3.10.4 Compounding of CM

[3.467, 3.470–3.472]

Blends. CM can be blended with other rubbers, which are crosslinkable by means of peroxides. It is, for instance, blended with EPDM to improve its low temperature flexibility and cure rate, or blended with NBR to improve its swelling resistance in oils or motor fuels. In amounts of about 5 phr, SBR is a useful coagent to improve the cure of CM.

Vulcanizing Agents. The conventional commercial CM grades are only crosslinked with peroxides, whose suitability depends on the particular manufacturing process for CM. Main considerations are scorch safety, and the specified cure temperature. Coagents in the peroxide cure are, as with EVM, triallyl cyanurate, as well as triallyl mellitate, trimethylpropane trimethacrylate [EDMA], or diallyl phthalate. The latter is also a useful plasticizer. More recently, thiadiazoles [3.465, 3.466] and triazine thiols [3.468] have been mentioned for peroxide-free crosslinking reactions.

Stabilizers and Protective Agents. Like any other chlorine-containing rubber, CM has to be protected against HCl cleavage by adding metal oxides to compounds. Especially well suited are MgO and Pb-compounds, as are epoxidized oils and epoxy resins. While MgO is inexpensive and a general purpose stabilizer, lead compounds give the best aging resistance, and the least water swell, but these compounds discolour the vulcanizate, and they are expensive and poisonous. ZnO and Zn-compounds cannot be used as stabilizers, since already in very small concentrations, they interfere with the stability of the vulcanizate.

Protective agents to improve the aging resistance are not generally required for CM. If particularly high demands for heat resistance are to be met, one could use TMQ or ADPA in these compounds.

Fillers. CM can accept high filler loadings, from about 30 to 200 phr. Fillers which are generally used by the rubber industry can be employed, namely blacks, silicas, and mineral fillers. To avoid interference with the peroxide crosslinking reaction, fillers with a pH of 7 or higher are preferred, such as calcium carbonate, silica, kaolin, and talc.

Softeners. The polarity of the rubber, and the possible interaction with the peroxide cure, should be considered when choosing softeners. This eliminates the use of mineral oils in higher quantities, and it restricts the choice to ester- and ether-based plasticizers, to those which do not react with peroxides. Dioctyl adipate and dioctyl sebacate are particularly well suited.

Process Aids. Because they process so well, CM compounds do not generally require process aids, obviously the flow properties can be markedly increased by the use of e.g. calcium soaps of saturated fatty acids [3.466a].

3.3.10.5 Processing of CM

CM can be processed without problems using conventional rubber machinery. Since CM is more thermoplastic than NR or SBR, all processing steps require a stricter temperature control.

3.3.10.6 Properties of CM Vulcanizates

[3.465a, 3.467, 3.470–3.472]

The properties of CM vulcanizates can be characterized as follows: good mechanical properties, low compression set (up to 150 °C), low brittleness temperature, very good dynamic fatigue, excellent aging, weathering and ozone resistance, good oil resistance, even at higher oil temperatures and for many oil blends [3.194], very good chemical resistance, including oxidizing chemicals, good flame resistance, and very good colour stability. Regarding flame resistance, it behaves like EVM (see page 103) or CR (see page 83).

3.3.10.7 Uses of CM

[3.467, 3.470–3.472, 3.474]

CM is especially recommended for those applications, in which there is a high demand for aging resistance in hot air, oils, chemicals, and a good ozone and weathering resistance, combined with a good flame resistance. The main uses are in the wire and cable industry, but these applications are threatened to some extent by oxygen-containing, chlorine-free polymers (see page 103).

3.3.10.8 Competitive Materials of CM

These are BIIR, CIIR, CO, CR, CSM, ECO, EVM, EPM, EPDM, and NBR.

3.3.11 Chlorosulfonated Polyethylene (CSM)

[3.290, 3.475–3.487]

3.3.11.1 General Background about CSM

The general property spectrum of CSM corresponds to that of CM, although CSM has in addition to chlorine sidegroups also chlorosulfonyl side groups in which the chlorine is less firmly bound. Therefore, CSM can be more readily crosslinked than CM. The structural formula of CSM is as follows:

$$\left[CH_2-\underset{\underset{Cl}{|}}{CH} \right]\left[CH_2-CH_2 \right]\left[\underset{\underset{SO_2Cl}{|}}{CH}-CH_2 \right]$$

Chloro-ethylene Ethylene Chlorosulfonyl ethylene Units

3.3.11.2 Production of CSM

CSM is produced by UV radiation of high pressure polyethylene in an inert chlorinated solvent at 70 to 75 °C in the presence of gaseous chlorine and sulphur dioxide. The mechanism is analogous to that of the Lee-Horn reaction [3.487]. CSM is produced on a large scale by Du Pont under the name of Hypalon [3.483] and for quite some time now also in Japan.

3.3.11.3 Structure of CSM and its Influence on Properties

Commercially available CSM grades contain 25 to 43% by weight of chlorine, and 0.8 to 1.5% by weight sulphur, so that, on average, a chlorine atom is attached to every seventh carbon atom, and a chlorosulfonyl group to every eighthy fifth carbon of the polymer chain [3.481]. The chlorine and sulphur are randomly distributed along the saturated polymer chain. The cure rate and cure state grow with increasing degrees of chlorination/sulfo-chlorination. Depending on the type of crosslinking agent, crosslinking takes place either through the chlorosulfonyl group or through the pendent chlorine on the chain. Due to its fully saturated polymer backbone, CSM is, like CM, very resistant to weathering, aging, ozone, and chemical degradation.

3.3.11.4 Compounding of CSM

CSM rubber grades with a low chlorine content of about 25% are best suited for applications that require an optimum heat resistance and/or optimum electrical resistivity. By contrast, those grades with higher chlorine contents of 35% have the optimum flame resistance.

Vulcanizing chemicals. Due to its higher reactivity, CSM is more easily crosslinked than CM. Polyvalent metal oxides, such as lead and magnesium oxides, react in the presence of small amounts of weak acids, like stearic acid, abietic acid, etc., and of accelerators or sulphur donors, like TMTD, DPTT and MBT. The crosslinks are bridges of $-SO_3-$ - Me $-O_3S-$. In special cases, polyfunctional alcohols, like pentaerythritol, are used as curing agents, together with bases, to form disulfonic acid ester bridges as crosslinks [1.12, 1.13, 3.484]. With special aliphatic diamines a rapid crosslinking reaction takes place, while aromatic diamines require higher curing temperatures.

Under certain conditions, CSM can also be crosslinked with peroxides, to form vulcanizates with a good heat resistance [3.477, 3.485, 3.486]. For this purpose, CSM is often blended with small amounts of EPDM or EPM [3.485]. Divalent basic metal oxides play a dual role in the compounds: they are acid acceptors, and they control the pH value. MgO is primarily used for general purpose compounds, while dibasic lead phthalate, for compounds with a low water swell. Polyols are added to improve the solubility of these metal compounds in the rubber formulation. Peroxide reactions are activated by coagents like TAC, TATM, or methylene bismaleimide (HVA-2) together with amine accelerators [3.476, 3.486].

Antioxidants are generally not needed. However, for an extremely high heat resistance at temperatures of 150 to 180 °C, synergistic combinations of thiodiethylene-bis-[3.5-ditert.butyl-4-hydroxyl]-hydrocinnamate (AD-1) and dilauryl-thiodipropionate (AD-2) are most beneficially used in peroxide cures.

Fillers, Plasticizers. The same fillers and plasticizers are used for CSM as for CM compounds.

3.3.11.5 Processing of CSM

CSM is more difficult to process than CR or CM.

3.3.11.6 Properties of CSM Vulcanizates

The spectrum of physical properties is very similar for CSM and CM vulcanizates.

3.3.11.7 Uses of CSM

For reasons given above, CSM is used in the same applications as CR and CM, namely in cables, hoses, moulded goods and membranes.

3.3.11.8 Competitive Materials of CSM

These are BIIR, CIIR, CM, CO, CR, ECO, EPM, EPDM, and NBR.

3.3.12 Acrylic Rubbers (ACM) and Ethylene-Acrylate-Copolymers (EAM) [3.488-3.541 a]

3.3.12.1 General Background about ACM

Through copolymerization of acrylic esters with monomers, which allow subsequent crosslinking reactions to take place, one obtains saturated and amorphous polymers which are very polar. As expected, these copolymers have excellent oil, heat, aging, and ozone resistance, and in this respect these ACM grades occupy an intermediate position between NBR and FKM. The acrylic esters are primarily ethyl acrylate and/or butyl or octyl acrylate, as well as ethylmethoxy or -ethyloxy acrylate.

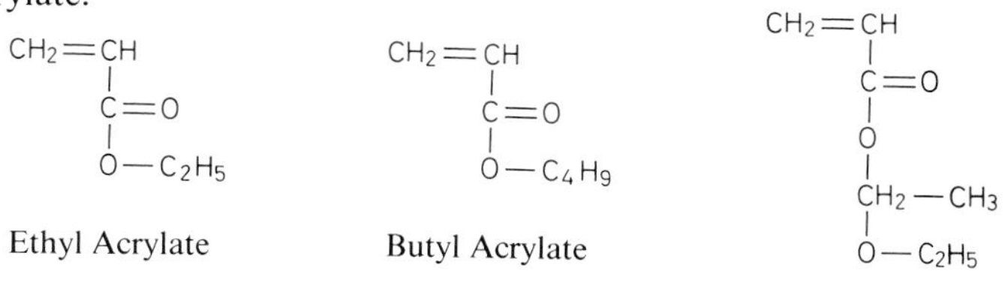

$$CH_2{=}CH-C({=}O)-O-C_2H_5$$

Ethyl Acrylate

$$CH_2{=}CH-C({=}O)-O-C_4H_9$$

Butyl Acrylate

$$CH_2{=}CH-C({=}O)-O-CH_2-CH_3 \quad (\text{with } O-C_2H_5 \text{ on } CH_2)$$

Ethylethoxy ethyl Acrylate

As vinyl comonomers, one can use those with [3.513–3.516] or without reactive chlorine groups. Recommended compounds are, for instance, 2-chloroethyl vinyl ether, vinylchloro acetate, chloromethyl acrylic acid, or its ethyl ester, glycidyl ethers [3.517–3.519], methylol compounds [3.520–3.522], imido esters [3.523, 3.524], hydroxy acrylates, such as β-hydroxy ethyl acrylate, carboxylic compounds, such as methacrylic acid, as well as alkylidene norbornene [3.525]. The latter three compounds can be used in combination with epoxy compounds. Examples of types of comonomers for ACM are:

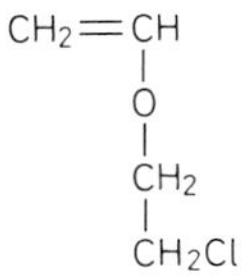

$CH_2{=}CH{-}O{-}CH_2{-}CH_2Cl$	$CH_2{=}CH{-}O{-}C({=}O){-}CH_2Cl$	$CH_2{=}CH{-}C({=}O){-}O{-}CH_2{-}Cl$
2 Chloroethyl Vinyl Ether	Vinyl Chloro Acetate	Chloromethyl Acrylate
$CH_2{=}CH{-}CH_2{-}O{-}CH_2{-}CH(O)CH_2$	$CH_2{=}CH{-}C({=}O){-}NH{-}CH_2OH$	$CH_2{=}CH{-}C({=}O){-}N$ (ring with N)
Allyl Glycidyl Ether	N-Methylol Acrylamide (for latex)	Acrylimide Derivative

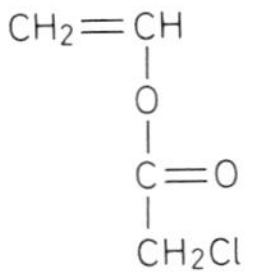

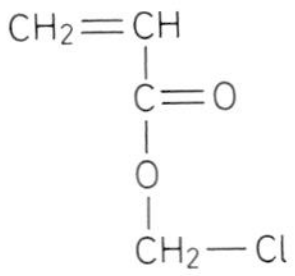

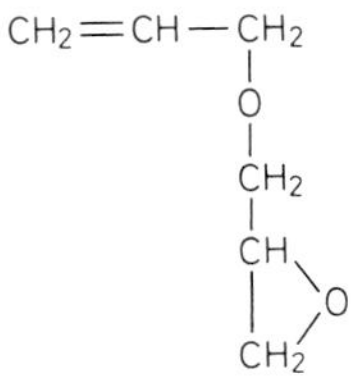

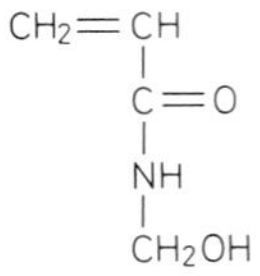

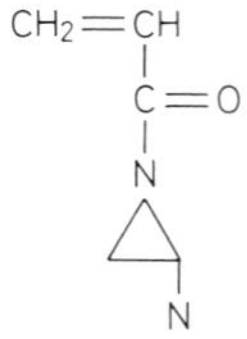

ACM was developed by Goodrich in the U.S.A. and has been produced on a large scale since 1948. Through modification of comonomers, products were developed with time, which gave great improvements in reactivity regarding cure rate and state of cure. More recently, Goodrich succeeded in the development of a new generation of ACM grades, which have such a high reactivity that it is no longer necessary (as it had been with conventional grades) to post cure the vulcanizates.

3.3.12.2 Production of ACM

[3.5, 3.491]

ACM is produced mainly by emulsion polymerization processes. For suspension polymerizations, which play a minor role, non-ionic water soluble dispersing agents and peroxides or azo compounds, which are soluble in the monomer, are used. Since the monomers are prone to hydrolysis, the emulsion polymerization has to be carried out in a basic medium at a pH of greater than 7. Therefore, long chain alkyl sulfates or -sulfonates are used here. The polymerization reaction is initiated by organic peroxides or azo compounds. Potassium persulfate or redox systems are also used. The additional use of chain modifiers is not necessary. The molecular weight of the polymer can be adjusted through the monomer/initiator ratio. The rubber is recovered from the latex using conventional methods, but the largest proportion, by far, of ACM, is used in latex form.

3.3.12.3 Structure of ACM and its Influence on Properties [3.15]

Influence of Acrylate. The choice of acrylate determines primarily the brittleness temperature, but also the swelling and heat resistance of the vulcanizate. Polyethyl acrylate has a higher polarity and gives, therefore, the best oil and heat resistance. However, because of its relatively high glass transition temperature T_g of $-21\,°C$, its vulcanizates have an insufficient low temperature flexibility. By contrast, polybutyl acrylate has a T_g of $-49\,°C$ [3.526], and its vulcanizates have, therefore, a good low temperature flexibility, but the oil and heat resistance are inferior. Better still is the low temperature flexibility of polyoctyl acrylate, but its oil and heat resistance are even further reduced. The use of alkoxy acrylates also results in improved low temperature flexibility [3.527]. A good compromise between the opposing properties is reached by copolymerization of ethyl- and butyl-acrylates.

Comonomers. The type of comonomer has little influence on the oil and heat resistance or low temperature flexibility, but it determines, instead, the cure behaviour, and with it, the physical properties of the vulcanizate. The early commercial grades, and some of today's chloroethyl vinyl ether and vinyl chloro acetate grades are characterized by the poor reactivity of the comonomers during polymerization, and poor reactivity of the halogen during crosslinking reactions. Also with several other comonomers, the goal of sufficient reactivities in polymerization and crosslinking reactions could not be reached satisfactorily. Only with the new generation of ACM grades, which, however, require special curing agents, was a significant progress made in satisfying both requirements. A different comonomer was used, whose identity has not been revealed [3.500, 3.503]. This newly developed copolymer gives high states of cure after relatively short cure times, with and without post vulcanization. This, in turn, results in distinctly improved mechanical properties and low compression set.

The reactive groups on the comonomer, namely chlorine, glycidil, or methylol groups, determine the choice of the vulcanization system. Chlorine groups require amines, sulphur, and accelerators, and/or combinations of metal soaps and sulphur. Glycidil groups enter the crosslinking reaction with compounds that split off NH_3-compounds, namely ammonium benzoate or dicarbonic acids [3.528]. Peroxides do not yield useful vulcanizates.

Contrary to copolymers with acrylamides, those with methylol acrylamide groups spontaneously crosslink by splitting off water, to form methylene-bisamide bridges between polymer chains. This reaction occurs when the copolymer is heated in the presence of phthalic anhydride, and it is exploited with latices in a great number of applications [1.7].

Saturation. Because they lack double bonds, ACM rubbers have the same characteristics as all other saturated polymers, namely excellent oxidation, ozone, and weather resistance. The resistance to chemicals is, however, limited, because of the possibility of saponification of ester groups.

3.3.12.4 Compounding of ACM [3.500, 3.502, 3.510, 3.529–3.535a]

Blends. ACM is rarely blended with other rubbers. Blends of ACM and FKM have been mentioned to optimize the properties of the individual blend components [3.534–3.535]. In addition, ECO has been proposed as blend partner to improve the low temperature flexibility of ACM [2.525]. These blends are rarely used.

Vulcanizing Agents. [3.500, 3.502, 3.506, 3.530] The early cure systems, consisting of diamines and Na-metasilicate, are not recommended any longer, because of poor processing behaviour and stickiness. The conventional chloroethyl vinyl ether ACM grades were mostly crosslinked with thiourea, and di- or polyamine cure systems. Examples are ETU/Pb_3O_4/hexamethylenediamine carbamate/basic Pb-phosphite and TETA or PEP/MBTS/sulphur. For vinylchloro acetate grades, one uses less reactive cure systems, which give less rise to corrosion, namely metal soaps and sulphur. These are today's most commonly used cure systems for ACM, although diamines are also used at times for these ACM grades. Crosslinking with metal soaps/sulphur can be inhibited by acids [1.12, 1.13], such as stearic acid, or the reaction can be accelerated by bases, such as MgO. The soap cation has a great influence on the vulcanization reaction. K-stearate is substantially more reactive than Na-stearate, particularly at cure temperatures below 175 °C. Sometimes, combinations of both stearates are used. Sulphur donors can also be used in the place of sulphur. Although glycidil-containing ACM grades are, at times, also crosslinked with metal soaps/sulphur, the preferred cure system for these grades is ammonium benzoate [3.528], or ammonium adipate, if a low compression set is required of the vulcanizate. The new ACM grades can also be cured with soap/sulphur and diamines, although these systems will not do them justice, particularly regarding the low compression set which can normally be achieved with vulcanizates of these new ACM grades. Instead, they should be crosslinked with soap/tertiary amines or soap/quaternary ammonium systems. For instance, a good balance of processing characteristics and vulcanizate properties is obtained when Na- or K-stearate and Diuron, which is a herbicide, are used as curatives. Combinations of accelerators, like DOTG/TMTM or DOTG/OTOS, as well as new thioketone accelerators give interesting vulcanizate properties. ACM without labile chlorine atoms, and with reactive carboxyl and epoxy groups instead, can be crosslinked with good results using OTBG and octadecyltrimethylammonium bromide (Arax B18, Bozetto) or new NPC system (Goodrich). ACM cannot be crosslinked with peroxides.

Stabilizers. As with all other chloride-containing polymers, the ACM grades with chlorine also require stabilizers which absorb HCl. However, this may not always be necessary when basic cure systems are used. Otherwise, ACM is stabilized using basic Pb-phosphite, and Pb-phthalate or Pb_3O_4. Other protective agents are normally not required with ACM vulcanizates, although for very demanding levels of heat resistance, one can add aromatic amines to compounds, namely p-phenylene diamine derivatives, ODPA, or organic phosphites.

Fillers. Gum vulcanizates or those compounded with non-reinforcing fillers have insufficient strength. Therefore, it is important to select the correct fillers. For good mechanical vulcanizate properties, active blacks or silicates are used. The fillers should be neutral or basic in order to avoid an interference with the basic vulcanization reaction. Of the blacks, N 326 and N 550, in particular, give a good balance of processibility and vulcanizate properties. Silicas, in combination with Al-silicates or silane-treated clays, are also recommended.

Softeners. Softeners are normally not used in ACM compounds, specifically with a view to the heat resistance and compression set of the vulcanizates. If, however, a particularly good low temperature flexibility is required, 5 to 10 phr of low-volatile adipic plasticizers, polyetheresters, alkylacrylopolyether alcohols, polyoxethylated nonylphenol, etc., are used. Softeners of low volatility and low extractability, are polyesters. To improve the tack of ACM compounds, coumarone indene resins and Koresin can be added.

Process Aids. ACM compounds are difficult to process. Therefore, it is necessary to use process aids, such as stearic acid, which retards, or zinc soaps, which accelerate, or fatty alcohol residues, octadecylamine, especially combinations of fatty alcohols with fatty acid soaps and best of all pentaerythritol tetrastearate [3.535 a], which do not influence the vulcanization.

3.3.12.5 Processing of ACM

[3.501, 3.504]

ACM compounds have a somewhat unusual rheological and vulcanization behaviour, which results in some processing difficulties. Yet, the compounds are mixed and processed in conventional rubber machinery and using conventional processes (see page 26). In addition to these, it is necessary in most cases, particularly if a low compression set is required, to post cure the vulcanizates for several hours at 150 °C. This is, however, not necessary any longer for a new generation ACM grades, although it is recommended.

3.3.12.6 Properties of ACM Vulcanizates

[3.488-3.492, 3.500]

Mechanical Properties. As with other specialty rubbers, the tensile properties of ACM vulcanizates do not reach the level of those from NR or NBR, but the tensile properties are sufficient for the usual requirements. The new generation ACM grades have, however, much better physical properties than the conventional grades, and this applies particularly to compression set, and the aging resistance after exposure for long times to high temperatures.

Aging, Heat, and Ozone Resistance. ACM grades, and particularly the new generation ones, and the ethylene/acrylate copolymers, can be used under certain conditions for 1000 hours at 160 to 170 °C. After exposure for 1000 hours to oil of 150 °C, the level of physical properties will not have changed noticeably. However, one has to remember that ACM vulcanizates soften at high temperatures to a greater extent than diene rubbers, such as NBR, because of their thermoplastic characteristics. ACM vulcanizates are also resistant to ozone degradation.

Swelling Resistance, Resistance to Chemicals. ACM vulcanizates are much more resistant to swelling in animal, vegetable, and mineral oils than all the rubbers discussed so far, and this applies also to swelling at high temperatures, where ACM is surpassed only by FKM. Unlike NBR, ACM is also resistant to most of the additives which are blended with technical oils [3.273]. Therefore, ACM has replaced NBR in many of its former applications. ACM vulcanizates are not resistant to motor fuels, and in this respect, they are surpassed by NBR, ECO and FKM. Finally, the resistance of ACM to chemicals is not particularly good.

Low Temperature Flexibility. ACM grades based on ethyl acrylate, compounded without plasticizers, have a brittleness temperature of −18 °C, which is not always low enough. Therefore, either additional plasticizers or an ethyl acrylate/butyl acrylate, or both, are used in many applications. There are no problems in reaching brittleness temperatures of −40 °C. Even with ethyl acrylate grades, one can reach dynamic brittleness temperatures of −38 °C with polyester ether plasticizers without impairing the heat resistance of the vulcanizate, but the compression set will be higher.

3.3.12.7 Uses of ACM

[3.492]

Over 90% of ACM is used in automotive or engineered products [3.252]. The main application is in shaft seals of all constructions, namely seals for crankshafts, automatic and differential transmissions, valves, or O-rings and oil hose. ACM is also used, at times, for special roll coverings. The relatively high price of ACM restricts its applications. ACM is threatened by FKM in some applications [3.252].

3.3.12.8 Competitive Materials of ACM

These are EAM, FKM, NBR, and Q.

3.3.12.9 Ethylene/Acrylate Copolymers (EAM)

[3.536–3.541 a]

In 1975, Du Pont introduced a new generate of acrylate rubbers under the name of Vamac, consisting of ethylene, methylacrylate, and a monomer with a carboxylic acid sidegroup. This acrylate rubber can be crosslinked with multivalent amines, and because of its molecular structure, it has a property spectrum between that of ACM and EVM.

While the heat resistance of EAM vulcanizates is better than that of ACM vulcanizates [3.541], the swelling resistance of EAM in hot oils corresponds more to that of butylacrylate vulcanizates [3.540]. The low temperature flexibility and elongation at break of EAM are considerably better than that of most acrylate rubbers, and it corresponds to that of polyoctyl acrylate vulcanizates. This is due to the improved chain flexibility of the ethylene sequences.

EAM is supplied as a rubber, and as a masterbatch with 20 phr of filler. For the compounding of the rubber, the usual rules for choosing fillers apply. The filler masterbatches can also be compounded with additional filler to obtain the desired hardness level.

EAM is vulcanized with capped diamines, such as hexamethylene diamine carbamate and DPG or DOTG, if faster cure rates are desired.

Process aids for EAM are octadecyl amine in combination with stearic acid, and a complex of organo phosphate (Gafac RL-210) and pentaerythritol tetrastearate. This prevents sticking of compounds to mill rolls.

Otherwise, EAM is compounded in the same way as ACM.

In spite of lower damping, vulcanizates of EAM are preferably used for static seals, damping devices, hose, but seldomly for crankshaft seals. Because it contains oxygen, EAM has intrinsically a relatively high oxygen index (LOI according to ASTM D2863), but with the aid of hydrated alumina, it is possible to develop EAM compounds with good flame resistance for halogen-free cable jackets. Because of their improved safety, these compounds can compete well with chlorine-containing cable jackets from CR, CM, and CSM (see page 83 and 105), but also with other oxygen-containing compounds from EVM (see page 103).

3.3.13 Epichlorohydrin Rubber (CO, ECO, and ETER)

[3.191, 3.204, 3.542–3.559b]

3.3.13.1 General Background about CO, ECO, and ETER

[3.542–3.556]

Through a ring-opening polymerization of epichlorohydrin, amorphous homopolymers, which have the structure of polyethylene ether with a chloromethyl sidegroup, are obtained [3.547, 3.557]. These polymers can be vulcanized through the chloromethyl group using a number of different crosslinking agents. These homopolymers, coded CO (-O- for ether), are extremely polar, and, therefore, swell resistant, and they have a relatively high glass transition temperature T_g. By copolymerizing epichlorohydrin with ethylene oxide, one obtains copolymers, ECO, with lower T_g temperatures [3.557]. Goodrich finally succeeded in a terpolymerization, where a diene monomer with unsaturation in the sidegroup could be incorporated in ECO. The exact nature of this diene monomer is not known, but it is likely allyl glycidil ether.

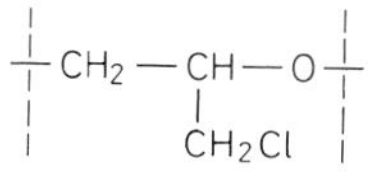

$$\left[CH_2-\underset{\substack{| \\ CH_2Cl}}{CH}-O \right] \qquad \left[CH_2-\underset{\substack{| \\ CH_2Cl}}{CH}-O \right]\left[CH_2-CH_2-O \right]$$

Polyepichlorohydrin CO 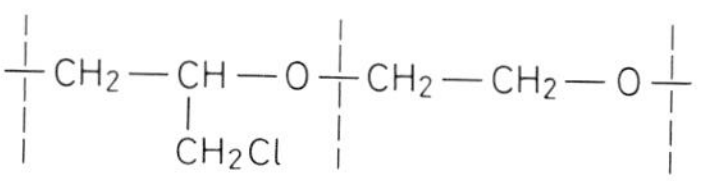 — Epichlorohydrin-Copolymer ECO — Ethylene oxide — Units

This terpolymer, also known as ETER, can be crosslinked through the double bond with sulphur but also with peroxides. Due to the high technical demands of the automotive industry, the importance of these rubbers has increased in recent years at the expense of NBR and CR [3.204, 3.252].

3.3.13.2 Production of CO and ECO

[3.547, 3.557]

Catalysts for the ring opening polymerization of epichlorohydrin are the same as for epoxy polymerizations, namely Al $(alkyl)_3$/water in a molar ratio of, for instance, 0.5 : 1. These catalyst systems can be modified with chelating agents, like acetyl acetone [3.558], to improve the catalyst activity. This is particularly important for copolymerizations. The polymerization reactions can be carried out in aliphatic, aromatic, or chlorinated hydrocarbon solvents, but also in ethers, and they proceed at room temperature at sufficiently high monomer conversion rates. In practice, higher polymerization temperatures are used to increase polymerization rates.

3.3.13.3 Structure of CO, ECO, and ETER, and its Influence on Properties

[3.5]

Concentration of Chloromethyl Groups. The homopolymer has the highest concentration of chloromethyl groups. Therefore, CO has, on the one hand, the highest vulcanization rates, the highest polarity, and with it, the best swelling and heat resistance, but on the other hand, the poorest low temperature flexibility of the vulcanizates. The permeability is also low and the flame resistance is very good of these vulcanizates. ECO, or the terpolymers, having fewer chloromethyl groups, represent a compromise between the optimum in swell and heat resistance, and the brittleness temperature. Because of the chlorine, the polymers tend to stick to mill rolls. This makes the processing difficult, unless processing aids are added to compounds.

Saturation. Since they have a fully saturated polymer backbone, CO and ECO are, like other saturated polymers, resistant to oxidative and ozone degradation. The ether-oxygen in the chain also contributes to this beneficial property.

Etherstructure. The polarity of the polyepichlorohydrin is also enhanced by the ether structure, compared with rubbers on ethylene basis with paraffinic and non-polar backbones.

Termonomers. As mentioned before, ETER contains a termonomer. Since the unsaturation resides in the sidegroups, the saturated backbone structure is, like EPDM, not interrupted, and therefore, there is no detrimental effect on the resistance to degradation. The termonomer also improves the cure reactivity, and permits higher states of cure.

Viscosity. The average molecular weight of CO is about 500,000 and above, which corresponds to Mooney viscosities ML 1+4, 100 °C of 45 to 70. At equimolar concentrations of epichlorohydrin and ethylene oxide, ECO has higher molecular weights of 10^6 and more, with correspondingly higher Mooney viscosities of 70 to 100.

Crystallinity. CO, ECO, and ETER are completely amorphous, and have no tendency to crystallize, like CR. As well, after storage at temperatures below 0 °C, the compounds or vulcanizates do not show any hardening due to crystallization, which makes it possible to use ECO at very low temperatures if very low compression set is required.

3.3.13.4 Compounding of CO, ECO, and ETER

[3.191, 3.202, 3.204, 3.544–3.546, 3.548–3.556, 3.559a, 3.559b]

Due to the fact, that polyepichlorohydrins are polyether rubbers, their compoundings follow some other rules compared with other rubbers in various respects. Therefore, the following remarks are somewhat more detailed.

Blends. CO and ECO or ETER are frequently used in blends with each other, mostly in the ratio of 40:60, in order to obtain an optimum balance of properties. Since the crosslinking reaction of the rubbers follows the same mechanism, a sufficient covulcanization of the polymer blend phases occurs. CO and ECO do not, however, occure with most of the other rubbers, therefore it is difficult to transfer their properties to other rubbers through blending. With the availability of the terpolymer, which, in addition to the conventional ETU/metaloxide vulcanization, can also be crosslinked with sulphur and accelerator as well as with peroxides, the opportunity for covulcanization of epichlorohydrin rubbers with other sulphur- or peroxide-curable rubbers exists. There seems to be a special interest in blending ETER with NBR [3.191] to improve the performance of NBR (see page 71), with NR, to improve its heat resistance, or with SBR, to improve its swelling resistance [3.545]. Most recently, ECO has been suggested as a blend partner to improve the low temperature brittleness of ACM [3.535].

Vulcanizing Agents. [3.191, 3.204, 3.544–3.546, 3.554]. The main, all purpose cure system for CO and ECO is ETU without sulphur. Properly chosen acid acceptors lead to the highest states of cure, the best mechanical properties, the best resistance to aging in hot air and oil [3.546], and the best resistance to dynamic fatigue, ozone, and hot compression set [3.545]. The three commercially available polyepichlorohydrins have a different cure reactivity. Therefore, they require different amounts of ETU, namely, the Goodrich grade only 1.0 phr, while the Herclor (now also Goodrich grades) and Japanese grades, 1.2 phr. Because they have a higher cure rate, sulphur, or especially DTDM, are used as cure inhibitors with the former Goodrich grades. This gives higher states of cure, but results also in an inferior heat aging and hot compression set. CBS at a level of 0.5 phr can also improve the scorch safety and the cure state, but it has the same disadvantages as sulphur or DTDM. The Herclor and Japanese grades do not require cure inhibitors, because they are intrinsically less reactive. ETU contributes to mould fouling. For this reason, and because it is undesirable on toxicological grounds, (although it is mostly masterbatched with ECO), other cure systems are also used, all of which have certain advantages and disadvantages. For instance, DETU gives lower states of cure, trimethylthiourea a poorer compression set, and capped amines or imidazoles slower cure rates [3.546]. The proprietory Echo cure systems of Hercules give the shortest cure times, but somewhat inferior mechanical properties and swelling resistance. In order to reduce mould fouling, smaller ETU concentrations should be used, or ETU together with capped amines, 1.3-propane diamine, or piperazinehexahydrate. With the Echo cure systems, mould fouling is also reduced.

CO and ECO cannot be crosslinked with peroxides. By contrast, ETER can also be cured with sulphur and accelerators or with peroxides, in addition to ETU [3.545, 3.550]. With ETER, the best heat resistance is also obtained when the vulcanizates have been cured with a sulphur-free ETU system.

ETER compounds with ETU have the shortest scorch time, the lowest degree of vulcanization, and the poorest adhesion to metal and textiles, and the vulcanizates have the best tensile and tear properties, average flex life and compression set, and the best heat, oil, and motor fuel resistance. ETER compounds with sulphur and accelerator exhibit the best processing properties and adhesion, as well as the best tear resistance and flex life. Their compression set, heat, and oil resistance are poorer, while the resistance to motor fuels is excellent with a somewhat inferior ozone resistance. With peroxides and coactivators, like ethyleneglycole dimethacrylate, or EDMA, very good processing properties are obtained, but the vulcanizates have poorer mechanical and dynamic properties. Peroxide cures with trimethylolpropane trimethacrylate (TRIM) give excessively scorchy compounds. The compression set, the heat and oil resistance, and the ozone resistance of peroxide-cured ETER vulcanizates are very good, as well as the cocuring with suitable blend partners, but the vulcanizates will swell more in motor fuels, because they have a lower state of cure.

Acid Acceptors, Cure Activators. [3.546] As with other chlorine-containing rubbers, CO, ECO, and ETER compounds also need to be formulated with acid acceptors, to absorb HCl. Standard acceptors are Pb_3O_4 and MgO, but also dibasic Pb-phosphite, dibasic Pb-phthalate, basic Pb-silicate, PbO, or Mg-, K-, and Na-stearates. It is important to add acid acceptors at the beginning of mixing cycles to compounds, using short mixing cycles and not too high dump temperatures. This reduces the corrosive action of compounds. The acid acceptors should also be added to the compound as a masterbatch, to ensure their beneficial action. This applies, in particular, to Pb_3O_4.

Pb-phthalate, Pb-phosphite, and basic Pb-silicate give vulcanizates with the best heat resistance, and especially also with respect to hot oil. If compounded in a proper mixing cycle, the less expensive Pb_3O_4 can also give very good tensile properties and very good long term performance of vulcanizates, when they are subjected to high temperatures [3.546]. In order to reduce the mould fouling caused by Pb_3O_4, and in order to achieve adequate cure rates, Pb_3O_4 is frequently used in combination with small anounts (2 phr) of MgO. MgO does not, however, easily disperse in compounds, and it also gives rise to sticking on mills. For this reason, a MgO brand that is coated with about 10% of paraffin wax is used. The addition of the basic MgO also improves the aging resistance and swelling properties of vulcanizates. In the place of Pb_3O_4/MgO, one can also use Pb-silicate, which, however, gives slower cure rates, but gives the best long term heat aging properties. The use of MgO by itself results in reduced heat aging properties [3.546], and zinc salts of unsaturated fatty acids, give a good balance of processibility and vulcanizate properties. However, for very demanding vulcanizate property specifications, these compounds should not be used. If ZnO is used alone in ETU crosslinking systems, the inherently good heat aging resistance of the vulcanizate vanishes [3.546]. With the Echo system, $BaCO_3$ is used in place of Pb_3O_4 as acid acceptor. This system practically does not give any mould fouling at all, but it leads to reduced mechanical properties, which, in most cases, cannot be tolerated.

In sulphur vulcanizations of ETER, ZnO has, of course, a dual role, that of acid acceptor and vulcanization activator.

Antioxidants. As protective agents [3.546], one uses mainly 2 phr or more Ni-dithiocarbamates. Ni-dibutyldithiocarbamate is even more effective as Ni-dimethyldithiocarbamate, but it is more soluble in oils and motor fuels, and is therefore leached out in service. Thus, in practice, the Ni-dimethyldithiocarbamate is preferred. In order to obtain a good balance of heat resistance for non-swollen and swollen vulcanizates (particularly after swelling), a combination of both Ni-compounds is mostly used, but in ETER these protective agents are only effective with ETU-crosslinked vulcanizates and not with those obtained by sulphur/accelerator or peroxides. For the latter vulcanizates, which have a good resistance to sour gas (see page 74), p-phenylene diamines (especially DNPD), BPH, MBI, and other antioxidants as well as UV-stabilizers, like Irganox 1010, 1024, or 1098, and Pb_2O_3 are used.

Fillers. The general rules for choosing fillers apply. With N-770 or N-990 black, one can achieve the highest filler loadings and the lowest compression set, but, at the same time, rather poor tensile properties. N-550 black gives a good compromise of both. If high elongations at break are required, N326 or N220 should be used. The gas permeation of the vulcanizates is already low, but it can be even further improved together with the gasohol resistance, if relatively high loadings of micro talc, such as Mistron Vapor, are used in the compound. This filler is more suitable than, for instance, mica.

Softeners. For reasons of heat resistance, softeners play a less important role in polyepichlorohydrin compounding than in NBR. Generally, only small amounts of high molecular softeners are used, which, in the case of peroxide cure of ETER, must not interfere with the crosslinking reaction. These softeners are, for instance, adipates, sometimes DOP, and polyesters.

Process Aids. It is advisable to use process aids like zinc soaps of unsaturated fatty acids, fatty alcohol residues, waxes, polyethylene, sorbitol monostearate, pentaerythritol tetrastearate etc. since they are reported to improve the milling behaviour of compounds and to facilitate filler dispersion. Also, octadecyl amine, fatty acids, polyethylene wax, and zinc soaps are good process aids as well. The best prevention of mill stick on mill rolls and best flow properties are achieved by the addition of zinc soaps of unsaturated fatty acids [3.559 a]. These do not influence the long term heat resistance of epichlorohydrin vulcanizates [3.546, 3.559 b]. Phthalic anhydride and nitrosodiphenyl amine act as chlorine acceptors in polyepichlorohydrins, and therefore reduce mould fouling caused by chlorine.

3.3.13.5 Processing of CO, ECO, and ETER

[3.543, 3.553]

Conventional processes and equipment are used. As with ACM, the optimum level of physical properties is often obtained only after a short post cure at 150 °C. CO and ECO by themselves do not stick to mill rolls and blades, but blends of both do. Therefore, it is necessary to use process aids in blend compounds. To eliminate the sticking problem, it is also advisable to start mixing on lukewarm mill rolls and keep compounds on rolls until they are cold. The lowest stick to mill rolls has been found in the presence of zinc soaps of unsaturated fatty acids which give at the same time the best flow [3.559 a].

3.3.13.6 Vulcanizate Properties of CO, ECO, and ETER

[3.204, 3.544–3.546, 3.550–3.553, 3.559 b]

Polyepichlorohydrin vulcanizates have an unusual and unique combination of properties. Beside FKM they have at the same time the best resistance to swelling in oils and motor fuels, a high heat resistance, the excellent ozone resistance, the least permeation to gases and motor fuels, very good damping properties, good flame resistance, and good low temperature flexibility.

Mechanical Properties. In relation to other specialty rubbers, the mechanical properties of polyepichlorohydrin vulcanizates are relatively high, although still somewhat lower than those of adequate diene rubber vulcanizates. Hardnesses ranging from about Shore A 50 to 90 can be obtained, although for most technical requirements they lie between 60 and 80. Compression set depends very much on the cure system, and very low values can be obtained. Polyethylene, sorbitol monostearate, pentaerythritol tetrastearate etc. are reported to improve the milling behaviour of compounds and to facilitate filler dispersion.

Heat Resistance. For 1000 hour exposures, the upper service temperature of unblended CO vulcanizates is about 150 °C, and that of ECO or ETER vulcanizates, about 135 °C.

Low Temperature Flexibility. The low temperature flexibility depends largely on the blend ratio of CO/ECO. The brittleness temperature of CO vulcanizates is about −23 °C, while that of ECO or ETER vulcanizates is below −40 °C. Intermediate values can be obtained by blending. The low temperature properties of polyepichlorohydrin are unique in that, contrary to other elastomers, including those with chlorine, the dynamic stiffening temperature and the brittleness temperature are approximately the same.

Swelling Resistance. Epichlorohydrin vulcanizates have an unusually good resistance to swelling in motor fuels, and are in this respect considerably superior to NBR vulcanizates. Even in aromatic fuels (Fuel C), the swell resistance is good. Of the epichlorohydrins, CO vulcanizates have the best swelling resistance, followed by those of ECO and ETER, with very little difference between that of the latter two. Compared with NBR, ECO and ETER have, in addition to the good swelling resistance, distinctly better low temperature flexibility, and higher heat resistance, together with a good ozone resistance. When peroxide cures are used, these good properties of ETER can also be transferred to NBR, when the two are blended [3.191].

The swelling resistance of epichlorohydrin vulcanizates in gasohol is somewhat inferior than that of NBR vulcanizates.

The permeation of motor fuels through epichlorohydrin is smaller by one order of magnitude than through NBR. This is important for fuel hoses. CO behaves best in this respect, and comparative figures are 100 mg alcoholfree fuel/m^2/day for NBR vulcanizates, less than 5 for CO, and about 10 to 20 for CO/ECO blends or ETER vulcanizates [3.204, 3.545, 3.552, 3.559]. Since 1981, strict rules on fuel permeation for all elastomeric automotive components have been enforced in the U.S.A. by the Environmental Protection Agency (EPA), and these can be complied with by using CO/ECO or CO/ETER blends. Compared with NBR vulcanizates, those from epichlorohydrin have approximately the same resistance to hydroperoxide-containing motor fuels (sour gas) (see page 74), which form in lead- and alcohol-free fuels or in fuel injection systems with suction pumps, but not in those with pressure pumps. Of the epichlorohydrins, ETER has the best resistance to "sour gas", particularly if the vulcanizates are compounded with suitable antioxidants. Peroxide-cured blends of NBR and ETER could conceivably be used as a compromise, if, in future, the requirements become even more demanding.

Damping Properties. ECO and ETER have remarkable damping properties and very much resemble NR vulcanizates in this respect.

After relatively short exposure to high temperatures, polyepichlorohydrin retains its damping properties better than NR, and therefore, it has already replaced NR in some high temperature-resistant damping pads for car engines mainly in USA.

Permeation. The resistance to permeation of gases or liquids is greater in CO vulcanizates than in the ones from IIR, but it is similar for ECO or ETER and IIR vulcanizates.

3.3.13.7 Uses of CO, ECO and ETER

Because of their well-balanced property profile, polyepichlorohydrins have gained in importance in recent years, and despite their higher price, have replaced NBR in many applications. The main use is in the automotive industry, for various seals, diaphragms, membranes, hoses (fuel and hot air hoses), heat resistant damping components, etc. In addition, the rubber is used for roll covers (paper rolls, printing rolls, copier rolls), various injection moulded rubber components, belts, and rubberized fabrics, to name a few. In future, ETER will also be increasingly used as blend partner with NBR, SBR, and NR [3.191].

3.3.13.8 Competitive Products of CO, ECO and ETER

Examples are, for instance, ACM, BIIR, CIIR, CM, CR, CSM, EVA, FKM, IIR, and NBR.

3.3.14 Polypropylene Oxide Rubber (PO and GPO)

[3.560–3.563]

As with epichlorohydrin, propylene oxide can be homopolymerized (PO) in a ring-opening reaction using $Zn(alkyl)_2$/water, or $Al(alkyl)_3$/water catalysts [3.560–3.562]. It can also be copolymerized with allylglycidil ether to give the copolymer GPO, which is the more important product of this class [3.563]. Since GPO has double bonds, it can be crosslinked with sulphur and accelerators. GPO has the following structure:

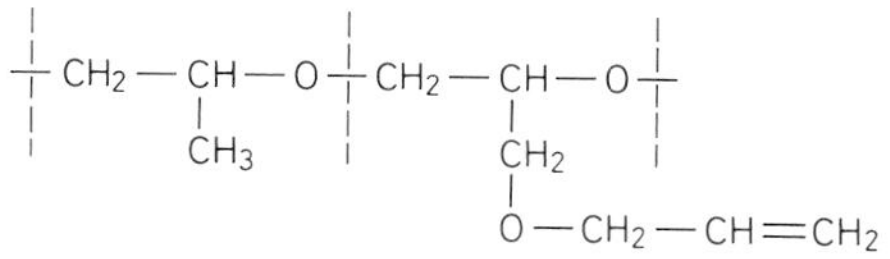

```
 |              |                  |
-|- CH2 — CH — O -|- CH2 — CH — O -|-
 |        |     |         |        |
         CH3             CH2
                          |
                          O — CH2 — CH=CH2
```

Propylene Oxide   Allylglycidil Ether   Units

The molecular weight of GPO is about 1×10^6 to 1.5×10^6, and it corresponds to Mooney viscosities ML 1+4 at 100 °C of about 75 to 80. Since it does not contain chlorine, the glass transition temperature of GPO is at −60 °C, lower than that of ECO.

GPO vulcanizates have a different property profile than sulphur vulcanizates of ETER. Although GPO has the same excellent heat and ozone resistance, the same high elasticity and good dynamic properties of ETER, it has a poorer swelling resistance in oils, and particularly in motor fuels. The low temperature flexibility of GPO is better, and surprisingly, the hydrolysis resistance and swelling resistance in water and alcohols is good. Since they do not contain chlorine, PO and GPO have a poor flame resistance.

The dynamic properties of GPO and NR are very similar, but GPO has the better heat resistance, and it is more expensive. Yet, GPO competes very little with NR, since at the low hardness levels, in which NR damping elements are normally used, GPO vulcanizates have relatively poor mechanical properties, in spite of the other advantages mentioned above.

3.3.15 Fluoroelastomers (FKM resp. FPM, formerly CFM)

[3.252, 3.281 a, 3.535, 3.538, 3.564–3.624]

3.3.15.1 General Background about FKM

[3.564–3.570]

Starting from the synthesis of polytetrafluoro ethylene in 1938, it became possible in 1956 to produce from vinylidenefluoride (VF_2) and chlorotrifluoro ethylene (CTFE) amorphous polymers with more than 60% fluorine and with a low glass transition temperature. These copolymers were termed CFM [3.605–3.609]. They have, however, been largely replaced by the newer, more heat resistant hydrofluoro elastomers (FKM according to ASTM; FPM according to ISO; here the abbreviation FKM is used), which can be obtained through co- or terpolymerization of the following monomers:

```
CF2=CF
     |
     CF3
```

Hexafluoro propylene (HFP) [3.610–3.613]

$CF_2{=}CF_2$	Tetrafluoro ethylene (TFE) [3.614]
$CHF{-}CF(CF_3)$	1-Hydropentafluoro propylene (HFPE) [3.615-3.617]
$CF_2{=}CF{-}O{-}CF_3$	Perfluoro(methylvinylether) (FMVE) [3.565]

By using a bulky monomer in the copolymerization, the chain symmetry and with it, the stiffness of the polyvinylidenefluoride chain, becomes smaller. While polyvinylidenefluoride has a melting temperature of −18 °C, the insertion of a comonomer into the polymer chain will reduce the glass transition temperature to about −40 °C.

In spite of its very high price, the demand for FKM has increased considerably, due to the exacting requirements of the automotive industry, and it has replaced NBR and Q in some of these applications. Therefore, the worldwide FKM production capacity has been enlarged dramatically in 1986 from 6500 to 12000 t per year [3.252].

3.3.15.2 Production of FKM

[3.565]

The preferred method of producing FKM is through emulsion polymerization, and by using peroxides as initiators at higher temperatures. The system is to be under high pressures, since fluoro olefins are gaseous at ambient pressures. The molecular weight of the polymers can be controlled with chain modifiers, such as chlorinated hydrocarbons, alkyl mercaptans [3.618], alkyl esters [3.619], halogens [3.620], or by adjusting the monomer/initiator ratio.

3.3.15.3 Structure of FKM and its Influence on Properties

[3.5]

Influence of Fluorine Monomer. The bond energy of the carbonfluorine bond is, at 442 KJ/mole, substantially higher than that of the carbon-hydrogen bond at 377 KJ/mole. It is also higher than the bond energy of carbon with other atoms, like chlorine. For this reason, and because of the shielding of the polymer chain by the larger fluorine substituents, as compared with non-substituted chains, FKM has a very high heat and chemical resistance [3.621]. Because of certain dehydro fluorination reactions, the fluorinated hydrocarbons are, however, reactive to some extent [3.621], and can therefore be crosslinked with a number of different cure systems, like diamines, dithiols, or aromatic diphenyl compounds, like bisphenol AF.

FKM grades with 65 to 67% fluorine can only be crosslinked with great difficulty, and they give very low states of cure.

In order to facilitate the crosslinking of FKM, the more recently developed FKM grades have a higher fluorine content of 68 to 71%, which gives, at the same time, a considerably better resistance to oils blended with amine or amine-donor stabilizers, and to methanol-containing motor fuels [3.576-3.579, 3.581, 3.583b, 3.584, 3.585, 3.586a, 3.587, 3.588]. To facilitate the crosslinking of FKM grades based on

VF_2, HFP and TFE, additional monomers with reactive groups are built into the polymer chain. Examples of such reactive groups are randomly distributed bromines, or terminal iodines. Perfluorated aromatic rings and nitrile groups have also been mentioned [3.584, 3.593].

The thermal and oxidative stability of the vulcanizates depends on the arrangement and the stability of the crosslinks. Thus, it can be expected that crosslinks via the bromine groups, distributed uniformly throughout the rubber matrix, give better vulcanizate properties than crosslinks via the terminal iodine groups only. In peroxide cures, these crosslinks do not form C–C bonds between FKM chain molecules. Instead, the bonds are formed via coactivators to produce hydrogen-containing bridges, whose thermal and oxidative stability is poorer than that of crosslinks that are formed when bisphenol A or bisphenol AF are used in the cure reaction. Peroxide cures also give FKM vulcanizates with poorer hot compression set, than do bisphenol AF cures. For these reasons, it is very important to choose the appropriate coactivators for peroxide cures. TAIC gives, for instance, a lower compression set than TAC [3.577]. Coactivators which give even better results than TAIC are said to be under development. FKM development grades based on FMVE with perfluoro aromatic rings can be crosslinked with bisphenol AF to give vulcanizates with an extremely high thermal and oxidative resistance [3.584, 3.623, 3.624]. Perfluoro-FKM grades with nitrile sidegroups can be crosslinked in the presence of organotin compounds to form at regularly used cure temperatures triazine structures of extremely high thermal stability. In this reaction, three polymer chains are connected via one triazine ring (see also Section 3.3.15.8, Fluorotriazene Elastomers, page 125). Since the crosslink bridges are free of hydrogens, they have nearly the same resistance as the polymer chain [3.584], which explains the extremely high oxidative and heat resistance. FKM grades with epoxy groups, which act as crosslinking sites, have also been discussed [3.583].

A new, but extremely expensive FKM grade, consisting of TFE and FMVE, ranks in its properties, between PTFE and the conventional FKM. This new grade, called Kalrez, is completely insoluble, but it can be compounded with fillers, and compression moulded. The vulcanizates are elastomers and have a similar chemical resistance as PTFE, and qualify for extended service at temperatures of over 280 °C [3.638].

A terpolymerization of VF_2, TFE, and FMVG [3.565] in a molar ratio of 75/10/15 results in FKM grades with considerably lower glass transition temperatures of −37 °C and brittleness temperatures of −50 °C. These values are about 10 °C lower than those of conventional FKM grades based on VF_2, HFP, and TFE, but they are much more expensive.

Saturation. FKM is fully saturated and it cannot be attacked by oxidative processes. Therefore, it is completely resistant to oxidation and ozone, and also to motor fuels with hydroperoxides (sour gas), but it can be attacked by some oil additives.

Viscosity. FKM is marketed in range of different viscosities.

Segmented and Thermoplastic FKM (TPE-FKM). These grades have been developed recently [3.582, 3.586], and they can be more readily processed.

3.3.15.4 Compounding of FKM

[3.571–3.575, 3.581, 3.593, 3.597, 3.601, 3.622]

Curing Agents. The first cure systems for FKM consisted of capped diamines, such as hexamethylenediamine carbamate, in combination with acid acceptors, like MgO, CaO, PbO, or other lead compounds. These cure systems offer, however, poor scorch safety, poorer vulcanizate properties and poorer compression set than the later bisphenol AF cure systems. On the other hand, the conventional amine systems produce vulcanizates with a good adhesion to metals, and a relatively good resistance to amine stabilizers which are blended with motor oils. Because of the better scorch safety and better compression set, one uses now primarily cure sytems based on bisphenol AF [3.593, 3.597] in combination with MgO and $Ca(OH)_2$ as acid acceptors. The resulting vulcanizates have good physical properties, high tensile strengths, low hot compression set, and release well from moulds. However, with the increased use of motor oils with amine donors as stabilizers, and of methanol-blended motor fuels, the use of bisphenol crosslinking systems has to be re-evaluated, because the vulcanizates soon form cracks after exposure to these oils at higher temperatures. In addition, FKM grades with low fluorine contents, that were cured with bisphenol, have a relatively poor swelling resistance in methanol-blended motor fuels.

FKM grades with higher fluorine contents can be crosslinked with peroxides. Peroxide cured vulcanizates are more resistant to amine stabilizers in blended oils and to methanol containing motor fuels, and therefore, peroxide cures are being used more extensively. The choice of peroxides and the required coreagents will be discussed on page 256ff. Peroxycarbamates appear to be particularly well suited to FKM cures [3.581]. Acid acceptors have to be used also with peroxide cures, and with high fluorine FKM grades, very good mechanical properties are obtained in addition to the already mentioned good resistance to blended oils and motor fuels. The cure systems tend to result in mould sticking of the vulcanizate, but this can be remedied to some extent by using carnauba wax.

Also mentioned above were the various developments in perfluoro FKM grades with reactive sidegroups [3.583, 3.584, 3.597], (see Section 3.3.15.3). They can be crosslinked with peroxides, although this does not always lead to an optimum environmental stability, or with special crosslinking agents, like bisphenol AF or organotin compounds, which give vulcanizates with the highest environmental stability [3.584]. This subject is by no means closed as yet, and since the different reactive groups of the polymers require specific crosslinking agents, the technical literature of the suppliers should be consulted for further details.

Stabilizers, Protective Agents. Basic compounds or metal oxides are required as HF-acceptors. For the most demanding heat resistance specifications, the choice of the HF-acceptors assumes particular importance. PbO is recommended if the vulcanizate is exposed to hot acids, and dibasic Pb-phosphite, together with ZnO, if the vulcanizate is exposed to steam or hot water. MgO and CaO give a superior vulcanizate performance in dry heat.

Fillers. To obtain good vulcanizate properties, the desired hardness, good processibility, and to reduce compound cost, one uses non-reinforcing blacks and mineral fillers. MT black gives a good balance of good processibility and physical properties. Since the compounds stiffen very soon after mixing, only relatively small amounts of filler (10–30 phr) can be used. Lowest compression set values are obtained with Austin Black in the compound.

Softeners. FKM is not compatible with conventional plasticizers. To improve the processibility of FKM compounds, one can use, instead carnauba wax, pentaerythritol tetrastearate or low molecular weight polymers which are chemically similar, namely those from VF_2/HFP.

3.3.15.5 Processing of FKM

[3.565, 3.580, 3.582, 3.586, 3.603]

The mixing and processing of FKM compounds is done using conventional methods of the rubber industry. An extensive post-curing of 24 hours at 200 to 260 °C is generally required. Only in this post cure is the crosslinking reaction completed, and the high cure states and good vulcanizate properties are obtained. The post cure is also important for obtaining low compression set.

3.3.15.6 Properties of FKM Vulcanizates

[3.565]

Mechanical Properties. The general level of mechanical properties of FKM vulcanizates is, as with most of the other specialty rubbers, distinctly lower than for conventional diene rubbers. The tensile strength depends greatly on the temperature, and it drops considerably at higher temperatures. The same applies to the hardness, which can be compounded for a range of Shore A 50 to 95, but preferably 70. A phenolic cure system results in a better hardness retention at high temperatures, and this cure system is also suitable for optimum compression set. FKM vulcanizates are not very elastic.

Heat and Aging Resistance. FKM has the best heat resistance of all rubbers, and continuous service for 1000 hour will be 220 °C and even a service life at 250 °C is possible. FKM vulcanizates also resist degradation from weathering and ozone.

Low Temperature Flexibility. The dynamic brittleness temperature of FKM is about −18 °C. More recently, FKM grades have been developed with an improved brittleness temperature of −40 °C.

Resistance to Swelling and Chemicals. FKM vulcanizates not only resist swelling in hot oils and aliphatic compounds, but also in aromatics and chlorinated hydrocarbons. They are also very resistant to most mineral acids, even acids at high concentrations. As already mentioned, FKM vulcanizates crosslinked with bisphenol AF are relatively easily attacked by amine stabilizers of blended oils. Peroxide-cured FKM vulcanizates with high fluorine contents are more stable in this respect. FKM vulcanizates also swell more in methanol, ketones, esters, and ethers, than in mineral oil, and motor fuels. The resistance to swelling improves with increasing fluorine contents, and, therefore, one chooses FKM grades with the highest possible fluorine content for applications with methanol-containing motor fuels. FKM vulcanizates have also a good resistant to motor fuels with hydroperoxides (sour gas), as well as to hydrogen sulphides, which are present in oil wells [3.205 a]. Hot hydrofluoric and chlorosulphonic acids attack FKM vulcanizates.

Permeation. The gas permeability of FKM vulcanizates is very low, and indeed, even lower than that of IIR vulcanizates.

3.3.15.7 Uses of FKM

Due to its interesting properties, FKM is used in special rubber products, such as shaft seals of motor cars, and components in aircrafts and rockets.

3.3.15.8 Additional Developments of Fluoro Elastomers

Tetrafluoroethylene-Propylene-Co-and Terpolymers (TFE/P). [3.625–3.634] Through alternating copolymerization of tetrafluoroethylene and propylene (TFE/P) [3.625], one obtains the following structure:

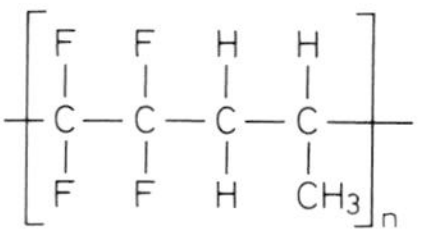

$$\left[-\overset{F}{\underset{F}{C}}-\overset{F}{\underset{F}{C}}-\overset{H}{\underset{H}{C}}-\overset{H}{\underset{CH_3}{C}}- \right]_n$$

Tetrafluoroethylene Propylene Units

This new generation of fluoro elastomers has unique properties, and it has been recently commercialized under the name of Aflas. A precise alternating arrangement of the comonomer molecules is important, since short propylene sequences impede the crosslinking, and reduce the heat resistance because of their thermoplastic nature. There are three TFE/P copolymer grades now available – a low molecular weight grade for extrusion, a higher molecular weight for compression moulding, and a very high molecular grade for highest vulcanizate performance levels, such as in oil drilling applications [3.627, 3.628]. More recently, a TFE/P-terpolymer has also been developed, in which the termonomer is a fluorinated vinyl compound. It can be readily crosslinked with peroxides through a vinyl sidegroup. This terpolymer has better low temperature flexibility [3.633]. Although TFE/P-grades can be crosslinked in nucleophilic reactions with dioles (e.g. bisphenol AF) and amines (e.g. hexamethylene diamine carbamate), most vulcanizates are crosslinked with peroxides in combination with TAIC as coagent, and $Ca(OH)_2$ as acid acceptor. This cure system gives a better all-round environmental resistance [3.633]. As fillers one uses small quantities of blacks (N550–N990), as well as aluminum silicates and hydrophobic fumed silica. The latter gives the best long-term resistance against amine corrosion inhibitors [3.627]. TFE/P vulcanizates can be obtained in hardness Shore A from 65 to 95, but mostly of 70. An extremely high hardness of up to Shore D 60 can also be obtained. At a hardness of Shore A 70, the tensile strength is about 17 MPa or even higher, at a maximum elongation of about 160%. The glass transition temperature of the copolymer is −3 °C, and that of the terpolymer is −13 °C. At very low temperatures, the vulcanizates become leathery, so that even at −54 °C, the components do not become brittle and shatter, but still remain functional [3.630]. TPE/P vulcanizates have a surprisingly good resistance even to concentrated sulphuric, nitric, hydrofluoric, and hydrochloric acids, concentrated caustic alkalis, hydrogen sulphide, steam (of 200 °C), alcohol, and mineral oils. The resistance to these reagents is at times even higher than that of FKM vulcanizates. By contrast, the resistance of TPE/P vulcanizates to motor fuels, benzene, and chlorinated hydrocarbons is poorer than that of FKM vulcanizates. The gas permeability of TPE/P vulcanizates is very low, and their resistance to hot air is exceptional. Even after 1500 hours at 200 °C, no loss of tensile strength was observed, and after 1000 hours at 230 °C, 70% of the original tensile strength was still retained. TPE/P vulcanizates also have good electrical properties, and are good insulators even at very high temperatures. Therefore, they are of interest to the cable and wire industry. The main applications of TFE/P are in parts for oil drilling components, [3.626, 3.628], to provide corrosion resistance of chemical equipment and transmission components [3.629, 3.630, 3.634], and in the aerospace, automotive, or electrical industry [3.627, 3.631 a].

Nitroso Elastomers (AFMU). [3.635] The alternating copolymerization of tetrafluoroethylene with perfluoronitroso-methane, together with small amounts of perfluoronitrosobutyric acid as reactive termonomer, results in an elastomer AFMU of the following structure:

$$-CF_2-CF_2-N(CF_3)-O-CF_2-CF_2-N(CF_2-CF_2-COOH)-O-$$

Tetrafluoro-ethylene | Perfluoro-nitrosomethane | Perfluoro-nitrosobutyric acid | Units

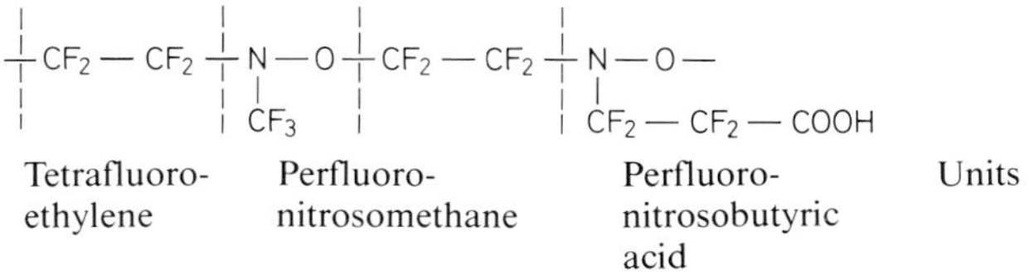

This elastomer has a very low glass transition temperature of −50 °C. However, above 175 °C, AFMU is not resistant to strong bases and oxygen. Crosslinking reactions with metal oxides and epoxides produce vulcanizates with excellent resistance to strong acids and oxidizing agents, and the vulcanizates do not burn, even in pure oxygen. These very expensive elastomers have been especially developed for the aerospace industry.

Fluorotriazine Elastomers. [3.635] Perfluoroalkyltriazines are obtained by reacting perfluoroalkyl-dinitriles and ammonia. The resulting poly(imidoamidine) is stable and at higher temperatures, it crosslinks. The fluorotriazine elastomers have the following structure:

$$\text{triazine ring } (C_3N_3) \text{ substituted at each C with } \sim(CF_2)_n\sim$$

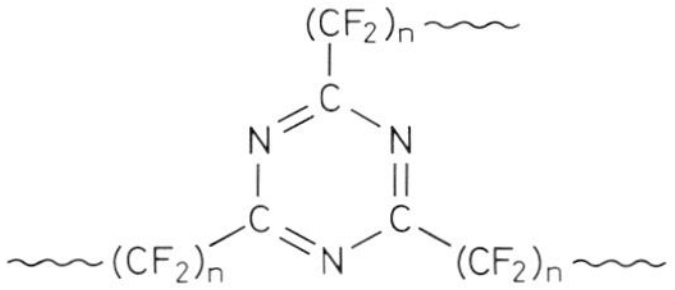

Like nitroso elastomers, the fluorotriazine elastomers are also resistant to acids and oxidizing agents, but they are attacked by bases.

Poly-(Fluoroalkoxyphosphazene) Elastomers (PNF) [3.636–3.637 c] have the following structure:

$$\left[-N=P(O-CH_2-CF_3)(O-CH_2-CF_2-CH_3)-\right]_n$$

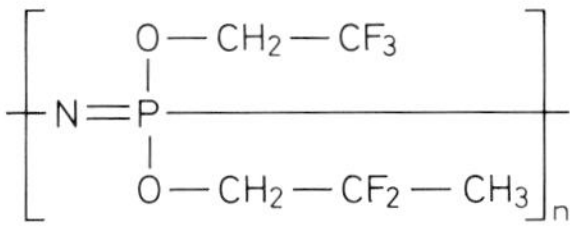

A new type of rubber based on fluorophosphonitrilic derivatives (PNF) has been brought on to the market in 1987, and this product too is in competition with FKM. Although PNF elastomers have a maximum service temperature which, at about 175 °C (or 200 °C for intermittent exposure applications), is 50 to 75 °C lower than that for FKM, they have excellent low temperature flexibility, with a T_g value of −65 °C. The swelling resistance is also very good; in fuels at room temperature it is appreciably better than that of FKM elastomers, and at higher temperatures (fuel injection temperatures) it is considerably better. The swelling resistance in aromatic solvents and in some chlorinated hydrocarbons is also better than that of FKM elastomers. The mechanical properties, and the compression set at 150 °C, are also good.

For this new type of rubber it is again too early to guess at likely substitution trends, as it has not yet been fully proved. In many of its properties PNF occupies a position intermediate between FVMQ and FKM (see page 135) [3.637 a–3.637 c].

TFE-FMVE Elastomers. [3.638] As already mentioned on page 121, the copolymerizations of TFE and FMVE, under the trade name Kalrez, give extremely stable but also very expensive elastomers.

Fluorosilicone Elastomers (FMQ). These elastomers will be discussed together with silicone elastomers (see page 135).

3.3.16 Polynorbornene (PNR)

[3.639–3.642]

General Background and Production. Polynorbornene is obtained by the *Diels-Alder* addition of ethylene to cyclopentadiene, and the subsequent ring-opening polymerization of the norbornene, to give the following structure:

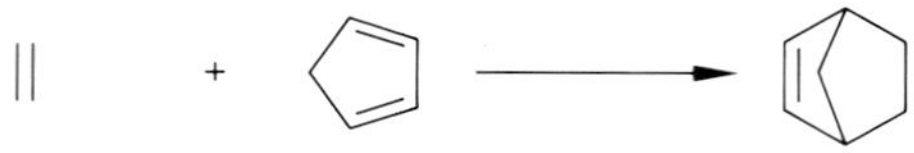

Ethylene + Dicyclopentadiene → Norbornene

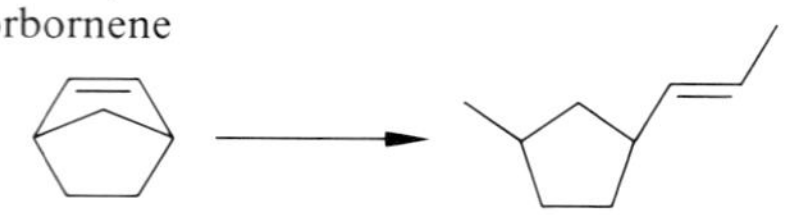

Polynorbornene

Structure and Properties. The polymerization process yields very high molecular weights with virtually no crosslinking or gelation taking place. Since the polynorbornene retains a double bond, it can be crosslinked with sulphur and accelerators. The glass transition temperature of PNR is +35 °C, and it can therefore be classed as a low-melting plastomer, but with the addition of conventional oils and ether plasticizers, the T_g value can be reduced to −60 °C. To facilitate the incorporation of oil, PNR is supplied in powder form or already as oil masterbatch. OE-PNR can be processed like conventional rubbers, and the high molecular weight and the cyclopentane ring in the polymer chain result in PNR vulcanizates with high tensile properties, even at high oil extensions.

Compounding. Here, the choice of the correct plasticizer is of special importance. Paraffinic oils and esters give particularly good low temperature properties. For high oil extensions (200 phr and higher), without loss of mechanical properties, compatible aromatic oils should be used. The kind of oil influences the damping properties; paraffinic oils give low and aromatic oils high damping. Naphthenic oils are in between. Vulcanizates from compounds without fillers have an insufficient level of physical properties, and the choice of fillers follows the same rules as for most other rubbers. For reinforcing blacks or silicas, smaller amounts of filler are used to avoid an excessive increase in viscosity and processing problems. Vulcanizing agents are sulphur and accelerators, such as thiazol/thiuram/dithiocarbamate, thiazol/guanidine/thiourea/dithiocarbamate, ZnO and stearic acid. The accelerator combination can be rather complex and larger amounts are required. Peroxide cures give a better heat resistance and better compression set. This makes the application of aliphatic oils neccessary.

Processing. PNR can be processed by conventional extrusion and compression moulding methods. For PNR powders, the usual powder mixing techniques are used.

Properties of Vulcanizates. PNR vulcanizates are characterized by high tensile properties even at very low hardness levels of, for instance, Shore A 15.

Uses of PNR. The rubber is used for seals, profiles, roll covers, bellows, damping elements, and in a wide range of Shore-hardnesses for vibration and noise reduction in equipment and vehicles. By blending PNR with NR one can influence the damping characteristics of NR vulcanizates.

3.3.17 Other Polymers

3.3.17.1 Special Butadiene Copolymers

General Background. Although butadiene copolymerizes with a great number of diene monomers other than styrene or acrylonitrile, none of these copolymers have found significant markets. For this not only the physical properties of the rubbers are important factors, but so are the availability and cost of the raw materials. Of the monomers which were closely investigated in copolymerizations with butadiene, only the acrylates, methacrylates, and vinylpyridine have gained some importance as termonomers for SBR and NBR latices. Other monomers, like N-dialkylacrylamide, styrene with an halogenated aromatic ring, vinylmethylketone, isopropenylmethylketone, and dimethylvinylethinyl carbinol have been thoroughly evaluated, but none give sufficiently attractive polymers to replace either styrene or acrylonitrile, which are more readily available.

The polymerizability of over 100 vinyl compounds with butadiene and the properties of the resulting copolymers has been discussed in several publications [3.643, 3.644].

Vinylpyridine Rubber. Copolymers of butadiene with 15 to 25% of 2-methyl-5-vinylpyridine play an important role as latex dips for rubber-textile adhesives. The rubber itself can be crosslinked with benzotrichloride and chloranil, to form quaternized bridges. Compared to sulphur bonds, these quaternized bonds give a better resistance to solvents, a good abrasion resistance, and low hysteresis [3.645]. The crosslinking reactions are interesting, from a theoretical point of view, but are not exploited in practical applications. The reaction with benzotrichloride or chloranil gives a lot of environmental problems.

Piperylene Rubber. With the greater availability of piperylene from C_5-cuts of refinery streams, there has been some activity in piperylene copolymerizations. For instance, a transbutadiene-piperylene copolymer has recently been developed [3.646, 3.647], but is not commercially available.

3.3.17.2 Special Isoprene Copolymers

[3.648]

Numerous copolymers of isoprene with other monomers have been investigated in addition to copolymers with isobutylene (IIR and dervatives see page 88), with acrylonitrile and with styrene as block-copolymers (SIS see page 150).

More recent attempts are the alternating copolymerizations of ethylene and isoprene to develop a new stereorubber [3.649], which is, however, not yet commercially available.

3.3.17.3 Dimethylbutadiene Polymers (Methyl Rubber)

[1.59, 1.60, 3.650–3.652]

Polydimethylbutadiene, or methyl rubber, was the first synthetic rubber that was produced on a large-scale. Due to World War I, it temporarily played an important role from 1916 to 1918. But compared to NR, it had insufficient properties, and after 1919, it disappeared from the market. By using modern coordination catalysts, it is possible to produce a methyl rubber which, although it has a higher glass transition temperature and greater rolling resistance, could find some use in tread compounds of winter tires, because of its improved traction.

3.3.17.4 Special Polymers from Ring-Opening Polymerizations

Trans-Polypentenamer (TPA). [3.653, 3.654] Several years ago the ring-opening polymerization of cyclopentene led to the development of a new elastomeric raw material, whose vulcanization properties are very similar to those of polybutadiene.

Refining C_5 streams to obtain isoprene, always gives a certain yield of cyclopentene and cyclopentadiene, which can be used for the production of the new rubber – the cyclopentadiene after partial hydrogenation.

The polymerization of cyclopentene is carried out in chlorinated or aromatic hydrocarbons using tungsten catalysts, which are stable in solution, together with alkylaluminum compounds. The ring-opening polymerization mechanism gives stereospecific polymers.

Polymers with greater than 85% trans-, and less than 15% cis-configuration have a glass transition temperature of $-90\,°C$, which corresponds to that of BR with 95% cis-1.4 structure.

Even high molecular weight TPA can be processed, and it accepts large amounts of black and oil and it has an excellent green strength, comparable with NR. Even at low concentrations of curatives, one obtains highly elastic, abrasion- and aging-resistant vulcanizates, whose properties are very similar to those of BR or BR/SBR blends.

TPA is not commercially available.

Trans-Polyoctenamer (TOR). [3.655, 3.656] Recently, a trans-polyoctenamer (TOR) has been produced by the ring-opening polymerization of cyclo-octene. In blends of 10 to 30 phr, it improves the extrudability of other diene rubbers, and it also facilitates the blending of rubbers through mutual compatibilization.

3.3.17.5 Alkylenesulphide Rubbers (ASR)

The structure of ASR resembles that of PO, whose oxygen atoms have been replaced by sulphur atoms:

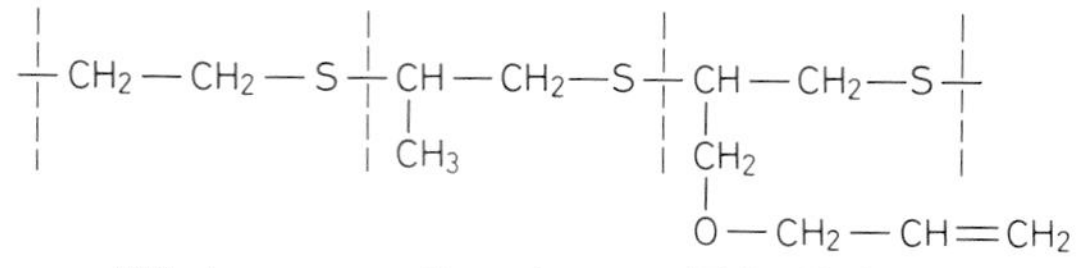

```
 |                |               |                |
-+-CH2—CH2—S-+-CH—CH2—S-+-CH—CH2—S-+-
 |                |  |            |  |             |
 |                |  CH3          |  CH2           |
                                     |
                                     O—CH2—CH=CH2
    Ethylene        Propylene       Thioalkyl-          Units
    Sulphide        Sulphide        glycidilether
```

The sulphur atom gives ASR similar properties to those exhibited by TM, without reaching the swelling resistance of the latter.

ASR does not quite fill the present gap in modern rubber technology, that is, the availability of a reasonably priced elastomer with good processing characteristics, and whose vulcanizates have a good swelling resistance in aromatic and chlorinated hydrocarbons or ketones, as well as good tensile properties.

Ethylene thioglycol ether with the abbreviation ST (analoguous to OT) is not on the market.

3.3.17.6 Crosslinkable Polyethylene (X-LPE)

[3.656a, 3.657, 3.658]

In spite of its crystallinity at room temperature, crosslinked polyethylene, X-LPE, has to be classed by definition as synthetic rubber, because the crosslinked polymer has elastomeric properties at room temperature, albeit combined with a relatively high hardness. This classification is not only for theoretical, but for practical reasons in particular. In the cable industry, X-LPE has replaced IIR in some applications.

The difficulties with processing X-LPE are connected with the relatively rapid dissociation of peroxides at processing temperatures. With the availability of more stable peroxides, and with an increased experience in processing these materials, processing problems have been largely solved.

3.3.17.7 Recently Developed Novel Elastomers

A number of very novel polymers [3.659] have been recently developed, which are, however, not commercially available. These are, for instance, polyoxazolidone elastomers [3.660], polymers from new telechelics [3.660a], and 2-isopropenylnaphthalene [3.661], elastomers from liquid polymercaptane resins [3.662], or new ionomeric elastomers [3.663–3.665]

3.4 Polycondensation and Polyaddition Products

3.4.1 Polysiloxane, Silicone Rubber (Q)

[3.666–3.721]

3.4.1.1 General Background about Q

[3.666–3.676]

The main polymer chain of Q does not consist of hydrocarbons, but of silicium and oxygen atoms. This structure creates unique properties in Q. The most important grades are dimethyl polysiloxane (MQ) and dimethylphenyl-polysiloxane (PMQ), both of which can contain small amounts of a termonomer with vinyl groups (VMQ and PVMQ).

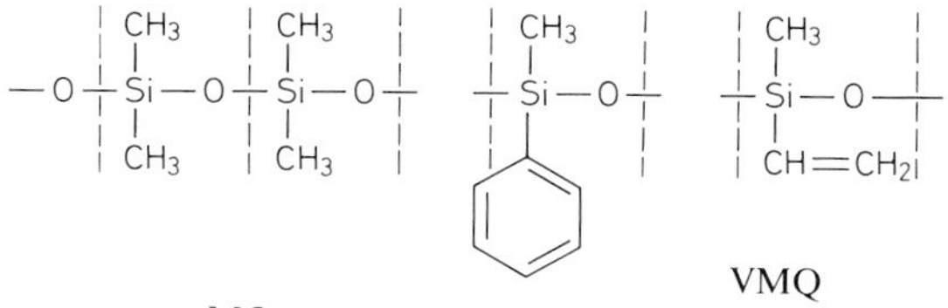

$$-O-[Si(CH_3)_2]-O-[Si(CH_3)_2]-O- \qquad -[Si(CH_3)(C_6H_5)-O]- \qquad -[Si(CH_3)(CH{=}CH_2)-O]-$$

MQ PMQ VMQ Termonomer Units

MQ was developed in 1942 by General Electric and Dow Corning, and was commercialized in 1945. The physical properties of the early vulcanizates were poor, compared with those of today's grades.

3.4.1.2 Production of Q

[3.666, 3.669–3.672, 3.674–3.678]

The starting material for the production of MQ is dimethyldichlorosilane, which is hydrolized in the presence of acid and condenses, after splitting off water, to yield a mixture of straight-chain and cyclic oligodimethyl siloxanes (rings with 3–5 siloxane units). These intermediates are further condensed in a secondary reaction step using acidic or basic catalysts and higher temperatures to produce high molecular weight polymers. Since the starting material contains trichloromethylsiloxane, the polymer becomes more or less branched or crosslinked, depending on its concentration. High degrees of branching can adversely affect the processibility of compounds and the strength properties of vulcanizates [3.709]. In order to avoid branching or crosslinking, modern processes for MQ employ either very pure dimethyldichlorosilane, or very pure cyclotetrasiloxane. The latter is condensed at higher temperatures of 200 °C with strong proton or Lewis acids (e.g. sulphuric acid, hydrochloric acid, or $FeCl_3$, BF_3, $SnCl_4$, phosphonitrilic chloride etc.), in the presence of small amounts of water, or with basic, anhydrous catalysts (e.g. KOH, alkalisilanolates, siloxanolates, quaternary ammonium- or phosphonium hydroxide).

The average molecule weight can be adjusted with modifiers, such as hexamethyl disiloxane. For the production of special Q grades, one also uses vinyl-methyl-dichlorosilane or phenyl-methyl-dichlorosilane as co- or termonomers. The polymerization reaction is terminated with monochlorosilane.

Since the polycondensation catalysts can also initiate a depolymerization reaction, they have to be deactivated through neutralization after the polycondensation reaction is completed.

3.4.1.3 Structure of Q and its Influence on Properties

[3.5]

Molecular Weight and Molecular Weight Distribution. High temperature vulcanizable MQ grades with an average molecular weight of about 300,000 to 700,000 represent a good balance between good processibility and a high level of vulcanizate properties. The molecular weights of room temperature vulcanizable grades have, of course, to be much lower at about 10,000 to 100,000, since these polymers need to be pourable or spreadable.

MQ has a broad molecular weight distribution. During processing and as finished vulcanizate, the polymer contains still relatively volatile siloxanes, which can lead to shrinkage [3.699]. It is, therefore, important to rid the polymer of low molecular weight components before use, and this can be done in hot internal mixers. Most commercial grades are, however, already degassed.

Chain Interaction, Influence of Substituents. Compared with most organic rubbers, the cohesive energy density is at 225 kJ/mole very low, and therefore there is only a small interaction force between chains. Thus, linear MQ grades have very low viscosities, and their viscosity depends very little on temperature. Furthermore, the polymer is highly compressible, it is very permeable to gases, and it has a high chain flexibility, which results in an extremely low glass transition temperature. The bond energy of a Si-O bond is at 373 kJ/mole higher than that of a C-C bond at 343 kJ/mole. Therefore, the polysiloxane chain is thermally and oxidatively much more stable than organic hydrocarbon chains.

In addition, the methyl sidegroups are also thermally stable, since they contain only primary hydrocarbon atoms. The strength of the siloxane bonds is determined by the substitutes of the sidegroups on the chain. Methyl groups reinforce Si-O bonds because of their electron-donating character. Phenyl groups, being electron acceptors, weaken the siloxane bonds [3.710]. The phenyl groups also contribute to chain flexibility, and therefore polysiloxanes with phenyl groups have a particularly low T_g of about $-120\,^{\circ}C$. Methyl groups polarize the siloxane groups, which can contribute to their hydrolizability. Phenyl groups cause the opposite, namely a greater resistance to hydrolysis. Because oxidation or hydrolysis can start under the influence of heat, PMQ grades have the optimum heat resistance.

Polarity. Q vulcanizates are not resistant to aliphatic, aromatic, and chlorinated hydrocarbons. They are to a certain extent swell resistant in paraffinic oils. Because of their polarity, the vulcanizates are also attacked by acids and bases.

Cure Activity. Crosslinking reactions induced by free radicals take place in the sidegroups of the polymer chain. For MQ, this occurs through dehydrogenation of two methyl groups and the formation of an ethylene bridge. Because of the greater chemical stability of the methyl groups in comparison with vinyl groups, MQ can be crosslinked less readily than VMQ. The presence of trace amounts of platinum enhances the reactivity.

Blockiness. Polysiloxane grades, which consist of alternating dimethylsiloxane and tetra-p-disilylphenylene blocks, have the properties of thermoplastic elastomers.

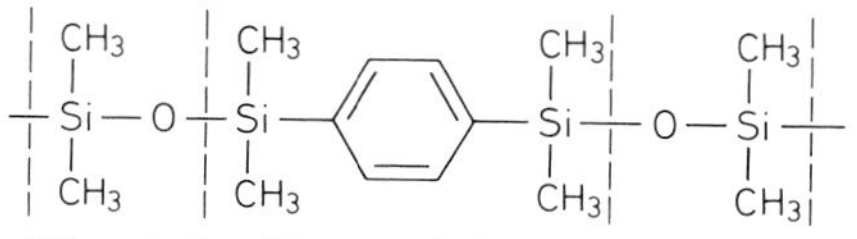

$$-[Si(CH_3)_2-O]-[Si(CH_3)_2-C_6H_4-Si(CH_3)_2-O]-$$

Dimethyl Siloxane — Tetramethyl-p-disilylphenylene — Units

Since the disilylphenylene blocks crystallize and thus form physical crosslinks (see page 159), they act like reinforcing fillers, to give Q, even without vulcanization, a high tensile strength. However, these crosslinks are thermally labile [3.711-3.714]. The dimethylsiloxane/bisphenol-A-polycarbonate block copolymers should also be mentioned in this context [3.707-3.708, 3.711-3.714].

3.4.1.4 Compounding of Q

[1.14]

Vulcanizing Agents. [1.12, 1.13, 3.681] MQ is a highly viscous liquid and it attains elastomeric properties only through crosslinking.

Since MQ is fully saturated, only peroxides can be used for hot vulcanization processes, and in order to obtain a good dispersion of the peroxides in the low viscosity compounds, it is advisable to mix them with silicone oils first. During mixing and further processing, compound temperatures will remain, as a rule, low. Therefore, peroxides with low decomposition temperatures, such as dibenzoyl peroxide or bis-(2.4-dichloro-benzoyl) peroxide can be used for the crosslinking of MQ. The latter can already be used at cure temperatures of as low as 100 °C in press cures, or in steam or hot air vulcanizations at ambient pressures without formation of porosity in the vulcanizate. For this peroxide the mixing and storage temperatures of the compound have to be about 40 to 45 °C. For higher processing temperatures, more stable peroxides are required. VMQ normally has a small number of double bonds, but they are not sufficient for a sulphur vulcanization. Therefore, VMQ (PVMQ) is also crosslinked with peroxides. The vinyl component accelerates the peroxide cure, and therefore lower amounts of crosslinking agent can be used, and higher states of cure are achievable. The peroxide cures do not require additional coagents.

It is of academic interest, that VMQ with higher degrees of unsaturation can be cured with sulphur and accelerators. However, this gives vulcanizates with such poor heat aging resistance, that they are useless for practical purposes [3.715].

Room temperature curable dihydroxypolysiloxanes can be cured at ambient temperatures with multifunctional alkoxysilanes or with alkoxysilane, together with catalytic amines or metal compounds, such as dibutyl-Sn-diacetate, dibutyl-Sn-dilaurate, dioctyl-Sn-maleinate, and Sn(II)-octoate [3.678, 3.682, 3.691]. The crosslinking takes place via the terminal reactive groups of the polymer. With two-component cure systems, the individual components are brought together only just prior to mixing. One component systems already contain the reactive ingredients and the cure reaction is triggered by the humidity of the air.

Fillers. [3.679, 3.680, 3.682, 3.684, 3.686, 3.687, 3.690] Gum vulcanizates from Q have practically no tensile strength at all. Therefore, it is necessary to compound Q with reinforcing fillers, and chemically related silica fillers are particularly well suited. Q-masterbatches with reinforcing fillers can be bought commercially. These facilitate the processing of compounds and the incorporation of additional quantities of fillers.

For optimum reinforcement, fumed silicas [3.680, 3.687] with specific surface areas (BET) of 50 to 400 m^2/g are mostly used. However, mixed compounds with fumed silicas harden during storage and have to be plasticized prior to further processing. Precipitated silicas do not harden mixed compounds during storage as much [3.684, 3.690], but give a lower heat resistance, particularly in closed systems, because they contain water and other electrolytes. Therefore, besides fumed silicas, non-reinforcing diatomic earths, quartz powder of fine particle size, zirconium silicate, or calcined clay are preferred. In the area of filler reinforcement, considerable progress has been made in recent years through surface treatment of the fillers [3.687].

Carbon black is used only in a few applications to produce electrically conductive vulcanizates from VMQ.

Stabilizers. In order to improve the good heat resistance even further, inorganic pigments like Fe_2O_3, TiO_2, or CdO are frequently used.

Softeners. For Q vulcanizates with a hardness of less than Shore A 50, it is necessary to use chemically similar plasticizers, besides filler loadings appropriate for a 50 to 60 hardness range. These are highly viscous silicone oils or gum Q. A liquid 1.2-polybutadiene has also been suggested as a cocurable plasticizer [3.683].

3.4.1.5 Processing of Q

[1.14]

Because of the low viscosity of the rubber, it is difficult to disperse larger amounts of fine particle size fillers in its matrix. However, with filler masterbatches, the compounding proceeds along the same lines as with other SRs. It is important to work with very clean mixing and processing equipment. A peculiarity of Q-compounds is the hardening during storage, so that it is necessary in most cases to remill the compounds at great expense prior to further processing. If the filler composition is properly chosen, the amount of remilling can be reduced to a minimum. It is also necessary in many instances to postcure the vulcanizates for about 6 to 15 hours at 200 °C, to obtain the optimum heat resistance and low compression set. During postcure it is also important to optimize the circulation of fresh air, so that a premature hydrolytic decomposition will not take place during postcure. In this process, all volatiles are also expelled from the vulcanizate, thus minimizing shrinkage. More recently, the processing of liquid silicone grades has gained in importance [3.691–3.698].

3.4.1.6 Properties of Q Vulcanizates

[1.14, 3.671–3.676, 3.690 a, 3.699–3.706]

Mechanical Properties. The tensile properties, and particularly the impact resistance and the abrasion resistance, are considerably poorer than those of other elastomers. The relative vulnerability, particularly due to mechanical impact, is a real weakness of Q vulcanizates. One has to remember, however, that unlike other elastomers, the vulcanizate properties of Q change very little with increasing temperatures, so that the difference in properties with other rubbers becomes smaller at higher temperatures. Above approximately 150 °C Q elastomers show the best mechanical properties of all elastomers. Q vulcanizates can be obtained in hardnesses of Shore A 30 to 80, but those in the hardness range of 50 to 60 generally have the best mechanical properties. This applies particularly to the elastic properties and the compression set.

Dynamic Properties. The damping properties depend relatively strongly on the temperature, and under certain conditions, they fall, for instance, from about 30% at −35 °C to about 15% at 180 °C. By contrast, the spring constant remains at 75 MPa/cm relatively constant over this temperature range. Therefore, it is possible to design from Q vulcanizates damping elements, whose properties remain largely constant over the usual range of service temperatures.

Heat Resistance, Aging Resistance. Q vulcanizates survive a long-term exposure to hot air at 180 °C, and even at 250 °C, their elasticity remains intact after 1000 hours' exposure. The vulcanizates even endure a short exposure to temperatures of 300 to 400 °C (heat shocks). However, steam of 120 to 140 °C attacks and corrodes Q vulcanizates after a certain time, and therefore the vulcanizates should not be exposed to steam. The favourable characteristics regarding heat resistance apply generally to exposures in open systems. During aging, two reactions compete with each another: hydrolytic decomposition by intruding foreign substances for instance water and condensation reactions through liberated water and subsequent crosslinking through oxidative processes. Under normal heat aging conditions, both competing reactions are well balanced, which explains the excellent heat resistance. However, when oxygen is excluded, that is, in closed systems, the equilibrium is shifted towards the hydrolytic decomposition reactions, and the heat resistance is thus reduced.

Q vulcanizates are extraordinarily resistant to aging, weathering, and ozone, and they can even be used for hoses to convey ozone.

Radiation Resistance. Compared with other elastomers, Q vulcanizates are quite resistant to radiation. They can, for instance, be exposed to radiation doses of up to 10 Mrad, with a loss of only 25% in the maximum elongation [3.702, 3.716]. PVMQ is particularly resistant to radiation.

Low Temperature Flexibility. The low temperature properties are also excellent. MQ and VMQ vulcanizates generally only harden below −50 °C and become brittle. It is even possible to prepare PVMQ vulcanizates which are flexible even at −100 °C. The low temperature flexibility depends very much on the vulcanizate hardness, and the optimum performance is in the hardness range of Shore 50 to 60, where the mechanical properties are also best. It should be remembered as well, that the good low temperature properties are not due to the addition of plasticizer, but rather, to the nature of the elastomer itself. Therefore, the good low temperature flexibility is not achieved at the expense of the heat resistance, as with other rubbers, and the same components can be used at either high or low temperatures without dimensional changes.

Swelling Resistance, Resistance to Chemicals. The volume swell of Q in oil compares generally with that of CR. Q vulcanizates are usually resistant to aliphatic motor and transmission oils, but naphthenic oils swell the vulcanizates to a greater extent. They should not be exposed to aromatic oil, particularly at temperatures of above 140 °C. As with ACM, oil additives attack Q vulcanizates to a lesser degree than NBR. The amount of swelling is, of course, also determined by the viscosity of the oil. In general, Q vulcanizates are resistant to heat transfer agents, like Askarels, but are not resistant to motor fuels, chlorinated hydrocarbons, esters, ketones, and ethers.

The resistance of Q vulcanizates to water and chemicals is determined by the possible hydrolysis of the siloxane molecules under extreme conditions. While the vulcanizates can still be exposed to boiling water, they are destroyed by steam of 130 to 140 °C. Alkalis and acids also attack the Q vulcanizates, and the chemically similar silicone oils swell the vulcanizates to some extent [3.701, 3.705].

Permeation. Because of their large molar volume of 95 cm^3/mol, Q vulcanizates have much higher gas and liquid permeation rates than other elastomers. The amount of permeation depends, of course, on the properties of a gas or liquid, and on the ambient conditions, such as temperature and pressure gradient. In general, the permeation rates in Q vulcanizates are roughly 100 times greater than those in IIR or NBR.

Flammability. The ignition temperature of Q vulcanizates is 400 °C, and on burning, a silica lattice forms, which is, unlike other polymers, an electrical insulator. Therefore, control wires which are insulated with Q vulcanizates still remain functional, even after exposure for a short time to fires.

Electrical Properties. The electrical properties of Q vulcanizates correspond at room temperature to those of the best elastomeric insulators, and they largely retain these properties at use temperatures of up to 180 °C, allowing the application of higher electrical ratings. Electrical conductive compounds are available as well [3.690 a].

Abhesive Properties. Q vulcanizates are abhesive and hydrophobic. Therefore, they exhibit some interesting qualities, in that they do not adhere to sticky surfaces, nor do they adhere to ice, which makes them applicable for refrigerator profiles.

3.4.1.7 Uses of Q

[1.14]

Q is an expensive specialty rubber which, like FKM, is only used in those applications where conventional organic elastomers fail. Q vulcanizates are mainly used for their high heat resistance and extreme low temperature flexibility, while maintaining a largely constant level of properties over a wide range of temperatures.

Their good weatherability, good resistance to ozone, aging, and UV radiation at high altitudes [3.702, 3.716], their good electrical insulating and abhesive surface properties, and their physiological compatibility are also exploited in many applications. Q vulcanizates are used in the electrical, electronic [3.703], aerospace [3.688], automotive, mechanical equipment, lighting, cable, and textile industries, as well as for pharmaceutical and medical components, for components in contact with food, or for rubber components with magnetic [3.704] and optically pure properties [3.679, 3.684].

Room temperature vulcanizable parts of MQ are used in such applications as elastic caulkings in the construction industry, and liquid RTL types for injection or transfer moulds and castings [3.682, 3.691, 3.696].

3.4.1.8 Competitive Materials of Q

These are ACM, EAM, FKM, AFMU, and TM.

3.4.1.9 Fluorosilicone Rubber (FMQ, FVMQ)

[3.565, 3.717–3.721]

The condensation reaction of methyl-trifluoropropyl siloxane gives a polysiloxane with a fluoropropyl sidegroup on a silicon-oxygen polymer backbone. These rubbers are known as fluorosilicone rubbers (FMQ resp. FVMQ), and they have the following structure:

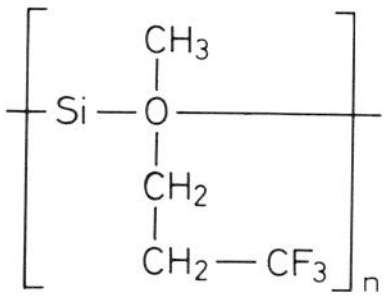

```
 [      CH3              ]
 |      |                |
-|-Si — O ———————————————|-
 |      |                |
 |      CH2              |
 |      |                |
 [      CH2 — CF3        ]n
```

FMQ [3.721]

The introduction of the fluoronated hydrocarbon moiety increases the polarity of FVMQ considerably over that of VMQ, with the result that FVMQ vulcanizates have a considerably better resistance to oils, motor fuels, and chemicals [3.721]. Even after exposure for 1000 hours to high-test motor fuels or methanol-containing fuels, the tensile strength of the vulcanizates has retained about 80 to 90% of its original value, which is surprisingly high [3.721]. FVMQ vulcanizates also swell very little in alcohols, but somewhat more in esters and chlorinated hydrocarbons. The brittleness temperature is, at −65 to −70 °C, higher than for VMQ [3.565], but this is still very much within the requirements of the automotive industry, which specifies brittleness temperatures of below −40 °C. Thus, FVMQ combines the good swelling resistance of fluoro elastomers with the good low temperature flexibility of silicone rubbers. The long-term heat resistance of FVMQ elastomers is somewhat poorer than that of VMQ elastomers owing to their tendency to decompose with loss of vinylidene fluoride. Nevertheless the maximum service temperature is still above 200 °C. Because of its polarity FVMQ is susceptible to hydrolysis by acids and bases. FVMQ elastomers have appreciably lower tensile strength, tear propagation strength and resistance to abrasion than those made of PNF (see page 125) and more especially of FKM (see page 123), but the elongation at break is considerably greater. FVMQ elastomers also have advantages over PNF elastomers with regard to moisture sensitivity and mechanical relaxation behaviour.

Owing to its inferior tensile strength, tear propagation resistance and abrasion resistance properties, the share of the market of FVMQ has remained small compared with that of FKM, despite considerable efforts by the manufacturers to increase its use. The development of PNF is likely to limit the potential for further growth in the use of FVMQ [3.637 a].

3.4.2 Polysulfide Rubbers (TM)

[3.722–3.732]

3.4.2.1 General Background about TM

The oldest examples of polycondensation products with elastomeric properties are the polysulfides (TM) marketed by the Thiokol Corporation. Despite their many shortcomings, TMs are still being used today. This is due to their excellent swelling resistance. The first grades were marketed as early as 1930 under the name of Thiokol A [3.726, 3.727]. The IG Farbenindustrie also marketed a grade under Perduren, but this was withdrawn again in 1945 [3.728].

Instead of a pure organic hydrocarbon backbone, TM has carbon-sulphur chains, which can have different endgroups and can, therefore, react differently.

$$\left[-CH_2-CH_2- \right]\left[-\underset{\underset{S}{\|}}{S}-\underset{\underset{S}{\|}}{S}- \right]\left[-CH_2-CH_2- \right]$$

Ethylene Polysulfide Units

TM is commercially available either as high molecular weight rubber or as low molecular weight pastes or fluids.

Ethylene thioglycol ethers (ASR resp. ST) are familiar to TM (see page 128), but they are not on the market.

3.4.2.2 Production of TM

[3.723]

TM is produced in an aqueous phase at 60 °C by the polycondensation of aliphatic dihalides (e.g. 1.2-dichloro-ethane) and alkali polysulfides (e.g. sodium tetrasulfide).

$$n\,ClRCl + n\,Na_2S_x \longrightarrow -(RS_x)_n- + 2n\,NaCl$$

(R = alkyl; X = 2 or greater)

In the presence of a dispersing agent, such as Mg or Ba-hydroxide, the reaction product is obtained as suspension with a particle size of about 15 μm. Because of their high specific gravity, the relatively large particles precipitate rapidly, facilitating the recovery of the polymer [3.729].

The polycondensation reaction yields high molecular weight products with terminal hydroxy groups [3.730], which have been formed by the hydrolysis of the alkyl halide groups in the basic reaction medium. Through a reductive scission of the polysulfide groups in aqueous dispersion with NaSH, Na_2SO_3, or other reducing agents, one obtains TM grades with terminal thiol groups (Thiokol LP).

$$-R-S-S-R- + NaSH \longrightarrow -RSH + RSNa + S$$

By varying the organic halide and the polysulfide, a great variety of reaction products can be obtained, whose properties can be tailor-made to suit specific requirements. This results in a wide range of commercial products.

3.4.2.3 Structure of TM and its Influence on Properties

[3.5]

Influence of Monomers. TM, which is produced from ethylene chloride and Na_2S_2, is a hard, plastic composition, while TM grades from the disulfide of dichloroethylformaldehydacetate and Na_2S_2 have rubbery properties, even at low temperatures. Co-condensates of both grades have an intermediate property spectrum (Thiokol FA).

Endgroups. The endgroups (i. e. -OH, -Cl, -SH) determine largely the cure properties of TM grades. TM with terminal SH groups can be cured with oxidizing agents like oxygen, in the presence of PbO_2, p-quinone dioxime, Co-salts, or peroxides, to give high molecular weight crosslinked networks, even from low molecular weight or liquid polymers. If these TM grades are properly compounded, they crosslink at ambient temperatures. TM grades with hydroxy or chlorine endgroups are mostly cured with ZnO. This acts as a chainextender by combining endgroups.

Molecular Weight. TM grades with molecular weights of 500,000 and higher are used as hot-vulcanizable rubbers. Room temperature vulcanizable (RTV) grades are pastes or liquids, and they have considerably lower molecular weights, with the latter about 2000 to 4000.

Sulphur Content. The sulphur content determines to a great extent the vulcanizate properties, and particularly the swelling resistance.

3.4.2.4 Compounding of TM

[3.724, 3.725]

Vulcanizing Agents. The cure systems for TM are very simple. The use of ZnO by itself is already sufficient, and increasing amounts of ZnO of up to 10 phr increase the state of cure. If required at all, one can use 0.5–1.0 phr of sulphur as accelerator.

Fillers. Carbon blacks are used as reinforcing agents and to reduce compound costs. The most widely used grades are N550 or N770. ZnO of up to 10 phr, or TiO_2, $BaSO_4$, and lithopone are used as light fillers.

Softeners. In concentrations of up to 0.3 phr, MTBS can serve as softener for TM grades, since, if added during the mixing process, it has a masticating effect, and thus gives smooth banding of the rubber on mills. This effect can be enhanced, if in addition to MBTS, DPG (0.1 phr) is used. To improve the inherently poor building tack, one can add about 5 phr of coumaron resins to TM compounds.

Process Aids. The use of stearic acid is recommended to reduce sticking of compounds to mill rolls, and to aid in the dispersion of fillers.

3.4.2.5 Processing of TM

[3.725]

TM is processed in conventional rubber machinery, but the mixing and further processing of TM can be difficult, and it often requires special procedures.

3.4.2.6 Properties of TM Vulcanizates

[3.725]

The raison-dêtre of TMs is the good resistance of their vulcanizates to aromatic and chlorinated hydrocarbons, and to ketones. In this respect, TM vulcanizates are superior to all other rubbers, although they have a somewhat lower level of mechanical properties, particularly compression set. TM vulcanizates also have a good weather and ozone resistance, and the rubber is relatively expensive. While the odour of the early TM grades was almost intolerable, the present grades have been greatly improved in this respect.

3.4.2.7 Uses of TM

[3.724]

Hot-Vulcanizable TM is mainly used for solvent (mostly aromatic) and oil resistant seals, moulded products and hose.

Pasty and Liquid TM is primarily used for caulking, since there is very little shrinkage on hardening and aging. Because of their good weather and ozone resistance, these caulkings are specifically used by the construction industry [3.731, 3.732].

3.4.2.8 Competitive Materials of TM

TM competes with AU, CO, ECO, ETER, NBR, and Q.

3.4.3 Polyester and Polyether Rubber (OT)

The polycondensation of esters or ethers yields polyesters resp. polyethers, which can either be crosslinked, or used as thermoplastic elastomers.

3.4.3.1 Paraplex (OT)

[3.733]

Paraplex X-100, which was developed by the Bell Telephone Laboratories from 1933 to 1942, is a polyether from ethylene glycol, 1.2-propylene glycol, and sebacic, adipic, or some maleic acid (3 weight %). After mixing with fillers, it can be crosslinked with benzoyl peroxides. Paraplex S200 had more unsaturation, and could therefore be crosslinked with sulphur and accelerator. These products are not used any longer, although low molecular weight ethers of similar composition are marketed as plasticizers, which are difficult to extract from the vulcanizates.

3.4.3.2 Norepol

[3.734]

Norepol of the North Regional Research Laboratories (U.S.A.), is a polyether of dimerized, unsaturated fatty acids and ethylene glycol, which can be crosslinked with sulphur and accelerators. Before the availability of GR-S, these products were used in limited quantities in the U.S.A., but today they are no longer being used.

3.4.3.3 Polyesters Based on Polyethylene Terephthalate and Polyethylene Glycol (Polyetheresters) (TPE-E)

[3.735–3.741]

Polyesters based on polyethylene terephthalate and polyethylene glycol are thermoplastic (with a melting point of approximately 200 °C), highly elastic and hard polymers. These products will be described in Section 3.5.2.1.3. "Thermoplastic Elastomers (TPE)", page 152.

3.4.3.4 Polyether Amides (TPE-A)

[3.742–3.744]

In recent years, several other polyester or polyether rubbers have been developed, which are mostly processed like thermoplastics. Of these, the copolymers of polyetheramides, and blocky amides, are of interest as TPE, see Section 3.5.2.1.4, page 153. Competitive materials are polyether esters and TPU.

3.4.4 Polyurethane Elastomers (PUR, AU, EU, TPU)

[3.745–3.786]

3.4.4.1 General Background about Polyurethane Elastomers (PUR)

[3.745–3.763]

Polyurethane elastomers, or PURs, have become an important class of organic materials for many technological applications. Their development sprung from the fundamental work of *O. Bayer* and his coworkers [3.745, 3.746] on diisocyanate polyaddition reactions.

By reacting a great variety of low or higher molecular weight compounds with diisocyanates, it is possible to obtain polymers with a corresponding variety of chemical structures and physical properties. With the introduction of PUR, it became possible, like thermoplastic polyetheresters, to close the then-existing gap between rubbery and highly elastic polymers and tough or hard polymers, like thermoplastics and duromers. Another attractive feature of polyurethane chemistry is the opportunity to tailor-make materials to specific property requirements, simply by changing the chemical components, their relative concentrations, and the condensation process conditions.

Liquid and highly viscous reactants allow energy-saving processing methods such as reaction injection moulding (RIM) [3.758–3.762], which has led to many novel applications. However, except for referring to the relevant literature, this subject cannot be dealt with in greater detail in this book.

In addition, it is possible to produce by means of PU-technology polymers, which can be processed and subsequently cured using conventional mixing and processing methods of the rubber industry. In the context of this book, these polymers are referred to as polyurethane elastomers. As a rule, they are based on polyester polyurethane prepolymers, (coded AU according to ASTM), and their main use is as millable gums [3.756, 3.757]. Polyether polyurethane prepolymers (EU) [3.763], are hardly at all available as millable gums.

The third main category of polyurethanes consists of high-molecular weight thermoplastic grades, also called TPU [3.764–3.777], which consist of hard segment domains, embedded in elastomeric soft phases (see page 151). The converter can buy TPUs as such, or he can produce them in-house from prepolymers. He can then process the materials by conventional methods, like calendering, extrusion, or injection moulding [3.766]. TPUs do not need to be chemically crosslinked, but this could be done by adding appropriate curing agents to compounds.

The polyurethane elastomers have the characteristics of elastomers - high tensile properties at high ultimate elongations, high hardness, and usually a high heat resistance. These properties are listed in reference texts [3.747-3.755], and therefore, differences in properties of individual AU grades will not be discussed in greater detail here.

Nor will some more recent PUR developments [3.747-3.786], like foams [3.754] be mentioned here, except for literature references [3.747-3.755]. The subject of foam rubbers is very specialized, ranging from open to closed-cell foam structures, unicellular products, moulded, structural foams, etc., and different materials or application techniques. This subject would justify a separate book altogether.

The following remarks are therefore restricted to the application of AU, which is the polyurethane rubber in the strictest sense, as well as to TPU (see page 151).

Polyester polyisocyanate prepolymers (AU), which are compounded and processed by conventional methods can be crosslinked with isocyanates or peroxides. Polyester grades which have double bonds can also be crosslinked with sulphur and accelerators, but this plays a less important role.

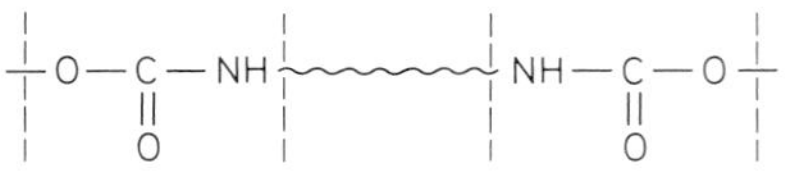

$$-O-\underset{\underset{O}{\|}}{C}-NH\sim\sim\sim\sim NH-\underset{\underset{O}{\|}}{C}-O-$$

Polyester Urethane Units

The crosslinking of polyurethanes is facilitated by incorporating reactive moieties in the polymer molecule, such as:

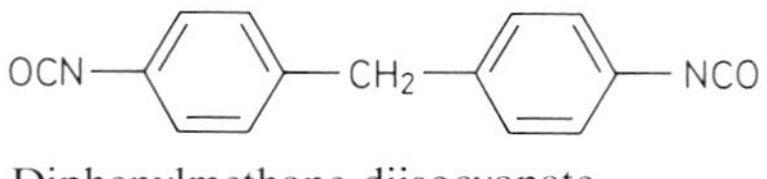

$$OCN-C_6H_4-CH_2-C_6H_4-NCO$$

Diphenylmethane diisocyanate

$$HO-CH_2-CHOH-CH_2-O-CH_2-CH{=}CH_2$$

Glycerol monoallylether

The crosslinking sites are the active hydrogen atoms of the urethane group, active methylene groups, or double bonds, which can react in a sulphur cure.

The great variety of starting materials results in a wide range of AU grades. This, naturally, leads to differences in processing, curing, and also in properties of the vulcanizates.

3.4.4.2 Properties of AU

[1.14, 3.756, 3.757]

The properties of AU depend greatly on the type of crosslinks.

Viscosity. AU grades which can be crosslinked with isocyanates (AU-I grades) are relatively soft, with Mooney viscosities between 14 and 25, while peroxide-crosslinkable grades (AU-P grades) have conventional Mooney viscosities of about 55.

Storage Life. This is good in AU. AU-I grades should be stored in a cool and dry environment.

Solubility. AU can only be dissolved in very few solvents. Only dimethylformamide completely dissolves AU-I grades, while AU-P grades also dissolve in ketones and tetrahydrofuran.

Hydrolysis Resistance. AU grades differ in their resistance to hydrolysis, and resistant or non-resistant grades are commercially available. EU grades are more resistant in this respect [3.763].

3.4.4.3 Structure of AU

[3.753, 3.756]

The structure of polyurethane elastomers is dealt with in Section 3.4.4.9 (see page 145).

3.4.4.4 Compounding of AU

[1.14, 3.756, 3.757]

Vulcanizing Agents. Isocyanates and peroxides are the main vulcanizing agents for AU, while sulphur vulcanization, which is possible with some AU grades, plays only a minor role. AU-I grades are commonly crosslinked with toluylene diisocyanate (TDI), and for products with a high hardness one uses higher TDI concentrations together with hydroquinonedioxyethyl ether. Organic lead salts act as accelerators. Although, in principle, it is also possible to use other diisocyanates, they do not, as a rule, offer any advantages over TDI as crosslinkers. Instead, most of them have serious shortcomings.

TDI can be readily dispersed in compounds, it has a good storage life in these compounds after mixing, and yet, it gives fast cure rates. The physical properties of TDI vulcanizates are also much better than vulcanizates from other diisocyanates.

Only stable peroxides qualify for a peroxide cure of AU-P grades, because a sufficient scorch safety is required for the mixing and subsequent processing steps. The choice of peroxides is also governed by the well known rules (see page 256ff).

If particularly high cure states are required, special coactivators, like triallylcyanurate (TAC) have to be used. They do not affect the potlife and processing safety of compounds significantly.

Peroxide cures cannot be accelerated, and the compounds must not contain sulphur or sulphur compounds, since these interfere with the peroxide cure.

Protective Agents Against Aging and Ozone. These are generally not required in AU compounds.

Protective Agents Against Hydrolysis. Hydrolysis inhibitors are important, particularly for hydrolysis-prone AU grades. Polycarbodiimides are successful for this purpose, and they also improve the resistance of AU vulcanizates to lubricants, hot air, and weathering, particularly in tropical climates.

Fillers. AU-I grades are frequently processed without filler, and in this instance, the hardness of the vulcanizates has to be adjusted through the amount of TDI and the coreagent, hydroquinonedioxyethyl ether. The hard isocyanate segments by themselves give reinforcement, a high hardness, and a good tensile strength. And yet, fillers with varying degrees of reinforcement can be used with AU-I grades. Already small amounts of reinforcing blacks or fumed silica can improve considerably the hardness and tear resistance of vulcanizates. Although fillers can be readily incorporated, the amount of filler that can be used is limited, because the mixing temperature has to be kept low due to the presence of free diisocyanate. In this way, premature scorch is avoided. Semi- or non-reinforcing fillers are used to modify the processibility of compounds, and to reduce compound cost. Moisture in fillers can adversely affect the isocyanate cure, and also the aging resistance of vulcanizates.

AU-P grades are compounded mostly with reinforcing or semi-reinforcing fillers to obtain optimum vulcanizate properties. Although N330 and N550 blacks are most frequently used, their choice regarding compound processibility and vulcanizate properties is governed by the same rules as for other rubbers. Silica fillers can also be used.

Softeners. The low swelling of AU-I grades in many solvents also means that the rubber is incompatible with most plasticizers. Therefore, they are hardly used at all. In special cases, the vulcanizates can be softened with dibenzyl ether, which results in lower processing temperatures, and thus improves the scorchlife in injection moulding or calendering processes.

AU-P grades can be plasticized with small amounts of phthalates and polyadipates to reduce the vulcanizate hardness.

3.4.4.5 Processing of AU

[1.14, 3.756, 3.757]

In general, AU compounds are mixed and processed using conventional rubber machinery and following generally accepted rules. With AU-I grades, however, there can be processing problems, and they should be treated like any other scorchy compound. Moisture and water should not come into contact with AU-I compounds, nor can open steam vulcanizations be used. Humidity, cure temperature, and cure time have an influence on the vulcanizate properties [3.782–3.784], and bimodal cures also change the property spectrum [3.785].

3.4.4.6 Properties of AU Vulcanizates

[1.14, 3.756, 3.757]

Mechanical Properties. The tensile strength of AU-I vulcanizates reaches as high as 40 MPa, and it is thus usually higher than that of other rubber vulcanizates. The hardness is generally high, and it ranges for AU-I grades from 70 to 99 Shore A. And yet, the AU-I vulcanizates exhibit a relatively high degree of elasticity at every hardness level. The abrasion resistance is excellent and better than that of other rubber vulcanizates, in spite of the high hardness of the AU-I vulcanizates. On the other hand, the compression set at higher temperatures is relatively high. AU-P vulcanizates, which are mostly used in the hardness range of Shore A 60 to 80, have poorer tensile properties and a poorer abrasion resistance than AU-I vulcanizates.

Heat Resistance, Aging Resistance. AU vulcanizates withstand a long-term exposure at 75–90 °C or even higher.

In keeping with their chemical nature, AU vulcanizates can be attacked by hydrolysis due to hot water, steam, acids, or bases, as well as by some lubricants, by heat, and particularly by longer exposure to tropical climates.

Vulcanizates from the hydrolysis-resistant AU-P grades can be used for one year at 65 °C.

The aging, weathering, and ozone resistance of AU vulcanizates is excellent. Products from AU are not attacked by oxygen nor ozone.

Low Temperature Flexibility. AU vulcanizates have relatively good low temperature properties. The dynamic brittleness temperatures are mostly between −22 and −35 °C.

Swelling Resistance. AU swells only very little in aliphatic and many other solvents. Highly polar solvents, like chlorinated hydrocarbons, aromatics, esters, and ketones swell AU vulcanizates to a greater extent, but in contact with these solvents, and with high-octane motor fuels, AU vulcanizates perform much better than many other rubber vulcanizates. If high vulcanizate hardnesses can be tolerated in specific applications, AU-I vulcanizates from high TDI and hydroquinone-dioxyethyl ether can be used, since the degree of swelling is inversely proportional to the modulus of the vulcanizates.

Permeation. The gas permeability of AU vulcanizates is of the same magnitude as for IIR vulcanizates.

3.4.4.7 Uses of AU

[1.14, 3.756, 3.757]

Products made from AU are particularly used in the automotive and mechanical engineering industries for seals, elastic-, shock-absorbing or damping members [3.757], for power transmission elements, for flexible joints, for suspensions and for supports with high abrasion resistance. In these applications, the chemical and physical performance characteristics are exploited, namely a combination of good weather and solvent swell resistance, high abrasion resistance, and elasticity even at high hardness levels, and good low temperature properties. However, in all applications, one has to consider the potential hydrolysis, the limited heat resistance, and the relatively high price of the rubbers.

3.4.4.8 Competitive Materials of AU

These are NBR, TM, polyesters from polyethylene terephthalate and from polyetheramides

3.4.4.9 Thermoplastic Polyurethane (TPU)

[3.753, 3.764–3.777]

Polyurethane elastomers are segmented copolymers, consisting of flexible long chain sequences [3.753]. The soft segments, formed by the addition of diols to diisocyanates, act as the elastic components when the polymer is mechanically stressed, while the hard segments from the reaction of diisocyanates with chain extenders can be compared to reinforcing fillers in conventional rubber compounds. There is, however, a difference between the conventional reinforcing fillers and the hard segments of AUs, since the latter are chemically combined with the elastomeric polymer molecule. Another distinguishing feature of reinforcing fillers is the ability of highly polar hard segments to form intermolecular associations. These physical crosslinks, partly in form of hydrogen bonds, are highly effective as long as the intermolecular associations remain thermally stable. This special morphology results in unusually high tensile strengths of polyurethanes, also called PUR or TPU. Not every PUR grade has a well developed segmented structure to qualify it as a truly thermoplastic grade. To do that it has to form specific morphological superstructures [3.753, 3.776, 3.777], and the physical crosslinks of the hard-segment domains have to remain intact at sufficiently high temperatures to give thermo-mechanical properties equivalent to those of conventional rubber vulcanizates. The physical crosslinks are thermally reversible. Above a critical temperature range, TPU can be processed like conventional thermoplastics, and below this critical temperature down to the glass transition temperature, the TPU behave like rubbers. The properties of TPU will be described in Section 3.5.2.1.2 (see page 151).

3.5 Thermoplastic Elastomers (TPE)

[3.787–3.850]

3.5.1 General Background about TPE and Definition resp. Classification

[3.787–3.809 g]

3.5.1.1 Problems

[1.44, 1.48, 3.795 a–3.795 c, 3.806 a]

In the sixties, the definition of high-polymer materials on the basis of their deformation mechanics (or thermodynamic behaviour), i.e. on the basis of the potential freedom of movement available to their molecule chains, led to the familiar three-way division into thermoplastics, elastomers and thermosetting plastics (resp. duromer) [1.44, 1.48, 3.795 a–3.795 c, 3.806 a]. During this period of standardisation, the class of thermoplastic elastomers (TPE) scarcely had any significance in technological terms. This was why TPEs which rank between thermoplastics and elastomers, did not feature in the definition of terms at this time.

In the meantime, thermoplastic elastomers of various compositions have become an indispensable group of materials and steadily development of new materials can be regarded. This is why it is becoming increasingly necessary to have definitions or universally acceptable classifications to make this increasingly confusing field more transparent and make it possible to establish whether finished products are made of elastomers or of TPE. Naturally, a number of different classifications have been proposed, but these are not generally based on the necessary physical or technological criteria and are even contradictory in parts. This is the reason why a pragmatic classification has been proposed [3.795 a].

3.5.1.2 Structure of TPEs

[3.795 a]

TPEs are polymers which, in the ideal case, combine the service properties of elastomers with the processing properties of thermoplastics. This combination of properties can be obtained through the simultaneous presence of soft, elastic segments that have a high extensibility and a low glass transition temperature (T_g value) and hard, crystallizable segments which have a lower extensibility, a high T_g value and are susceptible to association (crosslinking) (see Figure 3.6). The hard and soft segments must be thermodynamically incompatible with each other so that they do not penetrate each other but act as individual phases.

These different segments may be present either in the same molecule, as macromolecule segments, or in the form of a microheterogeneous phase distribution of thermoplastics in plasticizer (or polymer plasticizers). This means that TPE properties

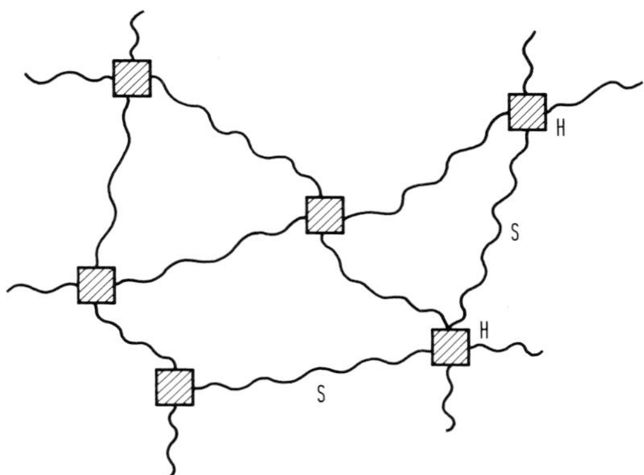

Figure 3.6 Idealized representation of thermoplastic elastomers with soft and elastic segments (S) and rigid, hard segments (H)

are characterized less by their chemical composition but more by their morphological behavior. The concept of TPE thus describes a state of aggregation of the material which is marked by the thermal lability of the crosslink points, arranged in a statistical configuration between the elastomer elements. These labile crosslink points are generally of a physical nature; they can, however, also be of a chemical nature e.g. hydrogenbonds.

The ratio of soft and hard segments determines inter alia the hardness and the modulus of elasticity and also comparable properties of the TPE. The chemical nature of the soft segments has an influence on elastic behaviour and the low-temperature flexibility, whilst the hard segments, which act as crosslink points, determine the heat resistance, the strength and the swelling behaviour. TPE research is endeavouring to find hard, crystalline segments with the highest possible melting point, which can be combined with segments that have low glass transition temperatures in order to guarantee a correspondingly high heat distortion temperature combined with low temperature flexibility in the resultant TPE.

3.5.1.3 Classification of TPEs

[3.795 a]

According to the definition given on page 4, the limiting structures of macromolecular compounds are exemplified by plastomers and duromers (thermosetting plastic). The plastomers are non-crosslinked compounds, which are used in a range of temperatures below their melting and above their glass transition temperature T_g. In the melting range, the crystallinity which gives coherence to the plastomer vanishes, and thus the plastomer can be easily deformed. By contrast, the tightly crosslinked duromers can, ideally, remain rigid, even at high temperatures.

The elastomers occupy an intermediate position between plastomers and duromers, since they have relatively weak crosslinked structures. They are, therefore, characterized by a high extendability, and their T_g value is below their application temperature. Furthermore, under ideal conditions the elastomers are not permanently deformed at higher temperatures, when the sress is removed.

There are, of course, transitions between the main structures represented by plastomers, elastomers, and duromers, and thermoplastic elastomers (TPE) occupy one of these.

If the DSC curve of a covalently crosslinked elastomer (Figure 3.7; above) which has only one glass transition and no melting range (pointing to a single-phase struc-

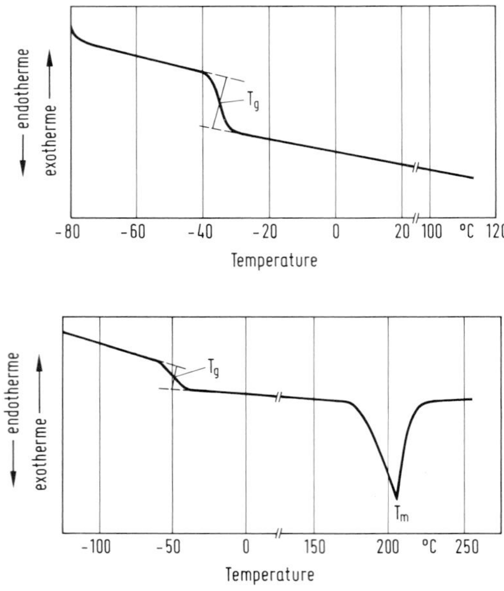

Figure 3.7 DSC curves (scheme) of crosslinked elastomer (above) and of TPE (below, example of a thermoplastic copolyester) [3.830a]

ture) is compared with the DSC curve of a TPE (see Figure 3.7 below) then, in the latter case, two clear phase transitions are evident, i. e. the glass transition T_g of the elastic phase and the melting range T_m of the crystalline phase. Similar DSC curves are obtained from all block copolymers, which show their two-phase structure. The various TPE types, however, differ considerably from each other in terms of the temperatures at which the transitions occur and the intensity of the transitions, such that the TPEs can be characterised on this basis.

In the same way as elastomers rank between thermoplastics and duromers in terms of their material state, TPEs can be placed in the gap between thermoplastics and elastomers (see Figure 1.2, page 5). If the temperature function of the modulus of elasticity of a TPE, for instance, is observed, then it will be seen that the soft segments thaw in the glass transition range T_g which is generally below room temperature. They then become highly elastic. If the temperature is increased still further, the hard segments of a TPE will reach a melt temperature T_m, in contrast to the thermostable covalent crosslinks of elastomers, whereby the modulus of elasticity falls to zero at T_m. The service temperature of a TPE generally ought to be some 30 °C lower than T_m. According to the deformation mechanics definitions for thermoplastics and elastomers, in the ideal case, soft TPEs satisfy the definition of elastomers at service temperatures, i. e. from T_g to (T_m - 30 °C), and the definition of thermoplastics in the melting range and at temperatures above this [3.809 f]. In the ideal case, soft TPEs are delimited vis-à-vis thermoplastics through phase structures which have the same effect as wide-mesh entropy-elastic crosslinks and lead to pronounced, almost fully reversible deformation at application temperatures at low levels of force. They differ from elastomers in that their macromolecules are no longer structurally fixed above critical temperatures, but display macro-Brownian movement, or in that the temperature and time function of the modulus of elasticity falls to zero. This relatively simple deformation mechanics definition of a TPE is essentially only applicable to a few soft grades of TPE which are close to the elastomers on account of their deformation mechanics properties. The majority of harder and more rigid TPE types are a long way from fulfilling this ideal.

Other criteria for drawing the boundary between TPEs and elastomers can be derived from the dependence of mechanical properties on the stressing temperature. The closer the stressing temperature of a TPE comes to its softening temperature, the greater the deviations will be in the properties of the TPEs in relation to a comparable elastomer. It is obviously possible to apply different properties for this, such as hardness, tensile strength and compression set, etc. When the properties are plotted versus the temperature, corresponding differences result in the curve profile in each case, these being different for elastomer and TPEs and also for TPEs with dissimilar softening ranges. These observations can be applied for a classification of TPEs or to draw the boundary between TPEs and elastomers. They permit the maximum service temperature to be established in each case. For instance the dependence of elongation set on strain can provide very sensitively information on the dissimilar behaviour of TPEs and elastomers. Even when elastomers are strained several hundred percent, there is only a small amount of residual deformation after the load has been removed. The TPEs are in between elastomers and thermoplastics, as is shown in the schematic diagram in Figure 3.8. The TPEs towards the bottom of the curve scatter range are closer to elastomers and those towards the top are closer to thermoplastics. Theoretically a complete continuum exists between the behaviour of elastomers and thermoplastics. Also the solution behavior shows differences between elastomers and TPEs. Covalently crosslinked elastomers are unsoluble in solvents in contrast to TPEs.

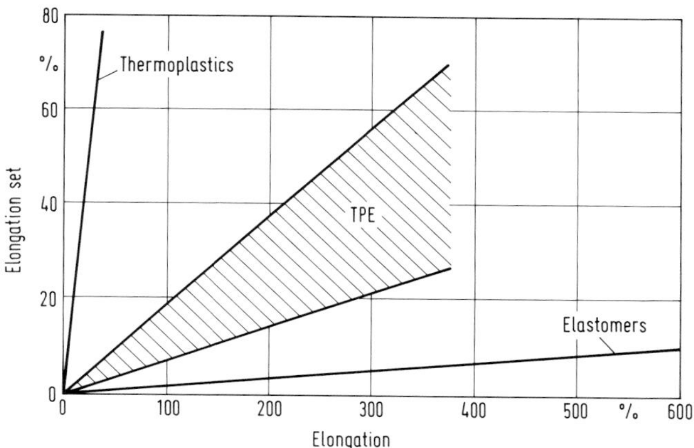

Figure 3.8 Elongation set of elastomers, TPEs and thermoplastics versus elongation [3.795a]

Since TPEs cover a continuous hardness range from approximately Shore A of 30 to Shore D of 75, i.e. from "soft and rubberlike" to "hard as a polyamide" (see Figure 3.9), this means that a uniform definition for all TPE material classes on the basis of deformation mechanics considerations is scarcely possible.

Pragmatic considerations thus take on increased importance when it comes to a classification e.g. regarding the hardness. Any hardness boundary, of course, will represent an arbitrary means of classification. If it is considered, however, that by

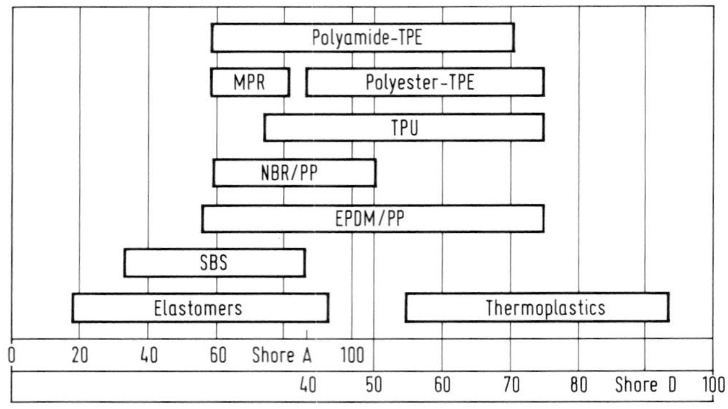

Figure 3.9 Comparison of hardness of TPEs versus elastomers and thermoplastics [3.795a]

far the majority of elastomer applications extend to hardness of some Shore A of 85 or Shore D of 35 then it would be feasible to draw a practice-oriented boundary between the TPE types that are closer to elastomers and those that are closer to thermoplastics.

If it is considered that rigid polymers only experience a correspondingly low level of deformation for a given level of stressing, then a new proposal takes on significance, whereby TPEs would be divided into three classes [3.795 a] leaving out the low-priced, non-engineering TPE grades (such as soft PVC, EVA, polyisobutylene and the like), which are not included in the definition considerations here. These would be the classes of:

- Blends of rubbers with thermoplastics, blended TPEs (e.g. EPDM/PP),
- soft block copolymers as "multi-purpose TPEs" (e.g. SBS) as well as other types of soft TPEs (e.g. also soft elastomeric alloys of EPDM/PP, NBR/PP etc.),
- hard block copolymers as "engineering TPEs" (e.g. thermoplastic polyurethanes, copolyesters, polyetheramides and also hard elastomeric alloys)

Whilst the first class, including the EPDM/PP grades is often closer to elastified thermoplastics than to TPEs in the medium and higher hardness range, the second class covers multi-purpose TPEs, including styrene triblock copolymers, and soft TPEs of other material classes - products which are close to elastomers in terms of service temperature range and which may be defined in the same way as these. The third class, the "engineering TPEs" follows the pattern already set in plastics with the designation of "engineering plastics". If the elastomeric alloys will be devided into soft and hard types, then remain only two groups for the classification of TPEs, the soft multipurpose TPEs and the harder engineering TPEs. Elastomers with thermally reversible labile crosslinks play a minor role on the market and are, therefore not regarded in this context.

3.5.2 Types of Thermoplastic Elastomers

In Sections 2.4.1.3 about NR (see page 14), 2.4.3.1 about guayule (see page 18), 3.3.2.3 about SBR (see page 62), 3.3.3.3 about NBR (see page 70), 3.3.8.3 about EPDM (see page 96), 3.4.3.3 about polyethyleneterephthalate polyethyleneglycole copolymers (see page 139), 3.4.3.4 about polyether polyamides (see page 139) and 3.4.4.9 about TPU (see page 143), the basic morphological requirements for TPEs

Table 3.4: TPE Classes[1] [3.795 a]

TPE Classes	Hardness (Shore)	Examples (Trade name/manufacturer)	Polymer segments		Service temperature limit °C
			Soft	Hard	
Styrene grades					
SBS, SIS (TPE-S)	30 A to 85 A (35 D)	Kraton D/Shell Cariflex TR/Shell Solprene/Phillips	Butadiene or isoprene	Styrene	– 70 to + 60 (80)
SEBS		Kraton G/Shell[2])	Ethylene butylene	Styrene	– 70 to + 60 (80)
Polyurethanes (TPE-U = TPU)	75 A to 75 D	Desmopan/Bayer Estane/Goodrich[3]) Pellethane/Upjohn[3])	Ester glycols or ether glycols	Isocyanate chain prolongers, H-bonds	– 40 to + 120
Polyetheresters (TPE-E)	85 A (35 D) to 72 D	Arnitel/Akzo Hytrel/Du Pont	Alkylene glycol	Alkylene terephthalate	– 50 to + 150
Polyetheramides (TPE-A)	60 A to 70 D	Pebax/Atochem	Ether diols	Amides	– 40 to + 80
Elastomeric alloys					
EPDM/PP (TPE-O = TPO)	55 A to 75 D	Levaflex/Bayer Santopren/Monsanto	Crosslinked EPDM	Propylene or Propylene/Ethylene	– 50 to + 100 (120)
NR/PP (TPE-NR)		Thermoplastic natural rubber/MRPRA	Crosslinked NR	Propylene or Propylene/Ethylene	– 50 to + 100
NBR/PP (TPE-NBR)	60 A to 50 D	Geolast/Monsanto	Crosslinked NBR	Propylene or Propylene/Ethylene	– 40 to + 120
Chloroolefine base	60 A to 80 A	Alcryn/Du Pont	N. N.	Chlorinated Polyolefine	– 40 to + 120

[1] A number of the most well-known classes (without the most recent development products) giving a few examples of each.
[2] Blended with thermoplastics also available as Elaxar or Kraton/PN or blended with polysiloxaneas C-Flex.
[3] Also blended with ABS.

Table 3.5: A Number of Abbreviations in Standard Use for Thermoplastic Elastomers [3.795 a]

Abbreviation	Groups for which used
TPE, TPR [3.804][1])	General designation of group
TPO, thermoplastics	Thermoplastic elastomers based on olefins
SBS, SIS, SBC, Y-SBR, Y-IR [3.804]	Styrene triblock copolymers
TP-NR, Y-NR	Thermoplastic natural elastomer
TP-NBR, Y-NBR	Thermoplastic nitrile elastomer
TP-FKM	Thermoplastic fluoro elastomer
TP-Q	Thermoplastic silicone elastomer
TPU	Thermoplastic polyurethanes
CPE, CPA [3.804]	Copolymer polyetheresters (Y-BPO, e.g. Hytrel)
PEBA	Polyether block amides
MPR [3.796 a][1])	Melt processable rubber, proposed for Alcryn

[1] The combination of an abbreviation with the letter R should be avoided, since the definition of the concept "rubber" does not apply in the case of TPEs [3.795 c] and is only valid in areas where the term "elastomer" is used.

have already been discussed. There are three major classes of thermoplastic elastomers:

1. block or segmented copolymers (styrene grades, polyurethanes, polyetheresters and polyetheramides)
2. elastomer-thermoplastic blends
3. elastomers with thermally reversible labile crosslinks. Some of the main types of groups 1 and 2 have been listed in Table 3.4 and a lot of abbreviations for TPEs have been listed in Table 3.5.

3.5.2.1 Block or Segmented Copolymers

[3.169, 3.582, 3.586, 3.707, 3.708, 3.710-3.714, 3.735-3.738, 3.740, 3.742, 3.764-3.775, 3.785, 3.786, 3.795 a, 3.795 b, 3.804, 3.809 g-3.819, 3.834, 3.835, 3.839-3.842, 3.844]

3.5.2.1.1 Styrene Triblock Copolymers (Y-SBR, resp. SBS; Y-IR, resp. SIS)

[3.795 a, 3.795 b, 3.809 g-3.819]

Styrene triblock copolymers (often generally abbreviated TPE-S) are made of styrene blocks - generally 200 to 500 styrene units - between which polydiene or polyene segments are intercalacted - as a rule, 700 to 1500 monomer units. The styrene blocks form crystallites (hard blocks), which theoretically act as physical crosslinks up to their melting point of approximately 100 °C. Since these crosslink points already lose some of their binding power a fair way below their melting point, however, they can only support loads at up to temperatures of 60 to 80 °C in technical terms. The soft poly(di)ene segments are generally made up of polybutadiene (SBS) or polyisoprene (SIS) blocks or, in order to avoid any susceptibility to oxidation through remaining double bonds, of butylene-ethylene units (SEBS).
The poly(di)ene segments lead to the material's good *low-temperature flexibility,* with softening points around or below -70 °C. This means that styrene triblock copolymers of the SBS type can theoretically (but not under practical applications) be applied in a temperature range from approximately -70 °C to a maximum of +100 °C.

In a *service temperature* range up to approximately +60 °C, styrene triblock copolymers in hardnesses of some Shore A of 60 behave like real elastomers. When it comes to the production of finished parts, they are cheap, readily processed, without waste, easily coloured and, a factor of prime importance, suitable for processing on rubber processing machines. Against this come the drawbacks of reduced *elasticity* and *dynamic properties,* a more pronounced softening gradient with increasing temperatures - giving poor *strength properties* as of temperatures as low as 60 °C, for instance, and, above all, a higher *compression set.* The *chemical resistance* is also lower. This then eliminates most applications which place stringent requirements on the elastomer character of a material, such as in the field of elastic force transmission, elastic temperature-resistant couplings, vibration insulation, sealing and protection against corrosion etc. [3.804]. TPEs of this type, by contrast, are used to a large extent in the adhesives sector as molten adhesives and bitumen modifiers and have extensively replaced NR and SBR, and also PVC and leather, in applications such as the shoe and sole sector. In 1986, world consumption totalled approximately 96500 t, with annual growth rates of approximately 7% [3.809 c].

For some time now, commission blending companies have also been blending triblock copolymers with thermoplastics, such as EVA and PP, thereby shifting the heat distortion temperature towards higher temperatures, e.g. as high as +120 °C, achieving a considerably more constant modulus of elasticity and hardness etc. at higher temperatures as well as an increased *ozone resistance.* An increase is also achieved in hardness and stiffness, however, at the same time. A further key field of application is the addition of small quantities of styrene triblock copolymers to thermoplastics in order to elastify these and make them impact-resistant. In theory, it is possible to employ any mixing ratio for styrene triblock copolymer and thermoplastics. In terms of properties, these polymer blends rank somewhere between TPEs and elastified thermoplastics, depending on the amount of triblock copolymer they contain.

3.5.2.1.2 Thermoplastic Polyurethanes (TPE-U, resp. TPU)

[3.753, 3.764-3.777, 3.795 a, 3.795 b, 3.809 g, 3.839]

As described on page 143 for such cases in which segmented structures with specific morphological superstructures are present in PUR types [3.753, 3.776, 3.777], these types can be processed like conventional thermoplastics. Those are TPUs, which behave below the critical temperature down to the glass transition temperature like rubbers.

The TPU grades cover a wide range of hardnesses, from Shore A of 75 to Shore D of 75 (or Rockwell R of 90). The grades of lower hardness of up to about Shore A of 85 (resp. Shore D of 40) compete with conventional elastomers, while the harder TPU grades of above Shore D of 55 compete with thermoplastics. The grades between Shore D 40 to 55 close the gap between rubbers and thermoplastics. High *rebound* values of 80% can be obtained, despite the high hardness levels, and TPUs retain their rebound between 10 °C and 100 °C much better than conventional rubber vulcanizates. This is important for shock absorbers. The *abrasion resistance* of TPUs is excellent, and therefore, the materials are used for baffles, roll and wheel covers, bearing pads, etc. At equivalent hardnesses, TPUs have a much better load bearing capacity than other conventional rubbers. The *impact resistance* of hard TPU parts is much better than that of most plastics. The tensile strengths at 30 to 45 MPa with *elongations at break* between 200 and 400% are also considerably higher than for other rubbers. TPU components too are very *resistant to cracking* after repeated bending, and the crack resistance of components can be greatly increased, if the gauge is reduced. The same applies, of course, to other rubbers. Because of the strength and toughness of TPUs, parts made from it can often be designed in thin gauges. Standard TPUs also have very good *low temperature properties*. Below −18 °C they start to stiffen, but do not become brittle even at temperatures of as low as −60 °C, or −80 °C for some special formulations. On the high temperature side, a relatively good *heat resistance* exists up to 120 °C, and the hardness level drops only by one Shore A unit after aging. However, the behaviour of TPU components should be tested in each instance, before they are approved for service temperatures of above 90 °C. TPUs are also very *resistant to aging*, weathering, oxygen, ozone and radiation (in pigmented compounds), as well as to oils, motor fuels, hydraulic fluids, and many different solvents. Therefore, TPUs are widely used in the automotive industry. On the other hand, TPU parts *swell* more or less in aromatic and polar hydrocarbons. The resistance of TPUs to long-term exposure to water is good. For instance, parts submersed in water of 50 °C for 12 months show only a loss of 5 MPa in tensile strength from the original 35 MPa. Because of their good *electrical properties*, TPUs are successfully used in pourable and encapsulating compositions for electrical applications of up to 100 kV and temperatures of up to 100 °C.

The fields of application of TPU correspond to those of AU-I vulcanizates (see page 142). They compete with TPE-Es and TPE-As.

3.5.2.1.3 Thermoplastic Copolyesters (TPE-E)

[3.735–3.741, 3.795 a, 3.809 c, 3.809 g, 3.844]

Copolymers based on alkylene terephthalate and alkylene glycol blocks are hard, thermoplastic and highly elastic polymers [3.736 a, 3.736 b, 3.795 a, 3.795 b, 3.809 a, 3.809 c, 3.809 g, 3.830 a]. They are best known under the trade names Hytrel (Du Pont) and Arnitel (Akzo). The structural formula of Hytrel is given below:

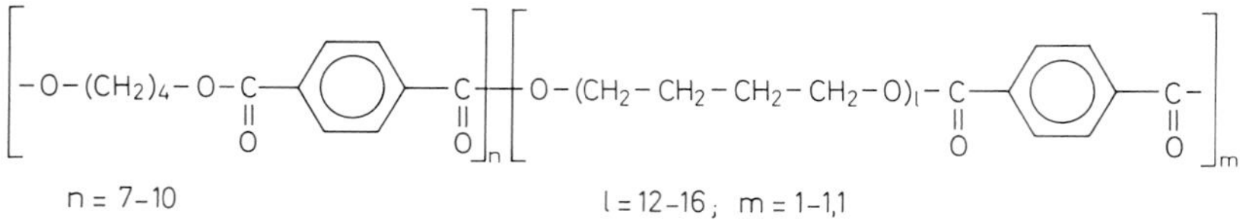

$$\left[-O-(CH_2)_4-O-\underset{\underset{O}{\|}}{C}-C_6H_4-\underset{\underset{O}{\|}}{C}-\right]_n\left[O-(CH_2-CH_2-CH_2-CH_2-O)_l-\underset{\underset{O}{\|}}{C}-C_6H_4-\underset{\underset{O}{\|}}{C}-\right]_m$$

$n = 7–10$ $l = 12–16;\ m = 1–1{,}1$

Hard Segment Soft Segment

Thermoplastic Copolyester (TPE-E)

The morphology of this thermoplastic copolyester resembles that of styrene triblock copolymers. The hard and crystallizing chain segments are here formed by oligobutylene terephthalate segments consisting of 7 to 10 monomers (corresponding to segment molecular weights of around 1700 to 2500), and have a melting point of around 200 °C. The soft and flexible segments, on the other hand, consist of oligobutylene glycol ether with e.g. 12 to 16 butylene ether monomer units (corresponding to segment molecular weights of about 1000–1400). The oligoester and oligoether units are linked with each other by ester groups. This is why oligoester-oligoether block polymers are usually referred to, for simplicity's sake, as thermoplastic copolyesters.

The glass transition temperature T_g of thermoplastic copolyesters is around −50 °C. Since they melt at around 200 °C, this means that they theoretically can be used at temperatures between −50 °C and +200 °C, an extremely wide service temperature range for thermoplastic polyesters. In actual practice, however, the operating temperature range is from −40 °C to +150 °C. Depending on the molecule length of the hard and soft segments and the ratio of hard and soft segments (e.g. from approximately 8 to 1), it is possible to make softer (e.g. Shore D of 35) and more flexible as well as harder (e.g. Shore D of 72) and thus more rigid grades of TPE-E.

Thermoplastic copolyesters are growing in popularity for many applications because of their high mechanical strength and heat resistance, their non-swelling properties in contact with oils, aliphatic and aromatic hydrocarbons, alcohols, ketones, esters, hydraulic fluids, etc., as well as their flexibility and high flexural modulus. This is confirmed by the annual growth rate of 20% [3.809 c]. Thermoplastic copolyesters can be processed only on plastics processing equipment because of their high melting points. Thermoplastic copolyesters are in competition mainly with thermoplastic polyurethanes and polyether amides.

Because of their hardness and elasticity, most thermoplastic copolyesters and other hard thermoplastic elastomers (often referred to as engineering TPEs [3.795 a] are used for applications where traditional elastomers or soft TPEs would hardly be suitable because of their relatively low elastic modulus and poor strength. On the one hand, TPEs are the ideal complement for engineering plastics because of their elasticity, and on the other they are a welcome addition to the covalently cross-linked elastomers or soft TPEs. Since thermoplastic copolymers are often fifteen times stronger than elastomers and soft TPEs, volume and, of course, weight can be reduced (in extreme cases) by a factor of 15 – although under practical conditions approximately 2 to 6. However, the elongation of these engineering TPE-Es up to which there is still spontaneous and almost completely reversible relaxation is only 7–25% [3.795 a].

When engineering plastics are replaced by engineering TPE-Es in order to take advantage of the latter's elasticity, concessions have to be made with regard to rigidity and strength. The modulus of elasticity in bending of engineering TPE-Es, for example, is up to about 500 N/mm^2, whereas engineering plastics can show values of more than 14000 N/mm^2 and elastomers a mere 25 N/mm^2. From the elastomer user's point of view, engineering TPE-Es do not show sufficient elastic recovery after major deformation. If an engineering TPE-Es is subjected to elongations of no more than 7 to 25% (depending on grade), no design modifications will be necessary, although the greater hardness will, of course, have to be taken into consideration. If, on the other hand, elongations are likely to be higher, design changes will have to be made.

A great deal of information is available about the use of thermoplastic copolyesters in place of other materials such as metals, plastics, leather, elastomers, etc. [3.795 a]. When thermoplastic copolyester are used as replacements for traditional elastomers, resulting in considerable weight savings, component design will have to be modified to ensure that the critical limiting elongation is not exceeded.
TPE-Es compete with TPUs and TPE-As.

3.5.2.1.4 Polyether/Polyamide Block Copolymers (PEBA, resp. TPE-A)

[3.742–3.744, 3.809 d, 3.809 e, 3.809 g, 3.834–3.836, 3.840]

The same applies to polyether/polyamide block copolymers (TPE-As) as to polyether-polyester block copolymers. Whilst the polyamide blocks constitute the hard segments, which obtain their strength through a high density of aromatic groups and amide groups, the polyether blocks are soft and elastic. The service temperature of TPE-A ranges from −40 to +80 °C [3.809 d], i.e. the temperature range is considerably narrower than for the thermoplastic copolyesters. The hardness of the commercial products currently available ranges from 60 Shore A to 75 Shore D, i.e. the soft grades extend further into the classic elastomer range than is the case with the thermoplastic polyurethanes and copolyesters, which constitute the chief rivals to TPE-As.

All TPE-As have, in general, the same property profile. They contain no plasticizer, the deMattia resistance at room temperature (without groove) with 60 million cycles is excellent (100000 with groove), the hysteresis is low and the tear resistance high. Furthermore they have excellent noise absorption properties, low wear, good resistance against chemicals. Beside this, their properties have similarity with those of thermoplastic polyesters. Recently new types with fiber reinforcement came on the market.

TPE-As must be processed on thermoplastic processing machinery and have good processibility. The processing cycles are up to 25% shorter than with soft TPE types like SBS. This is important for commerical reasons. Compared with vulcanized rubber the production rate of finished parts is reported to be eight times higher with TPE-As [3.809 a]. Because of the advantages TPE-As have at the time being a high market growth of approximately 20% [3.795 b].

3.5.2.2 Blends of Elastomers and Thermoplastics

[2.16 c, 2.51-2.52, 2.73, 2.74, 3.234-3.235 a, 3.809 g, 3.820-3.833 d]

3.5.2.2.1 General Background about Elastomeric Alloys

[3.795 a, 3.795 b]

Thermoplastics that contain plasticizer (such as plasticized PVC) can be regarded as a type of precursor or cheap version of TPE, although these are really elastified or impact-modified thermoplastics. Similar, but technically more effective results are obtained when thermoplastics are blended with non-crosslinked rubber [3.863]. They can have the same effect as macromolecular plasticizers but, by definition, they are not yet elastomers [3.795 c]. This is the case, for instance, with pairs of polymers and blend ratios in which the remaining crystal structures of the thermoplastics guarantee the strength and the still non-crosslinked rubber segments provide the elastic properties. These must therefore be regarded as elastified or impact-modified thermoplastics. Such blends do not contain any elastomer, by definition, and hence cannot be strictly classified as TPE.

The elastifying effect becomes even more pronounced when rubber/thermoplastic blends of this type have the rubber phase dynamically crosslinked in situ, making the mix into a true thermoplastic/elastomer blend, or when rubber molecules are grafted on to thermoplastic molecules, or when styrene triblock copolymers are blended with thermoplastics. Blends of these types are also called elastomeric alloys [3.821 a] or true types of TPE made of elastomers and thermoplastics.

Elastomeric alloys, such as those made up of thermoplastics and crosslinked elastomers, whose properties depend on the level of crosslinking of the elastomer segments, display different behaviour. There are many possibilities when it comes to manufacturing alloys of this type. A number of the more frequently cited polymer pairs in the literature will be listed by way of example: NR/PP [2.51]; EPDM/PP, EPDM/PS, EPDM/SAN, EPDM/PA 6, EPDM/PA 66 [3.822 a]; EVA/PS, EVA/SAN, EVA/PA 11 [3.822 a]; EVA/PVDC [3.821 a]; NBR/PP, NBR/PS, NBR/SAN, NBR/PA 6 to 9 [3.182, 3.822 a, 3.833 c]; NBR/PVC [3.235, 3.235 a]. Recently also blends of SBS (SIS) and PP came on the market. These obviously lead to very different classes of TPE – a number of which are listed in Table 3.4 (see page 149). This also explains the broad range of properties found in TPE based on elastomeric alloys, particularly when consideration is given to the type of bonding, i.e. grafting, covalent crosslinking and ionomer crosslinking [3.826, 3.827]. The class of elastomeric alloys thus covers a particularly wide hardness range (see Figure 3.8, page 147).

In the soft range (Shore A hardnesses of 55 to 85), elastomeric alloys display good elastic properties and are capable of replacing covalent crosslinked elastomers within certain temperature limits in the same way as styrene triblock copolymers. At higher hardness levels they fill the gap between elastomers and thermoplastics, in competition with other classes of TPE, such as thermoplastic polyurethanes, copolyesters and polyetheramides. At the top end of the hardness scale they consitute impact-modified thermoplastics.

The service temperatures of these TPE classes are obviously conditioned by the T_g values of the elastomers used and the melting points of the thermoplastics. At present there is no indication that TPE of this type can be used at temperatures higher than 100 to 120 °C. The high melt temperature calls for processing on plastics processing machinery.

3.5.2.2.2 EPDM/PP-Blends (TPE-O, resp. TPO)

[3.795 b, 3.820–3.824, 3.826–3.833 b]

Blends of elastomers and thermoplastics, particularly EPDM and PP, have been manufactured since the start of the seventies in Europe and Japan. These are used inter alia for bumpers on automobiles and for cable sheathing. Since it is possible to blend EPDM and PP, for example, in any ratio, there is theoretically a continuous spectrum from elastified PP to EPDM reinforced with thermoplastics. Where the PP component predominantes, which is by far the chief field of application, PP constitutes the continuous phase with uniformly and finely dispersed rubber segments [3.795]. With a very high EPDM component, by contrast, the structure reverses and the material is a PP-reinforced EPDM. The properties of blends containing a large quantity of PP – the impact-resistant PP grades which are available under a large number of brand names – depend on the quantity of EPDM and on the uniformity and size of the microheterogeneous rubber phase. They have a relatively low level of elasticity (above all, they have virtually no entropy elasticity), display a low elongation at break and high values for compression set [3.806 a]. They have only a low-level elastomeric character in the service temperature range, making it difficult to classify them as TPEs in the true sense of the term.

In contrast, through dynamic crosslinking in situ of EPDM with PP or PP/PE-blends during mixing [3.822 a, 3.832], elastomeric alloys (TPOs) can be made with rubber-elastic behavior over a wide range of temperatures (−40 °C to +125 °C [3.833 a, 3.833 b]). The rubber-elastic behavior of such elastomeric alloys is quite different from that of non-crosslinked EPDM/PP blends [3.833 b]. The morphology of in situ crosslinked polymer blends differs from those of non-crosslinked ones [3.832]. The distribution of rubber particles is much finer and, therefore, the heterogenius phase distribution is much better for the TPOs. The properties of the fully crosslinked elastomer particles do not change even at processing temperatures up to 180–225 °C [3.832]. In contrast, the melt temperature of the applied polyolefines influences the processing and application temperatures. The maximum application temperature can be expected to be in the range between 100 °C and 125 °C. The high melt temperatures of TPOs call for processing (e.g. calandering, extrusion, injection molding etc.) thermoplastics processing machines.

In general, the properties of each individual TPO type depend, apart from EPDM and the polyolefine types, also on the blend ratios, the degree of crosslinking and kind of compounding of EPDM phases, and, of course, of the microheterogenious phase distribution of EPDM in polyolefines. The well-known TPO sortiments from various producers e.g. [3.833 a, 3.833 b] vary in each case in wide hardness ranges from approximately Shore A of 55 up to Shore D of 75 with good elastic behavior of the softer ranges. These products can replace covalently crosslinked elastomers in a lot of applications and they fill the gap between elastomers and thermoplastics (see Figure 3.8). They compete with TPUs, thermoplastic copolyesters and with TPE-As. In the highest hardness range they compete also with impact resistant thermoplastics.

The UV and ozone resistance of TPOs is excellent and they are resistant in most inorganic chemicals as well as in a lot of polar organic solvents like break liquids [3.826, 3.827], but not in aliphatic and in aromatic solvents, in which they have great volume swell. TPOs also have good abrasion resistance and the electrical insulation properties are excellent. The compression set resistance between low minus degrees up to approximately 70 °C is low and it will be enlarged above 100 °C rapidly, coming at 125 °C near to 100%. The stress crack propagation as well as the migration of plasticizers and oils is low [3.833 a].

Under the additional application of primers finished TPO parts can be varnished with normal laquer systems. Because of the low density of TPOs, the short processing times under avoidance of vulcanization processes, low rate of waste etc. TPO parts can be produced cheaper than those made out of EPDM [3.833 b]. Obviously the TPO types are twice as expensive als EPDM. Cost savings up to 25% are reasonable. By 1986, some 48000 t of TPOs, based on elastomeric alloys, were consumed worldwide with an annual growth rate of approximately 7% (see Table 3.5).

3.5.2.2.3 NR/PP-Blends (TPE-NR)

[2.16 c, 2.51-2.52, 2.73, 2.74]

In the same way as described above compounded NR can also be blended with PP or PP/PE blends. As in situ crosslinking by peroxides during mixing cycles [2.51] results in thermoplastic NR (TPE-NR). Blends of Guayule rubber with PP are also possible [2.73, 2.74].

Because the olefinic component TPE-NR is stiffer than normal NR elastomers and the stiffness increases of course with increasing polyolefine share. The hardness, the modulus of elasticity and the tensile strength increase too. The higher stiffness limits the low temperature applicability to approximately −30 °C and the application under heat is limited to approximately +70 °C. At higher temperatures the compression set increases rapidly and the strength properties will be reduced drastically. TPE-NR has a high tensile strength at room temperature of approximately 20 MPa in flow direction and 17 MPa in the opposite direction [2.51]. The wheather and ozone resistance of TPE-NR is much better than that of NR-elastomers and the dynamic properties are also reported to be quite good. The electrical insulation properties are excellent. The finished TPE-NR parts can be varnished with normal lacker systems.

The applicability of TPE-NR (−30 °C up to +70 °C) is limited compared with that of NR elastomers. But because of the high stiffness TPE-NR can also be applied in such areas in which higher hardness is required, for instance in car body parts, shoes or boats parts etc. beside cables and a lot of other applications [2.51].

For commercial calculations the same arguments as for TPOs are valid (see Section 3.5.2.2.2). TPE-NR can be processed on rubber processing machinery, which is of advantage for the rubber industry. Therefore, TPE-NR will be an attractive material for technical as well as for commercial reasons. The products are too new yet for having a measurable market share.

3.5.2.2.4 NBR/PP-Blends (TPE-NBR)

[3.182, 3.795 b, 3.822 a, 3.833 c]

In the same way as described for TPOs, compounded NBR can also be blended with PP or PP/PE blends. An in situ crosslinking during the mixing cycle [3.822 a] results in thermoplastic NBR (TPE-NBR). Microheterogenious phase distribution can be obtained by crosslinked NBR in PP, which can be processed like other TPEs [3.833 c]. TPE-NBR is rubber-elastic in the application temperature range (−20 °C up to 100 °C), and the hardness range ranks on Shore A of 80 to Shore D of 50.

The good swelling resistance of TPE-NBR in automotive oils and fuels is combined with an excellent ozone resistance, contrary to covalently crosslinked NBR. Also the acid and alkali resistance is good. The compression set resistance is acceptable even at temperatures of 100 °C. Therefore, there are some possibilities to exchange covalent crosslinked NBR against the cheaper processable TPE-NBR for the same applications. Cost savings up to 50%, based on a finished part, have been mentioned for TPE-NBR versus covalently crosslinked NBR elastomers [3.833 c].

3.5.2.2.5 NBR/PVC Thermoplastic Blends

[3.235, 3.235 a]

By blending NBR grades (mainly X-NBR) with PVC, another kind ot thermoplastic NBR can be prepared [3.235, 3.235 a]. With increasing NBR share the blend gets stiffer. The hardness of the described products ranges from Shore A of 67 up to 75 with tensile strength of 10 up to 15 and elongation of break of 300% up to 470%. The properties vary with the kind of NBR grade, their molecular weight, acrylonitrile content, carboxylating rate, curing rate, blend ratio etc. These products seem to be experimental at the time being.

3.5.2.2.6 Thermoplastic Elastomers Based on Halogen Containing Polyolefins (Alcryn)

[3.829 a, 3.830 a, 3.833 d]

Because of the unknown constitution no abbreviation is possible for this class of TPEs. Therefore, the trade name Alcryn has to be used here.

Alcryn, which Du Pont started marketing in 1985, is based on a blend of an unnamed elastomer in chlorinated polyolefins. Its morphological structure is related to TPO [3.829 a, 3.830 a], but it does not show a sharp melting point; it behaves like an amorphous material which begins slowly to soften at over 135 °C. Alcryn can be processed by all the usual thermoplastic methods, including extrusion, injection molding, calendering, blow molding and thermoforming. Slightly modified rubber processing equipment which is capable of reaching the required temperatures and shear conditions can also be used, but when processing Alcryn, its different rheological properties must be kept in mind. Not only a temperature of 150–170 °C, but also sufficiently high shear are needed. Even if the temperature is over 170 °C, but at the same time shear is insufficient, flow properties are not adequate. For this reason the machines used for extrusion or injection molding must have an appropriate cylinder temperature and have to be equipped with screws which are long enough (e.g. L/D ratios of > 10:1; the optimum being 20:1), have shallow flights of < 2,5 mm and high compression ratios of roughly 2:1 to 4:1 (the optimum being 2,5:1). Raising the cylinder temperature is less likely to improve melt flow than raising the processing speed. Calenders for Alcryn should be equipped with rolls that can be heated to 135–165 °C. Before it is calendered, Alcryn must be homogenized first in an internal mixer and/or on the rolls to produce dollies or milled strips of material which are then fed to the calender. Because of the high processing speed and the small amount of scrap and production of rejects, Alcryn offers similar advantages to those obtained with TPOs.

Alcryn is a soft TPE with a Shore A hardness of 60–80 and low processing temperatures of 150 °C to about 190 °C. Because of the chlorine-containing base polymer, the product has outstanding oil resistance which is as good as that of covalently crosslinked NBR with medium acrylonitrile content and of TPE-NBR. As an illustration, the change in volume of Alcryn after seven days' immersion in ASTM #3 oil at 100 °C is 7–10%, and most of the mechanical properties showed little change after this test. Fuel resistance is, however, less good. The brittleness temperature of −42 °C to −46 °C is extremely attractive for an oil-resistant material.

The maximum service temperature of Alcryn – like that of all TPEs – is, of course, limited by the long-term heat resistance as expressed by the high-temperature compression set. After 22 hours at 100 °C, Alcryn has a compression set of 53 to 57%, which is still acceptable for TPEs, whilst at 121 °C the values are between 60 and 67% [3.830a]. Its compression set is, therefore, better than that for certain comparable TPO grades. Although the tensile strength is 12 to 15 MPa, which is lower than that of TPE-NR, but comparable with that of TPE-NBR, it is fully adequate for most practical applications. The tear resistance at room temperature is also satisfactory. The material is UV-resistant, mechanical properties remaining practically unchanged even after 20 days' irradiation. It is also resistant to ozone.

This thermoplastic elastomer is favored for tubing, industrial parts and for automotive applications, films, gaskets, shoe components, cable insulations, etc. The product is still too new for consumption statistics, although it may safely be assumed that it will become increasingly important because of its useful all-round properties, such as oil and ozone resistance and good low-temperature flexibility. At present more than 90% of the material is being processed by the plastics industry.

3.5.2.3 Elastomers with Thermally Labile Crosslinks

[3.845–3.850]

Another class of thermoplastic elastomers are those with thermally labile chemical crosslinks, which are capable of reforming after cooling down. These types of crosslinks have been observed first with polyurethane elastomers. There have been several attempts to produce such elastomers by introducing specific thermally labile crosslinks into NBR. The following compounds have been proposed for this purpose: 4-vinylpyridine or diaminomethacrylate crosslinked with metal salts [3.845], metal carboxylates, metal thiocarboxylates, metal alkoxides, metal mercaptides, metal dithiocarbamates, metal dithiophosphates [3.846], or polyisothiocyanates

[3.847]. Another interesting method of producing thermoplasticity is the copolymerization with acrylates or methacrylates, which have secondary or tertiary amine substituted hydrocarbons of chain length C_2 to C_4 in the molecule (e.g. diethylaminoethyl acrylate) [3.848–3.850]. Thermally labile crosslinks are then generated by the reaction with α, α'-dichloro-p-xylole. Thermoplastic NBRs of this type can be reversibly deformed several times at 175 °C over a time span of 5 minutes, but they are not yet commercially available.

3.5.2.4 Other Thermoplastic Elastomers

There is scarcely a major international rubber or plastics event today which does not see the presentation or introduction onto the market of a new sequence copolymer or new rubber-plastic blends or other products which can be used as TPE. The composition of these products, which are frequently (and in many cases, wrongly) denoted TPE remains confidential, thereby making classification difficult. A number of products of known composition are still under development [3.834–3.844] or have not yet really found a market – these include disilylphenyl-bisphenol A-polycarbonate block copolymers or polycarbonate-polydimethylsiloxane block copolymers or aminoalkylterminated polysiloxane segment polymers (thermoplastic silicone elastomers) [3.707, 3.708, 3.841] or polycarbonate-tetrafluoroethylene sequence polymers (thermoplastic fluoroelastomers) [3.582, 3.586]. Tetrafluoroethylene segments are being used in an attempt to introduce hard segments with a high melting temperature which will allow the thermal application range to be extended upwards. It can also be mentioned that attempts have recently been made to achieve improvements in the interpenetrating networks of polymer blends through the inclusion of polysiloxanes.

3.5.3 Service Temperatures of TPE

The range of service temperatures of TPEs is determined by the melting temperature of the hard segments or thermoplastic matrix, or by the thermal stability of the thermally labile crosslinks. For instance, the styrene blocks soften at 70 to 80 °C, with the result that products from SBS can only be exposed to temperatures of up to 65 °C. The upper service temperature of TPU can generally be 90 °C, and in many instances, even 120 °C. Polyethyleneterephthalate blocks melt only at about 200 °C, and therefore it is possible to use thermoplastic copolyesters at temperatures of up to 165 °C. PP blends with crosslinked EPDM should not be used above 100 °C, and in exceptional cases, 120 °C, while the upper service temperature limit of NBR/PP blends is 120 °C. In general, the upper temperature limits are determined by the specifictions for compression or tensile set, and it is therefore advisable to establish this by experimentation.

In the design of thermoplastic elastomers, it is also important that the melting temperature of the hard blocks is considerably below the decomposition temperature of the soft blocks.

Table 3.6 gives a survey of some commercially available thermoplastic elastomers and those which have been proposed, although it is not claimed that the list is complete.

Table 3.6: Some Examples of Thermoplastic Elastomers with Hard and Soft Segments

Hard Segment	Soft Segment	Type	Reference
Styrene	Butadiene Isoprene	SBS (Y-SBR) SIS (Y-IR)	[3.169, 3.810–3.819] (see page 62 and 150)
Methylstyrene Butylstyrene	Butadiene Isoprene		[3.816] [3.814]
Propylene	Ethylene	Sequential EPDM	[3.454] (see page 95)
Disilylphenyl Bisphenol A polycarbonate	Dimethyl Siloxane	Sequential Q	[3.707, 3.708, 3.711–3.714] (see page 131)
Aminoalkyl	Dimethyl Siloxane	Thermoplastic Q	[3.841–3.842]
Polycarbonate	Tetrafluoro Ethylene	Sequential FKM	[3.582, 3.586] (see page 159)
Polyethylene Terephthalate	Polyethylene Glycole	Thermoplastic Copolyester	[3.735–3.738, 3.809 c, 3.844] (see page 139 and 152)
Polyamide	Polyester	Thermoplastic Polyesteramide	[3.740, 3.742, 3.809 d, 3.809 e, 3.834, 3.835, 3.837, 3.840] (see page 139 and 153)
Urethane	Polyester Butadiene	TPU	[3.764–3.775, 3.785–3.786] [3.839]
Urethane/Urea	Polyester		[3.844]
Polypropylene	EPDM	Thermoplastic EPDM	[3.820–3.833] (see page 96 and 155)
	NR	Thermoplastic NR/Y-NR	[2.52, 2.53] (see page 14 and 156)
	Guayule	Thermoplastic NR/Y-NR	[2.73, 2.74] (see page 18 and 156)
Polypropylene (or other)	NBR	Thermoplastic NBR (Y-NBR)	[3.234, 3.822 a] (see page 70 and 157)
PVC	NBR	Thermoplastic NBR Polyblend	[3.235, 3.235 a] (see page 71 and 157)

3.6 A Comparison of Properties and Uses of Different Rubbers

3.6.1 Comparative Properties

[3.5, 3.851–3.852 f]

When the automotive, machine and cable industries, in particular, lay down their material specifications for elastomer components, stringent requirements are generally placed on a whole range of properties. A number of the key requirements are a high heat resistance on account of the operating conditions (operating temperature), volume increase in and chemical resistance to operating media (e.g. fuels, oils, water, coolants and chemicals), flexibility at low temperatures, ozone resistance, compression set, tensile strength, tear and abrasion resistance. Fulfilling an individual property laid down in the specifications is not generally difficult. It is when several properties have to be achieved at once, and achieving one property during compounding causes other similarly required properties to deteriorate, that it becomes problematical.

In order to give a clearly comprehensible overview of the best properties that can be achieved in each case, pairings are presented for a series of relevant elastomers, these being properties that are frequently set against each other (Figures 3.10–3.13). These data are based on the overview in Table 3.7, with the figures in the Table being taken from [3.851] and [3.851 a]. Where figures did not tally or where no figures were available, values were modified or supplemented in line with other indications given in the literature and on the basis of the author's own experience. These maximum properties are not always all attainable at the same time. The figures are also not absolute figures – they will frequently vary within an individual material class as a function of the grade involved and the type of compounding. The further to the left elastomers appear in the four Figures, the better their individual properties. In Figures 3.10, 3.11 and 3.13, properties are also all the better the higher they occur up the Figure, and in Figure 3.12, the lower their position in the Figure. In addition, it must be stressed that this information covers only material properties which are exceedingly complex to transpose to the performance of a moulded part (with its individual shaping, different wall thickness, potential reinforcement or sandwich construction etc.) and cannot always be worked out in advance. These are, after all, viscoelastic materials whose properties are a function of the loading parameters and the test conditions [3.851 a]. The properties cannot always be compared with the values given in tables and material data sheets on the basis of different test methods. This shows just how difficult it is to determine material boundary properties in the case of elastomers, and this is further complicated through the wide scope that exists for individual compounding measures. Despite this, Figures 3.10 to 3.13 permit a rapid overview of the probable suitability of individual elastomers for certain types of load and should be a sort of guide. For guide comparison the main properties of a number of elastomers are summarized also in Table 3.8.

The following fundamental considerations cover the performance limits of elastomers together with a number of property comparisons.

Table 3.7 Comparison of elastomers properties[1,2] [3.852]

Abbreviations	Elastomer grade	glass transition temperature T_G °C	approximate figures for low temperatures T_R °C	tensile strength[3]	resistance to tear propagation[3]	resistance of abrasion[3]
ACM	Polyacrylate rubber	−22 to −40	−10 to −20	M	M	M
AU	Polyester urethane rubber	−35	−22	H	H	H
BIIR[4]	Brominated butyl rubber	−66	−38	M	M/H	M
BR	Cis-1,4-polybutadiene rubber	−112	−72	M	L/M	VH
CIIR[4]	Chlorinated butyl rubber	−66	−38	M	M/H	M
CM	Chlorinated polyethylene	−25	−12	M/H	M	M
CO	Polyepichlorohydrin	−26	−10	M	M	M
CR	Polychloroprene	−45	−25	H	H	M/H
CSM	Sulfonated polyethylene	−25	−10	M/H	M/H	M
EAM	Ethylene-acrylate rubber	−40	−20	M	M	M
ECO	Epichlorohydrin ethylenoxide copolymer	−45	−25	M	M	M
EPDM, S	Ethylene-propylene-terpolymer, sulphur crosslinked	−55	−35	M	M	M
EP(D)M, P	Ethylene-propylene-copolymer, peroxide crosslinked	−55	−35	L/M	L	L
EU	Polyether-urethane rubber	−55	−35	H	H	H
EVM	Ethylene-vinylacetate copolymer	−30	−18	M/H	M	M
FKM	Fluororubber	−18 to −50	−10 to −35	M/H	M	M
FVMQ	Fluorosilicone rubber	−70	−45	L	L	L
H-NBR	Hydrogenated nitrile rubber	−30	−18	M/H	M	H
IIR[4]	Butyl rubber	−66	−38	M	M/H	M
MVQ	Dimethyl polysiloxane containing vinyl	−120	−85	L	L	L
NBR	Nitrile rubber					
low ACN content		−45	−28	M/H	M	H
medium ACN content		−34	−20	M/H	M	H
high ACN content		−20	−10	M/H	M	H
NR (IR)	Natural rubber (synthetic polyisoprene)	−72	−45	VH	VH	M/H
OT	Polyethylene glycol	−50	−30	L	L	L
PNF	Polyfluorophosphazene	−66	−42	M	L	L
PNR	Polynorbornene	+25		M	M	M
SBR	Styrene-butadiene rubber	−50	−28	H	H	H
X-NBR	Carboxyl groups containing NBR	−30	−18	H	M	VH

[1] Values for the different properties refer to typical examples. Some elastomers are available in a broad variety of products, in addition elastomer properties are determined in a certain extent by the compounding process; therefore there exists no distinct property values for a class of elastomers.

ozone resistance[3]	compression set at −20 °C %	compression set at room temperature %	compression set at +120 °C %	heat resistance after 5 h °C	heat resistance after 70 h	heat resistance after 1000 h °C	working temperature °C	swelling after 70 h in ASTM oil 3 %	swelling after 70 h, RT, in fuel C %
H	25	5	10	240	180	150	170	25 (150 °C)	65
H	25	7	70	170	100	70	75	40 (100 °C)	
M	12	10	60	200	160	130	150	> 140 (70 °C)	
L				170	100	75	90	> 140 (70 °C)	
M	12	10	60	200	160	130	150	> 140 (70 °C)	
H				180	160	140	150	80 (150 °C)	75
H			20	240	170	140	150	5 (150 °C)	10
M	50	10	30	180	130	100	125	80 (100 °C)	
H				200	140	130	150	80 (150 °C)	
H				240			175	50 (150 °C)	
H			20	220	150	130	135	10 (150 °C)	30
H	20	8	50	200	170	130	140	> 140 (70 °C)	
H	20	4	10	220	180	140	150	> 140 (70 °C)	
H	25	7	70	170	100	70	75	40 (100 °C)	
H	95	40	4	200	160	140	160	80 (150 °C)	
VH	50	18	20	> 300	280	220	250	2 (150 °C)	5
VH			30	> 300	220	200	215	2 (150 °C)	20
H			30	230	180	150	160	15 (150 °C)	65
M	12	10	60	200	160	130	150	> 140 (70 °C)	
VH	10	2	3	> 300	275	180	225	50 (150 °C)	
L	40	8	45	170	140	110	125	25 (100 °C)	45
L	45	8	50	180	145	115	125	10 (100 °C)	35
L	45	8	55	190	150	120	125	5 (100 °C)	25
L	15	8	70	150	120	90	100	> 140 (70 °C)	
H				170	120	60	100	10 (70 °C)	
H			30				175	10 (150 °C)	15
M							100	> 140 (70 °C)	
L				195	130	100	110	> 140 (70 °C)	
L			60	170	140	110	120	5 (100 °C)	20

[2] The values correspond to informations in [3.851] and [3.851 a], some are added or changed according to own experiences or other literature sources. In some cases no comparative informations were available.

[3] L: low; M: medium; H: high; VH: very high

[4] Fo BIIR, CIIR and IIR see Figs. 3.10–3.13, page 165–168 ((X)IIR)

Table 3.8: Comparative Vulcanizate Properties of NR and Other SR's*

Properties	Type of Rubber																
	NR	IR	SBR	BR	NBR	ACM	CR	ECO	CSM	FKM	(X)IIRs	EPDM	EAM	PVMQ	TM	SBS	AU
Tensile strength, gum	1	2	5	6	5	6	3	4	5	5	4	5	5	6	6	3	1
Tensile strength, with reinforcing fillers	1	2	2	4	2	3	2	3	3	3	3	3	3	4	4	1	1
Maximum elongation	1	1	2	3	2	4	2	3	3	3	2	3	3	4	4	1	2
Abrasion resistance with reinforcing fillers	4	4	3	1	2	4	3	3	3	4	4	3	2	5	5	5	1
Tear resistance	2	2	3	5	3	3	2	3	3	4	3	3	3	5	4	3	1
Rebound	2	2	3	1	3	4	3	3	4	5	6	3	3	3	5	4	3
Low temperature flexibility	2	2	3	2	3	5	3	3	5	5	2	2	4	1	4	2	4
Heat resistance	5	5	4	4	3	2	3	2	3	1	3	2	2	1	5	6	5
Oxidative resistance	4	4	3	2	3	2	2	1	2	1	2	1	1	1	1	5	1
UV resistance	4	4	3	3	3	2	2	1	2	1	2	1	1	1	1	5	1
Weather and ozone resistance	4	4	4	3	3	2	2	1	2	1	2	1	1	1	1	5	1
Oil resistance	6	6	5	6	1	1	2	1	2	1	6	4	4	1	1	6	1
Motor fuel resistance	6	6	6	6	2	3	3	1	2	1	6	5	5	6	1	6	1
Acid resistance	3	3	3	3	4	5	2	2	2	1	2	1	3	5	6	2	6
Alkali resistance	3	3	3	3	4	5	2	2	2	4	2	1	3	5	6	2	6
Flame resistance	6	6	6	6	6	6	2	2	3	3	6	6	6	6	6	6	6
Electrical resistivity	1	1	2	2	5	5	4	4	4	4	2	2	3	1	4	2	4
Gas permeation	5	5	4	4	2	3	3	1	3	3	1	4	2	6	1	4	1
Compression set − 40 °C	3	3	3	3	3	5	5	5	6	6	5	4	6	3	5	4	5
+ 20 °C	2	2	3	3	2	3	3	2	5	4	4	3	5	2	4	3	3
+100 °C	6	6	5	5	3	5	4	2	6	3	2	2	1	1	4	6	5

* 1 = excellent; 6 = insufficient.

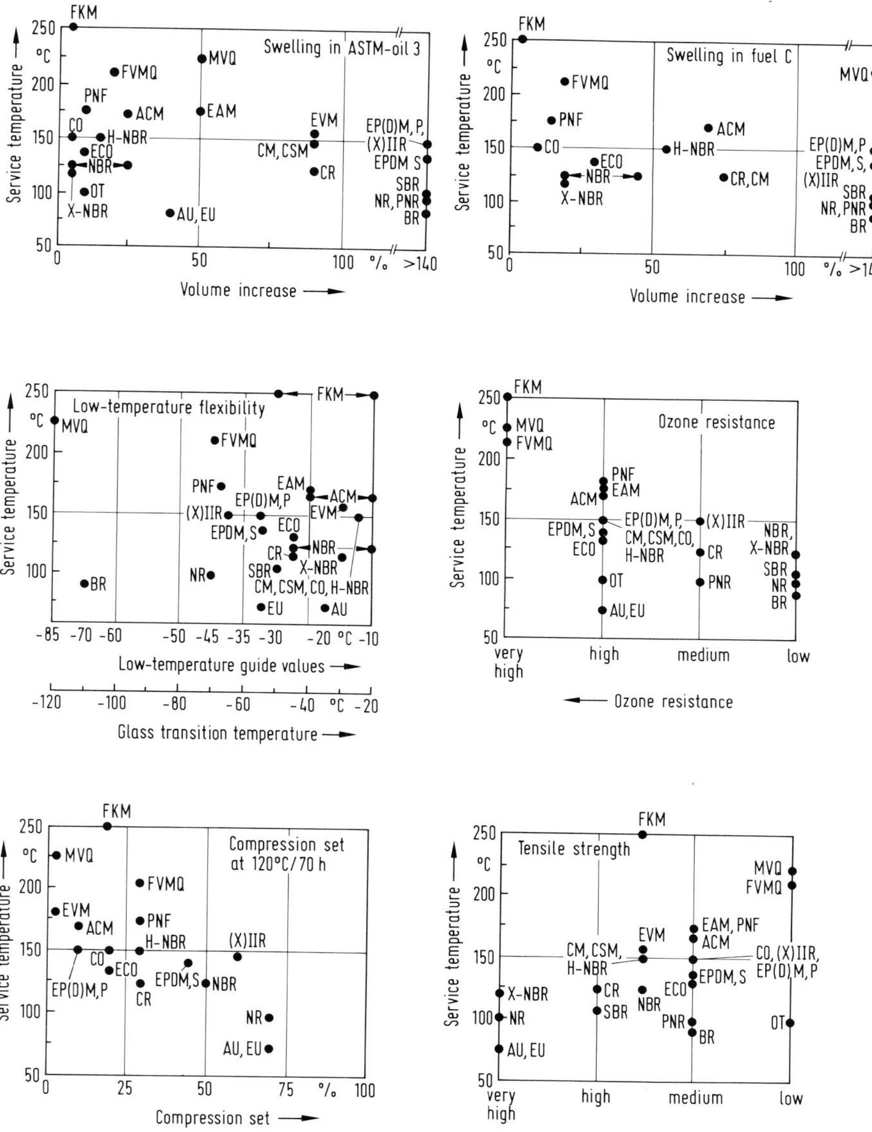

Figure 3.10 Correlation between service temperature and a number of important properties in engineering applications for different elastomers (for abbreviations see Table 3.7) [3.852]

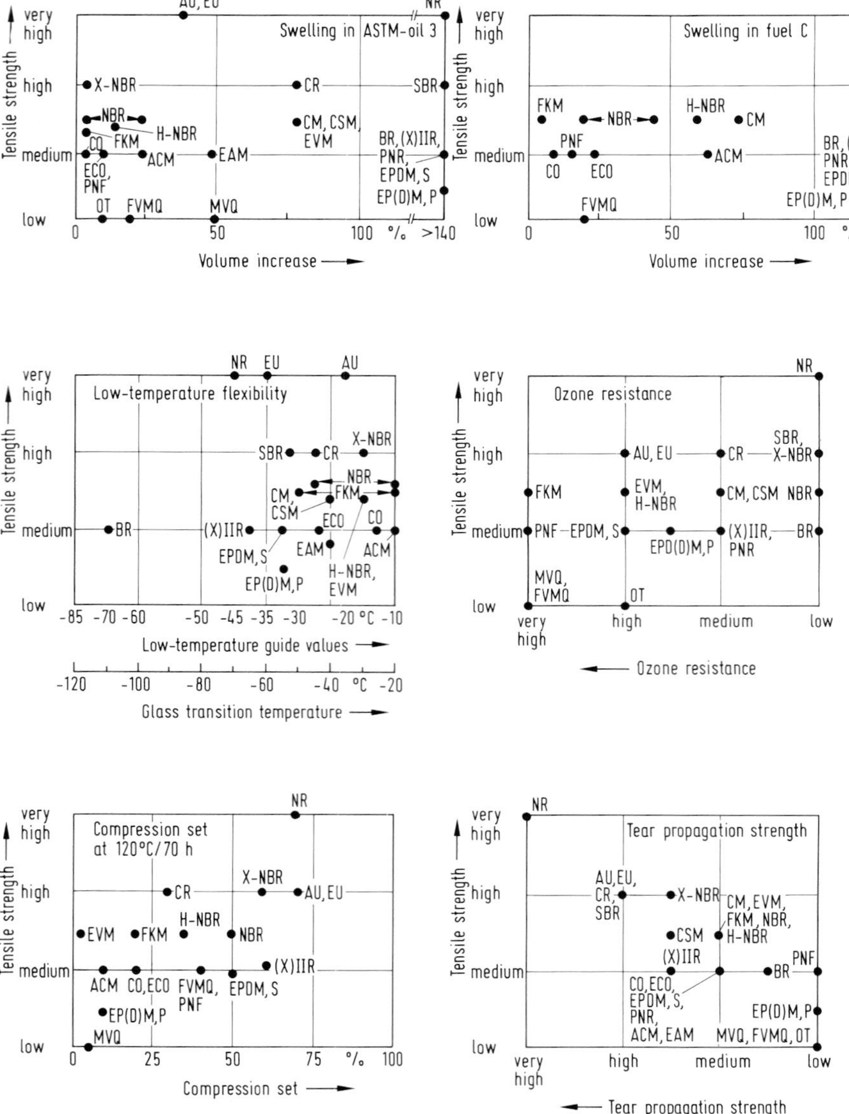

Figure 3.11 Correlation between tensile strength and a number of important properties for engineering applications for different elastomers (for abbreviations see Table 3.7) [3.852]

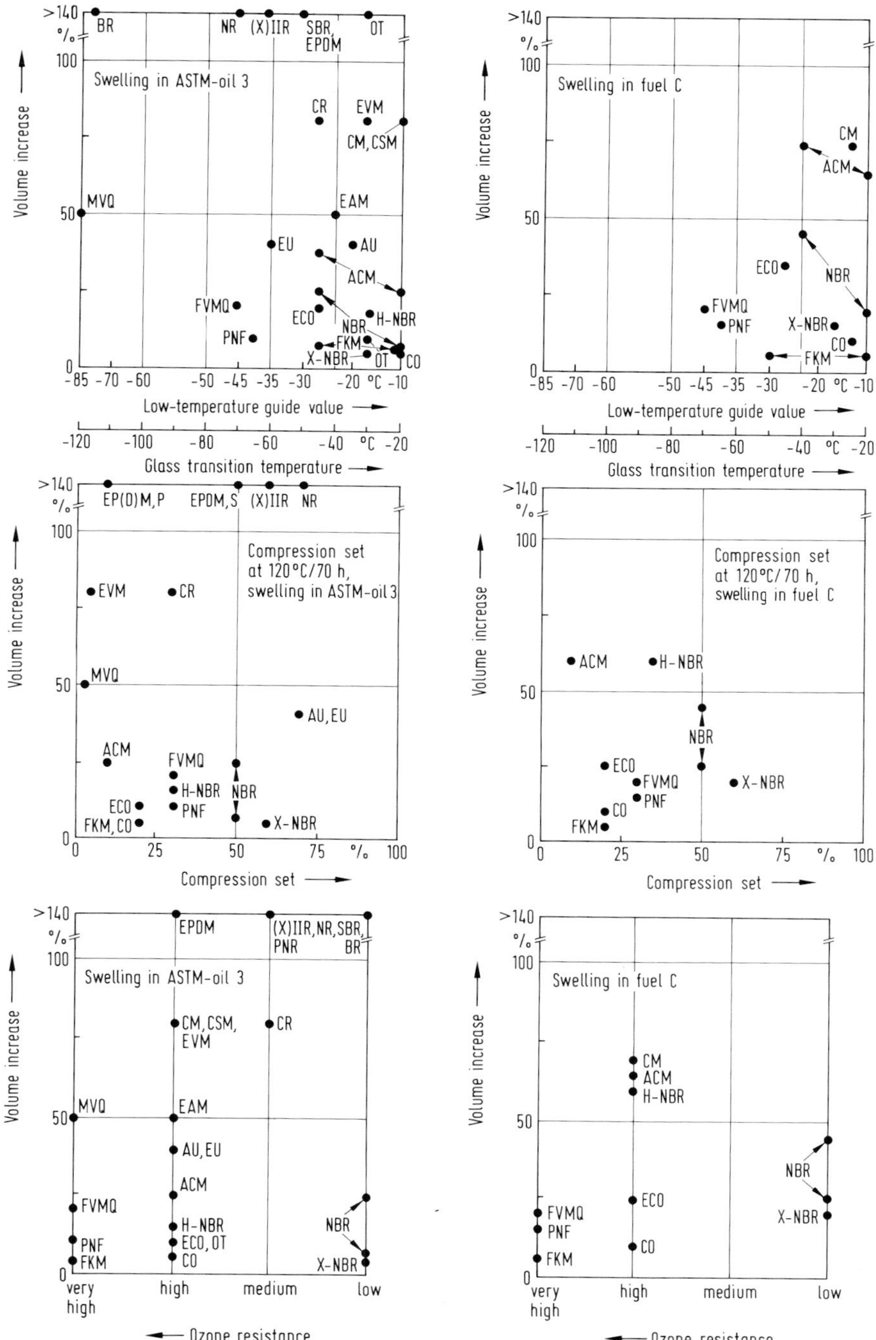

Figure 3.12 Correlation between volume increase and a number of important properties for engineering applications for different elastomers (for abbreviations see Table 3.7) [3.852]

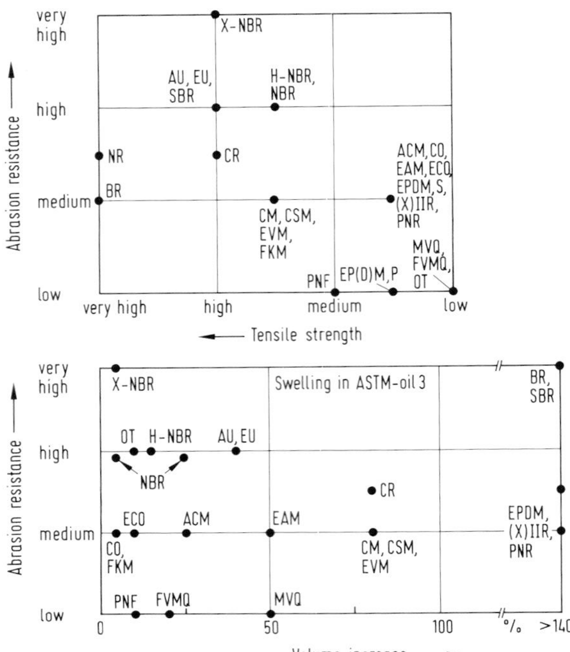

Figure 3.13 Correlation between abrasion resistance and tensile strength and volume increase for different elastomers (for abbreviation see Table 3.7) [3.852]

3.6.2 Performance Limits of Properties of Elastomers

[3.851, 3.851 a, 3.852]

3.6.2.1 Maximum Service Temperatures

The maximum permitted service temperature of an elastomer cannot be taken in isolation without its service conditions. Parameters such as load duration, continuous temperature or intermittent temperature cycles, action of media (oil, air, air containing ozone, anaerobic conditions etc.) and load (static or dynamic), play a decisive role. Under strictly anaerobic conditions, which scarcely occur in practical application and without additional mechanical load, the dissociation temperature of the weakest point in the elastomer network represents the maximum limit temperature (see Table 3.9).

Standard ageing processes take place under aerobic conditions, whereby oxidative processes display a clearly lower activation energy than anaerobic processes. This means that degradation commences at considerably lower temperatures than for thermal decomposition processes (see Table 3.10). Whilst elastomers display levels of heat resistance in nitrogen-fluxed, i.e. air-free oil, which come close to the thermal decomposition temperatures [3.852 a] (applications that can scarcely be performed in practice), elastomers in oils not fluxed in nitrogen, i.e. the standard service conditions, show a clearly lower level of heat resistance, although this is nonetheless still considerably higher than with applications in air. The heat resistance of a large number of elastomers is reduced still further in air containing

Table 3.9. Dissociation Temperature (T_{diss}) and Activation Energy (E_{diss}) of some typical chemical bonds [3.851 a]

Chemical Bond		T_{diss} in °C	E_{diss} in kJ/mol
1.	$-CF_2 \vdots CF_2-$	500	400
2.	$-Si-O \vdots Si-$	500	400
3.	$-CH_2 \vdots Ph$	420	380
4.	$-CH_2 \vdots CH_2-$	400	320
5.	$-CH_2 \vdots CH_2-C=C-$	390	300
6.	$-C-S \vdots S-C-$	~380	300
7.	$CH_2 \vdots O-$	345	330
8.	$CH_2 \vdots CH-CO$	330	280
9.	$-C \vdots S_x-C-$	~160	120

Table 3.10. Heat Resistance of Some Engeneering Elastomers (Classification after ISO/TR 8461, aerobic condition, Method ISO 4632/1 3days, (Retention of properties)) [3.851 a]

Heat resistant up to	Elastomer
100 °C	AU/EU, NR (IR), OT, SBR, PNR
125 °C	CR, NBR, X-NBR
150 °C	CO, ECO, EP(D)M, EVM, CM, CSM, (X)-IIR, H-NBR
175 °C	ACM, EAM, PNF
200 °C	FVMQ
225 °C	MVQ
250 °C	FKM

ozone. The heat resistance of an elastomer under the action of oil is determined not only by the presence of air oxygen but also by the nature and quantity of the oil additives already mentioned, which drastically reduce heat resistance and can lead to rapid destruction at much lower temperatures already [3.211, 3.273, 3.601].

The period of time for which the heat acts also, of course, plays the key role when it comes to the upper temperature limit required in practice. The cable industry, for instance, requires the capacity to withstand 20000 hours' continuous load, which naturally leads to a correspondingly low limit temperature being specified. The 1000 hours' continuous thermal loading frequently laid down by the automotive industry is disputed, since in practical service, relatively short periods of thermal load alternate with lower temperatures and recovery periods at ambient temperature (intermittent temperature cycles). Hence, the rather vaguely defined concept of service temperature is frequently used, being based on long-term experience (see Table 3.7 and Figure 3.10), and this is generally somewhat higher than the temperature for the 1000 hour heat resistance. For the practical application of a rubber moulding, consideration must also be given to the wall thickness, since the process of ageing-through, in so far as this is associated with crack formation, is dependent on diffusion processes and hence on a time function. Since there is no uniform definition of the term heat resistance, different figures are often found in the literature for the same classes of elastomer.

Clearly the heat resistance can be improved to a considerable extent by compounding as it is described in the particular chapters on elastomers.

3.6.2.2 Behaviour at Low Temperatures

The lowest temperature range for the application of elastomers is limited by the mobility of the chain segments. If low temperatures act for a sufficiently long period of time, then segments may become rigid and undergo transition to the glass state. This glass transition temperature T_g is the lowest figure for flexibility at low temperatures and the most important characteristic value in scientific terms. This indicates the temperature at which a product ceases to display an elastomeric character. Since the assessment of flexibility at low temperatures depends to a considerable extent on the testing time [3.852 b], a shorter loading or testing time can lead to higher temperature limit values. These temperature/time relationships can be quantified on an approximate basis according to *Williams, Landel* and *Ferry*, using the so-called WLF equation [3.851 a].

Correlations between the glass transition temperature T_g and the low temperature guide value T_R, which denotes the temperature to be allocated to the turning point of the modulus of elasticity or the maximum damping value, therefore, are possible. If the T_g values of different elastomers are plotted against their T_R values, then this gives a straight-line progression which corresponds in empirical terms to the relationship in the WLF equation:

$$T_g = 1.3\ T_R - 11\ (°C).$$

When indicating material data for elastomers, the T_g value is frequently used, although in most cases T_R is given. The appropriate sections of Figures 3.10 to 3.12 thus show both parameters on the abscissa. A certain degree of inaccuracy is, however, inevitable in this allocation.

The low temperature brittleness test (see Table 3.11) which is similarly applied in a large number of cases (brittleness temperature T_S) is an impact test with an even shorter loading and testing time than the T_R test. When this value is measured, the limit temperature of elastomers comes out even higher.

Low-temperature flexibility can also be shifted considerably towards lower temperatures through compounding measures (particularly through plasticizers, although these limit heat resistance) [3.852 b] and the field of application of elastomers can be extended in this way (see the particular chapters on elastomers).

Table 3.11. Brittleness Temperature of Some Engeneering Elastomers (Classification after ISO/TR 8461) [3.851 a]

Low Temperature Flexibility	Rubber
−75 °C	Q
−55 °C	NR, IR, BR, CR, SBR, (X)IIR, EP(D)M, CM, CSM, FVMQ, PNF
−40 °C	ECO, NBR, EP(D)M, CSM, FKM, AU, EU
−25 °C	ACM, NBR, OT, FKM
−10 °C	ACM, CO, FKM, TM, NBR

3.6.2.3 Resistance to Liquid Media

Elastomers, as crosslinked products, are insoluble in liquids, as far as the crosslinks are not destroyed. They show, however, a greater or lesser volume increase, the extent of which is a function of compatibility factors, and this increases with time until an equilibrium swelling rate has been attained. This absorption of liquid is associated with three-dimensional expansion of the network structure. The attendant loosening of the cohesive bonds causes a large number of mechanical properties to deteriorate, e.g. tensile strength, tear propagation strength and hardness. This process is theoretically reversible. If no material extraction takes place by virtue of the swelling process, then the original property state is obtained after full-scale reversal of the swelling. This seldom occurs, however, since the swelling media, e.g. stabilizers, antioxidants and plasticizers, are partially dissolved out. This causes changes in ageing properties and hardness etc. after swelling and subsequent reversal of the swelling. For this reason, chemicals that are highly difficult to extract are used and a low level of swelling is the target.

The swelling behaviour of elastomers in liquid media is determined by the difference between the cohesion energy densities or the solubility parameters that can be derived as the square root of these [3.182, 3.191]. The maximum thermodynamic compatibility of two materials, and hence their penetration, occurs when both have similar solubility parameters. The more dissimilar the solubility parameters become, i.e. the greater the difference between the elastomer and the swelling medium, the less the volume increase that can be expected. Since the solubility parameter of hexane at $\delta_1=7.3$ or super-grade petrol at $\delta_1=8.1$ is more or less the same as that of EPDM, for instance, at $\delta_2=7.0$, or NR at $\delta_2=8.0$, this produces the familiar high levels of swell in the elastomers made from them. Polar NBR elastomers with δ_2 values from 9.5 to 10.4 are more resistant to swelling in non-polar aliphatic compounds. Rubber with even greater polarity, such as AU, by contrast, is even more resistant to non-polar solvents. Polar rubber, however, swells extensively in polar solvents, unpolar rubbers, such as NR, IIR or EPDM as a rule displays better swell resistance.

To assess the swelling behaviour of thick-walled articles requires not only the actual media resistance but also knowledge of the time laws of the diffusion processes. Whilst the edge zones of the elastomers swell up rapidly, the swell front only moves slowly into the inside, so that the mechanical performance of the moulded part can be retained for a long period of time. The viscosity of the swelling medium naturally plays a key role here. In the case of higher-viscosity oils, for instance, years can pass before a 5 mm elastomer layer is completely penetrated [3.852 c]. In the case of fuels, the process is naturally faster. A liquid sheathing on the outer layer of a rubber part is not always rated negatively, moreover – it can in some cases serve as a protective layer against aggressive environmental influences.

It is understandable that swell resistance can also be influenced to a pronounced extent by compounding, and particularly through the type and quantity of the filler selected, although also through the degree of crosslinking. Readily extractable oils in the rubber blend can lead to an apparantly lower level of swelling in the case of swelling due to fuels or oils. This is due to the fact that part of the fuel or oil uptake is compensated for by oil extraction. The diffusion or permeation through a rubber moulding can be reduced through barrier-forming layer fillers, for instance.

A survey of the comparative chemical resistance of elastomers is given in reference [3.852 d] and [3.852 e].

3.6.2.4 Deformation Properties

The visco-elastic properties of elastomers are described in Section 6.2.1.1, page 469.

Different behaviour is observed when it comes to the deformation of elastomers as a function of the method of investigation used: stress relaxation, i. e. a reduction in stress with constant deformation (such as when a sealing force eases off); creep (i. e. an increase in deformation with constant stress); retention of starting properties and permanent deformation (compression set, set properties) after compressive shear or tensile or loading.

Ideally elastic elastomers show no relaxation or creep properties under ideal conditions at all and no permanent deformation. In practice, however, these phenomena always occur to a greater or lesser extent. They are attributable to the true viscoelastic properties, to inevitable gaps in the network structure and to interactions between the polymer matrix and the blend components. The complex relationships between the physical relaxation and creep processes are covered in detail in [3.851 a].

Relaxation and creep phenomena are due largely to the nature of loading. With compressive, shear and tensile loading they generally increase in a ratio of 1 : 1.3 : 1.5 [3.852 f]. At higher loading temperatures, non-differentiable chemical influences are superimposed on the physico-rheological influences in individual cases. This makes it more difficult to judge the suitability of elastomers for applications at high temperatures.

The visco-elastic behaviour of elastomers under static loading is generally characterised by the compression set under specified temperature, time and compression conditions. The percentage of permanent deformation is taken as a measure of set behaviour. In more seldom cases, the stress relaxation under constant deformation is measured, although this requires a greater outlay. Both the compression set and the stress relaxation [3.216, 3.219] are largely determined by the testing temperature (see Figure 3.14) [3.216, 3.219–3.222, 3.851 a]. Whilst at low temperatures the stiffening of the elastomer matrix through freezing plays the chief role, at higher temperatures it is the chemo-rheological processes that predominate.

Since elastomer mouldings used in engineering applications are generally stressed by high temperatures, Figures 3.10–3.12 show the compression set at 120 °C. This already represents a very high level of stressing for a large number of engineering

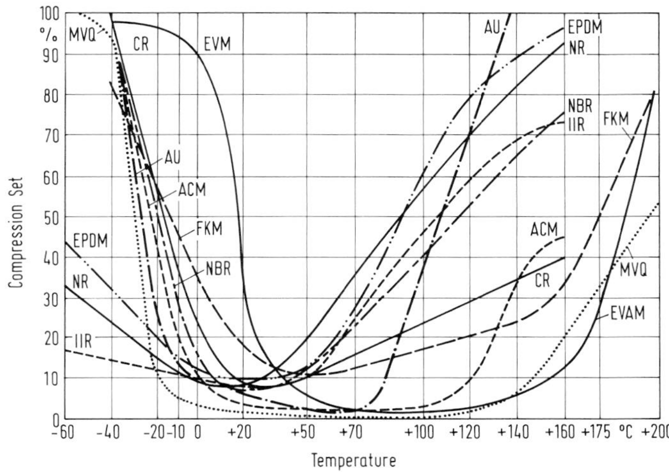

Figure 3.14 Compression set of some elastomers versus temperature

elastomers, and hence not all the elastomers are taken into account. Compression set and, in a similar manner, stress relaxation, can be influenced to a considerable extent by the crosslinking structure [3.220, 3.221] and the compounding (choice of crosslinking systems, fillers etc.) as well as the vulcanisation time [3.220–3.222].

3.6.2.5 Strength Properties

The strength properties in general are described in Section 6.2.2.1, page 474.

The highly complex problems of tear processes and their dependence on defects, testpiece shape (e.g. curvature gradients), tear energy and testing temperature (e.g. tear rate) are covered in [3.851 a]. Elastomers generally have a considerably lower tensile strength than plastics. With most synthetic elastomers, the tensile strength required in practice is only obtained through correct compounding, i.e. first and foremost through suitable fillers. The tensile strength data given in Figures 3.11 and 3.13 correspond to the values that can be obtained in the most favourable case at room temperature.

The tensile strength of elastomers generally falls as the temperature and loading duration increases. For this reason, the tensile strength of an elastomer established at room temperature (see Table 3.12) does not always permit its behaviour at higher temperatures to be predicted. Different elastomers display considerable differences; the more thermoplastic a rubber, the more its strength properties will decrease as the temperature rises. This means that the microstructure of the polymers and the type of crosslinking are highly important. Mention has already been made of the fact that for instance silicone elastomers, which have a relatively low tensile strength at room temperature, have the highest tensile strength of all elastomers at temperatures above 150 °C. For a lot of engineering applications, however, the maximum attainable tensile strength does not play a key role, since elastomers are frequently only used under pressure or in a slightly stressed state.

The tear resistance (see Table 3.13) is of greater importance. This is frequently fairly low, with the exception of a number of crystallising elastomers. A higher tear resistance of synthetic elastomers can only be achieved with selected crosslinking systems and active fillers [3.852 e].

Table 3.12. Tensile Strength of Some Amorphous and Crystallizable as well as Black-Filled Engeneering Elastomers [3.851 a]

Rubber unfilled	Tensile Strength N/mm^2	Rubber filled with Carbon Black	Tensile Strength
amorphous		BR	20
BR	3	NR	22
SBR	6	IR	22
NBR	7	CR	22
EP(D)M	7	SBR	20
ACM	2	NBR	20
EVAM	4	(X)IIR	18
(E)CO	6	EP(D)M	18
OT	3	CSM	20
MVQ	3	EVAM	15
		ACM	15
crystallizable		FKM	15
NR	20	(E)CO	15
IR	20	OT	10
CR	20	VMQ[1]	10
(X)IIR	14	[1] $+ SiO_2$	
CM	10		
CSM	18		
FKM	12		

[1] Instead of carbon black silica

Table 3.13. Limits of Tear Resistance of Some Elastomers, W (N/mm), Logarithmic Graph, Unifilled above, Filled with Carbon Black below the Line [3.851 a].

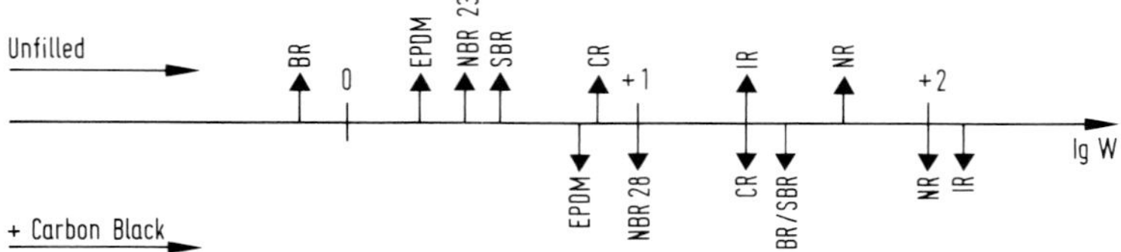

3.6.3 Polymer Blends

[2.16 c, 3.853–3.868]

In many instances, two (or more) rubbers are blended to modify the properties of the blend components [3.191, 3.234, 3.269, 3.413–3.425, 3.534, 3.535, 3.545, 3.656, 3.853–3.860 a]. In order to obtain a good covulcanization of the blend partners, it is important that the rubbers in the blend are as thermodynamically compatible as possible [3.191, 3.269, 3.853–3.860 a]. While blend combinations of relatively compatible rubbers like NR, IR, BR, and SBR are used on a large scale for tire production, little use is made of blending less compatible rubbers, although this could offer great benefits [3.234, 3.534, 3.535]. For these blends, one has to use special dispersing aids to obtain an intimate dispersion of the two heterogeneous microphases [3.269, 3.655, 3.656].

Rubbers are also blended with thermoplastics on a large scale, either to modify the properties of the rubber (e.g. NBR with PVC, see page 71 and 157), [3.235–3.240], PP with EPDM or PA with NBR to obtain thermoplastic rubbers, see page 70 and 157), [3.820–3.833], or to modify the properties of thermoplastics to make them more impact resistant [3.861–3.868].

3.6.4 Uses of Different Elastomers

For a quick survey the uses of a number of elastomers are summarized in Table 3.14. The comparison here is summarized tabular instead of detailed discussions. More details can be seen in the particular sections on elastomers.

Table 3.14: Some Major Uses of Different Rubbers (+ = major; (+) = occasional use)

Uses	Rubber								
	NR	IR	SBR	BR	NBR	ACM	CR	ECO	CSM
Passenger car tires									
Tread			+	+					
Carcass	+		+	+			(+)		(+)
Truck tires									
Tread	+	+	(+)	+					
Carcass	+	+							
Belting									
Conveyor belts			+		(+)		+		
V-belts	+		+				+		
Suspension elements	+	+	+	+	+		+	(+)	
Hose									
Fuel hose					+		+	+	
Milking machine hose					+		(+)		
Heating and cooling hose					+		+		
Oil and grease resistant hose					+	(+)	+	+	
Chemical resistant hose			+				+		+
Others			+		+		+	+	
Seals									
Profiles			+				+		
Shaft seals					+	+			
Heat resistant seals					(+)	+		+	(+)
Oil resistant seals					+	+	(+)	+	(+)
Other seals			+						
Food and pharmaceutical products	+	+	+		+		+		+
Fatty foods					+		+		
Nipples	+	+							+
Others	+	+	+		+		+		+
Sanitary rubber products and balloons	+	+					+		+
Rubberized fabrics							+	+	+
Gloves	+		+		+		+		+
Cables			+		(+)		+		+
Shoes and soling			+		+		+		
Latex products	+		+		+	+	+		

continued on next page

Table 3.14 (continued): Some Major Uses of Different Rubbers
(+ = major; (+) = occasional use)

Uses	Rubber								
	FKM	(X)IIR	EPDM	EAM	Q	AU	TM	SBS	CM
Passenger car tires									
Tread									
Carcass		+	(+)						
Truck tires									
Tread		+							
Carcass									
Belting									
Conveyor belts		(+)							
V-belts									
Suspension elements		(+)			(+)	+			
Hose									
Fuel hose	+								
Milking machine hose			(+)						
Heating and cooling hose		(+)	+						
Oil and grease resistant hose		(+)							
Chemical resistant hose		+	+	(+)				(+)	
Others		+	+						
Seals									
Profiles		+	+		+			(+)	
Shaft seals	+				+				
Heat resistant seals	+	+		+	+				+
Oil resistant seals	+			(+)	(+)	+	+		(+)
Other seals									
Food and pharmaceutical products		+	+	+	+	+		+	
Fatty foods		+	+		+				
Nipples								+	
Others		+	(+)	+	+	(+)		+	
Sanitary rubber products and balloons									
Rubberized fabrics		+	+	+	(+)	+			
Gloves									
Cables		+	+	(+)	+				+
Shoes and soling						+		+	
Latex products									

3.7 References on SR

3.7.1 General References on SR

(compare also Ref. [1.1]-[1.7], [1.14]-[1.21], [1.23]-[1.28], [1.66], [2.8])

[3.1] *Alari, G., Piazza, S.:* Entwicklungsrichtungen für Synthesekautschuke unter dem Gesichtspunkt der Verfügbarkeit von Feedstocks und Monomeren. GAK *36* (1983), p. 20.

[3.2] *Allen, P. W., Lindley, P. B., Payne, A. R.:* Use of Rubber in Engineering. MacLaren, London, 1967 (see also Refs. [2.95-2.97; 2.125-2.128; 2.136-2.139]).

[3.3] *Becker, W., Graulich, W.:* Synthetischer Kautschuk. In: *Houwink, R. (Ed.):* Chemie und Technologie der Kunststoffe, Vol. 2. Verlag Geest u. Portig, Leipzig, 1956, pp. 148-225.

[3.4] *Blackley, D. C.:* Synthetic Rubbers, their Chemistry and Technology. Elsevier Applied Science Publ., Barking, 1983.

[3.5] *Casper, R., Witte, J., Kuth, G.:* Synthetischer Kautschuk. In: Ulmann's Encyklopädie der technischen Chemie, 4th Ed., Vol. 13. Verlag Chemie, Weinheim, 1977.

[3.6] *Comyn, J.:* Polymer Permeability. Elsevier Applied Science Publ., Barking, 1985.

[3.7] *Davay, A. B., Payne, A. R.:* Rubber in Engineering Practice. MacLaren, London, 1964.

[3.8] *Gohl, W. (Ed.):* Elastomere Konstruktions- und Dichtungs-Werkstoffe. Lexika-Verlag, Grafenau, 1975.

[3.9] *Gröne, H.:* Synthesekautschuk - Gegenwart und Zukunft. KGK *31* (1978), p. 9.

[3.10] *Gröne, H.:* The Rubber Industry in Western Europe, Prospects for the next Decade. Paper at 124th ACS-Meeting, Rubber Div., Oct. 25-28, 1983, Houston, TX.

[3.11] *Hepburn, C., Reynolds, R. J. W.:* Elastomers, Criteria for Engineering Design. Elsevier Applied Science Publ., Barking, 1979.

[3.12] *Jenkins, A. D.:* Polymer Science.North Holland Publ. Comp., Vol. 2, 1972.

[3.13] *Kennedy, J. P., Törnquist, E. G. M.:* Polymer Chemistry of Synthetic Elastomers. Interscience Publ., New York, 1968, 1969.

[3.14] *Logemann, H.:* in: *Houben-Weyl:* Methoden der organischen Chemie, Vol. XIV/1. G. Thieme Verlag, Stuttgart, 1961, p. 703.

[3.15] *Naunton, W. J. S.:* The Applied Science of Rubber. Arnold Publ., London, 1961.

[3.16] *Purdie, H.:* Synthetic Rubber 1991, Managing for Profiles. Paper at 124th ACS-Meeting, Rubber-Div., Oct. 25-28, 1983, Houston, TX.

[3.17] *Stern, H. J.:* Rubber - Natural and Synthetic, 2nd Ed. MacLaren, London, Elsevier, Amsterdam, 1967.

[3.18] *Vogelzang, E. J. W., Rhoad, M. J.:* The Synthetic Rubber Industry to Global Economic Change. Paper at 124th ACS-Meeting, Rubber-Div., Oct. 25-28, 1983, Houston, TX.

[3.19] *Whitby, G. S.:* Synthetic Rubber. J. Wiley & Sons, New York; Chapman & Hall, 1954.

[3.20] *Wilson, A. D., Prosser, H. J. (Eds.):* Developments in Ionic Polymers. Elsevier Applied Science Publ., Barking, 1983.

[3.21] Not from Trees Alone, 2nd Ed. Publ.: BASRM, London, 1970.

3.7.2 References on Polymerization

3.7.2.1 General References on Polymerization

[3.22] *Howard, R. N.:* Development in Polymerization, Elsevier Applied Science Publ., Barking, Vol. 1, 1979; Vol. 2, 1979; Vol. 3, 1982.

[3.23] *Morton, M.:* Basic Principles of the Polymerization Process in Synthetic Rubber Production. Paper at 124th ACS-Meeting, Rubber-Div., Oct. 25-28, 1983, Houston, TX.

[3.24] *Odian, G.:* Principles of Polymerization, 2nd Ed. J. Wiley & Sons, New York, 1981.

3.7.2.2 References on Emulsion Polymerization

[3.25] *Blackley, D. C.:* Emulsion Polymerization, Theory and Practice. Elsevier Applied Science Publ., Barking, 1975.

[3.26] *Bovey, F. A., Kolthoff, I. M., Medalia, A. S., Mehan, E. J.:* Emulsion Polymerization. Interscience Publ., New York, 1955.

[3.27] *Poehlein, G. W., Doughery, D. J.:* Continuous Emulsion Polymerization. RCT *50* (1977), Review.
[3.28] *Ugelstad, J., Hansen, F. K.:* Kinetics and Mechanisms of Emulsion Polymerization. RCT *49* (1976), Review; *50* (1977), Addendum.
[3.29] *Uraneck,* C. A.: Molecular Weight Control of Elastomers Prepared by Emulsion Polymerization. RCT *49* (1976), Review.

3.7.2.3 References on Ionic Chain Polymerization

General Overview

[3.30] *Furukawa, J., Korayashi, K.:* Alternating Copolymerization. RCT *51* (1978), Review.
[3.31] *Iving, K. J. (Ed.):* Ring-Opening Polymerization, Elsevier Applied Science Publ., Barking, 1984, Vol. 1, 1984; Vol. 2, 1984; Vol. 3, in Prepar.
[3.32] *Kaiser, E. T., Kevan, L.:* Radical Ions. Interscience Publ., New York, 1968.
[3.33] *Nakahama, N.:* Reactive Block Copolymer. IRC '85, Oct. 15–18, 1985, Kyoto, Proc. p. 80.
[3.34] *Szware, M.:* Ions and Ion Pairs. Acc. Chem. Res. *2* (1969), p. 87.
[3.35] *Szware, M.:* Carbanions Living Polymers and Electron Transfer Processes. Interscience Publ., New York, 1968.
[3.36] *Tazura, S.:* Photosensilized Charge-Transfer Polymerization. Adv. Polym. Sci. *6* (1969), p. 321.
[3.37] *Tung, L. H., Lo, G. Y. S., Griggs, J. A.:* Block-Copolymer Preparation via Combined Anionic and Free Radical Polymerization. Gordon Res. Conf. of Elast., July 18–22, 1983, New London, NH.

Cationic Polymerization

[3.38] *Kennedy, J. P., Langer, A. W.:* Recent Advances in Cationic Polymerization. Fortschr. Hochpolym. Forsch. *3* (1964), p. 508.
[3.39] *Kennedy, J. P.:* Cationic Polymerization of Olefins, Critical Inventory. J. Wiley & Sons, New York, 1975.
[3.40] *Plesch, P. H.:* The Chemistry of Cationic Polymerization. Pergamon Press, London, 1963.
[3.41] *Plesch, P. H.:* Cationic Polymerization. In: *Robb, J. C., Peaker, F. W. (Eds.):* Progress in High Polymers, Vol. III, Heywood Publ., London, 1968, p. 137.

Anionic Polymerization

[3.42] *Halasa, A. F.:* Recent Advances in Anionic Polymerization. RCT *54* (1981), Review, p. 627; RCT *55* (1982), p. 253.
[3.43] *Morton, M., Fenfers, L. J.:* Homogeneous Anionic Polymerization of Unsaturated Monomers. Macromol. Revs. *2* (1967), p. 71.
[3.44] *Morton, M., Fenfers, L. J.:* Anionic Polymerization of Vinylmonomers. RCT *48* (1975), Review.
[3.45] *Mulvaney, J. E., Overberger, C. G., Schiller, A. M.:* Anionic Polymerization. Adv. Polym. Sci. *3* (1961), p. 106.
[3.46] *Young, R. N.:* Chain Transfer Reactions in the Anionic Polymerization of Dienes. Gordon Res. Conf. of Elastom., July 14–18, 1980, New London, NH.

Coordination Polymerization

[3.47] *Boos jr., J.:* The Nature of the Active Site in the Ziegler-Type Catalyst. Macromol. Revs. *2* (1967), p. 115.
[3.48] *Gaylord, G., Mark, H. F.:* Linear and Stereoregular Addition Polymers, Polymerization with Controlled Propagation. Interscience Publ., New York, 1958.
[3.49] *Henrici-Olivé, G., Olivé, S.:* Koordinative Polymerisation an löslichen Übergangsmetall-Katalysatoren. Adv. Polym. Sci. *6* (1969), p. 421.
[3.50] *Henrici-Olivé, G., Olivé, S.:* Coordination and Catalysis. Verlag Chemie, Weinheim, 1977, pp. 186, 210.

[3.51] *Hertler, W. R., Sogah, D. Y., Webster, O. W.:* Synthesis of Elastomers by Group Transfer Polymerization. Paper 1, 125th ACS-Meeting, Rubber-Div., May 8–11, 1984, Indianapolis, IN.

[3.52] *Horne jr., S. E.:* Polymerization of Diene Monomers by Ziegler Type Catalysis. RCT *53* (1980), p. G71.

[3.53] *Hsieh, H. L., Yeh, H. C.:* Polymerization of Butadiene and Isoprene with Lanthanide Catalysts, Characterization and Properties of Homopolymers and Copolymers. RCT *58* (1985), p. 117.

[3.54] *Natta, G.:* Angew. Chem. *68* (1956), p. 393.

[3.55] *Porvi, L., Gallazzi, M. C., Bianchi, F., Giarrusso, A.:* Some Characteristics of Neodynium Catalysts for Polymerization of Diolefins. Paper 2, 125th ACS-Meeting, Rubber-Div., May 8–11, 1984, Indianapolis, IN.

[3.56] *Reich, L., Schindler, A.:* Polymerization by Organometallic Compounds. Interscience Publ., New York, 1966.

[3.57] *Teyssié, Ph., et al.:* Stereospecific Polymerization of Dialetines by π-Allylic Coordination Complexes in Coordination Polymerization. A Memorial to K. Ziegler. Academic Press, New York, 1975.

3.7.2.4 References on Solution Polymerization

[3.58] *Beckmann, G., Engel, E.:* Zur Technik der Lösungspolymerisation. Chem. Ing. Techn. *38* (1966), p. 1025.

[3.59] *Engel, E., Schäfer, J., Kiepert, K. M.:* KGK *17* (1964), p. 702.

3.7.2.5 References on Suspension Polymerization (see also [3.5])

[3.60] *Hofmann, W.:* Crosslinking Agents in Ethylene-Propylene Rubber. Review in Progr. Rubb. Plast. Technol. *1* (1985) 2, pp. 18–50 (see also ref. [3.400]).

3.7.3 References on Structure of Polymers

3.7.3.1 References on Determination of Structure

[3.61] *Schneider, P.:* In: *Houben Weyl:* Methoden der organischen Chemie, Vol. XIV/2. Verlag G. Thieme, Stuttgart, 1961.

[3.62] *Sommer, N.:* KGK *28* (1975), p. 131.

3.7.3.2 References on the Influence of Structure of Polymers

[3.63] *Flory, P. J.:* Molecular Configuration in Bulc Polymers. RCT *48* (1975), Review.

[3.64] *Hamed, G. R., Shieh, C.-H.:* Flow Criterion for Elastomer Tack. RCT *55* (1982), p. 1469.

[3.65] *Hamed, G. R., Shieh, C.-H.:* Autohesion and Cohesion of Uncrosslinked Elastomers. Paper 15, 127th ACS-Meeting, Rubber-Div., Apr. 23–26, 1985, Los Angeles, CA.

[3.66] *Hamed, G. R., Song, J. H.:* Anisotropy Induced in an Uncrosslinked Elastomer via Large Strain Deformation. RCT *58* (1985), p. 407.

[3.67] *Koenig, J. L.:* Chemical Microstructure of Polymer Chains. J. Wiley & Sons, New York, 1980.

[3.67a] *Koszterszitz, G.:* Eigenklebrigkeit und Tack von unvernetzten Kautschuken. KGK *35* (1982), p. 11.

[3.68] *Llorente, M. A., Mark, J. E.:* The Effect of Chain Length Distribution on Elastomeric Properties. RCT *53* (1980), p. 988.

[3.69] *Markert, J.:* Einfluß der Molekülstruktur auf Misch- und Verformungseigenschaften einiger Synthesekautschuke. KGK *34* (1981), p. 260.

[3.70] *Nakajima, N., Harrell, E. R.:* Test of Strain-Time Corresponding Principle with Gel-containing Elastomers. Paper 85, 127th ACS-Meeting, Rubber-Div., Apr. 23–26, 1985, Los Angeles, CA.

[3.71] *Pearson, D.:* Rheological Behavior of Branched Elastomers. Gordon Res. Conf. of Elastom., July 18–22, 1983, New London, NH.

[3.72] *Plazek, D.J.:* The Effect of Long Chain Branching on Viscoelastic Behavior. Gordon Res. Conf. of Elastom., July 14–18, 1980, New London, NH.
[3.73] *Rhee, C.K., Andries, J.C.:* Factors which Influence Autohesion of Elastomers. RCT *54* (1981), p.101.
[3.74] *Schutz, D.N., Turner, S.R., Golub, M.A.:* Recent Advances in the Chemical Modification of Unsaturated Polymers. RCT *55* (1982), p.809.
[3.75] *Wool, R.P.:* Molecular Aspects of Tack. RCT *57* (1984), p.307.
[3.76] *Yamagushi, K., Okumura, J., Yokoyama, T.:* Molecular Design of Rubber by Graft Modification. IRC '85, Oct. 15–18, 1985, Kyoto, Proc. p.393.

3.7.4 References on Different Types of Rubber

3.7.4.1 References on BR

Synthesis

[3.77] *Adams, H.E., et al.:* The Impact of Lithium Initiators on the Preparation of Synthetic Rubbers. RCT *45* (1982), p.1252.
[3.78] *Anderson, W.S.:* Polymerization of 1,3-Butadiene on Cobalt- and Nickel-Halogenides. J. Polym. Sci. *5A-1* (1967), p.429.
[3.79] *Bawn, C.E.:* Polymerization of Butadiene by Soluble Ziegler-Natta-Catalyst. Rubber Plast. Age *46* (1965), p.510.
[3.80] *Childers, C.W.:* Cationic Coordination Catalysts of Polybutadiene with High cis-1.4-Content. J. Am. Chem. Soc. *85* (1963), p.229.
[3.81] *Duck, E.W.:* Recent Developments in Organometallic Solution Polymerization Catalysts. Paper at 13th IISRP Annual Meeting, June 21, 1972, Munich.
[3.82] *Guo-Jin, X., Jing-Li, L., Shen-Pu, Y.:* A Study of the BuLi-Toluene Butadiene Polymerization System. Paper 46, 123rd ACS-Meeting, Rubber-Div., May 10–12, 1983, Toronto, Ontario.
[3.83] *Hsieh, H.L.:* Alkyllithium Polymerization Catalyst. J. Polym. Sci. *31* (1965), pp.153, 163, 173, 181, 191.
[3.84] *van de Kemp, F.P.:* Organoaluminium/Cobalt Butadiene Polymerization Catalyst. Makromol. Chem. *93* (1966), p.202.
[3.85] *Kuzma, L.J.:* Polybutadiene and Polyisoprene Rubbers. Paper at 124th ACS-Meeting, Rubber-Div., Oct. 25–28, 1983, Houston, TX.
[3.86] *Stephenson, L.M.:* The Nature of Stereochemical Control in Metal Catalyzed Butadiene Polymerization. Gordon Res. Conf. of Elastom., July 18–22, 1983, New London, NH.
[3.87] *Takao, H., Imai, A.:* Polymer Design for Lower Rolling Resistance Rubber. IRC '85, Oct. 15–18, 1985, Kyoto, Proc. p.465.
[3.88] *Weber, H., et al.:* Die stereoregulierte Homo- und Copolymerization der Butadiene. Makromol. *101* (1967), p.320.

Structure and Properties

[3.89] *Churchod, J.:* The Physico Chemical Characteristics of Commercial Elastomers – Molecular Dimensions and Distribution Functions of Polybutadiene. RCT *43* (1970), p.1367.
[3.90] *Hamed, G.R., Shieh, C.-H.:* Tack and Related Properties of Isopropyl Azodicarboxylate Modified Polybutadiene. RCT *57* (1984, p.227.
[3.91] *Hamed, G.R.:* Tack and Green Strength of Elastomeric Material. RCT *54* (1981), p.576.
[3.92] *Imai, A.:* Molecular Design for Improvements of Rolling Resistance of Polybutadiene Rubbers. Gordon Res. Conf. of Elastom., July 15–18, 1985, New London, NH.
[3.93] *Kraus, G., Gruver, J.T.:* Rheological Properties of Cis-Polybutadiene. J. Appl. Polym. Sci. *9* (1965), p.739.
[3.94] *Nordsiek, K.H.:* Entwicklung und Bedeutung spezieller Homopolymerisate des Butadiens. KGK *25* (1972), p.87.

[3.95] *Nordsiek, K. H.:* Model Studies for the Development of an Ideal Tire Tread Rubber. Paper 48, 125th ACS-Meeting, Rubber-Div., May 8-11, 1984, Indianapolis, IN.
[3.96] *Nordsiek, K. H.:* The Integral Rubber Concept, an Approach to the Ideal Tire Tread Rubber. KGK *38* (1985), p. 178.
[3.97] *Nordsiek, K. H., Kiepert, K. M.:* Die charakteristischen Merkmale der Vinyl-Butadiene und ihre molekulare Deutung. KGK *35* (1982), p. 371.
[3.97a] *Nordsiek, K. H.:* Rubber Microstructure and Reversion. Rubbercon '87, June 1-5, 1987, Harrogate, GB, Proceed. A15.
[3.98] *Reichenbach, D.:* cis-trans-Umlagerung von cis-1.4-Polybutadien bei der Vulkanisation durch Schwefel. KGK *18* (1965), p. 213.
[3.99] *Ureda, A., Watanabe, H., Akita, S.:* Structure and Properties of Newly Developed High Vinyl Polybutadiene Rubbers. IRC '85, Oct. 15-18, 1985, Kyoto, Proc. p. 199.
[3.100] *Winter, J., Tomic, M., Silver, H.:* Use of 1.2-Syndiotactic Polybutadiene as Compatible Compounding Film. Paper 50, 120th ACS-Meeting, Rubber-Div., Oct. 13-16, 1981, Cleveland, OH.
[3.101] *Yamagushi, K., Hayashi, J., Asano, Y.:* Morphology of Syndiotactic 1.2-Polybutadiene in Vinyl-cis-Polybutadiene Rubber. IRC '85, Oct. 15-18, 1985, Kyoto, Proc. p. 193.

Compounding, Vulcanization, Processing

[3.102] *Bachin, W. W.:* Vulcanization-Characteristics of Polybutadiene. Ind. Engng. Chem. Prod. Res. Dev. *4* (1965), p. 15.
[3.103] *Cregg jr., E. C., Lattimer, R. P.:* Polybutadiene Vulcanization, Chemical Structure from Sulfur Donor Vulcanization of an Accurate Model. RCT *57* (1984), p. 1056.
[3.104] *Dräxler, A.:* Die Wirkung extrem hoher Verarbeitungstemperaturen auf kaltpolymerisiertes SBR und Polybutadien. KGK *17* (1964), p. 71.
[3.105] *Nordsiek, K. H., Vohwinkel, K.:* Das Verhalten von Schwefel in Polybutadien. KGK *18* (1965), p. 566.
[3.106] *Sims, D.:* Butadiene Polymer. J. IRI *1* (1967), p. 200.
[3.107] *Svetlik, J. F., Ross, E. F.:* Heat Stability of cis-Polybutadiene Compounds. Rubber Age *96* (1965), p. 570.

Low Molecular Weight and Liquid BR (see also Ref. [3.683])

[3.108] *Heitz, W.:* Oligobutadiene mit Ester-und Carbonatendgruppen. ikt '80, Sept. 23-26, 1980, Nürnberg, West Germany.
[3.109] *Jérôme, R., Broze, G.:* Viscoelastic Properties of a Class of Carboxylated Rubber, The Carboxylato-Telechelic Polydienes. RCT *58* (1985), p. 223.
[3.110] *Koshute, M. A.:* Bromine Terminated Liquid Polybutadiene as a Solventless Room Temperature Adhesive System. RCT *53* (1980), p. 285.
[3.111] *Lee, T. P. C., et al.:* The Potential and Limitations of Liquid Rubber Technology. KGK *31* (1978), p. 723.
[3.112] *ter Meulen, B. H., Murphy, W. T., Drake, R. S.:* Gießbare Elastomere auf der Basis von flüssigen Polymeren. GAK *31* (1978), p. 384.
[3.113] *Luxion, A.:* The Preparation, Modification and Application of Non-functional Liquid Polybutadiene. RCT *54* (1981), p. 596, Review.
[3.114] *Quack, G.:* Chemical Modification and Application of Low Molecular Weight Polybutadienes. Gordon Res. Conf. of Elastom., July 14-18, 1980, New London, NH.
[3.114a] *Drake, R. E.:* Flüssige 1.2-Polybutadiene als Coagentien für EPDM. GAK *35* (1982), p. 180.

Special References

[3.115] GB 848065, 1956, Phillips Petroleum Co.
[3.116] DAS 1112834, 1960, Firestone.
[3.117] DT 1165864, 1190441, 1962, DT 1242371, 1965, Bayer.
[3.118] DT 1128143, 1958, Goodrich.
[3.119] DAS 1213120, 1960, Bridgestone.

[3.119a] *Marwede, G., Stollfuß, B.:* Actual State of Butadiene Rubber for Tire Application. Rubbercon '87, June 1–5, 1987, Harrogate, GB, Proceed. B31
[3.120] DAS 1 087 809, 1956, Firestone.
[3.121] *van de Kamp, F. P.:* Makromol. Chem. *93* (1966), p. 202.
[3.122] *Grippin, M.:* Ind. Engng. Chem. Proc. Res. Dev. *4* (1965), p. 160.
[3.123] *Furukawa, J.:* Pure Appl. Chem. *42* (1975), p. 495.
[3.124] *Nordsiek, K. H.:* Spezialkautschuk auf Basis von 1.4- und 3.4-Polydienen. KGK *35* (1985), p. 1032.
[3.125] *Oberster, A. E., et al.:* Makromol. Chem. *29/30* (1973), p. 291.
[3.126] *Morton, A. A., et al.:* J. Am. Chem. Soc. *69* (1947), p. 950.
[3.127] DT 1 128 666, 1960, Bayer.
[3.128] *Hsieh, H. L.:* RCT *49* (1976), p. 1305.
[3.129] FR 1 449 382, 1 449 383, 1 456 282, 1964, FR 1 468 239, 1965, Chem. Werke Hüls.
[3.130] *Ring, W., Cantow, H. J.:* Makromol. Chem. *89* (1965), p. 138.
[3.131] DT 1 126 794, 1963, DDS 1 495 734, 1963, Bayer.
[3.132] *Nützel, K., Lange, H.:* Angew. Makromol. Chem. *14* (1970 (1970), p. 131.
[3.133] DT 1 570 099, 1965, Bayer.
[3.134] *Engel, E. F.:* IISRP-Annual Meeting, 1972, Munich.

3.7.4.2 References on SBR

Synthesis

[3.135] *Hamielec, A.:* Production of SBR, Dynamic Modeling of a Continuous Reactor Train. Paper 40, 123rd ACS-Meeting, Rubber-Div., May 10–12, 1983, Toronto, Ontario.
[3.136] *Hiraoka, M.:* Recent Advances in the Polymerization and Properties of Alternating Copolymers. Gordon Res. Conf. of Elastomers. July 19–23, 1982, New London, NH.
[3.137] *Hofmann, W.:* Mischpolymerisation. In: *Boström, S. (Ed.):* Kautschuk-Handbuch, Vol. 1. Verlag Berliner Union, Stuttgart, 1959, pp. 330–394.
[3.138] *Kolinek, E.:* Emulsion SBR Techology. 124th ACS-Meeting. Rubber-Div., Oct. 25–28, 1983, Houston, TX.
[3.139] *Luijk, P.:* The Versatility of the Lithium-Alkyl-Initiated Copolymerization of Styrene and Butadiene. KGK *34* (1981), p. 191.
[3.140] *Miller, J. R.:* Mechanisms of Retardation by Butadiene Dimer in SBR Polymerization. Paper 88, 124th ACS-Meeting, Rubber-Div., Oct. 25–28, 1983, Houston, TX.
[3.141] *Nuyken, O., Weidner, R.:* Synthese von azogruppenhaltigen Butadien-Styrol-Polymerisaten und deren Selbstvernetzung. ikt '85, June 24–27, 1985, Stuttgart, West Germany.
[3.142] *Oshima, O., Tsutsumi, F., Sakakihara, M.:* Solution Polymerized SBR-terminated with Tin Compound. I. Polymer Design. IRC '85, Oct. 15–18, 1985, Kyoto, Proceed. p. 178 (see also Ref. [3.149]).

Structure

[3.143] *Nordsiek, K. H., Kiepert, K. M.:* New Structures of Butadiene-Styrene Copolymers. IRC '79, Oct. 3–6, 1979. Venice, Italy, Proceed. p. 960.
[3.144] *Tanaka, Y., Sato, H., Saito, K., Miyashila, K.:* Determination of Sequence Distribution in Styrene Butadiene Copolymer. RCT *54* (1981), p. 685.

Compounding, Vulcanization, Uses

[3.145] *Briggs, G. J., Wei, Y. K., Holmes, J. M., Walker, J.:* Reduced Rolling Resistance with SBR. Paper 14, 120th ACS-Meeting, Rubber-Div., Oct. 13–16, 1981, Cleveland, OH.
[3.146] *Bhowmick, A. K., Gent, A. N.:* Effect of Interfacial Bonding on the Self-Adhesion of SBR and Neoprene. RCT *57* (1984), p. 216.
[3.147] *Dräxler, A.:* Einige Spezialanwendungen von kaltpolymerisiertem SBR. Gummi u. Asbest *14* (1961), p. 726.
[3.148] *Edwards, D. C., Sato, K.:* Interaction of Silica with Functionalized SBR. RCT *53* (1980), p. 66.

[3.149] *Fujimaki, T., Ogowa, M., Yamagushi, S., Tomita, S., Okuyama, M.:* Solution Polymerized SBR – Terminated with Tin Compound. II. Tire Application. IRC '85, Oct. 15–18, 1985, Kyoto, Proc. p. 184 (see also Ref. [3.142]).
[3.150] *Nakajima, N., Harrell, E.:* Effect of Extending Oil on Viscosity Behavior of Elastomers. RCT *56* (1983), p. 784.
[3.151] *Sato, K.:* Ionic Crosslinking of Carboxylated SBR. RCT *56* (1983), p. 942.

Special References

[3.152] DRP 570980, 1929, I.G. Farbenindustrie.
[3.153] DRP 891025, 1939, I.G. Farbenindustrie.
[3.154] US 2776276, 1953, Hercules Powder.
[3.155] US 2716107, 1953, US Rubber.
[3.156] DRP 753991, 1937, I.G. Farbenindustrie.
[3.157] US 3148225, 1962, Pennwalt.
[3.158] *Rostler, F. S.:* Rubber Age *69* (1951), p. 559.
[3.159] *Harrington, H. D., et al.:* India Rubber World *124* (1951), pp. 435, 571.
[3.160] *Rostler, F. S., White, R. M.:* Ind. Engng. Chem. *47* (1955), p. 1069.
[3.161] *Taft, W. K.:* Ind. Engng. Chem. *47* (1955), p. 1077.
[3.162] *White, D. W.:* Rubber J. *149* (1967) 9, p. 42.
[3.163] DT 1280545, 1961, Chem. Werke Hüls.
[3.164] *Willis, J. M., Barbin, W. W.:* Rubber Age *100* (1968), pp. 7, 53.
[3.165] *Zelinski, R., Childers, C. W.:* RCT *41* (1968), p. 161.
[3.166] *Hsieh, H. L., Glaze, W. H.:* RCT *43* (1970), p. 22.
[3.167] *Hoffmann, M., et al.:* KGK *22* (1969), p. 691.
[3.168] *Kuntz, I.:* J. Polym. Sci. *54* (1961), p. 569.
[3.169] US 3265765, 1962, Shell.
[3.170] *Engel, E. F.:* Rubber Age, March 1973, p. 25.
[3.171] *Cooper, Nash:* Rubber Age, May 1972, p. 55.
[3.172] *Railsbeck, Howard, Stumpe:* Rubber Age, April 1974, p. 46.
[3.173] *Railsbeck, Zelinski:* KGK *25* (1972), p. 254.
[3.174] *Alliger, Weissert:* IRC '66, 1966, Paris.
[3.175] *Duck, Locke:* J. IRI *2* (1968), p. 223.
[3.176] *Satake, Sone, Hamada, Hayakawa:* J. IRI *5* (1971), p. 104.

3.7.4.3 References on NBR

General Overview

[3.177] *Bertram, H. H.:* Developments in Acrylonitrile Butadiene Rubber (NBR) and Future Prospects. In: *Whelan, A., Lee, K. S. (Eds.):* Developments in Rubber Technology, 2nd Ed. Elsevier Applied Science Publ., Barking, 1981, pp. 51–85.
[3.178] *Bertram, H. H.:* 50 Jahre Perbunan N. Bayer-Mitt. f. d. Gummi-Ind. *53* (1981), pp. 3–12.
[3.179] *Dunn, J. R., Coulthard, D. C., Pfisterer, H. A.:* Advances in Nitrile Rubber Technology. RCT 51 (1978), pp. 389–405.
[3.180] *Hofmann, W.:* Nitrile Rubber. RCT *36* (1963) 5, pp. 1–262.
[3.181] *Hofmann, W.:* Nitrilkautschuk. Verlag Berliner Union, Stuttgart, 1965.
[3.182] *Hofmann, W.:* Recent Developments in Nitrile Rubber Technology. Progr. Rubber Technol. *46* (1984), pp. 43–84;
Neuere Entwicklungen auf dem Gebiet der Nitrilkautschuktechnologie – ein Überblick. KGK *37* (1984), pp. 753–769.
[3.183] *Morrell, J. P.:* Nitrile Elastomers. In: *Babbit, R. O.:* The Vanderbilt Handbook. Vanderbilt, 1978, pp. 169–187.

Synthesis

[3.184] Ref. [1.60], pp. 116-119, pp. 121-127; Ref. [3.180], pp. 46-106; Ref. [3.181], pp. 51-137.
[3.185] *Healy, J. C.:* The Nitrile Rubber. Paper at 124th ACS-Meeting, Rubber-Div., Oct. 25-28, 1983, Houston, TX.
[3.186] *Morton, M.:* Science and Technology of Rubber Polymerization. I. Nippon Gomu Kyakaishi *55* (1982), pp. 309-321.

Structure

[3.187] Ref. [3.180], pp. 107-126; Ref. [3.181], pp. 154-180.
[3.188] *Nakajima, N., Harrell, E. R.:* Viscoelastic Characterization of Long Branching and Gel in Butadiene-Acrylonitrile Copolymer Elastomers. RCT *53* (1980), p. 14.
[3.189] *Nakajima, N., Harrell, E. R., Seil, D. A., Jorgenen, A. H.:* Characterization of Butadiene-Acrylonitrile Copolymers, Relationship between Molecular Parameters and Properties. Paper 62, 124th ACS-Meeting, Rubber-Div., Oct. 25-28, 1983, Houston, TX.

Solubility Parameter

[3.190] *Bauer, R. F., Hale, P. T.:* Solubility Parameter Spectroscopy. Gordon Res. Conf. of Elastom., July 18-22, 1983, New London, NH.
[3.191] *Hofmann, W., Verschut, C.:* Covulkanisation von Kautschukgemischen aus diengruppenhaltigem Epichlorhydrinkautschuk (ETER) mit anderen Dienkautschuken. KGK *35* (1982), p. 95.
[3.192] *Neppel, A.:* Prediction of Equilibrium Swelling in Fuel and Lubricating Ingredients. Paper 72, 126th ACS-Meeting, Rubber-Div., Oct. 23-26, 1984, Denver, CO.
[3.193] *Orwole, R. A.:* The Polymer Solvent Interaction Parameters. RCT *50* (1977), Review.
[3.194] *Pausch, J. R., Carman, C. J., Pappas, L. G.:* A Molecular Interpretation of Hydrocarbon Solvent Resistance in Nitrile and Chlorinated Rubbers. Paper 2, 120th ACS-Meeting, Rubber-Div., Oct. 13-16, 1981, Cleveland, OH.

Resistance to Swelling in Solvents

[3.195] *Abu-Isa, I. A.:* Elastomer-Gasoline Blends Interactions. I. Effects of Methanol/Gasoline Mixtures on Elastomers. RCT *56* (1983), p. 135; II. Effects of Ethanol/Gasoline and Methyl-t-butylether/Gasoline Mixtures on Elastomers. RCT *56* (1983), p. 169.
[3.196] *Cheng, C. W.:* Effects of Gasohol and Alcohol on Elastomeric Material. 116th ACS-Meeting, Rubber-Div., Oct. 18, 1979, Detroit, MI.
[3.197] *Cheng, C. W., Bayer, K.:* Die Verträglichkeit von Gasohol und Alkoholen mit Elastomeren. KGK *35* (1982), p. 837.
[3.198] *Dunn, J. R.:* Heat and Fuel Resistance of NBR and NBR/PVC. PRI-Conf., April 10, 1981, London, Proceed. pp. 5-18.
[3.199] *Dunn, J. R., Pfisterer, H. A., Ridland, J. J.:* NBR Vulcanizates, Resistant to High Temperatures and Sour Gasoline. RCT *52* (1979), p. 331.
Gegen hohe Temperaturen und "saure" Motorenkraftstoffe beständige Nitrilkautschukqualitäten. GAK *33* (1980), p. 296.
[3.200] *Dunn, J. R., Sandrap, J. P.:* Improved Sour Fuel and Fuel Permeability Resistance in NBR Vulcanizates. SRC '79, Apr. 2-3, 1979, Copenhagen, Denmark.
[3.201] *Dunn, J. R., Vara, R. G.:* Oil Resistant Elastomers for Hose Applications. RCT *56* (1983), p. 557, Review.
[3.201a] *Dunn, J. R., Vara, R. G.:* Fuel Resistance and Fuel Permeability of NBR and NBR Blends. Paper 16, 128th ACS-Meeting, Rubber-Div., Oct. 1-4, 1985, Cleveland, OH.
[3.202] *Friberg, G.:* Making Rubbers for the Future Fuels. Europ. Rubber J. *182* (1980), 10, p. 25.
Kraftstoffe der Zukunft und ihre Wirkung auf Gummiteile im Auto. GAK *33* (1980), p. 156.
[3.203] *Hertz jr., D. L.:* Sour Hydrocarbons, The Elastomer Challenge. Paper 34, 117th ACS-Meeting, Rubber-Div., May 20-23, 1980, Las Vegas, NV.

[3.204] *Hofmann, W.:* Einige Aspekte über den Einfluß des Mischungsaufbaus von Nitril- und Epichlorhydrinkautschuk-Vulkanisaten gegen die Kraftstoffe der Zukunft. KGK *34* (1981), pp. 1017–1022.

[3.205] *Killgoar, P. C. jr., Lemieux, M. A.:* Improvements in the Oxidized Gasoline Resistance of NBR Elastomers. RCT *56* (1983), p. 853.

[3.205a] *Krumeich, P.:* Polymere Dichtungswerkstoffe. Resch-Verlag, Gräfelfing/Munich, 1988.

[3.206] *Pfisterer, H. A., Dunn, J. R.:* New Factors Affecting the Performance of Automotive Fuel Hose. RCT *53* (1980), p. 357.

[3.207] *Pfisterer, H. A., Dunn, J. R., Vucov, R.:* The Use of a Screening Test for Assessing the Hydrogen Sulfide Resistance of Elastomers. RCT *56* (1983), p. 418.

[3.208] *Mueller, W. J.:* Effect of Substituted Aromatic Gasoline Components on Properties of a Nitrile Hose Core. Paper 3, 117th ACS-Meeting, Rubber-Div., May 20–23, 1980, Las Vegas, NV.

[3.209] *Piazza, S., Santarelli, G., Pasarini, N.:* Fuels of Today and Fuels of the Future in Automobile Applicability Limits of Nitrile Rubber used for Technical Articles in Contact with Fuels. IRC '79, Oct. 3–6, 1979, Venice, Italy, Proceed. p. 451.
Nitrilkautschuk – Grenzen der Einsatzfähigkeit für Gummiartikel im Kontakt mit Kraftstoffen. GAK *33* (1980), p. 802.
Nitrile Rubber Articles in Contact with Automotive Fules. Ind. Gomma *24* (1980) 1, p. 37.

[3.210] *Trexler, H. E.:* The Effect of Oxidized Fuels on Polymers. RCT *54* (1981), p. 155.

[3.211] *Watkins, M. J., Derringer, G. C.:* Effects of Oil Field Corrosion Inhibitors on Nitrile Elastomers. Paper 89, 127th ACS-Meeting, Rubber-Div., Apr. 23–26, 1985, L. A., CA.

Heat Resistance

[3.212] *Byrne, P. S., Coulthard, D. C., Chalmers, D. C., Ridland, J. J.:* Optimization of Nitrile Rubbers and Nitrile Rubber Compounds to Obtain Maximum Heat Resistance. SAE-Paper 730537, SAE-Meeting, May 1973, Detroit, MI.

[3.213] *Coulthard, D. C., Gunter, W. D.:* New Compounding Approaches to Heat Resistant NBR. J. Elastomer Plast. *9* (1977), p. 131.

[3.214] *Dunn, J. R., Byrne, P. S., Coulthard, D. C.:* Heat Resistant Nitrile Rubber. 14th Annual Meeting of IISRP, May 1973, San Francisco, CA.

[3.215] *Dunn, J. R., Timar, J.:* Vulcanizates for High Temperature Service. Quebec Rubber and Plastic Group, Dec. 15, 1978, Montreal, Quebec.

[3.216] *Dunn, J. R.:* Compounding NBR for High Temperature Applications. PRI '84, March 12–16, 1984, Birmingham.
Rubber World *190* (1984) 3, p. 16.
Rezepturaufbau von NBR-Qualitäten für den Einsatz bei hohen Temperaturen. GAK *38* (1985), p. 275.

[3.217] *Walter, G.:* Elastomers in the Automotive Industry. RCT *49* (1976), p. 775.
Elastomere in der Automobilindustrie. GAK *28* (1975), p. 306.

[3.217a] *Walter, G.:* Kunststoffe und Elastomere in Kraftfahrzeugen. W. Kohlhammer-Verlag, Stuttgart 1985.

[3.217b] Nitrile Rubber Aging in Various Environments. Bibliography 6. Publ.: RCT.

Compression Set

[3.218] *Chang, D. M.:* Investigation of the Structure – Property Relationship of Improved Low Compression Set Nitrile Rubbers. RCT *54* (1980), p. 170.

[3.219] *Dunn, J. R.:* The Effect of Polymer and Compound Variables on the Sealing Force and Compression Set of NBR. Paper 42, 124th ACS-Meeting, Rubber-Div., Oct. 25–28, 1983, Houston, TX.

[3.220] *Hofmann, W.:* Optimierung des Wärmecompression-Set von schwefel-vulkanisiertem Nitrilkautschuk. KGK *36* (1983), p. 1044.

[3.221] *Hofmann, W.:* Influence of Crosslinking on Permanent Set Especially Compression Set of Nitrile Rubber at Elevated Temperatures. PRI-Conf., Oct. 27–29, 1982, London, Proceed. p. 91. Plast. Rubb. Proc. Appl. *4* (1984), p. 191.

[3.222] *Jahn, H.J.:* The Compression Set Behavior of Some Elastomers in Comparison with Natural Rubber. Rubber Plast. Age *46* (1965), p. 1028.

[3.222a] *Jahn, H.J., Betram, H.H.:* 102nd ACS-Meeting, Rubber-Div., 1972, Cincinatti, OH.

[3.223] *Jahn, H.J.:* Die Abhängigkeit des Compression Set von Elastomeren vom Mischungsaufbau. KGK *21* (1968), p. 469.

[3.224] *Lowman, M.:* The Effect of Oxygen on Compression Set. Paper 50, 117th ACS-Meeting, Rubber-Div., May 20-23, 1980, Las Vegas, NV.

NBR with Bound Antioxidant

[3.225] *Ajiboye, O., Scott, G.:* Mechanism of Antioxidant Action. Stabilization of Nitrile Rubber by Interpenetrating Networks based on Thiol Adducts. Polym. Degrad. Stabil. *4* (1982), p. 397. 2. Synthesis of Bound Antioxidant Concentrates in Nitrile Rubber. Polym. Degradat. Stabil. *4* (1982), p. 415.

[3.226] *Horvath, J.W., Bush, J.L.:* Exploring the Potential of Bound Antioxidant Nitrile in Automotive. Paper at SAE-Meeting, Sept. 26-30, 1977, Detroit, MI.

[3.227] *Horvath, J.W.:* Bound Antioxidant Stabilized NBR in Automotive Applications. Elastomerics *111* (1979) 8, p. 19.

[3.228] *Thomas, D.K.:* Antioxidants in Rubber Aged in Air and in Hydraulic Fluids. In: *Scott, G. (Ed.):* Development in Rubber Stabilization. Elsevier Appl. Sci. Publ., Barking, 1979, pp. 137-166.

Carboxylated NBR, X-NBR

[3.229] *Chakraborty, S.K.:* Effect of the Curing System on the Technical Properties and the Network Structure of Carboxylated Nitrile Rubber. KGK *36* (1983), p. 461.

[3.230] *Chakraborty, S.K., De, S.K.:* Silica and Clay Reinforced Carboxylated Nitrile Rubber, Vulcanized by a Mixed Crosslinking System. RCT *55* (1982), p. 990.

[3.231] *Sato, K., Blackshaw, G.C.:* Dynamic Properties of Carboxylated Nitrile Rubber. IRC '85, Oct. 15-18, 1985, Kyoto, Proc. p. 377.

[3.232] *Shaheen, F.G., Grimm, D.C.:* Carboxylated Nitrile Rubber. Paper at 127th ACS-Meeting, Rubber-Div., Apr. 23-26, 1985, Los Angeles, CA.

[3.232a] *Starmer, P.H.:* Effect of Metal Oxides on the Properties of Carboxylic Nitrile Rubber Vulcanizates. Rubbercon '87, June 1-5, 1987, Harrogate, GB, Proceed. A11.

[3.233] *Weir, R.J., Burkey, R.C.:* Carboxylierter Nitrilkautschuk, Eigenschaften und Anwendungen. GAK *36* (1983), p. 268.

Thermoplastic NBR

[3.234] *Coran, A.Y., Patel, R.:* NBR-Nylon Thermoplastic Elastomeric Compositions. RCT *53* (1980), p. 781.
Thermoplastische Vulkanisate aus verschiedenen Kunststoff-Verschnitten. KGK *35* (1982), p. 194.
Thermopalastische Verschnitte aus Nitrilkautschuk und Polyolefinen. GAK *37* (1984), p. 378.

[3.235] *Schwarz, H.F., Bley, J.W.F., Bleyie, P.:* Design of Alloys of PVC with NBR Polymers to Produce Thermoplastic Elastomers. Paper 90, 127th ACS-Meeting, Rubber-Div., Apr. 23-26, 1985, Los Angeles, CA.

[3.235a] *Schwarz, H.F., Bley, J.W.F.:* Design of Alloys of PVC with NBR Copolymers to Produce Thermoplastic Elastomers. Paper 93, 128th ACS-Meeting, Rubber-Div., Oct. 1-4, 1985, Cleveland, OH.

NBR/PVC Blends (see also Refs. [3.235, 3.235a])

[3.236] *Almond, C.J.:* Mixtures of Nitrile Rubber and Vinyl Polymers. Trans IRI *37* (1961), 3, p. 85.

[3.237] *Hofmann, W.:* NBR/PVC-Blends. Ind. Plast. Mod. 1961, p. 37.

[3.238] *Bittel, P.A.:* Processibility Testing of NBR/PVC Blends. Paper 38, 117th ACS-Meeting, Rubber-Div., May 20-23, 1980, Las Vegas, NV.

[3.239] *Schwarz, H. F.:* Carboxylated Nitrile/PVC Fluxed Blends for Premium Quality Products. Paper 7, 117th ACS-Meeting, Rubber-Div., May 20–23, 1980, Las Vegas, NV.
[3.240] *Vasilin-Oprea, C., Papa, M.:* Mechanochemical Reactions of Polyvinylchloride, XI. Optimum of Mechanical Properties of PVC by Blending with Nitrile Rubber. Coll. Polym. Sci. *260* (1982), p. 570.

Other Types of Nitrile Rubber

[3.241] *Eldred, R.:* Plasticization of Nitrile Elastomers by In Situ Grafted Acrylate Monomers. RCT *57* (1984), p. 320.
[3.242] *Weinstein, A. H.:* Elastomeric Tetramethylene-Ethylethylene-Acrylonitrile Copolymers. RCT *57* (1984), p. 203.

Hydrogenated NBR (H-NBR, ENM, HSN)

[3.242a] *Dinges, K.:* Possibilities and Limitations in the Synthesis of New Rubbers from Conventional Polymers. Paper 10, 20th Annual IISRP-Conference, May 1979, Vancouver, Brit. Columb.
[3.243] *Dunn, J. R., Blackshaw, G. C., Bradford, W. G., Coulthard, D. C.:* The Basic Properties of Highly Saturated Nitrile Rubbers. IRC '85, Oct. 15–18, 1985, Kyoto, Proc. p. 223.
[3.244] *Hashimoto, K., Watanabe, N., Yoshioka, A.:* Highly Saturated Nitrile Elastomer, A New High Temperature Chemical Resistant Elastomer. Paper 43, 124th ACS-Meeting, Rubber-Div., Oct. 25–28, 1983, Houston, TX.
[3.245] *Hashimoto, K., Watanabe, N., Oyama, M., Todani, Y.:* Zetpol, ein hochgesättigter Nitrilkautschuk. GAK *37* (1984), p. 602.
[3.246] *Hashimoto, K., Todani, Y., Oyama, M., Watanabe, N.:* Highly Saturated Nitrile Elastomers. Paper at 127th ACS-Meeting, Rubber-Div., Apr. 23–26, 1985, Los Angeles, CA.
[3.246a] *Hashimoto, K., Todani, Y., Oyama, M., Watanabe, N.:* Highly Saturated Nitrile Elastomer – Automotive Application. Paper 5, 128th ACS-Meeting, Rubber-Div., Oct. 1–4, 1985, Cleveland, OH.
[3.247] *Kube, Y.:* Highly Saturated Nitrile Elastomer. IRC '85, Oct. 15–18, 1985, Kyoto, Proceed. p. 32.
[3.248] *Mirza, J., Thörmer, J., Buding, H.:* Therban, A New Elastomer with Good Heat and Swelling Resistance. Paper 87, 124th ACS-Meeting, Rubber-Div., Oct. 25–28, 1983, Houston, TX; Paper at PRI-Conference '84, March 12–14, 1984, Birmingham.
[3.248a] *Mirza, J.:* Therban – A Special Elastomer for Requirements in Oilfields, Automotive and Industrial Applications. Paper 92, 128th ACS-Meeting, Rubber-Div., Oct. 1–4, 1985, Cleveland, OH.
[3.249] *Miyabayishi, T., Takemura, Y., Sakabe, N.:* A New Type Elastomer with Improved Heat, Ozone and Sour Gas Resistance. Paper 30, 125th ACS-Meeting, Rubber-Div., May 8–11, 1984, Indianapolis, IN.
[3.250] *Thörmer, J., Mirza, J.:* Eine neue Polymerklasse mit hohem mechanischem Werteniveau und besonderer Beständigkeit gegen aggressive Medien. ikt '85, June 24–27, 1985, Stuttgart, West Germany.

Special References

[3.251] DRP 658 172, 1930, IG Farbenindustrie.
[3.252] *Hofmann, W.:* Elastomere Werkstoffe für öl- und wärmebeständige Dichtungen – eine Trendanalyse. GAK *39* (1986), p. 511.
[3.252a] *Hofmann, W.:* Elastomere für technische Anwendungen. Kunstst. *74* (1984), p. 611.
[3.252b] *Hofmann, W.:* Der Natur- und Synthesekautschukmarkt der Welt. Kunstst. *76* (1986), p. 1150.
[3.252c] *Stephenson, W. A.:* Marketing Elastomers in the 1990s – Specialty Elastomers. Paper 51, 128th ACS-Meeting, Rubber-Div., Oct. 1–4, 1985, Cleveland, OH.
[3.252d] *Tessmer, W.:* Synthetic Rubber – Worldwide Outlook and Issues. Paper 48, 128th ACS-Meeting, Rubber-Div., Oct. 1–4, 1985, Cleveland, OH.

[3.253] Krynac, Oil Resistant Rubber. Ringbook, Publ.: Polysar.
[3.254] *Wolff, W.:* Plaste u. Kautschuk *15* (1968), p. 869.
[3.255] Ref. [3.180], pp. 50–58; Ref. [3.181], pp. 58–67.
[3.256] DB 864151, 1949, Bayer.
[3.257] US 2669550, 1950, Goodrich.
[3.258] DB 955901, 1953, Bayer.
[3.259] DB 950498, 1954, Bayer.
[3.260] US 2849426, 1954, Firestone.
[3.261] *Miller, V.A.:* Rév. gén. Caoutch. *34* (1957), p. 577.
[3.262] *Cooper, W., Bird, T.B.:* Rubber World *A 36* (1957) 1, p. 78; Ind. Engng. Chem. *50* (1958), p. 771.
[3.263] *Soos, I., Bikiks, I., Bisztriczky, J.:* Verschneiden von Epoxidharzen mit speziellen Kautschuken. KGK *34* (1981), p. 353.
[3.264] *Alliger, G., Sjothun, I.J.:* Vulcanization of Elastomers. Reinold Publ., Corp., New York, 1964.
[3.265] *Das, C.K., Millns, W.:* The Effect of Compounding Ingredients on the Peroxide cure of NBR. KGK *35* (1982), p. 402.
[3.266] *Abrams, W.J.:* Improving Ozone Resistance of Nitrile Rubber. Rubber Plastics Age *43* (1962), p. 451.
[3.267] *Haverland, A., Hofmann, W.:* Neuartige Bewertungskriterien für Verarbeitungs-Hilfsmittel. KGK *38* (1985), p. 1012. Paper at PRI-Conference '84, March 12–14, 1984, Birmingham.
[3.268] *Haverland, A., Hofmann, W.:* Ausgewählte Verarbeitungshilfsmittel in NBR und CR. DOG-Kontakt 30, Publ.: DOG Deutsche Oelfabrik, Hamburg, 1985.
[3.269] *Radke, R., Böttcher, E.:* Verbesserte Gleichförmigkeit in Mischungen aus Elastomerverschnitten. GAK *30* (1977), p. 72. Int. Polym. Sci. Technol. 5 (1978) 1, p. T/107.
[3.270] *Minnerly, H.E.:* Processing of Nitrile Rubber. Rubber World *152* (1965) 1, p. 76.
[3.271] *Bhowmick, A.K., De, S.K.:* Effect of Curing Temperature on the Technical Properties of Nitrile Rubber and Carboxylated Rubber. RCT *53* (1980), p. 107.
[3.272] *Zhang, S.W.:* Investigation of Abrasion of Nitrile Rubber. RCT *57* (1984), p. 769.
[3.273] *Bertram, H.H., Brand, D.:* RCT *45* (1972), p. 1224.
[3.274] *Hutchings, T.G.:* Permeation of Automotive Fuels – Laboratory Test-Methods. Paper at 126th ACS-Meeting, Rubber-Div., Oct. 23–26, 1984, Denver, CO.
[3.275] *Del Vecchio, R.J., Krakowski, F.J., McKenzie, G.T.:* Dynamic Property Variation and Analysis in NBR. Paper 83, 127th ACS-Meeting, Rubber-Div., Apr. 23–26, 1985, Los Angeles, CA.
[3.276] USP 3882094, 1975, Firestone.
[3.277] USP 3993855, 1976, Firestone.
[3.278] Jap. P. 67/47897, 1967, Bridgestone.
[3.279] Jap. P. 78/39744, 1978, Nippon Zeon.
[3.280] FP 2421923, 1979, Nippon Zeon.
[3.281] DP 2539132, 1977, Bayer AG; BP 1558491, 1980, Bayer AG; *Oppelt, D., Schuster, H., Thörmer, J., Braden, R.*
[3.281a] *Hofmann, W.:* Eigenschaften und Anwendung ausgewählter Elastomerklassen. Kunststoffe *78* (1988), pp. 132–141
[3.282] Therban, hydrierter Nitrilkautschuk, Leistungsprofil. Publ. KA 30 592, Bayer AG, Leverkusen, October 1987
[3.283] Tornac, The Basis of Highly Saturated Nitrile Rubbers. Publ.: Polysar, 1985.
[3.284] *Schultz, D.L.:* High Performance Elastomers for Automotive and Oil-Field Applications. Paper 23, 120th ACS-Meeting, Rubber-Div., Oct. 13–16, 1981, Cleveland, OH.
[3.285] *Nakao, K., Yamanaka, K.:* Effect of Addition of Liquid Amine Terminated Nitrile Rubber on Structure, Mechanical Properties and Bond Strength of Epoxi-Adhesives. IRC '85, Oct. 15–18, 1985, Kyoto, Proc. p. 905.

3.7.4.4 References on CR

General Overview

[3.286] *Hargreaves, C.A., Thompson, D.C.:* Neoprene. Encyclopedia of Polymer Technology and Science. Ref. [1.20].

[3.287] *Johnson, P.R.:* Polychloroprene Rubber. RCT *49* (1976), p.650.

[3.288] *Kirchhof, F.:* 2-Chlorbutadien-Polymerisate. In: *Boström, S. (Ed.):* Kautschuk-Handbuch, Vol.1. Verlag Berliner Union, Stuttgart 1959, p.286.

[3.289] *Knox, R.E.:* Neoprene, the First High Performance Elastomer. Paper 22, 120th ACS-Meeting, Rubber-Div., Oct.13–16, 1981, Cleveland, OH.

[3.290] *Mervin, J.K.:* Neoprene and Hypalon Rubbers. Paper at 124th ACS-Meeting, Rubber-Div., Oct.25–28, 1983, Houston, TX.

[3.291] *Stevenson, A.C.:* Neue Entwicklungen auf dem Gebiet der Polychloropren-Polymeren. GAK *19* (1966), p.144.

Synthesis and Structure

[3.292] *Dinges, K., Casper, R., Obrecht, W., Wendling, P.:* Sulfur Modified Polychloroprene, Influence of Reaction Conditions on Polymerization Kinetics and Polymer Composition. IRC '85, Oct.15–18, Kyoto, Proc. p.211.

[3.293] *Hrabák, F., Webr, J.:* Über die Dimerisierung und die thermische Polymerisation des Polychloroprens. Makromol. Chem. *104* (1967), p.275.

[3.294] *Mead, W.T.:* Molecular Fracture in Polychloroprene. RCT *53* (1980), p.245.

[3.295] *Miyata, Y., Matsunaga, S., Mitani, M.:* Structure and Properties of Polychloroprene with Different Modifications. IRC '85, Oct.15–18, 1985, Kyoto, Proc. p.217.

Particulate and Liquid CR

[3.296] *Hayashi, T., Ariyoshi, T., Sakanaka, Y.:* Some Characteristic of Polychloroprene Powders. IRC '85, Oct.15–18, 1985, Kyoto, Proc. p.784.

[3.297] *O'Neal, H.R.:* Chemistry and Applications of Liquid Chloroprene Polymers Containing Functional Groups. Paper 8, 117th ACS-Meeting, Rubber-Div., May 20–23, 1980, Las Vegas, NV.

Vulcanization

[3.298] *Abdel-Barry, E.M., Ghanem, N.A., Yehia, A.A., Younan, A.F.:* Effect of Thiadiazole Derivatives as Accelerators for Polychloroprene. Paper at IRC '79, Oct.3–6, 1979, Venice, Italy, Proc. p.150.

[3.299] *Behr, S., Rohde, E.:* Vernetzungssysteme für die Hochtemperatur-Vulkanisation von Polychloropren. KGK *32* (1979), p.492.

[3.300] *Eholzer, U., Kempermann, Th.:* Ein neuer Polychloroprenbeschleuniger als Ersatzprodukt für Ethylenthioharnstoff. KGK *33* (1980), p.696.

[3.301] *Fath, M.A., de Rudder, J.:* Die jüngste Entwicklung in den Vulkanisationssystemen für NBR, EPDM und CR. KGK *32* (1979), p.582.

[3.302] *Kato, H., Fujita, H.:* Development of Synergistic Curing Systems for Polychloroprene. RCT *55* (1982), p.949.

[3.303] *Kovacic, P.:* Neoprene-Vulcanization. Ind. Engng. Chem. *47* (1955), p.1090.

[3.304] *Sklernarz, R.:* Neuer Vulkanisationsbeschleuniger für CR. GAK *33* (1980), p.224.

Processing

[3.305] *Baument, J.C.:* Neoprene Processing. Rubber J. *146* (1964), 2, p.34, 3, p.50.

[3.306] *Murray, R.M., Thompson, D.C.:* Die Neoprene. Publ.: DuPont, 1965.

Flame Resistance and Combustion Products

[3.307] *Einbrodt, H.J., Hesse, H.:* Über die Toxizität der Brand- und Schwelgase bei Kabelbränden. GAK *36* (1983), p.648; *37* (1984), p.65.

[3.308] *Glaerer, J.:* Flammhemmung und Bedingungen der Feuerausbreitung bei elektrischen Kabeln. SRC '83, May 19–20, 1983, Bergen, Norway.

[3.309] *Fabris, H.J., Sommer, J.G.:* Flammability of Elastomeric Materials. RCT *50* (1977), Review.

[3.310] *Rohde, E.:* Isolier- und Mantelwerkstoffe aus Elastomeren. In: Kabel- und isolierte Leitungen. VDI-Verlag, Düsseldorf, 1984, pp. 33–60.

Properties

[3.311] *Bhowmick, A.K., Geni, N.N.:* Strength of Neoprene Compounds and the Effect of Salt Solutions. RCT *56* (1983), p. 845.

[3.312] *Hamed, G.R., Liu, F.-C.:* The Bonding of Polychloroprene to Brass, Rate and Temperature Effects. RCT *57* (1984), p. 1036.

[3.313] *Handvet, C., Morin, M.:* Influence de la température sur les propriétés du polychloroprène. Rev. Gén. Caoutch. *42* (1965), p. 395.

[3.314] *Madhoud, M., Abdel-Barry, E.M., Laundy, S.N., Darwish, N.A.:* Swelling and Stress-Strain Properties of Filled Neoprene Rubber. Paper 39, 118th ACS-Meeting, Rubber-Div., Oct. 7–10, 1980, Detroit, MI.

[3.315] *Menges, G., Haack, W.:* Vorausberechnung der mechanischen Eigenschaften von Natur- und Chloroprenkautschuken sowie Ermittlung des Verformungsverhaltens von Elastomerbauteilen. KGK *36* (1983), p. 973.

[3.316] *Subramaniam, N.S., Pinto, J.G., Hirsch, A.E.:* Dynamic Properties of Polychloroprene Elastomer. Paper 32, 117th ACS-Meeting, Rubber-Div., May 20–23, 1980, Las Vegas, NV; KGK *35* (1982), p. 23.

Uses

[3.317] *Mirza, J.:* Polychloroprene in Power Transmission Belting. Paper 71, 125th ACS-Meeting, Rubber-Div., May 8–11, 1984, Indianapolis, IN; KGK *38* (1985), p. 261.

[3.318] *Rohde, E.:* Aufbau dynamisch hochbelastbarer Polychloropren-Vulkanisate. KGK *34* (1981), p. 466.

Special References

[3.319] *Carothers, W.H., et al.:* J. Am. Chem. Soc. *53* (1931), p. 4203.

[3.320] *Michel, W.E.:* J. Polym. Sci. *8* (1952), p. 583.

[3.321] *Klebaanskii, A.L., et al.:* J. Polym. Sci. *30* (1958), p. 763.

[3.321a] *Göbel, W.:* 25 Jahre Baypren – ein historischer Abriß. KGK *35* (1982), p. 942.

[3.321b] *Rohde, E.:* 25 Jahre Baypren – Der Vielzweckspezialkautschuk und seine Anwendung im Gummi- und Kabelsektor. KGK *35* (1982), p. 946.

[3.321c] *Hummel, K., Siapkas, S.:* Radikalische Vernetzung von Mischungen aus Polychloropren und Poly(3-hydroxybuttersäure). KGK *34* (1981), p. 549.

3.7.4.5 References on IR

General Overview (see also Refs. [3.648, 3.649])

[3.322] *Schönberg, E., Marsh, H.E., Walters, S.J., Saleman, W.M.:* Polyisoprene. RCT *52* (1979), Review.

[3.323] *Osterhof, H.J.:* Isoprene und Polyisoprene Products. Trans. IRI *41* (1965), 2, p. T53.

Synthesis and Structure

[3.323a] *Al-Jarray, M.M.F., Al-Madfai, S.F.F.:* Synthesis and Characterization of New Isoprene Copolymers. Paper 85, 128th ACS-Meeting, Rubber-Div., Oct. 1–4, 1985, Cleveland, OH.

[3.324] *François, B., et al.:* Étude de la polymerization stéréospecifique de l'isoprene par les composés organolithiens. J. Polym. Sci. *4C* (1963), p. 375.

[3.325] *Hoffmann, M., Unbehend, M.:* Vulkanisationsstruktur, Relaxation und Reißfestigkeit von Polyisoprenen. Makromol. Chem. *58* (1962), p. 104.

[3.326] *Kuntz, I.:* Polymerization of Isoprene with n-, iso-, sec.- and tert.-Butyllithium. J. Polym. Sci. *2A* (1964), p. 2827.

[3.327] *Sato, H., Tanaka, Y.:* ^{1}H-NMR Study of Polyisoprenes. RCT *53* (1980), p. 305.

[3.328] *Wang, F., Liu, Y., Zhang, X.:* Structure and Properties of Polyisoprene Synthesized with Rare Earth Based Catalyst Systems. IRC '85, Oct. 15–18, 1985, Kyoto, Proc. p. 27.

[3.329] *Wasfold, D.J., Bywater, S.:* Anionic Polymerization of Isoprene. RCT *38* (1965), p. 627.

Properties

[3.330] *Bhowmick, A.K., Kuo, C.C., Manzur, A., MacArthur, A., McIntyre, D.:* Properties of Cis- and Trans-Isoprene Blend. Paper 59, 125th ACS-Meeting, Rubber-Div., May 8–11, 1984, Indianapolis, IN.

[3.331] *Houskus, J.:* Modification and Dielectric Properties of Polyisoprene. RCT *56* (1983) p. 718.

[3.332] *Kow, C., Hadjichristidis, N., Morton, M., Fetters, L.J.:* Glass Transition Behavior of Polyisoprene: The Influence of Molecular Weight, Terminal Hydroxy Groups, Microstructure and Chain Branching. RCT *55* (1982), p. 245.

[3.333] *Kuo, C.C., McIntyre, D.:* Blends of cis-Polyisoprene with Trans-Polyisoprene. Paper 43, 118th ACS-Meeting, Rubber-Div., Oct. 7–10, 1980, Detroit, MI.

[3.334] *Widmaier, J.M., Meyer, G.C.:* Glass Transition Temperature of Anionic Polyisoprene. RCT *54* (1981), p. 940.

[3.335] *Yano, S.:* Changes in the Dynamic Modulus During Thermal Degradation of Polyisoprene Vulcanizates. RCT *53* (1980), p. 944.

[3.336] *Yano, S.:* Photo-Oxidation of an IR-Vulcanizate. RCT *54* (1981), p. 1.

Special References

[3.337] DRP 250690, 1909, Farb.fabr. vorm. Friedr. Bayer Co.

[3.338] US 3114743, 1954, Goodrich Gulf.

[3.339] US 3285901, 1955, Firestone.

[3.340] US 3208988, 1955, Firestone.

[3.341] *Harries, C.D.:* Ann. *383* (1911), p. 213; *395* (1912), p. 220. Investigations on the natural and synthetic rubber types. Berlin, 1919.

[3.342] FR 1139418, 1955, Goodrich Gulf.

[3.343] DOS 1720772, 1968, Bayer.

[3.344] US 3047559, 1956, Goodyear.

[3.345] GB 870010, 1958, Goodrich Gulf.

[3.346] GB 880998, 1957, Dunlop.

[3.347] GB 992189, 1023853, 1963, Goodyear.

[3.348] US 3467641, 1963, Goodyear.

[3.349] DT 2720752, 2720763, 1967, Bayer.

[3.350] *Scott, K. W., et al.:* Rubber Plast. Age *42* (1961), p. 175.

[3.351] *Bruzzone, M. et al,:* Rubber Plast. Age *46* (1965), p. 278.

3.7.4.6 References on IIR

General Overview

[3.352] *Hofmann, W.:* Butylkautschuk und andere elastische Vinylpolymerisate. In: *Boström, S. (Ed.):* Kautschuk-Handbuch, Vol. 1. Verlag Berliner Union, Stuttgart, 1959, p. 395.

[3.353] *Kennedy, J.P.:* Polyisobutylene and Butyl Rubber. In: *Kennedy, J.P., Törnquist, E.G.M. (Eds.):* Polymer Chemistry of Synthetic Elastomers, Vol. 1. Interscience Publ., New York, 1968, p. 291.

[3.354] *Waddell, H.H., et al.:* Highly Unsaturated Butyl Rubber. Rubber World *196* (1962) 5, p. 57.

[3.355] *Wilson, G.J.:* Butyl and Halobutyl Rubbers. Paper at 124th ACS-Meeting, Rubber-Div., Oct.25–28, 1983, Houston, TX.

[3.356] *Zapp, R.L., Hons, P.:* Butyl and Chlorobutyl Rubber. In: *Morton, M. (Ed.):* Rubber Technology, 2nd Ed. Van Nostrand Reinold Publ., New York, 1973, p.249.

Compounding

[3.357] *Ford, W.E., et al.:* Effect of Carbon Black on Butyl Rubber. Rubber Age *94* (1964), p.738.

[3.358] *Gessler, R.M.:* The Reinforcement of Butyl with Carbon Black. Rubber Age *94* (1964), p.598.

[3.359] *Laue, E.W.:* Der Aufbau von Mischungen auf Basis von Butyl- und Chlor-Butyl-Kautschuk. KGK *22* (1969), pp.565, 648.

Vulcanization

[3.360] *Gau, L.M., Chew, C.H.:* Quinoid Curing of Butyl and Natural Rubbers. RCT *56* (1983), p.883.

[3.361] *Ghatge, N.D., Maldar, N.N.:* Vulcanization of Butyl Rubber: Curative Effects of 2-Pentadecylbenzoquinonedioxine. RCT *54* (1981), p.692.

[3.362] *Smith, W.C.:* in: *Alliger, G., Sjothum, I.J. (Eds.):* Vulcanization of Elastomers (Chapter 7). Van Nostrand Reinold Publ., New York, 1964.

[3.363] *Tarney, P.O., et al.:* The Vulcanization of Butyl Rubber with Phenole Formaldehyde Derivatives. RCT *33* (1959), p.229.

Properties

[3.364] *Haxo jr., H.F.:* Permeability of Polymeric Membrane Liners. Paper at 126th ACS-Meeting, Rubber-Div., Oct.23–26, 1984, Denver, CO.

[3.365] *Ferry, J.D., Fitzgerald, E.R.:* Dynamic Mechanical Properties of a Carbon Black-Loaded Butyl-Rubber Vulcanizate and a Carbon Black-Loaded Polyisobutylene. RCT *55* (1982), p.1403.

[3.366] *Khastgir, D., Ghoshal, PK., Das, C.K.:* Dielectric Behaviour and the Peroxide Vulcanization of the Cross-linked Butyl (XL-50) Rubber. KGK *36* (1983), p.277.

[3.367] *Trexler, H.E.:* The Permeability Resistance of Polymers to Air Conditioning Refrigerants and Water. RCT *56* (1983), p.105.

[3.368] *Wilkes, G.L., Bagrodia, S., Taut, M., Kennedy, J.P.:* Properties of New Polyisobutylene Based Elastomers. Paper 3, 127th ACS-Meeting, Rubber-Div., Apr.23–26, 1985, Los Angeles, CA.

[3.369] *Wilkes, G.L.:* Properties of New Polyisobutylene Based Ionomers. Gordon Res. Conf. of Elastomers, July 16–20, 1984, New London, NH.

Uses

[3.370] *Booth, P.A., et al.:* Selection of Butyl Compounds for Antivibration. Rubber Plastics Age *46* (1965), p.173.

[3.371] *von Hellens, W.:* Inner Liners for High Performance Tires. Paper 50, 125th ACS-Meeting, Rubber-Div., May 8–11, 1984, Indianapolis, IN.

[3.372] *Fusco, J.V., Gardner, I.J., Gursky, L.:* Conjugated Diene Butyl as an Elastomeric Additive in Fiber Glass Reinforced Polyester Composites. Paper 45, 118th ACS-Meeting, Rubber-Div., Oct.7–10, 1980, Detroit, MI.

[3.373] *Kumbhani, K.J.:* Specialty Applications of Butyl Rubber. Paper at 127th ACS-Meeting, Rubber-Div., Apr.23–26, 1985, Los Angeles, CA.

[3.374] *Smith, W.S.:* Butyl – The Original Water Saver Elastomer. Paper at 117th ACS-Meeting, Rubber-Div., May 20–23, 1980, Las Vegas, NV.

[3.375] *Strong, A.G.:* The Deterioration of Rubber and Plastics Linings on Outdoor Exposure – Factors Influencing Their Longlivity. Paper at 117th ACS-Meeting, Rubber-Div., May 20–23, 1980, Las Vegas, NV.

[3.376] *Strong, A. G.:* Features of Adhesive Jointing Systems for Rubber Roofing Based on Butyl, EPDM or Blends. Paper 60, 121st ACS-Meeting, Rubber-Div., May 4–7, 1982, Philadelphia, PA.

[3.377] *Strong, A. G.:* Performance Aspects of Polymer Membranes with Long Term Exposure. SRC '83, May 19–20, 1983, Bergen, Norway.

Special References

[3.378] *Thomas, K.:* India Rubber World *130* (1954), p. 203.

[3.379] USP 2356128, 1937, Jasco Inc.

[3.380] *Cesca, S., et al.:* RCT *49* (1976), p. 937.

[3.381] *Kennedy, J. P., Gillham, J. K.:* Fortschr. Hochpolym. Forsch. *10* (1972), p. 1.

[3.382] *v. Amerongen, G. J.:* J. Polym. Sci. *5* (1950), p. 307.

3.7.4.7 References on CIIR and BIIR

General Overview

[3.383] *Baldwin, F. P., et al.:* Rubber Plast. Age *42* (1961), p. 500.

[3.384] *Blackshaw, G. C.:* Bromobutyl Rubber. In: *Babbit, R. O. (Ed.):* The Vanderbilt Handbook. Publ.: Vanderbilt, 1978, pp. 102–132.

[3.385] *Condon, J. R.:* Chlorobutyl Rubber. In: *Babbit, R. O. (Ed.):* The Vanderbilt Handbook. Publ.: Vanderbilt, 1978, pp. 133–136.

[3.386] *Fusco, J. V.:* Chlorobutyl Development. Paper 13, 121st ACS-Meeting, Rubber-Div., May 4–7, 1982, Philadelphia, PA.

[3.387] *Hopkins, H., Jones, R. H., Walker, J.:* Bromobutyl and Chlorobutyl: A Comparison of their Chemistry, Properties and Uses. IRC '85, Oct. 15–18, 1985, Kyoto, Proc. p. 205.

Synthesis

[3.388] *van Tongerloo, A., Vucov, R.:* Butyl Rubber-Halogenation Mechanisms. Paper at IRC '79, Oct. 3–6, 1979, Venice, Italy, Proc. p. 70.

[3.389] *Walker, J., et al.:* Rubber Age *108* (1976), p. 33.

Compounding

[3.390] *Banks, S. A., et al.:* Compounds of Chlorobutyl with other Rubbers for Transportation Applications. Rubber Age *94* (1964), p. 923.

[3.391] *Dudley, R. H., Wallace, A. J.:* Compounding of Chlorobutyl Rubber for Heat Resistance. Rubber World *152* (1965) 2, p. 66.

[3.392] *Timar, J., Edwards, W. S.:* Compounding Bromobutyl for Heat Resistance. RCT *52* (1979), p. 319.

Vulcanization

[3.393] *Kuntz, I., Zapp, R. L., Pancirov, R. J.:* The Chemistry of the Zinc Oxide Cure of Halobutyl. RCT *57* (1984), p. 813.

[3.394] *Vucov, R.:* The Halogenation of Butyl Rubber and the Zinc Oxide Crosslinking Chemistry and Halogenated Derivatives. RCT *57* (1984), pp. 275, 284.

[3.395] *Vucov, R., Wilson, G. J.:* Crosslinking Efficiencies of Some Halobutyl Curing Reactions. Paper 73, 126th ACS-Meeting, Rubber-Div., Oct. 23–26, 1984, Denver, CO.

[3.396] *Yamada, A., Shiokaramatsu, Y., Kohjiya, S.:* Moisture-Curable Rubber Preapared from Halogen Containing Polymers and 3-Aminopropyltriethoxysilane. IRC '85, Oct. 15–18, Kyoto, Proc. p. 268.

Properties and Uses

[3.397] *Varrell, D.:* Aspects of Halobutyl Innerliner Performance and Technology. Hungarian Rubber Conference, Oct. 1982, Budapest, Hungary.

[3.398] *Young, D. G., Doyle, M. C.:* Fatigue Crack Propagation on Halobutyl Elastomer Blends for Tire Sidewalls. IRC '85, Oct. 15–18, 1985, Kyoto, Proc. p. 469.

3.7.4.8 References on EPM and EPDM

General Overview

[3.399] *Baldwin, F. P., VerStrate, G.:* Polyolefin Elastomers, Based on Ethylene and Propylene. RCT *45* (1972), pp. 709–781.

[3.400] *Hofmann, W.:* Vernetzungsmittel in Ethylen-Propylenkautschuk. KGK *40* (1987), pp. 308–332 (see also Ref. [3.60]).

[3.401] *Natta, G., et al.:* Ethylene-Propylene-Rubbers. In: *Kennedy, J. R., Törnquist, E. G. M. (Eds.):* Polymer Chemistry of Synthetic Elastomers, Vol. 2. Interscience Publ., New York, 1969, p. 679.

[3.402] *Samuels, M. E., Wirth, K. H.:* Propylene Rubbers. In: *Babbit, R. O. (Ed.):* The Vanderbilt Handbook. Publ.: Vanderbilt, 1978, pp. 147–168.

Synthesis

[3.403] *Corbelli, L., Milani, F., Fabri, R.:* New Ethylene-Propylene Elastomers Produced with Highly Active Catalysts. KGK *34* (1981), p. 11.

[3.404] *Balli, P.:* High Yield Catalyst for Polyolefins and EPM/EPDM Rubber Synthesis: A New Era in Process and Product Development. Gordon Res. Conf. of Elastomers, July 18–22, 1983, New London, NH.

[3.405] *Kaminsky, W.:* Elastomers by Copolymerization of α-Olefins and Diolefins with Soluble Ziegler Catalyst. Gordon Res. Conf. of Elastomers, July 19–23, 1982, New London, NH.

Structure

[3.406] *Beardslay, K. P., Ho, C. G.:* Rheological Properties as Related to Structure for EPDM-Polymers. Paper 75, 122nd ACS-Meeting, Rubber Div., Oct. 5–7, 1982, Chicago, MI.

[3.407] *Gilbert, R. D., Theil, M. H., Fornes, R. E.:* Ethylene Propylene Block Copolymers. Paper 32, 123rd ACS-Meeting, Rubber-Div., May 10–12, 1983, Toronto, Ontario.

[3.408] *Kresge, E. N., Cozewith, C., VerStrate, G.:* Long Chain Branching and Gel in EPDM. Paper 12, 125th ACS-Meeting, Rubber-Div., May 8–11, 1984, Indianapolis, IN.

[3.409] *Kühne, J. K., Kautt, J.:* Verzweigung langer Polymerketten und Gelbildung in Ethylen-Propylen-Elastomeren. KGK *37* (1984), p. 101.

[3.410] *Mark, J. E., Riande, E., Scholtens, B. J. R.:* Crystallization in Stretched and Unstretched EPDM Elastomers. Paper 34, 123rd ACS-Meeting, Rubber-Div., May 10–12, 1983, Toronto, Ontario.

[3.411] *Scholtens, B. J. R.:* The Effect of Variations in Chain Connectivity on Strain Induced Crystallization in Unvulcanized EPDM Elastomers. RCT *57* (1984), p. 703.

[3.412] *Starkweather, H. W.:* Order and Disorder in Copolymers of Ethylene and Propylene. Paper 57, 123rd ACS-Meeting, Rubber-Div., May 10–13, 1983, Toronto, Ontario.

Blends

[3.413] *Coran, A. Y., Patel, R.:* Rubber-Thermoplastic Compositions. I. EPDM-Polypropylene Thermoplastic Vulcanizates. RCT *53* (1980), p. 141.

[3.413a] *Coran, A. Y.:* Blends of Dissimilar Rubbers – Cure Rate Incompability. Rubbercon '87, June 1–5, 1987, Harrogate, GB, Proceed. A32.

[3.414] *Coulthard, D. C., Ritchie, K., Walker, J.:* Blending in Specialties for Specific Improvements. Europ. Rubber J., June 1977, p. 28.

[3.415] *Das, C. K.:* Effect of E/P Ratio on the Rheological Behavior of the Crosslinked Polyethylene and EPDM-Blends. IRC '85, Oct. 15–18, 1985, Kyoto, Proc. p. 419.

[3.416] *Das, C. K.:* Effects of Sulfur on the Peroxide Cure of NBR/EPDM Blends. KGK *35* (1982), p. 753.

[3.417] *Hamed, G. R.:* Morphology-Property Relationships for EPDM-Polybutadiene Blends. RCT *55* (1982), p. 151.

[3.418] *O'Mahoney, J. F.:* EPDM in Polymer Blends (EPDM/IR). Rubber Age, March 1970, p. 47.

[3.419] *Mitchell, J.M.:* Nitrile-EPDM-Blends, Fundamental Considerations for Development of Key Properties. J. Elastomers Plast. *9* (1977), p.329.
[3.420] *Mitchell, J.M.:* Verschnitte aus Nitrilkautschuk und EPDM. GAK *30* (1977), p.498,
[3.421] *Mitra, A., Millns, W., Das, C.K.:* Effect of Ethylene Propylene Ratio in the Peroxide Cure of NBR/EPDM Blends in the Presence of Sulfur. KGK *37* (1984), p.862.
[3.422] *Palla, H., Schnetger, J., Hochwald, H.:* Zur Wirkungsweise von Ethylen-Dien-Terpolymeren in Verschnitten mit Isoprenkautschuk bei Ozonangriff. Bayer-Mitt. f. d. Gummiind. *50* (1978), p.13.
[3.423] *Rehner, D., Wei, P.E.:* Heterogenität und Vernetzung von Kautschukverschnitten. Paper at IRC '65, Munich, West Germany, Proc. (1967), p.121; RCT *42* (1969), p.985.
[3.424] *Woods, M.E., Mass, T.R.:* in: *Platzer, N.A.J. (Ed.):* Copolymers, Polyblends and Composites. Adv. Chem. Ser. 142, Publ.: ACS, Washington, D.C., 1975, p.386.
[3.425] *Woods, M.E., Davidson, J.A.:* Fundamental Considerations for the Covulcanization of Elastomer Blends. II. Lead Oxide-Activated Cures of NBR-EPDM Blends. RCT *49* (1976), p.112.

Compounding and Processing (see also Ref. [3.114a])

[3.426] *Keller, R.C.:* Advances in Ethylene-Propylene Elastomer Compounding for Hose Applications. Paper 36, 122nd ACS-Meeting, Rubber-Div., Oct.5–7, 1982, Chicago, MI.
[3.427] *Shiga, S., Futura, M.:* Processibility of EPR in an Internal Mixer. II. Morphological Changes of Carbon Black Agglomerates During Mixing. RCT *58* (1985), p.1.
[3.427a] *Nordermeer, J.W.M.:* Verarbeitbarkeit von EPDM in Abhängigkeit von der Molekülstruktur. DKG-Meeting, Oct.23, 1987, Würzburg, West Germany.
[3.427b] *Nordermeer, J.W.M., Wilms, M.:* Processibility of EPDM in Internal Mixers. IRC '86, June 2–6, 1986, Göteborg, Sweden.
[3.428] *Wang, C.S.:* Processing Parameters of Continuous Microwave Heating of Ethylene Propylene Terpolymer. RCT *57* (1984), p.134.

Vulcanization

[3.429] *Baldwin, F.P.:* The Influence of Accelerators on EPDM Vulcanizate Structure. RCT *45* (1972), pp.1348–1365.
[3.430] *van den Berg, J.H.M., et al.:* Model Vulcanization of EPDM Compounds. I. Structure Determination of Vulcanization Products from Ethylidene Norbornane. RCT *57* (1984), p.265; II. Influence of Temperature and Time on the Vulcanization Products for Ethylidene Norbornane. RCT *57* (1984), p.725; III. Influence of Vulcanization System on the Vulcanization Products from Ethylidene Norbornane. RCT *58* (1985), p.58.
[3.431] *Booß, H.J.:* Zur peroxidischen Vernetzung von Ethylenhomo- und -copolymeren in Gegenwart von Antioxidantien und Füllstoffen. KGK *37* (1984), p.207.
[3.431a] *Dunn, J.R.:* Compounding Ethylene-Propylene Elastomers for High Temperature Aging Resistance. Rubbercon '87, June 1–5, 1987, Harrogate, GB, Proceed. B 24.
[3.432] *Gentschera, P., Teutscher, H.:* Flüssiges Dibrompolybutadien als Beschleuniger der Vulkanisation von EPDM. GAK *34* (1981), p.316.
[3.433] *Drake, R.E.:* Liquid 1.2-Polybutadiene Resins as Co-Agents for EPDM. Paper 1, 118th ACS-Meeting, Rubber-Div., Oct.7–10, 1980, Detroit, MI.
[3.434] *Hofmann, W.:* New Highly-Efficient Non-Blooming Accelerator Systems for Sulfure Cure of EPDM. IRC '85, Oct.15–18, 1985, Kyoto, Proc. p.44; GAK *39* (1986), pp.422–429.
[3.435] *Lederer, I.D.A., Kear, K.E., Kuhls, G.H.:* Diffusion of Curatives. RCT *55* (1982), p.1482.
[3.436] *van Meerbeek, A.:* Development of Curing Systems for Low Compression Set in EPDM. KGK *36* (1983), p.1062.
[3.437] *Recchnite, A.D., Dimeler, G.R.:* Petroleum Extender Oils for Reduced Peroxide Consumption in Compounded EPDM. Paper 36, 123rd ACS-Meeting, Rubber-Div., May 10–12, 1983, Toronto, Ontario.

[3.438] *Stella, G.:* EPDM Cure Systems for Continuous Vulcanization. KGK *34* (1981), p. 357.
[3.438a] *Stella, G.:* The Evolving Technology of EPDM. Rubbercon '87, June 1-5, 1987, Harrogate, GB, Proceed. B 33
[3.439] *Vidal, A.:* EPDM-Photocrosslinking and Characterization of Model-Networks. Gordon Res. Conf. of Elastomers, July 18-22, 1983, New London, NH.
[3.440] *Voigt, H. U.:* Über das Vernetzen von Polyolefinen. KGK *34* (1981), p. 197.

Properties and Uses

[3.441] *Allen, R. D.:* Improving the Heat Resistance of EPDM. Paper 65, 120th ACS-Meeting, Rubber-Div., Oct. 13-16, 1981, Cleveland, OH.
[3.442] *Allen, R. D.:* EPDM-Qualitäten mit verbessertem Einsatzverhalten bei hohen Temperaturen. GAK *36* (1983), p. 534.
[3.443] *Cassedy, P. E., Aminabhavi, T. M., Brunson, J. C.:* Water Permeation through Elastomer Laminates. I. Neoprene-EPDM. RCT *56* (1983), p. 357.
[3.444] *Crepeau, A. E.:* EPDM Flashing-Single Ply Membrane Roofing. Paper 61, 121st ACS-Meeting, Rubber-Div., May 4-7, 1982, Philadelphia, PA.
[3.445] *Fithian, L. E.:* EPDM Waterproofing Membrane - 15 Years Later. Paper 77, 126th ACS-Meeting, Rubber-Div., Oct. 23-26, 1984, Denver, CO.
[3.445a] *van Gunst, C. A., Paulen, H. J. G., Wolters, E.:* Eigenklebrigkeit von EPDM-Mischungen. KGK *28* (1975), pp. 714-720.
[3.446] *Hemmer, E., Jesse, H.:* Untersuchungen über das thermische Alterungsverhalten von ausgewählten Isolierstoffen für Kabel und Leitungen. GAK *34* (1981), pp. 64, 386.
[3.447] *Kato, H., Adachi, H., Fujita, H.:* Innovation in Flame- und Thermal-Resistive EPDM Formulations. RCT *56* (1983), p. 287.
[3.448] *Keller, R. C.:* Ethylene-Propylene Elastomers for Specialty Applications. Paper at 127th ACS-Meeting, Rubber-Div., Apr. 23-26, 1985, Los Angeles, CA.
[3.449] *Lal, J., Sandstrom, P. H., Senyek, M. L.:* High Flex, Ozone Resistant Polymers of High α-Olefins. Paper 4, 127th ACS-Meeting, Rubber-Div., Apr. 23-26, 1985, Los Angeles, CA.
[3.450] *Leibu, H. J.:* Adhesion of EPDM to Various Substrates with Emphasis on Textile Fibers. Paper at 119th ACS-Meeting, Rubber-Div., June 2-5, 1981, Minneapolis, MN.
[3.450a] *Pizzo, A., Milani, F.:* Ethylene-Propylene Elastomers in the Production of Composite Seals for the Automotive Industry. Paper 20, 128th ACS-Meeting, Rubber-Div., Oct. 1-4, 1985, Cleveland, OH.
[3.450b] *Shulman, C. B.:* Unique Features of High Unsaturated EPDM Polymers. Paper 90, 128th ACS-Meeting, Rubber-Div., Oct. 1-4, 1985, Cleveland, OH.
[3.450c] *Endstra, W., Seeberger, D.:* Vernetzung von Polymeren-Beschleuniger oder Peroxide? DKG-Meeting, Nov. 7, 1985, Hamburg, KGK *39* (1986), pp. 929-935.
[3.451] *Spenadel, L.:* Heat Aging Performance of Ethylene-Propylene Elastomers in Electrical Insulation Compounds. RCT *56* (1983), p. 113.
[3.452] *Strong, A. G.:* An Assessment of Key Parameters for Polymeric Roofing and a New Approach to Compounding EPDM for this Application. Paper 82, 124th ACS-Meeting, Rubber-Div., Oct. 25-28, 1983, Houston, TX.

Special References

[3.453] USP 3300459, 1955, Montecatini.
[3.454] *Kerrut, G.:* KGK *26* (1973), p. 341.
[3.455] *Eholzer, U., Kempermann, Th., Morche, K.:* Schnellheizende Vulkanisationssysteme für EN-EPDM-Kautschuk. GAK *28* (1975), p. 646.
[3.456] *Haverland, A., Hofmann, A.:* Ausgewählte Verarbeitungshilfsmittel in EPDM. DOG-Kontakt *29*, Publ.: DOG Deutsche Oelfabrik, Hamburg, 1985.

3.7.4.9 References on EVM

[3.457] *El Aasser, M. S., Vanderhoff, J. W.:* Emulsion Polymerization of Vinylacetate. Elsevier Appl. Sci. Publ., Barking, 1981.
[3.457a] *Alberts, H., Bartl, H., Kuhn, R., Morbitzer, L.:* Copfropfpolymere aus Celluloseestern und Ethylen-Vinylacetat-Copolymeren. KGK *38* (1985), p. 689.
[3.458] *Bartl, H., Peter, J.:* Kautschuk u. Gummi *14* (1961), p. WT 23.
[3.459] DAS 1126614, 1957, Bayer.
[3.460] FR 1225704, 1959, Bayer.
[3.461] *Goebel, W., et al.:* Kunststoffe *55* (1964), p. 329.
[3.462] *Bartl, H., Hardt, D.:* Angew. Chem. *77* (1965), p. 512.
[3.463] US 3358054, 1962, Bayer.
[3.464] *Wardig, C.:* Halogenfreier Kabelmantel mit gutem Brandschutzverhalten auf Basis Levapren 500. KGK *35* (1982), p. 115.

3.7.4.10 References on CM

[3.465] *Barnes, C., Sylvest, R. T.:* Vulcanization of Chlorinated Polyethylene without Peroxides. Paper 2, 118th ACS-Meeting, Rubber-Div., Oct. 7–10, 1980, Detroit, MI; GAK *36* (1983), pp. 150, 290.
[3.465a] *Bügel, H., Rohde, E., Wardig, G.:* Chloriertes Polyethylen und sein Einsatz in der Kabelindustrie. KGK *34* (1981), p. 551.
[3.466] *Davis jr., W. H., Flynn, J. H.:* CPE-Thiadizole Cure System Studies – Chemistry and Dispersion. Paper 24, 127th ACS-Meeting, Rubber-Div., Apr. 23–26, 1985, Los Angeles, CA.
[3.466a] Laborbericht Nr. 4278, January 1988, Publ.: DOG Deutsche Oelfabrik, Hamburg.
[3.467] *Johnson, J. B.:* Chlorinated Polyethylene Rubber. In: *Babbit, R. O. (Ed.):* The Vanderbilt Handbook. Publ.: Vanderbilt, 1978, pp. 295–299.
[3.468] *Mori, K., Nakamura, Y.:* Crosslinking of Halogen-Containing Rubbers with Triazine Thiols. RCT *57* (1984), p. 34.
[3.469] *Oates, W. G., Richards, R. B.:* Trans. Faraday Soc. *42A* (1946), p. 197.
[3.470] *Rohde, E.:* Chlorinated Polyethylene, Advantages on its Application in the Elastomer Sector. IRC '79, Oct. 3–6, 1979, Venice, Italy, Proc. p. 254.
[3.471] *Rohde, E.:* Bayer CM-Eigenschaften, Grundzüge des Mischungsaufbaues und Anwendungen. KGK *32* (1979), p. 304.
[3.472] *Rohde, E.:* Chloriertes Polyethylen für technische Gummiwaren. KGK *35* (1982), p. 478.
[3.473] *Rose, J. C., Coffey, R. J.:* Chlorinated Polyethylene for Wire and Cable. Paper 51, 121st ACS-Meeting, Rubber-Div., May 4–7, 1982, Philadelphia, PA.
[3.474] *Sylvest, R. T., Barnes, C., Warren, N. E., Worsley, W. R.:* Vulcanization of Chlorinated Polyethylene. Paper 15, 120th ACS-Meeting, Rubber-Div., Oct. 13–16, 1981, Cleveland, OH.

3.7.4.11 References on CSM

[3.475] *Baseden, G. A.:* A New Easy Processing Chlorosulfonated Polyethylene. Paper 94, 124th ACS-Meeting, Rubber-Div., Oct. 25–28, 1983, Houston, TX.
[3.476] *Dupuis, I. C.:* Chlorosulfonated Polyethylene. In: *Babbit, R. O. (Ed.):* The Vanderbilt Handbook. Publ.: Vanderbilt, 1978, pp. 300–307.
[3.477] *Dupuis, I. C.:* Composition of Hypalon Chlorosulfonated Polyethylene Synthetic Rubber for Higher Heat, Oil and Set Resistance. IRC '82 (Eurocaoutchouc), June 2–4, 1982, Paris; KGK *36* (1983), p. 353.
[3.478] *Guggenberger, S., Baseden, G. A.:* Hypalon Synthetic Rubber – A Specialty Elastomer with High Performance Properties. Paper at 127th ACS-Meeting, Rubber-Div., Apr. 23–26, 1985, Los Angeles, CA.
[3.479] *Koga, M., Fujii, S., Ishizuka, Y., Tabata, I.:* New Types of Chlorosulfonated Elastomers Having Improved Low Temperature and Dynamic Properties. IRC '85, Oct. 15–18, 1985, Kyoto, Proc. p. 796.

[3.480] *Maynard, J. T., Johnson, P. R.:* RCT *36* (1963), pp. 963–974.
[3.481] *Schlicht, R.:* Kautschuk u. Gummi *10* (1957), p. WT 66.
[3.482] *Scott, L. K.:* in: Physical Chemistry, College Outline Series, Chapt. 15, pp. 159–164.
[3.483] *Smook, M. A., et al.:* India Rubber Wld. *123* (1953), p. 348.
[3.484] *Stevenson, A. L.:* in: *Alliger, G., Sjothun, I. J. (Eds.):* Vulcanization of Elastomers. Van Nostrand Reinold Publ., New York, 1964, pp. 273–279.
[3.485] *Viadya, U. I.:* New Developments in Peroxide Cured Compositions Based on Hypalon Chlorosulfonated Polyethylene. Paper 53, 121st ACS-Meeting, Rubber-Div., May 4–7, 1982, Philadelphia, PA; GAK *38* (1985), p. 308.
[3.486] *Williams, J. E. A.:* Crosslinking Systems for Hypalon Chlorosulfonated Polyethylene (CSM). IRC '82 (Eurocaoutchouc), June 2–4, 1982, Paris.
[3.487] USP 2046090, 1933, Ch. L. Horn.

3.7.4.12 References on ACM and EAM

General Overview on ACM

[3.488] *Miles, D. C., Briston, J. H.:* Polymer Technology. Temple Press Books, London, 1965, pp. 216–227.
[3.489] *Morrill, J. P.:* Nitrile and Polyacrylate Rubber. In: *Morton, M. (Ed.):* Rubber Technology. Van Nostrand Reinold Co., New York, 1973, pp. 302–321.
[3.490] *Nielsen, L. E.:* Mechanical Properties of Polymers. Van Nostrand Reinold Co., New York, 1962.
[3.491] *Tucker, H. A., Jorgensen, A. H.:* Acrylic Elastomers. In: *Kennedy, J. P., Törnquist, E. G. M. (Eds.):* Polymer Chemistry of Synthetic Elastomers, Vol. 1. Interscience Publ., New York, 1968, p. 253.
[3.492] *Wolf, R. F., de Marco, R. D.:* Polyacrylic Rubber. In: *Babbit, R. O. (Ed.):* The Vanderbilt Handbook. Publ.: Vanderbilt, 1978, p. 188–206.
[3.493] *Beaulin, A. H., Bartmann, B., Sparks, W. J.:* Recent Advances in Acrylic Hot Melt Pressure Sensitive Adhesion, Technology. Paper 4, 126th ACS-Meeting, Rubber-Div., Oct. 23–26, 1984, Denver, CO.
[3.494] *Bleyie, P.:* Fortschritte in der Polyacrylattechnologie. GAK *39* (1977), p. 528.
[3.495] *Eldred, R. J.:* Plastizication by in Situ Grafted Acrylates. II. Effect of Graft Structures. RCT *58* (1985), p. 146.
[3.496] *Enyo, H., Zen, S., Takemura, Y.:* A New Type Sulfur Curable Acrylic Rubber. IRC '85, Oct. 15–18, 1985, Kyoto, Proc. p. 790.
[3.497] *Fukumori, T.:* Sulfur Curable Acrylic Elastomers. Paper 41, 123rd ACS-Meeting, Rubber-Div., May 10–12, 1983, Toronto, Ontario.
[3.498] *DelGato, J.:* Low Temperature Acrylic Debut. Rubber Wld. *152* (1965), 1, p. 95.
[3.499] *Giannetti, E., Mazzochi, R.:* Ammonium Salt Catalyzed Crosslinking Mechanism of Acrylic Rubbers. RCT *56* (1983), p. 21.
[3.500] *Hofmann, W.:* Polyacrylic Elastomers. PRI-Conf., April 10, 1981, London, Proc. pp. 32–67; KGK *35* (1982), p. 378.
[3.501] *Holly, H. W., et al.:* Increased Versatility for Acrylic Elastomers. Rubber Age Jan. 1965.
[3.502] *Lauretii, E., Mezzera, F., Santarelli, G., Spelta, A. L.:* Europrene AR, eine Gruppe von hochtemperatur- und ölbeständigen Spezialpolymeren. GAK *38* (1985), p. 296.
[3.503] *DeMarco, R. D.:* Polyacrylat-Elastomere einer neuen Generation. GAK *32* (1979), p. 588.
[3.504] *Mendelsohn, M. A.:* Acrylatkautschuk. Kautschuk u. Gummi *18* (1965), pp. 303, 788.
[3.505] *Rim, Y. S.:* A New High Temperature and Oil Resistant Elastomer Based on EPDM/ACM-Graft. Paper 86, 124th ACS-Meeting, Rubber-Div., Oct. 25–28, 1983, Houston, TX.
[3.506] *Saxon, R., Daniel, J. H.:* Crosslinking Reactions of Acrylic Polymers. J. Appl. Polymer Science *8* (1964), p. 352.
[3.507] *Scheer, E.:* Polyacrylate Elastomer Review. Paper at 127th ACS-Meeting, Rubber-Div., Apr. 23–26, 1985, Los Angeles, CA.

[3.508] *Starmer, P.H., Jorgenson, A.H.:* An Improved Acrylic Rubber. Rubber World *151* (1964) 4, p.78.

[3.509] *Stuesse, J.:* Selfcuring Acrylics. Rubber World *150* (1964) 2, p.78.

[3.510] *Trexler, H.E., Ileka, G.A.:* Compounding Acrylic Rubber for Minimum Corrosion. Rubber Age *98* (1966) 4, p.69.

[3.511] *Vial, T.M.:* Recent Developments in Acrylic Elastomers. RCT *44* (1971), p.344.

[3.512] *Weinstein, A.H.:* Elastomeric Diene Polyhalophenyl Acrylic Copolymers with Intrinsic Flame Retardance. RCT *54* (1981), p.767.

Special References on ACM

[3.513] US 2492170, 1945, US-Government, W.C. Mast et al.

[3.514] US 2568659, 1949, Goodrich.

[3.515] DOS 1910105, 1970, Bayer.

[3.516] DOS 1938038, 1970, Bayer.

[3.517] US 3312677, 1967, Thiokol Corp.

[3.518] US 3335118, 1967, Thiokol Corp.

[3.519] B.P. 1175545, 1969, Polymer Corp.

[3.520] DAS 1207629, 1965, Bayer.

[3.521] US 3315012, 1967, Goodrich, P.H. Starmer, J. Steusse.

[3.522] DOS 1808485, 1969, Goodrich.

[3.523] US 3448094, 1969, Baker Chem. Co.

[3.524] US 3475388, 1969, Dow Chem. Co.

[3.525] DOS 2358112, 1972, Am. Cyanamid.

[3.526] *Lewis, O.G.:* Physical Constants of Linear Homopolymers. Springer Verlag, Berlin, 1968.

[3.527] USP 3488331, 1968, Goodrich.

[3.528] USP 3317491, 1964, Thiocol Chemical Corp.

[3.529] *Jones, B.D.:* KGK *22* (1969), p.722.

[3.530] *Kändler, I., Peschk, G., Wöss, H.P.:* Angew. Makromol. Chem. *29/30* (1973), p.241.

[3.531] *Mast, W.C., Fischer, C.H.:* Ind. Engng. Chem. *41* (1949), p.790.

[3.532] *Seeger, N.V., et al.:* Ind. Engng. Chem. *45* (1953), p.2538.

[3.533] *Semegen, S.T., Wakelin, J.H.:* Rubber Age *71* (1952), p.57.

[3.534] *Antal, I.:* Verschnitt aus Polyacrylatkautschuk und anderen ölbeständigen Elastomeren. GAK *31* (1978), p.628; Int. Polym. Sci. Technol. *5* (1978) 2, p.T22.

[3.535] *Stanescu, C.:* Einfluß des Verschneidens von Acrylatkautschuken mit Epichlorhydrin- und Fluorkautschuken auf die physikalischen und chemischen Eigenschaften von Vulkanisaten. KGK *32* (1979), p.647.

[3.535a] Laborbericht Nr. 4270, September 1987, Publ.: DOG Deutsche Oelfabrik, Hamburg.

References on EAM

[3.536] *Colbert, G.P., Byam, J.D., Hagman, J.F.:* Designing Injection Moulding Systems for Vamac. Paper 11, 118th ACS-Meeting, Rubber-Div., Oct. 7–10, 1980, Detroit, MI.

[3.537] *Crary, J.A.:* Vamac Ethylene Acrylic Elastomer: Balanced Properties for Demanding Automotive Applications. Paper at 118th ACS-Meeting, Rubber-Div., Oct. 7–10, 1980, Detroit, MI.

[3.537a] *Hagman, J.F., Fuller, R.E., Witsiepe, W.K., Greene, R.N., Lewis, K.J.:* Ethylene/Acrylic Elastomer – A New Class of Heat and Oil Resistant Rubber. Rubber Age, May 1976.

[3.538] *Hübsch, D.:* Viton and Vamac in the Automotive Industry. SCR '79, Apr. 2–3, 1979, Copenhagen, Proceed. p.1.

[3.539] *Hirsch, E.A., Boyce, R.J.:* Dynamische Eigenschaften des neuen wärmebeständigen Ethylen-Acrylat-Elastomers. GAK *31* (1978), p.394.

[3.540] *Mayers, R.J., Seil, D.A.:* Effects of Long Term Fluid Immersions on Properties of Polyacrylic and Ethylene/Acrylic Elastomers. Paper 4, 120th ACS-Meeting, Rubber-Div., Oct. 13–16, 1981, Cleveland, OH.

[3.541] *Murrey, R.M., et al.:* Ethylene Acrylic Rubber Technical Developments. IRC '79, Oct. 3-6, 1979, Venice, Italy, Proceed. p. 291.

[3.541a] *Ochiltree, B.C., Warhurst, D.M.:* The Influence of Water on the Properties of Alumina Trihydrate Filled Ethylene-Acrylic Rubber Vulcanizates. Rubbercon '87, June 1-5, 1987, Harrogate, GB, Proceed. B 45.

3.7.4.13 References on CO, ECO, ETER

General Overview

[3.542] *Beier, E.:* Elastomere mit Durchstehvermögen - Epichlorhydrin-Kautschuke, Seifen, Öle, Fette, Wachse *106* (1980), p. 593.

[3.543] *Collins, E.A., Oetzel, J.T.:* Rheological Behavior of Elastomers. I. Melt Viscosity Characteristics of Uncured Epichlorohydrin Rubber. RCT *42* (1969), p. 790.

[3.544] *Ehrend, H.H.:* Zur Vernetzung von Epichlorhydrin-Elastomeren. KGK *38* (1985), p. 186.

[3.545] *Hofmann, W., Verschut, C.:* Epichlorhydrin-Terpolymerisat (ETER), ein neuer schwefel- u. peroxidvernetzbarer Kautschuktyp. GAK *33* (1980), p. 590; *34* (1981), pp. 24, 136.

[3.546] *Hofmann, W.:* Langzeitalterungsverhalten von Epichlorhydrinkautschukvulkanisaten. GAK *35* (1982), p. 563.

[4.547] *Ledwith, A., Fitzsimmond, C.:* Elastomers from Cyclic Ethers. In: *Kennedy, J.P., Törnquist, E.G.M. (Eds.):* Polymer Chemistry of Synthetic Elastomers, Vol. 1. Interscience Publ., New York, 1968, p. 377.

[3.548] *Mori, K., Nakamura, Y.:* Improvements of Sour Gasoline Resistance for Epichlorohydrine Rubber. RCT *57* (1984), p. 665.

[3.549] *Nakamura, Y.:* Polyether Elastomers - Compounding Studies for Improvement of Thermal Stability, Ozone Resistance and Air Retention Properties. IRC '85, Oct. 15-18, 1985, Kyoto, Proc. p. 802.

[3.550] *Oetzel, J.T.:* The Long-Term Heat Resistance of Acrylate and Epichlorohydrin Polymers. Paper at 107th ACS-Meeting, Rubber-Div., Apr. 30, 1975, Toronto, Ontario.

[3.551] *Oetzel, J.T., Scheer, E.N.:* Hydrin 400, an Improved Epichlorohydrin Elastomer. Paper at 113rd ACS-Meeting, Rubber-Div., May 2-5, 1978, Quebec.

[3.552] *Rijnders, R.F.R.T.:* Compounding Against Alcohol Containing Fuels. Paper at SRC '80, May 9, 1980, Rönneby, Sweden.

[3.553] *Scheer, E.:* Epichlorohydrin Elastomers. In: *Babbit, R.O. (Ed.):* The Vanderbilt Handbook. Publ.: Vanderbilt, 1978, pp. 275-294.

[3.554] *Schuette, W., Ehrend, H.H.:* Crosslinking of Epichlorohydrin Elastomers. Paper 32, 125th ACS-Meeting, Rubber-Div., May 8-11, 1984, Indianapolis, IN.

[3.555] *Vandenberg, E.J.:* Discovery and Development of Epichlorohydrin Elastomers. Paper 1, 121st ACS-Meeting, Rubber-Div., May 4-7, 1982, Philadelphia, PA.

[3.556] *Zemnickas, R.:* Hydrin 400, an Improved Epichlorohydrin Elastomer. Paper at SRC '78, June 8-9, 1978, Helsinki, Finland.

Special References

[3.557] *Vandenberg, E.J.:* Rubber Plast. Age *46* (1965), p. 1134.

[3.558] *Vandenberg, E.J.:* J. Polym. Sci. *47* (1960), p. 486.

[3.559] Polyepichlorohydrin Fuel Line Hose. Paper at the Detroit Rubber Group of the ACS, Oct. 18, 1979.

[3.559a] DOG-Lab.-Mitt. 4236, 1986, Publ.: DOG Deutsche Oelfabrik, Hamburg.

[3.559b] DOG-Lab.-Mitt. 4250, 1986, Publ.: DOG Deutsche Oelfabrik, Hamburg.

3.7.4.14 References on PO and GPO

[3.560] *Booth, C.:* Polymer *5* (1964), p. 479.

[3.561] *Foll, G.E.:* SCI-Monogr. *26* (1967), p. 103.

[3.562] *Furukawa, J., Saegusa, T.:* Polymerization of Aldehydes and Oxides. Interscience Publ., New York, 1963.

[3.563] *Gruber, E.E., et al.:* Ind. Engng. Chem. Prod. Res. Div. *3* (1964), p. 194.

3.7.4.15 References on FKM

General Overview

[3.564] *Albin, L. D.:* Current Trends in Fluoroelastomer Developments. RCT *55* (1982), p. 902; GAK *36* (1983), p. 450.

[3.565] *Arnold, R. G., Barney, A. L., Thompson, D. C.:* Fluoroelastomers. RCT *46* (1973), pp. 619–652, Review.

[3.566] *Kosmala, J. L., Tuckner, P. F.:* Fluoroelastomers: Polymers, Properties and Applications. Paper at 127th ACS-Meeting, Rubber-Div., Apr. 23–26, 1985, Los Angeles, CA.

[3.567] *Paciorek, I.:* Fluoropolymers. In: *Wall, L. A. (Ed.):* High Polymers, Vol. XXV. J. Wiley Interscience Publ., New York, 1972.

[3.568] *Procop, R. A.:* Fluorocarbon Elastomer Development, Past, Presence, and Future. IRC '85, Oct. 15–18, 1985, Kyoto, Proc. p. 38.

[3.569] *Stivers, D. A.:* Fluoroelastomers. In: *Babbit, R. O. (Ed.):* The Vanderbilt Handbook. Publ.: Vanderbilt, 1978, pp. 244–258.

[3.570] *Stivers, D. A.:* Fluorocarbon Rubbers. In: *Morton, M. (Ed.):* Rubber Technology, 2nd Ed. Van Nostrand Reinold Co., 1973, pp. 407–493.

Compounding

[3.571] *Mitchell, J., Itoh, K., Wada, T.:* High Performance Elastomeric Composition. Paper 23, 127th ACS-Meeting, Rubber-Div., Apr. 23–26, 1985, Los Angeles, CA.

[3.572] *Moggi, G., Civillo, G., Giunchi, G.:* Fluoroelastomers, The Effect of Polymeric Additions on Curing Behavior. IRC '85, Oct. 15–18, 1985, Kyoto, Proc. p. 838.

[3.573] *Novitskaya, S., Dontsov, A., Akimov, A.:* Some Regularities of Fluoroelastomers and Filler Interaction. IRC '85, Oct. 15–18, 1985, Kyoto, Proc. p. 328.

[3.574] *Stevens, R. D.:* Compounding Fluoroelastomers for Resistance to Extrusion at High Temperatures and Pressures. KGK *37* (1984), p. 770.

[3.575] *West, A. C., Ray, T. W.:* Fluoroelastomer Polymers Compounding and Testing for Downhole Environments. Paper 24, 117th ACS-Meeting, Rubber-Div., May 20–23, 1980, Las Vegas, NV.

New FKM Grades and Peroxide Cure

[3.576] *Albin, L. D.:* A New Peroxide Curable Fluoroelastomer for Use in Blends for Automotive Applications. Paper 5, 118th ACS-Meeting, Rubber-Div., Oct. 7–10, 1980, Detroit, MI.

[3.577] *Apotheker, D., Finley, J. B., Krusic, P. J., Logothetis, A. L.:* Curing of Fluoroelastomers by Peroxides. RCT *55* (1982), p. 1004.

[3.578] *Bauerle, J. G., Finley, J. B.:* A New Processing Fluoroelastomer Having Improved Fluid Resistance. Paper 4, 117th ACS-Meeting, Rubber-Div., May 20–23, 1980, Las Vegas, NV; GAK *34* (1981), p. 363.

[3.579] *van Cleef, A.:* Two New Peroxide Curable High Performance Elastomers. ikt '85, June 24–27, 1985, Stuttgart, West Germany.

[3.580] *Grossmann, R., Geri, S., Lagana, C.:* A New Easy Processing Fluoroelastomer. Paper 30, 124th ACS-Meeting, Rubber-Div., Oct. 25–28, 1983, Houston, TX.

[3.581] *Hepburn, C., Ogunniyi, D. S.:* Hexamethylene-N,N′-bis(tert. butyl peroxycarbamate) as a Curing Agent for Fluoroelastomers. IRC '85, Oct. 15–18, 1985, Kyoto, Proc. p. 287.

[3.582] *Ishiwari, K., Sakakura, A., Yutaka, S., Yagi, T.:* Some Properties of Segmented Fluoroelastomers. IRC '85, Oct. 15–18, 1985, Kyoto, Proc. p. 407.

[3.583] *Kojima, G., Morozumi, M., Wachland, H., Hisasue, M.:* Vulcanization and Vulcanizate Properties of a Fluoroelastomer Containing Epoxy Groups as Cure Sides. RCT *54* (1981), p. 779.

[3.583a] *Laguana, C., Monza, E., Geri, S.:* Terpolymer Fluoroelastomers with Various Ratios of Monomers. Paper 94, 128th ACS-Meeting, Rubber-Div., Oct. 1–4, 1985, Cleveland, OH.

[3.583b] *MacLaughlin, W.S.:* Fluoroelastomer Developments to Meet Demanding Automotive Requirements. Paper 21, 128th ACS-Meeting, Rubber-Div., Oct. 1-4, 1985, Cleveland, OH.

[3.584] *Logothetis, A.L.:* Developments in Perfluorocarbon Elastomers. IRC '85, Oct. 15-18, 1985, Kyoto, Proc. p. 73.

[3.585] *Morita, S., Yutani, Y., Tomoda, M., Oka, M.:* A High-Performance Fluorocarbon Elastomer. IRC '85, Oct. 15-18, 1985, Kyoto, Proc. p. 826.

[3.586] *Oka, M., Tomoda, M., Kawachi, S., Tatemoto, M.:* Thermoplastic Fluoroelastomers. IRC '85, Oct. 15-18, 1985, Kyoto, Proc. p. 832.

[3.586a] *Sohlo, A.M., Brullo, R.A.:* New Considerations in the Selection of Fluorocarbon Elastomers for Automotive Seal Applications. Paper 17, 128th ACS-Meeting, Rubber-Div., Oct. 1-4, 1985, Cleveland, OH.

[3.587] *Stevens, R.D., Pugh, T.L., Tabb, D.L., Bauerle, J.G.:* New Peroxide-Curable Fluoroelastomer Developments. IRC '85, Oct. 15-18, 1985, Kyoto, Proc. p. 342.

[3.588] *Stevens, R.D., Pugh, T.L., Tabb, D.L.:* New Peroxide-Curable Fluoroelastomer Developments. Paper 21, 127th ACS-Meeting, Rubber-Div., Apr. 23-26, 1985, Los Angeles, CA.

[3.589] *Tabb, D.L., Finlay, J.B.:* A New Gelled Fluoroelastomer. RCT *55* (1982), p. 1152.

Properties and Uses

[3.590] *Abu-Isa, I.A., Trexler, H.E.:* Mechanism of Degradation of Fluorocarbon Elastomers in Engine Oil. RCT *58* (1985), p. 326.

[3.591] *Arcella, V.:* Fluoroelastomers and New Oils. ikt '85, June 24-27, 1985, Stuttgart, West Germany.

[3.592] *Baddorf, C.R.:* Specialty Elastomers for Hose Applications. Paper 37, 122nd ACS-Meeting, Rubber-Div., Oct. 5-7, 1982, Chicago, MI.

[3.593] *Brown, J.H.:* Fluorocarbon Elastomers for High Temperature and Chemical Sealing Applications. IRC '79, Oct. 3-6, 1979, Venice, Italy, Proc. p. 333.

[3.594] *Campbell, R.R., Stivers, D.A., Kolb, R.E.:* Fluoroelastomer Applications for Pollution Control in the Automotive, Petrochemical and Electric Power Industries. RCT *55* (1982), p. 1137.

[3.595] *Eddy, J.D.:* New Fluoroelastomers with Outstanding Resistance to Harsh Fluids. Paper 32, 120th ACS-Meeting, Rubber-Div., Oct. 13-16, 1981, Cleveland, OH.

[3.596] *Field, S.D., Nersasian, A.:* Swelling of Fluorocarbon Elastomers in Synthetic Ester Lubricants. Paper 31, 125th ACS-Meeting, Rubber-Div., May 8-11, 1984, Indianapolis, IN.

[3.597] *Finlay, J.B., Moran, A.L., Logothetis, A.L.:* Curing Fluoroelastomers with Bisphenols for Optimum Properties. IRC '79, Oct. 3-6, 1979, Venice, Italy, Proc. p. 93.

[3.598] *Kosmala, J.L., Sohlo, A.M., Spoo, B.H.:* Fluoroelastomer Performance in Oil-field Environments. Paper 64, 124th ACS-Meeting, Rubber-Div., Oct. 25-28, 1983, Houston, TX.

[3.599] *Neppel, A., v. Kuzenko, M., Guttenberger, J.:* Swelling of Fluoroelastomers in Synthetic Lubricants. RCT *56* (1983), p. 12.

[3.600] *Pugh, T.L.:* Evaluation for Oil Field Service. KGK *37* (1984), p. 854.

[3.601] *Streit, G., Dunse, S.:* Fluorelastomere, Vernetzungssysteme und Wechselwirkung mit Motorenölen. KGK *38* (1985), p. 471.

[3.602] *Worm, A.T., Brullo, R.A., Kosmala, J.L.:* A High Fluorine-Containing Tetrapolymer for Harsh Chemical Environment. Paper 39, 123rd ACS-Meeting, Rubber-Div., May 10-12, 1983, Toronto, Ont.

[3.603] *Spoo, B.H.:* Injection Moulding of High Performance Fluoroelastomers. Paper at 105th ACS-Meeting, Rubber-Div., 1974.

[3.604] *Plazek, D.J., Choy, I.-C., Kelly, F.N., v. Meerwall, E., Su, L.:* Viscoelasticity and Tearing Energy of Fluorinated Hydrocarbon Elastomers. RCT *56* (1983), p. 866.

Special References on FKM

[3.605] *Conroy, M. E., et al.:* Rubber Age *76* (1955), p. 543.
[3.606] *Griffis, C. B., Montermoso, I. C.:* Rubber Age *77* (1955), p. 559.
[3.607] *Jackson, W. C., Hale, D.:* Rubber Age *77* (1955), p. 865.
[3.608] *Headrick, R. E.:* WADC-Techn. Rep. 1955, p. 55.
[3.609] G. P. 742907, 742908, 1953, Kellog, Co.
[3.610] *Dixon, S., et al.:* Ind. Engng. Chem. *49* (1957), p. 1687.
[3.611] *Rugg, S. I., Sterenson, A. C.:* Rubber Age *82* (1957), p. 102.
[3.612] US 3051677, 1962, DuPont.
[3.613] Chem. Engng. News *34* (1956), p. 4881.
[3.614] US 2968649, 1961, DuPont.
[3.615] US 3331823, 1967, Montecatini-Edison.
[3.616] US 3335106, 1967, Montecatini-Edison.
[3.617] *Miglierina, A., Ceccato, G.:* Fourth Int. Syn. Rubber Smp., 1969, No. 2, 65.
[3.618] US 3058818, 1962, Minnesota Mining.
[3.619] US 3069401, 1962, DuPont.
[3.620] US 3467636, DuPont.
[3.621] *Tatlow, J. C.:* Rubber Plast. Age *39* (1958), p. 33.
[3.622] *Schmiegel, W. W.:* KGK *31* (1978), p. 139.
[3.623] *Kalb, G. H., Khan, A. A., Quarles, R. W., Barney, A. L.:* Advances in Chemistry Series, No. 129, 1973, pp. 13–26.
[3.624] *Barney, A. L., Kalb, G. H., Khan, A. A.:* RCT *44* (1971), p. 660.

References on Tetrafluoroethylene and Propylene (TFE/P)

[3.625] *Hull, D.:* Aflas – A Unique Elastomer Based on a Regularly Alternating Structure of Tetrafluoroethylene and Propylene. Paper 26, 120th ACS-Meeting, Rubber-Div., Oct. 13–16, 1981, Cleveland, OH.
[3.626] *Hull, D.:* Copolymer of Fluoroethylene and Propylene Finds Oil Field Applications. Elastomeric, July 1982, p. 27.
[3.627] *Hull, D.:* Recent Developments Expand Aflas Elastomer Performance. Rubber World *186* (1982) 3, p. 31.
[3.628] *Hull, D.:* Oilfield Media Profile of Tetrafluoroethylene-Propylene Copolymer. Paper 63, 124th ACS-Meeting, Rubber-Div., Oct. 25–28, 1983, Houston, TX.
[3.629] *Hull, D.:* New Elastomer to Consider for Hot, Corrosive Service – Tetrafluoroethylene-Propylene Copolymer. Paper 50, 123rd ACS-Meeting, Rubber-Div., May 10–12, 1983, Toronto, Ont.
[3.630] *Hull, D.:* Tetrafluoroethylene-Propylene-Copolymer – A New High Temperature and Chemically Resistant Elastomer. Paper 4, PRI-Conference '84, March 12, 1984, Birmingham; KGK *38* (1985), p. 480.
[3.631] *Hull, D.:* Recent Technology Developments Concerning Tetrafluoroethylene-Propylene-Copolymer. Paper 20, 127th ACS-Meeting, Rubber-Div., Apr. 23–26, 1985, Los Angeles, CA.
[3.631a] *Hull, D., Kojima, G., Wachi, H.:* New Type of Fluoroelastomer (Tetrafluoroethylene/Propylene Copolymer) Provide Improved Resistance to Some Automotive Media. Paper 18, 128th ACS-Meeting, Rubber-Div., Oct. 1–4, 1985, Cleveland, OH.
[3.632] *Kojima, G., Yamabe, M., Wachi, H., Kodama, S.:* A New Fluoroelastomer with Outstanding Chemical Resistance and Improved Low Temperature Properties. Paper 95, 124th ACS-Meeting, Rubber-Div., Oct. 25–28, 1983, Houston, TX.
[3.633] *Kojima, G., Wachi, H.:* A New Tetrafluoroethylene-Propylene Based Fluoroelastomer with Improved Temperature Properties. IRC '85, Oct. 15–18, 1985, Kyoto, Proc. p. 242.
[3.634] *Morozumi, M.:* New Tetrafluoroethylene-Propylene Copolymer Lining Material. Paper 96, 124th ACS-Meeting, Rubber-Div., Oct. 25–28, 1983, Houston, TX.

References on Other Exotic Fluor Containing Elastomers

[3.635] Ref. [3.567], pp. 175, 267.
[3.636] *Brooks, J. T.:* Poly(Fluoroalkoxyphosphazene) Elastomers, Performance Profile. Paper 19, 127th ACS-Meeting, Rubber-Div., Apr. 23–26, 1985, Los Angeles, CA.
[3.637] *Lohr, D. F., Beckmann, J. A.:* PNF Phosphonitrilic Fluoroelastomer: Properties and Applications. Paper 34, 120th ACS-Meeting, Rubber-Div., Oct. 13–16, 1981, Cleveland, OH.
[3.637a] *Bjork, F.; Stenberg, S.:* Mechanical Properties of PNF Rubber of Sealing Application. Rubbercon '87, June 1–5, 1987, Harrogate, GB, Proceed. B 37.
[3.637b] *Penton, H. R.:* Polyphosphazenes-Semiorganic Elastomers for Specialty Applications. KGK *39* (1986), pp. 301–304.
[3.637c] *Bocks, J. T.:* Leistungsprofil der Poly(fluoralkoxyphosphazen)-Elastomere. GAK *39* (1986), pp. 374–376.
[3.638] *Barney, A. L.:* A High Performance Fluorocarbon Elastomer. Unpublished Works of DuPont; Kalrez, Perfluorelastomerteile. Bulletin of DuPont.

3.7.4.16 References on PNR

[3.639] Rev. Gén. Caoutch. 52 (1975), p. 71.
[3.640] *de Dellion, P.:* Current Applications of Polynorbornene Elastomers. IRC '79, Oct. 3–6, 1979, Venice, Italy, Proc. p. 349.
[3.641] *de Dellion, P., Manger, H. D.:* Vulkanisate mit niedrigen Härten aus Norbornenkautschuk. GAK *30* (1977), p. 518.
[3.642] *Manger, H. D.:* Polynorbornen – Eigenschaften und Anwendungen. KGK *32* (1979), p. 572.

3.7.4.17 References on Other Polymers

General Overview on Special Butadiene Copolymers

[3.643] *Starkweather, H. W., et al.:* Ind. Engng. Chem. *39* (1947), p. 210.
[3.644] *Bachmann, G. B., et al.:* Ind. Engng. Chem. *43* (1951), p. 997.

Vinylpyridine Rubber

[3.645] *Svetlik, J. F., et al.:* Ind. Engng. Chem. *48* (1956), p. 1084.

Piperylene Rubber

[3.646] *Carbonaro, A., Gargani, L., Sorta, E., Bruzzone, M.:* Trans Butadiene Piperylene Elastomers – Preparation and Structural Properties. IRC '79, Oct. 3–6, 1979, Venice, Italy, Proc. p. 312.
[3.647] *Lauretti, E., Santarelli, G., Canidio, A., Gargani, L.:* Trans Butadiene Piperylene Elastomers – Properties and Applications. IRC '79, Oct. 3–6, 1979, Venice, Italy, Proc. p. 322.

Special Isoprene Copolymers

[3.648] *Meeker, T. R.:* Specialty Isoprene-Based Polymers – Derivatives from the General Purpose Polymers. Paper 91, 127th ACS-Meeting, Rubber-Div., Apr. 23–26, 1985, Los Angeles, CA.
[3.649] *Wieder, W., Witte, J.:* Alternierende Copolymerisation von Ethylen und Isopren, ein neuer Stereokautschuk. KGK *36* (1983), p. 748.

Dimethylbutadiene Polymers (Methyl Rubber)

[3.650] *Hofmann, F.:* Chemiker-Ztg. *60* (1963), p. 693.
[3.651] *Kondakow, J.:* J. Prakt. Chem. *62* (1900), p. 172; 64 (1901), p. 109.
[3.652] *Holt, A.:* Angew. Chem. *27* (1914), p. 153.

Special Polymers by Ringopening Polymerization

[3.653] *Günther, P., et al.:* Angew. Makromol. Chem. *14* (1970), p. 87.
[3.654] *Haas, F. K.:* Properties of a 1.5-trans-Polypentenamer Produced by Polymerization Through Ring Cleavage of Cyclopentene. Paper at 97th ACS-Meeting, Rubber-Div., Spring 1970.
[3.655] *Dräxler, A.:* Trans-Polyoctenamer. KGK *34* (1981), p. 185.
[3.656] *Dräxler, A.:* Die Stellung der Polyoctenamere unter den technisch genutzten Kautschuktypen. KGK *36* (1983), p. 1037.

Crosslinkable Polyethylene (X-LPE)

[3.656a] *Dorn, M.:* Fortschritte auf dem Gebiet der PE-Vernetzung mit organischen Peroxiden. GAK *35* (1982), p. 808.
[3.657] *Martens, S. C.:* Chemically Crosslinked Polyethylene. In: *Babbit, R. O. (Ed.):* The Vanderbilt Handbook. Publ.: Vanderbilt, 1978, pp. 308–318.
[3.658] *Voigt, H. U.:* Grundlagen und Methoden der chemischen Vernetzung von Polyethylen. In: Kabel und isolierte Leitungen. Publ.: VDI-Verlag, Düsseldorf, 1984.

New Polymers

[3.659] *Aggarwal, S. L.:* Recent Developments, and a Look into the Future, of Synthetic Rubbers of Controlled Molecular Structure. Rubbercon '87, June 1–5, 1987, Harrogate, GB, Proceed. A 14.
[3.660] *Kitayama, M., Iseda, Y., Odaka, F., Auzai, S., Irako, K.:* Synthesis and Properties of Polyoxazolidone Elastomers from Diepoxides and Diisocyanates. RCT *53* (1980), p. 1.
[3.660a] *Kennedy, J. P., Faust, R.:* New Rubbery Polymers and Telechelic Prepolymers by Living Cationic Polymerisations. Rubbercon '87, June 1–5, 1987, Harrogate, GB, Proceed. A 8.
[3.661] *Schulz, R. C.:* Neue Polymere, ausgehend von 2-Isopropenylnaphthalin. ikt '85, June 24–27, 1985, Stuttgart, West Germany.
[3.662] *Lamp, C. M., Lindstrom, M. R.:* Elastomers from Liquid Polymercaptan Resin: The Effect of Curing Agent and Physical Properties. Paper 22, 118th ACS-Meeting, Rubber-Div., Oct. 7–10, 1980, Detroit, MI.
[3.663] *MacKnight, W. J., Lundberg, R. D.:* Elastomeric Ionomers. RCT *57* (1984), p. 652, Review.
[3.664] *MacKnight, W. J., Lundberg, R. D.:* Ionic Elastomers. IRC '85, Oct. 15–18, 1985, Kyoto, Proc. p. 106.
[3.665] *Kohjiyo, S., Hashimoto, T., Yamashita, S., Irie, M.:* Synthesis and Properties of Elastomeric Ionenes: Novel Hydrophilic Elastomers. IRC '85, Oct. 15–18, 1985, Kyoto, Proc. p. 481.

3.7.4.18 References on Q

General Overview

[3.666] *Gibbon, B.:* Silicone Rubbers. Paper at 127th ACS-Meeting, Rubber-Div., Apr. 23–26, 1985, Los Angeles, CA.
[3.667] *McGregor:* Silicones and Their Uses. McGraw-Hill Book Co., New York.
[3.668] *Kosfeld, R., Heß, M.:* Physikalisch-chemische Charakterisierung von Siliconkautschuken. KGK *36* (1983), p. 750.
[3.669] *Lewis, F. M.:* The Chemistry of Silicone Elastomers. In: *Kennedy, J. P., Törnquist, E. G. M. (Eds.):* Polymer Chemistry of Synthetic Elastomers, Vol. 2. Interscience Publ., New York, 1969, pp. 767–804.
[3.670] *Meals, R. N., Lewis, F. M.:* Silicones. Van Nostrand Reinold Publ., New York.
[3.671] *Noble, M. G.:* Silicone Elastomers. In: *Babbit, R. O. (Ed.):* The Vanderbilt Handbook. Publ.: Vanderbilt, 1978, pp. 216–232.
[3.672] *Noll, W.:* Chemie und Technologie der Silicone, 2nd Ed. Verlag Chemie, Weinheim, 1968.

[3.673] *Polmanteer, K.E.:* Current Perspectives on Silicone Rubber Technology. RCT *54* (1981), p.1051.
[3.674] *Razzano, J.:* Silicone Elastomers: Preparation and Performance. Paper at 124th ACS-Meeting, Rubber-Div., Oct.25–28, 1983, Houston, TX.
[3.675] *Rochow, E.G.:* Einführung in die Chemie der Silikone. Verlag Chemie, Weinheim, 1952.
[3.676] *Warrick, E.L., Pierce, O.R., Polmanteer, K.E., Saam, J.C.:* Silicone Elastomer Developments. RCT *52* (1979), Review.

Synthesis

[3.677] *Graiver, D., Huebner, D.J., Saam, J.C.:* Emulsion Polymerized Polydimethylsiloxane. RCT *56* (1983), p.918.
[3.678] *Saam, J.C., Graiver, D., Baile, M.:* Room-Temperature-Cured Polydimethylsiloxane Elastomers from Aqueous Dispersion. RCT *54* (1981), p.976.

Compounding, Fillers, Vulcanization

[3.679] *Baile, M.D.:* Optically Transparent Silicone Elastomers. Paper 60, 127th ACS-Meeting, Rubber-Div., Apr.23–26, 1985, Los Angeles, CA.
[3.680] *Brennan, J.J.:* Cab-O-Sil Fumed Silica in Silicones. Paper 2, 119th ACS-Meeting, Rubber-Div., June 2–5, 1981, Minnesota, MN.
[3.681] *Caprino, J.C.:* New Curing System for a Tough Silicone Rubber. Paper 4, 118th ACS-Meeting, Rubber-Div., Oct.7–10, 1980, Detroit, MI; GAK *35* (1982), p.118.
[3.682] *Cochrane, H., Lin, C.S.:* The Effect of Fumed Silica in RTL-Silicone Rubber Sealants. Paper 58, 127th ACS-Meeting, Rubber-Div., Apr.23–26, 1985, Los Angeles, CA.
[3.683] *Drake, R.E.:* 1.2-Polybutadiene as Additives to Silicone Elastomers. Paper 6, 119th ACS-Meeting, Rubber-Div., June 2–5, 1981, Minneapolis, MN.
[3.684] *Lutz, M.A., et al.:* Novel Wet-Process Silica Prepared from Alkyl Silicates; I. Synthesis, Paper 61; II. Performance in Reinforcing Silicone Elastomers, Paper 62; Use in Silicone Elastomers for Optical Applications, Paper 63, 127th ACS-Meeting, Rubber-Div., Apr.23–26, 1985, Los Angeles, CA.
[3.685] *Mark, J.E., Andrady, A.L.:* Model Networks of End-Linked Polydimethylsiloxane Chains. RCT *54* (1981), p.368.
[3.686] *Mark, J.E., Ning, Y.P.:* Ethylamine and Ammonia as Catalysts in the In-Situ Preparation of Silica in Silicone Networks. Paper 59, 127th ACS-Meeting, Rubber-Div., Apr.23–26, 1985, Los Angeles, CA.
[3.687] *Maxson, M.T., Lee, C.L.:* Effects of Fumed Silica Treated with Functional Disilazanes on Silicone Elastomer Properties. Paper 3, 119th ACS-Meeting, Rubber-Div., June 2–5, 1981, Minneapolis, MN.
[3.688] *Tanabe, T.M., Yoshioka, E.N., Anisman, A.M.:* Composite Silicone Rubber for the Space Telescope Program. Paper 8, 122nd ACS-Meeting, Rubber-Div., Oct.5–7, 1982, Chicago, MI.
[3.689] *Vick, S.C., Fairhurst, D., Sovio, A.:* Organosilicon-Chemikalien in mit Tonerdetrihydrat gefüllten Polyolefinen. GAK *37* (1984), p.336.
[3.690] *Wagner, M.P.:* Precipated Silicas in Silicone Rubber. Paper 4, 119th ACS-Meeting, Rubber-Div., June 2–5, 1981, Minneapolis, MN.
[3.690a] *Wolfer, D.:* Elektrisch leitfähiger Silicongummi – ein moderner Werkstoff mit vielen Anwendungsmöglichkeiten. KGK *34* (1981), p.640.

Liquid Silicones and their Processing

[3.691] *de Beers, M.D.:* A New Family of RTV-Silicone Sealants: The Octoates. Paper 9, 119th ACS-Meeting, Rubber-Div., June 2–5, 1981, Minneapolis, MN.
[3.691a] *Cush, J.R.:* Flüssiger Siliconkautschuk – eine vielseitige Alternative. GAK *35* (1982), p.666.
[3.692] *Huber, A.:* Flüssigsiliconverarbeitung, Verfahrenstechnik oder Maschinentechnik? Wo liegt die Problematik? KGK *38* (1985), p.191.

[3.693] *Laghi, A.A.:* Recent Developments for Liquid Injection Moulding of Silicone Elastomers. Paper 24, 118th ACS-Meeting, Rubber-Div., Oct. 7–10, 1980, Detroit, MI.
[3.694] *Macosko, C. W., Lee, L. J.:* Temperature and Cure Profiles in Liquid Injection Molding of Silicone Rubber. Paper 11, 119th ACS-Meeting, Rubber-Div., June 2–5, 1981, Minneapolis, MN.
[3.695] *Macosko, C. W., Lee, L. J.:* Heat Transfer and Property Development in Liquid Silicone Rubber Molding. RCT *58* (1985), p. 436.
[3.696] *Morrow, W. J.:* New Trends in Silicone Liquid Elastomers. Paper 12, 119th ACS-Meeting, Rubber-Div., June 2–5, 1981, Minneapolis, MN; KGK *35* (1982), p. 585.
[3.697] *Romig, Ch.A.:* Automatische Fertigung von Siliconkautschukteilen. GAK *38* (1985), p. 408.
[3.698] *Weise, C.:* HTV-Flüssigsiliconkautschuke, eine neue Materialgeneration. KGK *35* (1982), p. 111.

Properties and Uses

[3.699] *Dams, M. J., Murry, M. C.:* Silicone Rubber, Improvements in Heat Resistance. KGK *34* (1981), p. 15.
[3.700] *Gottlieb, M., Macosko, C. W., Lepsch, T. C.:* Stress-Strain Behavior of Randomly Crosslinked Polydimethylsiloxane Networks. RCT *55* (1982), p. 1108.
[3.701] *Henry, A. W.:* High Temperature Degradation of Silicone Rubber Compounds in Silicone Oil Environment. RCT *56* (1983), p. 83.
[3.702] *Mozisek:* Änderung der Eigenschaften von Kautschuken und Vulkanisaten durch ionisierende Bestrahlung. KGK *26* (1979), p. 92.
[3.703] *Murray, M. C.:* New Technology Improves Heat-Aging Stability of Silicone Rubber for Wire and Cable Insulation. Paper 7, 119th ACS-Meeting, Rubber-Div., June 2–5, 1981, Minneapolis, MN.
[3.704] *Stefcova, P., Schatz, M.:* Magnetic Silicone Rubbers. RCT *56* (1983), p. 322.
[3.705] *Tazawa, T.:* Swelling Behavior of Silicone Rubber in Silicone Oil. IRC '85, Oct. 15–18, 1985, Kyoto, Proc. p. 928.
[3.706] *Vallés, E. M., Rost, E. J.:* Small Strain Modulus of Model Trifunctional Polydimethylsiloxane Networks. RCT *57* (1984), p. 55.

Block and Segmented Polymers

[3.707] *Tang, S. H., Meinecke, E.A., Riffle, J. S., McGrath, J. E.:* Structure-Property Relationships of Perfectly Alternating Polycarbonate-Polydimethylsiloxane Block Copolymers. RCT *57* (1984), p. 184.
[3.708] *Tyagi, D., et al.:* Novel Segmented Elastomers Based on Aminoalkyl Terminated Polysiloxanes, Synthesis and Characterization. Paper 47, 123rd ACS-Meeting, Rubber-Div., May 10–12, 1983, Toronto, Ont.

Special References on Q

[3.709] *Warrik, E. L.:* RCT *49* (1977), p. 909.
[3.710] *Weit, C. E., et al.:* RCT *24* (1951), p. 366.
[3.711] *Merker, R. L., et al.:* J. Polym. Sci. *2* (1964), p. 31.
[3.712] US 3051684, 1962, M. Morton, A. Remhausen.
[3.713] US 3483270, 1969, E. E. Bostick.
[3.714] US 3665052, 1972, J. C. Saam, F. W. G. Fearon.
[3.715] *Polmanteer, K. E.:* Rubber Age *78* (1956), p. 83.
[3.716] *Harrington, R.:* Rubber Age *81* (1957), p. 971; *82* (1957), p. 461; *83* (1958), pp. 472, 1003; *86* (1960), p. 816.

References on Fluorosilicone Grades

[3.717] *Bush, R. B.:* Fluorosilicones-Compounding for Extended Life. Paper 35, 120th ACS-Meeting, Rubber-Div., Oct. 13–16, 1981, Cleveland, OH; GAK *36* (1983), p. 258.

[3.718] *Dams, M.J.:* Recent Developments and Improvements in Silicone and Fluorosilicone High Temperature Vulcanizing (HTV) Rubbers. ikt '85, June 24–27, 1985, Stuttgart, West Germany; KGK *38* (1985), p. 1109.

[3.718a] *Dams, M.J.:* Fluorosilicone Rubber – the Temperature and Oil Resistant Elastomer. Rubbercon '87, June 1–5, 1987, Harrogate, GB, Proceed. B 27.

[3.719] *Fiedler, L.D.:* Fluorosilicone Elastomers in the Automotive Industry. Paper at 118th ACS-Meeting, Rubber-Div., Oct. 7–10, 1980, Detroit, MI.

[3.719a] *Lynn, M.M., de Smedt, C., Groofaert, W., Kolb, R.E.:* Fluoroelastomers for Use in Aggressive Automotive Environments. Rubbercon '87, June 1–5, 1987, Harrogate, GB, Proceed. B 26.

[3.720] *Maxson, M.T., Lee, C.L.:* Fluorosilicone Liquid Rubber. Paper 18, 127th ACS-Meeting, Rubber-Div., Apr. 23–26, 1985, Los Angeles, CA.

[3.721] *Monroe, C.M.:* Die Leistung von Fluorsilicongummi in Treibstoffen. KGK *35* (1982), p. 667.

3.7.4.19 References on TM

General Overview

[3.722] *Flanders, S.K.:* Polysulfide Rubbers. Paper at 127th ACS-Meeting, Rubber-Div., Apr. 23–26, 1985, Los Angeles, CA.

[3.723] *Gobran, R.N., Berenbaum, M.B.:* Polysulfide and Monosulfide Elastomers. In: *Kennedy, J.P., Törnquist, E.G.M. (Eds.):* Polymer Chemistry of Synthetic Elastomers, Vol. 2. Interscience Publ., New York, 1969, pp. 805–842.

[3.724] *Schulman, M.A., Schultheis, J.J.:* Polysulfide Polymers. In: *Babbit, R.O. (Ed.):* The Vanderbilt Handbook. Publ.: Vanderbilt, 1978, pp. 207–215.

[3.725] *Paneck, J.R.:* Polysulfide Rubbers. In: *Morton, M. (Ed.):* Rubber Technology, 2nd Ed. Van Nostrand Reinold Co., New York, 1973, pp. 349–376.

Special References

[3.726] CH 127540, 1926, J. Baer.

[3.727] GB 302270, 1927, J.C. Patrik.

[3.728] DRP 670140, 1935, I.G. Farbenindustrie.

[3.729] *Hockenberger, L.:* Chem. Ing. Techn. *16* (1964), p. 1046.

[3.730] *Fettes, E.M., et al.:* Ind. Engng. Chem. *42* (1950), p. 2217; *46* (1954), p. 1539.

[3.731] *Ghatge, N.D., Vernekar, S.P., Lonikar, S.V.:* Polysulfide Sealants. RCT *54* (1981), p. 197.

[3.732] *Usmani, A.M., Chartoff, R.P., Warner, W.M., Butler, J.M., Salyer, I.O., Miller, D.E.:* Interfacial Considerations in Polysulfide Sealant Bonding. RCT *54* (1981), p. 1086.

3.7.4.20 References on Polyester and Polyether Elastomers

Polyesters and Polyethers

[3.733] *Harper, D.H.:* Trans. IRI *24* (1947), p. 181.

[3.734] *Cowan, C.J.:* Ind. Engng. Chem. *41* (1949), p. 1647.

[3.735] *Brizzolara, D.F.:* Hytrel Polyester Elastomer – High Performance Thermoplastic Elastomers. Paper 70, 127th ACS-Meeting, Rubber-Div., Apr. 23–26, 1985, Los Angeles, CA.

[3.736] *Coleman, M.L.:* Hytrel Polyester Elastomers. Paper 5, 121st ACS-Meeting, Rubber-Div., May 4–7, 1982, Philadelphia, PA.

[3.736a] *Hoeschele, G.K., Witsiepe, W.K.:* Polyetherester Block-Copolymere, eine Gruppe neuartiger thermoplastischer Elastomere. Angew. Makromol. Chem. *29/30* (1973), p. 267.

[3.736b] *Hoeschele, G.K.:* Über die Synthese von Polyetherester-Block-Copolymeren. Chimia *28* (1974) 9, p. 544.

[3.737] *Kane, R.P.:* Thermoplastic Copolyesters. In: *Babbit, R.O. (Ed.):* The Vanderbilt Handbook. Publ.: Vanderbilt, 1978, pp. 233–240.

[3.738] *Wells, S.C.:* Polyester Thermoplastic Elastomer. In: *Walker, B.M. (Ed.):* Handbook of Thermoplastic Elastomers. Van Nostrand Reinold Co., 1979, pp.103–215.
[3.739] *Alma, D.:* Les elastomères thermoplastiques de Polyester. Europlastique '86, Apr.22–23, 1986, Paris.
[3.740] *Souffie, R.D., Graff, R.S.:* Segmented Polyether Ester Copolymers – Thermoplastic Elastomers. Paper 19, 121st ACS-Meeting, Rubber-Div., May 4–7, 1982, Philadelphia, PA.
[3.741] *Souffie, R.D.:* Thermoplastische Elastomere auf Basis Polyether-Ester-Copolymeren. KGK *36* (1983), p.445.

Polyether Amides

[3.742] *Barot, P., Goletto, J.:* Les copolymères blocks amide-etheramide. Europlastique '86, Apr.22–23, 1986, Paris.
[3.743] *Biggi, A., della Fortuna, G., Peregio, G., Zofferi, L., Donato, S.:* Structure and Mechanical Properties on Regularly Alternated Polyesteramide. KGK *34* (1981), p.349.
[3.744] *Deleens, G.:* Propriété et applications des polyethers à block amide. Europlastique '86, Apr.22–23, 1986, Paris.

3.7.4.21 References on Polyurethanes

General Overview

[3.745] *Bayer, O.:* Das Diisocyanat-Polyadditionsverfahren (Polyurethane). 2. Mitteilung über Polyurethane. Angew. Chem. *A59* (1947), pp.257–272.
[3.746] *Bayer, O.:* Das Diisocyanat-Polyadditionsverfahren – Historische Entwicklung und Grundlagen. In: *Vieweg, R. (Ed.):* Kunststoff-Handbuch, Vol.7. Carl Hanser Verlag, München, 1963, pp.7–48.
[3.747] *Buist, J.M.:* Developments in Polyurethanes. Elsevier Applied Science Publ., Barking, 1978.
[3.748] *Fabris, H.J.:* Urethane Elastomers. Paper at 124th ACS-Meeting, Rubber-Div., Oct.25–28, 1983, Houston, TX.
[3.749] *Hepburn, C.:* Polyurethane Elastomers. Elsevier Applied Science Publ., Barking, 1982.
[3.750] *Saunders, J.H.:* Polyurethane Elastomers. In: *Kennedy, J.P., Törnquist, E.G.M. (Eds.):* Polymer Chemistry of Synthetic Elastomers, Vol.2. Interscience Publ., New York, 1969, pp.727–765.
[3.751] *Saunders, J.H., Frisch, K.C.:* Polyurethanes, Chemistry and Technology, I. Chemistry, High Polymers, Vol.XVI, 1962; II. Technology, High Polymers, Vol.XXIII, 1969. Interscience Publ., J. Wiley & Sons, New York.
[3.752] *Smith, V.:* Polyurethane Polymers. Paper at 127th ACS-Meeting, Rubber-Div., Apr.23–26, 1985, Los Angeles, CA.
[3.753] *Timm, Th.:* Derzeitige Erkenntnisse über physikalische und chemische Vorgänge bei der thermischen und thermo-oxidativen Beanspruchung von Polyurethanelastomeren. KGK *35* (1982), pp.568–584; *36* (1983), pp.257–268; *37* (1984), pp.933–944.
[3.754] *Woods, G.:* Flexible Polyurethane Foams, Chemistry and Technology. Elsevier Applied Science Publ., Barking, 1982.
[3.755] *Wright, P., Cumming, A.P.C.:* Solid Polyurethane Elastomer. MacLaren Sons Ltd., London, 1969.

General References on AU

[3.756] *Kallert, W.:* Neue Entwicklungen auf dem Gebiet der Chemie und Technologie der walzbaren Polyurethane. KGK *19* (1966), p.363.
[3.757] *Kleimann, H.:* Urepan – ein Urethankautschuk für die Gummiindustrie. KGK *83* (1983), p.175.

References on PUR

[3.758] *O'Connor, J.M., Sessions, W.J., Lickei, D.L.:* One Component Heat Curable Liquid Urethane Elastomers. RCT *55* (1982), p. 88.

[3.759] *Frisch, K.C.:* Recent Developments in Urethane Elastomers and Reaction Injection Moulded (RIM) Elastomers. RCT *53* (1980), p. 126.

[3.760] *Lee, L.J.:* Polyurethane Reaction Injection Moulding: Process, Materials and Properties. RCT *53* (1980), pp. 542-599, Review.

[3.761] *Lewis, G.D.:* RIM Urethane Elastomers - Application for the 80's. Paper 18, 119th ACS-Meeting, Rubber-Div., June 2-5, 1981, Minneapolis, MN; KGK *35* (1982), p. 200.

[3.762] *Peled, J.:* Polyurethane Cast Tires for Military Applications. Paper 68, 125th ACS-Meeting, Rubber-Div., May 8-11, 1984, Indianapolis, IN.

References on EU

[3.763] *Pechold, E., Pruckmayer, G., Robinson, I.M.:* A New Investigation of Tetrahydrofuran Based Polyether Glycols in Urethane Elastomers. RCT *53* (1980), p. 1032.

References on TPU

[3.764] *Boenig, H.V.:* Morphological Aspects on Polyurethane Properties. Paper 26, 118th ACS-Meeting, Rubber-Div., Oct. 7-10, 1980, Detroit, MI.

[3.765] *Cooper, S.:* Morphology and Properties of Polyurethane Block Polymers. Paper 16, 119th ACS-Meeting, Rubber-Div., June 2-5, 1981, Minneapolis, MN.

[3.766] *Cowell, R.D.:* Thermoplastic Polyurethane Elastomers - Chemistry, Properties and Processing for the 80's. Paper 17, 121st ACS-Meeting, Rubber-Div., May 4-7, 1982, Philadelphia, PA.

[3.767] *Doll, W., Könczol, L.:* Untersuchungen zum Spannungs-Dehnungsverhalten segmentierter Polyurethane. ikt '85, June 24-27, 1985, Stuttgart, West Germany.

[3.768] *Eisenbach, C.D.:* Segmented Polyurethanes - Synthesis and Properties of Elastomers with Monodisperse Segments. Gordon Res. Conf. of Elastomers, July 16-20, 1984, New London, NH.

[3.769] *Günther, C., Baumgartner, M., Eisenbach, C.D.:* Struktur und Eigenschaften von segmentierten Polyurethan-Elastomeren definierter Primärstruktur. ikt '85, June 24-27, 1985, Stuttgart, West Germany.

[3.769a] *Hoppe, H.-G.:* New Aspects in the Development and Application of Thermoplastic Polyurethane. Rubbercon '87, June 1-5, 1987, Harrogate, GB, Proceed. A 35.

[3.770] *Ikeda, Y., Kohjiya, S., Yamashita, S., Yamamoto, N.:* Synthesis of Novel Segmented Polyether Urethane Urea and its Application to Biomedical Uses. IRC '85, Oct. 15-18, 1985, Kyoto, Proc. p. 475.

[3.771] *McKnight, W.J.:* Segmented Polyurethanes with Monodisperse Hard Segments. Gordon Res. Conf. of Elastomers. July 15-19, 1985, New London, NH.

[3.772] *Koberstein, J.:* Microdomain Structure and Phase Mixing in Polyurethanes. Gordon Res. Conf. of Elastomers, July 15-19, 1985, New London, NH.

[3.773] *Lee, B.:* Synthesis and Characterization of Segmented Urea-Polyether Urethanes via Tertiary Alcohol Chain Extenders. Paper 6, 127th ACS-Meeting, Rubber-Div., Apr. 23-26, 1985, Los Angeles, CA.

[3.773a] *Petrovic, Z.S., Budinski-Semendic, J.:* Study of the Effect of Soft Segment Length and Concentration on Properties of Polyetherurethanes. I. The Effect on Physical and Morphological Properties. RCT *58* (1985), p. 685; II. The Effect on Mechanical Properties. RCT *58* (1985), p. 701.

[3.774] *Nefzger, H., Eisenbach, C.D.:* Synthese und Eigenschaften von thermoplastischen Polyurethan-Elastomeren ohne Wasserstoffbrückenbindung. ikt '85, June 24-27, 1985, Stuttgart, West Germany.

[3.775] *Schollenberg, C.S., Dinbergs, K., Stewart, F.D.:* Thermoplastic Elastomer Melt Polymerization Study. RCT *55* (1982), p. 137.

[3.776] *Wolkenbreit, S.:* Thermoplastic Polyurethane Elastomers. In: *Walker, B.M. (Ed.):* Handbook of Thermoplastic Elastomers. Van Nostrand Reinold Co., New York, 1979, pp. 216–246.

[3.777] *Wegner, G.:* Chemische Struktur, Überstruktur und Eigenschaften von thermoplastischen Elastomeren. KGK *31* (1978), p. 67.

References on New PUR Grades

[3.778] *Brunelle, C.M., McKnight, W.J.:* Thermal Transition and Relaxation Behavior of Polybutadien Polyurethanes on 2.6-Toluene-Diisocyanate. RCT *55* (1982), p. 1413.

[3.779] *Byrne, C.A., Sloan, J.M., Mack, D.P.:* A Study of Aliphatic Polyurethane Elastomers Prepared from Diisocyanate Isomer Mixtures. Paper 13, 127th ACS-Meeting, Rubber-Div., Apr. 23–26, 1985, Los Angeles, CA.

[3.780] *Cohen, A.:* Novel Urethane Elastomer Systems. Paper 23, 118th ACS-Meeting, Rubber-Div., Oct. 7–10, 1980, Detroit, MI.

[3.781] *Cooper, S.L.:* Morphology and Properties of Polyurethane Ionomers. Gordon Res. Conf. of Elastomers, July 19–23, 1982, New London, NH.

Special References

[3.782] *Kozabiewicz, K., Wlazlo, A.:* Studies of Moisture-cured Polyurethane Rubber, Modified with Coal Tar Pitch. IRC '85, Oct. 15–18, 1985, Kyoto, Proceed. p. 399.

[3.783] *Kimball, M.E., Fielding-Russel, G.S.:* Effect of Cure Temperature on Urethane Networks. RCT *53* (1980), p. 936.

[3.784] *Mack, J.E., Sung, P.H.:* Chain Extension Studies Relevant to the Completeness of Endlinking in Elastomeric Polyurethane Networks. RCT *55* (1982), p. 1464.

[3.785] *Kim, C.S., Bottaro, J., Farzan, M., Ahmad, J.:* Improvements in the Stress/Strain Behavior of Urethane Rubbers by Bimodal Network Formation. Paper 92, 127th ACS-Meeting, Rubber-Div., Apr. 23–26, 1985, Los Angeles, CA.

[3.786] *Mendelsohn, M.A., Navish, jr. F.W., Kim, D.:* Characteristics of a Series of Energy Absorbing Polyurethane Elastomers. Paper 45, 127th ACS-Meeting, Rubber-Div., Apr. 23–26, 1985, Los Angeles, CA.

3.7.4.22 References on Thermoplastic Elastomers (TPE)

General Overview

[3.787] *Allport, D.C., Janes, W.H.:* Block Copolymers. Elsevier Applied Science Publ., Barking, 1973.

[3.788] *Bull, A.L., v. Henten, K.:* Thermoplastic Elastomers, The Utilisation of Structural Parameters in Application Development. IRC '79, Oct. 3–6, 1979, Venice, Italy, Proc. p. 262.

[3.789] *Dreyfuss, P., Fettes, L., Hansen, D.R.:* Elastomeric Block Polymers. RCT *53* (1980), pp. 728–771; *54* (1981), p. 181.

[3.790] *Fithian, L.F.:* Where TPE's Do and Don't Fit Versus Conventional Rubbers. KGK *36* (1983), p. 448.

[3.791] *Folkes, M.J.:* Processing, Structure and Properties of Block-Copolymers. Elsevier Applied Science Publ., Barking, 1985.

[3.792] *Goodman, I.:* Development in Block Copolymers, I. Elsevier Applied Science Publ., Barking, 1982.

[3.793] *McGrath, J.E.:* Overview of Synthesis – Structure – Properties Relationships for Thermoplastic Elastomers. Paper 16, 121st ACS-Meeting, Rubber-Div., May 4–7, 1982, Philadelphia, PA.

[3.794] *Hashimoto, T.:* Order-to-Disorder Transition in Block Copolymers. Gordon Res. Conf. of Elastomers, July 19–23, 1982, New London, NH.

[3.795] *van Henten, K.:* Thermoplastic Elastomers – Multipurpose Rubbers. KGK *31* (1978), p. 426.

[3.795a] *Hofmann, W.:* Thermoplastische Elastomere, Stoffklassen, Versuch einer Klassifikation. Kunststoffe *77* (1987), pp. 767–776.

[3.795b] *Hofmann, W.:* Bewertung von thermoplastischen Elastomeren. GAK *40* (1987), pp. 650–659.

[3.795c] Ref. [1.16], Chapter „Der Gummi- und Elastomerbegriff", pp. 39–43.

[3.796] *Holden, G.:* Thermoplastic Elastomers. Paper 54, 127th ACS-Meeting, Rubber-Div., Apr. 23–26, 1985, Los Angeles, CA.

[3.797] *Kawai, H., Hashimoto, T., Miyoshi, K., Uno, H., Fujimura, M.:* Microdomain Structure and Some Related Properties of Block Copolymers. RCT *54* (1981), p. 1011.

[3.798] *Legge, N. R.:* Thermoplastic Elastomers Based on Three-Block Copolymers – A Successful Innovation. Paper 3, 121st ACS-Meeting, Rubber-Div., May 4–7, 1982, Philadelphia, PA.

[3.799] *Legge, N. R.:* Thermoplastic Elastomers – The Future. Paper 73, 127th ACS-Meeting, Rubber-Div., Apr. 23–26, 1985, Los Angeles, CA.

[3.800] *Morton, M.:* Structure-Property Relation in Amorphous and Crystallizable ABA Triblock-Copolymers. Paper 58, 123rd ACS-Meeting, Rubber-Div., May 10–12, 1983, Toronto, Ontario.

[3.800a] *O'Connor, G. E.:* Thermoplastic Elastomers – Opportunity or Threat. Paper 52, 128th ACS-Meeting, Rubber-Div., Oct. 1–4, 1985, Cleveland, OH.

[3.801] *Porter, L. S., Meinecke, E. A.:* Influence of Compression upon the Shear Properties of Bonded Rubber Blocks. RCT *53* (1980), p. 1133.

[3.802] *Quirk, R. P.:* New Block Polymers for Higher Temperature Applications. Gordon Res. Conf. of Elastomers. July 16–20, 1984, New London, NH.

[3.803] *Rader, C. P.:* Thermoplastic Elastomers. Paper at 127th ACS-Meeting, Rubber-Div., Apr. 23–26, 1985, Los Angeles, CA.

[3.804] *Schäfer, H. O.:* Thermoplastische Elastomere, Chancen oder Gefahren für die kautschukverarbeitende Industrie. KGK *36* (1983), p. 180.

[3.804a] *Scheele, W.:* Disskusionsbeitrag zu der Arbeit von *Th. Timm* [3.806a]. Kautschuk u. Gummi *14* (1961), pp. WT 392–394.

[3.805] *Simpson, B. J.:* Thermoplastic Elastomers. In: *Babbit, R. O. (Ed.):* The Vanderbilt Handbook. Publ.: Vanderbilt, 1978, pp. 238–240.

[3.806] *Teyssie, Ph.:* Further Developments in the Synthesis of Block Copolymers and their Application to Emulsions and Blends. Gordon Res. Conf. of Elastomers. July 15–19, 1985, New London, NH.

[3.806a] *Timm, Th.:* Organische hochpolymere Werkstoffe: Plastomere – Elastomere – Duromere, das Eigenschaftsbild hochpolymerer Werkstoffe als Grundlage einer Einteilung und Definition. Kautschuk u. Gummi *14* (1961), pp. WT 233–247 (see also Ref. [1.44, 1.48, 3.804a]).

[3.807] *Walker, B. M.:* Handbook of Thermoplastic Elastomers. Van Nostrand Reinold Co., New York, 1979.

[3.808] *Walker, B. M.:* Thermoplastic Elastomers – Overview and Trends. Paper 7, 121st ACS-Meeting, Rubber-Div., May 4–7, 1982, Philadelphia. PA.

[3.809] *Wallace, J. G., Abell, W. R., Hagman, J. F.:* Melt Processible Rubbers – A New Concept for the Rubber Industry. Paper 69, 127th ACS-Meeting, Rubber-Div., Apr. 23–26, 1985, Los Angeles, CA.

[3.809a] *Wright, M. A., Rader, C. P., Hamblin, N.:* Thermoplastic Elastomers in Transportation. Paper 19, 128th ACS-Meeting, Rubber-Div., Oct. 1–4, 1985, Cleveland, OH.

[3.809b] Thermoplastic Elastomers. Bibliography 3, Publ.: RCT.

[3.809c] Thermoplastic Polyester Elastomers. Akzo Plastics Information Nr. 22.000.64.

[3.809d] PEBAX – Eigenschaften und Verarbeitung. Brochure of Atochem, Düsseldorf.

[3.809e] Polyether-Block-Amide (PEBA) – eine neue Generation thermoplastischer Elastomere. KGK *34* (1981), p. 1048.

[3.809f] Thermoplastische Elastomere – Eine Kontroverse über eine gebräuchlich gewordene Terminologie. K-Plastic-Kautschuk-Zeitung, No. 366, April 21, 1988, p. 16.

[3.809g] Thermoplastische Elastomere (TPE) – eine Alternative. Seminar, June 23–24, 1988, Süddeutsches Kunststoffzentrum Würzburg, West Germany, Proceedings.

References on Butadiene and Isoprene Blockcopolymers (SBS, SIS)

[3.810] *Baker, M. L.:* Thermoplastic Elastomers from Dienes. Paper at 124th ACS-Meeting, Rubber-Div., Oct. 25–28, 1983, Houston, TX.

[3.811] *St. Clair, D. J.:* Rubber-Styrene Block Copolymers in Adhesives. RCT *55* (1982), p. 208.

[3.812] *Class, J. B.:* The Effect of Modifying on the Dynamic Properties of Styrene Block Polymers. Paper 56, 127th ACS-Meeting, Rubber-Div., Apr. 23–26, 1985, Los Angeles, CA.

[3.812a] *Flodin, P.:* Multiblock Copolymers as Elastomers. IRC '86, June 2–6, 1986, Göteborg, Sweden.

[3.813] *Holden, G.:* Current Applications of Styrene Block Copolymer Rubbers. KGK *36* (1983), p. 356.

[3.814] *Hoover, J. T., Ward, T. C., McGrath, J. E.:* Influence of Hydrogenation on the Chemical Structure – Physical Property Relationship of Star Block Polymers, Based on Tertiary Butylstyrene-Isoprene and Divinylbenzene. Paper 55, 127th ACS-Meeting, Rubber-Div., Apr. 23.–26, 1985, Los Angeles, CA.

[3.814a] *Hsiene, G. H., Wu, J. L., Yang, J. M.:* Surface Hydrogellation of Styrene-Butadiene Styrene Block Copolymers for Biomaterials. Paper 33, 128th ACS-Meeting, Rubber-Div., Oct. 1–4, 1985, Cleveland, OH.

[3.815] *Kraus, G.:* Modification of Asphalt by Block Polymers of Butadiene and Styrene. RCT *55* (1982), p. 1389.

[3.815a] *Morton, M.:* Thermoplastic Elastomers from Triblock Copolymers. IRC '86, June 2–6, 1986, Göteborg, Sweden.

[3.816] *Mistrali, F., Gargai, L.:* Long Term Elastic Properties of α-Methylstyrene-Diene Teleblock Copolymers. KGK *37* (1984), p. 377.

[3.817] *Recchutte, A. D., Dimeter, G. R.:* The Effect of Petroleum Extender Oil Types on the Properties of a Compounded Radical Styrene Block Thermoplastic. RCT *55* (1982), p. 1437.

[3.818] *Sogah, D. Y., Webster, O. W., Hertler, W. R.:* Synthesis of Block Copolymers by Group Transfer Polymerization. Paper 71, 127th ACS-Meeting, Rubber-Div., Apr. 23–26, 1985, Los Angeles, CA.

[3.819] *Yang, R. M., Meinecke, E. A.:* Dynamic Properties of SBS Block Copolymers. RCT *53* (1980), p. 1124.

TPE-E see Refs. [3.735–3.741]

TPE-A see Refs. [3.742–3.744]

TPE-U see Refs. [3.764–3.777]

References on Polyolefine-modified Elastomers

[3.820] *v. Bassewitz, K., zur Nedden, K.:* Elastomer-Polyolefin-Blends, Neuere Erkenntnisse über den Zusammenhang zwischen Phasenaufbau und anwendungstechnischen Eigenschaften. KGK *38* (1985), p. 42.

[3.821] *Blackley, D. C.:* Dynamic Mechanical Properties of Polyolefin Blend Thermoplastic Rubbers. Gordon Res. Conf. of Elastomers, July 19–23, 1982, New London, NH.

[3.821a] *O'Connor, G. E.:* Thermoplastic Elastomers – Opportunity or Threat? KGK *39* (1986), pp. 695–696.

[3.821b] *O'Connor, G. E., Fath, M. A.:* Thermoplastic Elastomers. Rubber World (1982), 12, (1982) 1.

[3.822] *Coran, A. Y., Patel, R.:* Rubber-Thermoplastic Compositions. I. EPDM-Polypropylen Thermoplastic Vulcanizates. RCT *53* (1980), p. 141; II. NBR-Nylon-Thermoplastic Compositions. RCT *53* (1980), p. 781; III. Predicting Elastic Moduli of Melt Mixed Rubber-Plastic Blends. RCT *54* (1981), p. 91; IV. Thermoplastic Vulcanizates from Various Rubber-Plastic-Combinations. RCT *54* (1981), p. 892. VII. Chlorinated Polyethylene Rubber-Nylon Composition. RCT *56* (1983), p. 210; VIII. Nitrile Rubber-Polyolefin Blends with Technological Compatibilization. GAK *37* (1984), p. 378.

[3.822a] *Coran, A. Y., Patel, R.:* Thermoplastische Vulkanisate aus verschiedenen Kautschuk-Kunststoffverschnitten. KGK *35* (1982), p. 194.
[3.823] *Coran, A. Y., Patel, R. P., Williams, D.:* Rubber Thermoplastic Compositions, V. Selecting Polymers for Thermoplastic Vulcanizates. RCT *55* (1982), pp. 116, 536; VI. The Swelling of Vulcanized Rubber-Plastic Composition in Fluids. RCT *55* (1982), p. 1063; IX. Blends of Dissimilar Rubbers and Plastics with Technological Compatibilization. Paper 68, 127th ACS-Meeting, Rubber-Div., Apr. 23-26, 1985, Los Angeles, CA.
[3.824] *Coran, A. Y.:* Useful Elastomeric Materials Based on Rubber-Thermoplastic Compositions. IRC '85, Oct. 15-18, 1985, Kyoto, Proc. p. 92.
[3.825] *Daalmans, H.:* Thermoplastic Rubber. IRC '79, Oct. 5-8, 1979, Venice, Italy, Proc. p. 281.
[3.826] *Danesi, S., Balzani, L.:* Composition of Thermoplastic Polyolefin. ICR '79, Oct. 5-8, 1979, Venice, Italy, Proc. p. 272.
[3.827] *Danesi, S., Gavagnani, E.:* Olefinic Thermoplastic Elastomers. KGK *37* (1984), p. 195.
[3.828] *Goettler, L. A., Richwine, J. R., Wille, F. J.:* The Rheology and Processing of Olefine-Based Thermoplastic Vulcanizates. RCT *55* (1982), p. 1448.
[3.829] *Ho, C. C., Kontos, E. G.:* Processability of Thermoplastic Polyolefin Elastomers - Sheet Vacuum Forming. RCT *56* (1983), p. 68.
[3.829a] *Hofmann, W., Koch, R.:* Neue Entwicklungen auf dem Gebiet des thermoplastischen Elastomer Alcryn. KGK *41* (1988), pp. 888-894.
[3.829b] *Karger-Koesis, J., Kozma, B., Schoeber, M.:* Tauroprene - A New Versatile Polyolefinic Thermoplastic Rubber. KGK *38* (1985), p. 614.
[3.830] *Kear, K. E., Rader, C. P.:* Fatigue Resistance of EPDM-Polypropylene Thermoplastic Vulcanizates. Paper 28, 126th ACS-Meeting, Rubber-Div., Oct. 23-26, 1984, Denver, CO.
[3.830a] *Koch, R.:* Neue Entwicklungen bei thermoplastischen Elastomeren. KGK *39* (1986), pp. 804-809.
[3.831] *Mitchell, J.:* Thermoplastic Elastomers from Olefins. Paper at 124th ACS-Meeting, Rubber-Div., Oct. 25-28, 1983, Houston, TX.
[3.832] *Stemper, J., Dufour, D. L., van Issum, E.:* Santoprene-Kautschuk, eine Herausforderung für die Gummiindustrie. KGK *37* (1984), p. 13.
[3.833] *Volforá, E., Zerzaň, J., Pelzbauer, Z.:* Einfluß der Wärmebehandlung auf die Eigenschaften von Polypropylen-EPDM-Mischungen. KGK *37* (1984), p. 291.
[3.833a] Levaflex EP, Thermoplastisches Polyolefin Elastomer. Brochure of Bayer AG, 1983.
[3.833b] Santoprene, Thermoplastischer Kautschuk. Brochure of Monsanto Europe S. A.
[3.833c] Geolast, Thermoplastic Elastomer. Brochure of Monsanto, Brussels, 1986.
[3.833d] Alcryn, eine neue Familie elastischer Thermoplaste. GAK *38* (1985), p. 294.

References on Other Thermoplastic Elastomers

[3.834] *Arcozzi, A., Biggi, A., Da Re, G.:* Elastoplastics Materials Based on Regularly Alternated Polyether/Esteramide. IRC '79, Oct. 3-6, 1979, Venice, Italy, Proc. p. 244.
[3.835] *della Fortunata, G., Melis, A., Peregio, G., Vitali, R., Zotteri, L.:* Morphology and Mechanical Properties of New Elastoplastics Based on Alternated Polyesteramide. IRC '79, Oct. 3-6, 1979, Venice, Italy, Proc. p. 229.
[3.836] *Gerger, W. P.:* Hydrogenated Block Copolymers. KGK *37* (1984), p. 284.
[3.837] *Hedrick, R. M.:* Nylon Block Copolymer Elastomers. Gordon Res. Conf. of Elastomers, July 16-20, 1984, New London, NH; Paper 16, 127th ACS-Meeting, Rubber-Div., Apr. 23-26, 1985, Los Angeles, CA.
[3.838] *Kennedy, J. P.:* Macromolecular Engineering by Carbocationic Techniques, New Polymers that Containing Elastomeric Sequences. RCT *56* (1983), p. 639.
[3.839] *McKnight, W. J.:* Structure and Properties of Polybutadiene Containing Polyurethanes. Paper 18, 121st ACS-Meeting, Rubber-Div., May 4-7, 1982, Philadelphia, PA.
[3.840] *Onder, K., Chen, A. T., Nelb, R. G., Sounders, K. G.:* A New High Temperature Co-Poly(esteramide) Thermoplastic Elastomer. Paper 2, 128th ACS-Meeting, Rubber-Div., Oct. 1-4, 1985, Cleveland, OH.

[3.841] *Tang, S. H., Meinecke, E. A., Riffle, J. S., McGrath, J. E.:* Structure. Property Studies on a Series of Polycarbonate-Polydimethylsiloxane-Block-Copolymers. RCT *53* (1980), p. 1160.

[3.842] *Tyagi, D., et al.:* Novel Segmented Elastomers Based on Aminoalkyl Terminated Polysiloxanes, Synthesis and Characterization. Paper 47, 123rd ACS-Meeting, Rubber-Div., May 10–12, 1983, Toronto, Ontario.

[3.843] *Wilkes, G. L., Badrogia, S., Kennedy, J. P.:* Structure/Property Studies of a Model Thermoplastic Ionomer Based on Polyisobutylene. Paper 13, 125th ACS-Meeting, Rubber-Div., May 8–11, 1984, Indianapolis, IN.

[3.844] *Yamamoto, N., Yamashita, I., Tanaka, K., Hayashi, K.:* Structure/Property Relationship in Biomedical Polyurethane Ureas based on PEO-PPO-PEO-Block-Copolyethers. IRC '85, Oct. 15–18, 1985, Kyoto, Proceed. p. 413.

Special References on Thermoplastic NBR

[3.845] Brit. P. 1339653, 1973, Aquitaine Total Organico.

[3.846] USP 3950313, 1976, Monsanto Chemicals, P. J. S. Bain, W. R. Foster, A. J. Neale.

[3.847] USP 3888964, 1975, Monsanto Chemicals, P. J. S. Bain, E. P. McCall.

[3.848] Can. P. 185313, 1973; 185359, 1973, Polysar, E. Lasis, E. J. Buckler.

[3.849] DBP 2613050, 1976, Polysar, E. Lasis, E. J. Buckler.

[3.850] Can. P. 1014295, 1977, Polysar, E. Lasis, E. J. Buckler, J. R. Dunn.

3.7.4.23 Comparative Properties and Rubber Blends

Comparative Properties

[3.851] *Dunn, J. R.:* Performance Limit of Elastomers. KGK *38* (1985), p. 611.

[3.851a] *Timm, Th.:* Die physikalischen Leistungsgrenzen von Elastomeren. KGK *39* (1986), p. 15.

[3.852] *Hofmann, W.:* Technische Elastomere. Kunststoffe *77* (1987), pp. 1057–1064.

[3.852a] *Dunn, R. J., Pfisterer, H.-A.:* Using Simulated End-use Conditions in assessing the high Temperature Performance of Oil Resistant Vulcanizates. J. Elastomer Plast. *9* (1977), p. 193.

[3.852b] *Engelmann, E.:* Kälteverhalten von Vulkanisaten, Einfluß des Mischungsaufbaues und Wahl der Prüfmethode. KGK *25* (1972), pp. 538–546.

[3.852c] *Lewis, P. H.:* Laboratory Testings of Rubber Durability. Polymer Testings *1* (1980), pp. 167–189.

[3.852d] SIS-Handbuch 131: Gummiquellungen. Gentner-Verlag, Stuttgart, 1979.

[3.852e] Chemical Resistance Data. Publ.: RAPRA, 1969.

[3.852f] Ref. [2.84a], pp. 22, 109.

References on Rubber Blends

[3.853] *Bhowmick, A. K., De, S. K.:* Effect of Curing Temperature and Curing System on Structure-Property Relations of Rubber Blends. RCT *53* (1980); p. 960.

[3.853a] *Bhowmick, A. K.:* Studies on Polymer Blends – Polymer – Solvent Interactions and Mooney-Rivlin Constants. KGK *34* (1981), p. 558.

[3.854] *Bohn, L.:* Incompatibility and Phase Formation in Solid Polymer Mixtures and Graft and Block Copolymers. RCT *41* (1968), p. 495.

[3.855] *Corish, P. J.:* Fundamental Studies of Rubber Blends. RCT *40* (1967), p. 324.

[3.856] *Corish, P. J., Powell, B. D. W.:* Elastomer Blends. RCT *47* (1974), p. 481.

[3.857] *Dschagarowa, E., Doikowa, S., Wassilew, A.:* Kinetik der Vulkanisation in Mischungen aus zwei Kautschuken. KGK *34* (1981), p. 116.

[3.858] *Gardiner, J. B.:* Studies in the Morphology and Vulcanization of Gum Rubber Blends. RCT *43* (1970), p. 370.

[3.859] *Inoue, T., Shomura, F., Ougizawa, T., Miyasaka, K.:* Covulcanization of Polymer Blends. Paper 4, 127th ACS-Meeting, Rubber-Div., Apr. 23–26, 1985, Los Angeles, CA.

[3.860] *Krause, S.:* Polymer Compatibility. J. Macromol. Sci., Revs. Macromol. Chem. *C7* (1972), 2, p. 251.

[3.860a] *Struckmeyer, H. F., Hofmann, W.:* Covulkanisate für High-Tech-Applikationen. GAK 42 (1989), in preparation.

References on Blends of Rubbers and Plastics

[3.861] *Adams, G. C.:* Understanding Toughness Measurements. Paper 30, 121st ACS-Meeting, Rubber-Div., May 4–7, 1982, Philadelphia, PA.

[3.862] *Bucknall, C. B.:* The Effective Use of Rubbers as Toughening Agents for Plastics. Paper 29, 121st ACS-Meeting, Rubber-Div., May 4–7, 1982, Philadelphia, PA.

[3.863] *Dinges, K.:* Die Rolle des Elastomeren in Elastomer-modifizierten Kunststoffen. KGK *32* (1979), p. 748.

[3.864] *Min, K., Plochocki, A. P., White, J. L., Fellers, J. A.:* Rheological Behavior, Phase Morphology and Processing of Polymer Blends. Paper 31, 121st ACS-Meeting, Rubber-Div., May 4–7, 1982, Philadelphia, PA.

[3.865] *Sperling, L. H.:* A Comparison of Interpenetrating Polymer Networks and Graft Copolymers as Materials for Rubber Toughened Plastics. Paper 32, 121st ACS-Meeting, Rubber-Div., May 4–7, 1982, Philadelphia, PA.

[3.866] *Takemori, M. T., Morelli, T. A.:* Fatigue Fracture Mechanisms in Rubber Modified Blends of PPO and PS. Paper 35, 121st ACS-Meeting, Rubber-Div., May 4–7, 1982, Philadelphia, PA.

[3.867] *Wu, S.:* Energy Dissipation Mechanisms in Impact Fracture of Rubber Toughened Semicrystalline Plastics. Paper 33, 121st ACS-Meeting, Rubber-Div., May 4–7, 1982, Philadelphia, PA.

[3.868] *Yee, A. F., Pearson, R. A.:* Studies on Toughening Mechanisms in Rubber Modified Epoxies. Paper 34, 121st ACS-Meeting, Rubber-Div., May 4–7, 1982, Philadelphia, PA.

4 Rubber Chemicals and Additives

[4.1–4.28 a]

The wide range of possible applications of elastomers is partially due to the fact that they can be compounded with numerous chemicals and additives, e.g., mastication aids, vulcanization chemicals, chemicals protecting against aging, fillers, softeners, blowing agents, etc.

In principle the nature of the elastomer determines the basic properties of the products manufactured from it. However, these properties may be significantly modified by the kind and amount of the compounding ingredients used. On the other hand the chemicals and fillers influence the mixing and processing behavior of the rubber compounds and make their vulcanization possible and permit a wide range modification of the vulcanizate properties and their adaption to a multitude of requirements. The more these opportunities available to the compounder are utilized, the easier it is to meet the often conflicting requirements placed upon a compound.

Because of the effort to arrive at dustfree processing operations, more and more chemicals in coated, pasted and granulated form have been available [4.16–4.28 a]. These efforts are desirable because of toxicological considerations and should be continuted [4.18].

4.1 Mastication and Peptizers

[4.29–4.36]

4.1.1 The Principle of Mastication

4.1.1.1 Processing Viscosity

The elastomer has to have a certain viscosity in order to incorporate fillers and other compounding ingredients easily and to disperse them uniformly. An optimized viscosity during the preparation of compounds from elastomer blends is especially important. A good dispersion of the compounding ingredients is only possible after a homogeneous dispersion of the macromolecules has been reached, when polymer blends are used. The correct viscosity adjustment is also important in order to obtain desired processing properties, e.g., for extrusion or calendering processes as well as for injection molding and other processing operations. Also, an especially low viscosity has to be obtained for the preparation of solutions of rubber compounds and low density foams.

A viscosity reduction – especially a minor one – can often be obtained by the application of process aids, e.g., stearic acid, and especially Zn-soaps, Ca-soaps of fatty acids and also processing plasticizers. However, if large viscosity reductions are required – as is the case especially for NR – the compound has to be masticated.

4.1.1.2 Mastication Without Peptizers

During the mastication process high shear forces are created because of the constant mechanical deformations and the high viscosity of the elastomers – especially at low temperatures – that the polymer chains are broken. The higher these shear forces are, the higher this effect is. A recombination of the fractured molecule ends is prevented because of their reaction with oxygen. Without presence of oxygen, mastication would not be possible. The consequence of mastication is a reduction of the average molecular weight, i.e., a degradation which results in a reduction of the viscosity.

This mechanical mastication process was invented in 1819 by *Hancock* using spiked rolls. The process is dependent on temperature with a negative temperature coefficient. With increasing temperature the elastomer softens because of its thermoplasticity and thus absorbs less mechanical energy because the polymer molecules can flow with respect to each other more easily (see Fig. 4.1). The result of this is an increasingly less effective mechanical degradation which vanishes altogether at temperatures of about 120–130 °C because of excessive softening.

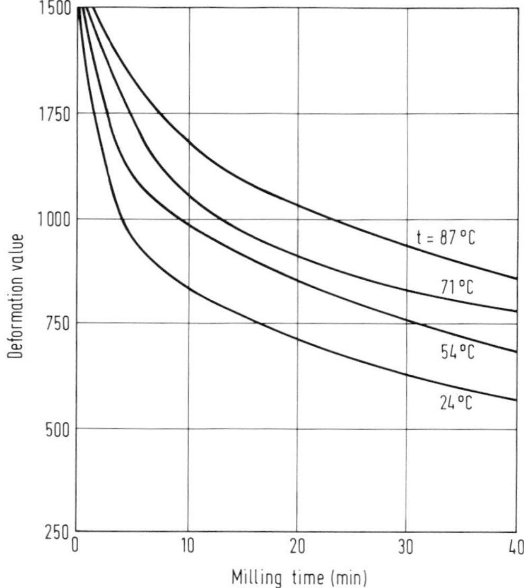

Figure 4.1 Influence of the mastication time on the viscosity of NR at various temperatures

However, simultaneously another degradation process with a positive temperature coefficient comes into play, the oxidative mastication process. With further increases in temperature, chain scission of the polymer molecules due to oxidative attack increases and thus the viscosity reduction accelerates.

Observing the degradation of an elastomer as a function of temperature reveals a maximum in the viscosity-temperature curve at 120–130 °C because of the two mechanisms involved (see Fig. 4.2).

The mechanical mastication effect is further reduced exponentially with time (negative time coefficient, see Fig. 4.1) because the shear forces decrease with time because of the softening of the elastomer caused by the mastication itself. In contrast, the oxidative mastication is ideally independent of the viscosity which results in a linear degradation with time.

For a mechanical mastication the problematic requirement has to be met that the temperature of the elastomer be kept as low as possible in order to obtain high shear forces. Nevertheless, during the mastication of 40 kg smoked sheets on a 60″ mill (2500 × 650 mm roll dimensions) the temperature rises above 80° because of frictional losses even though the rolls are fully cooled. At this temperature the

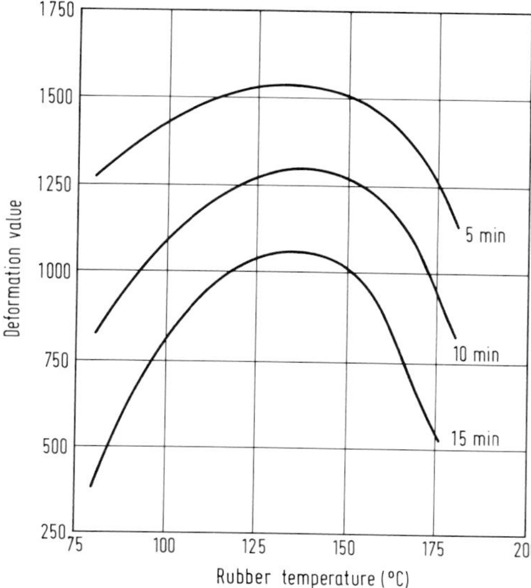

Figure 4.2 Mastication rate of NR as a function of temperature and time

mechanical mastication is not optimal anymore. Thus, already after 5–10 min milling time a marked reduction in the mastication effect is noticeable. Therefore, relatively long mastication times are necessary using this technique or it may even be required to perform several mastication processes of the same batch with cooling periods in between (see page 359).

4.1.1.3 Mastication With Peptizers

The consequence of these considerations is the preferred application of the oxidative mastication where the mastication effect increases with increasing temperature. This technique was only made possible by the application of ingredients which catalyze the oxidation process, the so-called peptizers. They enhance the formation of the primary radicals necessary for chain scission. Thus the oxidative mastication is greatly accelerated and starts already at much lower temperatures (see Fig. 4.3).

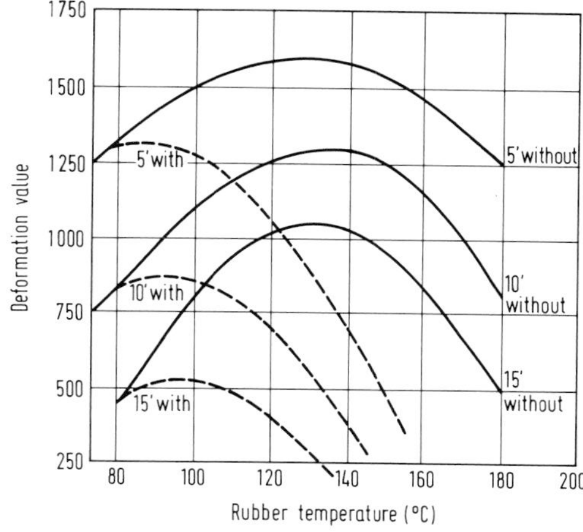

Figure 4.3 Mastication rate of NR as a function of the elastomer temperature with and without 0.1 phr ZnPCTP

It is important that the oxidation catalysts be de-activated during the preparation of the compound so that they will not promote any degradation of the compound or the vulcanizate during later temperature or aging influences.

The effect of the peptizer is – in the ideal case – independent of the viscosity so that the mastication curves do not exhibit an asymptotic but approx. a linear form. After a deactivation of the peptizer the viscosity should then remain constant.

The catalytic influence of substances upon mastication was already discovered in the 1930s. To date numerous substances were developed, e.g., β-Thionaphthol, Xylylmercaptan and its Zn salt, 2,3,5-Trichlorthiophenol, 9-Mercaptoanthracen, 0,0′-Dibenzamidodiphenyldisulfide, Pentachlorothiophenol (PCTP) and its Zn-salt (Zn PCTP). Many of these are not fully satisfactory (toxicology, distribution in the elastomer, staining, price, specifity of effectiveness, effective temperature and effectiveness). Most of these have only a very limited technological importance today.

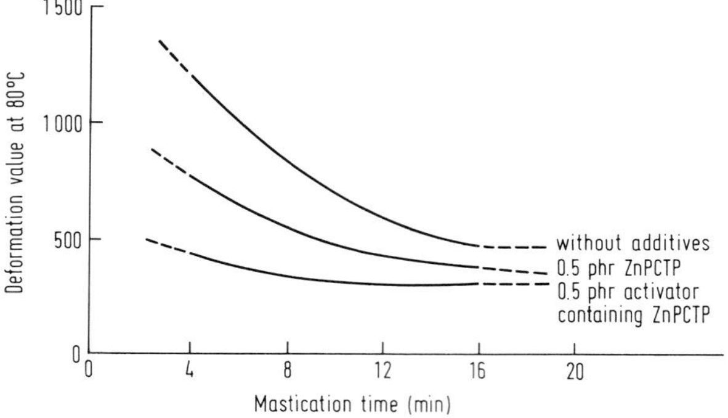

Figure 4.4 Mastication of NR at 80 °C elastomer temperature with and without 0.5 phr ZnPCTP or activated ZnPCTP, respectively

Only very recently, potent activators based on inner-complex salts have been developed. For example, compounds of tetraazo porphyrines or phthalocyanines with metals like Co, Cu, Mn and Fe and other complexes of Fe, V and other metals were used. Thus it is possible to perform the oxidative mastication at a low temperature similar to mastication on an open roller mill (Fig. 4.4). Furthermore, these activators can be used with NR as well as with SR. They are useful in very small quantities and are applied on an inert support for ease of handling and weighing.

4.1.2 Peptizing Agents

[1.14]

4.1.2.1 Peptizing Agents Without Activators

Pentachlorothiophenol (PCTP) and its zinc salt (Zn PCTP), and dibenzamidediphenyl disulfide among others were at one time frequently used as peptizing agents, while PCTP was best used in SR such as SBR and NBR, Zn PCTP was preferred for NR, respectively IR. Since the development of activators containing ZnPCTP as well as activators containing Zn-soaps, the use of the non-activated peptizers has decreased drastically.

4.1.2.2 Activator Containing Peptizers

Activated Sulfur Compounds. The activator containing products, especially those based on ZnPCTP have the advantage over the non-activated products that they can be used over a wide temperature range, for example 80–180 °C on the open mill as in an internal mixer and for NR and SBR. Because of their high potency they can drastically shorten the mastication time, increase mastication capacity as well as lower labor cost. Since their activity is high at temperature levels reached automatically in the mixing process, cooling water or heating energy can be saved. The oxidation-catalytic action of the activators can be inhibited by sulfur, antioxidants or active fillers. When inactive fillers are added, the peptizer remains active and therefore mixing can begin during the mastication process.

One of the problems with such products is the difficulty in distributing extremely small portions in big quantities of rubber in the usually used short mixing times, which causes, therefore, often insufficient uniformity of mastication effect in the rubber. Another problem is the possibility of formation of toxic dioxines during mastication. These are the reasons why in some cases pentachlorthiophenol-free mastication aids are preferred.

Activator Containing Zn-Soaps [2.121]. Compared to activated Zn PCTP, activated Zn-soaps have the additional advantage that they are soluble in rubber and are readily mixed and uniformly dispersed. The consequence is rapid bonding and completely homogeneous mastication that makes smooth further processing possible. Furthermore, a drastic lowering of the viscosity is obtained when the Zn-soap is added to the mastication process, which results in further energy and time savings. The Zn-soap addition also improves the flow properties of the rubber compounds [2.120] which in turn facilitates processing and increases production. The application of process aids can therefore be reduced or completely avoided.

4.2 Vulcanization and Vulcanization Chemicals

4.2.1 The Vulcanization Process

[1.7, 1.12, 1.13, 2.3, 3.264, 4.37–4.77]

4.2.1.1 General Observations

The Concept of Vulcanization. Vulcanization is the conversion of rubber molecules into a network by the formation of crosslinks. Vulcanizing agents are necessary for the crosslink formation (see page 229 ff). These vulcanizing agents are mostly sulfur or peroxide and sometime other special vulcanizing agents or high energy radiation (see page 403 ff). As long as the molecules are not tied to each other, they can move

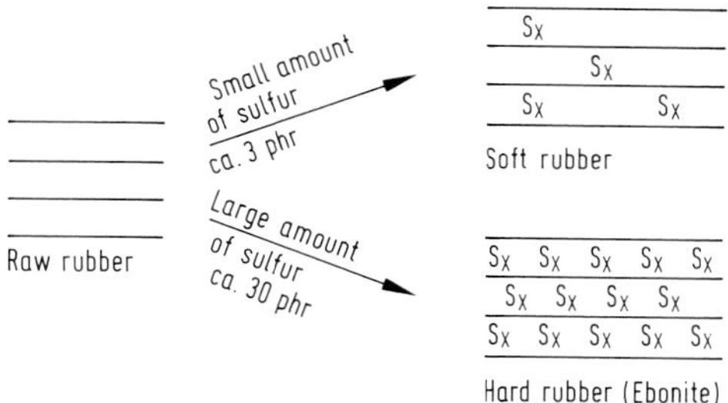

Figure 4.5 Uncrosslinked and crosslinked rubber

more or less freely, especially at elevated temperatures (macro-Brownian motion); at that point the material is plastic. It exhibits mechanical and thermodynamic irreversible flow (see Fig. 1.1, page 4). By crosslinking, the rubber changes from the thermoplastic to the elastic state. As more crosslinks are formed, the vulcanizate becomes tighter and the forces (stress forces) necessary to achieve a given deformation, increase (Fig. 4.5).

Degree of Vulcanization. The number of crosslinks formed depends on the amount of vulcanization agents, its activity and the reaction time. One calls it the degree of vulcanization or crosslink density. In sulfur vulcanization, the most commonly used, various types of crosslinks are formed depending on the quantity and activity of the other additives, particularly accelerators. These crosslinks can be anything between monosulfidic to polysulfidic (Fig. 4.6). The resulting properties of the vulcanized rubber depend a great deal on the number and type of crosslinks. The type and quantities of fillers and plasticizers, etc. can have an even stronger effect on the properties of the vulcanizate than the crosslink density.

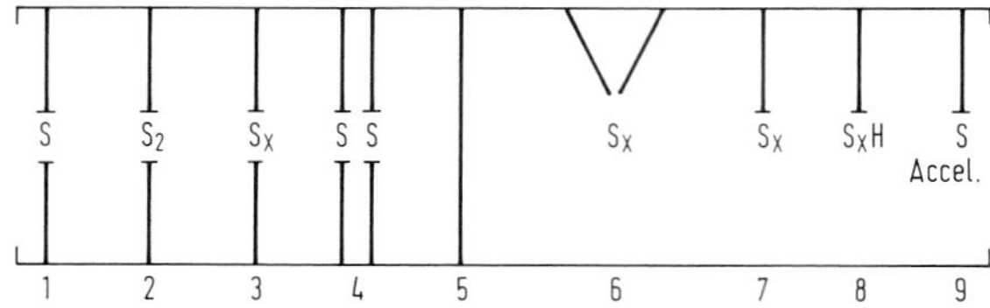

Figure 4.6 Different crosslink structures: monosulfidic (1), disulfidic (2), polysulfidic (3), when $X \geq 3$, vicinal (4), C-C-crosslinking (5), chain modifications, cyclic sulfur structures (6), sulfur chains (7), thiol groups (8), bound accelerator residues (9)

Stages of Vulcanization (Curing Stages). The stress-vulcanization time curve can be divided into several phases: The ratio of pre-vulcanization at the beginning of vulcanization, the under-vulcanization, the optimum vulcanization and the over-vulcanization (over cure) which typically leads to "reversion" [4.67–4.77] in NR and IR (Fig. 4.7, see also page 427).

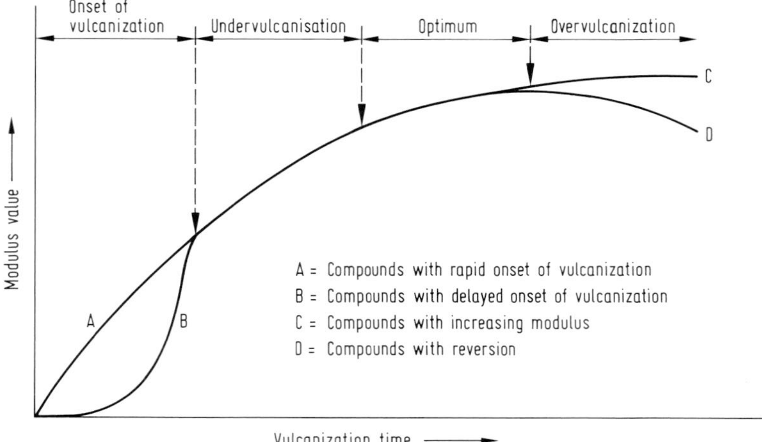

Figure 4.7 Stages of vulcanization

Pre-vulcanization is the time during which tight curing (thickening) is observed, during that period a rubber compound still flows uninhibited, as for example, in compression molding. Depending on the choice of vulcanization chemicals, the start of vulcanization of a rubber compound can be rapid or retarded. In a compression molding process it is necessary for the rubber compound to maintain a more or less prolonged flow in order to fill all cavity spaces and for all entrapped air to escape. In an open cure very fast cure initiation is needed. The tightening caused by crosslinking by which an undesirable deformation is prevented is opposed by the softening caused by the heat of vulcanization.

A rapid vulcanization initiation is undesirable in most cases, since it interferes with safe processing of the compounds and ultimately is the cause of vulcanization already occurring during mixing and processing. The pre-vulcanized compound can no longer be extruded or calendered.

In the under-cure phase, most technological properties of rubber are not yet fully developed. It is therefore usually necessary to vulcanize to the optimum cure stage (maximum stress values). Since all technological properties do not reach their optimum value simultaneously, it is necessary to compromise with a light over- or undercure.

On prolonged heating of SR-types no decrease in the stress/heating time curve is observed (broad vulcanization plateau) and even a small increase in stress after the optimum cure occurs (increasing stress value characteristics). With NR one observes after passing the vulcanization optimum (depending on the choice of vulcanization systems), a more or less rapid decrease in crosslinking (reversion) that is accompanied by a lowering of the mechanical properties. Depending on how rapid the reversion occurs, one speaks of short or broad plateaus in the stress-time curve of a vulcanizate, etc. The width of the plateau is a measure of the heat stability since it indicates the influence of the heat of vulcanization on the stress value of the vulcanizate.

4.2.1.2 Change of Properties of Elastomers Depending on the Degree of Vulcanization

[4.54–4.66]

Mechanical Properties. Since, by definition, vulcanization is the process of converting the gum-elastic raw material into the rubber-elastic end product, the ultimate properties of the latter depend on the course of the vulcanization and the choice of the vulcanization chemicals. As long as the single rubber molecules are not tied to each other, the stress values are low and approach zero. The stress value of a vulcanizate represents the number of crosslinks and is therefore proportional to the degree of vulcanization as crosslink density at low elongation. The following equation correlates the *stress value* P and crosslink density M_c^{-1} (stress = f (crosslink density))

$$P = \delta RTA_o^{-1}M_c^{-1}\,(l - l^{-2})$$

In this equation, δ is the density of the rubber, R is the gas constant, T the absolute temperature, A_o the cross sectional area of the test specimen before extension and M_c the average molecular weight between crosslinks as a reciprocal measure of crosslink density and l the elongation (deformation) (see Fig. 4.8). According to this equation, the stress at a certain elongation depends on the reciprocal value of the molecular weight of the rubber molecule between two crosslinks at constant den-

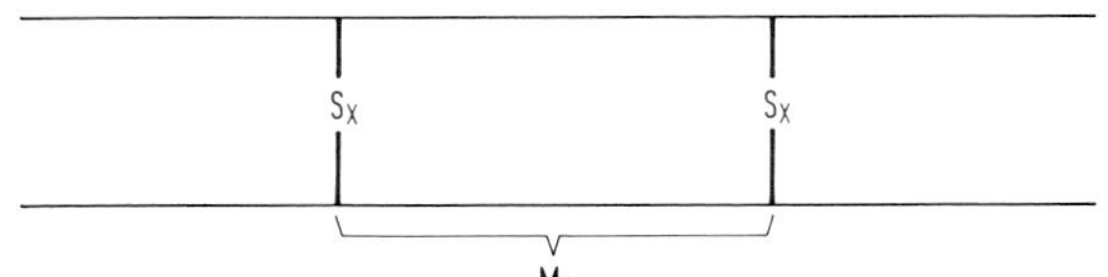

Figure 4.8 Average molecular weight M_c of the rubber molecules in between two crosslink sites

sity, temperature and dimensions of the test specimen and therefore the number of crosslinks. The denser the network, the shorter are the molecular segments between crosslinks and the higher the stress values. On the other hand one can deduce the crosslink density of an ideal vulcanizate from the stress value. The stress values are largely independent of the chemical composition of the macromolecules or the nature of the crosslinks.

The *hardness* increases analogous to the stress value with increasing crosslink density until it has reached the Hard Rubber (Ebonite) stage (see Fig. 4.5, page 221). This has to be expected if one remembers that hardness and stress values are determined by similar principles (see page 475). While stress values and hardness increase with the increasing number of crosslinks, the tensile strength at first increases to an optimum value and then readily decreases (overcuring) with increasing crosslink density. With even higher crosslink density, tensile strength of some rubbers again increases until it reaches the high strength of "Hard Rubber" [1.12, 1.13].

Elongation at break decreases with increasing crosslink density. It approaches asymptotically very low values at high crosslink densities.

Also the *permanent deformation* (permanent set) (for example, permanent set on elongation or compression set on compressions at room temperature) decreases with increasing crosslink density to an optimum value. It is therefore necessary to achieve an optimum crosslink density for products that have to show extremely low set [3.220]. Frequently it is inversely proportional to the elasticity.

Optimum *resistance to tearing* (tear strength) is usually reached somewhat earlier than the stress value. This means that slightly undercured compounds show highest tear strength. At high crosslink densities, especially in overcured rubber, the tear strength rapidly decreases.

Rubber *elasticity* (entropic elasticity) stems from the movements of chain segments (Micro-Brownian motion). Using a small force, large reversible deformations occur. One deals here with straightening and de-entangling of the macromolecules into a statistically improbable position (lower entropy). Elasticity increases with increasing Micro-Brownian motion with increased crosslinking to an optimum value. In the ideal plastic state the macromolecules remain in their new position after extension. As a consequence of "fixing" the macromolecules after having been removed from their starting position, they have the tendency to return to that location. With increasing number of crosslinks, this tendency becomes strong. At high crosslink density, the segments of the macromolecules become immobile, the system becomes stiffer and the elasticity decreases. The correlation between elasticity and crosslink density is minute to that existing between stress and crosslink density. This means that rubber elasticity depends primarily on the number of crosslinks.
Elasticity = f (crosslink density)

$$E = \frac{1}{2}\,\delta R T M_c^{-1}\,(l_1^2 + l_2^2 + l_3^2 - 3)$$

In this equation, E is the modulus of elasticity, δ the density of the rubber, R the gas constant, T the absolute temperature, M_c the average molecular weight between crosslinks as a reciprocal measure of crosslink density and l_1, l_2, l_3 the elongations along the three coordinates. Accordingly, the elasticity within narrow limits is proportional to the crosslink density.

Where crosslink density becomes too high, i.e., on overcuring, rigid structures supercede entropic elasticity, called "energy elasticity" (or steel elasticity), as for example in the "hard rubber" (Ebonite) range.

During elastic recovery this in reality never reaches 100% (otherwise it would represent a "Perpetual Motion" of the first type) a smaller or larger portion of energy is lost as heat because of friction of molecules, depending on the amount of the elasticity. In vulcanizates of high elasticity, the loss of energy is small and the heat build-up during dynamic stress is small.

Low Temperature Flexibility. The low temperature flexibility of a vulcanizate is primarily dependent on the glass transition temperature (T_g) of the rubber and the types and amounts of plasticizers used. However, rubber elasticity has a significant bearing on low temperature flexibility. With decreasing temperature the elasticity decreases until it changes into steel-elasticity at the freezing point (T_g) see (Fig. 1.3, page 5); elasticity goes through a minimum during this process. At that temperature, a vulcanizate ceases to be rubber-elastic. With increasing crosslinking to an optimum value of micro-Brownian mobility and elasticity, low temperature flexibility is also improved (compare also Section 3.6.2.2, page 170).

Swelling. Non-crosslinked rubber (gum rubber) can swell in certain compatible solvents till it loses all cohesive strength and goes into solution. A swelling equilibrium does not occur in this process. The necessary condition for this process is the higher osmotic pressure that the solvent exacts on the polymer, compared to the forces that hold the molecules of the high polymer together. When the forces that hold the macromolecules together are increased through crosslinks, we could assume that the high polymer will not dissolve any longer but will swell to a smaller or larger extent. With increasing crosslinking, swelling becomes less.

A correlation between equilibrium swelling and crosslink density can be established similar to that found between crosslink density and stress value. The so called *Flory-Rehner* equation describes it as follows:

$$-\ln[(1-V_2)\,V_2+\gamma V_2^2]=\left(\frac{V_1}{\overline{V}_2\,M_c}\right)\left(1-\frac{2M_c}{M}\right)\left(V_2\cdot{}^1\!/\!_3-\frac{V_2}{2}\right)$$

In this equation, V_2, the volume fraction of rubber in the swollen network, V_1 the molar volume of solvent, $\overline{V}_2$ the specific volume of the polymer, γ the Huggins solubility parameter (an interaction constant between solvent and polymer), M the molecular weight of the polymer (before crosslinking), M_c the molecular weight between two crosslinks. The solubility parameter is not important in this context [3.190–3.194]. It allows conclusions with respect to the thermodynamic compatibility of polymer and solvent (compare also Section 3.6.2.3, page 171).

The extent of swelling is only weakly influenced by the degree of cross-linking. It is mainly determined by the chemical nature of the polymer and the solvent.

Gas permeability. Diffusion of gasses through elastomers is lowered with increasing crosslink density because of a slight interaction with the rubber. That is the reason for obtaining somewhat lower gas permeability in well crosslinked rubber than in those with lower crosslink density. The differences are minor in soft rubber and cannot even be compared with those caused by different fillers.

Heat stability. Heat stability shows very little dependence on crosslink density compared to stress values or elasticity. The heat stability depends to a larger extent on the chemistry of the crosslinks that are formed during curing and the chemical constitution of the high polymers. Among other things, heat stability of the vulcanizate depends on the bond energies of the chemical bonds formed at the crosslink site (compare also Section 3.6.2.1, page 168).

Summary. With increased crosslink density, the following properties of the vulcanizate are affected *strongly:*
- stress values, tensile strength and elongation
- rebound elasticity at elevated temperature
- dynamic damping at elevated temperature
- resistance to cut growth
- permanent set, compression set
- fatigue resistance
- abrasion resistance
- dynamic heat build-up
- swelling stability

less strongly affected:
- abrasion resistance
- gas permeability
- rebound elasticity at room temperature
- dynamic damping at room temperature
- low temperature flexibility
- resistivity

Not all properties are optimized to the same degree of vulcanization as the stress value. The following properties are optimized at

a slight undercurve:
- abrasion resistance
- crack growth resistance
- dynamic tear resistance

the vulcanization optimum:
- tensile strength
- aging resistance

a slight overcure:
- rebound elasticity
- permanent set, compression set
- wear resistance, heat build-up
- dynamic damping
- swelling stability
- low temperature flexibility.

When strong overcuring occurs, especially if there is a tendency for reversion, most properties suffer, especially toughness, cut-resistance and aging stability. Which degree of cure to choose has to be based frequently on a compromise.

It has to be emphasized that the property changes discussed here at constant composition of the compound occur only because of changes in the degree of vulcanization. The fact that these properties may be affected more drastically by changes in the composition of the coumpounds is not being discussed here.

4.2.1.3 Effect of Crosslink Structure on Vulcanizate Properties

[4.57]

Type of crosslink structure. Depending on the vulcanization system used, for example, the amounts of sulfur, different crosslink structures are obtained. In high sulfur systems *(conventional vulcanization)* polysulfide (C-S_x-C, X > 2) links are formed. In *semi-efficient vulcanization* (semi-EV) disulfide crosslinks (C-S-S-C) are formed; with very low sulfur and in sulfurless thiuram vulcanization, *efficient vulcanization,* (EV), even in the presence of sulfur donors, monosulfidic and disulfidic (C-S-C, C-S-S-C) crosslinks predominate. The sulfurfree vulcanization, for example, with peroxides, practically yields only C-C crosslinks.

The free mobility of chain segments of the macromolecules (micro-Brownian movement) depends on their relative distance, and therefore on the length of the crosslinks. Therefore the type of crosslink structure will influence the property spectrum of the vulcanizate. The larger the crosslinks (larger × in the C-S_x-C structure) the larger the possible displacement during mechanical or thermal strain on the vulcanizate. Since the differences in chain distance are usually not very large, the consequences are modest. Nevertheless, one can confirm certain differences between vulcanizates with C-C crosslinks, in which the chains are rigidly connected and those with longer, mobile crosslinks (polysulfidic crosslinks). Even larger differences are observed with, for example, long chain urethane crosslinking (see page 261).

Not only the number of sulfur atoms between two cross-link sites but also sulfur atoms which are bound intermolecularly, Thiole groups or sulfur containing accelerator remains (see Fig. 4.6, page 222) are important for the properties of the vulcanizates. The medium number of bound sulfur atoms, referred to as the number of total cross-link sites can be determined by *Moore's* Efficiency Parameter, E [2.3]. This parameter accounts for all sulfur atoms which are ineffective for the cross-link formation. The larger the parameter E, the larger is the number of sulfur atoms bound per cross-link site. A *Moore* Efficiency Parameter of 7–12 as it occurs in conventional vulcanization systems [3.220, 3.221, 4.60] indicates that numerous sulfur links will be broken at high temperatures and are then available to form new cross-links (revulcanization). Poor heat stability, reversion and permanent set will result. E values of approximately 3 (Thiuram-EV) or of 1 (Cadmate system) [3.220, 3.221, 4.60] indicate, in contrast, considerably higher thermal stability. Even though a sulfur-less TMTD vulcanization is characterized by a mono-sulfidic cross-linking system, its efficiency parameter is around 2–3. This unexpectedly high value is due to the bound dithiocarbamate residues [4.63] which cannot be split under normal application conditions and act as bound antioxidants. The excellent thermal stability of TMTD efficient vulcanization systems is explained by this.

Mechanical Properties. The mechanical properties of NR are affected by the cross-link structure more than those of SR (at a certain cross-link density) even though this is modest. In general the *tensile strength and the elongation at break* are higher for polysulfidic links compared to low sulfur compounds. The *tear strength* is often lower for sulfur free compounds. The *elastic behavior* at room temperature improves somewhat, as expected, with increasingly longer crosslinks due to the increased free mobility of the chain segments; the *permanent set* improves with decreasing values of X in C-S_x-C links. These improvements are especially seen at higher temperatures where the increasing thermal resistance plays an additional role. The vulcanizate structure does, however, not influence the *abrasion resistance* significantly.

Aging Behavior and Permanent Set. Since the sulfur links exhibit increasing bond energies with decreasing sulfur content, they influence to a large extent the aging of vulcanizates, where specially the *thermal stability* has to be mentioned.

Vulcanizates with high bond strength crosslinks (C-C bonds) are superior in this respect than those with lower bond strength (C-S_x-C bonds). This is the reason why vulcanizates with shorter crosslinks (semi-EV, EV and peroxide systems) have generally better heat stability than those with polysulfidic crosslinks (conventional vulcanization).

Therefore the sulfur/accelerator ratio is especially important for the thermal stability of the vulcanizates. A temperature stability not achieved by the proper adjustment of the sulfur dose cannot be corrected by any other means, not even the addition of antidegradents. Low sulfur vulcanization systems also have a beneficial effect on the *tendency to reversion* [4.67–4.77] (see page 23, 232, 247).

The higher bond strength of shorter crosslinks is of importance for the high temperature *compression set* in addition to the heat stability of the vulcanizate; low values of x in C-S_x-C generally give better permanent set values (see page 231). EV and peroxide systems furnish therefore generally lower compression set values, especially at elevated temperatures (see page 258).

No correlation between *static ozone stability* and crosslink type has been established.

Dynamic Properties. One observes a sometimes significant influence of crosslink structure on the dynamic properties of vulcanizates. This is especially the case with *dynamic damping* which is better in vulcanizates with polysulfidic bonds, i.e. gives lower values with shorter sulfur links. This response is similar to the observed improved elasticity and corresponds to the theoretical expectation that micro-Brownian motion can occur more easily when the distance between polymer molecules is longer. However, damping elements often do not only require good dynamic properties but also good temperature stability. Therefore, in order to optimize both properties, mono-to-disulfide crosslink structures (EV or semi EV) are employed.

The *fatigue cut growth* becomes larger in vulcanizates with shorter crosslinks. High sulfur systems give clearly better fatigue resistance than those which are sulfur free. DeMattia fatigue as well as normal fatigue show distinct improvement with longer crosslinks. Cut growth is also slowed with increasing X in C-S_x-C. Because of the higher stability of low sulfur systems, the situation can be reversed in short order (for example in tires after a short test run). This is more apparent in NR than in SR types.

In destructive dynamic test methods where the load is increased to ultimate destruction (degradation with heat build-up), the effect of crosslink structure is difficult to predict. Aside from energy balance considerations, for which long chain crosslinks look favorable, reversion stability has also to be considered. The latter is, of course, better in low sulfur vulcanizates. The result of these two factors is not always clear. It has been frequently shown that in tests with extremely high degradation effects, for example the Goodrich test or the San-Joe flexometer, reversion stability is dominant and better values are observed with short sulfur chain crosslinked vulcanizates.

There is some indication that the *rolling coefficient of friction* depends on the crosslink structure. Vulcanizates containing low amounts of sulfur frequently give a slightly lower coefficient of rolling friction and therefore lower fuel consumption in tire applications, than high sulfur vulcanizates. For this reason and because of reversion resistance, low sulfur tire compounds have recently be used.

Summary. The following properties are optimized in NR vulcanizates with *short crosslinks* (C-C or C-S-C links):

- permanent set, compression set at elevated temperatures
- reversion stability
- heat stability
- degradation resistance (for vulcanizates with reversion)

longer crosslinks (C-S_x-C links):

- tensile strength
- elasticity
- permanent set at low temperature
- dynamic damping
- fatigue resistance
- coefficient of rolling friction (higher)

no significant influence:

- ozone stability
- abrasion resistance.

4.2.2 Sulfur and Sulfur Containing Vulcanizing Agents

[1.12, 1.13]

4.2.2.1 General Observations about Sulfur and Sulfur Containing Vulcanization Agents

[4.78–4.94]

Vulcanization can be achieved by using high energy radiation without any vulcanization chemicals (see page 403).

As practiced today, the type of materials necessary for vulcanization depends primarily on the type of rubber to be vulcanized. The diene-rubbers, whose polymer molecules still contain unsaturation, like NR, SBR, NBR, etc. can be vulcanized with sulfur and usually with peroxides. Sulfur vulcanization is preferred for several reasons [3.434, 4.99]

- because of the easier adjustment of the balance between the onset of vulcanization and the vulcanization plateau
- the higher flexibility during compounding
- the possibility of air heating
- the better mechanical properties
- the possibility to control the length of the crosslinks and
- because of economic reasons.

However, as a rule peroxide crosslinking has the following advantages so that it achieved considerable technological importance (see page 256):

- better heat stability
- better reversion resistance
- lower high temperature compression set
- absence of sulfur avoids corrosion in cable metals.

For the preferred type of sulfur vulcanization, additional materials are necessary for the activation of the sulfur, such as vulcanization accelerators and activators and frequently also retarders. Optimization of a vulcanization systems involves

- processing safety,
- cure time and temperature,
- vulcanization type and
- achievement of the desired technological properties necessitates a delicate balance of all components.

Saturated polymers can not be sulfur vulcanized. Other materials have to be used. The ones mostly used are mainly peroxides, quinonedioximes, metal oxides, amines, certain resins and isocyanates that will be treated in a different section (see Section 4.2.4, page 255).

4.2.2.2 Sulfur

Vulcanization Sulfur The most important vulcanization agent for rubber is the so-called vulcanization sulfur. For the preparation of soft rubber goods, dosages of about 0.25–5.0 phr are used; for hard rubber compounds (ebonite) the sulfur content is raised to 25.0–40.00 phr. The range between 5.0 and 25.0 phr of sulfur is not of importance for most applications (except for example, for floor coverings and some roll covers); rubber articles prepared within those limits are in the so-called leathery region, have poor strength and elastic properties.

The sulfur quantity used for the preparation of soft rubber differs greatly with the amount of accelerators used and the demand on properties of the vulcanizate. For accelerator free NR compounds, hardly used in today's technology, relatively high sulfur levels are necessary (for example 5 phr). This type of vulcanization leads, aside from polysulfidic sulfur with intermolecular crosslinks, to many side reactions that do not lead to crosslinking (for example intramolecular cyclic structures with high efficiency parameter) [3.220, 3.221, 4.60]. In the presence of vulcanization accelerators, these side reactions are surpressed and, depending on activity and amount of accelerators used, the average number of sulfur atoms per crosslink is decreased. For that reason less sulfur is used with increasing amount of accelerator. Higher amounts of sulfur are necessary for less active accelerators, i.e., bases like guanidine, than for highly active acceleration, i.e., sulfenamides.

For *conventional*, most frequently used systems, one uses for example roughly 1.5–2.5 phr sulfur with 1.0–0.5 phr accelerator. When accelerator dosage is increased (for example 1.5–2.5 phr), the sulfur content has to be lowered (for example 1.2–0.5 phr) to achieve the same crosslink density; the consequence is formation of crosslinks with less sulfur content. Such semi-efficient (semi-EV) systems produce, as expected, heat- and reversion-resistant vulcanizates.

Because the high accelerator content leads to a more effective use of the sulfur and thus to shorter sulfur links as well as to a reduction of that amount of sulfur which is not used for crosslinking (a decrease of the *Moore* efficiency parameter E) (see page 227). When the amount of accelerators is drastically increased (e.g. 6 phr CBS), small amounts of sulfur are sufficient to obtain a satisfactory degree of crosslinking. When the amount of sulfur is reduced even further, the use of accelerators which are at the same time sulfur donors (e.g. TMTD) is called for. For example, when 2.5–3.5 phr TMTD are used, vulcanization takes place with only very small amounts of sulfur (e.g., 0.2 phr), or even in the absence of all elemental sulfur. For these *semi-EV* or *EV compounds* the sulfur donors without any accelerator activity also play a certain important role (see Section 4.2.2.4, page 231).

A so-called *Efficient*-vulcanization (EV) leads to mono- and di-sulfidic crosslink structures. The consequence is very good heat- and reversion-stability, respectively a low high temperature compression set.

For the vulcanization of NR one uses usually a slightly higher sulfur dosage and somewhat less accelerator than for SR. Among the SR- types for those with very low unsaturation, for example IIR or EPDM, larger amounts of sulfur and accelerator are needed than for classical diene rubbers.

Sulfur Purity. Sulfur that is suitable for vulcanization should be SO_2 free and be at least 95% pure (ash content <0.5%). Highly uniform dispersion is necessary for a uniform vulcanization product and the resultant good rubber property spectrum. When small amounts of sulfur are used as in cases where sulfur is difficult to disperse in the rubber compound, frequently sulfur containing dispersing agents or sulfur pastes are used.

Insoluble Sulfur. Normal vulcanization sulfur tends to bloom to the surface during storage of compounded rubber; this makes subsequent adhesion or welding steps more difficult. Bloom can be avoided by using so-called insoluble sulfur (sulfur with a 60–95% CS_2 insolube fraction) that can not bloom because it is also insoluble in the rubber. To achieve uniform dispersion its particle size must be lower than that of normal vulcanization sulfur. Furthermore, insoluble sulfur is frequently used as a rubber dispersion or as a paste.

Insoluble sulfur is not stable – it slowly converts to the normal soluble form, at room temperature faster at elevated temperatures. During the mixing process and processing of the rubber compound one has to be careful not to raise the temperature above 120 °C.

Colloidal sulfur. Due to the fact that this type of sulfur contains very fine particle size, it plays an important role in latex compounds, but it is of lower importance in solid rubber and will not be described in detail in this book [1.16].

4.2.2.3 Selenium and Tellurium

Selenium and tellurium can be used instead of sulfur for rubber vulcanization. These products act somewhat weaker compared to sulfur. Because of their price and toxicity they are not of practical importance.

4.2.2.4 Sulfur Donors

[4.79–4.94]

Aside from elemental sulfur, compounds that contain sulfur in a heat-labile form can also be used. They liberate sulfur at the vulcanization temperature. Sulfur donors can be subdivided into those that exhibit an accelerator activity and can be substituted directly for sulfur, without drastic change of the vulcanization characteristics, and those that are simultaniously vulcanization accelerators. Products of the first type are for example dithiodimorpholine, DTDM(I) and caprolactamdisulfide, N,N′-dithio bis-(hexahydro-2H-azepinone), CLD(II). For sulfur donors that are at the same time vulcanization accelerators, the vulcanization system has to be properly modified, thus, for example 2-morpholino-dithio-benzothiazole, MBSS (III), dipentamethylene thiuramtetrasulfide, DPTT (IV), N-oxydiethytlene dithiocarbamyl-N′-oxydiethylene sulfenamide OTOS (V) as well as tetramethyl thiuramdisulfide, TMTD (VI), whose main function is that of an accelerator (see page 246ff).

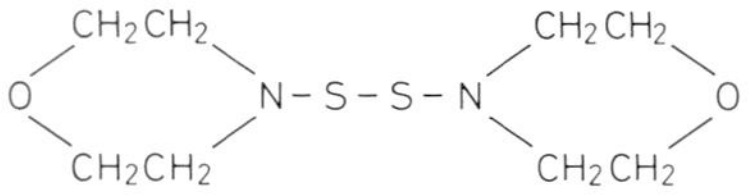

DTDM, molecular weight 236, active sulfur 13.6 Mol. % *

CLD, molecular weight: 288, active sulfur 11.1% *

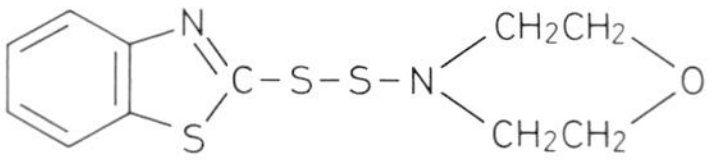

MBSS, molecular weight: 284, active sulfur: 11.3 Mol % *

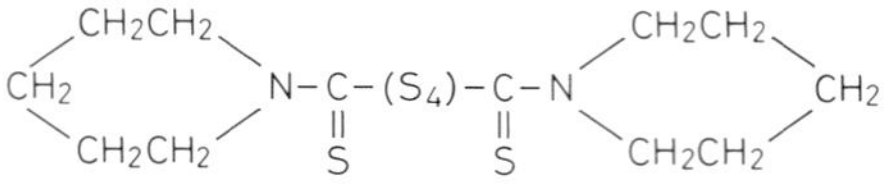

DPTT, molecular weight: 384, active sulfur: 16.6% **

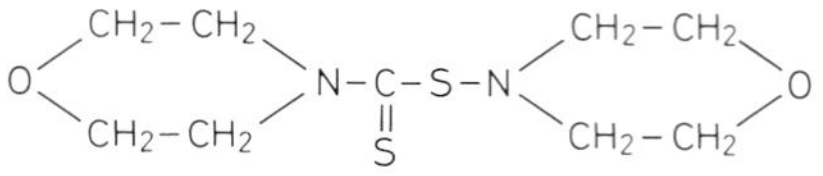

OTOS, molecular weight: 248, active sulfur: 12.9% *

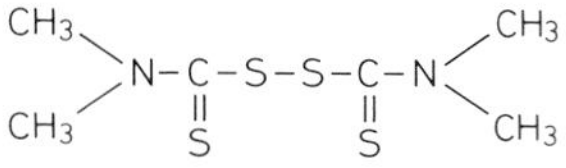

TMTD, molecular weight: 240, active sulfur: 13.3% *.

Of these products, DPTT has the highest sulfur content. Theoretically, four sulfur atoms could be given up by DPTT, however, usually only two become activated and are incorporated as mono- or disulfides. When DPTT is used it should be possible that small amounts of tri- and tetrasulfide crosslink may be incorporated. DTDM, MBSS, CLD and TMTD have only two sulfur atoms that can be activated for crosslinking and can lead to mono- or disulfidic bonds. In comparison, OTOS contains only one sulfur atom, that can be activated and therefore its use in EV-systems leads strictly to monosulfidic crosslinks.

When equal weights are used, DPTT forms the most crosslinks compared to other sulfur donors and gives therefor a very high crosslink density. Since these crosslinks frequently contain multiple sulfur atoms, the lowest heat stability is obtained. In comparison, OTOS forms crosslinks of highest heat and reversion stability. The other sulfur donors give products of intermittent heat stability, of the latter group, DTDM gives the highest crosslink density.

The important role of the sulfur donors lies in the semi-EV and EV systems. A reduction of elemental sulfur requires, as already mentioned (see page 230) a drastic increase in the amounts of accelerator when the crosslink density shall be kept constant and thus also the modulus and other mechanical properties. If the sulfur content is less than 0.5 phr, even very high doses of those accelerators which are not sulfur donors cannot compensate [4.82]. A decrease of crosslink density (seen in a decrease in modulus) is the result. In such low sulfur systems the application of sulfur donors is essential, where TMTD plays the most important role as accelerator and sulfur donor. A sulfur substitution by using TMTD or other accelerators with sulfur donating capabilities cannot be accomplished without a complete change of

* Refers to one available S atom (monosulfidic structure)

** Refers to two available S atoms (disulfide structure)

the vulcanization characteristics. Practically, this means that completely different vulcanization systems are required as the sulfur content is lowered. If no change of the original cure system is desired, sulfur donors with accelerating efforts (e.g., DTDM) have to be used. The amount of sulfur donors required to replace the elemental sulfur are in relation to the existing sulfur/accelerator ratios. For DTDM the following guidelines apply [4.82]: For the substitution of each weight equivalent of sulfur the equivalent amounts of DTDM are: in conventional systems 0.5–1.0, 1.0–2.0 in semi-EV systems and 2.0–5.0 in EV systems. These equivalent quantities are required in order to obtain the same crosslink density, i.e., the same modulus.

Since sulfur donors are used generally with accelerators and combined with small amounts of sulfur, their individual action can not be observed. They frequently give synergistic action in practical compounds, whereby the full crosslinking potential of the sulfur donors can be achieved. The strongest synergist among the sulfur donors is probably OTOS.

One has to add as sulfur donor bis-[3-triethoxysilylpropyl)tetrasulfide [4.77] that is used as an activator for fillers (see page 289) as well as thiokols [4.84] (see page 136ff). They can be used instead of sulfur, in small amounts for the vulcanization of NR and SR. A portion of the bound sulfur in thiokols becomes labile on heating and is available for rubber vulcanization. Thiokols are not used in practice as sulfur donors. Factice with high sulfur content has to be considered also when a sulfur balance is made.

One of the oldest vulcanizing agents is sulfurdichloride which was already discovered in 1846. It makes vulcanization of thin NR articles possible at room temperature or better at slightly elevated temperatures (cold vulcanization). It is of little importance because of poor aging properties of the resultant vulcanizates. Applications are described on page 452.

4.2.3 Vulcanization Accelerators

4.2.3.1 General Observation and Classification of Accelerators

[1.12–1.14, 4.3, 4.95–4.113]

General Observations. Sulfur, by itself, is a slow vulcanizing agent. Large amounts of sulfur are necessary, high temperatures and long heating periods and one obtains an unsatisfactory crosslinking efficiency with unsatisfactory strength and aging properties. Only with vulcanication accelerators the quality corresponding to today's level of technology can be achieved. The multiplicity of vulcanization effects demanded can not be achieved with one universal substance, a large number of diverse materials is necessary.

There are many theories as to the chemical action of vulcanization accelerators [1.12, 1.13, 2.3, 4.49, 4.50] that will not be discussed here in detail. Almost all accelerators need metal oxides for the development of their full activity, ZnO usually being the best additive.

Classification. Since a high number of commercial products are available, accelerators are best classified according to their chemical structure. There are also a number of specialty products that are difficult to classify.

In addition there are several diverse methods of preparation that follow primarily the trend to dust-free application: coated powder, pastes and granules. Many products contain only 80% or less of active ingredient. Because of their improved dispersibility many of them can be used like 100% active compounds without decreased effectiveness [5.54].

It is not possible within the framework of this discussion to mention all commercially available products, therefore, only the following important classes, respectively products will be enumerated.

One can differentiate principally between inorganic and organic vulcanization accelerators. Since the inorganic products are hardly anything other than activators, they can be treated briefly. The organic accelerators are of major importance.

The following facts are the basis for the importance of the organic accelerators:

- They increase the rate of the crosslinking reaction with sulfur considerably. This making for shorter, economical curing times and at the same time, production under milder conditions, which improves aging stability of the rubber product.
- An adjusted dosage and combination of different accelerators promotes a broad adjustment of rates of onset and vulcanization stages.
- By combining two or more accelerators one can observe in many cases a positive synergistic effect (secondary accelerator effect) [4.103, 4.104], by which one can obtain more than additive, sometime very favorable, results.
- By adding accelerators one can lower the sulfur content necessary to achieve optimum vulcanizate properties. Sulfur is used more efficiently by the use of organic accelerators for the formation of crosslinks and, as has been mentioned before, fewer side reactions occur.
- This leads to a drastically improved aging stability of the rubber product.
- A further important consequence of the lower sulfur dosage is a flattening of the stress/heating time curve. That brings about a longer period at the maximum (plateau effect) and less danger of over-vulcanization and particularly of reversion.
- Lowering of vulcanization temperature permits the use of organic dyes while previously only inorganic pigment could be used; this permits multiple coloring. With certain accelerators, transparent articles can be produced.

Since organic accelerators frequently have long and complicated chemical names, an association, initiated by the author, of the most important rubber chemical producers of the world *(WTR)* has attempted to develop a *system* of *abbrevation.* This system proposed by WTR includes also sulfur donors, retarders, activators, antiaging materials, blowing agents, and others, but by no means all materials.

The following list contains the WTR-proposed abbreviations. With materials where no WTR-proposal exists, commonly used abbreviations or some proposed by the author were used.

The most important organic vulcanization accelerators can be summarized in the following classes (alphabetic list of abbreviations see Section 8.2, page 564).

		WTR Number
Mercapto-accelerators		
2-Mercaptobenzothiazole	MBT	23
Zinc-2-mercaptobenzothiazole	ZMBT	23b
Dibenzothiazyl disulfide	MBTS	23d
Sulfenamide-accelerators		
N-cyclohexyl-2-benzothiazylsulfenamide	CBS	19
N-tert-butyl-2-benzothiazylsulfenamide	TBBS	21

		WTR Number
2-Benzothiazyl-N-sulfenemorpholide	MBS	22
N,N-dicyclohexyl-2-benzothiazylsulfenamide	DCBS	20
Thiuram accelerators		
Tetramethylthiuram disulfide	TMTD	46
Tetramethylthiuram monosulfide	TMTM	47
Tetraethylthiuram disulfide	TETD	48
Dimethyldiphenylthiuram disulfide	MPTD	
Dipentamethylenthiuram tetrasulfide	DPTT	68
Dithiocarbamate accelerators		
Zinc dimethyldithiocarbamate	ZDMC	36
Zinc diethyldithiocarbamate	ZDEC	38
Zinc dibutyldithiocarbamate	ZDBC	40
Zinc pentamethylenedithiocarbamate	Z5MC	
Zinc ethylphenyldithiocarbamate	ZEPC	44
Zinc dibenzyldithiocarbamate	ZBEC	
Piperidine pentamethylenedithiocarbamate	PPC	
Sodium dimethyldithiocarbamate	NaDMC	
Sodium dibutyldithiocarbamate	NaDBC	
Selenium dimethyldithiocarbamate	SeDMC	
Tellurium dimethyldithiocarbamate	TeDMC	
Lead dimethyldithiocarbamate	PbDMC	
Cadmium dimethyldithiocarbamate	CdDMC	
Cadmium pentamethylenedithiocarbamate	Cd5MC	
Copper dimethyldithiocarbamate	CuDMC	
Copper dibutyldithiocarbamate	CuDBC	
Bismuth dimethyldithiocarbamate	BiDMC	
Dithiocarbamylsulfenamide		
N-oxydiethylenedithiocarbamyl-N′-oxydiethylene sulfenamide	OTOS	
Xanthate Accelerators		
Zinc isopropylxanthate	ZIX	
Zinc butylxanthate	ZBX	
Sodium isopropylxanthate	NaIX	
Guanidine Accelerators		
Diphenylguanidine	DPG	27
Di-o-tolylguanidine	DOTG	28
o-Tolylbiguanidine	OTBG	
Amine Accelerators		
Butyraldehydeaniline	BAA	
Tricrotonylidenetetramine	TCT	
Hexamethylenetetramine	HEXA	
Polyethylenepolyamines	PEP	
Cyclohexylethylamine	CEA	
Dibutylamine	DBA	

(continued on next page)

		WTR Number
Thiourea Accelerators		
N,N′-ethylenethiourea (=2-mercaptoimidazoline)	ETU	31
N,N-diphenylthiourea (=thiocarbanilide)	DPTU	30
N,N′-diethylthiourea	DETU	
Dithiophosphate Accelerators		
Zinc dibutyldithiophosphate	ZDBP	
Copper diisopropyldithiophosphate	CuIDP	
Sulfur Donors		
2-Benzothiazole-N-morpholyldisulfide	MBSS	23E
Dimorpholine disulfide	DTDM	67
Vulcanization Retarders		
Cyclohexylthiophthalimide	CTP	
Phthalic anhydride	PTA	56
Benzoic acid	BES	
Salicylic acid	SCS	55
N-nitrosodiphenylamine	NDPA	

The numbers are WTR proposals (WTR= Working Group Toxicology of Rubber Auxiliaries [4.28a], Avenue de Tervuren 270–272, Boite 3, B-1150 Brussels, Belgium, Europe)

4.2.3.2 Inorganic Vulcanization Accelerators

[4.114, 4.115]

Soon after the discovery of vulcanization it was found that addition of magnesium oxide, calcium hydroxide, lead oxide or antimony tri- and pentasulfide shortened the vulcanization time. At the same time it was found that the until then customary high sulfur charge could be reduced and that products with better mechanical properties and longer durability would be produced.

Inorganic accelerators mentioned above are rarely used by themselves ever since organic accelerators were introduced. Sometimes they are used in addition to zinc oxide as additional activators for organic accelerators. Mixtures of inorganic accelerators and high sulfur charges are solely used sometimes for thick walled, voluminous products, such as large textile- and papermill roll covers.

4.2.3.3 Thiazole Accelerators

[1.14, 4.116–4.131 a]

Thiazoles are the accelerator class of highest economic importance. 2-Mercaptobenzothiazole and its derivatives, that can be subdivided into mercapto accelerators and benzothiazole sulfenamide accelerators, belong to the accelerators most often used, their use gives a multiplicity of vulcanization effects and the best level of vulcanizate properties. An estimated 80% of all vulcanization accelerators used are thiazoles. In addition the other accelerator classes, such as guanidines, thiurams and dithiocarbamates are frequently used in conjunction with thiazoles. Since the especially interesting sulfenamide accelerators have a separate classification, we chose in the following discussions the subdivision of "mercapto accelerator".

4.2.3.3.1 Mercapto Accelerators

[1.14]

Materials. The basic compound among the mercapto accelerators is 2-mercaptobenzothiazole (MBT)

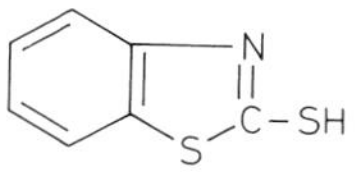

MBT

Dibenzothiazyl disulfide (MBTS) as well as zinc-2-mercaptobenzothiazole (ZMBT) are derived from the former:

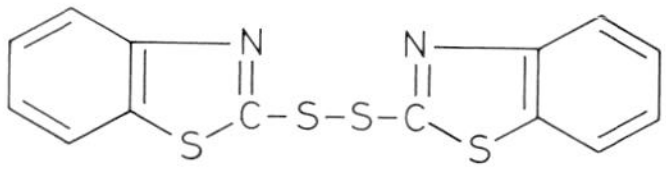

MBTS

The main products in this class of accelerators, MBT and MBTS are used broadly in many types of rubber and may be called "all purpose" accelerators. While vulcanization begins rapidly with MBT, MBTS gives a slightly delayed start. This is based on the fact that MBTS has to decompose thermally into MBT fragments before vulcanization begins. With ZMBT one obtains an intermediate onset of vulcanization, this accelerator is mainly used in latex mixtures.

Efficiency and Combinations. The mercapto accelerators are very efficient and confer good processing safety to the rubber compounds (MBTS better than MBT), intermediate vulcanization rates (MBT higher than MBTS), a broad vulcanization plateau and to the vulcanizate very good aging resistance. Mercapto accelerators may be used alone to give a relatively low crosslink density. They are used preferably in combination with other accelerators, for example, guanidines, thiuram or dithiocarbamate accelerators, thioureas, dithiophosphates and the like, combinations with basic accelerators, for example, guanidines, act synergistically and experience a secondary acceleration and activation. These combinations cause faster vulcanization than each product separately (secondary acceleration) and a considerably higher crosslink density (activation) which is positive for the general property spectrum of the vulcanizate. Therefore, prepared combinations are commercially available and are of considerable technical and economic importance, for example, MBTS/DPG, MBTS/DOTG, MBT/HEXA. Combinations with, for example, thiurams, like MBT/TMTD, are used primarily in slow vulcanizing rubbers, such as IIR as well as EPDM. The mercapto accelerators alone in low sulfur compounds (semi-EV-systems) are not very effective, but in combination with bases, MBT and MBTS become very effective in semi-EV-systems.

Mercapto accelerators in combination with thiurams, respectively dithiocarbamates give vulcanizates with excellent aging properties; basic activator systems have a detrimental influence on the length of the curing plateau and reversion stability in reversion prone vulcanizates. Optimum heat stability is therefore not obtained with basic combination mercapto accelerators.

Activation. For development of the full activity of mercapto accelerators addition of ZnO is necessary, and in the absence of basic secondary acelerators, also stearic acid. Both can be replaced by Zn-soaps. Fatty acids because of better dispersion also improve the activity of basic activated mercapto accelerators.

Retardation. The start of vulcanization in compounds containing only mercapto accelerators is little affected by the presence of retarders. When base activated mercapto accelerators are used, the flow period can be regulated by variation in stearic acid amounts. More intensive retardation can be obtained by addition of, for example, benzoic acid, salicylic acid and phthalic anhydride; these additives also prolong the total curing time. One can slow down a compound that starts curing rapidly by the additional use of benzothiazolesulfenamide accelerators or OTOS.

Properties of the Vulcanizates. Since a relatively low degree of vulcanization is obtained by using uncombined mercapto accelerators, the strength and elastic properties are not at an optimum in such vulcanizates. They become excellent by using additional basic accelerators, but as has been mentioned earlier, the plateau becomes shorter as a consequence. Mercapto accelerators do not stain the vulcanizate, but they turn yellow easily.

Application. Mercapto accelerators are basically suitable for all *types of vulcanization* (press vulcanization, hot air vulcanization, steam vulcanization, salt bath and high frequency vulcanization or other continuous methods); in some cases secondary acceleration must be used. In the case of hot air vulcanization the use of MBTS and its combination gives often a vulcanization start that is too slow. By partial replacement of MBTS by the faster MBT a broader flow period can be adjusted from delayed to fairly rapid onset of vulcanization. This capability to adjust the beginning of vulcanization by minor adjustments in dosage of accelerator combinations is particularly useful in production of hot air heated products. Conversely, mixtures of MBT+guanidine are not sufficiently safe for the above production method. Partial replacement of MBT by MBTS in similar combinations increases processing safety considerably, the slower vulcanization finish can be tolerated in the hot air vulcanization process. For hot air cured products, combinations of MBT+MBTS+DPG (or DOTG) without or with little stearic acid are frequently used. In addition, the combined application of MBT+ZDEC as well as TMTD is also useful.

The accelerators of mercapto classification are for example, used with the following *types of rubber:* NR, IR, BR, SBR and NBR. MBTS is used as retarder for CR. For the vulcanization of polymers with low unsaturation like IIR and EPDM, mixtures of mercapto accelerators and thiurams and/or dithiocarbamates are applied (see page 89, 97). In principle, similar combinations in corresponding dosages may also be used for diene rubbers. For the quinone dioxime cure of IIR, MBTS beside lead dioxide, read lead, thiuram and others are used as activators. In mixtures of thiocols, MBTS does not function as vulcanization accelerator but rather as chemical degradant. In the vulcanization of CSM with metal oxides additional organic vulcanization accelerators are used. Among these, thiurams have been applied in as well as MBT and MBTS.

The mercapto accelerators are used in a multitude of *rubber products,* for example, molded products (seals, stoppers), hoses and profiles, conveyor belts and transport rolls, shoe products (including soles and heels), cable, bicycle and automobile tires, foamed products and many others. Mercapto accelerators (uncombined) are also used with rubber thread production. ZMBT is mostly used in the production of latex articles, whereby the accelerating as well as the sensibilitizing properties of this compound are used. For products that necessitate a long flow period (complicated shaped press molded parts), MBTS+bases are suitable. This is, however, generally the domain of the sulfenamide accelerator. MBT+dithiocarbamates is important for example for rapidly heated shoe articles. MBT+thiurams (respectively dithiocarbamates) are especially used for products made of IIR and EPDM.

The vulcanizates have a bitter taste; mercapto accelerators are therefore only used in very small quantities for products that come in contact with food and the like.

4.2.3.3.2 Benzothiazole Sulfenamide Accelerators

[1.14]

Products The benzothiazole sulfenamide accelerators are also derived from 2-mercapto-benzothiazole, since an amine is oxidatively bound to the mercapto sulfur. The accelerator differ according to the type of amine used. Typical sulfenamide accelerators are the following.

CBS:	$R_1 = H$, $R_2 =$ Cyclohexyl
TBBS:	$R_1 = H$, $R_2 =$ tert. Butyl
MBS:	$R_1 + R_2 =$ Oxydimethylene ($-CH_2CH_2-O-CH_2CH_2-$)
DCBS:	$R_1, R_2 =$ Dicyclohexyl
5BS:	$R_1 = H$, $R_2 =$ tert. Amyl

In this subclass we are dealing with a "molecular combination" of mercapto accelerators and bases. The accelerators become active as the amines are split off during vulcanization. The base activates 2-mercaptobenzothiazole as it is formed.

Activity and Combinations. Consequently the benzothiazole sulfenamides produce a retarded vulcanization start compared to the mercapto accelerators MBT and MBTS and therefore improved processing safety. The thermal stability of the sulfenamide depends on the substituants on the nitrogen atom, that produces in turn various degrees of retardation. The processing safety increases from CBS to TBBS, MBS and is highest with DCBS. Because of the base activation of the 2-mercaptobenzothiazyl residue formed after the decomposition of the sulfenamide, vulcanization proceeds rapidly after the retarded start. Vulcanization behavior of the sulfenamides resembles the base activated mercapto accelerators, except for the prolonged flow time. The vulcanization curve of sulfenamide mixtures approach the ideal for molded products, long flow time, rapid vulcanization finish. The vulcanization speed can be synergistically influenced by addition of secondary accelerators, especially thiurams and dithiocarbamates. Further activation with basic accelerators is no longer very effective. If one adds sulfenamides to fast-acting accelerators respectively, accelerator combination, the favorable flow time/cure time relationship of the sulfenamides is also obtained. The sulfenamides, in combination with fast accelerators, can act as retarders.

At equal levels, TBBS gives the highest crosslink density (stress value) among the sulfenamides, somewhat lower values are obtained with CBS and MBS, while DCBS presents the end of the series. Altogether, these differences are unimportant. All sulfenamides give higher crosslink density than the simple mercapto accelerators. The vulcanization plateau observed for example in NR compounds, is fairly broad with sulfenamides, less favorable than with uncombined or thiuram or dithiocarbamate combined mercapto accelerators. Strong over-vulcanization has to be avoided with sulfenamides. Combination of dithiocarbamates or thiurams with sulfenamides improves somewhat the reversion tendency of those vulcanizates that are reversion prone, but do not reach the very high level reached with dithiocarbamyl sulfenamide (see page 248).

Analogous to thiurams and opposite to most other accelerators, the sulfenamides are suitable for "Semi-Efficient" and "Efficient" vulcanization. These vulcanization systems contain, beside 0.3–1.0 phr sulfur, either only sulfenamides (up to 6 phr) or better yet, a combination of 1.5–3.0 phr sulfenamides and 0.2–1.0 phr thiurams or dithiocarbamates. These systems combine several advantages. In comparison with pure thiuram vulcanization, they give rapid heating time, better flow time/heating time ratio, very broad vulcanization plateau, and very good aging resistence and low compression set of the vulcanizate. As disadvantages one can list somewhat lower elasticity respectively higher damping of the vulcanizate and frequently lower resistance to fatigue crack growth. Among the sulfenamides, MBS and TBBS are most suitable for EV-systems. The thermal instability of the benzothiazole sulfenamide bond is the reason for a limited shelf-life of these materials. Depending on the heat history of a given compound [2.123] and its storage time, undesirable bond changes can occur in the benzothiazole sulfenamides. The consequence is that the expected processing safety and the retardation of the vulcanization start are reduced. After mixing in high efficiency mixers and several day storage, Mooney scorch values can drop to half the potentially attainable value. In this respect dithiocarbamyl sulfenamides behave much better. CBS, TBBS and 5BS don't split off secondary amines, which may form nitrosoamines.

Activation. Addition of ZnO is necessary for the activation of the benzothiazole sulfenamides. In compounds with CBS addition of stearic acid is recommended to increase stress values. Addition of stearic acid is necessary in compounds containing TBBS, MBS or DCBS to achieve higher crosslink densities. Instead ZnO and stearic acid, Zn-soap may be used.

Retardation. Even though benzothiazole sulfenamide accelerators retard the start of the vulcanization, the addition of vulcanization retarders is necessary in many practical applications. Acidic compounding ingredients, for example, stearic acid and particularly benzoic acid, salicylic acid and phthalic anhydride prolong the flow period. At the same time, the total cure time is also prolonged. Retarders of the type N-cyclohexyl-thiophthalimide (see page 264) have special activity regarding correlating retardations with total cure time and are therefore important products.

Sulfenamide accelerators may be used, as has been mentioned before, as vulcanization retarders for faster vulcanization accelerators.

Properties of the Vulcanizates. Because of the high crosslink density achieved with the benzothiazole sulfenamide, very good strength and elastic properties (damping, heat build-up) are obtained. Fatigue resistance of the vulcanizates is also good. However, the aging resistance of the conventional vulcanization system without the addition of antioxidants is not optimal. The sulfenamide accelerators do not cause staining but do cause yellowing of the vulcanizate.

Application. The *vulcanization process* of choice for sulfenamides is press curing (compression-, injection- and transfer molding). They are also usable for continuous vulcanization in steam, fluid bed, salt bath, and high frequency cure but a second accelerator such as thiuram or dithiocarbamate is usually added. When used in continuous or discontinuous steam vulcanization, a special property of sulfenamides has to be remembered. Under the influence of heat and humidity, they decompose immediately and produce a more rapid onset and finished cure [4.128].

Retardation of the vulcanization start is reduced. This shortening of the flow time is very desirable in case of steam vulcanization, since deformation is thus prevented. DCBS is not suitable for steam cure. These materials are not suitable for hot air cures because of retardation of the vulcanization onset.

The following *rubbers* are for example vulcanized with sulfenamides and sulfur: NR, IR, BR, SBR and NBR. Sulfenamides are not at all or less suitable for CR, in some combinations it is also used in EPDM compounds.

As is the case with mercapto accelerators, the members of this category have many areas of application in the production of *rubber articles*. Based on their special property spectrum they are predominantly, but not exclusively used in the tire industry. Further applications are in technical articles that undergo high dynamic stress (conveyor belts, bumpers, elastic couplings) respectively, general technical articles (gaskets, hoses, profiles, sleeves, stoppers etc.). The shoe and cable industry also use considerable amounts of sulfenamides. They are not used in the latex industry. Because of their bitter taste they are not used with parts that have contact with food.

4.2.3.4 Triazine Accelerators

[4.132–4.134]

Some years ago a completely new class of accelerators was developed, the triazine accelerators, that are technically closely related to the thiazole accelerators. An example are the aminomercaptotriazines.

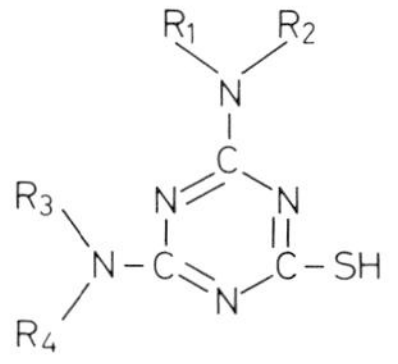

Base Compound

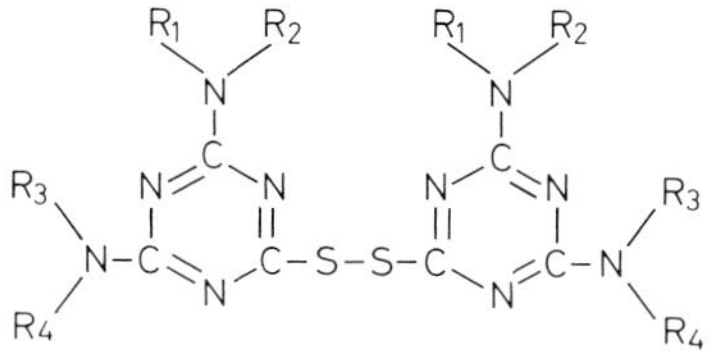

Disulfide

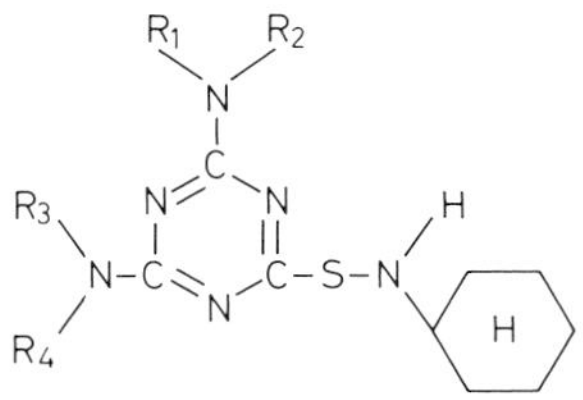

Sulfenamide

They are more effective than the thiazole accelerators, and therefore can be used in smaller amounts. As sole accelerators or in combination with thiazole accelerators they produce highly reversion resistant vulcanizates among those that are reversion prone. They are used in large articles such as farm tractor tires. Furthermore, triazine accelerators do not cause yellowing of the vulcanizate, that may be useful in production of transparent shoe soles. Triazine accelerators are being used very little.

4.2.3.5 Dithiocarbamate Accelerators

[1.12–1.14, 4.3, 4.135–4.139]

These products known in the rubber industry as ultra accelerators are among the first organic vulcanization accelerators [4.135]. They have even today considerable technological importance. One differentiates between zinc-, ammonium-, and other metal dithiocarbamates; technically they only differ in scorch time behavior, with respect to their behavior during manufacturing, the metal- and the ammoniumdithiocarbamate differ. The zinc- and other metaldithiocarbamates are used in the solid rubber and latex section. The ammonium dithiocarbamates primarily for solutions and latex articles and are treated separately for that reason. Selenium-, lead-, bismuth and copper dithiocarbamates are special products and are rarely used. Dithiocarbamates as well as thiurams are highly polar accelerators and tend to bloom in nonpolar rubbers like NR and especially EPDM [3.400]. The highest blooming rate because of lowest solubility will be found with dimethyl derivatives. The longer the alkyl chains in these accelerators are, the better their solubility in unpolar rubbers is and, therefore, the higher their tolerable dosages without blooming are. Dithiocarbamates need ZnO activation [4.138, 4.139].

Vulcanization speed of the dithiocarbamates is so high that processing safety is not sufficient when they are used alone in solid rubber. Therefore they are used in these applications in most cases in combination with slower acting accelerators. In latex compounds, where no elevated temperatures are used during processing, the dithiocarbamates are well suited as sole accelerators.

4.2.3.5.1 Zinc Dithiocarbamates

[1.14]

Products. The base component of zinc dithiocarbamate accelerators is the zinc dimethyldithiocarbamate (ZDMC)

$$\left[(CH_3)_2N-C(=S)-S-\right]_2 Zn$$

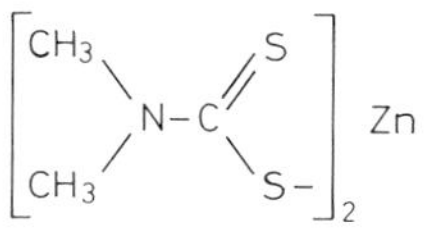

Other dithiocarbamates are derived by replacing the methyl by other groups. The most important zinc dithiocarbamate accelerators are: zinc dimethyldithiocarbamate (ZDMC) zinc diethyldithiocarbamate (ZDEC), zinc dibutyldithiocarbamate (ZDBC), zinc pentamethylenedithiocarbamate (Z5MC), zinc ethylphenyldithiocarbamate (ZEPC) and zinc dibenzyldithiocarbamate (ZBEC).

Activity and Combinations. The beginning of vulcanization of the zinc dithiocarbamates as sole accelerators is so fast that sulfur containing compounds on addition of these products vulcanize at low temperatures, for example room temperature. This is especially the case if one adds basic activators, for example CEA. The rapid start of vulcanization necessitates special precautions during processing, such as addition of the accelerator shortly before further processing of the compound.

The scorch time respectively processing safety in solid rubber compounds increases in the following sequence: ZDBC, ZDEC, ZEPC, ZDMC, Z5MC, ZBEC. Changes in this sequence may occur depending on the contents of the mixture. In latex compounds, Z5MC shows the shortest scorch time (vulcanization start); following in decreasing rate (longer scorch time) sequence ZBEC, ZDBC, ZDEC and ZDMC. An especially high rate of vulcanization in latex compounds is obtained by mixtures of ZDEC and Z5MC (approximately 1:1). For optimum vulcanization time, the same sequence applies generally. Zinc dithiocarbamates cause a steep increase in stress value up to a high crosslink density level that is only slightly flatter than with thiuram accelerators, but they give a shorter vulcanization plateau. For that reason in case of, for example, NR-compounds, only low vulcanization temperatures are required and recommended.

Basically, dithiocarbamates can be applied in low sulfur vulcanization. Better with respect to crosslink density, flow time/heating time relationship, etc. are thiurams, especially in combination with sulfenamides; for that reason beside in EPDM, dithiocarbamates are rarely used in this connection.

Addition of bases increases the crosslink density of compounds containing zinc dithiocarbamates, i.e., it represents an activation. Scorch time and optimum vulcanization times of compounds with zinc dithiocarbamates can also be further shortened by adding basic accelerators; this represents a secondary acceleration effect, which is not suitable in most applications because of processing limitations. Exemptions are self-vulcanizing mixtures and adhesive solutions, where combinations of ZEPC and CEA proved to be especially suitable in practice. Short time hot air vulcanization can be achieved (at 80 °C) in the presence of CEA vapor (vapor vulcanization) if zinc dithiocarbamate and sulfur are present in the compound. By a combination with thiuram or thiazole accelerators the vulcanization rate with zinc dithiocarbamate can be reduced and processing safety increased.

Activation. Addition of ZnO is necessary in compounds with zinc dithiocarbamates. Addition of stearic acid is also beneficial for vulcanization properties. Instead of these two, zinc soaps also can be used.

Retardation. When zinc dithiocarbamates are used, retardation becomes important. It is usually obtained by the use of other accelerators, but also by using acid retarders, as for example benzoic acid, salicylic acid or phthalic anhydride.

Vulcanizate Properties. Strength and elastic properties of zinc dithiocarbamate accelerated vulcanizates are excellent. To improve aging, especially with NR and IR, addition of compounds with good aging resistance and additionally low cure temperatures are useful. Zinc dithiocarbamate does not cause any changes in color or discoloration on light exposure.

Applications. Zinc dithiocarbamates alone are only usable in open cures because of their short scorch time (start of vulcanization). Together with slower accelerators they are useful for all *types of vulcanization*, i.e., for press-, steam-, and hot air vulcanization. They can also be used for transfer- and injection molding and other methods of conditious vulcanization. When used as sole accelerator in NR compounds the temperature should not exceed 135 °C because of reversion considerations.

In SBR and similar compounds the temperature can be a little higher. When used as secondary accelerators, they affect scorch time strongly with increasing dosage and temperature.

The zinc dithiocarbamates are suitable for most *types of rubbers*, especially all diene rubbers, including IIR and EPDM. CR is an exception.

They are used for the production of a multitude of *rubber articles* and have a multitude of applications in the solid rubber sector, where they are used alone or together with other accelerators for a variety of products. They are used in the shoe- and cable area, for surgical and hygiene applications, for parts in the food industry, articles in the pharmaceutical and cosmetic industry, rubber coated fabrics, dipped goods and self vulcanizing mixtures and solutions. Because of the recent discussion of carcinogenicity of nitrosoamines which are formed out of secondary amines and split off during vulcanization, a competition exists between dithiocarbamates and dithiophosphates (see page 254) [4.18a].

In the latex field, zinc dithiocarbamates as water insoluble accelerators are frequently used in combination with water soluble products for the production of practically all articles, even rubber thread and foam rubber.

4.2.3.5.2 Ammonium Dithiocarbamates

[1.14]

Ammonium dithiocarbamates are water soluble accelerators and preferred for use in the latex sector. A typical accelerator in this class is piperidyl ammonium-piperidyl dithiocarbamate (PPC)

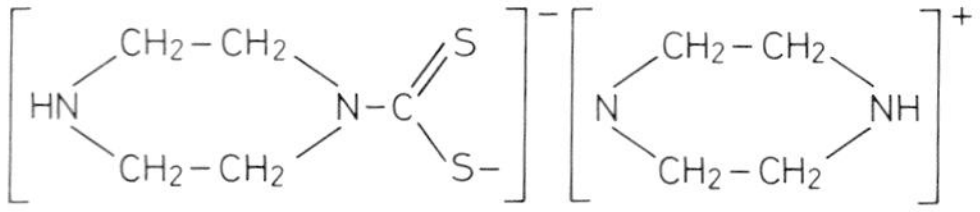

$$\left[HN \begin{matrix} CH_2-CH_2 \\ \\ CH_2-CH_2 \end{matrix} N-C \begin{matrix} S \\ \\ S- \end{matrix} \right]^- \left[N \begin{matrix} CH_2-CH_2 \\ \\ CH_2-CH_2 \end{matrix} NH \right]^+$$

Products of this type accelerate vulcanization extremely fast and can hardly be mixed into solid rubber without danger of scorching. They and the xanthates are the fastest known accelerators. This high cure rate can be utilized in the latex sector, where hardly any heating through processing takes place. Frequently, combinations with water insoluble zinc dithiocarbamates are used, where a synergistic effect is observed. With exception of the higher vulcanization rate and the special latex use, the same applies to ammonium dithiocarbamates as was explained for zinc dithiocarbamates (see page 242).

4.2.3.5.3 Sodium Dithiocarbamates

In addition to ammonium dithiocarbamates, sodium dithiocarbamates are also sometimes used as water soluble latex accelerators. The base compound of this class of accelerators is sodium dimethyldithiocarbamate (NaDMC).

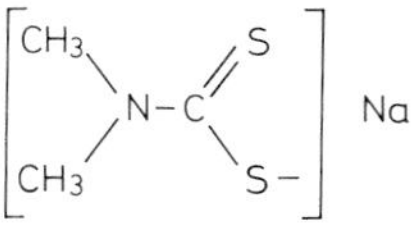

$$\left[\begin{matrix} CH_3 \\ \\ CH_3 \end{matrix} N-C \begin{matrix} S \\ \\ S- \end{matrix} \right] Na$$

The vulcanization rate of these accelerators is a little lower than with ammonium dithiocarbamate. Otherwise the same applies as mentioned for these materials, or their Zn-derivatives.

4.3.2.5.4 Selenium and Tellurium Dithiocarbamates

[4.3]

Selenium or tellurium dithiocarbamates have structures analogous to zinc dithiocarbamate. The base compound of these classes of accelerators are selenium or tellurium salts of dithiocarbamates (SeDMC, resp. TeDMC)

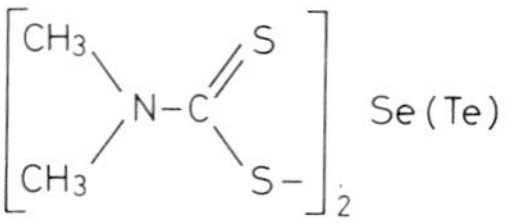

$$\left[\begin{matrix} CH_3 \\ \\ CH_3 \end{matrix} N-C \begin{matrix} S \\ \\ S- \end{matrix} \right]_2 Se\,(Te)$$

The selenium and telluriam salts of dithiocarbamic acid are high priced and are therefore only used for special applications. Primarily they are used as accelerators or secondary accelerators for IIR, CSM or EPDM for MBT/TMTD combinations, where they act as strong activators. In particular tellurium dithiocarbamate gives very high vulcanization rates, whereby the processing properties and storage stability of the blends may be adversely affected. They are used to some extent in continuous vulcanization, for example, in production of cables and heat stable articles. For normal diene rubber blends, selenium or tellurium dithiocarbamates are hardly ever used. Because of their toxicological properties the compounds are rarely employed.

4.2.3.5.5 Lead, Cadmium, Copper and Bismuth Dithiocarbamates

[4.3]

Products. These accelerators are structured similarly as the zinc dithiocarbamates, lead-, cadmium-, copper- respectively bismuth dithiocarbamates (PbDMC, CdDMC, Cd5MC, CuDMC, BiDMC) and have the following structure:

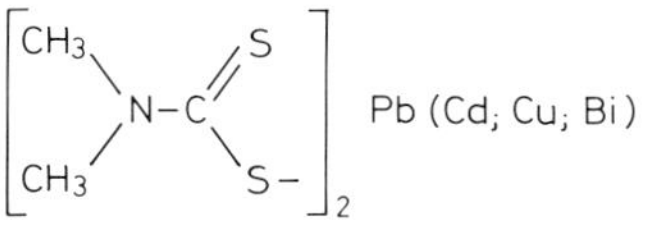

$$\left[\begin{matrix} CH_3 \\ CH_3 \end{matrix} \!\!>\! N-C \!\begin{matrix} \!/\!\!/ S \\ \backslash S- \end{matrix}\right]_2 \; Pb\,(Cd;\, Cu;\, Bi)$$

Cadmium pentamethylenedithiocarbamate (Cd5MC) and copper dibutyldithiocarbamate (CuDBC) also have some technical applications.

Efficiency and Combinations. These substances also exhibit high vulcanization speeds. Begin of the vulcanization is somewhat slower with lead- and cadmium-dithiocarbamates than with zinc dithiocarbamate. Especially Cd5MC shows a retarted onset of vulcanization (longer scorch time). BiDMC gives especially high vulcanization rates. Because of the high vulcanization speeds, processing safety of such compounds is poor, similar to the use of zinc dithiocarbamates. CuDBC is of some interest in accelerator combinations in EPDM because of its highest activity; it leads to high crosslinking density and to low heat compression set.

By combination with MBTS, processing safety is much improved. If vulcanization temperatures are not too high (below approx. 125 °C) the vulcanization plateau of NR vulcanizates is good. The main characteristics of the accelerators is the steeper Arrhenius activation energy curve compared to zinc dithiocarbamates. Thus the vulcanization rate versus temperature increases more rapidly (steeper curve) than is the case with zinc dithiocarbamates; this can be important in continuous vulcanization.

MBT does not activate these mutual salts. These dithiocarbamates serve as secondary accelerators for products of the mercapto class. CdDMC, CuDMC and BiDMC among others are recommended as secondary accelerators for the high temperature activation of sulfenamides in SBR compounds. Because of the high vulcanization rate, BiDMC is used for continuous vulcanization of cable covers in the steam pipe (CV-vulcanization). CuDMC gives a brown discoloration in light colored vulcanizates, but not CuDBC.

Activation. The lead-, cadmium-, copper- and bismuth salts of dithiocarbamic acid need addition of zinc oxide or cadmium oxide for full development of their activity. Fatty acids are beneficial similar to the other dithiocarbamates. Zn soaps may also be used.

Application. These accelerators are used to some extent for cable and molded parts of SBR, IIR, EPDM and also of NR. To obtain highly heat stable NBR vulcanizates CdDMC with CdO activators is recommended [3.213]. Similar applications are questioned because of toxicological reasons [3.181, 3.182] (see page 71).

Nickel dithiocarbamates are in contrast to the substances mentioned, not vulcanization accelerators but light and aging protectors for CR, CSM, CO, ECO and ETER against the effect of heat.

4.2.3.6 Xanthate Accelerators

[4.3, 4.140-3.142]

Materials. Xanthate accelerators are derivatives of xanthic acid and are usually alkali or zinc salts. One of the best known accelerators of this class is zinc isopropylxanthate (ZIX)

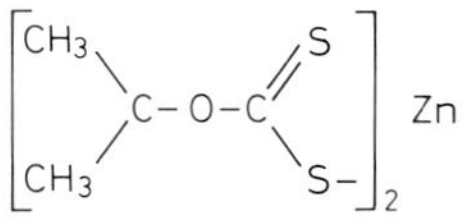

$$\left[\begin{matrix} CH_3 \\ \quad \diagdown \\ \quad C-O-C \\ \quad \diagup \\ CH_3 \end{matrix} \begin{matrix} \diagup\!\!\diagup S \\ \\ \diagdown S- \end{matrix} \right]_2 Zn$$

Among the known xanthate accelerators are zinc butylxanthate (ZBX) and the water soluble sodium isopropylxanthate (NaIX).

Activity. The xanthate accelerators are among the fastest acting ultraaccelerators. They are faster acting than the ammonium salt of dithiocarbamic acids. They therefore are only used in special cases for solid rubber. They have some use in benzene solutions in spite of their short storage stability (pot life).

In latex application, NaIX is slightly faster than the corresponding zinc derivative. A combination of the two accelerators (water soluble and water insoluble) vulcanizes somewhat faster than the separate products, similar to the dithiocarbamates. With these very fast accelerators this difference is difficult to establish.

Because of the great vulcanization speed the vulcanization plateau in vulcanizates prepared with these accelerators is very short; low cure temperatures (80 °C-110 °C) are therefore preferred.

Activation. Similar to the dithiocarbamates, the xanthates also need ZnO to achieve their full activity.

Application. The main application for xanthate accelerators is in the latex field mainly in the production of foam rubber.

4.2.3.7 Thiuram Accelerators

[1.12-1.14, 2.3, 4.3, 4.79-4.94, 4.143-4.152].

Products. The thiuram accelerators are derived from the dithiocarbamates by dimerization of the dithiocarbamate molecules. The base compound of this class is the tetramethylthiuram disulfide (TMTD) [4.143, 4.144]

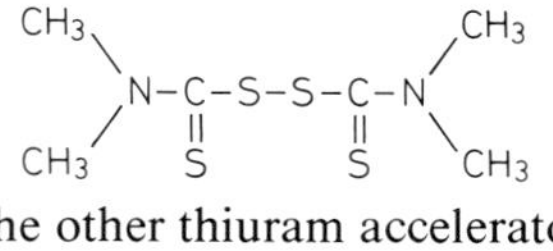

$$(CH_3)_2N-\underset{\underset{S}{\|}}{C}-S-S-\underset{\underset{S}{\|}}{C}-N(CH_3)_2$$

The other thiuram accelerators are derived from the base compound by variation of the alkyl group and the sulfur content. The other best known thiuram accelerators are tetramethylthiuram monosulfide (TMTM) [4.145, 4.146], tetraethylthiuram disulfide (TETD), dimethyl diphenylthiuram disulfide (MPTD) and dipentamethylenethiuram tetrasulfide (DPTT) [4.147] (compare page 231).

Efficiency and Combination. Thiuram accelerators have to be decomposed into the corresponding dithiocarbamates to begin their own activity. Therefore, the onset of vulcanization is slower than with the corresponding dithiocarbamates. The thiurams may, like no other class of accelerators, be used within wide limits of sulfur/accelerator ratios. For conventional vulcanization systems (1.2-2.0 phr sulfur) one uses 1.5-0.3 phr, for semi-EV systems (0.5-1.2 phr sulfur) 2.0-1.5 phr and for EV systems (0–0.2 phr sulfur) for example 3.0–2.5 phr thiuram accelerator. Because of this and the varied effects resulting from combination with other accelerators, thiurams are of special importance. In combination with rapid accelerators, for example dithiocarbamates and xanthates, thiurams have a retarding effect, without negatively changing the slope of the vulcanization curve on the crosslink density. In combination with thiazole accelerators on the other hand, they function as activating secondary accelerators in that the flow time/curing time relationship is somewhat improved because of the slightly faster vulcanization start [4.152]. During vulcanization secondary amines are split off. They can form nitrosoamines under environmental conditions, which seem to be human carcinogenics. TMTM as a monosulfidic thiuram type forms less secondary amines than TMTD [4.151].

In *conventional systems,* the vulcanization starting and finishing rates increase in the sequence DPTT, TETD, MPTD, TMTM, TMTD. Since the vulcanization in any case proceeds rapidly, these materials are classified as ultra accelerators. The thiurams, similar to the sulfenamides and dithiocarbamates, show a steep stress value curve and rapidly achieve the vulcanization optimum. When mixing only thiuram accelerators, a short vulcanization plateau is obtained, that necessitates with for example NR, low vulcanization temperatures (for example 135° C). When thiurams are employed as secondary accelerators, especially favorable flow times/cure time ratios are obtained and a broader plateau, depending on the primary accelerator. TMTM is mostly the material of choice in that case.

By combining thiurams with sulfenamides, vulcanization start is retarded without influencing the total cure time appreciably [4.152]. These combinations make a rapid complete cure possible with surprisingly high scorch safety. Furthermore, high crosslink densities are obtained.

Combinations of thiurams with mercapto accelerators, especially MBT are important for less active SR types, for example IIR, EPDM, CSM, ETER, since they show synergistic action.

In *semi-EV systems,* thiuram accelerators are frequently combined with sulfenamides. In these systems, TMTM gives a somewhat better flow time/cure time relation than for example TMTD; the stress curve is slightly steeper. In equal amounts, TMTD gives a higher crosslink density than TMTM. Vulcanization with less sulfur presents a broader vulcanization plateau combined with good reversion stability with reversion-prone vulcanizates, and better aging- and heat stability. Similar systems may be activated with basic accelerators, for example DPG, DOTG, HEXA and especially with thioureas, like ETU, DETU, etc.

In *sulfur-less EV-systems* [4.49, 4.50, 4.55, 4.65, 4.72, 4.92] only di- or oligosulfide thiuram accelerators may be used. They act there as sulfur donors, since even without the use of free sulfur, sulfur links are formed [2.3]. During vulcanization with TMTD no free sulfur is formed; that can be demonstrated by nonblackening of a silver mirror. The most important material for that is TMTD. However, this material gives a relatively rapid start of vulcanization with a flat flow time/cure time relation. While TETD and DPTT give better processing safety than TMTD, they are not much used. Combining TMTD with sulfenamides a steeper flow time/cure time relationship can be obtained with better scorch safety. For that reason combinations with sulfenamides are important. The EV-vulcanization with thiurams can also be activated by adding thioureas [4.150] like ETU, DPTU and others. Other typical secondary accelerators have little activity. In equal dosage, DPTT, because of its higher sulfur content gives considerably higher crosslink densities than TMTD combined with lower reversion tendency and heat stability. For that reason DPTT is used in EV-vulcanization in small amounts mainly as additional sulfur donor. In sulfurless, so called " thiuram vulcanization" one obtains an excellent vulcanization plateau, low reversion tendency and good heat stability.

Activation. Thiuram accelerators need ZnO to develop their activity. The addition of fatty acids, like stearic acid is useful. Zn-soaps can be used instead of them.

Retardation. To delay the onset of vulcanization, mainly thiazole accelerators are used. In conventional systems, scorch safety can also be improved with classical vulcanization retarders, whereby the vulcanization time is correspondingly prolonged. In EV-systems most retarders do not work, for that reason, sulfenamides are preferably used to obtain a desired flow time/cure time relationship.

Vulcanizate Properties. Using in combined thiurams in *conventional systems*, good strength (tensile)- and elastic properties are obtained because of the high crosslink density. Reversion- and heat stability are limited. In *EV-systems*, reversion- and heat stability are optimized while tensile- and elastic properties are poorer. By using *semi-EV systems* a compromise between both effects can be obtained.

Thiurams do not discolor the vulcanizate nor do they cause color change on exposure to light.

Application. The thiurams are suitable for all *vulcanization processes* (press, steam, hot air). They are also used in transfer- and injection molding and other methods of continuous vulcanization.

The thiurams are suitable primarily for the vulcanization of the following *rubber types:* NR, IR, BR, SBR and NBR. Besides, they are also important for rubbers with low unsaturation like EPDM, ETER or IIR. In use are TMTD, respectively TMTM in combination with mercapto accelerators (MBT, ZMBT, MBTS, and in some cases also sulfenamides like CBS). For the vulcanization of CR, thioura derivatives are generally used. However, if a more consistent vulcanization start is desired, a combination of TMTM + DOTG and sulfur is recommended. Thiurams are used beside MBTS in the vulcanization of CSM.

From the above described property spectrums, the manifold areas of application for thiuram accelerators in *rubber products* can be explained. One important area of application is in heat stable articles of all kinds, like conveyor belts, hoses, gaskets, sleeves, etc. Furthermore, articles should be mentioned which must be physiologically clean, for example items in contact with food or surgical rubber pads. A further field is in rubber articles that need to be vulcanized very rapidly. Finally one can find thiurams used as secondary accelerators for cable, technical articles, sometimes for tires. Sulfurless EV-systems are only considered for articles that do not contain sulfurs, as for example, cable isolators, spot-light gaskets and the like. TMTD is also suitable in 3.0–5.0 phr levels as a rapid accelerator for hard rubber, either in solid rubber or based on latex.

For the vulcanization of latex mixtures, vulcanization speed of the thiurams is usually too slow; usually dithiocarbamates are preferred. By the addition of thioureas, thiuram vulcanization can be accelerated to the point where it can be applied in the latex field.

4.2.3.8 Dithiocarbamylsulfenamide Accelerators

[3.219–3.221, 3.400, 3.434, 4.153–4.161]

Products. Benzothiazyl sulfenamide shows weakness in regard to reversion stability and therefore is used in combination with thiurams (respectively dithiocarbamate derivatives) which also improve the flow time/cure time relationship; based on these facts it was obvious that a molecular combination of dithiocarbamate- and sulfenamide structure should be developed [4.157, 4.161]. This development by Goodrich leads to the youngest and especially interesting class of accelerators, the dithiocarbamyl- sulfenamides. They have the basic structure

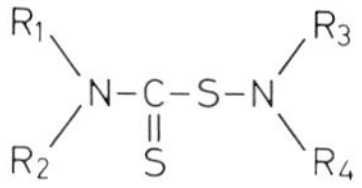

$$\begin{matrix} R_1 \\ R_2 \end{matrix} \rangle N-\underset{\underset{S}{\|}}{C}-S-N \langle \begin{matrix} R_3 \\ R_4 \end{matrix}$$

The first commercial product of this class is N-oxydiethylenedithiocarbamyl-N′-oxydiethylene-sulfenamide (OTOS)

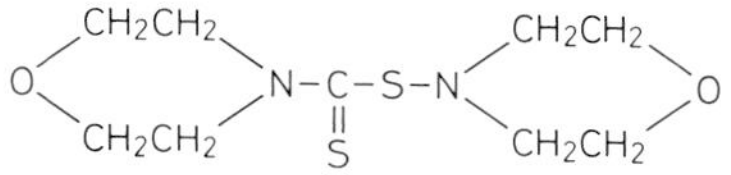

$$O \langle \begin{matrix} CH_2CH_2 \\ CH_2CH_2 \end{matrix} \rangle N-\underset{\underset{S}{\|}}{C}-S-N \langle \begin{matrix} CH_2CH_2 \\ CH_2CH_2 \end{matrix} \rangle O$$

A further product developed from this class is the N-oxydiethylenedithiocarbamyl-N′-tert. butylsulfenamide (OTTBS) [4.158 a].

Efficiency and Combinations. OTOS acts like a molecular combination of a sulfenamide and dithiocarbamate. This means that an exceedingly steep flow time/cure time relationship develops. After a strongly retarded start of vulcanization, that roughly equals that of TBBS, vulcanization proceeds very rapidly to a very high crosslink density, that is higher than with benzothiazylsulfenamides using equal amounts. Based on the high efficiency of OTOS and on the fact that this material can furnish monosulfidic sulfur for the crosslinking reaction, on the average sulfur-poorer crosslinks are formed than from benzothiazole sulfenamides. Consequently an especially broad vulcanization plateau is formed and high reversion and heat stability obtained. The high temperature compression set values of the vulcanizates are especially low [3.219–3.221].

OTOS belongs to the strongest synergists among vulcanization accelerators. In the presence of thiazole accelerators [4.153–4.156, 4.158–4.160], especially MBS and MBTS, its vulcanization properties get fully developed. This goes fully for NR, where only these combinations are recommended. In SR, for example, SBR, SBR/BR blends and the like, OTOS as sole accelerators can be used. In EPDM OTOS is mainly combined with dithiocarbamates and thiurams.

OTOS has better chemical stability than benzothiazole sulfenamides; for that reason a once established scorch time is preserved independent of heat history or storage time of the blends. For that reason one prefers for some SR-mixtures uncombined OTOS over a mixture with benzothiazole sulfenamides.

Since OTOS acts as accelerator and sulfur donor similar to TMTD, it can be used like the latter in broad sulfur to accelerator ratios. In semi-EV systems, the reversion and aging stability is fully developed. OTOS is also used advantageously in EV-systems, when for example the benzothiazole sulfenamide is completely or partly replaced by OTOS. Thus it is possible to obtain very low compression set values [2.108]. For a sulfur less vulcanization without TMTD addition, OTOS is too slow.

Activation. ZnO is necessary for OTOS activation, similar to the dithiocarbamates and benzothiazole sulfenamides. Fatty acids, like stearic acid, are beneficial. Instead of both, Zn-soaps may be used.

Retardation. The cure rate depends on the accelerator combination used. Vulcanization is retarded by combination with MBT. Otherwise a further retardation can be achieved with procedures similar to those for benzothiazole sulfenamides.

In EPDM compounds a certain retardation in scorch can be achieved by replacing sulfur partially by sulfur donors. There a typical scorch retardation with subsequent rapid vulcanization rate can be obtained because of the synergistic OTOS/sulfur donor action in EPDM.

Properties of the Vulcanizates. Because of its high efficiency, equal amounts of OTOS give correspondingly higher crosslink density than the benzothiazole sulfenamides. An equalization is possible by lowering the acceleration or sulfur content up to 30%. In the first case economic advantage is achieved with roughly equal properties while in the second case better heat stability and permanent set are obtained. When OTOS is used, heat stability, compression set and heat build-up are almost always improved. OTOS and especially OTTBS form drastically reduced secondary amines during vulcanization [4.161 a], which seems to be important concerning the actual discussion of the toxicity of nitrosoamines.

Applications. OTOS analogous to benzothiazole sulfenamides is preferred in compression molding, injection molding and transfer molding processes. It is particularly suitable for the following rubbers: NR, SBR, NBR, EPDM and BIIR. Its most important application is in the following rubber products: tires (especially radial-passenger car treads and innerliners), conveyor belts, heat resistant products with low compression set, for example engine mounts, gaskets, window profiles as well as heat stable NBR- and EPDM- parts.

4.2.3.9 Guanidine Accelerators

[1.12-1.14, 4.3, 4.162-4.166]

Products. The best known guanidine accelerators are diphenylguanidine (DPG) [4.162-4.165] as well as di-o-tolylguanidine (DOTG) and o-tolylbiguanide (OTBG) [4.166]. Of these three products, DPG is of greatest economic importance. This material as well as DOTG is only rarely used as a single accelerator; the guanidines serve mainly as secondary accelerators.

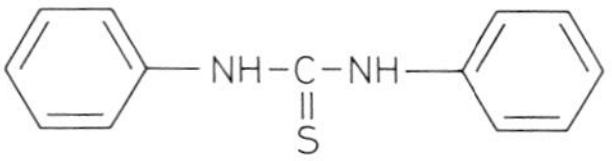

C_6H_5–NH–C(=S)–NH–C_6H_5 DPG

Efficiency and Combinations. When used as a single accelerator, the guanidines have a long scorch time and long vulcanization time, have an unfavorable vulcanization plateau and their use with some exceptions is uneconomical, the products having insufficient age resistance. In combination especially with mercapto accelerators they show synergistic action (see page 237). Herein lies the main importance of guanidines. As activating secondary accelerators they often increase the vulcanization rate and crosslink density of the primary accelerator drastically. This secondary accelerator action is best developed with mercapto accelerators. With thiurams and dithiocarbamates as well as to a lesser extent with sulfenamides, synergistic activity is also observed. In triple combinations, for example sulfenamide + dithiocarbamate (thiuram) + guanidine, considerable activation is also observed.

Activation. Addition of ZnO is necessary if DPG and DOTG are used as sole accelerators. In compounds with un-combined OTBG, ZnO can be left out; the accelerator only becomes fully active if ZnO is added. Stearic acid retards compounds containing guanidine accelerators. Dosages above 1.0 phr give furthermore poorer mechanical properties. When using guanidines as secondary accelerators, the activation rules of the primary accelerators apply: that necessitates addition of ZnO. In these systems, fatty acids generally give some retardation. Instead of ZnO and fatty acids, Zn-soaps can also be used.

Retardation. Guanidines, since they are bases, are generally retarded by lowering the P_H value, e.g. by addition of acids.

Properties of the Vulcanizates. Vulcanizates prepared with uncombined guanidine accelerators often show good tensile properties and elasticity when high sulfur (approx. 3 phr) and low stearic acid (less than 1 phr) charges are used. This is valid especially in blends with (acidic) silica fillers. Because of polysulfidic crosslinks, age resistance and compression set are sometimes insufficient. Properties of vulcanizates with accelerator systems using guanidines as secondary accelerators depend primarily on the primary accelerator used. The guanidines may have a positive influence because of an increased crosslink density and a negative influence because of a shortened vulcanization plateau.

Vulcanizates with guanidines discolor on light exposure. This is naturally also the case for combinations with guanidines depending on the quantity employed. DOTG is better than DGP with respect to discoloration. HEXA as a non-discoloring secondary accelerator is sometimes preferred to guanidines, it even acts to lighten the color of a product. For that reason a partial replacement of DPG by HEXA in activated accelerator systems can compensate for the slight discoloration of the DPG.

Applications. Guanidines, especially in many combinations can find application in all types of *vulcanization processes*. It is obvious that blends adjusted for slow vulcanization (for example with guanidine as sole accelerator) can not be used for open air or continuous vulcanization processes.

Looking at the applicability of guanidines in various *rubber types*, a different picture appears. In diene rubbers like NR, IR, BR, SBR, and NBR they are often combined with thiazole-, thiuram- and dithiocarbamate accelerators. They are less useful (even in combination) in IIR and EPDM. In CR, usually cyclic or open thioureas are used as accelerators, if one seeks higher scorch resistance, combinations of for example DOTG/TMTM/sulfur are used. In ACM guanidines, especially OTBG (see page 110) may be used as basic compounds for crosslinking.

Guanidines as sole accelerators are only used in a limited number of *rubber products*. Because of their slow cure, they are of use primarily in thick-walled rubber goods, like for example, rolls. The recipes contain frequently other basic products in addition to guanidines, for example HEXA. Guanidines are often important for compounds containing large amounts of silica. OTBG is particularly suitable for repair-compounds and beyond that – in specialty applications – for compounds containing strongly acid components. OTBG is also used in eraser compounds and chrome leather waste molds (beside polyethylene polyamines PEP).

Combination of guanidines and other accelerators, especially mercapto accelerators, are used in large numbers of technical rubber goods like injection molded parts, etc. Furthermore, some tire applications (e.g. cycle tires), cable covers and insulation, rubber shoe parts, soles and heels and rubber coated parts should be mentioned. They are generally not suited for production of parts used in contact with food because of their odor and bitter taste.

4.2.3.10 Aldehyde-Amine Accelerators

[1.12–1.14, 4.3, 4.167–4.176]

Products summarized in this category show great variation in their activity and are less important than accelerators treated thus far. They partly belong to the relatively fast acting accelerators such as condensation products of butyraldehyde and aniline (BAA) [4.170] or heptaldehyde with aniline [4.171] and partly to the weak accelerators, like hexamethylenetetramine (HEXA) [4.168], tricrotonylidinetetramine (TCT) [4.169], formaldehyde-p-toluidine, resp.-aniline.

4.2.3.10.1 Butyraldehydeaniline (BAA)

[1.14]

This is a very rapid accelerator that leads to high crosslink densities. Because of the very high elastic properties it is sometimes used for rubber springs. Because of its high scorch tendency it also plays a small role as a secondary accelerator.

$N{=}CH{-}CH_2{-}CH_2{-}CH_3$

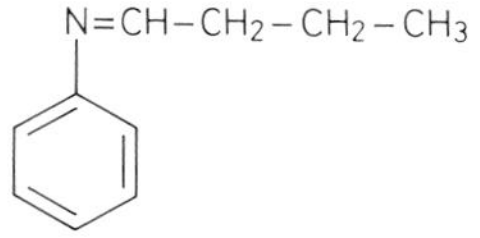

4.2.3.10.2 Hexamethylenetetramine (HEXA)

[1.14]

It has practically no application as sole accelerator. As single accelerator it is even slower than guanidines. It has some importance as activating, non-discoloring secondary accelerator, that is used instead or in conjunction with guanidines (see page 250).

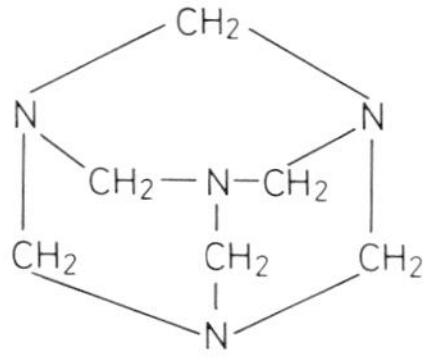

CH_2
N N
$CH_2{-}N{-}CH_2$
CH_2 CH_2 CH_2
N

4.2.3.10.3 Tricrotonylidinetetramine (TCT)

[1.14]

This is a special accelerator for the vulcanization of hard rubber. This accelerator acts very slowly; it is practically never used as a single accelerator.

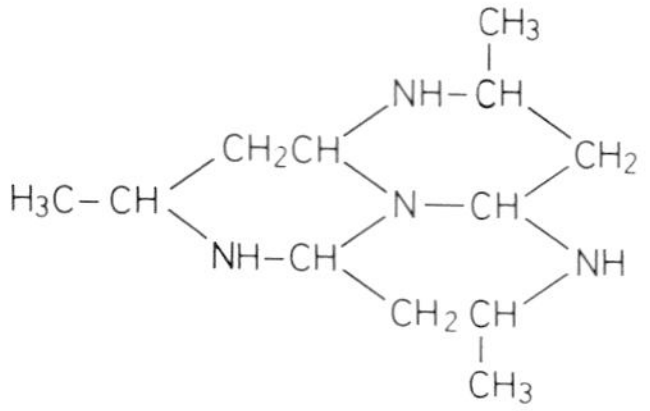

CH_3
$NH{-}CH$
CH_2CH CH_2
$H_3C{-}CH$ $N{-}CH$
$NH{-}CH$ NH
$CH_2\,CH$
CH_3

4.2.3.11 Other Amine Accelerators

[1.12–1.14, 4.3, 4.172–4.176]

Secondary amines, such as *dibutyl* (DBA)-, *dibenzyl-* or *cyclohexylethylamine* (CEA) are generally speaking weak accelerators and are rarely used alone, but practically only as activating secondary accelerators. The amines that were first described as organic accelerators [4.172–4.176] like aniline, piperidine, etc. are too slow and are no longer being used.

Cyclohexylamine (CEA) and dibutylamine (DBA) have limited importance for the activation of dithiocarbamate accelerators for example in the so called vapor vulcanization (see page 243). In the presence of sulfur and zinc dithiocarbamate, CEA resp, DBA can be used like sulfur chloride (see page 452).

Polyethylenepolyamines (PEP) are relatively rapid accelerators; they play a role as activators for silica fillers and in special cases, for example in the presence of acid compounding ingredients, sometimes are used together with OTBG (see page 251).

4.2.3.12 Thiourea Accelerators

[4.150, 4.177–4.183]

Products. Thioureas are special products used in the vulcanization of CR, epichlorohydrines and EPDM. They are only used in special cases in classical diene rubbers.

The oldest accelerator in this class that used to be used to any extent is diphenylthiourea, also known as thiocarbanilide (DPTU) [1.177]

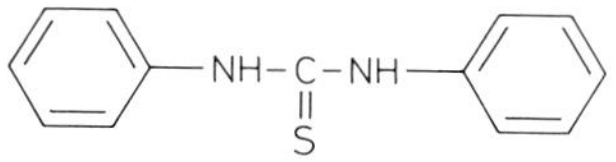

$$C_6H_5-NH-\underset{\underset{S}{\|}}{C}-NH-C_6H_5$$

Other known materials in this classification are the especially active ethylenethiourea (ETU), also called 2-mercaptoimidazoline,

$$\begin{matrix} CH_2-NH \\ | \\ CH_2-NH \end{matrix}\!\!>C=S \longleftrightarrow \begin{matrix} CH_2-NH \\ | \\ CH_2-N \end{matrix}\!\!\gg C-SH$$

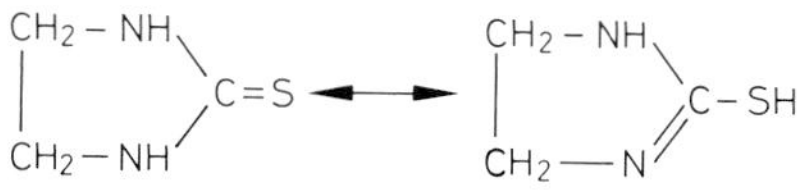

Ethylenethiourea 2-Mercaptoimidazoline

diethylthiourea (DETU) and dibutylthiourea (DBTU). The antioxidant 2-mercaptobenzimadazole (MBI, see page 273) as thiourea derivative has some accelerator activity in CR compounds [4.182]

Efficiency and Combinations. ETU and DETU are the materials of choice in CR and epichlorohydrin compounds among all the known thiourea derivatives because of their high crosslinking activity and relatively favorable flow time/cure time relationship. DPTU can be used theoretically for the compression molding of CR; in general it is not used in practice because of its low crosslinking ability and unfavorable flow time/cure time ratio. However, DPTU is used in self-vulcanizing blends and solutions based on CR, in combination with other accelerators such as PEP or BAA as well as an additional accelerator for EPDM. Thiourea accelerators show crosslinking activity in CR-latex vulcanizates. In latex, DPTU is superior to ETU.

ETU (with or without sulfur) and also DETU acts more rapidly as accelerator for CR in respect to scorch and vulcanization rate than the also used combination TMTM + DOTG + sulfur; the former generally gives higher crosslink density and a higher level of physical properties and is especially suitable for high temperature vulcanization of CR [4.178]. The second combination will only be used if difficulties in scorch resistance exist with thiourea, or if these can't be used for other reasons.

In EV-vulcanization of diene rubbers with thiurams, thiourea accelerators exhibit a clearly activating secondary accelerator action in contrast to most typical secondary accelerators (see page 247) [4.150]. In EPDM, ETU shows a strong activating activity, for example in compounds containing CaO it can overcompensate for the CaO-caused retardation. Especially in slow vulcanizations of the DCP-EPDM type, ETU is of special value [4.97] (see page 96). ETU is also the best accelerator for compounds based on polyepichlorohydrins; with it highest heat stability is obtained (see page 114).

Activation. Addition of ZnO is necessary for the activation of thioureas; in CR compounds it is usually used with MgO; stearic acid is usually not necessary. Zn-soaps can also be used as activators.

Retardation. Compounds containing ETU are retarded by MPTD, sulfenamides and especially MBTS. Vulcanization retarders are not active.

Properties of the Vulcanizates. When ETU is employed instead of thiuram/guanidine combinations in CR compounds, a property spectrum based on high crosslink density is obtained. Furthermore no discoloration is observed in the presence of Cu, Mn and other heavy metals.

Applications. ETU and DETU can be used for all *processes* that are applied in CR and epichlorohydrin rubber technology, beginning with press- and hot steam vulcanization to the modern processes of injection- and transfer molding, the salt bath- and UHF-vulcanization, continuous vulcanization in steam, and others.

DPTU in combination with basic products like PEP or BAA is mainly used for the self-vulcanization or very fast vulcanization of CR. It is suitable for all vulcanization methods for diene rubbers as activator or secondary accelerator. ETU and DPTU may be used as additional accelerators for EPDM in all vulcanization processes used for these rubbers.

ETU and DETU are suitable as single accelerators for CR and epichlorohydrin type rubbers. In other diene *rubbers* they only act as activator and secondary accelerators in sulfur-poor or sulfurless systems and furthermore as additional accelerators for EPDM. DPTU is not used alone. Combinations with basic accelerators are used in self-vulcanizing or very rapid curing blends of CR. In other diene rubbers, DPTU acts mainly as an activator. In EPDM it is also used as secondary accelerator. In the latex area, these accelerators are only used with CR.

Thiourea derivatives are mainly used for technical *rubber goods*, cables, films, rubber coatings, solutions and similar articles. The thioureas are not used in products that come in contact with food because of toxicological reasons. Because of the latter they are threatened as vulcanization accelerators; a search is on for suitable substitutes (see page 81, 115, 255) [4.179–4.183].

4.2.3.13 Dithiophosphate Accelerators

[4.184, 4.185]

Products. The base compound of the dithiophosphate accelerators is zinc-dibutyl-dithiophosphate (ZDBP)

$$\left[(C_4H_9O)_2P(=S)S^- \right]_2 Zn$$

Copper-diisopropyldithiophosphate (CuIDP) and ammoniumdithiophosphates are in use. Among these materials, ZDBP has a slower scorch time than CuIDP. Generally the dithiophosphates are a little slower than dithiocarbamates.

Efficiency and Combinations. Since the dithiophosphate accelerators are chemically closely related to dithiocarbamates [4.185], the former can be substituted in diene rubbers for the latter [4.184]. Dithiophosphates are, however, clearly slower and pricewise less attractive and have therefore not been able to penetrate this field.

In EPDM compounds, where dithiophosphates are preferred, they are partly substituted for dithiocarbamates to overcome their bloom limits and beside this to overcome the formation of toxic nitrosoamines [4.18 a] (see page 243). There they show a weak synergistic effect. Because of price consideration, this application is limited.

Activation. ZnO is necessary for the activation of dithiophosphates similar to the dithiocarbamates in addition to stearic acid. Zn soaps are also used.

Retardation. In EPDM, retardation is usually not desired. By substituting some of the sulfur by sulfur donors a certain retarded scorch can be achieved.

Properties of the Vulcanizates. Properties that can be achieved with dithiophosphates are very similar to those with dithiocarbamates. Generally reversion stability in conventional systems is improved.

Applications. The dithiophosphates are used mostly for all applications in EPDM. They are rarely used alone but usually, especially in EPDM, as accelerator combinations. The most common combination is with thiazole-, thiuram-, and dithiocarbamate accelerators. The dithiophosphates can here in all cases be substituted for the dithiocarbamates.

While ZDBP has no discolorating tendency, CuIDP does discolor and leads to discoloration of light-colored adjacent surfaces on PVC, coatings, etc.

4.2.3.14 Other Vulcanization Accelerators

Very recently, new accelerators have been developed to replace ETU. We deal here mainly with, for example, oxadiazine-, thiadiazine-, thiadiazole-, and thiazolidine derivatives [4.179–4.183] (see page 81, 115). Other new accelerators have also been suggested [4.186–4.189].

Beyond the best known accelerator classes mentioned here, many other substances are available for special applications, for example for ACM, CSM, FKM (see page 122), etc., that were mentioned in connection with the specific types of elastomers (see there).

Many commercial products are special combinations of the accelerators mentioned or are derived from the accelerator classes already treated by other substituants.

Accelerator and sulfur donor systems (quite often very complex) are necessary for the slow vulcanizing EPDM. Synergisms have to be exploited. Therefore, accelerator mixtures developed with the special know-how of their manufacturers are used with good success [3.400, 3.434, 4.189] (see page 97).

4.2.4 Crosslinking Agents without Sulfur

[1.12, 1.13]

The vulcanization of elastomers can also be carried out by radicals in addition to the ionic reactions with sulfur and accelerators. Here, the initiating step is the formation of polymer chain radicals which can be created in a number of ways. Also the application of high energy radiation can be employed as well and has gained increased importance lately. This procedure can be applied in principle without the use of any vulcanization chemicals [3.400]. Because of its technological importance, radiation crosslinking will not be covered in this chapter, but in chapter [5.3.3.6] (see page 403 ff) instead.

4.2.4.1 Peroxides

[1.12, 1.13, 3.400, 4.3, 4.190–4.223]

Crosslinking with peroxides has been known for a long time [4.190] but gained importance only with the development of the saturated synthetic rubbers, like EVM, EPM, CM, Q, etc. In the meantime, one has also recognized their action in NR and the classic diene types of rubber, SBR and NBR. Based on the heat stability that can be obtained by peroxide vulcanization of diene rubbers, especially of NBR but also of EPDM, this type of vulcanization has achieved a great importance.

The temperature of decomposition (half-life) of the peroxide is the main determinant for scorch temperature and cure rate. For that reason, the composition of the peroxide determines its usefulness as a vulcanizing agent.

Products. The formula of all peroxides can be derived from hydrogen peroxide; the single substituted products are the hydroperoxides; the double substituted products are the peroxides:

H-O-O-H Hydrogenperoxides
R-O-O-H Hydroperoxides
R-O-O-R Peroxides

Hydroperoxides do not lead to the formation of crosslinks, but can interfere with the crosslinking capacity of other peroxides, respectively can degrade polymer molecules. They are therefore not important in the crosslinking of rubber; they are important in the radical initiation during polymerization (see page 42). They frequently initiate aging reactions (see page 265).

For the crosslinking of rubber, only peroxides can practically be used, which are stable and not dangerous during the usual handling, and that, on the other hand, decompose sufficiently fast at customary curing temperatures. For that purpose, peroxides with tertiary carbon atoms are suitable. Peroxides bonded to primary and secondary carbon atoms are less stable. One can distinguish two groups of organic peroxides that are stable enough for rubber processing:

	aliphatic for example:	aromatic for example:
Peroxides *with carboxy groups*	$CH_3-\underset{\substack{\|\\O}}{C}-O-O-\underset{\substack{\|\\O}}{C}-CH_3$ Diacetylperoxide	$C_6H_5-\underset{\substack{\|\\O}}{C}-O-O-\underset{\substack{\|\\O}}{C}-C_6H_5$ Dibenzoylperoxide
Peroxides *without carboxy groups*	$CH_3-\underset{CH_3}{\overset{CH_3}{C}}-O-O-\underset{CH_3}{\overset{CH_3}{C}}-CH_3$ Di-tert.butylperoxide	$C_6H_5-\underset{CH_3}{\overset{CH_3}{C}}-O-O-\underset{CH_3}{\overset{CH_3}{C}}-C_6H_5$ Dicumylperoxide

Almost all peroxides can be derived from these basic types, that are of interest in curing of rubbers. Peroxides containing more than one peroxy group are called *polymeric or polyvalent peroxides*, for example

```
      CH3       CH3                 CH3       CH3
      |         |                   |         |
CH3 - C - O - O - C - CH2 - CH2 - C - O - O - C - CH3
      |         |                   |         |
      CH3       CH3                 CH3       CH3
```

2.5-Bis-(tert.butylperoxy)-2.5-dimethyl hexane

or

```
      CH3       CH3                CH3       CH3
      |         |                  |         |
CH3 - C - O - O - C —(C6H4)— C - O - O - C - CH3
      |         |                  |         |
      CH3       CH3                CH3       CH3
```

1.4-Bis-(tert.butylperoxyisopropyl)-benzene.

In addition to the *symmetrical peroxides, unsymmetric (mixed) peroxides* are also being used, as for example, tert.-butyl-perbenzoate, tert.-butylcumylperoxide, and also mixed polymeric peroxides.

The peroxides in the different classes have the following characteristics:

Carboxy group peroxides:

- low sensitivity to acids
- low temperature of decomposition
- high sensitivity to oxygen; therefore, cure problems in the presence of carbon black.

Peroxides without carboxy groups:

- sensitivity to acids, aliphatics better than aromatics
- higher temperature of decomposition
- lower sensitivity to oxygen than peroxides with carboxy groups.

For the curing of elastomers those peroxides are preferred that form the following radicals [4.191]:

```
         CH3                        CH3
         |                          |
(C6H5)— C - O*     (C6H5)*    CH3 - C - O*     CH3*
         |                          |
         CH3                        CH3
cumyloxy-          phenyl     tert.butyloxy-   methyl-radical
```

Efficiency. Decomposition of peroxides can occur under the influence of the following factors: by heat, by light or high energy radiation, or reactions with other materials. Preferably the decomposition occurs at the peroxy groups. In covalent peroxides, decomposition can occur homolytically into peroxide radicals or heterolytically into ions. For the peroxide crosslinking of rubber a homolytic decomposition is presumed. This occurs without the influence of other compounds, for example ideally in the gas phase; it can be negatively affected by other compounding ingredients, that can lead to reduced radical yield and thereby influence the crosslink density. When symmetrical peroxides are used, two similarly active radicals are formed on homolytic cleavage, that can initiate the crosslinking reaction equally. With mixed peroxides, two radicals of unequal reactivity are formed. At medium curing temperatures (approx. 150 °C) the more active radical acts as a crosslinker, while the other one remains essentially inactive. A lower crosslink density follows. At increasing temperatures (approx. 180–190 °C) the activities of the two radicals equalize more and more, whereby the theoretical crosslink density is approached. When unsymmetric peroxides are used a larger amount must be used at medium curing temperature or normal peroxide dosage be used to cure at higher temperatures.

A peroxide that decomposes at low temperatures would be desirable because of rapid vulcanization and high production rate. These advantages however are countered by decreased processing safety, because curing practically starts when the peroxide begins to decompose. Because of production problems and high reject production rates, peroxides with short half-lives can frequently not be used. Instead, it is necessary to have the stability of the peroxide used conform to the desired scorch time (start of vulcanization) and with it the desired processing safety. The stability of the peroxide is of course important for the choice of the vulcanization temperature. Peroxides with acid groups, for example diaroylperoxides, decompose at considerably lower temperatures than the dialkyl-, alkylaryl- or diarylperoxides. For that reason, compounds with dibenzoylperoxide can only be heated to 45 °C without scorching (that can be realized with Q compounds). Compounds with dicumylperoxides can tolerate approx. 110 °C without scorching danger. The stability of the peroxide also determines the maximum curing temperature that one can use; it shouldn't exceed about 130 °C for dibenzoylperoxide and 170 °C for dicumylperoxide. The radical yield and the crosslink density depend very much on the temperature.

Depending on the energy necessary for the dehydration of the rubber, for any given energy content of the peroxide radical, a crosslink or a decomposition of the polymer chain can result [4.204]. One can not draw conclusions from the behavior of a peroxide in a given rubber to the behavior of the same peroxide in a different rubber. Different behavior can also be expected in blends. While Q, EVM, EPM, CM and AU vulcanize well with peroxides, and NR and NBR can obtain a high crosslink density with peroxides, vulcanization of SBR and BR is a problem. IIR can not be vulcanized with peroxides, but is decomposed instead under the influence of peroxides. With most peroxides, oxygen should be excluded [4.206]. For that reason, hot-air heating (also UHF-heating) can not be used. An exception is for example 2.4-dichlorobenzoylperoxide.

The advantages and disadvantages of peroxide vulcanization are [3.400]:

Advantageous are:
- scorchfree storage of compounds
- the rapid vulcanization at high temperatures
- possibility to apply high vulcanization temperatures without reversion
- good balance between processing properties and crosslink density
- simple formulation
- low compression set even at high temperatures
- good electrical properties, no copper corrosion
- good high temperature stability
- no discoloration
- no bloom
- easy covulcanization with other polymers, also with reactive plasticizers

Disadvantageous are, however:
- limited compounding because of reaction of compounding ingredients with peroxides (e.g. antioxidants, plasticizers, resins etc.)
- sensibility of vulcanization reactions to oxygen (e.g. on UHF vulcanization)
- difficult adjustment of scorch time/plateau time relation
- at low temperatures long vulcanization times
- mostly lower tensile strength, lower tear strength
- lower abrasion resistance
- higher swelling
- frequently disturbing odors and toxicity of fragments
- mostly higher costs.

Accelerators. In contrast to sulfur vulcanization, accelerating additives (other than coactivators) are not known. Acceleration is only possible within limits by an increased temperature. For that reason it is important to select peroxides with a certain half-life dependent on the vulcanization conditions. These are mentioned for example [3.400, 3.401], resp. [4.208].

Activation. It is not possible to activate peroxides analogous to sulfur vulcanization with *metal oxides*, like ZnO, or stearic acid.
The degree of crosslinking achieved in peroxide vulcanization depends primarily on the type and amount of peroxide, respectively the radical yield as well as the reactivity of the rubber. The radical yield and with it the crosslink density is lowered by all materials that interact with peroxides without forming crosslinks (for example antioxidants and most plasticizers, with exception of highly paraffinic mineral oils, and the like [3.431, 3.437, 4.199–4.201]). Crosslink density can be considerably increased by using *coactivators* [3.458, 4.211–4.223]. Here we are dealing with polyvalent compounds (for example di- or triallyl compounds, maleic acid or reactive acrylic derivatives) that produce several consecutive reactions with one peroxydic initiation. They get involved in the crosslinking scheme and make an increased crosslink yield possible, that influences the stress values and hardness and the total vulcanization property spectrum. Different coactivators are recommended for different rubbers. For the Q-crosslinking none are necessary, while for EVM [3.458] and CM [3.470] highly active coactivators are used.

Those used for that purpose are for example triallylcyanurate (TAC) and -isocyanurate (TAIC) [3.458, 4.216], triallylphosphate (TAP), triallyltrimellitate (TAM) [3.458], diallylphthalate (DAP, also a plasticizer), m-phenylene-bis-maleimide [4.215], ethyleneglycoldimethacrylate (EDMA), trimethylolpropane trimethacrylate (TPTA), 1,3-butyleneglycol dimethacrylate, and others [4.217]. Also, vinylsilane and titanate coupling agents should be mentioned [4.214, 4.480, 4.480a]. The triallyl compounds are preferably used in spite of their high price for slow crosslinking polymers, like EVM and CM, while the cheaper methacrylates are preferred in many other cases, such as NBR and ETER. As acrylate monomers they act as plasticizers like DAP for the unvulcanized mixture and permit high hardness vulcanizates, as for example peroxide cured NBR. TPTA gives a much shorter scorch time (sometimes scorching occurs during mixing) than EDMA; for that reason the latter is preferred in some cases in spite of its lower degree of coactivation. Small amounts of sulfur can also be used as coactivators [3.400, 4.211–4.213], which usually creates a strong odor, but higher tear strength.

Retardation. Recently, N-nitrosodiphenylamine [4.205] has been suggested to retard peroxide crosslinking [4.205], which is considered questionable because of its toxicity.

Properties of the Vulcanizates. In peroxide vulcanization one obtains usually low ultimate tensile *properties* (especially low *tear resistance*) and lower *elasticity*, as well as poorer *dynamic properties* than in sulfur vulcanization. *Swelling resistance* is usually lower. *Heat stability* and hot *compression set* are generally especially good in peroxide cures. They are even better than sulfur-less or sulfur donor EV-systems (for example thiuram vulcanization).
A number of peroxides, especially dicumylperoxide and tert.-butylcumyl peroxide cause, after the vulcanization, an unpleasant *odor*. The decomposition products can be vaporized in polymer Q and other heat stable elastomers by post-vulcanization hot air treatment (tempering). The dialkylperoxides, particularly 2,5-bis-(tert.butylperoxy)-2,5-dimethylhexane and several new products, produce a low level odor

compared to the other products. In many cases it is necessary to remove decomposition products by tempering because of their effect on *aging*. In polymer Q, decomposition products formed from peroxides with acidic groups catalyze hydrolytic depolymerization, causing poor stability of Q-vulcanizates on aging in closed systems. With peroxides without acidic groups, the danger for this occurrance is much lower.

Applications. Peroxides are preferred for the crosslinking of saturated *rubbers*, such as AU, CM, EPM, EVM, Q, X-LPE (crosslinkable polyethylene) and so on, as well as halogenated IIR. Because of the high heat stability inherent in peroxide crosslinking, peroxides are also applied in sulfur curable rubbers, such as NR, NBR, EPDM and ETER.

Preferred *vulcanization processes* are compression molding, injection-, respectively transfer-molding and the LCM-process. Steam-tube vulcanization is possible. Hot air vulcanization as well as the UHF process are used in special cases.

Peroxides are used in diene rubbers for *products* where high heat stability and/or low compression set are demanded, for example for cable covers (because sulfur-free compounds are necessary to avoid copper corrosion), gaskets, heat stable spring components, like motor mounts, building profiles, and the like.

4.2.4.2 Quinonedioximes

[1.12–1.14, 3.360–3.362, 3.400, 4.224–4.232]

General. p-Benzoquinonedioxime (CDO) as well as its dibenzoyl derivative (dibenzo-CDO) are crosslinkers for many rubbers because of their free radical reactions. Best known and of greatest interest is their use in IIR because of the excellent heat- and steam stability that can be obtained [4.231]. They have been unimportant for the classic diene rubbers.

Activity. When CDO and dibenzo-CDO are used, vulcanization without sulfur can take place. They are mainly used for IIR types with low unsaturation where sulfur vulcanization is slow. Addition of sulfur increases the stress values and furthermore leads to a longer scorch time. Heat stability and compression set of the vulcanizates are poorer when sulfur is added.

By combining CDO with MBTS, lead oxide, ZnO and sulfur, high enough vulcanization rates are reached that they can even be used for continuous vulcanization. Dibenzo-CDO gives a somewhat longer scorch time than CDO without prolonging the total vulcanization time; when CDO is used oxydizing agents like MBTS or lead oxide are necessary. Even though CDO is highly active in EPDM, it is of no practical importance.

Activation. CDO crosslinking necessitates also addition of ZnO. Increasing amounts of ZnO in CDO compounds decrease scorch time, increase heat stability and stress values. The sames goes for dibenzo-CDO. Compounds without ZnO show better processing stability but the vulcanizates have insufficient mechanical properties. To obtain good heat stability with IIR vulcanizates with CDO, higher ZnO levels are recommended.

Increasing stearic acid produces lower mixing viscosity. While in lead oxide containing mixtures shorter scorch time and faster vulcanization is observed when stearic acid is used, this can not be found in MBTS containing compounds.

Properties of the Vulcanizates. Compounds with MBTS give, at optimum vulcanization, highest tensile strength, highest elongation at break and best age resistance; but also the highest compression set. Lead oxide containing products show at equal curing time the highest degree of vulcanization, furthermore the best elastic properties (especially at higher temperature) as well as lowest elongation at break and permanent set.

4.2.4.3 Polymethylolphenolic Resins

[1.12, 1.13, 3.363, 4.233–4.236]

The resin vulcanization, that has established itself to some degree in IIR-crosslinking, gives similar to the p-quinonedioxine vulcanization, excellent heat- and steam stability. For that purpose, partially reacted polymethylolphenolic resins are used, that finish their reaction during vulcanization and which are introduced as crosslinkers between the polymer molecules (see page 89).

Apart from the resins, special activators, for example $SnCl_2$, are necessary which has the disadvantage of being difficult to mix into the rubber, irritating to the mucous membranes and corrosive. For certain halogenated resins these activators are not necessary.

These vulcanizing agents are used at the level of 5–12 phr. In spite of the fact that this vulcanization is slow, one obtains surprisingly low compression set values.

While it is theoretically possible to resin-cure diene rubbers and EPDM, no practical use has been made of this process.

4.2.4.4 Di- and Triisocyanates, Urethane Crosslinkers

[2.110–2.113, 3.745, 3.746, 4.238]

Polyurethane prepolymers, that are used as AU can be crosslinked in addition to peroxides with specific diisocyanates (see page 141). For crosslinking of polyesters, primarily 2,4-tolylenediisocyanate (TDI) resp. methylene-bis-chloraniline (MOCA) are applied. By using an equivalent amount of hydroquinone dihydroxyethyl ether the high crosslink density and hardness can further be increased. Vulcanization rate can be increased by the addition of organic lead salts.

Diene rubbers can also be crosslinked with isocyanates, for example with triphenylmethanetriisocyanate, resp. tris-(p-isocyanatophenyl)-thiophosphate. This process is used mainly for rubber adhesives to obtain heat- and oil resistant adhesive joints and sometimes also for textile adhesion as well as metal adhesives (see page 315, 316).

A new class of compounds based on urethane reagents has recently been proposed for rubber vulcanization, that can preferably crosslink NR with urethane crosslinks. This leads to good reversion stability. The crosslinking agents are a blocked diphenylmethanediisocyanate, that dissociates thermally into two quinoneoxime molecules and one diphenylmethanediisocyanate. The quinone molecules are bound to the rubber via nitroso groups after tautomerization and shedding of phenolic side groups. Those react with the compounds that have diisocyanate groups to form crosslinks. The type of blocking agents influences the crosslinking activity of the urethane vulcanizing agents [4.237]. Urethane crosslinking plays a minor role only in hard NR compounds (compare page 23).

4.2.4.5 Other Crosslinking Agents

For crosslinking of FKM several polyamines have been developed, some of which are blocked to avoid spontaneous vulcanization initiation (see page 122). Some of them are also applicable for ACM (see page 110). Bisphenol AF is also utilized for the vulcanization of FKM (see page 122).

In addition to those mentioned, there are other crosslinking agents known, that play a very modest role in the rubber industry. For example 1, 3, 5 trinitrobenzene and m-dinitrobenzene in the presence of lead are vulcanizing agents. Furthermore epoxy resins [1.12, 1.13], polyfunctional amines, azo-compounds [4.238], metalloorganic and organo silicone compounds [1.12, 1.13], high energy radiation and silane crosslinking [4.239] should be mentioned.

4.2.5 Accelerator Activators

[1.12, 1.14, 4.241–4.242]

With organic vulcanization accelerators it is necessary to use organic or inorganic "activators" to achieve their full potential. ZnO is the most important of these additives. Apart from ZnO in special cases MgO (in CR) and Ca $(OH)_2$ are used. PbO respectively Pb_3O_4, that used to play an important role in accelerator-free compounds are used if a very low water swelling is required, and also in a number of specialty rubbers, for example ACM, CSM, CO, CR, ECO, ETER, and others.

The system rubber-sulfur-accelerator-ZnO is furthermore activated by the addition of fatty acids (stearic acid) as zinc stearate, zinc laurate, and the like. Similar activity is found with dibutylaminooleate, 1, 3-diphenylguanidinephthalate and amines, like mono-, di- and triethanolamine, mono- and dibutylamine, dibenzylamine, etc. Generally it can be stated that increasing the pH leads to activation of the vulcanization. The basic activators mentioned lead to improved strength properties of the vulcanizates and come to a shortening of the vulcanization time. The fatty acids and fatty acid salts give better processing and improved dispersion of fillers and chemicals which is also important for the total property spectrum of the vulcanizates; they often cause a longer scorch time.

4.2.6 Vulcanization Retarders

[1.12–1.14, 4.243–4.246]

4.2.6.1 General Considerations

Vulcanizate specifications often require accelerater systems, which may have unsufficient scorch safety. In such cases or when using short vulcanization times or high processing temperatures one often has to retard the onset of vulcanization (scorch time) to assure sufficient processing safety, that for example compression molded articles can flow into the mold cavities [4.247, 4.248].
Compounds that highly retard vulcanization are for example N-nitroso compounds of secondary aromatic amines such as N-nitrosodiphenylamine (NDPA) [4.250], which are not longer in use for toxicological reasons. Low volatility organic acids such as benzoic acid (BES), phthalic anhydride (PTA) and salicylic acid (SCS) [4.251] are important retarders. N-Chlorosuccinimide and nitroparaffines [4.249] have also been mentioned as retarders but have no practical importance. Recently certain sulfenic acid – and sulfonic derivatives [4.243–4.246, 4.252–4.257] have gained importance; they present practically a new generation of vulcanization retarders.

Fundamentally it should be contemplated, if a vulcanization retarder is really necessary. Because the choice of vulcanization accelerators makes it possible to achieve a retarded vulcanization start (longer scorch time), since with a number of accelerators, vulcanization is slower to start, for example MBTS and sulfenamide accelerators. Furthermore, a retarded vulcanization can be obtained by the proper combination of accelerators (primary- and secondary accelerators) by for example combining a fast vulcanization accelerator with a slow one that does not act synergistically. The secondary accelerators are in that case only used in very small quantities. Combinations with sulfur donors that release the sulfur slowly, are of some importance in this connection. The antioxidant MBI acts as retarder for most accelerators (see page 273). The action of dithiocarbamate accelerators and thiurams are also retarded by addition of PbO.

Fatty acids and fatty acid salts, like stearic acid and oleic acid, as well as zinc stearate and zinc laurate can have retarding effect on scorch time. These compounds are at the same time important as activators for the vulcanization.

4.2.6.2 N-Nitrosodiphenylamine (NDPA)

[1.14]

NDPA increases the processing safety of mixtures with sulfur and accelerators and prolongs the flow time during vulcanization. Since NDPA tends to discolorize severely it was only used for dark blends. NDPA has strongest activity in mixtures with sulfenamide- and guanidine accelerators. When thiuram-, dithiocarbamate-, and mercapto accelerators are used, the retarding effect is small or non existing. In peroxide crosslinking [4.205] NDPA has also a scorch prolonging effect. In sulfurless compounds with TMTD (EV-systems) and with aldehyde amines NDPA has practically no retardation activity.

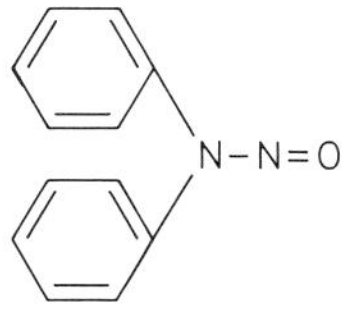

N-N=O

NDPA does not only retard the beginning of vulcanization, but the complete vulcanization time is prolonged. However the prolonged optimum vulcanization time is usually small compared to the gain in the scorch time (in black compounds with sulfenamide-, mercapto- and guanidine accelerators).

NDPA, that as nitrosamine is relatively harmless toxicologically, can function as nitro group transmitter (donor) in the presence of accelerators with secondary amino groups, for example with thiurams, dithiocarbamates, some sulfenamides etc. when some highly toxic aliphatic or heterocyclic nitrosamines may be formed. The use of NDPA is therefore problematic.

4.2.6.3 Phthalic Anhydride (PTA) and Benzoic Acid (BES)

[1.14]

In contrast to NDPA, PTA and BES show no tendency to discolorize and they are preferred in light-colored mixtures. In carbon black compounds they do act as retarders, the retarding effect is however lower. The acidic retarders are active in the presence of almost all accelerators, one exception is the EV-vulcanization with TMTD. For peroxide crosslinking, PTA and BES are not useful.

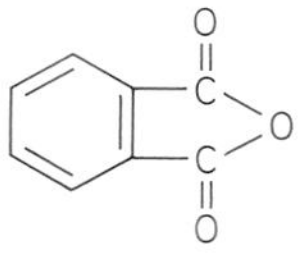

PTA 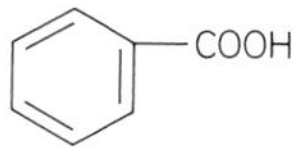 BES

The acidic vulcanization retarders cause not only a delayed scorch effect, but at the same time an almost similar vulcanization time delay; for that reason only low dosages (for example 0.2–0.5 phr) are used. In that respect, PTA is somewhat better than BES. PTA however has very limited solubility in rubber blends, while BES can be used in larger amounts. The latter (BES) has a softening action on unvulcanized blends and causes hardening of the vulcanizate.

4.2.6.4 Phthalimidesulfenamide

[4.243, 4.245, 2.246], respectively Sulfonamide Derivatives [4.244]

With products of this class the wish of the rubber technology can be fulfilled to delay scorch time without greatly prolonging the total cure time. When these compounds are used, the vulcanization curve undergoes practically a parallel transposition. These materials interact differently with different vulcanization accelerators; for that reason the choice of retarder depends on the individual conditions. Generally speaking, today CTP may be looked upon as a standard material that can be used in small dosage for highly effective activity. Its action in compounds with mercapto accelerators is less than with benzothiazole sulfenamides. CTP does not cause discoloration. One can speculate that products of this class of materials will replace older vulcanization retarders since they are universally applicable. In this area, other products are being developed [4.252–4.257].

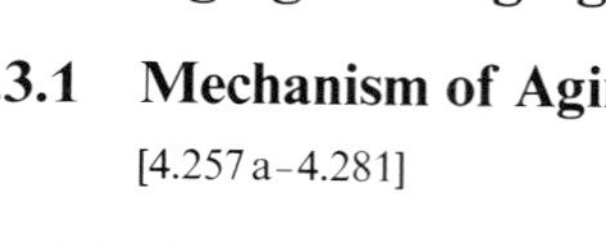

N-Cyclohexylthiophthalimide (CTP)

4.3 Aging and Aging Protectors (Antioxidants)

4.3.1 Mechanism of Aging

[4.257a–4.281]

4.3.1.1 General Considerations

[4.257a–4.275]

Aging is a collective term for changes in property of materials that occur on longer term storage without the action of chemicals that lead to partial or complete degradation. These changes can occur in form of degradation processes, enbrittlement-, rotting-, softening- and fatigue processes, static crack formation, and the like.

Uncured and cured rubber are especially prone to such aging effects. The unsaturated groups in diene rubbers make it possible to cure with sulfur, but at the same time present a sensitivity toward oxygen, ozone and other reactive substances. These reactions cause changes in the rubber. Since soft rubbers based on diene rubbers contain free double bonds, even after vulcanization they remain sensitive to the above agents. Higher temperatures make these effects more noticable. In the presence of oxidation catalysts (rubber poisons), like Cu- and Mn- compounds, these aging phenomena occur rapidly. When overvulcanization (reversion) occurs these effects become more apparent.

Since unreacted double bonds are present the possibility exists of further reaction with sulfur causing hardening (postvulcanization). In SR a continuation of the polymerization, resp. an intramolecular crosslinking called cyclization causes hardening and embrittlement of the material. Materials with hydrolizable bonds, for example polyester based elastomers can become useless through moisture, especially at elevated temperatures.

All these effects lead to various forms of destruction summarily called "aging". There is therefore no single process of "aging" but various forms of aging processes that exhibit different resulting end effects.

Basically one can differentiate the following aging processes:

- Oxydation at lower or higher temperatures (aging in the real sense)
- Oxydation accelerated by heavy metal compounds (rubber poison)
- Changes caused by heat in the presence of moisture (steam aging, hydrolysis)
- Oriented crack formation by dynamic stress (fatigue)
- Oriented crack formation by static ozone action (ozone crack formation)
- Random crack formation by high energy light and oxygen (crazing effect)
- Changes in surface luster (frosting phenomenon)
- Other processes

While the first three processes affect the total bulk of the rubber part, the other processes at first only affect changes on its surface.

4.3.1.2 Oxygen Aging. Aging in the Real Sense

[4.257a-475]

Diene rubber vulcanizates take up oxygen from the air during storage that is partially bound to the vulcanizate and partly given off as carbondioxide and water and other low molecular weight oxidation products.

Reactions occurring during the oxygen exposure are chain reactions and active radicals are the real reaction carriers. During oxidation the first products are hydroperoxides that again decompose into free radicals that start new chain reaction, that can also react with double bonds of the rubber (autocatalytic reaction).

At low temperature oxygen absorption is roughly linear to reaction time; at elevated temperature the originally linear reaction changes to an autocatalytic one. Relatively small amounts of bound oxygen lead to deep seated changes in the structure of the vulcanizate, not only on the surface but also in the bulk of the vulcanizate. Depending on the type of rubber, oxygen can

- cause molecular chain cleavage, whereby the molecular network is "loosened" (degradation, softening).
- cause crosslinking whereby a higher crosslink density is effected (cyclization, hardening)
- be bound chemically to the molecular chains without cleavage or crosslinking (indifferent action)

The net result of three concurrent reactions determines which is the preponderate change in the properties of the vulcanizate. While the first and second reactions cause deep seated structural changes of the vulcanizate, the third reaction is at first largely indifferent to aging.

Vulcanizates that are based on NR, IR and IIR undergo preferably cleavage reactions during the oxydation process; they generally become softer. During progressive aging, the crosslinking mechanism can dominate; completely oxidized NR is usually hard and brittle. On the other hand, vulcanizates prepared from SBR, NBR, CR, EPDM, ETER etc. undergo cyclization reactions that lead to hardening of the aged part. These products are when completely oxydized, hard and brittle.

Rubbers that do not contain diene structures, such as ACM, CM, CSM, CO, EAM, ECO, EPM, EVM, FKM, Q, and others are much less sensitive to oxydation than the diene rubbers.

4.3.1.3 Accelerated Oxidation in the Presence of Heavy Metal Compounds (Rubber-Poison Aging)

Many heavy metal compounds, as for example those of copper and manganese have a catalytic action on the oxidation of rubber blends and vulcanizates. In particular even traces (0.001 wt%) of Cu- and Mn-compounds in NR are able to accelerate the autoxidation of rubber and the vulcanizates. They are therefore called rubber poisons. NR and IR are especially affected by this action while most SR-types are less sensitive to these poisons.

Beside these rubber poisons, other heavy metal compounds, like Fe^{++} especially poisonous in SBR, as well as Co- and Ni-compounds also accelerate aging phenomena. These are effective in NR and IR in higher concentration than Cu and Mn.

For rubber poisons it is important that they are present in rubber soluble form. While for example metallic Cu or CuO exhibits only low aging accelerating activity, Cu-oleate is very agressive.

In the autoxidation of NR accelerated by rubber poisons and in vulcanizates prepared with them, competition between softening- and hardening reactions also occurs.

4.3.1.4 Heat Aging in the Absence of Oxygen

In the presence of heat, various reactions can take place in the absence of oxygen, for example in steam or immersed in oil, whose net-result determines the property changes in the vulcanizate:

- thermal decomposition of crosslinks, also hydrolysis of water sensitive structures (softening)
- continuation of inter- or intramolecular network formation (hardening)
- shifting of crosslinks without change in the total numbers (no resulting changes).

Aging in the absence of oxygen (in steam or oil) leads, in oxidation sensitive rubbers, to a slower structural change than in oxygen, which allows the application of higher temperatures. In hydrolyzable rubbers, i.e. AU, EVM, Q and others, steam aging proceeds faster because of additional breaking of, for example C-N-, C-O-, resp. Si-C-bonds.

4.3.1.5 Fatigue

[4.281]

When rubber is subjected to prolonged mechanical stress changes, for example periodic back and forth bending, cracks will slowly develop in the surface, they will grow until they can lead to a complete break of the article. The cracks develop perpendicular to stress direction. NR vulcanizates form these cracks relatively quickly, but they grow slowly. SBR shows slower start of crack formation, but once formed they grow faster. This is connected to the low tear resistance of SBR-vulcanizates. Higher temperatures and naturally also higher frequency of stress change accelerate crack formation. If higher ozone concentrations are also present dynamic crack formation is accelerated. It has not been clearly established if in the complete absence of ozone, crack formation can happen.

The fatigue crack resistance is of course not only dependant on the type of rubber but considerably on the crosslink density and type of crosslinking (higher crosslink-density and sulfur-rich crosslinking structures are preferred) (see page 226, 229).

4.3.1.6 Ozone Crack Formation

[4.276–4.281]

When double bond containing vulcanizates are exposed to weather conditions in a static extended mode, cracks appear slowly perpendicular to the direction of the applied stress, that grow slowly and lead ultimately to a break of the vulcanizate. This phenomenon is the static analogue to the crack formation under dynamic stress. Today the small amounts of ozone (O_3) in the atmosphere are usually looked upon as its cause. Without extension of the vulcanizates cracks are not formed; a critical extension has to be exceeded in any case before cracks show up, in NR that is less than 10%. With increasing extension, the number of cracks formed per unit area and time increase rapidly. The speed of ozone crack formation also depends strongly on the temperature and humidity of the air. Totally saturated elastomers are ozone resistant.

4.3.1.7 Crazing-Effect

When an unstressed vulcanizate is exposed to weathering, especially when exposed to prolonged sunlight, a system of small, unoriented connected cracks can develop on the surface. The surface appears then like a wrinkled orange or an elephant's skin; for that reason this effect is called elephant skin formation or "crazing" effect. The surface can, after prolonged irradiation, become brittle and the filler can chalk. The vulcanizate is in this case usually not destroyed. This effect is only found in light colored vulcanizates. Articles with carbon black filler and dyed articles, that absorb high energy radiation, do not show this effect.

4.3.1.8 Frosting Phenomenon

Another changed surface effect is the so-called "frosting" that is caused by the action of a warm, moist, ozone containing atmosphere on vulcanizates with light colored fillers; it appears as dulling of a previously glossy rubber surface. This type of aging has not yet been thoroughly explored.

4.3.2 Aging-, Fatigue- and Ozone Protective Agents

[1.14, 4.282–4.339]

4.3.2.1 General Observations and Classification of Antioxidants (Aging Protectors)

[4.282, 4.283, 4.306, 4.310]

The stability of vulcanizates against single or combined action of degrading agents is primarily determined by the type of rubber affected. Dienes, e.g., are attacked considerably more by oxygen and especially by ozone than saturated rubbers. The latter are, for that reason, quite stable toward oxygen and ozone. Dienes based on isoprene are less stable because of the electron donating influence of the methyl group than those based on butadiene [2.82]. The crosslinking system too, has a considerable influence on the aging- and heat stability of elastomers. EV-systems and sulfur-free crosslinking agents have a stronger influence on aging- and heat stability than the addition of antioxidants (aging resistors) (see p. 227). The type and amount of fillers used has considerable influence on the stability of elastomers.

In any given rubber compound the relevant degradation processes can be retarded by addition of chemicals that are summarily called aging protectors (antioxidants). Those antioxidants are added to rubber mixtures in amounts of 1–3 phr, and sometimes 5 phr and more. Thereby the rubber part is more or less protected against the influence of the aging condition. The degree of the protection depends primarily on the composition of the antioxidant.

Recently rubbers have been offered to the trade (for example NBR) that contain molecularly bound antioxidants [3.225–3.228, 4.295–4.299]. Own experiences showed that these do not have any advantage over rubbers without bound antioxidants to whom similar active additives had been added except that they are extraction proof; because they are not able to migrate, the surface region becomes impoverished.

There is no antioxidant that gives maximum protection against all previously named aging processes without causing discoloration. Rather one finds that each antioxidant has a certain "activity spectrum", i.e. differential protective action against mentioned processes, as well as discoloration behavior on illumination, that ranges from non-discoloring, weakly discoloring to strong brown-black coloring. The same goes for coloration of contact surfaces (staining), like in lacquers (contact discoloration).

Almost always, strongly discoloring antioxidants protect more effective than non-discoloring; otherwise strongly discoloring products would not be used. This rule is a rough approximation, in no way is the protective effect strictly parallel to the discoloration behavior.

The various commercial products are usually classified according to their discoloration tendency as well as their behavior in fatigue- and ozone attack. In contrast to the pure antioxidants, those materials effective against ozone are called antiozonants.

The phenylenediamines and phenols are the largest groups of aging protectors.

A number of aging protectors have recently fallen into disrepute because of their toxicological properties. [4.18]

Analogous to the vulcanization accelerators, the aging protectors have usually long and complicated chemical names; for that reason abbreviations are chosen that are partly suggested for international use by society of rubber chemical producers (WTR, see page 234 [4.18a]). Where WTR proposals have not been made, common usage or self-proposed abbreviations are used.

Abbreviations for aging protectors (antioxidants) classified according to their chemical composition

		WTR-Number
p-Phenylinediamine-Derivatives (strongly discoloring)		
N-Isopropyl-N′-phenyl-p-phenylenediamine	IPPD	1
N-(1,3-dimethylbutyl)-N′-phenyl-p-phenylenediamine	6PPD	2
N-N′-Bis-(1,4-dimethylpentyl)-p-phenylenediamine	77PD	3
N,N′-Bis-(1-ethyl-3-methylpentyl)-p-phenylenediamine	DOPD	
N,N′-Diphenyl-p-phenylenediamine	DPPD	4
N,N′-Ditolyl-p-phenylenediamine	DTPD	4a
N,N′-Di-β-naphthyl-p-phenylenediamine	DNPD	
Dihydroquinoline-Derivatives (strongly discoloring)		
6-Ethoxy-2,2,4-trimethyl-1,2-dihydroquinoline	ETMQ	6
2,2,4-Trimethyl-1,2-dihydroquinoline, polymerized	TMQ	7
Naphthylamine-Derivatives (strongly discoloring)		
Phenyl-α-naphthylamine	PAN	10
Phenyl-β-naphthylamine	PBN	11
Diphenylamine-Derivatives (strongly discoloring)		
Octylated diphenylamine	ODPA	8
Styrinated diphenylamine	SDPA	16a
Acetone/disphenylamine condensation product	ADPA	9
Benzimidazole-Derivatives (non-dicoloring)		
2-Mercaptobenzimidazole	MBI	12
Zinc-2-mercaptobenzimidazole	ZMBI	
Methyl-2-mercaptobenzimidazole	MMBI	12a
Zinc-2-methylmercaptobenzimidazole	ZMMBI	
Bisphenol-Derivatives (non-discoloring)		
2.2′-Methylene-bis-(4-methyl-6-tert.butylphenol)	BPH	14
2.2′-Methylene-bis-(4-methyl-6-cyclohexylphenol)	CPH	
2.2′-Isobutylidene-bis-(4-methyl-6-tert.butylphenol)	IBPH	
Monophenol Derivatives (non-discoloring)		
2,6-Di-tert.butyl-p-cresol	BHT	15
Alkylated phenol	APH	16b
Styrenated and alkylated phenol	SAPH	16a
Styrenated phenol	SPH	16
Other Materials (non-discoloring)		
Tris-nonylphenylphosphite	TNPP	17
Polycarbodiimide	PCD	
Benzofuran derivative	BD	
Enolether	EE	

Numbers are WTR proposals, compare page 234. Alphabetic list of abbreviations see Section 8.2, page 564.

4.3.2.2 Discoloring Protectors with Fatigue- and Ozone Protection (Antiozonants)

[1.14, 4.287, 4.299 a, 4.304, 4.305, 4.308, 4.310, 4.320–4.335]

The most effective compounds for ozone- and fatigue protection under static and dynamic stress are nitrogen substituted p-phenylenediamines

R–NH–C6H4–NH–R' p-Phenylenediamine derivatives

They increase the critical energy necessary to form ozone cracks under static conditions. Therefore, crack formation will start at higher extension. At the same time the crack growth rate under static and dynamic conditions will be reduced [4.324].

The effectiveness of the p-phenylenediamine depends on the type and size of the nitrogen substituent.

Symmetric N,N′-diaryl-p-phenylenediamines, for example, N,N′-di-β-naphthyl-p-phenylenediamines, DNPD, are excellent antioxidants but give little fatigue- and ozone protection [4.324]. Because of the large size of the substituents, DNPD can hardly migrate through the rubber and its migration to the surface, where it would be active as ozone protectant, is small. By reducing the size of the substituents (for example changing from the naphthyl- to tolyl groups, (DTPD) migration is accelerated, whereby the fatigue- and ozone protection improves [4.308]. By going to even smaller substituents, symmetric N,N′-dialkyl-p-phenylene-diamines, for example N,N′-bis(1,4-dimethylpentyl)-p-phenylenediamine (77PD) resp. the N,N′-bis-(1-ethyl-3-methylpentyl)-derivative, DOPD, migrates rapidly to the surface and has to be considered an antiozonant. When even smaller substituents are used, migration becomes so strong (ozone protection is even better) that the compound can not be used for dermatological reasons [4.18].

Unsymmetrical substituted N-alkyl-N′-aryl-p-phenylenediamines have the best properties under static and dynamic stress, when the alkyl substituents are isopropyl or isobutyl groups and the aryl substituent a phenyl group. The optimum effectiveness occurs here with the N-isopropyl-N′-phenyl-p-phenylenediamine IPPD. N-1,3-Dimethylbutyl-N′-phenyl-p-phenylenediamine, 6PPD, migrates a little less because of its larger alkyl substitutent; for that reason its effectiveness as fatigue- and ozone protector is less compared to IPPD. The material has the advantage of being less volatile and exhibits smaller water leaching effect that makes its effectiveness last longer. With longer alkyl substituents migration, volatility and water elution further decrease but also its effectiveness [4.300, 4.304, 4.305, 4.308, 4.323, 4.324].

The migration of the aging protectors depends naturally on the type of rubber used. In polar NBR for example, the p-phenylene-diamine derivatives migrate less than in SBR or NR. That is the reason why it is more difficult to protect NBR against ozone than the non-polar rubbers.

In addition to the p-phenylenediamines, that are the most important fatigue protectors on the market, 6-ethoxy-2.2.4-trimethyl-1,2-dihydroquinoline (ETMQ) plays a certain role. This material is however less effective for fatigue- and ozone protection than the p-phenylenediamines.

Phenyl-α-and -β-naphthylamines (PAN and PBN) used previously in NR-mixtures as fatigue protectors, have only low activity in SBR and BR and are therefore not used any longer for that purpose.

4.3.2.3 Discoloring Antioxidants with Fatigue- but without Ozone Protection

[1.14]

The already mentioned naphthylamine derivatives

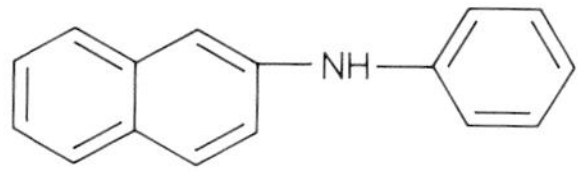

Phenyl-β-naphthylamine (PBN)

PAN and PBN are highly effective antioxidants, but have become much less important because of toxicological considerations. The substituted diphenylamines should be mainly mentioned in this classification.

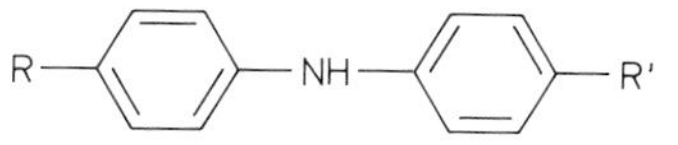

Diphenylamine derivatives

Those mainly used are octylated (ODPA), styrenated (SDPA) as well as acetonated (ADPA) derivatives. These are substances with good antioxidant- and heat protection activity. While these compounds are roughly equally effective in tire rubbers, ODPA turns out to be an exceptionally good heat protector in CR. Diphenylamine derivatives provide a certain amount of fatigue protection in NR- and IR-mixtures, but not as much as PAN or PBN; in SBR- and BR-vulcanizates, fatigue protection is very small.

4.3.2.4 Discoloring Antioxidants with little or no Fatigue- or Ozone-Protective Activity

[1.14]

For protection against fatigue and ozone it is necessary for the material to have sufficient ability to migrate through the vulcanizate. High molecular weight aging protectors and those with blocking substituents are not able to do this and are only useful as antioxidants. The best protection against oxidation is afforded by DNPD. But also a polymeric 2,2,4-trimethyl-1,2-dihydroquinoline (TMQ) has low mobility in the vulcanizate because of its molecular weight, for that reason it is eliminated as fatigue protector. With increasing molecular weight,

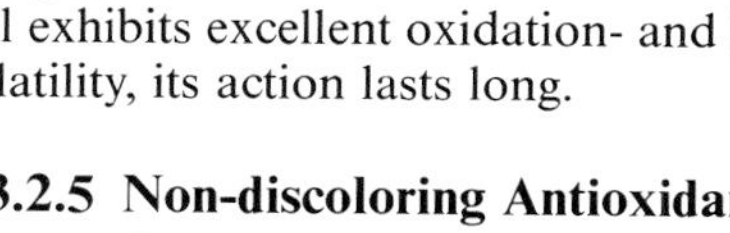

(n = ca. 3)

TMQ becomes less volatile and better suited for heat stable vulcanizates. This material exhibits excellent oxidation- and heat protection like DNPD. Because of its low volatility, its action lasts long.

4.3.2.5 Non-discoloring Antioxidants with Fatigue- or Ozone-Protective Activity

[1.13]

Aralkylated phenols, like for example styrenated phenol (SPH) as well as some benzofuran derivatives (BD) [4.238, 4.339] belong to the non-discoloring aging protectors, that exhibit some fatigue- or ozone protective action. SPH and BD afford roughly the same fatigue protection as ODPA and are therefore much less effective than for example p-phenylenediamines. As an anti-autoxidant, BD is superior to SPH. Since it is also less volatile, it is effective at higher temperatures.

BD has definite activity against ozone crack formation and can be called a chemically active, non-discoloring antiozonant. It is preferred in CR and CR-containing blends, nowadays in CM. SPH as well as BD show also activity against the "crazing effect" but are poorer in that regard than phenolic antioxidants.

4.3.2.6 Non-discoloring Antioxidants without Fatigue- or Ozone Protective Activity

[1.14, 4.285, 4.286, 4.288, 4.311–4.316]

Aging protectors in this class are mainly phenols, but also mercaptobenzimidazoles and their derivatives. Of the 2.4.6-substituated phenols, the 2,6-di-tert.butyl-p-cresol (BHT) is frequently used.

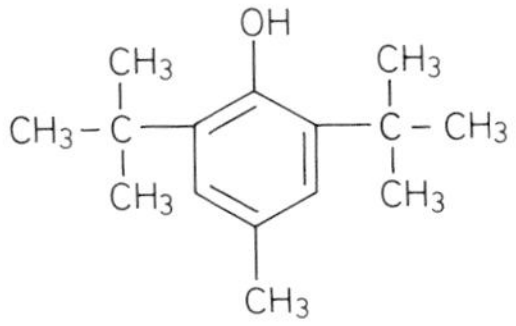

BHT

Because of its high volatility it only acts as antioxidant at relatively low temperatures. Better heat stability and sometimes better protective action is produced by for example 2,4-dimethyl-6-tert.butyl- 2,4-dimethyl-6-(α-methyl-cyclohexyl)- and 4-methoxymethyl-2,6-di-tert.butyl-phenol. A number of 2,4,6-substituted phenols with undisclosed structure are available commercially.

Among the bifunctional phenols, 2,2′-methylene-bis(4-methyl-6-tert.butyl-phenol) (BPH)

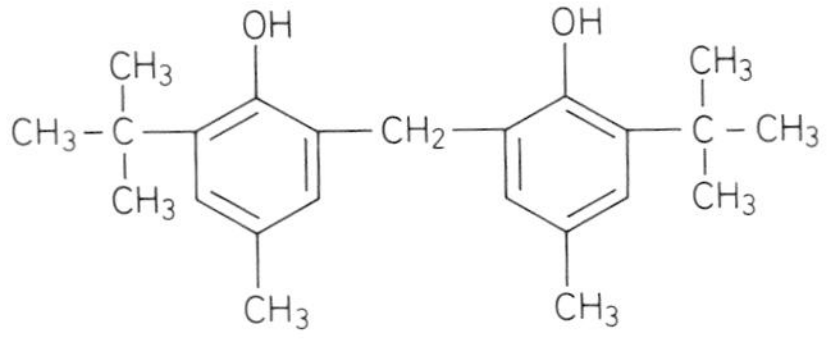

BPH

2,2′-methylene-bis(4-methyl-6-cyclohexyl-phenol) (CPH), 2,2′-isobutylidene-bis-(4,6-dimethyl-phenol) (IBPH) as well as 2,2′dicyclopentyl-bis(4-methyl-6-tert.butyl-phenol) (DBPH) give excellent protection against the action of oxygen. However, after prolonged light exposure, a certain amount of pink discoloration occurs, which is very small with IBPH. A lower discoloration tendency but also weaker activity is shown by the corresponding ethyl derivatives, for example 2,2′-methylene-bis-(4-ethyl-6-tert.butyl-phenol). 4,4′-Thio-bis-(3-methyl-6-tert.butyl-phenol) is used as antioxidant for rubber but also as stabilizer for low and high density polyethylene.

Trifunctional phenols that show very low volatility, for example tris-1,1,3-(2′-methyl-4′-hydroxy-5-tert.butyl-phenyl)-butane, β-(4-hydroxyl-3,5-di-tert.butyl-phenyl)-propionic acid octadecylester or 1,3,5-trimethyl-2,4,6-tris(3′, 5′-di-tert.butyl-4′-hydroxy-benzyl)-benzene are used mainly for rubber that is processed at higher temperatures [4.311–4.313].

Heterocyclic mercaptans, like mercaptobenzimidazole (MBI) its zinc salt (ZMBI) as well as the 4- or 5-methyl compounds (MMBI and its zinc salt ZMMBI) are moderately active, non-discoloring aging protectors.

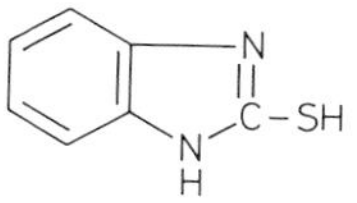

N
C–SH MBI
N
H

Since they are very active synergistically, they are rarely used alone. They increase, in combination with other antioxidants, their activity. ZMBI has some use as "sensibilizer" in latex processing.

The phenolic antioxidants present a certain amount of protection against rubber poisons. They are in that respect not as good as the p-phenylenediamines. MBI and MMBI have weaker activity than the phenols. If they are combined with a bisphenol like BPH, a synergistic action is obtained. The importance of this combination lies in its non-dis-colorating property.

Against the crazing-effect, the phenolic compounds, especially BPH and BHT should be mentioned for their activity.

The phenolic and heterocyclic, non-discoloring antioxidants have no activity against fatigue- and ozone cracks.

4.3.2.7 Non-discoloring Antiozonants without Aging Protection

[1.14, 4.336.–4.339]

While there is no special problem in producing dark, ozonestable rubber goods (carbon black and p-phenylenediamines), production of corresponding light-colored articles made of diene rubbers is only conditionally possible.

For that purpose, physically acting microcrystalline waxes, paraffine waxes, ozokerite, etc. are primarily used, they bloom and form a wax film on the surface that protects the rubber article from ozone. If that protective film is damaged, ozone resistance is reduced. For that reason waxes alone cannot be effective against dynamic crack formation. The type of wax that is being used (migration properties, elasticity, plyability, etc.) is very important [4.337]. The level at which the ozone protection wax is used depends on the wax type and the desired effect and lies between 1 to 3.5 phr and more. Even higher levels of some waxes would improve ozone stability further, but the improvement is in no relation to the increased cost.

Several benzofuran derivatives (BD) and enolethers (EE) act chemically and give good ozone resistance to light colored vulcanizates. While BD is only active in CR and CR-containing blends, EE can be used in other diene rubbers, such as NR, IR, BR, SBR and with less activity in NBR, especially synergistically in combination with microcrystalline waxes. EE is only active against ozone while BD is an active aging protector in addition to being an antiozonant (see page 272).

4.3.2.8 Aging Protectors with Hydrolysis Protection

Elastomers containing hydrolizable groups in the rubber matrix or in the side chains, like for example urethane- or ester groups, can be degraded or changed by moisture (steam/ or hydrolytic chemicals/ acids, bases). Products that counteract hydrolysis and therefore increase the durability of the polymer considerably even at higher temperatures, are for example polycarbodiimides (PCD).

4.3.3 Selection and Dosage of Antioxidants for Various Applications

4.3.3.1 Stabilization of SR (Oxidation Protection)

For the stabilization of unvulcanized rubber (oxidative stability, prevention of gel forming) one can use discoloring aging protectors (especially secondary aromatic amines, for example SPDA, PBN, p-phenylenediamine, and others) as well as non-discoloring (especially sterically hindered mono- or bisphenol derivatives, for example BHT, SPH and others); the amounts used depending on the activity of the substance and the desired protection range from 0.4 to about 1.25 phr (see also page 59).

Discoloring stabilizers are almost always superior to the non-discoloring in respect to heat stability and prevention of gelforming. There are also special stabilizers that are not aging protectors for vulcanizates, for example arylphosphite esters (for example trisphenyl-nonylphosphite, TNPP).

4.3.3.2 Antioxidants against Uncatalyzed Autoxidation of Vulcanizates

For the prevention of autoxidation of vulcanizates, amounts of 0.8–1,5 phr protective agents are used. Especially active products (p-phenylenediamines) give some protective action as low as 0.2 phr. On the other hand, completely non-discoloring chemicals (for example styrenated phenols) have to be added as much as 2 phr. The resulting protection is always more apparent in NR than in SR, partly because of the inherent better aging stability of SR and partly because the added stabilizer during production acts as aging protector.

Almost all antioxidants can prevent autoxidation, except the special antiozonants without aging protective action and the antihydrolytic agents. The strongest activity is exhibited by the p-phenylenediamine- and naphthylamine derivatives. They are strongly discoloring and can only be used for dark or black colored articles.

The weakly discoloring diphenylamine derivatives protect also very well. The bisphenols have the strongest activity among the non-discoloring materials, followed by the aralkylated phenols. Frequently a synergistic effect is obtained in combination with MBI.

4.3.3.3 Aging Protectors against Rubber Poison Accelerated Autoxidation [4.317–4.319]

As protective agent against the damaging action of rubber poisons, generally the same materials are used at the same level as for the uncatalyzed oxidation, that is in quantities of about 0.8.–1.5 phr; less is used for very active chemicals. There are however certain exceptions observed.

Deactivation of Cu- and Mn-compounds is a special property of a protecting agent that does not necessarily equal the uncatalyzed protective effect. For example, PAN and PBN show very little difference with respect to uncatalyzed aging, while the α-isomer is much preferred for protection against Cu and Mn. DNPD too has strong activity against rubber poisons. Certain combinations (for example MBI with phenols, especially sterically hindered bisphenols or secondary aromatic amines) show surprisingly high activity, even if each component by itself shows very little potency (synergistic action).

The influence of antioxidants against rubber poisons depends at least partially on a complex formation (chelation) of the damaging ion. In favor of this theory is the fact that simple complexing agents that have no aging protective activity, like ethylene diamine tetracetic acid, act as copper protectors. Also nickel dimethyldithiocarbamate (NiDMC) acts as special rubber-poison protective.

4.3.3.4 Aging Protectors Against Heat Aging

For the selection of antioxidants for heat resistant vulcanizates their volatility is important apart from their effectiveness. The volatility depends strongly on the molecular structure. The least volatile antioxidants generally yield the highest high temperature resistance. The least volatile antioxidant is DNPD [3.182]. Only 1.2% of this chemical has evaporated after 2 weeks when stored at 150 °C. This material is followed by α-methyl-SDPA (1.5%), TMQ (depending on the degree of polymerization 4–15%), ADPA (7.8%), ODPA (17.5%), DDPA (45–65%), and PBN (85%). In NBR using EV cure systems very good heat stability can be achieved with the least volatile products, i.e. DNPD, α-methyl-SDPA, TMQ or APDA by themselves or in combination with synergistic anti aging compounds such as MBI or ZMBI. However, CR vulcanizates can be effectively protected using products of the diphenylamine family such as ODPA. Sufficiently good heat aging is also obtained with other p-phenylenediamines as long as they have sufficiently high molecular weight so that they are not very volatile. It has already been mentioned that the choice of the correct cure system (EV system, peroxides or other sulfur free crosslinkers) is of extreme importance in order to obtain good heat aging properties (see page 227).

4.3.3.5 Fatigue Protection Agents

There is a smaller number of fatigue protectors than there are aging protectors. As a rule a fatigue protector is effective also against static autoxidation, while the opposite is not true. Varied work has been done on the specific mechanism of action in fatigue protection. The fatigue protection is to some extent parallel to ozone protection.

All highly active products are discoloring, normally the better they are, the more they discolor. The most effective fatigue protectors are arylalkyl substituted p-phenylenediamines.

The selection of products is guided by discussion in Section 4.3.2.2 (see page 270).

4.3.3.6 Antiozonants

The number of antiozonants (specifically) is even smaller than the fatigue protectors. Discoloring antiozonants are usually also fatigue protectors. Not every fatigue protector is an antiozonant. Antiozonants are with exception of the waxes, generally also protectors against oxygen and heat. On the other hand, not every antioxidant is effective against ozone crack formation.

The most important agents are p-phenylenediamines. Almost all presently known chemically active antiozonants are discoloring. Only very recently a few chemically acting non-discoloring antiozonants have been discovered that are benzofuran derivatives (BD) respectively enolethers (EE) [4.338–4.339]. They show optimum activity in combination with waxes (synergistic action). As materials that act by a physical mechanism, the waxes should be mentioned.

The choice of antiozonants depends on features discussed in Section 4.3.2.2, 4.3.2.5 and 4.3.2.7 (see page 270, 271, 273).

There is hardly anything known about the mechanism of antiozonant action. One only knows that the agent is used in a chemical reaction in the surface area of the

Table 4.1 Activity Spectrum of Aging Protectors[1] [4.1]

Protective Agent	I Autoxidation[2]	II Heat[3]	III Fatigue Crack Formation	IV Static Ozone Crack Formation	V Metal Poisons	VI Crazing	VII Discoloration	VIII Staining[4]	IX Permitted in contact with food[5]	X State of Aggregation
DNPD	1	1–2	6	6	1	3	2	1–2	no	solid
DTPD	2	2–3	2	3	2	–	5	4	no	solid
77PD	3–4	3–4	2	1	–	–	5	–	no	liquid
DOPD	3–4	3–4	2	1	–	–	5	–	no	liquid
IPPD	2	2–3	1	1–2	2	–	5–6	5	no	solid
6PPD	2	2–3	1–2	2	2	–	5–6	–	no	solid
ETMQ	2–3	3	2	3–4	–	–	5	4	no	liquid
PAN	2	2–3	2–3	6	2–3	–	5	4	no	solid
PBN	2	2–3	2–3	6	3–4	–	5	4	no	solid
ODPA	2–3	2[6]	3–4	6	3	6	1–2	1–2	no	solid
TMQ	2	1–2[7]	4–5	5	3–4	6	2	1–2	no	solid
SPH	3–4	3–4	4	6	–	2	0	0	yes	liquid
BD	3	3	3–4	6[8]	6	2	0–1	0	no	solid
BHT	3–4	4–5	6	6	4–5	1	0	0	yes	solid
BPH	2–3	3	6	6	3[9]	1	1	0	yes	solid
MBI	4[10]	3[11]	6	6	6[12]	6	0	0	no	solid
MMBI	4[10]	3[11]	6	6	6[12]	6	0	0	no	solid
EE	6	6	6	2[13]	6	6	0	0	no	liquid

[1] In column I–VI, 1 means best and 6 is worst; in column VII–VIII, 0 means no discoloration and 6 strongest discoloration; – means not tested or not important. [2] For NR or IR. [3] Does not apply to CR. [4] Rubber/Rubber. [5] Recommendation XXI, Utensiles made of NR and SR (German Public Health Publication 22 (1979), pg. 283. [6] In CR: 1. [7] In combination with MBI: 1. [8] Good protective in CR. [9] In combination with MBI:1. [10] In mixtures with dithiocarbamate accelerators. [11] In combination with IPPD or TMQI; depending on the mixture; in steam even without other aging protectors: 1–2. [12] In combination with BPH: 1. [13] Wax addition necessary (except in CR)

article either by ozone itself or by the oxidation products of the rubber. The protective action continues in proportion to the ability of the agent to diffuse from the bulk of the rubber surface part. Since certain waxes can aid the migration, their combinations have a certain importance.

4.3.3.7 Protective Agents against Crazing- and Frosting Effects

It is relatively easy to prevent the "crazing" effects. Since they only occur in light-colored vulcanizates, only non-discoloring or weakly discoloring preventive agents can be used. The quantities used lie between 0.5–2 phr. According to present observations, there is a clear parallel between protective action against uncatalyzed and light-catalyzed autoxidation. Protective agents are alkylated and aralkylated phenols. The yellow to brown coloration of the rubber (absorption of the blue part of light) also presents a protective effect.

Protection against the "frosting" phenomenon is a problem. To some extent, p,p′-diaminediphenylmethane acts against "Frosting". This action has not been accepted by everyone. Addition of paraffines as physical-action antiozonants is beneficial. Chemically active antiozonants may also have a positive effect. The best protection is obtained by proper choice of fillers and vulcanization accelerators.

4.3.3.8 Protective Agents against Steam Aging

When discussing protection against steam aging one has to differentiate between diene rubbers and rubber types that can be hydrolized. For diene rubber vulcanizates, the same protective agents are beneficial that also provide good heat protection, namely p-phenylenediamines, diphenylamine derivatives, TMQ especially in combination with MBI and its derivatives. Vulcanizates stable to reversion are here advantageous. For that purpose valid criteria are given in Section 4.3.2.2–4.2.2.4 (see page 270f).

For rubbers with groups that can be hydrolized, protective agents are suitable that react with steam and moisture faster than the rubber itself; that prevents chain- or pendant group cleavage. For that purpose are for example carbodiimides suitable (see chapter 4.3.2.8, page 273).

4.3.3.9 Summary

Table 4.1 gives an activity spectrum overview over the most important antioxidants (summary from [4.1]).

4.4 Reinforcement, Fillers and Pigments

The use of fillers is – aside from the vulcanization system – of utmost importance in order to obtain the desired properties of vulcanizates. One can produce soft, filler free NR vulcanizates of high tensile strength because of its stress crystallization (see page 24), but for most applications the use of fillers is more or less desirable or necessary. Fillers are certainly required, however, for SR vulcanizates. In the rarest cases are they used because of a desire to reduce compounds costs, rather their specific effect on the elastomer is the determining factor. The fillers should therefore most often be looked upon as quality enhancing materials, not as cost cutting ones. This is true both for the processing properties of the unvulcanized compound as well as for the properties of the vulcanizates.

4.4.1 The Principle of Reinforcement and its Determination

[1.17, 4.340–4.374]

4.4.1.1 Influence of the Fillers on the Vulcanizate Properties

Reinforcement is defined as the ability of fillers to increase the stiffness of unvulcanized compounds and to improve a variety of vulcanizate properties, e.g. tensile strength, abrasion resistance and tear resistance. At the same time the stress values and the hardness are generally increased and as a rule other properties such as elongation at break and rebound and other properties depending on these lowered.

The reinforcement effect of a filler shows up especially in its ability to change the viscosity of a compound and also the vulcanizate properties with increasing amount of filler loading. Those fillers which only lead to small increases in the viscosity of the compound and otherwise to an worsening of the mechanical properties of the vulcanizate, are not reinforcing; they are called non-reinforcing or inactive fillers. In contrast, reinforcing or also active fillers lead to dramatic increases in viscosity of the compound as well as to maxima of the tensile and the tear strength and the abrasion resistance with increasing amounts of filler loading. Fillers for which these property changing characteristics are only weakly developed are called semi-active fillers. Often the stress value is also used as a measure for active filler effectiveness; however, this judgement is controversial since the influence upon the stress value is strongly affected by the chemical nature of the fillers. For example, highly active silicas as judged by their tensile strength enhancement would be mistakenly judged as only semi-active based upon the stress value [4.346]. The complete property spectrum of the unvulcanized compounds and that of the vulcanizate has therefore to be evaluated in order to classify the activity of a filler correctly.

The reinforcement effect can be explained using stress-strain diagrams [4.350]. The tensile stress-strain curve of a filler reinforced vulcanizate is steeper up to a higher stress at break compared to an unfilled vulcanizate or one with inactive fillers. From this observation a reinforcing effect can be seen. The fracture energy, obtained from the integration of the area under the stress-strain curve is, however, often smaller for the filled vulcanizate than for the unfilled one [1.17]. thus, the reinforcing filler does not produce a reinforcement as far as the fracture energy is concerned. This example shall demonstrate that it is not meaningful to generalize about the reinforcement properties of fillers. Rather, they should always be evaluated in connection with certain mechanical properties, especially with the already mentioned tensile strengths, abrasion and tear resistances as well as to a lesser degree with the stress value. It must be stated furthermore, that hardly any filler will enhance all of these properties to the same optimal degree. The reinforcing effect of an active filler as well as the dosage required can be quite different for different elastomers. As an example, the activity of fillers in BR, SBR and NBR is often quite more pronounced because of their different structure and lack of strain crystallization than in NR and partially also in CR [4.355]. Their effect in these elastomers is still more pronounced because of the higher reactivity of these elastomers than for example in EPDM with its low reactivity. The variation in the filler effectiveness of NR and SR can be explained with the theory of over-stressed molecules [4.356–4.359]. During stressing of NR a partial parallel orientation of the molecules occurs, where those which are more oriented than others form stressinduced crystals. These are, compared to the molecules surrounding them which are not yet fully extended, over-extended and are responsible for strength enhancement. However, similar orientations with overstressed molecules can be obtained in SR types which do not exhibit stress orientation by affixing two filler particles inbetween two polymer molecules. This leads to analogous strength improvements.

4.4.1.2 Effects between Filler and Elastomer

Even though the reasons for reinforcement are not completely known inspite of numerous investigations [4.341, 4.343–4.345, 4.354], it has to be assumed, based upon the statements above, that it is caused by interreaction forces between the elastomer and the fillers. These lead to conditions which range from weak *van der Waal's* forces to chemical bonds [4.359, 4.360]. These adhesion forces are not distributed equally over the surface of the filler because the latter is energetically heterogeneous. On the one hand the chemical composition of the filler's surface, its special structure on the other hand are of importance. The formed bonds cause, of course, an increased deformation stiffness, because of less movability of the polymer chains. The reactivity of the elastomer affects, of course, the filler-elastomer interaction as well. Because of this one should refer to active fillers with regard to certain elastomers. While, for example silica fillers without coupling agents reinforce diene elastomers only moderately, they lead to high reinforcement effects in reactive SBR [3.148].

Also, clay types which are barely active in NBR can have strong reinforcement effects in X-NBR [3.230]. The differences in reinforcement between diene elastomers and EPDM were already pointed out.

The active centers of the filler surface can polarize the double bonds of the rubber molecules and can thus influence reactions. Fillers can have chemically or adsorptively bound functional groups on their surface, depending on their origin. On carbon black surfaces, for example, phenolic, hydroxyl, quinone, carboxyl, lactone, groups and reactive hydrogen bonds and others [4.360–4.363] as well as free radicals [4.364] have been formed which can react chemically with rubber molecules. From this results that on the one hand the surface structure and its active centers and on the other hand the size of the total surface area, and thus the particle size, are responsible for the reinforcement effects. The latter becomes understandable when one assumes that a larger surface area leads statistically also to the possibility of a larger number of absorption centers.

4.4.1.3 Mullins Effect

The so-called *Mullins* effect confirms that a wide spectrum of absorption forces exists, from weak absorption to chemical bonding between elastomer and fillers [4.365, 4.366]. When a vulcanizate filled with active fillers, is pre-stretched and then after a relaxation the tensile stress-strain curve is determined and it is compared to the stress-strain curve of the unfilled vulcanizate, a curve intercept results which approaches that of the unfilled sample. However, above this pre-stretch the curve corresponds again closely to the original stress-strain curve of the filled elastomer. This behavior can be explained by the fact that the forces expended for the pre-stretch destroyed the absorptive bonds. Therefore, the pre-stretched vulcanizate behaves up to this elongation similar to an unfilled one. Above the applied tensile stress the original stronger reinforcement attachments still are intact.

4.4.1.4 Primary and Secondary Structures

The magnitude of the surface is not the only quantity determining the reinforcement effect of a filler. For example, one can obtain only slight differences. Aside from the filler surface area, its structure plays an important role. Many fillers, among them especially the carbon blacks, have the ability to form chains from single particles which are not even destroyed by mechanical influences. These are called primary filler structures.

From these, because of adsorption forces, more or less large secondary structures can be formed by agglomeration which, however, can be destroyed by mechanical forces. The magnitude of the attractive forces between the filler particles is of importance for the reinforcement process also, since these forces have to be overcome during the deformation of the filler reinforced elastomer.

4.4.2 Determination of Filler Activity

4.4.2.1 Bound Rubber

The rubber fraction which is bound by the reinforcement effect to the filler surface can be determined as "bound rubber" which is also a measure for the degree of reinforcement. When an elastomer-filler mixture is extracted with a solvent, e.g. benzene, then the gel-like elastomer which is bound to the filler surface cannot be dissolved any more while the rest will go into solution. If one calls the total insoluble fraction FG, consisting rubber gel (G) and filler fraction (F), the fraction of bound rubber (rubber gel) can be determined by the relation:

$$\text{Bound Rubber} = \frac{\%\text{FG} - \%\text{F}}{\%\ \text{Total Rubber}} \cdot 100$$

The higher this value, the more rubber has been bound to the filler surface, i.e. the higher are its adsorption forces.

4.4.2.2 Determination of Particle Size

The particle diameter can, for example, be determined by electron microscopy where, for example, 1200 to 1500 single particles are measured under a magnification of 50–75000. From these data particle size distribution curves are plotted and the average particle diameter determined.

4.4.2.3 Determination of the Filler Surface

From the average particle diameter the *theoretical total surface area* may be calculated assuming a spherical particle shape. This, however, does not take into account the structure of the particles, or any possible porosity. In order to obtain this completely, the magnitude of the adsorption of gasses consisting of small atoms, e.g. Nitrogen, which penetrates the finest pores and crevices is measured. According to this *Nitrogen adsorption* method, developed by *Brunauer, Emmett* and *Teller* [4.367], one obtains the so-called BET-value, which defines the surface area of a filler in m^2/g Nitrogen. Generally one can assume that fillers with BET values from 0 to 10 m^2/g are inactive, from approximately 10 to 60 m^2/g partially active and from approximately 60 to 250 m^2/g active. From the ratio of the 0_{BET} determined by Nitrogen adsorption and the 0_{EM} determined by electron microscopy the so-called *roughness factor* (the "structure") of the filler particles can be determined: $R = 0_{BET}/0_{EM}$. This value is related to the activity of the filler.

The larger the inner surface area, the higher are the possibilities for reactions between the fillers and the rubber molecules, but the higher is also the possibility that compounding chemicals are adsorbed to the filler surfaces and are thus not available for the elastomer. This factor can be determined by measuring the *DPG-adsorption*.

The *BET method* measures also pores and crevices which are too small for the elastomer to penetrate. Therefore, adsorption measurements with other molecules are carried out also, e.g. with Iodine *(Iodione adsorption)* and with Bromine-hexadecyl-trimethylammonia *(CTAB method)*. The CTAB determines the surface area active for the polymer, since the micropores are not measured by this method.

Another filler specific *constant A* [4.353] is in direct relation to the polymer reinforcing surface of the fillers. It is extracted from the difference of the rebound of an unfilled (Ro) and filled (R) test compound:

$$A = \frac{Ro - R}{\frac{mF}{mP}}$$

where mF is the amount of filler and mP the amount of polymer (in g). The A-value approaches with increasing BET surface area a limiting value.

4.4.2.4 Determination of the Filler Structure

It is difficult to determine the attractive forces which lead to an attachment of a filler and thus the degree of attachment itself. However, by using the *Oil* or *Dibutylphthalate (DBP) Adsorption Method* the relative values of the filler attachment may be determined.

The *oil adsorption* method, where the amount of oil is determined which is sufficient to form from a certain amount of oil a paste, is relatively difficult to reproduce. Formerly it was carried out manually using linseed oil. Today, preferably DBP is used for this purpose and the measurement is carried out in a plastograph or in an adsorptometer especially developed for this purpose by Cabot. The cavities formed by the filler structure take up the oil or the DBP. The plastograph shows a torque maximum at that amount of added oil or DBP at which all crevices have been filled and the sample becomes a paste. The oil or DBP adsorption is expressed in ml oil or DBP per 100 g filler, e.g. 20 ml DBP per 100 g Carbon black. For carbon black the different levels of oil or DBP adsorption differentiate between three levels of agglomeration: low, normal and high structure types [4.368–4.370]. The larger the oil or DBP value is at constant carbon black surface, the stronger are the *van der Waal* forces between the Carbon black particles. These adsorption methods measure, of course, only the filler structure before the fillers are incorporated into a compound. Since the original filler structures degrade partially during mixing, the *DBP method* can, of course, not define those structures which are present after the mixing process [4.353]. In order to estimate the degradation of structure due to shear forces, the so-called *24M4-test* was developed for Carbon blacks. Carbon black is compressed using 24000 psi (170 MPa) in a special apparatus. Thereby the secondary structures are destroyed just as during the mixing process. After a repeated DBP determination and comparison to the original DBP adsorption value, the structural changes can be estimated which might occur during the mixing process.

Another significant method to determine the structure of Carbon black imbedded within a polymer network is the measurement of the so-called *α_F-value*. This value is measured by comparing the crosslinking isotherms of a filled and an unfilled SBR 1500 standard compound [4.371–4.373] using torque measurements in a rheometer. The α_F value is, as a first approximation, directly proportional to the structure of the Carbon black.

Carbon blacks can be classified using the α_F value and the A parameter and the expected properties can be calculated [4.353, 4.374].

Another method to determine the structure of Carbon blacks remaining in the elastomer is the measurement of the *electrical conductivity*.

4.4.2.5 Black-Tinting-Intensity

This procedure is only useful for carbon black and only for certain types of black. With the measurements obtained from the Nigrometer one may correlate the tinting intensity with particle size and with it, the activity.

4.4.2.6 pH-Values

The pH-value can give important clues for the vulcanization behavior and the possible adsorption of accelerator by the filler; this can upset the vulcanization process and result in poor properties of the vulcanizate.

4.4.2.7 Results of Testing

Even though the test results are important for the characterization of a filler, the pre-investigation does not allow a final judgement as to the behavior of the filler in rubber mixtures. It is for example possible that a highly active filler is so difficult to process, that it is not suitable as filler.

It is further important, that testing of a filler is not done in a fixed recipe, but that in each case acceleration is optimized. Investigations with fixed test recipes only lead to wrong conclusions and are therefore worthless. Physical pre-testing gives leads for reasonable structure of a recipe and purposeful accelerator dosage; the pre-testing can therefore save labor. The safest method for the determination of filler reinforcement is to measure the property change of rubber compounds or vulcanizates in a formulation which is similar to the one intended for end use.

4.4.3 Fillers

4.4.3.1 Classification of Fillers

The fillers are primarily classified as carbon blacks and light colored fillers. Among the light colored fillers chemical composition is primarily the basis for classification. For example one can list colloidal silica, calcium- and aluminium silicate, alumina gel, Kaoline, silica, talcum, chalk (calcium carbonate), metal oxide, like zinc oxide and metal carbonates.

With each class of fillers, different degrees of activity are present. Basically, most carbon blacks, colloidal silica and most small particle size silicates belong to the high- and medium activity fillers, while chalk belongs to the inactive fillers.

4.4.3.2 Carbon Blacks

[4.375–4.442]

4.4.3.2.1 General Considerations and Classification

The application of Carbon black in rubber compounds is over a hundred years old. Before 1872 only lamp black was utilized as a black pigment. It was manufactured in China by the deposition of oil flames onto china plates. After the discovery of the channel black in 1872 the lamp black, which was only used as an extender, was successively replaced by channel black. Even though the rubber reinforcement by channel black was already discovered in 1911, it took until 1940 before extensive scientific investigations of the mechanism of reinforcement were undertaken. Because of these development efforts the principles of the modern gas and oil furnace black manufacture were found, even though SRF black had been developed already in 1922. Since approximately 1950 the triumphal progression of the oil furnace black began and since approximately 1965 the variety of furnace blacks was extended to new special application areas and special properties. Since that time the manufacture of carbon black has been in hectic change. From the numerous literature of the last five years can be seen that the area of black production has not stabilized yet.

Aside from the flame, channel, gas and furnace blacks which are produced by incomplete combustion of oil, coal tar products and natural gas, the thermal, acetylene and arc blacks play only a minor role. The latter blacks are produced by thermal cracking of natural and coke gas, acetylene or low molecular weight hydrocarbon gasses. Lately, it was also attempted to produce carbon blacks from coal [4.427, 4.430, 4.431], graphite [4.428] and other raw materials.

Carbon blacks are largescale technical products. World production is roughly 2.5 million metric tons per year. It obtained this position thanks to its rubber reinforcing properties which is the basis of today's tire and rubber industry. With respect to reinforcing it has not been possible to completely replace carbon blacks with other materials.

According to the production process, the U.S. War Production Board in 1943 classified carbon blacks as follows:

"F" = Furnace Black
"C" = Channel Blacks
"T" = Thermal Blacks

Furnace black is today the most important of them. Channel blacks have practically disappeared from the rubber industry. Thermal blacks have in recent years been replaced by suitable furnace blacks because of economic and ecological factors, often with suitable changes in recipes. Reasons were increase in price of natural gas and costly air pollution control installations.

The old classification of carbon blacks was based on their historic names. The "High Abrasion Furnace" black (HAF) does not provide high abrasive resistance according to today's requirements and is hardly ever used any longer for tire treads. The "High Modulus Furnace" black (HMF) must be categorized today more likely as low modulus black. A new scheme introduced a few years ago by ASTM (see Table 4.2) tries to account for today's situation less ambiguously [4.442]. The type notations consist of a letter indicating rate of vulcanization ("N" for normal, "S" for slow) and three numbers, the first of which is an index of primary particle size.

The use-directed choice of carbon black types and amounts can be found in the various chapters on "Compounding" of the various types of rubber; it does not have to be discussed here in great detail.

4.4.3.2.2 Furnace Black

[4.436–4.442]

Preparation. The furnace process – the first continuous process for carbon black production – was introduced in 1922; it continued for 20 years with natural gas as feedstock and with SRF as the only product. Later, HMF and FF were added. In 1943 the Oil-Furnace process superceded the natural gas based process. Today all Furnace Blacks are produced from liquid aromatic feedstock, that originate from petroleum fractionation, coal tar distillates or ethylene crackers. Basically, the feedstock is pre-heated and burned in a reaction zone with insufficient air supply. The temperature and other conditions are regulated by burning in the reaction zone auxiliary gas or other secondary feedstocks. The reaction is quenched by a water spray and the black is separated from the steam/gas mixture in Zyclones or table filters and finally pelleted.

Structure. The classification of carbon blacks proceeds (see Table 4.2) according to their iodine absorption (an index as to the size and activity of the surface) and their structure that measure the extent of agglomeration and aggregation of primary particles to form chain- and grape-like structures; similar aggregates represent the smallest technically active units even after being mixed into the rubber, in spite of a certain amount of degradation. The primary particles could be regarded more as a hypothetic idea. The carbon black structure survives to some extent even in the vulcanizate and is characterized as the so-called αF-value. Low structure blacks have been available for at least 25 years. It has been possible to produce high-structure blacks since the middle 1960's. Differences in surface and structure show up clearly in the technological properties of the rubber. Higher iodine adsorption always goes with higher reinforcing action (higher "activity"). High black structure causes good dispersion in the mixture, higher compound viscosity, low die-swell with smooth extrudate surface and with higher modulus, higher hardness and better wear resistance of the vulcanizate. Low structure blacks give low dynamic heat build-up, high tensile and tear strength, good crack growth resistance on flexing and low stress values. A linear correlation exists between carbon black structure measured in the vulcanizate (αF-value) and the abrasion resistance when one keeps rubber-active surface (A-value) constant. Improvement in abrasion resistance are possible as long as the αF-value can be increased [4.353].

Improved Carbon Blacks. Since 1970 "improved" or "new technology" carbon blacks have been developed by international modification of production variables [4.436–4.441]. These blacks present almost as good processing- and vulcanization properties as conventional types of higher activity and price category; they have therefore gained a large portion of the market. N-375, N-339 and N-234 are the successful numbers of these new types of carbon blacks in the USA as well as in Europe. It does not seem impossible that conventional large-use products like N-220 (ISAF) and N-110 (SAF) will be completely replaced in a few years by "new technology"-blacks. Many in-between types like N-242 (ISAF-HS), N-219 (ISAF-LS) or N-440 (FF) are already today commercially unobtainable. There is a trend if not demand to rationalize among carbon blacks that are shipped in silo-can and containers to large-scale users. A choice of furnace blacks can be seen in [4.353], resp. in ASTM-D 1765 [4.442].

Table 4.2: Classification of Carbon Blacks according to ASTM-D 1765 [4.442], Typical Properties

ASTM Designation	Target Values					
	Iodine Adsorption No.,[a] D 1510 g/kg	DBP No., D 2414 $cm^3/100$ g	DBP No., Compressed Sample, D 3493, $cm^3/100$ g	CTAB, D 3765, m^2/g	Nitrogen Adsorption D 3037, m^2/g	Tint Strength D 3265
N110	145	113	98	126	143	124
N121	121	132	112	121	132	121
S212	...	85	82	119	117	115
N220	121	114	100	111	119	115
N231	121	92	86	108	117	117
N234	120	125	100	119	126	124
N242	121	124	106	111	125	116
N293	145	100	92	114	130	117
N299	108	124	105	104	108	113
S315	...	79	75	95	88	...
N326	82	72	69	83	84	112
N330	82	102	88	83	83	103
N332	84	101	90	...	...	118
N339	90	120	101	95	96	110
N347	90	124	100	88	90	103
N351	68	120	97	74	73	100
N358	84	150	112	88	87	99
N375	90	114	97	98	100	115
N472	250	178	114	145	270	...

(continued on next page)

Table 4.2: (continued) Classification of Carbon Blacks according to ASTM-D 1765 [4.442], Typical Properties

ASTM Designation	Target Values					
	Iodine Adsorption No.,[a] D 1510 g/kg	DBP No., D 2414 $cm^3/100$ g	DBP No., Compressed Sample, D 3493, $cm^3/100$ g	CTAB, D 3765, m^2/g	Nitrogen Adsorption D 3037, m^2/g	Tint Strength D 3265
N539	43	111	84	41	41	...
N550	43	121	88	42	42	...
N630	36	78	62	38	38	...
N642	36	64	62	37	37	...
N650	36	122	87	38	38	...
N660	36	90	75	35	35	...
N683	35	133	...	39	37	...
N754	24	58	57	29	...	...
N762	27	65	57	29	28	...
N765	31	115	86	33	31	...
N774	29	72	62	29	29	...
N787	30	80	74	32	30	...
N907	...	34	...	...	11	...
N908	...	34	...	...	...	...
N990	...	43	40	9	9	...
N991	...	35	38	8	7	...

Note 1: The iodine adsorption number and DBP number values represent target values. A target value is defined as an agreed upon value on which producers center their production process and users center their specifications. All other properties shown are averages of typical values supplied by several manufacturers.

Note 2: IRB data was obtained from tests performed during the certification of IRB No. 6 carbon black.

[a] In general, Test Method D 1510 can be used to estimate the surface area of furnace blacks but no channel, oxidized, and thermal blacks

Producers. The main carbon black producers are: Ashland, Cabot, Columbian Carbon, Continental, Degussa, Huber and Philips. Cabot is the biggest of them, followed by Degussa and Columbian Carbon.

4.4.3.2.3 Thermal Blacks

Thermal blacks are generally produced from natural gas in preheated chambers without air. They are inactive, improve the tensile strength of the vulcanizates very little, but give only moderate hardness at high loading, and good processing and dynamic properties. Thermal blacks have in recent years been in short supply and expensive, but recently Cancarb has increased its capacity enormously. Their use has been limited to special applications (highly loaded CR-parts; FKM; X-LPE, etc.).

4.4.3.2.4 Channel Blacks

Till the end of World War II the channel blacks were the most important reinforcement blacks. They have been completely replaced by the abovementioned furnace blacks that had been developed during the last years of the war. The furnace blacks in SBR give much better abrasion resistance than the comparable channel blacks. The channel blacks are more acidic (pH-value of about 5 compared to furnace black 6.5–10) than the other blacks. They therefore cause a more or less strong vulcanization retardation.

The channel blacks are prepared by partial combustion of gaseous hydrocarbons, mostly natural gas, through thousands of single burners, are deposited on cooled steel rings and scraped off and collected.

4.4.3.2.5 Other Carbon Blacks

Aside from the main classes of carbon blacks, one can mention also:

Acetylene Blacks, that are prepared by thermal decomposition of acetylene, are outstanding because of their high electric conductivity. They are advantageous for many applications where high conductivity is necessary, respectively where electrostatic charge must be avoided, for example rolls, tanker hoses, containers for powdered materials. They are frequently today replaced by conductive furnace blacks.

Flame Blacks, prepared from combustion of liquid fuels give at high loadings good processing properties with attractive dynamic properties. They are today increasingly exchanged for furnace blacks, especially those with high structure.

Electric Arc Carbon Blacks were byproducts from the acetylene production in the electric arc. They are not being produced any longer.

4.4.3.3 Non-Black Fillers

[4.443–4.475b]

4.4.3.3.1 Active Non-Black Fillers

[4.443–4.458]

Preparation: The highly active, light colored fillers are, chemically, silicas (silicic acids). They can be manufactured by two methods: Solution process [4.457, 4.458] or pyrogenic process (fumed silica) [4.457].

Those most important for the rubber industry are made by precipitation: Alkalisilicate solutions are acidified under controlled conditions. The precipitated silicic acid (silica) is washed and dried. Depending on the condition during preparation, the silica filler is more or less active. The products with highest activity are pure silicic acids (silicas) with large specific surfaces. Ca-silicates are a little less active but easier to process. Al-silicates have in this series the lowest activity.

In the preparation of colloidal silica by the pyrogenic process, silicone tetrachloride is reacted at high temperatures with hydrogen and oxygen:

$$SiCl_4 + 2\ H_2O \longrightarrow SiO_2 + 4\ HCl$$

The reaction products are quenched immediately after coming out of the burner. One obtains very finely divided silica that is important as filler for example for Q. For the normal types of rubber, pyrogenic silica is too active and too expensive.

Structure. As was done for carbon blacks, for the characterization of silica fillers the single particle size and the specific surface area is used. The smallest, physically observable single filler particles, so-called primary particles, have for pyrogenic (fumed) silica a particle diameter of approx. 15 μm, for precipitated silica approx. 15–20 μm. The surface forces of the small primary particles are so high that many thousand conglomerate to form the so-called secondary particles. There is no mixing technique known that develops high enough shear forces to destroy this secondary structure to disperse primary filler particles in the rubber. The knowledge of primary particle size is therefore quite unimportant for the use in rubber of the group of filler here under discussion.

The secondary particles of silica fillers form further agglomerates. They form net-like or chain-like structures, the socalled tertiary structures. Even though these tertiary structures are relatively stable, they get more or less degraded by shear forces during mixing. All means to increase shear forces during mixing improve therefore homogeneity of the filler dispersion. In plasticizer free mixtures based on high Mooney viscosity polymers good filler dispersion is obtained due to the high shear forces. If a low Mooney viscosity rubber is used and if in addition a large amount of plasticizer is necessary, the latter should be added if possible after the filler. In special cases, for example if granular filler is used, dispersion can still be a problem.

The specific surface area is frequently considered for the judging of a silica in addition to particle size. Determination of the surface area by nitrogen adsorption (BET method) is generally used. It must be remembered that the total surface of the filler is determined, even that portion of the surface that lies inside the secondary conglomerate and would not have activity in the rubber technology. Knowledge of the BET filler surface does not furnish a true indication of filler activity, if the fillers were prepared by different processes; it can however be used as production control constant of a certain filler type.

Adsorption of Fillers. All silicic acids (silicas) have strongly polar surface character. The centers of this polarity are mostly hydroxyl groups bound to silicon. The reactivity of the surface causes foreign substances to be adsorbed on the filler surfaces till they are saturated. The behavior of the filler and its effect on the rubber may be strongly influenced by that process.

All precipitated silicas and silicate fillers contain a certain amount of water from their preparation. Since the water content can clearly influence processing- and vulcanization properties it is customary that a constant amount is present during packaging by controlling the production process. During transportation and storage, this water content can change, i.e. the water content can vary depending on the moisture content of the surroundings. A monomolecular layer of water on the surface of the silica- resp. silicate particles remains even after the harshest drying methods.

With increasing water content the dispersion time of colloidal silica into the rubbers is somewhat prolonged; the scorch time and vulcanization time with conventional sulfur-accelerator-combination are prolonged and the color of transparent vulcanizates is made lighter. Certain variations in processing- and vulcanization properties are in most cases due to insufficient climate control of the filler before incorporation into the rubber.

The filler surface is capable not only to adsorb water but also compounds, particularly basic ones. If silica containing compounds are accelerated with basic products such as DPG or DOTG, a certain amount of accelerator is taken up by the filler and is not available as vulcanization accelerator. These compounds need a larger amount of basic additives to compensate, unless other measures are taken. Respective adsorption, water has a preferred status over other compounds. If a filler has taken up additional amounts of water during storage, less basic accelerator will be bound and a higher amount of accelerator is available for vulcanization and thereby fast vulcanization occurs at higher water content. During vulcanization with TMTD or peroxides these adsorption phenomena do not occur.

Filler activation [4.477–4.489]. Not only can silica fillers interact with water or accelerators, but also with polymers. Consequently with increasing filler content, respectively increasing filler activity the viscosity of the mixture increases and makes processing more difficult. To the extent that filler-polymer interaction is decreased, compound viscosity can be reduced. If materials are added to the mixtures that are adsorbed stronger on the filler than the rubber, softer mixtures are obtained that are processed more easily. Additives of this type are DPG or DOTG type accelerators, hydroxyl group containing compounds, like glycols, glycerol, etc. as well as almost all materials with basic nitrogen like for example triethanolamine and secondary amines. These additives are frequently in technical terminology called filler-activators since they act not only to improve processing but also to reduce accelerator absorption. Frequently they act as vulcanization activators that makes it difficult to differentiate the two effects.

It is peculiar that these activators, especially glycols, triethanolamine and secondary amines beyond the effects mentioned above, also cause degradation of the tertiary structure of the filler which may cause better dispersion of the filler. Especially with highly active silicas by the addition of small amounts of activators an increased viscosity may be observed because of better filler dispersion and the lower viscosity described previously can be obtained only when more activator is added, after passing a viscosity maximum.

Carbon blacks have reactive organic groups on the surface that furnish affinity to rubber. These reactive centers are missing in light-colored fillers. The previously mentioned filler activators, particularly silanes [4.77, 4.451, 4.460, 4.465–4.467, 4.477, 4.481–4.489], titanates [4.214, 4.478–4.480], zirconates [4.480, 4.480 a] and others can make the filler more reactive.

These materials, especially the silanes may, as silicic acid derivatives, be chemisorbed to the silica particle and the reactive organic groups increase affinity to the rubber. By the use of additional silanes, titanates, and the like, the activity of light colored fillers can be raised. For some time now, bis-(3-triethoxysilylpropyl)-tetrasulfide is commercially available, that functions not only as filler activator but also as crosslinker in conjunction with MPTD for obtaining reversion-stable NR-vulcanizates [4.77] (see page 233).

Customarily dosage of additives are calculated on the amount of rubber in use; filler-activators should be used in relation to the amount of filler used, for example for silicas, 8.0–10.0 parts per weight of polyethylene glycol or 4.0–6.0 parts per weight secondary amines or DOTG relative to the amount of filler, if mercapto accelerators are present. If the main accelerator is of the sulfenamide type, on the average the basic accelerator may be reduced by 20–30%. If the vulcanization is produced with TMTD without or with little sulfur, base addition is not necessary, since no accelerator absorption occurs, but glycols or silanes should be used in customary dosage to improve processing. When silicates are used because of their low activity the need for filler-activators is reduced.

The proper choice and dosage of light colored reinforcing fillers can be obtained from the various chapters on compounding for the various rubber types and does not have to be discussed here.

Pure Silicas. The pure silicas represent very active fillers. With comparable specific surface areas one obtains vulcanizates that, compared to reinforcing carbon blacks, show nearly equal tensile strength and tear resistance, while abrasion resistance is 15–20% lower. Strain values and hardness are usually lower. On the other hand, electrical behavior is somewhat better. Silicas with highest activity, because of their large surface area give mixtures of high viscosity that makes processing more difficult; that can be adjusted by the use of filler activators.

Silicates: *Calcium silicates,* that may be called semi-active fillers, give even at high loadings, soft and elastic vulcanizates. Because of their low activity, they process better than the silica fillers. The *aluminium silicates* are less active than the calcium silicates and do not produce properties achieved with silicas or calcium silicates.

4.4.3.3.2 Inactive Non-Black Fillers

[3.230, 4.451, 4.459–4.467]

Among the inexpensive inactive light colored fillers, chalk and Kaolin are used in large quantity to cheapen light colored as well as carbon black containing mixtures. When used in NR it is important that these fillers do not contain any rubber poisons, especially copper and manganese.

Chalk [4.462–4.465]. Among the different types of chalk one differentiates between milled, washed and precipitated products. It should be mentioned that it is possible under certain suitable conditions to precipitate very fine calcium carbonate that has very small particles and is a semi-reinforcing filler.

The different types of chalk mentioned above are different in color and also in their effect on processing, extrudability and vulcanization. These differences are not very extensive.

Kaolin [3.230, 4.451, 4.459–4.461]. Aside of inactive Kaolins like Kaolin G and china clay that are used mostly as extenders, there are also a number of semi-active Kaolins, for example Dixie Clay and Suprex Clay, that have BET values of 20–50 m^2/g and contribute some reinforcing.

Kieselerde [4.466]. The Kieselerde and Kieselgur types are inactive fillers that show different activity dependant on particle size.

Zinc Oxide [4.242]. The ZnO-types differ considerably with respect to their activity. Those prepared by the pyrogenic process (oxidation of zinc vapor) so-called zinc white types have low activity and those prepared by the wet-chemical process can have high activity.

The ZnO-types are primarily used at the 1–5 phr level as vulcanization activators. For that same purpose other metal oxides (lead oxide, antimony oxide, magnesium oxide) can be used but are applied only in exceptional cases.

ZnO in larger amounts has the lowest effect on elasticity loss of the vulcanizate of any filler; ZnO, for example active ZnO is occasionally used as filler for highly elastic vulcanizates of high hardness, for example spring elements, that could not be obtained with other fillers.

Aluminium oxyhydrate [3.308–3.310]. This material is used less for its filler effect but rather for its ability to split off water at high temperatures as a flame retardant.

Talcum and Micro Talcum [4.467]. By the application of certain fillers the permeation behavior of vulcanizates may be affected favorably [3.204]. In order to obtain high permeabilities large amounts of platelet micro talcum are required. Since micro talcum is an inactive filler the mechanical vulcanizate properties suffer. However, by utilizing silane treated micro talcum this property decrease may be limited.

4.4.4 Other Fillers, Short Fibers

[4.468–4.476]

Aside from those fillers already mentioned, a multitude of other products is being applied. For a long time and still today the application of re-ground rubber [4.476], i.e. finely ground vulcanizate waste, is being discussed. This may be an ecologically sensible approach [4.17, 4.18]; however, the application of such fillers always results in a worsening of the mechanical vulcanizate properties since ground rubber is an inactive filler [4.476].

Better results are obtained with anisotropic fillers, e.g. short fibers [4.468–4.475]. These orient during processing in the direction of flow, for example during extrusion in the extrusion direction and provide because of their anisotropy no reinforcement in the machine direction, however a strong one perpendicular to it. These fillers are gaining in importance. The following short chopped fibers are being utilized: Jute [4.469, 4.472], Silk [4.474, 4.475], Cotton, Rayon and other fibers [4.468, 4.471, 4.473]. Such fibers were also used to reinforce the tire carcass in order to eliminate experimentally the conventional cord construction. These attempts were not successful however.

4.4.5 Pigments

[1.14, 4.490–4.494]

4.4.5.1 General Considerations

Organic and inorganic pigments are used to color rubber compounds. Only those substances can be used that are insoluble in rubber and in solvents that usually come in contact with rubber articles (water, fats, organic solvents). They must not contain water solubilizing groups, like carboxyl or sulfonic acid groups, must be easily dispersible in rubber, sufficiently heat stable and insensitive to all vulcanization conditions, vulcanizing agents and other additives; they must give color tones that are light fast and independant of the pH of the compound. To prevent premature aging the pigments have to be free of Cu- and Mn-compounds. They must be insoluble in rubber to avoid blooming or bleeding. Most inorganic pigments are stable against bases and acids.

4.4.5.2 White Pigments

Today's most important white pigments for the rubber industry are the different types of titanium dioxide [4.492]. In spite of their relatively high price, they are still very economical for the rubber sector because of their great whitening ability. Therefore small amounts are sufficient. A further advantage of the titanium dioxide pigments is that they change the properties of the vulcanizates very little because of the small amounts used.

Lithopone [4.493] has relatively low whitening ability; for that reason, relatively large amounts have to be used. Since Lithopone acts as an inactive filler, it can cause a reduction in quality in high quality vulcanizates. For that reason, one prefers today as white pigment the more effective titanium dioxide. This is particularly true for highgrade rubber items.

Titanium Dioxide [1.14]. Titanium dioxide pigments appear commercially in two modifications, as anatas- and rutile types that differ in their crystal lattice; their physical properties are slightly different.

Vulcanizates tinted with titanium dioxide "anatas" show an outstanding white color (blueish white). Most "rutile" types produce a creme-colored white and they are not rubber compounds that tend to have a yellowish white tinge already. Rutile-types have however a 20 percent better covering ability than the anatas types. For that reason, rutile types with disturbed crystal lattice, with the desirable blueish-white color were developed. Rutile types in comparison to anatas types, confer increased light and weather resistance to vulcanizates in colored and white compounds.

Vulcanizates tinted with anatas-types exhibit somewhat larger decrease in mechanical properties after weatherometer testing than those tinted with rutile-types. They furthermore tend to crack more under stress. Nevertheless, anatas-types are used for articles where a pure white is important and the decrease in mechanical properties after weather exposure is not significant. In those products a small amount of chalking takes place after relatively weather exposure, that produces a purer white than when chalking-resistant types are post-treated. This so-called "self cleansing effect" is frequently used in practical applications, e.g. white sidewalls of tires. The frequently observed slight yellowing of light colored vulcanizates on exposure to the weather, does not appear when anatas types are used.

All white pigments having BET-values of 8–12 belong to the inactive fillers. Since titanium dioxides are frequently used higher than 20 phr, the inactivity of the filler plays a role.

4.4.5.3 Inorganic Color Pigments

[1.14, 4.494]

For the production of colored rubber articles, a number of inorganic pigments are used; they do not have the brilliance of the organic pigments, but have outstanding weathering properties, good chemical resistance and sometimes they are low priced. By mixing different color pigments, all coloristic effects can be obtained. The color pigments are also considered inactive fillers because of their small specific surface. They are always used in small amounts.

Iron Oxide Pigments. The iron oxide pigments are especially suitable to obtain subdued red, brown, beige and yellow shades. Since iron oxides frequently contain rubber poisons like Mn as impurities, it is important when they are chosen, to test if they are free of rubber poisons. Special iron oxide pigments for the rubber industry have been tested and their rubber poison content has been limited.

Chromium Oxide pigments. For the production of subdued green and yellow-green color tones, chromium oxide pigments are chosen. There two types that are practically free of rubber poisons are available to the rubber industry.

Cadmium Pigments. If rubber articles are supposed to have especially luminous color shades, the brilliant cadmium pigments are superior to the iron oxide and chromium oxide pigments. One obtains with them brilliant yellow, orange and red shades. However, cadmium containing products are restricted in some countries for toxicological reasons.

Ultramarine. Ultramarines are mainly produced for blue coloration.

4.4.5.4 Organic Color Pigments

In comparison to inorganic compounds, the organic coloring agents are more efficient, give brilliant color tones, but are less light resistant, have less covering ability, and are more expensive. The biggest portion of suitable compounds are azodyes, for example coupling products of diazotized o-chloroaniline with p-nitrophenyl-3-methyl-5-pyrazole, that is used for orange shades. Besides these, some other dye types are used, as for example alizarine dyes. For blue and green shades, the phthalocyanine dyes are used.

These materials are commercially available as pure powders or in paste form. The pastes usually contain pigments, that are for example prepared with special factice-types, where the pigment is present in almost colloidal dispersion. For application in latex, aqueous pastes with high pigment concentration are suitable as well as powders that contain dispersants that can easily be dispersed in water.

4.4.6 Organic Fillers

[1.14, 4.495–4.502]

4.4.6.1 Styrene Resins

Butadiene-styrene copolymers with 50–90 wt.%, preferably 85 wt.% styrene (styrene resins) have a room temperature hardening effect on rubber vulcanizates; the density of the mixture is not increased, as would be the case with fillers. The hardening is larger with increasing styrene content. With very high styrene content, the melt temperature gets too high and makes mixing a problem. Since styrene resins are thermoplastic they cause excellent processing characteristics. They are primarily used in NR and SBR, for example for hard and light soles. Since the styrene resins are not incorporated into the rubber network, they severely soften the vulcanizates at higher temperatures.

At high styrene content one obtains at room temperature hard rubber like, high impact vulcanizates. The amounts of styrene resins used are 5–60 phr and more, in shoe-sole mixtures between 10 and 30 phr.

4.4.6.2 Phenolic Resins

[4.498-4.502]

Incompletely condensed phenolic resins of the novolac type give NBR-based vulcanizates improved physical properties (see page 71). They increase hardness, tensile strength, tear- and abrasion-resistance as well as swelling resistance and heat stability of the vulcanizates. This effect is more pronounced as the acrylonitrile content of the rubber increases. Elongation and elasticity decrease with increasing dosage. The loss of elasticity may be partially aliviated by the addition of plasticizers. The rubber to metal bond is clearly improved by the addition of this type of phenolic resins.

These phenolic resins act as plasticizers for the rubber compounds. They improve filler incorporation and flow behavior of the compounds before vulcanization. They are for example used alone or in combination with carbon black or other fillers for oil- and gasoline resistant gaskets, hose, break linings and all rubber goods that need to have high hardness, tensile strength and resistance to swelling. Some of such resins can be used as curatives (see page 261).

4.4.6.3 Polyvinylchloride (PVC)

[4.499, 4.500]

PVC is also used as an NBR additive, where it serves as a hardening function and at the same time improves weathering resistance (see page 71). Up to a 50:50 proportion, PVC can be considered as hardening and reinforcing component for NBR; at higher contents, NBR has to be looked as a PVC polymer plasticizer.

4.5 Plasticization and Plasticizers, Process Aids and Factice

[4.503-4.570]

4.5.1 The Principle of Plasticization and Classification of Plasticizers

[4.503-4.509]

Besides fillers, plasticizers play the biggest quantitative role in building a rubber compound. The reasons for the use of plasticizers are manifold:

- Decrease of elastomer content by using high dosages of carbon black and plasticizer to lower the price of the blend, i.e. extending the rubber.
- Improvement in flow of the rubber compound and energy savings during processing; especially reduced energy peaks.
- Improved filler dispersion in the rubber compound.
- Improvement in processing and tackiness of the rubber compounds.
- Influence on the physical properties of the vulcanizate, especially its elongation and elasticity, especially at low temperatures, lowering of the glass transition temperature, elevating the electrical conductivity, increasing of flame protection etc.

SR is usually harder to process and less tacky than NR and needs larger additions of plasticizers. In comparison to NR, the larger amounts added to SR have less influence on the properties of the vulcanizate [4.506].

Rubber/Plasticizer Interaction. One can differentiate between two groups of plasticizers based on their interaction with the rubber, the primary and secondary plasticizers. Depending on their function, the deformation behavior of the rubber compound can be changed considerably. A quantitative treatment is different and only possible using thermodynamic considerations [3.180, 3.181, 3.400].

Those plasticizers that solubilize rubber, so called *"primary plasticizers"*, assist the micro- and macro *Brownian* motion of the polymer chains and thereby also the viscous flow. Since the plasticizers swell the rubber, they reduce the viscosity of the unvulcanized rubber compound considerably. They confer generally good elastic properties to the vulcanizate at low hardness levels. For different rubbers, different plasticizers can function as primary plasticizers, for example polar products in polar elastomers and non-polar ones in non-polar rubbers. For example, high molecular weight rubber molecules with dipolar character form intermolecular force fields; they combine with polar plasticizers like esters or ethers to form associations [4.508] that cause reduction of the force fields and thereby produce higher mobility of the chain segments. With polar plasticizers, sometimes called "elasticators", the elastic behavior of NBR-vulcanizates is improved. However, such plasticizers reduce strength and hardness of the vulcanizates considerably.

Plasticizers that solubilize very little or not at all are called *"secondary" plasticizers;* they act as lubricants between the rubber chain molecules, improve the formability, without any appreciable effect on the viscosity of the compounds. Mineral oils, paraffine or ozokerite belong to this group. They are rubber soluble at the processing temperature, however they readily exude at relatively low levels and thereby reduce the tackiness of the rubber compound.

Between these groups of plasticizers (primary and secondary) there are many transitions possible and it is impossible to draw a sharp deviding line between them.

The efficiency of the plasticizers of both groups depends on their chemical structure and their physical properties as well as on the type of rubber [4.508]. High boiling petroleum distillates are therefore primary plasticizers for NR and SBR; for compounds with NBR they are typically secondary plasticizers. On the other hand, coal/tar oils can be used as primary plasticizers for NR, SBR as well as NBR.

The *physical properties of the plasticizers* are important for practical applications. For high processing and application temperatures, high boiling point, low vapor pressure and chemical stability are necessary. The gel point is important for the elastic properties of the vulcanizate at low temperature, while viscosity at high levels influences especially the hardness. Plasticizers that react basic or acidic, can accelerate or decelerate the vulcanization. Unsaturated compounds react chemically with the sulfur that is needed for vulcanization.

Classification. Plasticizers are generally classified as mineral oils, natural products and synthetic plasticizers.

To extend particularly the non-polar rubbers, relatively cheap mineral oils are used. The choice of mineral oil type depends on its price, polymer type, and compatibility and has relatively little influence on the property spectrum of the vulcanizate.

Beyond that, a multitude of natural products can be used, like fatty acids, wool grease, vegetable oils, glue, rosin, as well as modified natural products, like factice, in order to improve the processing properties, tackiness or filler dispersion of the rubber compound.

The application of the more expensive synthetic plasticizers, that are usually PVC-plasticizers as well, is based on other considerations. Important are the improvement of low temperature flexibility, the elastic properties of the vulcanizate and the beneficial effect certain types have on tackiness of the compound or in the inflammability of the vulcanizate. Additional consequences in some cases are some lowering of hot air stability of the vulcanizate and some extraction by solvents, oils and fats. The wide variation in chemical structure one can achieve with synthetic plasticizers makes it possible to fulfill such demands. The compatibility of synthetic plasticizers especially with polar rubbers is generally better than with mineral oils.

Many plasticizers, especially aromatic mineral oils and ether types interfere with the peroxide crosslinking (see page 257) and cannot at all or only in small amounts be used in compounds that are to be peroxide cured.

4.5.2 Plasticizers and Process Aids

4.5.2.1 Mineral Oil Plasticizers

[4.517–4.525]

4.5.2.1.1 Mineral Oils

The mineral oil plasticizers are quantitatively the most important group since they are low-priced extenders and compatible with many types of rubber. Here too several subclasses have to be differentiated, that vary characteristically in relative content of aromatic, naphthenic and paraffinic hydrocarbons. To orientate oneself among the many products of varied origin and composition that are presented to the rubber industry, methods are used that are customary for mineral oil analysis. They permit a satisfactory characterization and classification and are equally suitable for quality control.

One of the most important methods of classification is the determination of the viscosity-density constant (VDK) that can be determined from the density and Sayboldt-viscosity of the mineral oil as well as its aniline point [1.14, 4.521–4.525]:

$$\mathrm{VDK} = \frac{\mathrm{D} - 0.24 - 0.022 \log (\mathrm{V_t} - 35.3)}{0.755}$$

where

$$\mathrm{D} = \frac{\text{density of the mineral oil @ 60 °F (15.6 °C)}}{\text{density of water @ 60 °F (15.6 °C)}}$$

V_t = Sayboldt-viscosity @ 210 °F (98.9 °C). (Conversion of kinematic viscosity into conventional units for the calculation is necessary).

$$\mathrm{VDK} = \frac{1196 - \text{Aniline point (°F)}}{1170}$$

The VDK-number allows an estimate of the properties of the mineral oil (Table 4.3).

Table 4.3: Relation between VDK and Composition of Mineral Oils

Type of Mineral Oil	Range of VDK
paraffinic	0.791–0.820
relatively naphthenic	0.821–0.850
naphthenic	0.851–0.900
relatively aromatic	0.901–0.950
aromatic	0.951–1.000
highly aromatic	1.001–1.050
extremely aromatic	> 1.050

Table 4.4: Correlation between Refraction Intercept (RI) and Composition of Mineral Oils

Type of Mineral Oil	RI-Range
paraffinic	< 1.048
naphthenic	1.048 – 1.065
aromatic	1.053 – 1.065
highly aromatic	> 1.065

The refractive index intercept, RI, is also important for the determination of the paraffinic, naphthenic, aromatic and highly aromatic contents (see Table 4.4).

$$RI = n_D - 0.5\ d_4^{20}$$

where n_D = refractive index at 20 °C

d_4^{20} = density of the mineral oil at 20 °C/density of water at 4 °C

By knowing VDK and RI, an accurate determination of the composition of the mineral oil is possible by the carbon-distribution analysis (so-called RI-VDK-method, Figure 4.9). The intercepts of lines of equal RI and VDK give the statistical percentage composition of the mineral oil. Example for Figure 4.9: A mineral oil

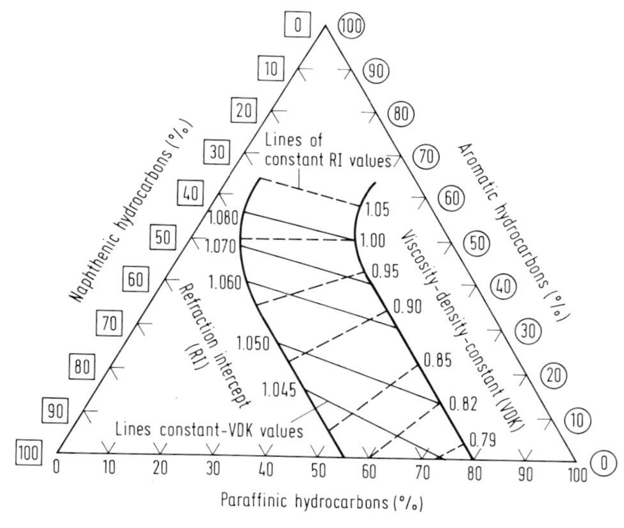

Figure 4.9 Diagram for the carbon distribution analysis (RI-VDK-Method) of mineral oil plasticizers

Table 4.5: General Properties of Mineral Oils

	Density	Storage Stability	Temperature Viscosity Relation	Aniline Point	Low Temp. Properties	Coloration Discoloration	Peroxide Cure
paraffinic	low	good	low	high	good	good	good
rel. naphthenic	↑	↑	↑	↑	↑	↑	↑
naphthenic	│	│	│	│	│	│	│
rel. aromatic	│	│	│	│	│	│
aromatic	↓	↓	↓	↓	↓	↓	↓
highly aromatic	high	poor	high	low	poor	strong	poor

Table 4.6: Compatibility of Mineral Oils with Different Types of Rubber

	NR	SBR	BR	NBR	CR	CSM	EPDM	IIR
paraffinic	+	+	+	–	–	–	+	+
rel. naphthenic	+	+	+	–	–	–	+	+
naphthenic	+	+	+	○	○	○	+	○
rel. aromatic	+	+	+	○	+	+	+	–
aromatic	+	+	+	+	+	+	○	–
highly aromatic	+	+	+	+	+	+	○	–

\+ Compatibility good, ○ conditionally compatible, – incompatible

plasticizer with RI = 1.050 and VDK = 0.90 contains 35% paraffinic, 40% naphthenic and 25% aromatic components. From the knowledge of its composition, one can derive approximately the properties of the plasticizer (see Table 4.5) and its compatibility with different types of rubber (see Table 4.6).

Generally, one can assume that polar plasticizers are compatible with polar rubbers and non-polar plasticizers with non-polar rubbers and that they should be applied correspondingly.

The amounts in which mineral oils are used in rubber compounds range between approximately 5–30 phr. In some cases, for example EPDM, larger amounts up to 100 phr are being used.

4.5.2.1.2 Paraffines and Ceresine

[1.3]

Above a chain length of C_{17} the normal unbranched paraffine hydrocarbons are solid. Paraffines with chains of 17 to 30 carbon atoms have freezing points of 50–62 °C. Higher molecular weight *paraffines* with chain length of 40–70 carbon atoms with melting points of 80–105 °C are partially produced synthetically and are called *"hard paraffines"*. Branched paraffines of equal molecular weight are softer, more pliable and more plastic than the normal ones; their freezing points are also lower. The branched paraffines are also called *"isoparaffines"*. When paraffines or isoparaffines contain alicyclic structures they are called *"ceresines"* or *"isoceresines"*. Ceresines and isoceresines have a very similar property spectrum. In technical mineral waxes, all three groups are generally mixed; they are difficult to separate. During the synthesis of paraffines similar mixtures are produced.

Mineral waxes are naturally found in places where at one time petroleum had percolated. These naturally found waxes are called *raw ozokerites*. They can be separated and purified to produce paraffines and particularly ceresine.

Paraffines are primarily made from petroleum. On fractional distillation of petroleum, one obtains among others, the fraction called axle- and cylinder oil.

From the *"axle oil"* fraction, *"paraffine mush"* separates, that can be used directly in rubber compounds. When this mush is dissolved in methylene chloride, paraffins crystallize out; by fractional extraction the high melting portions are purified from the low melting portions. The *"cylinder oil"* fraction contains *"petrolata"*, which can be found in old recipes and are frequently found even today in American recipes. By solvent treatment, finely crystalline paraffins can be obtained from petrolata.

While paraffines and ceresines with low freezing points have some use as plasticizers or process aids, those with high freezing points are used as ozone protection waxes (see page 273).

4.5.2.1.3 Cumarone- and Indene Resins

[1.3]

After treatment of the *"light oil"* fraction of petroleum with first dilute caustic (to remove phenols and cresols) and then with dilute sulfuric acid (to separate the pyridine bases) the soluble components can be polycondensed and resinified by treatment with concentrated sulfuric acid. In this way, *cumarone-* and *indene resins* are formed that can be distilled between 160–180 °C.

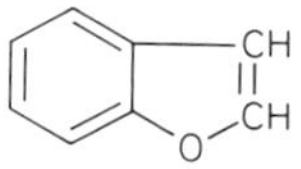

CH ‖ CH O — Cumarone

CH ‖ CH C H_2 — Indene

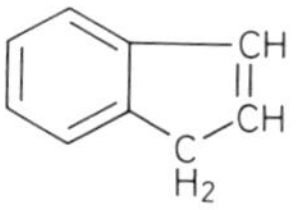

These resins are very useful plasticizers and tackifiers for NR, but especially for SR. Depending on the conditions during preparation and degree of condensation, different cumarone-, respectively indene-resins are obtained. These resins are used at 1 to 5 phr levels.

4.5.2.1.4 Petroleum Distillation Residues

[1.3]

When distilling petroleum, ground wax, oil shale, and the like there are high boiling and solid, meltable hydrocarbon residues (bituminous component). They contain relatively small amounts of mineral impurities and still oxygen containing compounds. From the residues, the following components can be obtained:

- In CS_2 soluble, saponifiable components, "*Montan* waxes".
- CS_2 partially soluble, incompletely saponifiable components, "*ozokerite, asphaltines* (silsonite)" etc.
- CS_2 insoluble and non-saponifiable component, *"bitumens, pitches"*.

Petroleum contains variable amounts of bituminous components. Largely paraffinic oils contain on distillation little bitumen or none at all. On the other hand, there are aromatic oils that contain large amounts of bitumen of highly variable composition. In the rubber industry, bitumen with softening points between 60 and 100 °C at the 5 phr level and higher are being still used as process aids, which lower the elastic properties of the rubber compounds considerably.

4.5.2.2 Fatty Acids, Fatty Acid Derivatives, Process Aids

[3.267–3.269, 3.456, 4.538–4.555]

Fatty Acids are primarily necessary, in small amounts, as vulcanization activators. Larger amounts would give a plasticizing effect, but affect tackiness negatively. Increasing amounts of fatty acids also reduce the rate of vulcanization. In most cases stearic acid is used, sometimes also palmitic acid.

Fatty acid salts are used as emulsifiers in the production of SR and therefore many SR-types already contain certain amount of fatty acids.

Metallic Soaps. Beside the fatty acids, fatty acid metal salts play an important role as process acids. Among them, the *Zn-soaps* [3.267, 3.268, 4.543, 4.543 a, 4.548 a,

4.548 b, 4.551 c] are especially important. Contrary to fatty acids, the salts can be used because of their higher solubility in higher dosage without danger of blooming whereby correspondingly better effects can be obtained.

Aside from being excellent lubricants, they facilitate and improve the preparation of the compound and further processing. They also lower the mixing and processing temperatures and help to save energy. The soaps also improve dispersion of the components of the rubber compounds, and permit the use of larger amounts of filler; they also prolong the scorch time without prolonging, in some cases even shortening, the vulcanization time [4.554]. In addition, one observes with NR a mastication effect, that can be increased by the use of a combination of Fe- and Mn-complexes [4.555] (see page 221). Because of the better filler dispersion and vulcanization activation, the property spectrums of the vulcanizates are improved. Because of all these effects, the Zn-soaps are among the most important process aids.

While *Zn-Stearate* is frequently used as a dusting medium because of its ability not to affect the tackiness of rubber compounds (see page 319), it has only a limited importance as a lubricant. *Zn-soaps of unsaturated fatty acids* are much more effective in this respect [3.267]. Since, of course the process aids (as the plasticizers) cannot be equally effective in all elastomers [3.267, 4.543] and since furthermore their effectiveness depends upon the shear rate during the processing operation [4.547], their effectiveness has to be considered very critically.

Zn-soaps of unsaturated fatty acids exhibit in *NBR* at all shear rates the strongest effect, followed by *fatty alcohol esters* [3.267, 3.268]. In *EPDM*, however, such Zn-soaps exhibit their effectiveness only at the lowest shear rates and *Ca-soaps of saturated fatty* acids are most effective at higher shear rate [3.267, 3.456]. In *CR*, on the other hand, the Zn ions contained in the Zn-soaps accelerate the crosslinking reactions already during processing. Therefore, the Zn soaps affect the processibility behavior negatively, especially at high shear rates. Fatty alcohol ester are the most effective substances [3.267, 3.268]. In *General purpose elastomers*, however, very good improvements of the processibility behavior can be obtained with Zn soaps [4.548]. The pure compounds lead, of course, generally to the largest effects while lesser effectiveness has to be expected when diluted products, including emulsion plasticizers, are utilized. *Ca soaps of saturated fatty acids* have proven to be especially effective lubricants in EPDM [3.267, 3.456] and they play an important role for this elastomer.

Na- und K-soaps have in that respect a more pronounced effect and are preferred if a rubber compound is to be made more alkaline, for example in amine cures. They are, for example, used in the ACM-vulcanization (see page 110).

Fatty Acid Esters and Fatty Alcohols. Aside from fatty acids and metal soaps as well as vegetable and animal fats and oils, increasing use of low priced mixtures of fatty acid esters and fatty alcohols are used because of their outstanding processing improvement activity (without providing mastication and vulcanization activation). Since they are free of Zn and water they can also be used advantageously in cases where the Zn-content of the Zn-soaps (for example during CR vulcanization) or the water in emulsion plasticizers (for example in pressure-less vulcanization) may interfere.

Pentaerythritol tetrastearate is also a process aid with a wide application range. It does not bloom and not interfere with peroxides and is, therefore, a product that is well used.

Emulsion Plasticizers. Many combinations of metal soaps, fatty acid esters and fatty alcohols in aqueous emulsion are on the market under the general expression "emulsion plasticizers" and are recommended as general purpose process aids. But in recent investigations it has been found that they can do a lot less than they are said to [3.267, 3.268, 4.543, 4.543 a]. Due to the fact that parts of the water will remain in the compound, bubble-free vulcanizing without pressure is difficult.

Polyethyleneglycol fatty alcohol ethers are used for instance as antistatic plasticizers especially for NBR (see page 303, 483) [1.14].

General purpose process aids do not exist [4.548 a].

4.5.2.3 Animal and Vegetable Fats, Oils and Resins

[1.12–1.14, 2.86, 3.146, 3.363, 4.495–4.502, 4.525 a–4.537]

Raw Wool Fat (Lanolin). Lanolin is a well liked plasticizer that facilitates mixing, extruding and calandering greatly and that improves the surface of the extruded or calandered part.

Fat Emulsions. Fat emulsions, also emulsion plasticizer types have been useful in practical applications; they exhibit a dispersing activity for both carbon black and light colored fillers. They aid a little the processing of compounds during extrusion and calander processing. They contain about 10–15% water. Water that is mixed into the rubber in this procedure has a positive effect on banding and cooling on the rubber mill. If a pressure-less vulcanization is used subsequently the water can cause considerable difficulties. The action on processibility (extrudability, injection molding etc.) is less.

Vegetable (Plant) Oils. Several vegetable (plant) oils lower the processing viscosity of the rubber compound and improve filler uptake. *Palmoil* is used for example. The influence of the unsaturation on aging of the vulcanizate must be considered. Palm oil melts at 30–40 °C and softens the rubber greatly. The same considerations apply to *soyabean oil.*

When soyabeans are processed, a yellowish brown, pasty material is obtained, *Rubberine Gel,* that consists of a mixture of fatty acid esters that contain some lecithines. Rubberine Gel is used in place of stearic acid; it does not extend the scorch time as does stearic acid, but shortens it. Because of its hydrophilic character, it prevents mill-sticking of highly loaded compounds.

Tall oil is also used to some extent as plasticizer in the rubber industry.

Animal Glue. This is a fairly interesting material for the rubber industry. In order to mix it into rubber it must first be swelled. It is useful to start with glue-rubber premixes (master batches) in a 40:60 proportion. Similar batches can easily be prepared in an internal mixer. Swelling is not necessary with glues that contain some wax alcohol.

Glue acts as a plasticizer for rubber compounds and facilitates shaping. In steam vulcanization, it improves the green strength of uncured parts. Glue acts as a weakly reinforcing filler for the vulcanizate, i. e. it increases hardness and stress values.

Rosins (pine tars) are well liked pasticizers for NR. When using mercapto accelerators it has to be remembered that this additive can lead to increased scorchiness; this has to be considered when designing the compound. The acid number and content of volatile materials must not be too high.

Abietic acid derivatives are important emulsifiers in the production of E-SR-types and increase the tackiness. Even some types of rubber that are made in solution contain added rosin acids to destroy the activators, which again leads to an improved tackiness. Since the rosin acids are only partially removed during washing, preparation of the rubber for compounding, some of the residues are still present in the finished rubber articles. The amounts of rosin acid present in SR from its preparation are generally approx. 2–7 phr (see page 61). *Colophony-derivatives* play an important role as tackifiers. They are among the best tackifiers for NR.

Coal Tar Pitch and derivatives as well as blown bitumen residues (mineral rubber) that cause softening of the rubber compound and facilitate processing, and cause at the same time some hardening of the vulcanizate, are of no importance today.

Zewa Resins. Related to Tall oil is the Zewa-resin, which gives an exothermic reaction when mixed into NR. It has to be added, therefore, at the end of the mixing cycle. It hardens the vulcanizate and improves its strength. Its action is especially beneficial in highly loaded compounds, for example those used for soles, heel- and floorcover compounds. In nitrile rubber the Zewa resin serves as a process aid.

4.5.2.4 Synthetic Plasticizers

[1.12–1.14, 3.181, 3.182, 3.214, 3.216, 4.510–4.516]

4.5.2.4.1 General Considerations

[1.12, 1.13]

Because of their clearly higher prices, synthetic plasticizers are quantitatively less used than mineral oils. Because of the possible variations of their compositions and manifold specific activities, their number is very large. They are more important qualitatively to achieve certain properties in the vulcanizates, that have increased with the requirements of modern technology; they are furthermore necessary, for certain types of rubber that are not compatible with mineral oils. This plays an important role in polar rubbers like NBR and CR, that are barely compatible with low- or non-polar mineral oils. Synthetic plasticizers are in most cases primary plasticizers.

The addition of synthetic plasticizers on the two roll mill or in the interal mixer does not present any difficulty; to the contrary, they act as quasidispersing agents for the incorporation of the fillers and at the same time soften the compounds. Compounds that contain plasticizers show frequently improved tackiness because of their lower viscosity, improved extrusion behavior, whereby large temperature increases are avoided.

Generally, synthetic plasticizers do not influence the storage stability or scorchiness of the compounds. However, in the vulcanizate a lower shore hardness and poorer tensile properties are observed, depending on the amounts used with otherwise constant compound composition. Frequently, rebound and low temperature flexibility are improved, especially when using monomeric ester- and ether types. To achieve good high temperature stability, a careful selection of the plasticizer type is necessary, since especially low molecular products with high volatility are detrimental. For that purpose, higher molecular weight products with low volatility are preferable [3.181, 3.182, 3.214, 3.216].

No generalizations can be made about swelling stability of plasticizer containing vulcanizates, since organic solvents extract more or less plasticizers from the vulcanizate, by which a lower swelling appears to take place. Swelling and extraction act therefore in opposite directions. Only polymeric plasticizers migrate less from the vulcanizate.

The synthetic plasticizers do not generally have any effect on discoloration of the vulcanizate. Exceptions are mentioned under the heading of specific products.

Comparison of a multitude of plasticizers for NBR can be made from [4.515].

4.5.2.4.2 Ether Plasticizers

[1.12–1.14]

Ethers or thioethers are especially well suited for NBR.

Dibenzylether used to be applied in larger amounts to improve processing of synthetic rubber but has lost its importance today largely because of its volatility.

Polyether and Polyether-Thioethers play today, depending on the type used, a role as antistatic agents (see page 301) or as very effective plasticizers for achieving highest elasticity and low temperature flexibility in rubber products made of NBR and CR.

Thioether-esters are important as very effective plasticizers to improve elasticity and low temperature behavior or rubber parts made of NBR and CR. Examples are methylene-bis-thioglycolic acid butylesters, thiobutyric acid butylesters, etc. Quantities used are between 5–30 phr. The optimum on elastification is reached at the highest dosage mentioned. These products are rather volatile and cannot be applied where very high heat stability is called for.

4.5.2.4.3 Ester Plasticizers

[1.12–1.14]

Phthalic Acid Esters are frequently used as cheap plasticizers to improve elasticity and low temperature flexibility, particularly for NBR and CR vulcanizates. The following are mainly considered: dibutyl-, dioctyl- (DOP), diisooctylphthalate (DIOP) and diisononylphthalate (DINP).

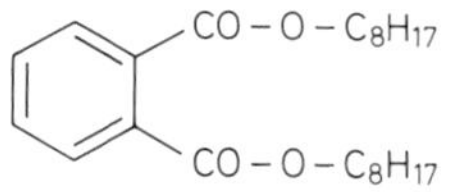

$CO-O-C_8H_{17}$

DOP

$CO-O-C_8H_{17}$

Higher molecular weight products are less interesting since low temperature flexibility becomes poorer with long chain phthalates. Quantities added are between 5–30 phr.

Adipic Acid- and Sebacic Acid Esters. For *adipic acid esters* e.g. dioctyl adipate (DOA) the same considerations hold as for phthalic acid esters. However, because of their price, use is only justified, if estreme improvement of elastic properties has to be achieved, for example for NBR vulcanizates.

Sebacic acid esters play a small role for the production of rubber articles because of price and technological reasons, in spite of the fact that dioctyl sebacate (DOS) gives very good low temperature behavior. The same goes for ester plasticizers based on *azelaic acid*. To achieve the best heat resistance, sebacates are preferred.

Trimellitates, e.g. triisooctyl-trimellitate (TIOTM) are plasticizers with extremely low volatility. They influence good low temperature behavior [4.510].

Phosporic Acid Esters. Corresponding to their use in plastics, phosphoric acid esters are also used in rubber because of their low flammability. They are primarily used in SR types that by themselves exhibit low flammability like CR and require addition of low flammability plasticizers. Primarily considered are tricresyl- and diphenylcresyl phosphate, rarely also esters or mixed esters of xylol or trioctylphosphate.

Dosage lies between 5–15 phr. Because chlorine containing plasticizers have become less important as protectants against combustion, the use of phosphoric acid esters has also diminished.

Other Ester Plasticizers: Because of the increasing importance of heat resistant NBR formulations [3.181, 3.182, 3.214, 3.216] less and less volatile products are utilized. However, their application involves compromises as far as low temperature flexibility is concerned. Because of reasons to obtain simultaneously high temperature performance and low temperature flexibility these plasticizers are also often combined with volatile ones which have a stronger influence to lower the low temperature flexibility. As plasticizers with especially low volatily the following ones are used:

Polyester with different degrees of condensation (of which, of course, the ones with higher degree of condensation exhibit the lowest volatility but result in the least desirable low temperature flexibility (see Section 4.5.2.4.5), *citrates* (e.g. tributyl acetyl citrate), *ricinoleates* (e.g. butylacetylricinoleate), *octyl-iso-butyrate, tri-glycoldioctylate, tetra-hydro-furyl-octoate, butyl-carbitol-formaldehyde, penta-erythritol-ester* (see page 300) etc.

4.5.2.4.4 Chlorinated Hydrocarbons

Chlorinated paraffines are used in rubber articles in amounts of 20 phr to lower flammability. When higher levels are used, the physical properties of the vulcanizates become poorer. In chlorine containing rubber types like CR, lower amounts are used and usually do not exceed 10 phr. Frequently, the chlorinated hydrocarbons are used in combination with antimonytrioxide. Also, Al-oxyhydrate, Zn-borate, Mg-carbonate or selenium are considered as additives. Aside from chloroparaffines, chlorinated diphenyl, chlorinated naphthaline and chlorinated rubber are added as flame retardants. Because of newer flame protection concepts, the use of chlorinated hydrocarbons has diminished.

4.5.2.4.5 Polycondensation- and Polymerization Products

Polyesters of Adipic- and Sebacic Acids and 1,2-Propyleneglycol have as liquid non-volatile and non-migrating plasticizer a certain importance for rubber parts made of NBR. The elastic properties of the vulcanizate at low temperatures are not much better than with monomeric esters.

Alkyd Resins. To make NBR compounds suitable for extrusion, alkyd resins can be used.

Polymerization Products from Croten Aldehyde are good dispersants for fillers.

Liquid BR, SBR, NBR and EPDM can be used for the respective solid rubbers as non-volatile and non-extractible plasticizers [3.108–3.114a, 3.181, 3.182, 3.433, 4.555a].

Tackifiers. Since synthetic rubber is generally much less tacky (sticky) than NR, the addition of resins is particularly important for its processing. By using tackifiers one achieves better adhesion and welding at the points of contact during the manufacturing process. These materials lend tackiness only to the unvulcanized rubber compound and at the same time they lower the viscosity and improve processing. The vulcanizates do not exhibit any tack and are relatively softened only slightly by the resins compared to plasticizers that improve elasticity [4.525a–4.537]. *Colophony* is usually not sufficiently compatible. That makes the use of synthetic resins attractive.

The most important tackifier, especially for NBR and CR, is a viscous *xylol-formaldehyde-resin,* as well as the highly active *koresin,* that allows the construction tack to increase with increasing dosage according to demand. Increased performance has been achieved by other hydrocarbon resins. Other synthetic resins, like *alkyd resins,* are also in use.

4.5.3 Factice

[4.556–4.570]

4.5.3.1 General Considerations about Factice and Classicification

Factices are unsaturated oils that have been reacted with sulfur or similar materials. The reactions that occur during "factication" correspond to a hard rubber vulcanization. One differentiates generally between brown, yellow and white factice.

The factice industry is almost as old as the rubber industry and many kinds of factice have been produced. They are based on a variety of raw materials and production processes as well as requirements for special products for SR-types and taylor-made products for certain customers. This explains that more than 60 different kinds are commercially available, even if some seven products could cover the broad area of applications.

For the production of factice vegetable and animal oils with an iodine number >80 are primarily used that have three or more double bonds per molecule and sulfur, sulfure chloride or other materials are used as crosslinkers.

Among the factice-types one differentiates according to the crosslinking procedure

- Sulfur Factice
- Hydrogensulfide Factice
- Sulfurchloride Factice
- Sulfur-free Factice

Further subdivisions are made according to the type of base oil used, for example rape seed oil, soybean oil, fish- or castor oil as well as to the pre-treatment, like polymerization, blowing, hardening and the like. For further subdivisions, low priced additives like mineral oil and fillers have to be considered, that do not enter the factice production process.

During factice production depending on selection of starting materials and reaction temperature, white, yellow, amber to dark brown types can be generated.

Depending on rubber technological requirements the following seven factice-types can be differentiated:

- white sulfur factice
- yellow sulfur factice
- brown sulfur factice
- white sulfur chloride factice, stabilized
- white sulfur chloride factice, not stabilized
- special factice for SR
- sulfur and chlorine-free special factice

The preparation is given in [4.557, 4.566]

4.5.3.2 Use and Properties

The former English name for factice is "rubber substitute" which indicates that factice was originally used as a substitute for the very expensive rubber. Today factice is used as additive to achieve certain special effects.

Advantages. Various factice-types serve various demands. They specifically serve the following purposes:

During *processing* of rubber compounds

- improvements of the green strength and dimensional stability during extrusion
- reduction of the calandering effect
- improvement of surface smoothness
- increasing plasticizer absorption
- improvement of handling of very soft compounds with high plasticizer content (for example printing roller covers, closed cell sponge-rubber compounds etc.)

In vulcanizates: factice

- gives a good feel (textile character)
- improves grinding (for example with rolls)
- improves fatigue crack resistance

In the cost analysis: factice

- saves energy during milling and mixing
- saves labor time by shortening mixing time
- increases extrusion speed
- increases calandering speed
- reduces the cost per volume by lowering the density

Selection. In selecting suitable factice types one has to consider the basic oils, preparation method and color as well as extender additives (because of lowering the factice effect). Sulfur factices and stabilized sulfur chloride factices (which frequently have to be used in large quantities) do not affect the action of the vulcanization accelerators. However, sulfurchloride factice retards vulcanization, and on heating, hydrogenchloride is given off.

Only through the use of strongly basic vulcanization accelerators in larger quantities, white unstabilized sulfurchloride factice can be used even if high vulcanization temperatures are to be applied, for example in the manufacture of erasers. Therefore, stabilized sulfurchloride factice is preferred.

Influence on properties. Tensile strength and compression set are more or less impaired by factice addition, dependent on factice type used. Factice reduces abrasion resistance. For that reason factice is used in large amounts in erasers, where abrasion is desired. This increased abrasion makes factice use in tire treads impossible. However, small amounts of factice are used in tire side walls, since the dynamic fatigue resistance is good in the presence of factice. Since good factice types do not show aging phenomena on prolonged storage, they tend to have a beneficial effect on aging of the rubber goods.

Application Areas. In practice they are used only in nontire applications.

Factice based on *rape seed oil* are preferred in NR, IR, SBR and similar allpurpose rubbers. They are also suitable for NBR and CR; for good swelling behavior the low-swell factices based on *castor oil* should be applied. Small amounts of factice (5–10 phr) are recommended to use in injection compounds to prevent surface flow faults (see page 422). All sulfur factices are not heat stable based on their sulfur crosslink structure. For that reason one uses for heat resistant articles factice types that have crosslinks other than sulfur. The following types can be used:

Oxygen-Factice. These factices are prepared by peroxide crosslinking of the fatty oils and they contain oxygen crosslinks instead of sulfur.

Isocyanate Factice, where isocyanates react with hydroxyl groups, for example in castor oil to form urethane crosslinks.

Epoxidized factices are being developed [4.562a].

Specialty Factices play an important role in other fields, for example as suede protectant or as additives for abrasives in metal polishes.

Liquid Sulfur- or *Sulfurchloride Factices* are suitable additives to improve lubricating- and cutting oils; furthermore to make pastes of dyes that are hard to disperse and other products for the rubber industry.

4.6 Blowing Agents

[1.14, 4.571–5.576]

4.6.1 General Considerations

For the production of sponge- (open cell), closed cell sponge (small, closed and open cells) or micro cellular rubber (small closed cells with thin cell walls) inorganic and organic blowing agents are used. Two processes are mainly used: the blowing process for the production of sponge-rubber and the expansion process for the production of micro cellular rubber (closed cell) (see page 440ff).

Blowing agents are substances that are worked into the rubber and are stable at room temperature, but decompose at higher temperature with gas release before or during vulcanization. The gas that is formed, nitrogen or carbon dioxide, gives rise to the formation of pores.

Inorganic blowing agents are difficult to work into the rubber, their dispersion is primarily poor, causing irregular pore structure. The storage stability of such mixtures, especially those containing ammonium bicarbonate, is not very high. For that reason, nitrogen releasing organic blowing agents were developed, that have better properties and have, in spite of their higher price, largely replaced the inorganic blowing agents. A good blowing agent should fulfill the following requirements:

- contain a large amount of releasable gas
- be toxicologically "clean"
- decomposition products should not have a bad odor
- should not discolor the vulcanizate
- should disperse well in the rubber
- should not influence the vulcanization and not worsen aging properties
- have a suitable decomposition range (varies for different uses)
- be low-priced

Since it is difficult to combine all these properties in one product, special blowing agents have been developed for different areas of application, that produce vulcanizates with optimum properties combined with good processing safety. For example, in some cases (hard micro cellular rubber soles) a high decomposition temperature is beneficial; in other cases (for example for sponge rubber production) a low decomposition temperature of the blowing agent is necessary.

4.6.2 Inorganic Blowing Agents

For the production of sponge rubber, sodium bicarbonate used to be preferred, partly in combination with a weak organic acid, like tartaric, stearic- or oleic acid. For improved dispersion, similar inorganic blowing agents were used as pastes, for example in mineral- or paraffine oil. The rubber or rubber compound must be masticated strongly and the viscosity of the raw rubber as well as the storage time and temperature, carefully controlled.

The low price of the inorganic blowing agents is more than made up for by the uniform pore type and size and processing safety achieved with organic blowing agents. One exception is sodium nitrite combined with ammonium chloride that is generally used as tablets in the production of hollow rubber objects, especially balls.

4.6.3 Organic Blowing Agents

Based on the reasons mentioned, organic blowing agents are mostly in use today. Since nitrogen diffuses slower through rubber cell walls than carbon dioxide or ammonia, blowing agents that give off nitrogen are most preferred today.

4.6.3.1 Azo Compounds

Diazoamino-Compounds belong to the oldest blowing agents. The best known blowing agent in this group is diazoaminobenzene

$$C_6H_5 - NH - N = N - NH - C_6H_5$$

The melting point of the technical grade is 90-96 °C (decomp.). It has an isonitrile-like odor depending on its purity. It develops under normal conditions 113 cm^3 nitrogen per g of material. Quantities of 2-3 phr necessary for the production of sponge rubber are still soluble in the rubber. Diazoaminobenzene is very effective and stable at mixing temperatures. During the decomposition, aside from nitrogen, a mixture of intensively yellow-brown colored materials is formed that also contain aniline. Light colored material like paper, textiles, plastics, linoleum, nitrocellulure lacquers, and others, discolor on contact with vulcanizates that contain diazoaminobenzene or its decomposition products. A further disadvantage is the skin irritation caused by compounds and vulcanizates containing diazoaminobenzene. For those reasons its usefulness is limited.

Azonitrile. During World War II, azonitrile was developed by IG Farbenindustrie as a nitrogen-releasing blowing agent for rubber and plastics. α,α'-Azo-bis-(isobutyronitrile)

$$(CH_3)_2C(CN) - N = N - C(CN)(CH_3)_2$$

The melting point of the technical grade is 103-104 °C (decomp.).

It is superior because of its high gas release (136 cm^3 nitrogen/g material under normal conditions) to the slower decomposing α,α'-azo-bis-(hexahydrobenzonitrile) (84 cm^3 N_2/g material, normal conditions). Because of its high efficiency, on addition of 2-4 phr of isobutyronitrile-derivative a soft sponge rubber expansion of 500-600% can be achieved. Furthermore, production of a very light porous hard rubber with a density of 0.065 g/m^3 is possible. The products remaining after decomposition, especially the tetramethyl succinic-dinitrile, have toxic properties and require special precautions during processing and use of the rubber articles. α,α'-Azo-bis-(hexahydrobenzonitrile) has the advantage that the decomposition products are less toxic, but its lower blowing capacity and slower decomposition have to be considered.

Due to the toxicological reasons of isonitrile derivatives and the technological disadvantages (slow decomposition and low gas production) of hexahydronitrile derivatives, these two materials are practically no longer used in the rubber industry.

Azodicarbonamide (ADC) [1.14] A further development of the azo-nitriles by IG-Farbenindustrie lead to the development of the azodicarbonic acid, and its derivative azodicarbonamide

$H_2N-CO-N=N-CO-NH_2$

which today is the most important substance of this group. Azodicarbonamide, a light colored powder, decomposes at approx. 215 °C and is today still a very interesting nitrogen releasing blowing agent that gives high processing- and storage safety for the production of odorless, cellular vulcanizates. It is especially suitable for articles that are cured at relatively high temperatures (for example closed cell sponge rubber profiles in LCM- or UHF-setups). Gas generation begins in rubber compounds at approx. 140 °C, it is completed only at vulcanization temperatures of 160 °C and above. For these reasons no gas loss occurs when the uncured rubber compounds containing azodicarbonamide are stored. Gas evolution is approx. 190 cm^3/g under normal conditions. Beside the large gas generation, the lack of odor during processing and in the vulcanizate as well as the non-discoloration are important.

By using so-called *"kickers"*, the decomposition temperature can be lowered by some 20–30 °C and the decomposition rate increased at constant temperature. Suitable substances are ZnO and Zn-salts, glycols (for example diethyleneglycol) and (less appropriate) urea as well as sulfinic acids. A number of commercial products are blowing agent/kicker mixtures with corresponding lower temperatures of decomposition.

By using the blowing agent process (e.g. for closed cell sponge) one obtains a uniform, visible pore structure. Expanded articles (for example light micro cellular rubber) show a uniform cell structure. Cell size is strongly influenced by the mold pressure and can practically vary from invisible to clearly discernable cells.

4.6.3.2 Hydrazine Derivatives

[1.14]

Hydrazine derivatives of organic sulfonic acids, which were developed by Bayer in 1949, represent a new generation of blowing agents. They belong to the most modern nitrogen releasing blowing agents.

Benzenesulfohydrazide (BSH). Among the hydrazides of aromatic sulfonic acid, benzenesulfohydrazide has proven to be an excellent important blowing agent.

$C_6H_5-SO_2-NH-NH_2$

Technical benzenesulfohydrazide is a colorless and odorless crystalline powder (decomposes at 95–100 °C) that is of unlimited storage stability under customary conditions. To facilitate dispersion in rubber it is preferably used as a paste, for example 27% in mineral oil. 1 g material develops 115–130 cm^3 of nitrogen under normal conditions.

Good dispersion and processing properties which are applicable on a large scale, excellent blowing action, absence of odor in sponge- and micro cellular rubber articles prepared from it, no discoloration in even the lightest color shades, are the excellent properties of this product. Since the degree of blowing also depends on the rubber compound remaining viscous long enough before the beginning of the vulcanization, it must be considered that addition of benzenesulfohydrazide activates vulcanization accelerators. In rubber compounds the decomposition of benzenesulfohydrazide begins at ca. 80–90 °C. The rate of decomposition increases with increasing vulcanization temperature. It takes ca. 10 min at 110 °C. Basic accelerators decrease the temperature of decomposition considerably [1.14, 4.573].

The decomposition of benzenesulfohydrazide proceeds to the corresponding sulfinic acid as nitrogen is given off that in turn disproportionates to give diphenyldisulfide and diphenyldisulfoxide. Fatty amides can be added to the compound as dispersing agents, to achieve defined pore sizes and uniformity. One can vulcanize in compression molds as well as in hot air and open steam.

Benzene-1,3-disulfohydrazide. Compounds that contain benzene-1,3-disulfohydrazide start gas evolution (approx. 166 cm^3/g nitrogen under normal conditions) at approx. 115 °C. This blowing agent can be used in hard, highly loaded compounds that reach higher temperatures during processing. The temperature of decomposition is lowered by compounds that react basic. Since pure benzene-1,3-disulfohydrazide is very energy rich and tends to explode, it is commercially available only in diluted form, for example as 50% paste in chloroparaffin. Due to the high temperature of decomposition it is very suitable for production of micro cellular rubber soles.

Diphenyloxide-4,4′-disulfohydrazide does not have to be diluted, in contrast to the above. The temperature of decomposition lies at ca. 160 °C (ca. 125 cm^3/g nitrogen, under normal conditions), but the decomposition starts already at 120 °C. The material is commercially available coated with approx. 3% mineral oil.

p-Toluenesulfonic Acid Hydrazide resembles the benzenesulfohydrazide technologically in its application. Its point of decomposition is approx. 105 °C (approx. 120 cm^3/g nitrogen under normal conditions). The material is commercially available as powder or paste (for example 80%).

The sulfohydrazides described above do not affect odor or taste of the vulcanizates. On opening of the vulcanization presses, no offensive odor is noticable, nor have the vulcanizates unfavorable physiological properties.

4.6.3.3 N-Nitroso Compounds

[1.14]

An other important group of blowing agents are N-nitroso derivatives of secondary amines and N-substituted amides. They also give off nitrogen.

N,N′-Dinitrosopentamethylenetetramine (DNPT, WTR-number 62, see page 234, 269) [4.28 a, 4.205]

```
      H2C — N — CH2
       |    |    |
ON – N     CH2  N – NO
       |    |    |
      H2C — N — CH2
```

a fine crystalline powder of light yellow color, has a decomposition point of approx. 205 °C. Because of its high energy content it can only be handled without danger in diluted form. Even in 80% concentration it releases exceptionally large quantities of gas (approx. 260 cm^3/g under normal conditions). Gas release begins in rubber compounds at approx. 120–125 °C. Complete blowing action occurs at the customary vulcanization temperatures. Because of the large amount of releasable gas, N,N′-dinitrosopentamethylenetetramine is very economical as a blowing agent.

DNPT is acid sensitive in contrast to the sulfohydrazides. Aside from acid components of the compound (e.g. benzoic acid), the hydroxyl group containing components (e.g. glycols) also lower the decomposition temperature. Small amounts of such compounds are frequently added for activation, i.e. acceleration of gas release.

For dilution, DNPT is mixed with chalk and mineral oil or with liquid polyisobutylene.

DNPT can only be used for porous rubber articles, where the characteristic odor of the decomposition products does not matter. Because of the large amount of generated gas it is a very economical blowing agent of worldwide use and is primarily used for production of porous soles of all qualities and of micro cellular rubber. It is less suited for sponge rubber because of the high decomposition temperature.

N,N′-Dimethyl-N,N′-Dinitrosophthalamide is a crystalline, yellow, odorless powder that decomposes at about 105 °C. It gives off approx. 180 cm^3/g nitrogen under normal conditions. In compounds decomposition begins already at 80 °C. Since it occurs very rapidly in the concentrated form, danger of explosion exists. For that reason production has recently ceased, since even the diluted product was not safe enough. The material is particularly suitable in Q-compounds.

4.6.4 Comparison of Important Blowing Agents

[1.14]

A comparison of suitable blowing agents for important fields of application is given in Table 4.7.

Table 4.7: Survey of Suitable Blowing Agents for Various Applications

	Benzene-sulfo-hydrazide	Benzene-1,3-disulfo-hydrazide	Dinitrosopen-tamethylene-tetramine	Azodicarbon amide
Blowing Process				
Sponge Rubber	very good	good	possible	not recommended
Slipper Soles	very good	very good	possible	not recommended
Closed Cell Sponge Rubber[1)]	very good	good	possible	possible good
Closed Cell Sponge Rubber[2)]	possible	good	very good	very good
Expansion Process				
Soft Micro Cellular Rubber[3)]	very good	good	very good	very good
Porous Soles	very good	good	very good	good
Hard Micro Cellular Soles[4)]	possible	very good	very good	good
Cellular Ebonite	very good	good	good	possible
Type of Pores	very fine to invisible	visible pores	very fine to invisible	very fine

[1)] Open air heating
[2)] Vulcanized in salt bath or in UHF-system
[3)] Density app. 0.3 g/cm^3
[4)] Leatherlike

4.7 Adhesion and Adhesives

[4.577–4.583]

4.7.1 Rubber-Textile-Adhesion and Adhesives

[4.584–4.628]

4.7.1.1 Rubber-Fabric-Adhesion

In many products of the rubber industry, like tires, V-belts, conveyor-belts, hose, etc. stringent requirements are set for the bond between rubber and the incorporated load carrier made of half- resp. fully synthetic fibers, metals and recently also glass fibers. That is particularly important in the dynamic use of the articles. It is only by using special adhesives that sufficient adhesion is obtained.

In some cases it is sufficient if one assures intimate contact between the surface of woven goods and the rubber as well as an extensive mechanical interaction of the rubber in the texture of the woven material. It is especially easy to achieve this with cotton, for which an impregnation with a dissolved rubber mixture or with latex can give satisfactory adhesion for many applications.

Up to the 1940s, cotton was the only load carrier available, until the half-synthetic rayon was developed. In the 1950s the polyamide Nylon 6.6 (USA) and Nylon 6 (Perlon, Germany) were added and in the beginning of the 1960s, polyester (USA, at first Good-year). In the beginning of the 1970s, Du Pont developed the aramide-fiber Kevlar (earlier Fiber B). Also a small amount of polyvinylalcohol fiber is in use (preferred in Japan). In addition to the organic fibers, Michelin developed already in 1936 steel cord and in the 1960s, Owens-Corning glassfiber as reinforcement members. All these half- or fully synthetic resp. inorganic fibers are monofilament yearns compared to cotton that is processed from spun fibers; the monofilaments do not have the capability to solidly anker rubber compounds. They have therefore a low affinity to the rubber compound and sufficient adhesion between textile (woven goods) and rubber can only be achieved by using special adhesives.

4.7.1.2 Adhesives Based on Resorcinol-Formaldehyde

[1.14]

Pre-impregnation. The resorcinol-formaldehyde resins, already developed in 1935, are the oldest adhesives to achieve rubber-textile adhesion; they are used in aqueous phase in conjunction with latex (RFL-Dip) and have found wide use [4.612]. With the use of new fibers and SR-types, RFL preparations were modified and improved. For example, for use with polyamide fibers a special latex based on butadiene-styrene-2-vinyl-pyridine (70:15:15 wt%), so called vinylpyridine latex (VP-latex) was developed [4.613]. It is today also used mixed with NR- resp. SBR-latex for rayon impregnation. For polyester resp. inorganic fibers, these aids are insufficient. They are partly in a pretreated form commercially available.

The RFL-dips are frequently prepared by the rubber fabricators from the single components. For example, the resorcinol and formaldehyde are mixed in a molar ratio of 1:1.5–2 in an alkaline medium. After approx. 6 hours storage at room temperature, the resin solution is added to NR-, SBR- or VP latex or mixtures of these latexes. A rather sensitive maturation process follows during 12–24 hour storage, that leads only by careful execusion to equal resin quality and equal adhesion levels in the final articles. It is particularly difficult to maintain constant temperature during the reaction and maturation, since the reaction of resorcinol and formaldehyde is a exothermic process. To avoid the problem, precondensed resorcinol-formaldehyde resins with optimum and constant degree of condensation are commercially available adhesives. They are also used in conjunction with latexes, especially in combination with NR-, SBR- and VP latexes [4.614]. On preparing the dipping solution, formaldehyde has to be added. A maturation process is not necessary. The resin content of the impregnating bath is related to the rubber content of the latex and is usually 20 wt%. The solids content of the RFL-dip has to be tailored to the textile, for rayon, for example, 10–15 wt%, for polyamide approx. 15–20 wt% and for polyester, 20 wt% or more.

By adding more adhesion promoting materials to the RFL-dip, adhesion of textile to rubber can be further improved. Among those, reaction products of triallylcyanurate (TAC) [4.615] resp. p-chlorophenol [4.616, 4.617] with resorcinol and formaldehyde should be mentioned.

The pre-treatment of cord- and fabrics with RFL dips necessitates special equipment for dip preparation as well as appropriate machinery for the coating and drying (see page 380f). Depending on the fabric used and the expected rubber compound, corresponding modified dipping (impregnating) processes are necessary. Special precautions have to be taken for storage of the impregnated fabric. The dip uptake of the RFL-dip adhesive should be for normal rayon up to 4–8 wt% of the textile weight. For heavier fabrics, dip pickup up to 15 wt% is necessary; for glass fiber cord it goes up to 25–30 wt%.

The pre-treatment of textiles plays still a considerable role in rubber factories that have a well functioning and already amortized dipping installation.

Adhesive Compounds. Many of the problems mentioned as well as the use of the dipping process disappear if the so called direct-adhesive process with adhesive compounds is used, which was developed by Bayer in 1965 [4.618] and independently by Degussa.

To make these adhesive rubber compounds, the addition of the following components to the rubber compound is necessary (RFK-system):

- Resorcinol
- Formaldehyde-Donor
- Silica reinforcing filler

As formaldehyde donors, hexamethylolmelamine ethyl ethers or hexamethylenetetramine are used. A combined action of all three components is absolutely necessary to achieve optimum adhesion. During the vulcanization process the resorcinol-formaldehyde resin is formed which acts as an adhesion promotor. The function of the active silica has up to now not been clarified. It certainly acts as a catalyst for the resin formation; here the humidity seems to play an important role.

A carefully balanced rubber compound composition is important with the RFK-system.

Primarily one deals with solid rubber compounds when using the RFK system. They can also be used in solution in an organic solvent for the pre-treatment of fabrics.

The process is universally applicable for all types of rubbers in combination with all commonly used textiles, for example cotton, rayon, polyamide, polyesters with special spin finishes (Diolen 1645, Trevira 715, Terylen 111H) including glass cord or glass fabric, as well as metals, especially steel cord (raw, brass or zinc coated steel cord). Even on bare steel (cut ends of brasserated steel) the RFK-system is active.

4.7.1.3 Impregnation with Isocyanates

[1.14]

If especially high adhesive strength is necessary, or if it is difficult to achieve sufficient adhesion with polyester fabrics, the fabrics can be impregnated with isocyanates that were developed in the 40's to 60's, for example triphenylmethanetriisocyanate as well as thionophosphoric acid-tris-(p-isocyanato-phenyl)-ester [4.618–4.622] (see page 261, 380). However, this does not only mean working with organic solvents instead of water, but also adjusting to the limited storage stability of the "dough" or impregnating solution (short pot-life). For that reason, the use of isocyanates in the rubber industry has steadily diminished. It is today only important in speciality applications, for example for impregnation of cords for open V-belts.

The triisocyanates mentioned above are especially active, while the diisocyanates, for example, diphenylmethane diisocyanate (MDI), are somewhat weaker.

4.7.1.4 Other Rubber-Fabric Adhesives

Since polyester and aramide fibers are difficult to adhere to rubber, they are commercially available pre-impregnated with a spin finish. One can use a two-bath dipping process. A pre-impregnation during fiber production, for example with phenol-blocked MDI in combination with a peroxide [4.623]. Similar esters with adhesive finish can then be treated with the RFL-dip described above [4.624]. Beside the phenolblocked MDI, epoxy resins are also used [4.625].

In order for glass cord to be finished "rubber friendly", a silane coating is put on the fiber. It is important for glass cord that each filament is separated completely and enclosed by the rubber matrix since they are destroyed by self abrasion [4.604, 4.626–4.628].

Steel cord is usually brass (see Section 4.7.2.2) or zinc coated before it is used, and becomes by that process "rubber-friendly".

All these pre-treated textiles can produce good adhesion by the use of RFL-dips or RFK systems. Later, rust can develop in steel cords. With steel cord, adhesion can be improved by Cobalt-salts, for example Co-naphthenate. Co-naphthenate containing compounds can be counted among the adhesive compounds for steel cord. With zinc coated steel cord, the addition of PbO can improve adhesion.

4.7.2 Rubber-Metal Adhesion and Adhesives

[4.629–4.674]

4.7.2.1 Rubber-Metal Adhesion

Because of the increased use of rubber in engineering elements of the automobile-, aircraft and machine industry, strong bonds between rubber and metal are needed. Since the adhesion of soft rubber to various metals is weak, good bonding agents for rubber-metal elements play an important role.

Earlier the problem was solved with an interlayer of hard rubber. Since a longer heating period is necessary because of the hard rubber, fewer applications with dynamic requirements and lower heat resistance of the bond, mostly other bonding agents are used today.

4.7.2.2 The Brass Process

[4.669]

One bonding agent [4.670] that gives a reasonably good bond between rubber and metal is a galvanic layer of brass. Advantageous is mainly the great stability of the bond toward heat and solvents. The process requires, compared to chemical adhesives, a large investment in processing machinery. The biggest difficulty with the brass process is the necessity to keep all variables in the galvanic bath constant. The brass process is advantageous, if the compound is applied to the metal part under high pressure (for example in the injection molding process). Under these conditions, chemical adhesives can be partially displaced and the adhesion be partly reduced. On steel cord for automobile tires steel cord with brass coating has proven very successful [4.652–4.668].

4.7.2.3 Chemical Adhesives

For many years mainly isocyanates, chlorinated rubber, cyclized rubber, rubber hydrochloride, epoxides, etc. were used that formed good adhesion between various rubber types and metals, without the brass layer.

For rubber-metal adhesion a clean preparation of the metallic surface is essential. This can be obtained by mechanical (for example sand blasting) or chemical treatment [4.635].

Isocyanates [1.14, 4.671, 4.672]. With isocyanates, especially triphenylmethanetriisocyanate and thionophosphoric acid-tris-(p-isocyanato-phenyl)-ester (see page 261) one obtains not only good adhesion, but the rubber to metal bond has especially good heat and swelling resistance. The great efficiency of the isocyanates keeps the cost of this process very low.

With isocyanates, compounds of almost all common types of NR and SR can give excellent adhesion to almost all customary heavy and light metals except bronze. Working with isocyanate does pose certain problems. The isocyanate coating is not only sensitive to moisture and steam but the composition of the rubber compound is important. Since isocyanates are very reactive, reaction with a number of rubber chemicals can lead to undesirable side reactions.

Halogenated Adhesives [4.641–4.673]. For these reasons adhesives with less problems have appeared in recent years that are based on compounded chlorinated and brominated NR- and SR-types. These adhesives consist of chlorinated rubber, rubber hydrochloride as well as polymerized 2-chlorobutadiene resp. 2,3-dichlorobutadiene that are after the polymerization sometimes further chlorinated. As crosslinking agents for example trinitrosobenzene and p-quinonedioxime with oxidizing agents are added, also fillers and solvents [4.674].

With these compounds one frequently gets more consistent results because of the fewer side reactions; the heat and swelling stability can however not be compared to that obtained with diisocyanates. When these materials are used, rusting below the adhesive bond has been observed.

Reaction Adhesives. For polar rubbers, like NBR and CR, reaction adhesives based on epoxides or polyurethanes can be applied. They are today of no importance.

4.7.3 Crosslinking Agents for Rubber Adhesive Solutions

[1.14, 4.675–4.688]

By adding crosslinking agents to rubber adhesive solutions one can improve the adhesive and cohesive strength of the adhesive film, as well as its heat and solvent stability, independent of the resins used.

As crosslinker one used to apply practically exclusively rapid vulcanization systems (auto-vulcanization). Because of the isocyanates, application of the vulcanizing systems has been greatly reduced. The cohesion is increased with these systems as well as increased adhesion to various materials is obtained. Adhesive solutions with crosslinkers have to be applied in two parts compared to blocked isocyanates or resin based solutions. Isocyanate containing adhesive solutions have proven themselves for adhesion to rubber, leather, textiles, wood, metal and plastics. These adhesives set up rapidly and make efficient work possible. They give high static and dynamic adhesion values, good heat stability and good swelling resistance with NBR and CR, even against PVC-plasticizers.

4.7.4 Influence of the Rubber Compound on Adhesion

When rubber is bonded to textiles, metals, rubber, plastics or other materials, the composition of the rubber compound is of importance in addition to the adhesive itself.

- Soft rubber compounds (for example below 50 shore A) are more difficult to bond than the hard ones.
- A possible exuding of plasticizers or bloom of sulfur, waxes or other products can lead to complete loss of adhesion.
- The vulcanization process is frequently also important in achieving a good bond [4.590, 4.653–4.665]. It is important to have a long scorch period, to give the bond time to form. Bond failure usually occurs for scorched rubber compounds. With strong under- or extreme over-vulcanization, a poor bond is frequently formed also [4.641].
- Certain antioxidants, like MBI, can weaken the bond or sometimes destroy it completely [4.590].
- All fillers can influence the bond [4.686, 4.687].
- Finally the rubber type also influences the bond [4.688]. Rubbers decrease in their "adhesiveness" in the following sequence: NBR > CR > NR > SBR > BR > EPDM > IIR.

For the production of rubber-metal parts it is also important that the mold design is such, that the flow field of the rubber compound does not move the adhesive. In the design of the parts, stress concentration in the adhesive interface should be avoided.

4.8 Latex Chemicals

[1.16]

Since latex is an aqueous colloidal system, water insoluble powders have to be added as dispersions because of their poor wetting ability; water insoluble liquids, have to be added as emulsions. Aside from the customary rubber chemicals, like sulfur, vulcanization accelerators, antioxidants, etc. the real latex-chemicals play an important role; they are dispersants, emulsifiers as well as stabilizers, crosslinking and foaming agents, foam stabilizers, defoaming agents, coagulants, conservation agents, etc.

These products will not be discussed in this book, because this deals only with solid rubber technology (see [1.16]).

4.9 Other Compounding Aids

4.9.1 Compound Components

4.9.1.1 Hardening Substances

[4.689–4.692]

For the production of various rubber goods, like soles, or porous soles, heels, floor coverings, etc., other additives beside inactive and active fillers as well as resins and high quantities of sulfur are used to increase the hardness of the vulcanizates. In the past, glue (perle glue) and shellac, experimentally also sodiumsilicate (waterglas) were added for that purpose. The hardening effect was often considerable. These materials were difficult to disperse.

To achieve a small increase in hardness, benzoic acid at the 3–6 phr level is used; in carbon black loaded compounds it softens the uncured compound and also acts as a weak scorch retarder. Styrene resins (see page 293) and phenoplasts (page 294) can also be considered. In NBR, the addition of PVC gives a hardening effect (page 294). Another way is the use of acrylate monomers, which have a softening effect on uncured NBR compounds and harden the vulcanizates (page 259).

4.9.1.2 Odor Improving Agents

[1.14]

In many cases, the typical rubber smell is undesirable, that is inherent in NR and SR and is increased by the rubber chemicals and vulcanization. It can be made weaker by prolonged storage of the rubber compound, tempering or steaming, but not completely removed. By adding odor-improving materials (odorants), the rubber odor can be made part of a new pleasant odor nuance or can be completely covered up (with a perfume). An odor nuance is for example leather odor for leather-like articles or textile odor for rubber sheetings.

4.9.1.3 Anti-Termite Agents

[4.693, 4.694]

With increasing industrialization of tropical and subtropical countries, the termite proof requirement for rubber articles like cables, gaskets, conveyor belts etc. has become more important. The pesticides to be used have to be vulcanizationproof. Certain phosphoric acid esters should be considered.

4.9.1.4 Antimicrobials

[4.695]

Antimicrobial properties may be important for articles like rubber footware, gloves, hygienic rubber goods, etc., to prevent spreading of pathogens or severely restrict them. The same goes for rubber articles that are used in homes or hospitals.

Suitable antimicrobial products are for example salicylaldehyde and dihydroxy-dichlorodiphenylmethane derivatives. Zinc dithiocarbamates and thiurams have a mild activity. Furylbenzimidazole which gave excellent experimental test results, proved too allergenic for general use.

4.9.2 Materials for Special Effects

[1.14]

4.9.2.1 Mold Release Agents

Problem. Press vulcanized articles tend to adhere and stick more or less tenatiously to the molds. When they are taken out of the mold, this can lead to damage of the vulcanized article and of the vulcanization mold. Suitable mold release agents are important for an efficient production of molded products. Through their use scrap is reduced, the appearance of the products improved, mold cleaning cost reduced and mold life increased.

Requirements for mold release agents vary with the production process, type and shape of the article as well as its posttreatment. An ideal release agent that would be optimal for all needs has not been developed yet.

Soap Solutions. For many years, simple soap solutions were used - and in some cases they are used even today. Some of the newer synthetic soap substitutes have excellent release properties. These products have the disadvantage that they slowly decompose at vulcanization temperatures causing impurities on the mold surface, that make a clean molding process impossible. For that reason a frequent cleaning of the mold is necessary that is not only time consuming, expensive and very difficult with complex molds, but frequently leads to premature scrapping of the sometimes very expensive molds.

Silicone Mold Release Agents [1.14]. By the use of suitable silicone mold release agents the disadvantages of the soap solutions can be avoided. Release agents that are based on silicone do not decompose because of their high temperature stability, and the vulcanized parts have a high gloss and handle smoothly. The silicones, in contrast to the purely organic release agents, make a subsequent coating, glueing or welding of the surfaces more difficult. Furthermore, the release capacity of silicone release agents is somewhat weaker than that of organic agents.

Combinations of Silicones and Soaps. An especially favorable compromise can be achieved by combining silicone oil emulsions with good organic release agents.

Fluorinated Polymers. Recently, fluorinated polymers have been available in spray cans, that are sprayed on the molds and then heat fused. Because of the evaporating solvents (toxicology) and difficulties in repairing damaged coatings, these release agents are not much used.

4.9.2.2 Powders and Release Agents to Avoid Adhesion Between Unvulcanized Compounds

To a large extent talcum, Zn-stearate and other powders are used to avoid adhesion between rubber compound intermediates, for example extruded articles. They have the known disadvantage of dust irritation. One has gone over largely to liquid release agents. A certain success has been obtained with soap solutions, especially with hard, heavily loaded compounds. With medium hard and especially soft compounds the soaps penetrate into the surface and the surface is depleated of the coating. By adding high molecular weight materials that are insoluble, for example cellulose derivatives to such a soap solution, a separating film remains on the surface that prevents sticking together during storage or vulcanization.

Zinc stearate as separation agent (powder agent) has the remarkable property that it prevents undesirable sticking of unvulcanized parts at room temperature, but does not avoid welding during vulcanization.

The release agents mentioned above or solutions of silicone oils, 0.5–1 wt% in benzene, carbon tetrachloride or ethylacetate are often used for example for easy removal of dipped articles from the forms. The vulcanizates obtain a uniform glass and improved appearance.

4.9.2.3 Products for Surface Treatment

[1.14, 4.696]

Problem. One obtains surface effects on molded rubber articles by the use of polished molds or mold coatings, that are suitable for many applications; these effects are hard to obtain on open air cured articles without a lacquer coating.

If one aims to improve the appearance of such articles or if molded goods have special gloss or "feel" requirement, it is necessary to give them a special coating, i.e. apply a lacquer coating.

Valvet looking surfaces can be established by the use of factice (see page 305).

Drying Oils. Suitable varnishes are for example drying oils, that were blown or sulfur treated. Sometimes they contain alkyl- and other synthetic resins. As base, polymerized vinylcompounds can be used, like polyacrylates, polyvinylacetate as well as plasticized ureaformaldehyde resins. Depending on its property, the lacquer is applied to the vulcanizate or to the rubber compound and the coating is formed by air drying or during vulcanization. The latter method has the advantage that a dry (non-sticky) film is formed that allows immediate packaging of the rubber product.

Polyurethane lacquer. Recently, lacquers made from polyesters and polyisocyanates have frequently been used. With stabilized isocyanates a storage stable solution is formed, that is applied to rubber compounds. Only at the vulcanization temperature free isocyanate groups are formed that react with the polyester to give an elastic, tightly adhering coating. Solutions with limited potlife made from non-stabilized polyisocyanates and polyesters can be applied to the rubber compound or to the vulcanizates as air drying lacquers [4.696].

Influence of Compound Composition on Lacquer Coatings. Rubber compound and lacquer have to be tuned to each other. The rubber compound must not contain paraffine, mineral oil or large amounts of plasticizers, that can concentrate at the surface and thereby decrease adhesion. The lacquer must not have components that interfere with the vulcanization. If the lacquer is applied to the vulcanizate and the vulcanization is taking place in a mold, mold coating, silicone oils, soaps or similar materials cannot be used. On the other hand a powder coating of zinc stearate interferes little with the drying or appearance of the lacquer coating. One has to be cautious with the use of antioxidants. Discoloring compounds dissolve in the solvents of the lacquer and diffuse in the film and discolor on light exposure.

Suitable lacquers do not only improve the looks of the rubber product, but protect it also from weathering. Lacquers can be used with compounds for solid rubber goods as well as latex products.

4.9.2.4 Reclaim Agents

[4.697]

In the production of rubber articles, rubber waste (scrap) is generated that can be re-used after a reclaim process. The importance of reclaim or waste depends closely on the respective price of new rubber.

When vulcanizates are heated for prolonged periods to about 200–250 °C they depolymerize and a material called reclaimed rubber is produced that is liquid-to paste-like and can be re-used. This process can be accelerated by the addition of chemicals. These materials are called reclaiming agents.

The inorganic materials used in reclaiming production waste, like caustic soda or sulfuric acid, are not really reclaiming agents, they are added in order to destroy textile fabrics contained in the waste product.

Reclaiming agents are thiophenols and disulfides that are also mastication aids (see page 220ff), also dixylyldisulfide or alkylated phenylsulfides which are preferred for reclaiming SR vulcanizates.

So called reclaiming oils, mineral oils in combination with resins as well as highly unsaturated oils of different origin and composition are used alone as well as in combination with the reclaiming agents just mentioned.

The higher the reclaiming temperature the stronger becomes the undesirable cyclization, especially with SR. Most SR vulcanizates have to be reclaimed at the lowest possible temperature in the shortest possible time.

The type of chemical additives depends on the type of rubber. For NR waste, practically all reclaiming agents can be used; they get blended at the 1–2 wt% level with ground waste and about 2 wt% resin oil.

For reclamation of SBR and NBR vulcanizates, substituted thiophenols o,o′-dibenzamidodiphenylsulfide or disulfides of isomeric xylylmercaptans are indicated. The ground waste is mixed with approx. 2–2.5 wt% of the reclaiming chemical, 3–5 wt% colophony and 5–10 wt% plasticizer, and then heated.

4.9.2.5 Solvents

[4.698]

For the solvents useful in the rubber industry the following properties are required aside from being good solvents (see Table 4.8) for the gum rubber and its compounds: favorable toxicological properties, possibility of recovery, boiling within certain boiling range limits and low flammability. Solubility of crude (uncured) rubber depends on how much it was pre-treated mechanically, i.e. by working on a rubber mill or chemically, for example by the use of a plasticizer. Similar considerations are valid for the respective rubber components (preparation of solutions see page 383).

Depending on concentration and degree of plastication of the rubber, solution of different viscosities are obtained. Table 4.8 gives information about properties of some solvents used in the rubber industry.

Table 4.8: Suitability of Some Solvents for Various Applications

	Bp °C	NR	SBR	IIR	CR	NBR[1]	NBR[2]	TM
p-Hexane	69	gs	gs	gs	ml	q	n	n
Gasoline	80–110	gs	gs	gs	ml	q	n	n
Crude Petroleum	160–250	gs	gs	gs	q	n	n	n
Mineral oil	–	l	mq	mq	q	l	n	n
Terpentine	160–180	gs	gs	gs	gs	l	n	l
Benzene	80	gs	gs	gs	gs	gs	q	gs
Toluene	111	gs	gs	gs	gs	gs	q	l
Xylene	138–142	gs	gs	gs	gs	gs	q	l
Styrene	145–146	gs	gs	gs	gs	gs	gs	gs
Tetraline	205	gs	gs	ms	gs	ms	mq	gs
Chloroform	61	gs	gs	ms	gs	gs	gs	gs
Carbontetrachloride	77	gs	gs	gs	gs	l	n	n
Chlorobenzene	132	gs	gs	ms	gs	gs	gs	ms
Diethylether	35	gs	gs	q	l	n	n	n
Dibenzylether	295–298	ms	ms	n	ms	ms	ms	gs
Acetone	56	n	n	n	l	gs	gs	n
Ethylmethylketone	81	l	l	q	gs	gs	gs	n
Cyclohexanone	155–157	gs	gs	q	gs	gs	ms	gs
Ethylacetate	77	l	n	l	ms	ms	ms	n
Butylacetate	127	gs	gs	l	gs	ms	ms	n
Dibutylphthalate	339	q	mq	n	ms	mq	qs	n
Dioctylphthalate	216	mq	mq	n	gs	q	n	n
Tricresylphosphate	295	n	l	n	ml	mq	mq	n
Pyridine	115	q	gs	n	gs	gs	gs	ms
Carbondisulfide	16	gs	gs	ms	gs	gs	ml	q

[1] With 28% acrylonitrile [2] With 38% acrylonitrile

gs = good solubility
l = limited swelling or softening
q = gel formation or unlimited swelling
n = no influence
ms = medium solubility
ml = medium limited swelling or softening
mq = medium gel formation or unlimited swelling

4.10 References on Rubber Chemicals and Additives

4.10.1 General References

Books, Lists of Substances (see also [1.2, 1.3, 1.6, 1.14, 1.17 1.18, 1.21])

[4.1] *Abele, M. et al.:* Kautschuk-Chemikalien und Zuschlagstoffe. In: Ullmanns Encyclopädie der technischen Chemie. Verlag Chemie, Weinheim, 14th Ed., Vol. 13, 1977, p. 640 ff.

[4.2] *Alphen, J. v.:* Rubber Chemicals. Ed.: *Turnhout, C. M. van,* TNO, Delft, D. Reidel Publ. Co., Dordrecht, Holland; Boston, MA, 1973.

[4.3] *Babbit, R. O.:* The Vanderbilt Handbook, Eigenverlag, 1978, p. 337–432.

[4.4] *Häggström, B., Kinnhagen, S., Laurell, L. G.:* Nordisk Gummiteknisk Handbook. Sveriges Gummitekniska Förening, Sunde, 7th Ed., 1983.

[4.5] ...: Rubber Blue Book. Publ.: Rubber World.

[4.6] ...: Rubber Red Book. Publ.: Elastomerics.

Trends

[4.7] *Buswell, A. G.:* Developments in Rubber Chemicals. Progr. Rubb. Technol. *40* (1977), p. 97.

[4.8] *Hofmann, W.:* Kautschuk-Zusatzstoffe in Elastomeren für technische Anwendungen, Kunstst. *74* (1984), p. 640.

[4.9] *Kaiser, E.:* Rubber Chemicals – Alive and Hopefully well Through 2000 AD. Paper 62, 128th ACS-Conf., Rubber Div., Oct. 1–4, 1985, Cleveland, OH.

[4.10] *Kempermann, Th.:* Trends im Einsatz von Kautschuk-Chemikalien in der Gummi-Industrie. KGK *31* (1978), p. 234.

[4.11] *Pyne, I. R.:* Compounding Ingredients. Progr. Rubb. Technol. *35* (1971), p. 54.

General References on Compounding (see also [1.9, 1.23, 2.8, 2.19, 2.66])

[4.12] ...: Rubber Compounding. 10 Lectures at the 121th Akron Rubber Group (1984), Annual Lectures Series, Publ.: ACS, Rubber Div.

Quality Assessment

[4.13] *Bertram, H. H.:* Qualität – Chancen und Grenzen – aus Sicht eines Rohstoffherstellers. KGK *38* (1985), p. 1093.

[4.14] *Bertrand, G.:* Modern Scientific Methods of Tests for the Quality of Rubber Chemicals. Paper 65, 128th ACS-Conf., Rubber Div., Oct. 1–4, 1985, Cleveland, OH.

[4.15] *Kempermann, Th., Lahousse, G.:* Kritische Betrachtungen zur Qualitätsbewertung und -kontrolle von Kautschuk-Chemikalien. KGK *39* (1986), p. 697.

[4.15 a] *Mersch, F.:* Status of ISO Normalisation Work for Raw Materials Used in the Rubber Industry. KGK *39* (1986), p. 528– 531.

[4.15 b] *Paffrath, W., Kempermann, Th.:* Qualitätskontrolle von Kautschuk-Chemikalien mit Hilfe der elektronischen Datenverarbeitung. KGK *36* (1983), p. 755.

[4.15 c] *Roebuck, H.:* Rationalization and Cost Saving for Quality Assurance. Rubbercon '87, June 1–5, 1987, Harrogate, GB, Proceed. A 27.

Toxicity and Dust-Free

[4.15 d] *Garnett, A. A.:* A Review of Predispersed Rubber Chemicals, Rubbercon '87, June 1–5, Harrogate, GB, Proceed. B 19.

[4.15 e] *Gupta, H.:* Nitrosoamine in der Gummiindustrie – Gefahr und Möglichkeiten der Vermeidung. GAK *39* (1986), pp. 6–7.

[4.16] *Hofmann, W.:* Gummiallergien. Bayer-Mitt. f. d. Gummi-Ind. *46* (1972), p. 39, Druckschrift der Bayer AG.

[4.17] *Hofmann, W.:* Ökologie in der Gummiindustrie. KGK *27* (1974), p. 487.

[4.18] *Hofmann, W.:* Umweltschutzprobleme bei der Anwendung von Synthesekautschuk und Kautschukchemikalien. GAK *27* (1974), pp. 624–632, 830–840.

[4.18 a] *Lohwasser, H.:* Nitrosamine in der Gummiindustrie. GAK *39* (1986), pp. 385–391.
[4.19] *Löser, E., Lohwasser, H., Schön, N.:* Toxikologische und arbeitshygienische Gesichtspunkte beim Umgang mit Kautschukchemikalien und polymeren Rohstoffen unter Berücksichtigung bestehender gesetzlicher Regelungen. ikt '80, Sept. 24–26, 1980, Nürnberg, West Germany.
[4.19 a] *Möbius, K.:* Formaldehyd-Report löst neue Debatte über Krebsrisiko aus. GAK *39* (1986), pp. 392–393.
[4.20] *Munn, A.:* Bladder Cancer and Carcinogenic Impurities in Rubber Additives. Rubber Ind. *8* Nr. 1 (1974), p. 19.
[4.21] *Schultheiß, E.:* Gummi und Ekzem. Württ. Verlag Editio Cantor, Aulendorf, 1959.
[4.22] *Spivey, A. M.:* Same Health Considerations in the Use of Chemicals in the Rubber Industry, SRC 79, April 2–3, 1979, Kopenhagen.
[4.23] ...: Manual about Toxicity and Safe Handling of Rubber Chemicals. Code of Praxis, Ed.: BRMRA, Birmingham, 1978.
[4.24] *Ehrend, H., Morche, K.:* Polymergebundene Kautschukchemikalien – ein Beitrag zur Auswahl optimaler Bindersysteme. KGK *38* (1985), p. 499.
[4.25] *Hofmann, W.:* Vermeidung der Staubemission in der Gummiindustrie. GAK *29* (1976), p. 852.
[4.26] *Magg, H., Kempermann, Th.:* Moderne Lieferformen von Kautschuk-Chemikalien. KGK *35* (1982), p. 1039.
[4.27] *Rijnders, R. F. R. T., Katzanevas, A.:* The Evolution of Various Presentation Forms of Rubber Accelerators. Int. Polym. sci. Technol. *6* No. 11 (1979), p. TI 84; GAK *32* (1979), p. 309.
[4.28] *Schuette, W., Ehrend, H., Morche, K.:* Polymer Bound Rubber Chemicals – some Aspects of Optimized Polymer Binders, Paper 31, 124th ACS-Conf., Rubber Div., Oct. 25–28, 1983, Houston, TX.
[4.28 a] ...: WTR-Rubber Chemicals, Safety Data and Handling Precautions, Ed.: WTR, Brussels.

4.10.2 References on Mastication and Mastication Aids

(see also [2.82, 2.121, 2.122])

[4.29] *Bristow, G. M.:* Mastication of Elastomers. Trans. IRI *38* (1962), pp. 29, 104.
[4.30] *Fries, H., Pandit, R. R.:* Mastication of Rubber. RCT *55* (1982), p. 309.
[4.31] *Kempermann, Th.:* Mechanischer und thermischer Abbau von Kautschuk. GAK *27* (1974), p. 566.
[4.32] *Lober, F.:* Peptisations- und Mastikationsmittel. In: *Boström, S.* (ed.): Kautschuk-Handbuch. Verlag Berliner Union, Stuttgart, Vol. 4, 1961, pp. 385–394.
[4.33] *Morton, M., Piirma, I., Stein, R. J., Meier, I. F.:* Cold Mastication of Diene Elastomers. 4th Rubb. Technol. Conf., May 1962, London; Pre print..
[4.34] *Redetzky, W.:* Developments in Peptizing Agents. IRI-Conf. on Recent Developments in Rubber Compounding. Manchester, 1969, Proc. p. 5.
[4.35] *Schneider, P.:* Mastizieren und Regenerieren mit chemischen Hilfsmitteln. Kautschuk u. Gummi *6* (1953), pp. WT 21, WT 48.
[4.36] *Stöcklin, P.:* Über den thermischen Abbau von Buna. Kautschuk *15* (1939), p. 1.

4.10.3 References on Vulcanization and Vulcanization Chemicals

4.10.3.1 References on Vulcanization

General References (see also [1.7, 1.12, 1.13, 2.3, 3.264])

[4.37] *Bateman, L., Moore, C. G.:* in: *N. Kharash (Ed.):* Organic Sulfur Compounds. Pergamon Press, New York, 1961.
[4.38] *Campbell, R. H., Wise, R. H.:* Vulcanization. RCT *37* (1964), pp. 635–649; 650–667.
[4.39] *Coran, A. Y.:* Vulcanization. RCT *37* (1964), pp. 668–672; 673–678; 679–688; 689–697; *38* (1965), pp. 1–14.

[4.39 a] *Coran, A. Y.:* New Curing System Components. IRC '86, June 2-6, 1986, Göteborg, Sweden, Proceed, p. 387.

[4.40] *Coran, A. Y.:* Science and Technology of Rubber. VII. Vulcanization, Nippon Gomu Kyokaishi; *56* (1983), pp. 232-257.

[4.40 a] *Kempermann, Th.:* Schwefelfreie Vulkanisationssysteme für Dienkautschuke - eine Übersicht. KGK *40* (1987), pp. 741-751.

[4.41] *Kirkham, M. C.:* Current Status of Elastomer Vulcanization. Progr. Rubb. Technol. *41* (1978), pp. 61-95.

[4.42] *Kuan, Th.:* Vulcanizate, Vulcanizate Structure and Properties - An Overview. Paper 51, 123rd ACS-Conf., Rubber-Div., May 10-12, 1983, Toronto, Ont.

[4.43] *Lautenschläger, F. K., Edwards, K.:* Model Compound Vulcanization. Part 1-5, RCT *52* (1979), pp. 213-231; 1030-1043; 1044-1049; 1050-1056; *53* (1980), pp. 27-47.

[4.44] *Lautenschläger, F. K., Zeeman, P.:* Simulierte Vulkanisation an Modellolefinen. GAK *33* (1980), p. 386.

[4.45] *Mark, H. F., Gaylord, N. G., Bikales, N. M. (Eds.):* Vulcanization, in Encyclopedia of Polymer Science and Technology, Interscience Publ., New York, Vol. 14, 1972, p. 740.

[4.45] *Matoba, Y:* Novel Crosslinking Systems for ECO, dkt '88, July 4-7, Nürnberg, West Germany.

[4.46] *Morita, E., Sullivan, A. B., Coran, A. Y.:* Vulcanization Chemistry, RCT *58* (1985), p. 284.

[4.47] *Porter, M.:* in: *Kharash, N. (Ed.):* Mechanisms of Reactions of Sulfur Compounds. Interscience Res. Foundation, Santa Monica, CA, Vol. 3, 1968, p. 145.

[4.48] *Porter, M.:* in: *Tobolski, A. V. (Ed.):* The Chemistry of Sulfides. Interscience Publ., New York, 1968, p. 165.

[4.49] *Scheele, W.:* Chemismus und Mechanismus der Vulkanisation. KGK *15* (1962), p. WT 482; Angew. Makromol. Chem. *16/17* (1971), p. 128; RCT *34* (1961), p. 1306.

[4.50] *Scheele, W., Bielstein, G.:* Die Vulkanisation des Naturkautschuks. Kautschuk u. Gummi *8* (1955), p. WT 251; see also p. WT 27.

[4.51] *Schnecko H.:* Bedeutung und Aufbaumöglichkeiten von Netzwerken. KGK *32* (1979), p. 297.

[4.52] *Shershnev, V. A.:* Vulcanization of Polydiene and Other Hydrocarbon Elastomers. RCT *55* (1982), p. 537, Review.

[4.52 a] *Sommer, J. G.:* Stabilized Curative Blends for Rubber, Rubbercon '87, June 1-5, 1987, Harrogate, GB, Proceed. B 22.

[4.52 b] *Vergnaud, J. M.:* Application of Modelling of Rubber Vulcanization by Taking into Account the Heat Evolved from Cure Reaction and Heat Transfer, Rubbercon '87, June 1-5, 1987, Harrogate, GB, Proceed. A 57.

[4.53] *Vukov, R., Wilson, G. J.:* Factors Affecting State of Cure in Low Unsaturated Elastomers. Gordon Res. Conf. Elastomers, Juli 18-22, 1983, New London, NH.

Kinetics, Vulcanization Structures, Influence on Properties

[4.54] *Amin, M., Osman, H., Abdel-Bary, E. M.:* Effect of Some Vulcanizing Systems on the Electrical Conductivity of Butadiene Rubber. KGK *35* (1982), p. 1049.

[4.55] *Campbell, D. C.:* Structural Characterisation of Vulcanizates. J. Appl. Polym. Sci. *14* (1970), p. 1400; *15* (1971), p. 2661; RCT *44* (1971), p. 771; *45* (1972), p. 1366.

[4.56] *Eckelmann, W., Reichenbach, D., Sempf, H.:* Über die Abhängigkeit der Eigenschaften von Vulkanisaten von Vulkanisationszeit und -temperatur. KGK *22* (1969), p. 5.

[4.57] *Hofmann, W.:* Einfluß der Vernetzungsart und -dichte auf das Eigenschaftsbild makromolekularer Stoffe. Paper at the Techn. Universität Berlin, Feb. 9, 1978.

[4.58] *Jurkowski, B., Kubis, J.:* The Determination of Kinetic Constants of the Crosslinking Reactions of the Rubber Compounds. KGK *38* (1985), p. 515.

[4.59] *Kempermann, Th.:* The Relationship between Heat Build-up and the Chemistry of Crosslinking Systems. RCT *55* (1982), p. 391.

[4.60] *Lee, T. C. P., Morrell, S. H.:* Network Changes in Nitrile Rubber at Elevated Temperatures. J. of IRI, Febr. 1963, p.27; RCT *46* (1973), p.483.

[4.61] *Mark, J. E.:* Experimental Determinations of Crosslinking Densities. RCT *55* (1982), p.762, Review.

[4.62] *Pal, D., Basu, D. K.:* Some Kinetic Aspects of Crosslinking of Natural Rubber. KGK *36* (1983), p.358.

[4.63] *Parks, C. R., Parker, D. C., Chapman, D. A., Cox, W. L.:* Pendent Accelerator Groups in Rubber Vulcanizates. RCT *43* (1970), p.572–587; *45* (1972), p.467.

[4.64] *Reichenbach, D., Eckelmann, W.:* Zur Konzentrationsabhängigkeit der Vulkanisation. KGK *24* (1971), p.443.

[4.65] *Saville, B., Watson, A. A.:* Structural Characterization of Sulfur Vulcanized Networks, RCT *40* (1967), p.100–148, Review.

[4.66] *Sharaf, M. A., Mark, J. E.:* The Effects of Crosslinking and Strain on the Glass Transition Temperature of Polymer Network. RCT *53* (1980), p.982.

Reversion

[4.67] *Aarts, A., Baker, K. M.:* The Application of New Analytical Techniques to Study the Vulcanization and Degradation of Rubber Compounds, KGK *37* (1984), p.497.

[4.68] *Baker, C. S. L., Swift, L.:* Vulkanisation von Kautschuk bei höheren Temperaturen. GAK *31* (1978), p.712.

[4.69] *Bhowmick, A. K., De, S. K.:* Effect of Curing Temperatures and Curing System on Network Structure and Technical Properties of Polybutadiene and Styrene-Butadiene Rubber. J. Appl. Polym. Sci. *26* (1980), pp.529–541.

[4.70] *Chen, C. H., Koening, J. L., Shelton, J. R., Colling, E. A.:* Characterization of the Reversion Process in Accelerated Sulfur Curing of Natural Rubber. RCT *54* (1981), p.734.

[4.71] *Davies, K. M.:* Practical Consequences of Vulcanizate Structure Changes at High Cure Temperatures. Plast. Rubb. Proc., Sept. 1977, p.87.

[4.72] *Dogadkin, B., Tarazova, Z. N.:* Vulcanization Structures and their Influence on the Heat Resistance and Fatigue of Rubber. RCT *27* (1954), p.883.

[4.73] *Kempermann, Th., Eholzer, U.:* Physikalische und chemische Aspekte der Reversion. KGK *34* (1981), p.722.

[4.74] *Kempermann, Th., Eholzer, U.:* Improvements of Reversion Resistance Through Post-Vulcanization Accelerator Systems, Paper 46, 128th ACS-Conf., Rubber Div., Oct. 1–4, 1985, Cleveland, OH.; see also: *Eholzer, U., Kempermann, Th.:* Verbesserung der Reversionsbeständigkeit durch nachvernetzende Beschleunigungssysteme. ikt '85, June 24–27, 1985, Stuttgart, KGK *38* (1985), p.710.

[4.75] *Morrison, N. J., Porter, M.:* Temperature of Intermediates and Crosslinks in Sulfur Vulcanization. RCT *57* (1984), p.63.

[4.76] *Soos, I., Nagy, K.:* Wirkung der Beschleunigung auf den Naturkautschuk-Abbau bei verschiedenen Temperaturen, GAK *33* (1980), p.608.

[4.77] *Wolff, S.:* A New Development for Reversion Stable Sulfur-Cured NR-Compounds, IRC 79, Oct. 3–6, 1979, Venice, Italy, Proc. p.1043.

4.10.3.2 References on Sulfur and Sulfur Donors

Insoluble Sulfur

[4.78] *Kearnan, J. E.:* Rheograph Studies of High Stability in Soluble Sulfur, Paper 113, 128th ACS-Conf., Rubber Div., Oct. 1–4, 1985, Cleveland, OH.

Sulfur Donors

[4.79] *Ascroft, K., Robinson, K. J., Stuckey, J. E.:* A Comparison Study of Vulcanization of Natural Rubber with Various Sulphur-Donor Systems. J. of IRI *3* (1969), p.159.

[4.80] *Bhowmick, A. K., De, S. E.:* Dithiodimorpholine-based Accelerator System in Tire Tread Compounds for High Temperature Vulcanization. RCT *52* (1979), p.985.

[4.81] *Bhowmick, A.K., De, S.E.:* Kinetics of Crosslinking and Network Changes in Natural Rubber Vulcanizates with a Dithiodimorpholine-based Accelerator System. RCT *53* (1980), p.1015.
[4.82] *Hofmann, W.:* Wirkung und Anwendung von Dithiodimorpholin (DTDM) als Schwefelspender. GAK *36* (1983), p.602; *37* (1984), pp.110, 179.
[4.83] *Hofmann, W.:* Deovulc M. DOG-Kontakt 28, Druckschrift der D.O.G. Deutsche Oelfabrik, Hamburg, West Germany, 1983.
[4.84] *Hoffmann, R.F., Schultheiß, J.J.:* Polysulfidic Polymers as Rubber Modifiers. Thiocol Techn. Report, October 1976.
[4.85] *Kempermann, Th., Eholzer, U.:* Einsatzmöglichkeiten von Schwefelspendern. GAK *26* (1973), p.272.
[4.86] *Lawrence, J.P.:* Efficient and Semi-Efficient Vulcanization Systems. Paper at the 110th ACS-Conf., Rubber Div., Oct.5-8, 1976, San Francisco, CA.
[4.87] *Lawrence, J.P.:* N-(Aminothio)imide Cure Modifiers. RCT *49* (1976), p.333.
[4.88] *Lloyd, D.G.:* Developments in Accelerators and Retarders. J.of PRI *38* (1975), p.77.
[4.89] *Mitra, A., Das, C.K., Willns, W:* Benzothiazyldithiomorpholide Vulcanization of the Natural Rubber at Elevated Temperatures. KGK *36* (1983), p.103.
[4.90] *Morita, E.:* S-N-Compounds as Delayed Action Chemicals in Vulcanization. RCT *53* (1980), pp.393-437, Review; pp. 1013.
[4.91] *Rodger, E.R.:* Vulkanisationssysteme. GAK *34* (1981), pp.124, 300.
[4.92] *Skinner, T.S.: Watson, A.A.:* EV-Systems for NR. Rubber Age *99* No.11 (1967), p.76; RCT *42* (1969), p.404.
[4.93] ...: Sulfasan R to Eliminate Bloom. Techn. Note 103/73, Bulletin of Monsanto.
[4.94] ...: EV and Semi-EV Curing Systems. Techn. Bulletin O/RC-8A, Bulletin of Monsanto.

4.10.3.3 Vulcanization Accelerators

General References (see also [1.12-1.14, 2.66, 2.108, 2.109, 3.102, 3.103, 3.105, 3.219-3.223, 3.301, 3.361, 3.429, 3.430, 3.434, 3.436, 3.438, 3.544, 3.545, 4.79-4.94])

[4.95] *Blokh, G.A.:* Organic Accelerators and Curing Systems for Elastomers. RAPRA-Transl. 1981.
[4.96] *Dibbo, A., Hammersley, D.A.:* Beschleunigungsysteme für wirtschaftliche Fertigung. KGK *22* (1969), p.379.
[4.97] *Eholzer, U., Kempermann, Th.:* Reaktions- und Folgeprodukte beim Einsatz von Beschleunigern. GAK *36* (1983), p.470.
[4.98] *Eholzer, U., Kempermann, Th., Morche, K.:* Schnellheizende Vulkanisationssysteme für EN-EPDM-Kautschuke. GAK *28* (1975), p. 646.
[4.99] *Endstra, W., Seeberger, W.:* Vernetzung von Polymeren - Beschleuniger oder Peroxide. Paper at the DKG-Bezirksgruppentagung, Nov.7, 1985, Hamburg, West Germany.
[4.100] *Hofmann, W.:* Vulkanisationshilfsmittel. In: *Boström, S. (Ed.):* Kautschuk-Handbuch. Verlag Berliner Union, Stuttgart, Vol.4, 1961, pp.281-352.
[4.101] *Kempermann, Th.:* Zusammenhang zwischen Konstitution und Wirkung bei Beschleunigern. GAK *30* (1977), pp.776-787; 868-877; *31* (1978), pp.247-258.
[4.102] *Kempermann, Th.:* Neue rasch vernetzende Vulkanisationssysteme für wärmebeständige Gummiartikel. KGK *20* (1967), p.126.
[4.103] *Kempermann, Th., Redetzky, W.:* Synergistische Effekte bei Vulkanisationsbeschleunigern. KGK *22* (1969), p.706.
[4.104] *Mathur, R.P., Mitra, A., Ghoshal, P.K., Das, C.K.:* The Effect of Binary Accelerator-Systems on the Sulfuration of Natural Rubber. KGK *36* (1983), p.1067.
[4.105] *Morita, E., Bonstany, K., D'Amico, J.J., Sullivan, A.B.:* Rubber Chemicals from Cyclic Amines. RCT *46* (1973), p.67.
[4.106] *Morita, E.:* Correlation Analysis of Curing Agents. RCT *57* (1984), p.744.

[4.107] *Parts, W. W.:* Vulcanization, its Activation and Acceleration. Paper at the 122nd Conf., Rubber Div., Oct. 5–7, 1982, Chicago, MI.
[4.108] *Saville, R. W.:* The Reactions of Amines and Sulfur with Olefins. J. Chem. Soc. 1958, p. 12 880; RCT *32* (1959), p. 577.
[4.109] *Sibley, R. L.:* Organic Sulfides as Vulcanizing Agents for Rubber. RCT *24* (1951), p. 211.

Migration and Distribution of Vulcanization Chemicals (see also [3.105, 3.434])

[4.110] *Grishin, B. S.:* Diffusion of Compounding Ingredients in Rubber. Paper 49, 124th ACS-Conf., Rubber Div., Oct. 25–28, 1983, Houston, TX.
[4.111] *Leblanc, J. L.:* Verbesserte Eigenschaften von Naturkautschukverschnitten mittels kontrollierter Interphasenverteilung der Vulkanisiermittel. KGK *36* (1983), p. 457.
[4.112] *Roebuck, H., Moult, B.:* Dispersion of Accelerator Particles. KGK *38* (1985), p. 510.
[4.113] *Vergnaud, J. M., Bonzou, J., Ferradou, C., Sadr, A.:* Regulation of Cure Conditions Preventing a Variation in the Distribution of Vulcanizing Agents Through Rubber Mass. Paper 68, 124th ACS-Conf., Rubber Div., Oct. 25–28, 1983, Houston, TX.

Inorganic Accelerators

[4.114] *Fisher, H. L., Davies, A. R.:* Ind. Engng. Chem. *40* (1948), p. 143.
[4.115] USP 3633 (1844), Ch. Goodyear.

Thiazole Accelerators (see also [1.12–1.14, 2.123, 4.90]

[4.116] *Bedford, C. W., Sebrell, L.:* Ind. Engng. Chem. *13* (1921), p. 1034; *14* (1922), p. 25; USP 1 566 687 (1921), Goodyear, *Bedford, C. W., Sebrell, L. B. (MBT).*
[4.117] *Sebrell, L. B., Boord, C. E.:* Ind. Engng. Chem. *15* (1923), p. 1009 *(MBT).*
[4.118] *Bruni, G., Romani, E.:* Ind. Rubb. J. *62* (1921), p. 18; India Rubber World *67* (1922), pp. 20, 94; Giorn. Chim. Ind. Appl. *3* (1921), p. 351 *(MBT).*
[4.119] *Sebrell, L. B., Boord, C. E.:* J. Am. Chem. Soc. *45* (1923), p. 2390 *(MBTS).*
[4.120] *Teppema, J., Sebrell, L. B.:* J. Am. Chem. Soc. *49* (1927), pp. 1748, 1779 *(MBTS).*
[4.121] DRP 573 570 (1931); DRP 587 608 (1931), IG Farbenindustrie, *Zaucker, E., Orthner, L., Bögemann, M.;* DRP 586 351 (1932), IG Farbenindustrie, *Zaucker, E. (ABS).*
[4.122] BP 517 451 (1938); USP 2 191 856 (1938); USP 2 191 857 (1938); DRP 855 564 (1938), Monsanto, *Harman, M. W. (CBS).*
[4.123] DRP 615 560 (1933), IG Farbenindustrie, *Tschunkur, E., Köhler, H. (MBS).*
[4.124] USP 2 730 526 (1951); USP 2 730 527 (1952); DRP 950 465 (1952); DRP 1 013 653 (1953); EP 713 496 (1952); EP 737 252 (1953), Cyanamid, *Kinstler, R. C.;* USP 2 758 991 (1953); DBP 1 012 914 (1959), Cyanamid, *Sullivan, F. A. V. (MBS).*
[4.125] USP 2 367 827 (1941), Firestone, *Smith, G. E. (TBBS).*
[4.126] USP 2 807 620 (1955), Monsanto, *Cooper, R. H., D'Amico, J. J. (TBBS).*
[4.127] DBP 1 052 478, 1 053 775 (1956), Bayer, *Freytag, H., Lober, F., Pohle, H.;* DBP 1 046 056 (1957), Bayer, *Lober, F., Freytag, H., Kracht, H. (DCBS).*
[4.127 a] *Banerjee, B.:* Influence of Carbon Black on Sulfenamide Acceleration System. KGK *39* (1986), pp. 521–523.
[4.128] *Chakravarty, S. N., Rajamani, A., Kapur, A. L.:* Effect of Moisture on Sulphenamide Accelerated NR Gum and Carbon Black Filled Compound Properties. KGK *34* (1981), p. 122.
[4.129] *Dogadkin, B. A., Feldshstein, M. S., Eitington, I. I., Pevznev, O. M.:* The Action of Some Heterocyclic Disulfides as Agents and Accelerators of Vulcanization. RCT *32* (1959), pp. 976–982; 983–991.
[4.130] *Fegade, N. B., Millns, W.:* Hydrobenzamide – An Excellent Booster for Thiazole Accelerated Sulfur Vulcanization of Natural Rubber. KGK *34* (1981), p. 1023.
[4.130 a] *Das, P. K., Datta, R. N., Basu, D. K.:* Cure Modification of NR Effected by Thioamines in the Presence of Dibenzothiazyl Disulfide. KGK *41* (1988), pp. 59–62.

[4.131] *Skinner, T. D.:* The CBS-Accelerated Sulfuration of Natural Rubber and Cis-1.4-Polybutadiene. RCT *45* (1972), p. 182.

[4.131 a] *Rayner, G. H., Leleu, E.:* Improved Productivity by New Accelerators, Sulfenamides of Mercaptobenzothiazole. Rubbercon '87, June 1–6, 1987, Harrogate, GB, Proceed. B. 23.

Triazine Accelerators (see also [1.12, 1.13])

[4.132] *Westlinning, H.:* KGK *23* (1970), p. 219; RCT *43* (1970), p. 1194.

[4.133] *Westlinning, H., Schwarze, W.:* Rubbercon '72, London, Proc. p. G4 1–9.

[4.134] *Ahne, W., et al.:* KGK *28* (1975), p. 135.

Dithiocarbamate Accelerators (see also [1.12–1.14, 4.3])

[4.135] DRP 280198 (1914), Farb. Bayer, *Gottlob, K., Bögemann, M.* (Amin-Dithiocarbamate).

[4.136] DRP 380774 (1919); F.P. 520477 (1920); EP 140387 (1920), Pirelli, *Bruni, G. (ZDMC).*

[4.137] *Bögemann, M.:* Angew. Chem. *51* (1938), p. 113. In: *Schwab (Ed.):* Handbuch der Katalyse. Vol. 7 II, 1943, p. 569.

[4.138] *Cranor, D. F.:* India Rubber J. *58* (1919), p. 1199.

[4.139] *Maximoff, A.:* Caoutch. et Gutta Percha *18* (1921), p. 10944, 10956.

Xanthogenate Accelerators (see also [4.3])

[4.140] *Ostromysslenski, I.:* Chem. Ztbl. I (1916), p. 703.

[4.141] USP 1440961 (1920), Naugatuck, *Cadwell, S. M.*.

[4.142] USP 1735701 (1928), Chemical Corp., *Whitby, G. S.*.

Thiuram Accelerators (see also [1.12–1.14, 2.3, 4.3, 4.55, 4.79–4.94])

[4.143] USP 1413172 (1920), Vanderbilt, *Lorentz, B. E. (TMTD).*

[4.144] USP 1440962 (1922), Naugatuck, *Cadwell, S. M. (TMTD).*

[4.145] USP 1440963 (1922), Naugatuck, *Cadwell, S. M. (TMTM).*

[4.146] USP 1788632 (1928), Du Pont, *Powers, D. H. (TMTM).*

[4.147] USP 2414014 (1943), Du Pont, *Cable, G. M., Richmond, J. L. (DPTT).*

[4.148] *Banerjee, B.:* Influence of Hydrofuramide on Thiuram Vulcanization of Natural Rubber in the Absence of Elemental Sulfur. KGK *32* (1979), p. 13.

[4.149] *Banerjee, B.:* Thiuram Vulcanization of Natural Rubber in the Presence of Amines, Application of Electron Spin Resonance Technique to Elucidate the Mechanism of Vulcanization. KGK *37* (1984), p. 21.

[4.150] *Beniska, J., Kysela, G., Staudner, E., Tuan, D. M.:* Contribution of Thiuram Vulcanization activated by Thiurea. IRC 85, Oct. 15–18, 1985, Kyoto, Proc. p. 891.

[4.150 a] *Beniska, J.:* Contribution to the Mechanism of the Activated Vulcanization in the Presence of TMTD. IRC '86, June 2–6, 1986, Göteborg, Sweden, Proceed. p. 490.

[4.151] *Kelm, J., Vogel, L., Groß, D.:* Untersuchung der Aminbildung bei der Thiuramvulkanisation mit Hilfe der Fluor-NMR-Spektrometrie. KGK *36* (1983), p. 274.

[4.152] *Kempermann, Th., Leibbrandt, F., Abele, M.:* Verzögerung der Thiuram-Vulkanisation. GAK *28* (1975), pp. 278, 316.

Dithiocarbamylsulfenamide Accelerators (see also [3.219–3.221, 3.400, 3.434])

[4.153] *Adhikari, B., Pal, D., Basu, D. K., Chandhuri, A. K.:* Studies on the Reaction between Thiocarbamylsulfenamides and Dibenzothiazyldisulfide. RCT *56* (1983), p. 327.

[4.154] *Das, M. M., Basu, D. K., Chandhuri, A. K.:* Effect of Thiocarbamylsulfenamide and Dibenzothiazyldisulfide in the Ageing of Natural Rubber Vulcanizates. KGK *36* (1983), p. 569.

[4.155] *Das, M.M., Datta, R.N., Basu, D.K., Chandhuri, A.K.:* Structural Characterization of Aged Natural Rubber Gum Vulcanizates by Using Thiocarbamylsulfenamide-Dibenzothiazyldisulfide Accelerator System. KGK *38* (1985), p.113.
[4.155 a] *Datta, R.N., Das, P.K., Basu, D.K.:* Effect of Zinc Dithiocarbamate on the Network Structure of NR Vulcanizates Formed in the EV Systems Containing Thiocarbamyl Sulfenamide and Dibenzothiazyl Disulfide. KGK *39* (1986), pp.1090–1093.
[4.156] *Hofmann, W.:* OTOS – ein neuer hochwirksamer Vulkanisationsbeschleuniger. GAK *32* (1979), pp.158, 318, 392; SRC 79, April 3–4, 1979, Kopenhagen.
[4.156 a] *Khamrai, A.K., Adhikar, B., Maiti, S., Maiti, M.M.:* New Accelerators of Vulcanization of Rubber, 5.Polymeric Thiocarbamoyl Sulfenamides. KGK *40* (1987), pp.826–828.
[4.157] *Krymowski, J.F., Taylor, R.D.:* Chemical Reactions between Thiocarbamyl Sulfenamides and Benzothiazylsulfenamide leading to Cure Synergisms. RCT *50* (1977), p.671.
[4.158] *Layer, R.W.:* Curing SBR with Thiocarbamyl Sulfenamide Accelerators. Paper 112, 128th ACS-Conf., Rubber Div., Oct.1–4, 1985, Cleveland, OH.
[4.158 a] *Layer, R.W., Chasar, D.W.:* Thiocarbamyl Sulfenamides: New Properties and Applications. dkt '88, July 4–7, 1988, Nürnberg, West Germany.
[4.159] *Moore, K.C.:* OTOS/MBT-Derivative Vulcanization Systems. Elastomerics, June 1978, p.36.
[4.160] *Pal, D., Adhikari, B., Basu, D.K., Chandhuri, A.K.:* Study of Cure Synergism of the Thiocarbamyl Sulfenamide-Dibenzothiazyldisulfide Accelerator System on the Vulcanization of Natural Rubber. RCT *56* (1983), p.827; The Kinetics of Crosslinking in Natural Rubber Gum Vulcanizates Containing A Thiocarbamyl Sulfenamide – Dibenzothiazyldisulfide Accelerator System. KGK *36* (1983), p.859.
[4.161] *Taylor, R.D.:* Effect of Alcyl Group Structure on Cure Characteristics of N,N-Dialkylthiocarbamyl-N′,N′-Dialkylsulfenamide Vulcanization Accelerators. RCT *47* (1974), p.406.

Guanidine Accelerators (see also [1.12–1.14, 4.3])

[4.162] USP 1411231 (1921), Dovan Chem. *Weiss, M.L.*
[4.163] USP 1721057 (1922), Du Pont, *Scott, W.*
[4.164] EP 253197 (1925), British Dyestuffs, *Turrell Cronshaw, C.J., Smith Naunton, W.J.*
[4.165] DRP 481994 (1929), IG Farbenind., *Meis, H. (DPG, DOTG).*
[4.166] EP 201885 (1923), Pirelli, *Bruni, G., Romani, E (OTBG)).*

Aldehydamine Accelerators (see also [1.12–1.14, 4.3])

[4.167] EP 7370 (1914), *Peachy, S.J.*
[4.168] USP 1365495 (1920), Du Pont, *Scott, W. (HEXA).*
[4.169] DRP 551805 (1929), IG Farbenind., *Weigel, Th. (TCT).*
[4.170] USP 1417970 (1920), Naugatuck, *Cadwell, S.M. (BAA).*
[4.171] USP 1780326 (1925); USP 1780334 (1926), Du Pont, *Williams, J., Burnett, W.B.*

Other Amine Accelerators (see also [1.12–1.14, 4.3])

[4.172] DRP 221310 (1908); DRP 243346 (1909), Farb. Bayer, *Oswald, Wo., Oswald, Wa.*
[4.173] DRP 265221 (1912), Farb. Bayer, *Hofmann, F. et al.*
[4.174] *Gottlob, K.:* Gummiztg. *30* (1916), pp.303, 326; *33* (1918), p.87.
[4.175] *Oenslager, G.:* Ind. Engng. Chem. *25* (1933), p.232.
[4.176] *Bögemann, M.:* Angew. Chem. *51* (1938), p.113.

Thiourea Accelerators (see also [1.14, 4.3])

[4.177] USP 1365495 (1920), Du Pont, *Scott, W.*
[4.178] *Behr, S., Rohde, E.:* Vernetzungssysteme für die Hochtemperatur-Vulkanisation von Polychloropren. KGK *23* (1979), p.492.

[4.179] *Beadle, H.C.:* A Non-Thiourea Neoprene Accelerator. Paper 31, 117th ACS-Conf. Rubber Div., May 20–23, 1980, Las Vegas, NV.

[4.180] *Eholzer, U., Kempermann, Th.:* Ein neuer Polychloroprenbeschleuniger als Ersatz für Ethylenthioharnstoff. KGK *33* (1980), p. 696.

[4.181] *Hadhoud, M.K. et al.:* Alternative Accelerators to Ethylene Thiourea for Loaded Chloroprene Rubber. Paper 7, 124th ACS-Conf., Rubber Div., Oct. 25–28, 1983, Houston, TX.

[4.182] *Pal, D. et al.:* New Accelerators for Vulcanization of Rubber, III. Synthesis, Characterization and Evaluation of Some 2-Mercaptobenzimidazole Derivatives. RCT *58* (1985), p. 713.

[4.183] *Sklenarz, R.:* Neuer Vulkanisationsbeschleuniger für CR. GAK *33* (1980), p. 224.

Dithiophosphate Accelerators (see also [4.98, 4.102])

[4.184] *Ehrend, H.:* Über die Wirkung von Dithiophosphaten in Vernetzungssystemen für Dienkautschuke. SGF 76, Kopenhagen, Denmark.

[4.185] *Pimblott, I.G. et al.:* Bis-(diisopropyl)-thiophosphorylsulfides in Vulcanization Reactions. J. Appl. Polym. Sci. *23* (1979), p. 3621.

Other Vulcanization Accelerators (see also [4.179–4.183])

[4.186] *Byers, J.T.:* New Vulcanizing Agent for Tire Compound. Paper 45, 128th ACS-Conf., Rubber Div., Oct. 1–4, 1985, Cleveland, OH.

[4.187] *Stieber, S.:* A New Delayed Action Accelerator for Tire. Paper 74, 126th ACS-Conf., Rubber Div., Oct. 23–26, 1984, Denver, CO.

[4.188] *Yamazaki, N., Nakahama, S., Zama, Y., Yamagushi, K.:* Polyethersulfides as Curing Agents. Paper 9, 124th ACS-Conf., Rubber Div., Oct. 25–28, 1983, Houston, TX.

[4.189] ...: Deovulc EG Beschleunigergemische – Eine Gruppe von Beschleunigergemischen für EPDM. DOG-Kontakt 26, Sept. 1981, Techn. Bulletin by Deutsche Oelfabrik, Hamburg, West Germany.

4.10.3.4 References on Peroxides

[4.190] *Ostromyslenski, I.:* J. Russ. Phys. Chem. *47* (1915), p. 1467; India Rubber J. *52* (1916), p. 470.

General References
(see also [1.12, 1.13, 3.400, 3.401, 3.485, 3.576, 3.577, 3.579, 3.587, 3.588, 4.3, 4.99])

[4.190 a] *Beyer, G.:* Bestimmung der Aktivierungsenergie des DCP-Zerfalls durch thermogravimetrische Analyse. GAK *40* (1987), p. 337–339.

[4.190 b] *Endstra, W.C.:* Vernetzung von Polymeren mittels multifunktionellen Peroxiden. dkt '88, July 4–7, 1988, Nürnberg, West Germany.

[4.190 c] *Endstra, W.C., Seeberger, D.:* Vernetzung von Polymeren mit Schwefel/Beschleunigern oder Peroxiden? KGK *39* (1986), p. 929–935.

[4.191] *Endstra, W.C.:* Vernetzung von Elastomeren und Thermoplasten durch organische Peroxide; March 1983, Paper at several Conferences of DKG-Bezirksgruppen, FRG.

[4.192] *Endstra, W.C.:* Peroxid-Vernetzung von Polymeren. KGK *32* (1979), p. 756.

[4.192 a] *Gleim, W., Oppermann, W., Rehage, G.:* Die elastischen Eigenschaften von peroxidisch vernetztem Gummi. KGK *39* (1986), pp. 516–520.

[4.192 b] *de Groot, J.J.:* Sicherheit der Vernetzungs-Peroxide, dkt '88, July 4–7, 1988, Nürnberg, West Germany.

[4.193] *Parakel, T.L., Purper, R.T.:* Organic Peroxides in Rubber Vulcanization. Techn. Bulletin Pennwalt; Ind. Gomma *22* No. 12 (1978), p. 45.

New Peroxides (see also [3.581])

[4.194] *Endstra, W.C.:* Peroxide for Microwave Curing of Polymers. Rubbercon 81, June 1981, Harrogate, GB, Caoutch. Plast. *60* No.630 (1983), p.35.
[4.195] *Hepburn, C.:* Bis-peroxycarbamates as Vulcanizing Agents for Rubber. PRI-Tagung, May 1982, London; KGK *37* (1984), p.390.
[4.196] *Kmiec, C.J., Kamath, V.R.:* Elastomer Crosslinking with Diperoxyketals. Rubber World, Oct. 1983, p.26.
[4.197] *Sacrini, E., Fontanelli, R., Fontana, A.:* New Organic Peroxides and their Use in Elastomer Curing. KGK *37* (1984), p.312.
[4.198] ...: Newest Peroxides, Peroxyketals. Mod. Plast. *60* No.4 (1983), p.96.

Influence of Compounding on Peroxides (see also [3.431, 3.437])

[4.199] *Bandyopadhyay, P.K., Banerjee, S.:* Effect of Diphenylguanidine on 2-(Morpholinodithio)-benzothiazole Accelerated Sulphur Vulcanization of NR Both in Presence and Absence of Dicumyl Peroxide. KGK *32* (1979), p.588.
[4.200] *Das, C.K., Millns, W.:* The Effect of Compounding Ingredients on the Peroxide Cure of NBR. KGK *35* (1982), p.402.
[4.201] ...: Process and Extender Oils for EPDM-Rubbers. Techn. Bulletin MP 617 by BP Oil International.

Selected References on Peroxides

[4.202] *Beyer, G.:* Kinetische Untersuchungen zur peroxidischen Vernetzung von Polymeren. GAK *38* (1985), p.368.
[4.203] *Brazier, D.W., Schwarz, N.N.:* The Cure of Elastomers by Dicumyl Peroxide as Observed in Different Scanning Calorimeter. Thermochim. Acta *39* No.1 (1980), p.7.
[4.204] *Capla, M., Barsig, E.:* Simultaneous Degradation and Crosslinking Effect of Dicumyl Peroxide on Ethylene Propylene Copolymer. Eur. Polym. J. *16* (1980), p.611.
[4.205] *Chow, Y.W., Knight, G.T.:* Peroxidische Vernetzungssysteme mit verzögerter Wirkung. GAK *31* (1978), p.716.
[4.206] *Groepper, J.:* Peroxidische Vulkanisation in Anwesenheit von Luftsauerstoff. KGK *36* (1983), p.466.
[4.207] *Honsberg, W.:* Improved Peroxide Cures for Chlorosulfonated Polyethylene. Paper 3, 124th ACS-Conf., Rubber Div., Oct. 25–28, 1983, Houston, TX.
[4.208] *Perkins, G.T.:* Peroxide Curing of Nordel. Techn. Bulletin ND-310, 2, Du Pont.
[4.209] ...: Peroxide Vulcanization of EPM and EPDM. Techn. Bulletin by Akzo, Düren, West Germany.
[4.210] ...: Di-Cup und Vul-Cup Peroxides Vulcanizing Ethylene-Propylene Elastomers. Technical Bulletin by Hercules, Wilmington.

Peroxide Co-activators (see also [3.400, 3.433, 3.458, 3.470, 3.683])

[4.211] *Das, C.K.:* Effect of Sulfur on the Peroxide Cure in NBR/EPDM-Blends. KGK *35* (1982), p.753.
[4.212] *Fujio, R., Kitayama, M., Katarka, N., Anzai, S.:* Effect of Sulfur on the Peroxide Cure of EPDM and Divinylbenzene Compounds. RCT *52* (1979), p.74.
[4.213] *Mitra, A., Das, C.K.:* Effect of Ethylene Propylene Ratio in the Peroxide Cure of NBR/EPDM Blends in the Presence of Sulfur. KGK *37* (1984), p.862.
[4.214] *Monte, S.J., Sugerman, G.:* The Effect of Titanate Coupling Agents. Paper at the 116th ACS-Conf., Rubber Div., Okt. 23–26, 1979, Cleveland, OH.
[4.215] *Okita, T., Nagasaki, N., Okamura, H., Ohashi, K.:* The Vulcanizing Mechanism of N,N′-m Phenylene-bis-maleimide and Vulcanizing Properties. IRC 85, Oct. 15–18, 1985, Kyoto, Proc. p.875.
[4.216] *Sinebryukhova, N.Y., Inshakova, L.M., Sovodin, G.A.:* Study of the Action of Triallylcyanurate as Coagent in Peroxide Vulcanization of Ethylen-Propylene Rubbers. Proiz. Shin. RTI i ATI, *1981* No.4, p.3.

[4.217] *Weatherstone, E., Younger, J.:* Methacrylate/Peroxide Vulcanization of Rubbers. IRC 79, Oct. 3-6, 1979, Venice, Italy, Proc. p. 1054.
[4.218] ...: Polyfunctional Methacrylic Monomers as Coagents in Peroxide Vulcanization. Techn. Bulletin, Sartomer.
[4.219] ...: Saret Crosslinking Plasticizers for Injection Moulding. Techn. Bulletin, Sartomer.
[4.220] ...: Effect of Coagents on Peroxide Cure. Bulletin ORC 110 D, Techn. Bulletin by Hercules. Wilmington.
[4.221] ...: Verschiedene Coagentien in peroxidisch vernetztem EPDM. Techn. Bulletin G.10/81, Lehmann u. Voss.
[4.222] ...: Use of Dicyclopentadiene Acrylate as a Coagent in Peroxy-Cured Elastomers. Res. Discl. *215* (1982), p. 78.
[4.223] ...: Triallyltrimellithate Crosslinking Agent. Mod. Plast. *11* No. 10 (1981), p. 76.

4.10.3.5 References on Chinondioxime (see also [1.12-1.14])

[4.224] *Fisher, H.:* Ind. Engng. Chem. *31* (1939), p. 1381.
[4.225] *Rehner, J., jr., Flory, P. J.:* Ind. Engng. Chem. *38* (1946), p. 500.
[4.226] *Haworth, I. P.:* Ind. Engng. Chem. *40* (1948), p. 2314.
[4.227] USP 2975153 (1958), Monsanto.
[4.228] *Schoenbeck, M. A.:* Rubber Age *92* (1962), p. 75.
[4.229] *Quirk, F. J., Minter, H. F.:* Rubber Age *99* (1967), p. 63.
[4.230] *Imoto, M. et al.:* Nippon Gomu Kyokaishi *41* (1968), p. 583.
[4.231] *Mironyk, V. P. et al.:* Kauch. i Rezina *1973* No. 3, p. 77.
[4.232] ...: QCD (Para-Quinone Dioxime), Techn. Bulletin DS 10-8003 D, Hughson Chemical Div.

4.10.3.6 References on Resin Cross-Linking (see also [1.12, 1.13, 3.363])

[4.233] *Giller, A.:* KGK *13* (1960), p. WT 288; *14* (1961), p. WT 201; *17* (1964), p. 174.
[4.234] *Giller, A.:* Phenolic Resins in Today's Rubber Industry. IRC 78, Kiev 1978, Preprint 17.
[4.235] *Fries, H., Esch, E., Kempermann, Th.:* Neues vulkanisationsfähiges Phenolharz für die Gummiindustrie mit breitem Anwendungsspektrum. KGK *32* (1979), p. 860.
[4.236] ...: Vulkanisationssysteme für Butylkautschuk. Bayer Mitt. f.d. Gummiind. *29*, pp. 34-57. Techn. Bulletin Bayer AG.

4.10.3.7 References on Other Accelerators (see also [2.110-2.113, 3.745, 3.746])

[4.237] *Shapkin, A. N. et al.:* Int. Polym. Sci. Technol. *5* No. 2 (1978), p. T 11.
[4.238] *Esser, H.:* Kautschuk u. Kunstst. *11* (1958), p. WT 5.
[4.239] *Wouters, G., Woods, F.:* Moisture Curable Silane Grafted Ethylene Propylene Elastomers. Rubbercon '81, Harrogate, GB, 1981.

4.10.3.8 References on Acceleration Activators (see also [1.12-1.14])

[4.240] *Ecker, R.:* Kautschuk u. Gummi *7* (1954), p. WT 96.
[4.241] *Starmer, P. H.:* Effect of Metaloxides on the Properties of Carboxilic Nitrile Rubber. Rubbercon '87, June 1-6, 1987, Harrogate, GB, Proceed. A 11.
[4.241 a] *Proycheva, A. G., Fomin, A. G., Andryakov, E. I., Romanova, T. V.:* Investigation of Interphase Transfer Catalyst as Activators of Sulphur Vulcanization of Diene Rubber. Rubbercon '87, June 1-6, 1987, Harrogate, GB, Proceed. A 17.
[4.242] ...: Use of Zinc Oxide in the Rubber Industry. Bibliography 2, Publ.: ACS.

4.10.3.9 References on Vulcanization Retarders (see also [1.12–1.14]

General References

[4.243] *Davies, K.M., Lloyd, D.G.:* Die Entwicklung von Vulkanisationsverzögerern. GAK *27* (1974), p. 92; Rev. Gén. Caoutch. *51* (1974), p. 217.

[4.244] *Kempermann, Th. et al.:* Neue Erkenntnisse auf dem Gebiet der Vulkanisationsverzögerer. GAK *25* (1972), p. 510; *28* (1975), pp. 278, 316.

[4.245] *Lloyd, D.G.:* Anwendungen eines geregelten Vulkanisationseinsatzes. KGK *31* (1978), p. 576.

[4.246] *Trivette, C.D. jr., Morita, E., Maender, O.W.:* Prevulcanization Inhibitors. RCT *50* (1977), Review.

Selected References

[4.247] BP 219 247 (1924), Naugatuck, *Cadwell, S.M.*

[4.248] USP 1 734 633 (1928), *Burage, A.C., Morse, H.B.*

[4.249] USP 1 871 037 (1930), Naugatuck.

[4.250] *Weber, C.O.:* B. *35* (1902), p. 1947.

[4.251] *Campbell, A.W.:* Ind. Engng. Chem. *33* (1941), p. 809.

New Developments

[4.252] *Hopper, R.J.:* N-Sulfonyl-Sulfilimines as Prematur Vulcanization Inhibitors. RCT *53* (1980), p. 1106.

[4.253] *Martin, M.A., Ibarra, L., González, L., Royo, J.:* Neuartige Verzögerer der Anvulkanisation von Kautschukmischungen. KGK *35* (1982), p. 1047.

[4.254] *Morita, E., Sullivan, A.B.:* Thioketals as Prevulcanization Inhibitors. RCT *54* (1981), p. 1132.

[4.255] *Morita, E.:* Linear Free Energy Relationship Involving the Properties of Prevulcanization Inhibitors. RCT *55* (1982), p. 352.

[4.256] *Sullivan, A.B., Davis, L.H., Maender, O.W.:* Reactivity Optimization of Prevulcanization Inhibitors. RCT *56* (1983), p. 1061.

[4.257] *Sullivan, A.B., Morita, E., Leib, R.:* New Developments in Prevulcanization. Paper 43, 128th ACS-Conf., Rubber Div., Oct. 1–4, 1985, Cleveland, OH.

4.10.4 References on Aging and Anti-Degradents (Antioxidants)

4.10.4.1 References on Aging

General References on Oxidative and Heat Aging (see also [2.82])

[4.257a] *Ab-Malek, K., Stevenson, A.:* Ageing of thick Rubber Blocks. Rubbercon '87, June 1–5, 1987, Harrogate, GB, Proceed. A 48.

[4.258] *Ahagon, A., Hiratsuga, Kanagawa:* Oxidative Aging of Black-Filled Elastomers. KGK *38* (1985), p. 505.

[4.259] *Allen, N.S.:* Degradation and Stabilization of Polyolefins. Elsevier Applied Science Publ., London, 1983.

[4.260] *Andrews, E.A. (Ed.):* Development in Polymer Fracture, Elsevier Applied Science Publ., London, 1979.

[4.261] *Bateman, L.:* Oxidation of Natural Rubber Hydrocarbon, Trans. IRI *26* (1950), p. 246.

[4.262] *Buist, I.M.:* Ageing and Weathering of Rubber. Heffer, W. sons, Cambridge, 1956.

[4.263] *Carpenter, A.:* Absorption of Oxygen by Rubbers. Ind. Engng. Chem. *39* (1947), p. 187.

[4.264] *Genskens, G.:* Degradation and Stabilisation of Polymers. Elsevier Applied Science Publ., London, 1975.

[4.265] *Grassie, N. (Ed.):* Developments in Polymer Degradation, Elsevier Applied Science Publ., London, 1st 1977; 2nd 1979; 3rd 1981; 4th 1982; 5th 1984.

[4.266] *Hashimoto, K., Miyahara, N., Yonehama, T.:* A Study of Thermal Behaviour of Sulfur Crosslinks in Vulcanized Natural Rubber by Pyrolysis Glass Capillary Gas Chromatography. IRC 85, Oct. 15-18, 1985, Kyoto, Proc. p. 608.
[4.267] *Hemmer, E., Jesse, H.:* Untersuchungen über das thermische Alterungsverhalten von ausgewählten Isolierstoffen für Kabel und Leitungen. GAK *34* (1981), pp. 503, 794.
[4.268] *Kelen, T.:* Polymer Degradation. van Nostrand Reinold Co., New York, 1982.
[4.269] *Kuzminskii, A. S., Leyland, B. N.:* Ageing and Stabilization of Polymers. Elsevier Applied Science Publ., London, 1971.
[4.270] *Leuchs, O.:* Wärmealterung von Naturgummi; 1. Das Gesetz von Arrhenius. KGK *35* (1982), p. 651; 2. Der Zerreißwertgang. KGK *35* (1982), p. 830; 3. Enthomogenisierung. KGK *35* (1982), p. 1023.
[4.271] *Mitsuhashi, K., Kurumija, H.:* The Study of Long Term Natural Ageing of Rubber Vulcanizates. IRC 85, Oct. 15-18, 1985. Kyoto, Proc. p. 135.
[4.272] *Murakami, K.:* Recent Studies on Chemorheology. IRC 85, Oct. 15-18, 1985, Kyoto, Proc. p. 129.
[4.273] *Schneider, P.:* Ein Überblick über die Alterung von Kautschuk und Kautschuk-Vulkanisaten. Kautschuk u. Gummi *6* (1953), p. WT 111.
[4.274] *Tobisch, K.:* Untersuchungen zum Alterungsverhalten hitzebeständiger Elastomerdichtungen. KGK *31* (1978), p. 917.
[4.275] *Wolf, A.:* Zum Alterungsverhalten von Dichtmaterialien unter verschiedenen Umwelteinflüssen. KGK *33* (1980), p. 930.

General References on Weathering and Ozone Attack (see also [4.262])

[4.276] *Amsden, L. S.:* Static Ozone Resistance and Treshold Strain. J. IRI *1* (1976), p. 214.
[4.277] *Andrews, E. H.:* Resistance to Ozone Cracking in Elastomer Blends. J. Appl. Polym. Sci. *10* (1966), p. 47.
[4.277 a] *Braun, D., Müller, I.:* Ozoninduzierter Abbau von Elastomeren. KGK *39* (1986), pp. 507-509.
[4.278] *Davies, A., Sims, D.:* Weathering of Polymers. Elsevier Applied Science Publ., London, 1983.
[4.279] *Keller, R. W.:* Oxidation and Ozonization of Rubber. RCT *58* (1985), p. 637, Review.
[4.280] *Kempermann, Th. et al.:* Über die kritische Dehnung als Maß für die Ozonfestigkeit von Vulkanisaten. KGK *18* (1965), p. 638.
[4.281] *Lake, G. I. et al.:* Ozone Cracking, Flex Cracking, Fatigue of Rubber. Rubb. J. *146* (1964), No. 10, p. 34; No. 11, p. 30.

4.10.4.2 References on Anti-Degradants (Antioxidants, Stabilizers)

General References (see also [4.259, 4.264, 4.269])

[4.282] *Kempermann, Th.:* Alterungsschutzmittel. In: *Boström, S. (Ed.):* Kautschuk-Handbuch. Verlag Berliner Union, Stuttgart, Vol. 4, 1961, p. 353.
[4.283] *Scott, G. (Ed.):* Developments in Polymer Stabilization. Elsevier Applied Science Publ., London, 1. 1979; 2. 1980; 3. 1980; 4. 1981; 5. 1982; 6. 1983; 7. 1984.

Mechanism of Antioxidant Activity

[4.283 a] *Abdel-Bary, E. M.:* Effect of Some Sulfanil Derivatives as Antioxidants in Natural Rubber. Rubbercon '87, June 1-5, 1987, Harrogate, GB, Proceed. A 10.
[4.284] *Ashworth, B. T., Hill, P.:* Some Observations on Unusual Antioxidant Effects. KGK *34* (1981), p. 18.
[4.285] *Bravar, M., Kempermann. Th., Ljubic, B.:* Zusammenhang zwischen chemischer Konstitution und Wirkung bei Alterungsschutzmitteln vom Typ monofunktionelles Phenol. KGK *36* (1983), p. 95.

[4.286] *Kempermann, Th.:* Zusammenhang zwischen chemischer Konstitution und Wirkung bei Alterungsschutzmitteln vom Typ Mehrkernphenol. Paper 25, 99th ACS-Conf., 1971, Cleveland, OH.

[4.287] *Kempermann, Th.:* Über den Zusammenhang zwischen Konstitution und Wirkung bei Rißschutzmitteln vom Typ p-Phenylendiamin. GAK *26* (1973), p. 90.

[4.288] *Kempermann, Th.:* Alterung von Polymeren und Wirkungsweise von Alterungsschutzmitteln. GAK *35* (1982), pp. 238, 402.

[4.288 a] *Levy, M.:* Interactions between Antidegradants and Compounding Ingredients in a Rubber Formulation. dkt '88, July 4–7, 1988, Nürnberg, West Germany.

[4.288 b] *Paul, T.K., Bhowmick, A.K., Samajar, N.C., Bhargava, P.S.:* Influence of Binary Antioxidant System on the Aging Properties of Natural Rubber. KGK *39* (1986), pp. 1192–1194.

[4.289] *Dean, P.R., Kuczkowski, J.A.:* Enhanced Polymer Oxidation Resistance Through the Use of Secondary Antioxidants. Paper 25, 127th ACS-Conf. Rubber Div., April 23–26, 1985, Los Angeles, CA.

[4.290] *Dean, P.R., Dessem, R.W., Kline, R.H., Kuczkowski, J.A.:* Persistance Factors Influencing Antioxidant Performance in NBR, Paper 15, 126th ACS-Conf., Rubber Div., Oct. 23–26, 1984, Denver, CO.

[4.291] *Miller, D.E., Dean, P.R., Kuczkowski, J.A.:* Effect of Persistant Primary Antioxidants Upon the Heat Resistance of High Performance Elastomers, Paper 26, 127th ACS-Conf., Rubber Div., April 23–26, 1985, Los Angeles, CA.

[4.292] *Nando, G.B., De, S.K.:* Effect of Peptizer, Antioxidant and Retarder on the Network Structure and Technical Properties of Natural Rubber. KGK *33* (1980), p. 920.

[4.293] *Nethsinghe, L.P., Scott, G.:* Mechanisms of Antioxidant Action: Complementary Chain-Braking Mechanisms in the Mechano-Stabilization of Rubbers. RCT *57* (1984), p. 918.

[4.294] *Uchiyama, Y.:* The Effect of an Antioxidant on the Wear of Rubber. IRC 85, Oct. 15–18, 1985, Kyoto, Proc. p. 627.

Bound Antioxidants (see also [3.225–3.228])

[4.295] *Dweik, H.S., Scott, G.:* Mechanisms of Antioxidant Action: Aromatic Nitroxyl-Radicals and their Derived Hydroxylamines as Antifatigue Agents for Natural Rubber. RCT *57* (1984), p. 735; The Antifatigue Mechanism of Nitrosoamines, RCT *57* (1984), p. 908.

[4.296] *Kuczkowski, J.A., Gillick, J.G.:* Polymer-Bound Antioxidants, RCT *57* (1984), p. 621, Review.

[4.297] *Scott, G.:* Neuere Entwicklungen bei an Kautschuk gebundenen Antioxidantien. GAK *31* (1978), p. 934.

[4.298] *Scott, G.:* Wege zur Herstellung alterungsbeständiger Gummisorten. GAK *36* (1983), p. 276.

[4.299] *Wolff, J.R., jr.:* Copolymerized Antioxidants in Polyether-Ester Elastomers, RCT *54* (1981), p. 988.

Staining Anti-Degradants

[4.299 a] *Aarts, A.J., de Coninck, De, Burhin, H., Orband, A.:* A Study of The Fate of PPD Antidegradants by New Chromatography and Spectroscopic Techniques. Rubbercon '87, June 1–5, 1987, Harrogate, GB, Proceed. P 8.

[4.300] *Chakravarty, S.N. et al.:* Loss of Antioxidants in Truck Tires, 1. Loss of Antioxidants during Mixing. KGK *36* (1983), p. 22; 2.1. *John, A.G. et al.:* Loss due Volatility. KGK *36* (1983), p. 363; 2.2. Loss due Leaching by Water and Migration, KGK *37* (1984), p. 115.

[4.301] *Dean, P.R., Kuczkowski, J.A.:* The Function and Selection of Antidegradants, Paper at 122nd ACN-Conf., Rubber Div., Oct. 5–7, 1982, Chicago, MI.

[4.301 a] *Fries, H.:* Schutzmittel für Langzeitwirkung in Kautschuk. GAK *40* (1987), pp. 238–259.

[4.301 b] *Kempermann, Th., Redetzky, W.:* New Investigations on the Long-Term Effect of Antidegradants. Rubbercon '87, June 1–5, 1987, Harrogate, GB, Proceed. P 18.
[4.302] *Lattimer, R. P., Hooser, E. R., Zakriski, P. M.:* Characterization of Aniline-Acetone Condensation Products by Liquid Chromatography and Mass Spectroscopy. RCT 53 (1980), p. 346.
[4.303] *Lederer, D. A. jr., Helt, W. F.:* Comparison of Effectiveness and Durability of Amine-Based Antidegradants. Paper 80, 124th ACS-Conf., Rubber Div., Oct. 25–28, 1983, Houston, TX.
[4.304] *Lorenz, O., Hanlena, F., Braun, B.:* Verbrauch von p-Phenylendiaminen bei der Alterung. KGK *38* (1985), p. 255.
[4.305] *Spirk, E., Stefanic, P., Orlitz, J., Collonge, J.:* Beständigkeit gegen Auslaugen von p-Phenylendiamin-Derivaten. GAK *34* (1981), p. 734.
[4.306] *Ranney, M. W.:* Antioxidants, Recent Developments. Chemical Technical Review No. 127, Noyes Data Corp., New York, 1979.
[4.307] *Otomo, S., Kobayashi, Y., Yamamoto, Y.:* Preparation and Evaluation as Reactive Antioxidants of 5-(3′,5′-dialkyl-4′-hydroxyphenyl)tetrazoles. IRC 85, Oct. 15–18, 1985, Kyoto, Proc. p. 568.
[4.308] *Widmer, H.:* Langzeitalterungsschutz von Gummiartikeln. DKG-Tagung Nov. 7, 1975, Wiesbaden, West Germany (see also [4.324]).
[4.309] *Yehia, A. A., Nouseir, M. H., Younan, A. F.:* New Antioxidants for Rubber. IRC 85, Oct. 15–18, 1985, Kyoto, Proc. p. 562.
[4.309 a] *Yehia, A. A., Younan, A. F., Ismail, M. N.:* Pyridazine Derivatives as Antifatigue Agents for NR Vulcanizates, Rubbercon '87, June 1–5, 1987, Harrogate, GB, Proceed, P. 5.
[4.310] . . .: Protective Materials for Rubber, Antioxidants, Antiozonants, Waxes. Rubber Age *77* (1955), p. 705.

Non-Staining Anti-Degradants (see also [4.306, 4.310])

[4.311] FP 1 263 659 (1960), ICI.
[4.312] EP 605 950 (1960), Shell.
[4.313] USP 3 285 855 (1965), Geigy.
[4.314] *Ambelang, I. C. et al.:* RCT *36* (1963), p. 149.
[4.315] *Schneider, P.:* Angew. Chem. *67* (1955), p. 61.
[4.316] *Williams, G. E.:* Trans. IRI *32* (1956), p. 43.

Anti-Degradants Against Heavy Metal Aging

[4.317] . . .: Gummi Ztg. *5* (1891), p. 7; *6* (1892), p. 7.
[4.318] *Villain, H. V.:* Rev. Gén. Caoutch. *26* (1949), p. 740.
[4.319] *Parks, C. R. et al.:* Ind. Engng. Chem. *42* (1950), p. 2552.

4.10.4.3 References on Antiozonants

Staining Antiozonants (see also [4.287, 4.300, 4.304, 4.305, 4.308, 4.310])

[4.320] *Bergmann, E. W. et al.:* Antiozonants for Diene Elastomers. Rubber World *148* No. 6 (1963), p. 61.
[4.321] *Brück, D., Königshofen, H., Ruetz, L.:* The Action of Antiozonants in Rubber. RCT *58* (1983), p. 728; KGK *38* (1985), p. 372.
[4.322] *Dibbo, A.:* Prüfung, Beschreibung und Einsatz von Ozonschutzmitteln. GAK *18* (1965), p. 120.
[4.323] *Kempermann, Th., Clamroth, R.:* Antiozonantien in ölverstrecktem Kautschuk. Kautschuk u. Gummi *15* (1962), p. WT 135 (see also p. WT 422).
[4.324] *Langner, G. P. et al.:* Über den Langzeiteffekt von Antiozonantien. KGK *32* (1979), p. 81 (see also [4.308]).
[4.325] *Lattimer, R. P. et al.:* Mechanisms of Ozonization of N,N′-di-(1-Methylheptyl)-p-phenylenediamine. RCT 53 (1980), p. 1170.

[4.326] *Lattimer, R.P., Hooser, F.R., Layer, R.W., Rhee, C.K.:* Mechanisms of Ozonization of N-(1,3.-Dimethylbutyl)-N'-Phenyl-p-phenylenediamine. RCT *56* (1983), p.431.
[4.327] *Lattimer, R.P., Layer, R.W., Rhee, C.K.:* Mechanisms of Antiozonant Protection: Antiozonant-Rubber Reactions During Ozone Exposure. RCT *57* (1984), p.1023.
[4.328] *Lattimer, R.P. et al.:* Mechanisms of Antiozonant Protection: Unextractable Nitrogen in Aged cis-Polybutadiene Vulcanizates. Paper 111, 128th ACS-Conf., Rubber Div., Oct.1-4, 1985, Cleveland, OH.
[4.329] *Lederer, D.A., Fath, M.A.:* Effects of Wax and Substituted p-Phenylenediamine Antiozonants in Rubber. RCT *54* (1981), p.415.
[4.330] *Lorenz, O., Parks, C.R.:* Mechanism of Antiozonant Action. RCT *36* (1963), p.194; 201; see also RCT *34* (1961), p.816.
[4.331] *Miller, D.E., Dessent, R.W., Kuczkowski, J.A.:* Long Term Antiozonant Protection of Tire Sidewalls. Paper 14, 126th ACS-Conf., Rubber Div., Oct.23-26, 1984, Denver, CO.
[4.332] *Porter, N.M. et al.:* Determination of p.-Phenylenediamine Antiozonant in Polychloroprene. RCT 57 (1984), p.801.
[4.333] *Scott, G.:* The Mechanism of Antifatigue Agents in Tyres, Paper 62, 125th ACS-Conf., Rubber-Div., May 8-11, 1984, Indianapolis, IN.
[4.334] *Scott, G.:* A Review of Recent Developments in the Mechanisms of Antifatigue Agents. RCT *58* (1985), p.269.
[4.335] *Tabaddor, F.:* Review of Fatigue in Rubber and Rubber Composites. Paper 21, 122nd ACS-Conf., Rubber-Div., Oct.5-7, 1982, Chicago, MI.

Non-Staining Antiozonants

[4.336] *Buist, I.M., Meyrink, T.I.:* Eur. Rubber J. *157* No.10 (1975), p.26.
[4.336a] *Lévy, M.:* The Activity of Phenolic Antioxidants in Rubber Application - Theory and Practice. KGK *40* (1987), pp.1043-1052.
[4.337] *Nah, S.H., Thomas, A.G.:* Migration and Blooming of Waxes to the Surface of Rubber Vulcanizates. RCT *54* (1981), p.25.
[4.338] DP 1620800 (1966), Bayer.
[4.339] DP 1693163, 1795646 (1967); 1917600 (1969), Bayer.

4.10.5 References on Reinforcement, Fillers and Pigments

4.10.5.1 References on Reinforcement

General References on Reinforcement

[4.340] *Bachmann, J.H. et al.:* Collection of References on Reinforcing Problems. RCT *32* (1959), p.1286.
[4.341] *Dannenberg, E.M.:* The Effects of Surface Chemicals Reactions on the Properties of Filler Reinforced Rubbers. RCT *48* (1975), Review.
[4.342] *Dannenberg, E.M.:* Filler Choice in the Rubber Industry. RCT *55* (1982), p.860, Review.
[4.343] *Dannenberg, E.M.:* Reinforcement of Rubber. Paper at the 125th ACS-Conf., Rubber Div., May 8-11, 1984, Indianapolis, IN.
[4.344] *Donnet, J.B., Vidal, A.:* Highlights of Elastomer-Filler Reinforcement, Paper 49, 125th ACS-Conf., Rubber Div., May 8-11, 1984, Indianapolis, IN.
[4.344a] *Donnet, J.B., Mang, M.J., Papirer, E., Vidal, A.:* Influence of Surface Treatment on the Reinforcement of Elastomers. KGK *39* (1986), pp.510-515.
[4.345] *Donnet, J.B.:* New Insights on the Mechanism of Filler Reinforcement of Elastomer. IRC 85, Oct.15-18, 1985, Kyoto, Proc. p.123; KGK *39* (1986), pp.1082-1083.
[4.346] *Fromandi, G., Ecker, R.:* Anwendungstechnische Versuche zur Charakterisierung des aktiven Verhaltens von Füllstoffen in Kautschuk. Kautschuk u. Gummi *5* (1952), p.WT 191.

[4.347] *Heitz, E.:* Füllstoffe als qualitätsverbessernde Modifikation. GAK *28* (1975), p. 268.
[4.348] *Kraus, G.:* Reinforcement of Elastomers, Interscience Publ., New York, 1965.
[4.349] *Parkinson, D.:* Reinforcement of Rubbers. IRI, London, 1957.
[4.350] *Soos, I.:* Charakterisierung des Verstärkungseffektes von Füllstoffen aufgrund der Auswertung der Spannungs-Deformationskurve. GAK *37* (1984), pp. 232, 300, 509.
[4.351] *Westlinning, K.:* Verstärkerfüllstoffe für Kautschuk. KGK *15* (1962), p. WT 475.
[4.352] *Westlinning, K.:* Verstärkung und Abrieb. KGK *20* (1967), p. 5.
[4.353] *Wolff, S.:* Füllstoffentwicklungen heute und morgen. KGK *32* (1979), p. 312.
[4.354] *Yamada, E., Inagaki, S., Okamoto, H., Furukawa, J.:* Theory of Reinforcement and New Active Filler, IRC 85, Oct. 15–18, 1985, Kyoto, Proc. p. 538.

Selected References on Reinforcement and its Determination

[4.355] *Harwood, J. A. C., Payne, A. R.:* RCT *43* (1970), pp. 6, 687.
[4.356] *Harwood, J. A. C., Payne, A. R.:* RCT *39* (1966), p. 1544.
[4.357] *Houwink, R.:* Kautschuk u. Gummi *5* (1952), p. WT 65.
[4.358] *Dannenberg, E. M., Brennan, J. I.:* RCT *39* (1966), p. 597.
[4.359] *Dannenberg, E. M.:* Trans. IRI *42* No. 2 (1966), p. 26.
[4.360] *Drogin, I., Messanger, T. H.:* 3rd Rubber Techn. Conf., Cambridge 1955, Proc. p. 585.
[4.361] *Rivin, D.:* RCT *44* (1971), p. 307.
[4.362] *Marvin, L. D., Wittington, E. L.:* RCT *41* (1968), p. 382.
[4.363] *Puel, R. B., Bansal, R. C.:* Rev. Gén. Caoutch. et Plast. *41* (1964), p. 445.
[4.364] *Donnet, J. B., Papierer, E.:* Rev. Gén. Caoutch. et Plast. *42* (1965), p. 389.
[4.365] *Mullins, L.:* J. Rubb. Res. *16* (1947), p. 275; Trans. IRI *32* (1956), p. 231.
[4.366] *Bueche, F.:* RCT *35* (1962), p. 259.
[4.367] *Brunauer, S., Emmett, P. H., Teller, E.:* J. Am. Chem. Soc. *60* (1938), p. 309.
[4.368] *Hess, W. M. et al.:* RCT *46* (1973), p. 1.
[4.369] *Kraus, G.:* RCT *44* (1971), p. 199.
[4.370] *Stacy, C. J. et al.:* RCT *48* (1975), p. 538.
[4.371] *Westlinning, H., Wolff, S.:* KGK *19* (1966), p. 470.
[4.372] *Wolff, S.:* KGK *23* (1970), p. 7.
[4.373] *Reichenbach, D.:* KGK *33* (1980), p. 349.
[4.374] *Timm, Th., Messerschmidt, W.:* KGK *27* (1974), pp. 83, 130.

4.10.5.2 References on Carbon Black (see also [4.453])

Reinforcement by Carbon Black and its Determination

[4.375] *Deviney, M. L., jr.:* Neuere Fortschritte bei der Anwendung der Radiochemie in der Kautschuk- und Rußforschung. KGK *25* (1972), pp. 51, 92.
[4.376] *Fowkes, F. M.:* Acid-Base Contribution to Adsorption on Fillers, RCT *57* (1984), p. 328.
[4.377] *Hess, W. M. et al.:* Morphological Characterization of Carbon Blacks in Elastomer Vulcanizates. Rubber Conf. 1973, Sept. 17, 1973, Prag, Paper A 17.
[4.378] *Hess, W. M., Donald, G. C.:* Morphology Analysis of Carbon Black, 76. Annual Meeting ASTM, June 24, 1973, Philadelphia; in: Rubber and Related Products, Philadelphia, 1973, pp. 3–18.
[4.379] *Hess, W. M., Donald, G. C.:* Improved Particle Size Measurement on Pigments for Rubber. RCT *56* (1983), p. 892.
[4.380] *Kraus, G., Janzen, I.:* Verbesserte physikalische Rußprüfung für Korrelation und Vorhersage des Verhaltens in Vulkanisaten. KGK *28* (1975), p. 253.
[4.381] *Pausch, J. B., McKalen, C. A.:* A Low-Cost, Rapid Method for Determining Carbon Black Surface Area. RCT *56* (1983), p. 440.
[4.382] *Medalia, A. I.:* Reinforcement by Carbon Black. Gordon Res. Conf. on Elastomers, July 15–19, 1985, New London, NH.
[4.383] *Meinecke, E. A.:* A Theory to Predict Extrudate Swell from Viscoelastic Dynamic Measurement. Gordon Res. Conf. on Elastomers, July, 1985, New London, NH.

[4.383 a] *Meinecke, E.A.:* Effect of Carbon Black on the Mechanical Properties of Elastomers. dkt '88, July, 4–7, 1988, Nürnberg, West Germany.

[4.384] *Nakajima, N., Harrell, E.R.:* Compaction as a Part of Mechanisms in Incorporating Carbon Black into Elastomer. Paper 74, 122nd ACS-Conf., Rubber Div., Oct. 5–7, 1982, Chicago, MI.

[4.385] *Rigbi, Z.:* Reinforcement of Rubber by Carbon Black. RCT *55* (1982), p. 1180.

[4.386] *Stevenson, L.G., Svanson, S.E.:* An NMR Investigation of the Interaction between Carbon Black and cis-Polyisoprene. RCT *53* (1980), p. 975.

[4.387] *Wolff, S.:* Abhängigkeit des Eigenschaftsbildes von OTR-Laufflächenmischungen von Vernetzungs- und Verstärkungsparametern. KGK *36* (1983), p. 968.

Selection and Characterization of Carbon Blacks

[4.388] *Brown, W.A., Patel, A.C.:* The Characterization of Carbon Blacks. Paper 78, 128th ACS-Conf., Rubber-Div., Oct. 1–4, 1985, Cleveland, OH.

[4.389] *Maire, U.:* Recent Experiences with Pneumatic Conveying of Carbon Black. Paper 31, 128th ACS-Conf., Rubber-Div., Oct. 1–4, 1985, Cleveland, OH.

[4.389 a] *Manley, T.R.:* The Determination of the Particle Size of Carbon Black. Rubbercon '87, June 1–5, 1987, Harrogate, GB, Proceed. P33.

[4.390] *Mc Neish, A.A., Swor, R.A.:* The Role of Carbon Black in High Performance Rubber Compounds. Paper 7, 125th ACS-Conf., Rubber Div., May 8–11, 1984, Indianapolis, IN.

[4.391] *Neish, A.A.:* Carbon Black Prospects in Rubber Products. Paper 63, 128th ACS-Conf., Rubber-Div., Oct. 1–4, 1984, Cleveland, OH.

[4.392] *Patterson, W.J.:* Selection of Carbon Black for Rubber Compounds. Paper at the 122nd ACS-Conf., Rubber-Div., Oct. 5–7, 1982, Chicago, MI.

[4.393] *Studebaker, M.L.:* Carbon Black, A Survey for Rubber Compounds. Phillips Chemical Co., Bull. P-10, 1954.

Dispersion Determination and Influence of Carbon Blacks on Properties

[4.394] *Aboytes, P., Marsh, P.:* Hysteresis Effect of Carbon Black Aggregate Size, Paper 7, 117th ACS-Conf., Rubber-Div., May 20–23, 1980, Las Vegas, NV.

[4.395] *Cotton, G.R.:* Mixing of Carbon Black with Rubber, I. Measurement of Dispersion Rate by Changes in Mixing Torque. Paper 38, 124th ACS-Conf., Rubber-Div., Oct. 25–28, 1983, Houston, TX.

[4.395 a] *Cotton, G.R., Murphy, L.J.:* Mixing of Carbon Black with Rubber – Analysis of SBR/BR-Blends. Rubbercon '87, June 1–5, 1987, Harrogate, GB, Proceed. A 23.

[4.396] *Janzen, J., Kraus, G.:* New Methods for Estimating Dispersibility of Carbon Blacks in Rubber. RCT *53* (1980), p. 48.

[4.397] *Hess, W.M., Swor, R.A., Micek, E.J.:* The Influence of Carbon Black, Mixing and Compounding Variables on Dispersion. RCT *57* (1984), p. 959.

[4.398] *Hess, W.M., Swor, R.A., Vegvari, P.C.:* The Influence of Carbon Black Distribution on Elastomer Blend Properties. KGK *38* (1985), p. 1114.

[4.398 a] *Leblanc, J.L., Swiderski, Z.:* Temperature-Pressure Induced Rubber-Carbon Black Interaction in Uncured Rubber Compounds. KGK *40* (1987), pp. 829–836.

[4.399] *Lin, S.S.:* Auger-Electron Spectroscopy Analysis of Dispersion in Rubber. Paper 12, 127th ACS-Conf., Rubber-Div., April 23–26, 1985, Los Angeles, CA.

[4.400] *Nakajima, N., Harrell, E.R.:* Contributions of Elastomer Behavior to Mechanisms of Carbon Black Dispersion. RCT *57* (1984), p. 153.

[4.401] *Patel, A.C., Brown, W.A.:* Carbon Black Structure and Viscoelastic Properties of Rubber Compounds. Paper 9, 127th ACS-Conf., Rubber-Div., April 23–26, 1985, Los Angeles, CA.

[4.401 a] *Patel, A.C., Brown, W.A.:* Rußstruktur und viskoelastische Eigenschaften von Kautschukmischungen und Vulkanisaten. 1. GAK *40* (1987), pp. 260–271; 2. GAK *40* (1987), pp. 327–334.

[4.402] *Sircar, A.K., Wells, J.L.:* Prediction of Hysteresis Properties from Reticulation Studies of Carbon Black Suspensions in Mineral Oil, RCT *56* (1983), p. 51.

[4.402a] *Vohwinkel, K.:* Augenblicklicher Stand und Entwicklungsrichtungen des Einsatzes von Ruß in Kautschuk. KGK *39* (1986), pp. 810–815.

[4.403] *West, J. R., Lovett, L. L., Keach, C. B.:* A New Familiy of Carbon Blacks with Mixing Energy Reduction and Enhanced Dispersion. Paper 37, 126th ACS-Conf., Rubber-Div., Oct. 23–26, 1984, Denver, CO.

Influence of Carbon Blacks on Mixing and Processing Behavior

[4.404] *Cotton, G. R.:* Mixing of Carbon Black with Rubber, II. Mechanisms of Carbon Black Incorporation, Paper 8, 127th ACS-Conf., Rubber-Div., April 23–26, 1985, Los Angeles, CA.

[4.405] *Nakajima, N.:* Mechanisms of Mixing Carbon Black into Elastomer and Wasted Energy in the Process. Paper at the IRC 85, Oct. 15–18, 1985, Kyoto, Proc. p. 112.

[4.406] *Wolff, S., Arnold, U.-E., Panenka, R.:* Verarbeitungseigenschaften von Rußen in SBR 1500, KGK *34* (1981), p. 110.

[4.407] *Wolff, S.:* Optimization of Carbon Black Mixing for Energy Savings. Paper 55, 124th ACS-Conf., Rubber-Div., Oct. 25–28, 1983, Houston, TX.

Influence of Carbon Blacks on Dynamic Properties

[4.408] *Isono, Y., Ferry, J. D.:* Stress Relaxation and Differential Dynamic Modulus of Carbon Black Filled Styrene Butadiene Rubber in Large Shearing Deformations. RCT *57* (1984), p. 925.

[4.409] *Hess, W. M., Vegvari, P. C., Swor, R. A.:* Carbon Black in NR/BR Blends for Truck Tires. RCT *58* (1985), p. 350.

[4.410] *Hirakawa, H.:* Hysteresis Loss Mechanism of Carbon Black Filled Rubber. KGK *38* (1985), p. 898.

[4.411] *Medalia, A. I.:* Effect of Carbon Black on Dynamic Properties of Rubber Vulcanizates. RCT *51* (1978), Review.

[4.412] *Meinecke, E. A., Taftaf, M. I.:* The Influence of Static Deformations and Carbon Black Loading on the Dynamic Properties of Elastomers. I. Extension, Paper 46, II. Compression, Paper 47, 124th ACS-Conf., Rubber-Div., Oct. 25–28, 1983, Houston, TX.

[4.413] *Trexter, H. E., Lee, M. C. H.:* Effect of Types of Carbon Black and Cure Conditions on Dynamic Mechanical Properties of Elastomers. Paper 46, 127th ACS-Conf., Rubber-Div., April 23–26, 1985, Los Angeles, CA.

[4.414] *Wilder, C. R., Haws, J. R., Cooper, W. T.:* Effects of Carbon Black Types of Treadwear of Radial and Bias Tires at Various Test Severities. RCT *54* (1981), p. 427.

[4.415] *Wolff, S., Tan, E. H., Panenka, R.:* Present Possibilities to Reduce Heat Generation of Tire Compounds, Paper 101, 128th ACS-Conf., Rubber-Div., Oct. 1–4, 1985, Cleveland, OH.

Influence of Carbon Black on Electrical Conductivity

[4.416] *Abdel-Bary, E. M., Amin, M., Hassan, H. H.:* Die elektrische Leitfähigkeit von mit HAF-Rußen gefüllten SBR während der Vulkanisation. GAK 34 (1981), p. 728.

[4.417] *Burton, L. C., Patt, J.:* Electrical Resistivity Relaxation of Carbon Black Filled Rubber. Paper 80, 125th ACS-Conf., Rubber-Div., May 8–11, 1984, Indianapolis, IN.

[4.418] *Hindmarsh, R. S., Norman, R. H., Gate, G. M.:* Extrusion of Rubbers with Reproducible Electrical Conductivity. Paper 70, 124th ACS-Conf., Rubber-Div., Oct. 25–28, 1983, Houston, TX.

[4.419] *Juengel, R. R.:* Carbon Black Selection for Conductive Rubber Compounds, Paper 78, 127th ACS-Conf., Rubber-Div., April 23–26, 1985, Los Angeles, CA.

[4.419a] *El-Mansky, M. K., Hassan, H. H.:* Elektrische Leitfähigkeit von mit HAF-Ruß gefüllten NR/SBR-Vulkanisaten. GAK *40* (1987), pp. 648–649.

[4.420] *Medalia, A. I.:* Electrical Conduction in Carbon Black Composites, Paper 95, 127th ACS-Conf., Rubber-Div., April 23–26, 1985, Los Angeles, CA.

[4.420a] *Poulaert, B., Probst, N.:* Thermal Conductivity of Carbon Black Loaded Polymers. KGK *39* (1986), pp. 102–107.

[4.420b] *Probst, N.:* Extra Conductive Carbon Black and its Multiple Facets in the Polymer Industry. Rubbercon '87, June 1–5, 1987, Harrogate, GB, Proceed. P26.

[4.421] *Swor, R.A., Harries, D.R., Lyon, F.:* Carbon Blacks for Electrical Conductivity Application. KGK *37* (1984), p. 198.

[4.422] *Toub, M.:* Semiconductive Silicone Rubber Properties and Applications, Paper 99, 127th ACS-Conf., Rubber-Div., April 23–26, 1985, Los Angeles, CA.

[4.423] *Voet, A.:* Temperature Effect of Electrical Resistivity of Carbon Black Filled Polymers. RCT *54* (1981), p. 42.

[4.424] *Wardell, G.F.:* Electrical Properties as Related to Rubber-Carbon Black Interaction. Paper 71, 124th ACS-Conf., Rubber-Div., Oct. 25–28, 1983, Houston, TX.

Influence of Carbon Black on Heat Conductivity

[4.425] *Hassan, H.H., El-Mansky, M.K.:* Wärmeleitfähigkeit von rußgefüllten NR/SBR-Verschnittqualitäten. GAK *37* (1984), p. 448.

[4.426] *Poulaert, B., Probst, N.:* Thermal Conductivity of Carbon Black Loaded Polymers. ikt '85, June 24–27, 1985, Stuttgart, West Germany.

New Types of Carbon Black

[4.427] *Baumann, H., Klein, J., Jüntgen, H.:* Kohlen als Füllstoffe für Kautschuke. KGK *35* (1982), p. 843.

[4.428] *Chartoff, R.P., Eriksen, E.H., Miller, D.E., Salyer, I.O.:* Enhanced Damping in Elastomers by Use of Platelet Graphit Fillers. Paper 48, 127th ACS-Conf., Rubber-Div., April 23–26, 1985, Los Angeles, CA.

[4.429] *Mark, J.M.:* In-Situ Filling of Elastomers. Gordon Res. Conf. on Elastomers, July 15–19, 1985, New London, NH.

[4.430] *Martin, J.W.:* Evaluation of Semireinforcing Carbon Black from Coal. Paper 30, 128th ACS-Conf., Rubber-Div., Oct. 1–4, 1985, Cleveland, OH.

[4.431] *Yan, D., Chen, W.-J., Savage, R.L., Meinecke, E.A.:* Characterization of Coal Carbon Black. Paper 29, 128th ACS-Conf., Rubber-Div., Oct. 1–4, 1985, Cleveland, OH.

Toxicity of Carbon Blacks

[4.432] *Nau, C.A., Neal, J., Stembridge, V.A.:* Arch. Ind. Health *17* (1958), p. 21; *18* (1959), p. 511; Arch. Environm. Health *1* (1960), pp. 512, 516; *4* (1962), p. 45.

[4.433] *Neal, J., Thorten, M., Nau, C.A.:* Arch. Environm. Health, *4* (1962), p. 46.

[4.434] *Ingalls, Th.:* Arch. Environm. Health *2* (1961), p. 429.

[4.435] ...: Mutagene und karzinogene Wirkung von Rußen, KGK *35* (1982), p. 555.

Selected References on Furnace Blacks

[4.436] *Boonstra, B.B. et al.:* RCT *51* (1974), p. 823.

[4.437] *Dannenberg, E.M.:* RCT *48* (1975), p. 410.

[4.438] *Medalia, A.I. et al.:* RCT 46 (1973), p. 1239.

[4.439] *Toussaint, H.E.:* 5th Int. Symp. f. Gummi, Gottwaldow, CSSR, September 1, 1975, Paper A 18.

[4.440] *Vohwinkel, K.:* SGF 70, Rönneby, May 28, 1970, Paper 37, Proc.

[4.441] *Wolff, S.:* KGK *27* (1974), p. 511; *Wolff, S. et al.:* KGK *28* (1975), p. 379.

[4.442] ...: Annual Book of ASTM Standards (1986), D 1765, pp. 434–435.

4.10.5.3 References on White Fillers

Reinforcing Silicate Fillers (see also [3.148]

[4.443] *Bomo, F., Morawski, J.C., Lami, P., Brachon, B.:* Quantitative Determination of the Morphology of Precipitated Silicas, Relationship between Morphology, Specific Area and Mechanical Properties of Filled Elastomers. Paper 77, 128th ACS-Conf., Rubber-Div., Oct. 1-4, 1985, Cleveland, OH.

[4.443 a] *Bühler, A.:* Aerosil R 202 als Thixotropierungsmittel für vorbeschleunigte Bisphenol-A-Polyesterharze und Vinylesterharze. KGK *40* (1987), pp. 1040-1042.

[4.444] *Dannenberg, E.M., Cotten, G.R.:* Rev. Gén. Caoutch. *51* (1974), p. 347.

[4.445] *Donnet, J.B., Wang, J., Papirer, E., Vidal, A.:* Influence of Surface Treatment on the Reinforcement of Elastomers with Silicas. ikt '85, July 24-27, 1985, Stuttgart, West Germany.

[4.446] *Ecker, R.:* Versuche zur Aktivierung heller Verstärkerfüllstoffe. Kautschuk u. Gummi *12* (1959), p. WT 351.

[4.446 a] *Ferch, H., Reisert, A., Bode, R.:* Aerosil R 972 als Füllstoff für Fluorelastomere. KGK *39* (1986), pp. 1084-1089.

[4.447] *Fetterman, M.Q.:* Utilizing the Unique Properties of Precipitated Silica in Design of High-Performance Rubber. Paper 8, 125th ACS-Conf., Rubber-Div., May 8-11, 1984, Indianapolis, IN.

[4.447 a] *Gonzalez, L.H., Iberra, C.R., Royo, J.M., Rodriguez, A.D., Chamorro, C.A.:* Sepiolithe - A New Inorganic Active Filler for the Rubber Industry. KGK *40* (1987), pp. 1053-1057.

[4.448] *Maxson, M.T., Lee, C.L.:* Effects of Fumed Silica Treated with Functional Disilazanes on Silicone Elastomer Properties. RCT *55* (1982), p. 233.

[4.449] *Morawski, J.C.:* Dispersion of Silica in Rubber, Relation with Mechanical Properties. Paper 67, 124th ACS-Conf., Rubber-Div., Oct. 25-28, 1983, Houston, TX.

[4.450] *Pal, P.K., De, S.K.:* Effect of Reinforcing Silica on Vulcanization, Network Structure, and Technical Properties of Natural Rubber. RCT *55* (1982), p. 1370.

[4.451] *Pal, P.K., De, S.K.:* Studies of Polymer-Filler Interaction, Network Structure, Physical Properties and Fracture of Silica and Clay Filled EPDM-Rubber in the Presence of a Silane Coupling Agent. RCT *56* (1983), p. 737.

[4.452] *Polmanteer, K.E., Lentz, C.W.:* Paper 7, 107th ACS-Conf., Rubber-Div., May 6, 1975, Cleveland, OH.

[4.453] *Shaw, I., Seeberger, E.F.:* GAK *27* (1974), p. 592.

[4.454] *Wagner, M.P.:* Reinforcing Silicas and Silicates. RCT *49* (1976), Review.

[4.455] *Wagner, M.P.:* The Role of Non Black Filler Reinforcing Materials in Rubber Compounding, Paper at the 122th ACS-Conf., Rubber-Div., Oct. 5-7, 1982, Chicago, MI.

[4.456] ...: White Reinforcing Fillers for Natural and Synthetic Rubbers. Publ.: Washington Chemical, Washington.

[4.457] DP 879 834, DP 900 339, 1951, Degussa.

[4.458] FP 1 064 230, 1952; 1 082 945, 1953, Columbian Southern. Chem. Corp.

Partially Active and Non-Active Fillers (see also [3.230, 4.451])

[4.459] *Hutchinson, J., Birchelli, J.D.:* Research on a New Reinforcing White Filler for Rubber. IRC 79, Oct. 3-6, 1979, Venice, Italy, Proc. p. 133.

[4.460] *Jeffs, D.J.:* Methods for Increasing the Efficiency of the Silane Treatment of Clays. IRC 79, Oct. 3-6, 1979, Venice, Italy, Proc. p. 850.

[4.461] *Kaeuffer, J.L.:* Clays in the Rubber Industry. KGK *35* (1982), p. 396.

[4.462] *Kishibe, M., Yokoyama, T.:* Effect of Bound Rubber on a Surface of Modificating Calcium Carbonate on Reinforcement of Filled Styrene-Butadiene-Rubber. IRC 85, Oct. 15-18, 1985, Kyoto, Proc. p. 487.

[4.463] *Korena, T., Tsukisata, R., Yamashita, S.:* Novel Calcium Carbonates modified with Azide Derivatives for Reinforcing Synthetic Rubber. IRC 85, Oct. 15-18, 1985, Kyoto, Proc. p. 492.

[4.464] *Nakatsuka, T., Kawasaki, H.:* End Group Effects in Polymer Modification of Calcium Carbonate Fillers. RCT *58* (1985), p. 107.
[4.465] *Skelhorn, D. A.:* A New Semi Reinforcing Filler for Rubber. Techn. Bulletin of ECC International.
[4.465 a] *Washabaugh, F. J.:* Einsatzverhalten von oberflächenmodifizierten Kaolinen in EPDM-Kautschuk. GAK *41* (1988), p. 120–127.
[4.466] *Wiedemann, P.:* Silanbehandelte mineralische Füllstoffe. KGK *38* (1985), pp. 377, 518.
[4.467] ...: Cyprubond, Technical Bulletin, Cyprus.

4.10.5.4 Short Fibers as Anisotropic Fillers

[4.468] *Abrate, S.:* The Mechanisms of Short-Fiber Reinforced Composites, A Review. Paper 28, 127th ACS-Conf., Rubber-Div., Oct. 23–26, 1985, Los Angeles, CA.
[4.469] *Chakraborty, S. C., Setua, D. K., De, S. K.:* Short Jute Fiber Reinforced Carboxy Nitrile Rubber. RCT *55* (1982), p. 1286.
[4.469 a] *Daojie, D.:* Structure and Properties of Natural Rubber-Short Cellulose Fibre Composite. IRC '86, June 2–6, 1986, Göteborg, Sweden, Proceed. p. 312.
[4.469 b] *Daojie, D.:* Dynamic Performance of Rubber Composite Reinforced with Short Fibre. Rubbercon '87, June 1–5, 1987, Harrogate, GB, Proceed. B 39.
[4.470] *Goettler, L. A., Chen, K. S.:* Short Fiber Reinforcement Elastomers. RCT *56* (1983), p. 619, Review.
[4.471] *Moghe, S. H.:* Short Fiber Reinforcement of Elastomers. Paper 20, 122th ACS-Conf., Rubber-Div., Oct. 5–7, 1982, Chicago, MI.
[4.472] *Murty, V. M., De, S. K.:* Short Jute Reinforced Rubber Composites. RCT *55* (1982), p. 287.
[4.472 a] *Nanda, G. B.:* Short Polyester – Fibre Reinforced Natural Rubber Composites. IRC '86, June 2–6, 1986, Göteborg, Sweden, Proceed. p. 541.
[4.472 b] *Nogushi, T.:* Dynamic Fatigue of Short Fibre – Rubber Composite under Compressive Stress. IRC '86, June 2–6, 1986, Göteborg. Sweden.
[4.472 c] *Nogushi, T., Mashimo, S., Nakajima, M., Yamagushi, Y, Ashiden, M.:* Stress Decay and Surface Temperature Distribution of Short Fibre-Runner Composites under Dynamic Fatigue. Rubbercon '87, June 1–5, 1987, Harrogate, GB, Proceed, A 50.
[4.473] *Utracki, L. A.:* The Shear and Elongation Flow of Polymeric Containing Anisomeric Filler Particles. Paper 19, 124th ACS-Conf., Oct. 25–28, 1983, Houston, TX.
[4.474] *Setua, D. K., De, S. K.:* Short Silk Reinforced Natural Rubber Composites. RCT *56* (1983), p. 808; *57* (1984), p. 351.
[4.475] *Setua, D. K.:* Tear and Tensile Properties of Short Silk Fiber Reinforced Styrene-Butadiene Rubber Composites. KGK *37* (1984), p. 962.
[4.475 a] *White, J. L., Czarnecki, L., Tanaka, H.:* Experimental Studies of the Influence of Particle and Fibre Reinforcement on the Reological Properties of Polymer Melts. RCT *53* (1980), pp. 823–835.
[4.475 b] ...: Technische Anwendung von Santowebb-Fasern in der Gummiindustrie – Ein Überblick. GAK *35* (1982), p. 522.

4.10.5.5 Rubber Regrind as Filler

[4.476] *Efferding, P.:* Gummimehl als Sekundärstoff für die Gummiindustrie. GAK *35* (1982), p. 348.

4.10.5.6 References on Filler Activators

(see also [4.77, 4.214, 4.451, 4.460, 4.465–4.467])

[4.477] *Cameron, G. M. et al.:* Eur. Rubber J. *156* Nr. 3 (1974), p. 37.
[4.477 a] *Debnath, S., Bhattacharya, A. K., Khastgir, D., De, S. K.:* Effects of Silane Coupling Agent on Network and Kinetics of Vulcanization of Mica – Filled Styrene – Butadiene Rubber. KGK *40* (1987), pp. 938–940.

[4.478] *Monte, S.J., Sugerman, G.:* Use of Titanate Coupling Agents in Elastomers. Paper 28, 122nd ACS-Conf., Rubber-Dir., Oct.5–7, 1982, Chicago, MI.

[4.479] *Monte, S.J., Sugerman, G.:* New Titanate Coupling Agents Designed to Eliminate Particulate Pretreatment. Paper 61, 125th ACS-Conf., Rubber-Div., May 8–11, 1984, Indianapolis, IN.

[4.480] *Monte, S.J., Sugerman, G.:* Neoalkoxy Titanate and Zirconate Coupling Agent Applications in Thermosets. Paper 30, 127th ACS-Conf., Rubber-Div., April 23–26, 1985, Los Angeles, CA.

[4.480 a] *Monte, S.J.:* Neoalkoxy-Titanate und -Zirkonate als Kupplungsmittel. GAK *39* (1986), pp. 112–117; pp. 242–249.

[4.481] *Pickwell, R.J.:* Bistriethoxysilylethyltoluenepolysulfide, a Scorch Resistant Silane Coupling Agent for Mineral-Filled Elastomers. RCT *56* (1983), p. 94.

[4.482] *Ranney, M.W. et al.:* KGK *26* (1973), p. 409; GAK *27* (1974), p. 600.

[4.483] *Vondrácek, P., Hradec, M., Chralovský, V., Khani, H.D.:* The Effect of the Structure of Sulfur-Containing Coupling Agents on their Activity in Silica-Filled SBR. RCT *57* (1984), p. 675.

[4.484] *Wolff, S.:* Non-Black Reinforcing Agents, Paper 6, 116th ACS-Conf., Rubber-Div., Oct. 23–26, 1979, Cleveland, OH.

[4.485] *Wolff, S.:* Si 230, ein neues Silan für die kautschukverarbeitende Industrie. KGK *32* (1980), p. 1000.

[4.486] *Wolff, S.:* Si 230 – A New Compounding Ingredient for Siliceous Filler-Loaded Halogen-Rubber Compounds. Paper 28, 119th ACS-Conf., Rubber-Div., June 2–5, 1981, Minneapolis, MN.

[4.487] *Wolff, S.:* Reinforcing and Vulcanization Effects of Silane Si 69 in Silica Filled Compounds. KGK *34* (1981), p. 280.

[4.488] *Wolff, S.:* Optimization of Silane-silica OTR Compounds, I. Variations of Mixing Temperature and Time During the Modification of Silica with Bis-(3-triethoxysilylpropyl)-tetrasulfide. RCT *55* (1982), p. 967.

[4.489] *Yamashita, S., Shigaraki, M., Orita, M., Nishimura, J.:* Vulcanization of 1-Chlorobutadiene – Butadiene Rubber by an Amino Silane Coupling Agent. IRC 79, Oct. 3–6, 1979, Venice, Italy, Proc. p. 1076.

4.10.5.7 References on Pigments

[4.490] *Ingle, G.W.:* Mod. Plastics *31* No. 11 (1954), p. 69.

[4.491] *Smith, D.A.:* Rubber J. *153* (1971), p. 19.

[4.492] *Jacobsen, E.E.:* Ind. Engng. Chem. *41* (1949), p. 523.

[4.493] *Becker, E.A.:* Lithopone. Verlag Berliner Union, Stuttgart, 1957.

[4.494] ...: Pigments in Rubber 1960–1970. Bibliography 112, Ed.: ACS, Rubber-Div., Library and Inf. Service, University of Akron.

4.10.5.8 References on Organic Fillers

[4.495] *LeBras, J.:* RCT *35* (1962), p. 1308.

[4.496] *Powers, P.O.:* RCT *36* (1963), p. 1542.

[4.497] *Green, K.M.:* Paper at the ACS-Conf., Southern Rubber Group, Feb. 22, 1974, Houston, TX.

[4.498] *Newberg, G.R.:* Rubber Age 62 (1948), p. 533.

[4.499] *Hofmann, W.:* Nitrilkautschuk. Verlag Berliner Union, Stuttgart, 1965, p. 304.

[4.500] *Hofmann, W.:* Industrie des Plastiques Modernes, 1961, p. 37.

[4.501] *Brown, G.L.:* Resins in Rubber. Clairtown Pa., Pennsilvania Industrial Chemical Corp. 1969.

[4.502] *Fries, H., Esch, E., Kempermann, Th.:* Neuartige härtende Phenolharze für die Gummiindustrie mit breitem Anwendungsspektrum. IRC 79, Oct. 3–6, 1979, Venice, Italy, Proc. p. 111; KGK *32* (1979), p. 860.

4.10.6 References on Plasticizers, Resins, Process Aids and Factice

4.10.6.1 References on Plasticizers

General References on Plasticization and Plasticizers

[4.503] *Dimeler, G.R.:* Plasticizers for Rubber and Related Polymers. Paper at the 122nd ACS-Conf., Rubber-Div., Oct. 5–7, 1982, Chicago, MI.
[4.504] *Hamed, G.R.:* Use of WLF Equation. Paper at the 123rd ACS-Conf., Rubber-Div., May 10–12, 1983, Toronto, Ont.
[4.505] *Ritchie, P.D.:* Plasticizers, Stabilizers, Fillers. Publ.: Plastics Institute, Liffe Books, London, 1972.
[4.506] *Ludwig, L.E. et al.:* India Rubber World *111* (1944), pp. 55, 180.
[4.507] *Stephens, H.L.:* Plasticizer Theory – An Overview Based on Work of Doolittle, Bueche, Ferry. Paper at the 123rd ACS-Conf., Rubber-Div., May 10–12, 1983, Toronto, Ont.
[4.508] *Stöcklin, P.:* Beitrag zur Frage der Weichmacherwirkung in Hochpolymeren. Kautschuk u. Gummi *2* (1949), p. 367; *3* (1950), pp. 45, 86, 199.
[4.509] ...: Weichmacher in der Gummiindustrie, Teil 1, Grundlagen und Analysemethoden. Ed.: WdK, Frankfurt, 1971, Grünes Buch No. 32.

Synthetic Plasticizers (see also [1.12–1.14, 3.182, 3.214, 3.216])

[4.510] *Dougherty, P.C., Cassis, F.A.:* Vinyl-Plasticizers from Trimellitic Anhydride. SPE-Journal, November 1962, p. 1387.
[4.511] *Hertz, D.L., jr.:* Phthalate Ester Plasticizer – A Compounding Study. Paper at the 123rd ACS-Conf., Rubber-Div., May 10–12, 1983, Toronto, Ont.
[4.512] *Hoppe, J.:* Synthetische Kohlenwasserstoff-Weichmacher und ihre Verträglichkeit mit Polymeren. GAK 34 (1981), p. 486.
[4.513] *Kumanova, B.K., Angelova, A.V., Voynova, S.Ch.:* Studies on the Reagent Composition During the Synthesis of the Plasticizer Dioctylphthalate. KGK *35* (1982), p. 495.
[4.514] *Matheson, D.R.:* Effect of Plasticizers on Nitrile Rubber. Paper at the 123rd ACS-Conf., Rubber-Div., May 10–12, 1983, Toronto, Ont.
[4.515] *Seto, K., Timar, J.:* Plasticizers in Krynac Rubbers. Techn. Bulletin by Polysar.
[4.516] *Whittington, W.H.:* Monomeric and Polymeric Ester (Polar) Plasticizers – A Review. Paper at the 123rd ACS-Conf., Rubber-Div., May 10–12, 1983, Toronto, Ont.

Mineral Oils

[4.517] *Corbin, H.E.:* Oil Use in Rubber Processing. Rubber Age *106* (1974), p. 49; Rubber India *26* (1974), p. 26.
[4.518] *Dimeler, G.R.:* Etude de la Volatilé des Huiles de Mise en Œuvre pour Caoutchouc et Influence sur le Compartement des Vulcanisats. Rév. Gén. Caoutch. *51* (1974), p. 91.
[4.518a] *Glenz, W.-D., Wommelsdorf, R.:* Die technischen Möglichkeiten der Mineralölindustrie zur Herstellung von Prozeßölen für die Kautschukindustrie. KGK *39* (1986), p. 816–822.
[4.519] *Godail, M.J.:* Plasticizing EPDM (Non Polar) Elastomers – A Review. Paper at the 123rd ACS-Conf., Rubber-Div., May 10–12, 1983, Toronto, Ont.
[4.520] *Hamilton, G.G.:* Can Mixtures of Aromatic and Paraffinic Oils Replace Naphthenics in Rubber Compounds. GAK *34* (1981), p. 570.
[4.521] *Holl, I., Coats, H.:* Ind. Engng. Chem. *20* (1928), p. 641.
[4.522] *Kurtz, S. et al.:* Engng. Chem. *48* (1950), p. 2233.
[4.523] *Robinson, R.A.:* Process Oils in Rubbers and PVC, Information Report 5947, RAPRA, Shawbury, GB, 1971.
[4.524] *Rostler, F.:* Rubber Age *69* (1951), p. 559; *70* (1952), p. 735; *72* (1952), p. 223.
[4.525] *Sweeney, I. et al.:* Rev. Gén. Caoutch. *34* (1957), p. 170.

4.10.6.2 References on Resins and Tackifiers
(see also [1.12-1.14, 2.86, 3.146, 3.363, 4.499, 4.502])

[4.525 a] *Abendroth, H., Umland, H., Schuster, R.A., Schmidt, R.:* Evaluations on the Mechanisms of Homogenizing Resins. Rubbercon '87, June 1-5, 1987, Harrogate, GB, Proceed. B 15.

[4.526] *Copley, B.C.:* Tackification of Natural Rubber/Styrene Butadiene Blends. RCT *55* (1982), p.416.

[4.527] *Class, J.B.:* The Effect of Low Molecular Weight Resins on the Viscoelastic Properties of Elastomers. Paper at the 123rd ACS-Conf., Rubber-Div., May 10-12, 1983, Toronto, Ont.

[4.528] *Fries, H.:* Entwicklung des Einsatzes von Harzen in der Gummiindustrie. GAK 38 (1985), p.454.

[4.529] *Giller, A.:* Rubber Plastics Age *45* (1964), p.1205.

[4.530] *Giller, A.:* Die Wirkung von Klebrigmachern. GAK *29* (1976), p.766.

[4.531] *Giller, A.:* Phenolic Resins in Todays Rubber Industry. Paper C 7, IRC 78, Kiev.

[4.532] *Ghatge, N.D., Shinde, B.M.:* The Synthesis of Isocyanate-Modified Novolak Resins for Use in Natural and Synthetic Rubber. RCT *53* (1980), p.239.

[4.533] *Guja, J., Bravar, M., Jelenčić, J.:* The Influence of the Type of Alkylphenols in Phenol-Formaldehyde Resins and of Activators on the Vulcanization of Butyl Rubber. KGK *38* (1985), p.28.

[4.533 a] *Härtel, E.:* Synthesis, Crosslinking and Characterization of Modified Epoxies. IRC '86, June 2-6, 1986, Göteborg, Sweden.

[4.533 b] *Leicht, E., Sattelmeyer, R.:* Phenolic Resins for Reinforcement, KGK *40* (1987), pp.126-129.

[4.533 c] *Nieberle, J., Paulus, G., Queins, H., Schöppl, H.:* Über den Einfluß von Phenol-Formaldehyd-Novolaken in Kautschuk. KGK *39* (1986), pp.108-114.

[4.534] *Shephard, A.F., Boiney, J.F.:* Modern Plastics *24* (1946), pp.154, 210, 212.

[4.534 a] *Souphantong, A., Cseh, M., Edelenyi, A.:* Structure and Properties of Phenolic Resin Cured EPDM. Rubbercon '87, June 1-5, 1987, Harrogate, GB, Proceed. p.15.

[4.535] *Spearman, B.P., Hutchinson, J.D.:* Rubber Age *106* No.8 (1974), p.41.

[4.536] *White, L.T.:* The Role of Tackifying Resins on the Tackification of Rubber Compounds. Paper 73, 124th ACS-Conf., Rubber-Div., Oct.25-28, 1983, Houston, TX.

[4.537] *Wolney, F.F., Lamb, J.:* Alkylphenol-Based Tackifiers - A Critical Review. Paper 74, 124th ACS-Conf., Rubber-Div., Oct.25-28, 1983, Houston, TX.

4.10.6.3 References on Process Aids (see also [3.267-3.269, 3.456])

[4.538] *Brichzin, D.:* Bewertung der Wirksamkeit von Verarbeitungsadditiven als Gleitmittel in Kautschukmischungen. KGK *36* (1983), p.451.

[4.539] *Crowther, B.G.:* Process Aids - A Review. Paper at the PRI-Conf. 84, March 12-15, 1984, Birmingham, GB.

[4.540] *Crowther, B.G.:* Process Aids - Are they Successful in Natural Rubber? Paper 110, 128th ACS-Conf., Rubber-Div., Oct.1-4, 1985, Cleveland, OH.

[4.541] *Drake, R.E., Labriola, J.M.:* Vulcanizable Peptizer and Process Aid for Natural, SBR and Polybutadiene Rubber. Paper 27, 127th ACS-Conf., Rubber-Div., April 23-26, 1985, Los Angeles, CA.

[4.542] *Fegade, N.B., Deshpande, N.M., Millns, W.:* Amides of Fatty Acids and Resin as Multifunctional Ingredients for Rubber, KGK *37* (1984), p.604.

[4.543] *Haverland, A., Hofmann, W.:* New Rating Techniques for Process Aids, Paper at the PRI-Conf. 84, March 12-15, 1984, Birmingham, GB.

[4.543 a] *Haverland, A., Hofmann, W.:* Neuartige Bewertungskriterien für Verarbeitungshilfsmittel. 3. Verbesserung des Verarbeitungsverhaltens von Mischungen ausgewählter Spezialkautschuke. dkt '88, July 4-7, 1988, Nürnberg, West Germany.

[4.544] *Hepburn, C.:* A Tensid Plasticizer to Increase Tensile Strength of Rubber, Paper at the SRC 83, May 19-20, 1983, Bergen, Norway.

[4.545] *Hepburn, C.:* Novel Multipurpose Additives. Europ. Rubber J., June 1984, p.23.

[4.546] *Hepburn, C., Mahd, M.S.:* Amine Bridged Amines which Function as Multipurpose Agents and Processing Aids in Polychloroprene Rubber. ikt 85, June 24–27, 1985, Stuttgart, West Germany; KGK *39* (1986), pp. 629–632.

[4.547] *Hofmann, W.:* Einige für die Beurteilung des Verarbeitungsverhaltens von Kautschukmischungen wichtige rheologische Gesetzmäßigkeiten. KGK *38* (1985), p. 777.

[4.548] *Hofmann, W., Haverland, A.:* New Rating Techniques for Process Aids, II. The Production Rate influenced by Process Aids in General Purpose Rubbers. IRC 86, June 2–6, 1986, Göteborg, Sweden.

[4.548 a] *Hofmann, W., Haverland, A.:* Neuartige Bewertungskriterien für Verarbeitungshilfsmittel in verschiedenen Kautschuken. Paper at Denmarks Gummiteknologiske Förening, Summer Meeting, Jund 2–3, 1988, Braedstrup, Denmark.

[4.548 b] *Hofmann, W., Haverland, A.:* Processing of General Purpose and Special Rubbers under Rheological Considerations. Rubbercon '88, Oct. 10–14, 1988, Sydney, Australia.

[4.549] *Klingensmith, W.H., Danilowicz, P.A., Larsen, L.C.:* Effective Use of Homogenizing Agents. Paper 52, 118th ACS-Conf., Rubber-Div., Oct. 7–10, 1980, Detroit, MI.

[4.550] *Klingensmith, W.H.:* The Effect of Process Aids in Rubber Compounding. Paper at the 122nd ACS-Conf., Rubber-Div., Oct. 7–10, 1982, Chicago, MI.

[4.551] *Larsen, L.C., Klingensmith, W.H.:* Overcoming Process in Problems of High Performance Vulcanizates using Processing Agents. Paper at the PRI-Conf. 84, March 12–15, 1984, Birmingham, GB.

[4.551 a] *Leblanc, J.L.:* Wirksamkeit von Verarbeitungshilfsmitteln in Kautschukmischungen. GAK *39* (1986), pp. 528–531.

[4.551 b] *Lloyd, D.G., Devaux, A., Leblanc, J.L.:* New Additives to Improve Processing of Rubber Compounds. Rubbercon '87, June 1–5, 1987, Harrogate, GB, Proceed. A 38.

[4.551 c] *Morche, K., Ehrend, H.:* Processing Aids for Tire Manufacturing. IRC '86, June 2–6, 1986, Göteborg, Sweden, Proceed. p. 514.

[4.552] *Nauwelaerts, H.E., Vermeulen, E., Worm, A.T.:* Processing Additives for Elastomers, A New Approach. KGK *34* (1981), p. 284.

[4.552 a] *Steger, L.:* Moderne Verarbeitungshilfsmittel für Synthesekautschuk. dkt '88, July 4–7, 1988, Nürnberg, West Germany.

[4.553] *Wolers, M., Klingensmith, W.H., Danilowicz, P.A., Howard, E.C.:* The Use of Processing Aids to Solve Tire Processing Problems. KGK *37* (1984), p. 17.

[4.554] ...: Dispergum – Zinkseifen, DOG-Kontakt No. 24, Techn. Bulletin of Deutsche Oelfabrik D.O.G., Hamburg, West Germany.

[4.555] ...: Dispergum 24 – ein hochwertiges Mastiziermittel. DOG-Kontakt No. 25, Techn. Bulletin of Deutsche Oelfabrik D.O.G., Hamburg, West Germany.

[4.555 a] ...: Flüssiges EPM/EPDM Trilene. Techn. Bulletin of Lehmann & Voss & Co., Hamburg, West Germany.

4.10.6.4 References on Factice

[4.556] *Berchelmann, F.:* Spezielle Einsatzmöglichkeiten für Faktis in Kautschuk-Mischungen. KGK *18* (1955), p. 577.

[4.557] *Erroll, F.J. (Hrsg.):* Symposium. Factice as an Aid to Productivity in the Rubber Industry. National College Rubb. Technol., 1962.

[4.558] *Flint, C.F.:* Factice, Relation of Structure to Properties. Proc. IRI *2* (1955), p. 151.

[4.559] *Flint, C.F.:* Factice in SBR-Compounds. Paper 6, Newton Heath Technical College. Nov. 8, 1961, Manchester.

[4.560] *Flint, C.F.:* Use of Factice in Butyl Rubber. Rubber J. Intern. Plastics *139* (1960), p. 490.

[4.561] *Flint, C.F.:* The Chemical Structure of Factice and its Behaviour at Vulcanization Reactions. J. IRI, June 1969, p. 110.

[4.562] *Harrison, I.B.:* Factice – Its Use and Function in Rubber Technology, Trans. IRI *28* (1952), p. 117.

[4.563] *Hofmann, W.:* Faktis. In: Ullmanns Encyclopädie der techn. Chemie. Verlag Chemie, Weinheim, 4th Ed., Vol. 13, 1977, p. 658.

[4.564] *Hofmann, W.:* Faktis. DOG-Kontakt No.27. Techn. Bulletin by Deutsche Oelfabrik D.O.G., Hamburg, West Germany.

[4.565] *Kirchhof, F.:* Über die praktische Anwendung von Faktis in der Kautschuk-Industrie. Gummi, Asbest *4* (1951), p.313.

[4.566] *Kirchhof, F.:* Neuere Ergebnisse über die Chemie der Faktisbildung und verwandter Reaktionen. Kautschuk u. Gummi *5* (1952), p.WT 115.

[4.567] *Kirchhof, F.:* Über den gegenwärtigen Stand der Chemie der Faktis-Bildung. Kautschuk u. Gummi *15* (1962), p.WT 168.

[4.568] *Kirchhof, F.:* Die Rolle des Faktis in der Technologie des Kautschuks. KGK *16* (1963), pp.201, 266, 431.

[4.569] *Lever, A.E.:* Factice - A Review of its Characteristics. India Rubber J. *120* (1951), p.820.

[4.570] *Webb, C.E.:* Sulphur Chloride Reactions in Relation to the Rubber Industry. Trans. IRI *27* (1951), p.279.

4.10.7 References on Blowing Agents

[4.571] *Eckelmann, W., Kaiser, G.:* Über den Zerfall von Benzosulfohydrazid in NR und in Mischungen. KGK *22* (1969), p.220.

[4.572] *Hunter, B.A.:* Chemical Blowing Agents. Rubber Age *108* No.2 (1976), p.19.

[4.573] *Lober, F.:* Entwicklung und Bedeutung von Treibmitteln bei der Herstellung von Schaumstoffen aus Kautschuk und Gummi. Angew. Chem. *64* (1952), p.65.

[4.574] *Mark, H.F., Gaylord, N.G., Bikales, N.M.:* Encyclopedia of Polymer Science. Interscience Publ. - J. Wiley Sons, New York, London, Vol.2, Blowing Agents, p.532.

[4.575] *Reed, R.A.:* The Chemistry of Modern Blowing Agents, Plastic Progress. British Plastic Convention. Kiffe & Sons, 1955, London.

[4.576] *Overbeck, W.:* Die Verfahren zur Herstellung von porösen Kautschukwaren. GAK *8* (1955), pp.560, 604, 686.

4.10.8 References on Adhesion and Adhesives (see also [4.525a-4.537])

4.10.8.1 References on Theory of Adhesion

[4.577] *Allen, K.W. (Ed.):* Adhesion. Elsevier Applied Science Publ., London, 1. 1977; 2. 1978; 3. 1979; 4. 1980; 5. 1981; 6. 1982.

[4.578] *Jensen, W.B.:* Lewis Acid-Base Interactions and Adhesion Theory. Paper at the 119th ACS-Conf., Rubber-Div., June 2-5, 1981, Minneapolis, MN.

[4.579] *Kelber, J.A.:* A New Method of Surface Characterization: Auger Line Shape Analysis. Paper at the 119th ACS-Conf., Rubber-Div., June 2-5, 1981, Minneapolis, MN.

[4.580] *Parks, C.R.:* Brass Powder in Rubber Vulcanizates, The Effect on Adhesion. RCT *55* (1982), p.1170.

[4.581] *Roberts, A.D.:* Adhesion Theory. Paper at the 119th ACS-Conf., Rubber-Div., June 2-5, 1981, Minneapolis, MN.

[4.582] *Westley, S.A., Gervase, N.J.:* Composite to Rubber Bonding, Vortrag, 119th ACS-Conf., Rubber-Div., June 2-5, 1981, Minneapolis, MI.

[4.583] *Wool, R.P.:* Molecular Aspects of Tack. Paper 56, 124th ACS-Conf., Rubber-Div., Oct.25-28, 1983, Houston, TX.

4.10.8.2 References on Rubber-Fabric Adhesion and Adhesives

General References on Rubber-Fabric Adhesion

[4.584] *Brodsky, G.I.:* Comprehensive Evaluation of Cord-To-Rubber Adhesion. Paper 101, 124th ACS-Conf., Rubber-Div., Oct. 25-28, 1983, Houston, TX.

[4.585] *Cembrola, R.J., Dudek, T.J.:* Cord/Rubber Material Properties. RCT *58* (1985), p.830.

[4.586] *Daan, H.A.:* The Special Requirements on the Adhesion of Textile Reinforcing Materials in Mechanical Goods. KGK *38* (1985), p.904.

[4.587] *Fielding-Russel, G.S., Livingston, D.I.:* Factors Affecting Cord-to-Rubber Adhesion by A Tire Cord Adhesion Test. RCT *53* (1980), p.950.

[4.587a] *Lievens, H.:* Effect of Aging on the Steel Cord-Rubber Interface. KGK *39* (1986), pp.122-126.

[4.588] *Salomon, T.S.:* An Overview of Cord Adhesion. Paper 57, 124th ACS-Conf., Rubber-Div., Oct.25-28, 1983, Houston, TX.

[4.589] *Wootton, D.B.:* The Adhesion of Textile Materials. Paper 58, 124th ACS-Conf., Rubber-Div., Oct.25-28, 1983, Houston, TX.

Rubber-Fabric Adhesives and Influence of Compounding on Adhesion

[4.590] *Albrecht, K.D.:* Untersuchungen über den Einfluß des Vulkanisationssystems auf die Gummi-Gewebe- und Gummi-Stahlcord-Haftung. KGK *25* (1972), p.531; RCT *46* (1973), p.981.

[4.591] *Ford, I.E.:* Observations on Adhesive Dip in Reyon Tire Cord. Trans. IRI *39* (1963), p.1.

[4.592] *Ingram, L.B.:* New Polyester Dip System Result in Improved Adhesion and Processing Cost Reduction. KGK *35* (1982), p.213.

[4.593] *Lattimer, M.B., Weber, C.D., Harat, Z.R.:* An Improved Adhesive System for Textile-Reinforced Rubber Products. RCT *58* (1985), p.383.

[4.594] *Morita, E.:* Reactions of Resin-Formers - Dry Bonding Rubber Systems. RCT *53* (1980), p.795.

[4.595] *Peerman, D.E.:* A New Terpolyamide for Bonding Textiles. KGK 35 (1982), p.398.

[4.595a] *Pokluda, I.:* Possibilities of Use of Univalent Phenol-Based Phenol-Formaldehyde Resins for Textile Dipping Cord. IRC '86, June 2-6, 1986, Göteborg, Sweden, Proceed. p.546.

[4.596] *Smely, Z., Mzourek, Z.:* Einige Verfahren zur Erhöhung der Festigkeit der Textil-Gummi-Bindung. Plaste u. Kautschuk *12* (1965), p.674.

[4.597] *Tomlinson, R.W.:* Improved Adhesion of EPDM Sulfur Vulcanizates to RFL-Treated Fabrics. RCT *55* (1982), p.1516.

Textiles (see also [4.468-4.475b])

[4.598] *Aitken, R.G. et al.:* Terylene Polyester Cord as Reinforcement of Tires, Rubber World 151 No.5 (1965), p.58.

[4.599] *Bradley, E.K.:* The Effect of Ozone and Humidity on Polyester to Rubber Adhesion. Paper 99, 124th ACS-Conf., Rubber-Div., Oct.25-28, 1983, Houston, TX.

[4.600] *Curley, I.B.:* The Case for Reyon. Rubber World 156 No.6 (1967), p.53.

[4.601] *Ebert, A.:* Entwicklung und Wettbewerbssituation der Chemiefasern für den technischen Einsatz. KGK *18* (1965), p.372.

[4.601a] *Heissler, H.:* Verstärkungsfasern in Hochleistungsverbundstoffen. GAK *40* (1987), pp.182-187.

[4.602] *Keefe, R.I., jr.:* Radial Truck Tire Aramid Reinforcement, Paper 45, 126th ACS-Conf., Rubber-Div., Oct.23-26, 1984, Denver, CO.

[4.603] *Kovac, F.I., McMillan, R.C.:* Polyester Tire - New Fibre Venture, Rubber World 153 No.5 (1965), p.83.

[4.603a] *Lang, B.:* Polyester as a Versatile Reinforcement. KGK *39* (1986), pp.115-121.

[4.604] *Marzocchi, A., Gagnon, R.K.:* Glass Fibre - Big Potential for Versatile Reinforcer. Rubber World 156 No.5 (1967), p.55.

[4.605] *Priest, M.H.:* Developments for Rubber and Plastics Reinforcement. Rubber and Plastics Age *46* (1965), p.491.

[4.606] *Pulvermacher, P., Pavlow, G.D.:* Hochfeste Fasern in der Kautschukindustrie. KGK *36* (1983), p.870.

[4.607] *Schröder, W.A., Puttman, I.B.:* Synthetic Fibers as Tire Cords, Rubber Age *99* (1967), p.72.

[4.608] *Seegel, V., McMahon, P.E.:* Eigenschaften der Kohlefaser mit verschiedenen Schlichten. GAK *37* (1984), p. 225.
[4.609] *Vanderbilt B.M., Clayton, R.E.:* Bonding of Fibres of Glass to Elastomers. Ind. Engng. Chem. Prod. Res. Dev. *4* (1965), p. 18.
[4.610] *Veen, A.M. van de:* Die Festigkeitsträger für den technischen Einsatz. KGK *32* (1979), p. 97.
[4.610 a] *Wagenmakers, J.C., Jansen, F.:* Garn- und Fertigprodukt auf dem Nichtreifen-Gummigebiet. KGK *34* (1981), p. 292.
[4.611] *Weening, W.E.:* Static and Dynamic Aspects of Aramid to Rubber Bonding, Paper 100, 124th ACS-Conf., Rubber-Div., Oct. 25–28, 1983, Houston, TX.
[4.611 a] *Weening, W.E.:* Aramid, Öffnung zu neuen Wegen bei Cord-Gummi-Compositen. KGK *37* (1984), p. 379.

Selected References on Rubber-Fabric-Adhesion

[4.612] US 2128229, 1935, Du Pont.
[4.613] US 2561215, 1945, Du Pont.
[4.614] US 2619445, 1949, General Tire & Rubber Ca.
[4.615] GB 1082531, 1963, ICI.
[4.616] DAS 1620816, 1966, ICI.
[4.617] *Mather, J.:* Br. Polym. J. *3* (1971), p. 58.
[4.618] DT 1301475, 1965 Bayer.
[4.619] *Meyrick, T.J., Watts, J.T.:* Trans. IRI *25* (1949), p. 150.
[4.620] DT 928252, 1942, Bayer.
[4.621] BE 668068, 1964, Bayer.
[4.622] FR 1366471, 1962, US Rubber Co.
[4.623] US 2994671, 1956, Du Pont.
[4.624] US 3307996, 1963, Du Pont.
[4.625] *Iyengar, Y.:* Paper at the ACS-Conf., May 10, 1976, San Francisco, CA.
[4.626] *Marzocchi, A., Jannarelli, A.E.:* Rubber World *158*, No. 6 (1968), p. 67.
[4.627] *Bartrug, N.G., Kolek, R.L.:* Adhes. Age *11*, No. 6 (1968), p. 2.
[4.628] FR 1459078, 1965, Owens Corning.

4.10.8.3 References on Rubber-Metal Adhesion and Adhesives

General References on Rubber-Metal Adhesion and Adhesives

[4.629] *Anthoine, G., Orband, A., Roebuck, H.:* Rubber Metal Adhesion. ikt 85, June 24–27, 1985, Stuttgart, West Germany.
[4.630] *Arklas, R.:* Theory and Practice of Silane Adhesion Promotors. Paper at the 119th ACS-Conf., Rubber-Div., June 2–5, 1981, Minneapolis, MN.
[4.631] *Bourrain, P., Morawski, J.C.:* Optimization of the Adhesion on the Rubber/Brass Interface by Means of a Silica Promoter. Paper 16, 126th ACS-Conf., Rubber-Div., Oct. 23–26, 1984, Denver, CO.
[4.632] *Buchan, S.:* Rubber to Metal Bonding. Crosby Lockwood, 1959.
[4.633] *Czerwinski, R.W.:* Surface Preparation and Quality Control-Key to Consistent Bond, Paper at the 119th ACS-Conf., Rubber-Div., June 2–5, 1981, Minneapolis, MN.
[4.634] *Hamed, G.R., Donatelli, T.:* Effect of Accelerator Type on Brass-Rubber Adhesion. RCT *56* (1983), p. 450.
[4.635] *Klement, G.:* Gummi/Metall-Bindung – Untersuchung der Grenzfläche Metall/Bindemittel. GAK *24* (1971), p. 430.
[4.636] *Klement, G., Scheer, H., Wirtz, W.:* Bindung von rußgefülltem Naturkautschuk an Stahl in Abhängigkeit von Vulkanisations-System und Wärmeübergang. KGK *24* (1971), p. 160.
[4.637] *Mori, K., Ohara, H., Nakamura, Y.:* Adhesion of NR to Triazine Thiols Treated Brass Plate during Vulcanization. IRC 85, Oct. 15–18, 1985, Kyoto, Proc. p. 899.

[4.638] *Mori, K., Sakakida, H., Nakamura, Y.:* Study on the Factors of Fixing between Metals and Vulcanizates. IRC 85, Oct. 15–18, 1985, Kyoto, Proc. p. 321.

[4.639] *Nitzsche, C. H.:* Haftung von Kautschuk an Metalle und Kunststoffe, Einflüsse durch Werkstoffe, deren Form und Oberflächenbehandlung. KGK *36* (1983), p. 572.

[4.640] *Özelli, R. N., Scheer, H.:* Untersuchungen über die Einwirkung organischer Medien und hoher Temperaturen auf die Gummi-Metall-Bindung. GAK *27* (1974), p. 612.

[4.641] *Özelli, R. N., Scheer, H.:* Untersuchungen der Bindung von in unterschiedlichen Verfahren hergestellten Gummi/Metall-Teilen. GAK *28* (1975), p. 512.

[4.642] *Özelli, R. N., Scheer, H.:* Abhängigkeit der Gummi/Metall-Bindung vom Vulkanisationssystem. GAK *32* (1979), p. 701.

[4.643] *Özelli, R. N.:* Entwicklung lösungsmittelfreier Gummi/Metall-Bindemittel unter Berücksichtigung der Umweltschutzgesetze. KGK *33* (1980), p. 260.

[4.644] *Özelli, R. N.:* Vergleichende Untersuchungen mit lösungsmittelhaltigen und lösungsmittelfreien Gummi/Metall-Bindemitteln. GAK *34* (1981), p. 653.

[4.645] *Özelli, R. N.:* Die Anwendung unterschiedlicher Chemosil-Produkte für spezielle Gebiete. Swiss Plastics *5* No. 9 (1983), p. 57.

[4.646] *Özelli, R. N.:* Untersuchungen an der Grenzfläche Gummi/Bindemittel. Swiss Plastics *6* No. 1–2 (1984), p. 27.

[4.646 a] *Özelli, R. N.:* Verbundteile bzw. Verbundstoffe in der Gummiindustrie. GAK *39* (1986), pp. 616–627.

[4.646 b] *Özelli, R. N.:* Das Polymer, wichtigster Parameter in Haftmechanismen. dkt '88, July 4–7, 1988, Nürnberg, West Germany.

[4.647] *Schneberger, G. L.:* Surface Preparation for Anhancing Adhesion, Paper at the 119th ACS-Conf., Rubber-Div., June 2–5, 1981, Minneapolis, MN.

[4.648] *Seitz, N., Schmid, R.:* Untersuchungen von Grenzflächenproblemen Gummi/Metall in der Schadensanalyse moderner Pkw-Schnittgürtelreifen – Erkenntnisse für die Betriebssicherheit. KGK *38* (1985), p. 1100.

[4.649] *Sexsmith, F. H., Weih, M. H., Siverling, C. E.:* Effect of Compounding Ingredients on Rubber-to-Metal Adhesion, I. Postvulcanization Bonding, Paper 97, 128th ACS-Conf., Rubber-Div., Oct. 1–4, 1985, Cleveland, OH.

[4.650] *Weening, W. E.:* Theorie und Praxis bei Untersuchungen über die Gummi-Metallhaftung. KGK *31* (1978), p. 227.

[4.650 a] *Weih, M. A., Siverling, C. E., Sexsmith, F. H.:* Einfluß von Mischungsbestandteilen auf die Bindung von Elastomeren an Metallen. GAK *40* (1987), pp. 422–433; pp. 558–561.

[4.651] . . .: Rubber-To-Metal Bonding. Bibliography 9, Publ. by RCT.

Rubber-Steel Cord Adhesion (see also [4.590])

[4.652] *Bourgois, L.:* Effect of the Adhesion System in Natural Rubber on Corrosion Rate of the Steel Cord Rubber Composite. Paper 41, 126th ACS-Conf., Rubber-Div., Oct. 23–26, 1984, Denver, CO.

[4.653] *Davies, J. R.:* Verbesserte Adhäsion von Kautschuk an verzinktem Stahlcord für Förderbänder. KGK *37* (1984), p. 493.

[4.654] *Ginffria, R., Marcelli, A.:* Configuration of the Brass on Brass-Plated Steel Wires in Tire Cords. RCT *55* (1982), p. 513.

[4.655] *Ishikawa, Y.:* Effects of Compound + Formulation on the Adhesion of Rubber to Brass-Plated Steel Cord. RCT *57* (1984), p. 855.

[4.656] *Ishikawa, Y., Kawakami, S.:* Effects of Salt Corrosion on the Adhesion of Brass Plated Steel Cord to Rubber. Paper 88, 127th ACS-Conf., Rubber-Div., April 23–26, 1985, Los Angeles, CA.

[4.657] *Lievens, H.:* Effects of Aging on the Steel Cord – Rubber Interface. ikt 85, June 24–27, 1985, Stuttgart, West Germany.

[4.658] *Lunn, A. C., Evans, R. E., Ong, J. L.:* The Adhesion of Rubber to Steel Tire Cord Under Cyclic Loading. Paper 20, 120th ACS-Conf., Rubber-Div., Oct. 13–16, 1981, Cleveland, OH.

[4.659] *Ooij, W.J. van:* Fundamental Aspects of Rubber Adhesion to Brass Plated Steel Tire Cord. RCT *52* (1979), Review.
[4.660] *Ooij, W.J. van, Weening, W.E., Murray, P.F.:* A New View on the Mechanism of Rubber Adhesion to Steel Cord. Paper *36,* 118th ACS-Conf., Rubber-Div., Oct. 7–10, 1980, Detroit, MI.
[4.661] *Ooij, W.J. van, Weening, W.E., Murray, P.F.:* Rubber Adhesion of Brass-Plated Steel Tire Cords: Fundamental Study of the Effects of Compound Formulation Variations on Adhesive Properties. RCT *54* (1981), p. 221.
[4.662] *Ooij, W.J. van:* Mechanism and Theories of Rubber Adhesion to Steel Tire Cords – An Overview. RCT *57* (1984), p. 421.
[4.663] *Ooij, W.J. van:* A Novel Class of Rubber to Steel Cord Adhesion Promoters. RCT *57* (1984), p. 686.
[4.663 a] *Orband, A., Anthoine, G., Roebuck, H.:* Verbesserte Haftung von Naturkautschuk an vermessingtem Stahlcord. KGK *39* (1986), pp. 37–42.
[4.664] *Peterson, A., Dietrick, M.L.:* Resorcinol Bonding Systems for Steel Cord Adhesion. Paper 77, 125th ACS-Conf., Rubber-Div., May 8–11, 1984, Indianapolis, IN.
[4.665] *Solomon, T.S.:* Systems for Tire Cord-Rubber Adhesion. RCT *58* (1985), p. 561, Review.
[4.666] *Tate, P.:* Maximize Steel-Cord Adhesion Using a Combined System of Cobalt and Resorcinal Formaldehyde Resin. Paper 21, 126th ACS-Conf., Rubber-Div., Oct. 23–26, 1984, Denver, CO.
[4.667] *Wagner, M.P., Hewitt, N.L.:* Compounding for Dynamic Adhesion of Rubber to Steel Cord, Paper 102, 124th ACS-Conf., Rubber-Div., Oct. 25–28, 1983, Houston, TX.
[4.668] *Wolf, G.H., Singenstroh, F.:* 1. Der Einfluß der Cordkonstruktion auf die Haftung Gummi/Stahl, KGK *34* (1981), p. 365; 2. Betrachtungen zur dynamischen Beanspruchung von Fördergurtseilen. KGK *38* (1985), p. 908.

Selected References on Rubber-Metal Adhesion

[4.669] *Gallagher, J. W.:* Adhesives Age 1968, p. 29.
[4.670] *Irrin, H., Cornell, W.H.:* Rubber World *132* (1955), p. 55.
[4.671] DT 928 252, 1942, Bayer.
[4.672] *Proske, W.:* Kautschuk u. Gummi, *6* (1954), p. WT 137.
[4.673] *Spearman, B.P., Hutchinson, I.D.:* GAK *28* (1975), pp. 519 ff.
[4.674] DT 1 143 017, 1958, Lord Manufacturing Co.

4.10.8.4 References on Rubber Cements (see also [4.525 a–4.537])

General References on Adhesion and Rubber Cements

[4.675] *Bhowmick, A.K., Gent, A.N.:* Effect of Interfacial Bonding on the Self-Adhesion of SBR and Neoprene. Paper 72, 124th ACS-Conf., Rubber-Div., Oct. 25–28, 1983, Houston, TX.
[4.676] *Dahlquist, C.A.:* An Overview of Adhesive Technology. KGK *38* (1985), p. 617.
[4.677] *Gent, A.N.:* The Role of Chemical Bonding in the Adhesion of Elastomers. RCT *55* (1982), p. 525.
[4.678] *Nakamura, Y.:* Coupling Effect of Triazine Thiols for the Interface in Rubber Composites. IRC 85, Oct. 15–18, 1985, Kyoto, Proc. p. 56.
[4.679] *Özelli, R.N.:* Klebstoffe und Klebeverfahren für Kunststoffe. VDI-Verlag 1974, pp. 123–137.
[4.680] *Özelli, R.N.:* Kleben von Elastomeren. GAK *29* (1976), p. 764.
[4.681] *Pocius, A.V.:* Elastomer Modification of Structural Adhesives. Paper 17, 126th ACS-Conf., Rubber-Div., Oct. 23–26, 1984, Denver, CO.
[4.682] *Wake, W.C.:* Adhesion and the Formulation of Adhesives. Elsevier Applied Science Publ., London, 1982.

Selected References on Rubber Cements

[4.683] *De Crease:* Rubber World *158* No. 1 (1958), p. 55.
[4.684] *Maesele, A., Debruyne, E.:* RCT *42* (1969), p. 613.
[4.685] *Ayerst, R. C., Rodger, E. R.:* RCT *45* (1972), p. 1497.
[4.686] *Hicks, A. E., Lyon, F.:* Adehes. Age 1969, p. 21.
[4.687] *Meyrick, T. J., Watts, J. T.:* Trans. Proc. IRI *13* (1966), p. 52.
[4.688] *Cox, D. R.:* Rubber J. 1963, p. 73.

4.10.9 References on Other Chemicals

[4.689] *Le Bras, J.:* RCT *35* (1962), p. 1308.
[4.689 a] *Brophy, J. F.:* Antimikrobielle Agentien für neue PUR-Produkte. GAK *40* (1987), pp. 135–138.
[4.690] *Powers, P. O.:* RCT *36* (1963), p. 1542.
[4.691] *Brown, G. L.:* Resins in Rubber. Clairtown Pa. Pennsylvania Industrial Chemical Corp. 1969, p. 101.
[4.692] *Green, K. M.:* ACS Div. Rubber Chem. Southern Rubber Group Meeting, Houston, Feb. 22, 1974, p. 9.
[4.693] *Hofmann, W.:* VDI-Nachrichten *19* No. 1 (1965), p. 2.
[4.694] *Hofmann, W.:* Indian Rubber Bull. *245* (1969), p. 5.
[4.695] *Hofmann, W.:* Kautschuk u. Gummi, Kunstst. *15* (1962), p. WT 501; Acta Medicotechnica *10*, 419 (1962); Rev. Gen. Caoutch. Plast. *41* (1964), p. 1119.
[4.696] *Hebermehl, R.:* Farbe u. Lack *280* (1955).
[4.696 a] *Ruhl, L.:* Trennmittelfreie Beschichtung von Spritz- und Pressformen. GAK *39* (1986), p. 250.
[4.697] *Schneider, P.:* Kautschuk u. Gummi *6* (1963), p. WT 21.
[4.698] *Sarbach, D. V., Garvey, B. S., jr.:* India Rubber World, *116* (1947), p. 798.

5 Processing of Elastomers*

[5.1–5.17]

The majority of rubber articles is manufactured from elastomeric gum which by being mixed with inorganic and organic substances before being vulcanized attains a variety of applications like no other polymer.

The usual additives (see Chapter 4, page 217 ff) are selected according to technical as well as economical considerations. While the technical considerations for articles which have to meet severe mechanical specifications such as car tires, conveyor belts, highly elastic mounts etc. are of highest importance, a lowering of the manufacturing price is desirable for rubber articles with less demanding service requirements and mass produced parts by extending the compound with substances such as reclaim, inert fillers, plasticizers, etc. By quality standards for an ever increasing number of articles, quality demands have been defined in most industrialized countries.

The elastomeric gum is mixed with the additives on heavy duty processing machines. Preliminary processing steps are used to obtain compound sheets, strips or pellets as intermediate material forms for the final processing by calendering, impregnation, coating, extrusion, molding, hand fabrication and dipping. In many cases the forming process is intimately interwoven with the always necessary vulcanization (e.g. in the case of molding). In other cases it is a separate production step, e.g. in the case of extrusion or calendering even though combined forming and continuous vulcanization systems are known for these processes also. For the manual manufacturing processes, a separation of the forming and vulcanization steps is always required.

5.1 Compound Preparation

[5.18–5.90]

The compound preparation which was formerly predominantly and then later only occasionally done on two roll mills is accomplished today predominantly in internal mixers. The mill is used mostly to sheet out and cool the rubber compound coming from the internal mixer. In some cases the final mixing process like the incorporation of sulfur and vulcanization accelerators as well as other minor components is done on the mill also.

In modern, large scale production facilities the material flow is extensively automated using micro processor controlled equipment which controls and monitors the whole mixing room (see Fig. 5.1) [5.18–5.26]. The elastomers which are delivered in the form of bales are reduced in size using bale cutters and choppers so that the required doses can be weighed. Today's mixing technology includes either one or multiple step mixing for which specific mixer types and system concepts are required. For multiple step production, first a preliminary compound (master batch) is prepared, which is then stored in the form of sheets or pellets. These are then brought to the final mixers together with the additional additives which flow from individual silos to the mixer, adjusted by micro processor controlled weighing equipment.

* The processing of latex is not covered in this book. See [1.16] regarding this subject matter.

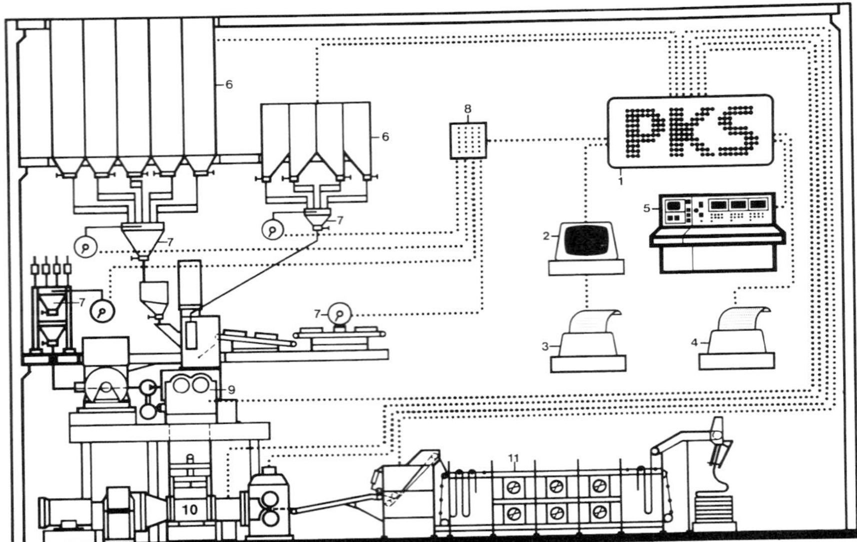

Figure 5.1 Complete mixing line, micro processor controlled
(Photo: Werner & Pfleiderer)
1 Process computer system PKS, 2 screen data entry, 3 data printer, 4 protocol and error printer, 5 control console for high efficiency internal mixer with control screen, 6 material storage (silos), 7 weighing of carbon blacks, additives, plasticizers and rubber, 8 weighing control, 9 high efficiency internal mixer, 10 forming extruder with roller die, 11 slab cooling line with wig-wag, cutting and stacking equipment

The principles which have to be followed when preparing compounds from NR and SR are presented in the chapters discussing the individual elastomer types and do not have to be discussed here in any detail.

NR is at temperatures which are considerably above the freezing point "frozen in" and is hard to cut. Therefore, it is heated in heating rooms resp. in ovens up to about 50 °C.

Reducing the Rubber into Small Pieces. NR and SR in most cases are delivered in bales of approx. 25 to 100 kg. They have to be cut down to smaller pieces by cutting machines or with knifes.

5.1.1 Operation on the Two Roll Mill

The compound preparation on the two roll mill is utilized primarily in small factories or for small size compounds. Additionally, mills have their importance as follow-up machines after internal mixers or as break-down and warm-up equipment in front of calenders or extruders. They also have great importance to masticate NR or SR as well as for finishing of rubber compounds. Therefore, they are still among the standard equipment in rubber factories. As mixing equipment, they are primarily used to prepare colored, tacky or very hard compounds.

5.1.1.1 Mixing Mills

The mixing mills are constructed from two horizontal parallel arranged *rolls* (see Fig. 5.2) made from hard castings which are supported through strong bearings, e.g. friction bearings, in the *mill frame* made from steel castings. The rolls run at different surface speeds (friction ratio) against each other.

Figure 5.2 Mixing and sheeting mill (450 mm diam., 1100 mm length) (Photo: Troester)

For *heating* or *cooling* purposes the rolls are hollow where it is important that the wall thickness is uniform in order to obtain an even surface temperature. In contrast to plastics processing, the rolls are seldom heated and then mostly only for start-up of production. Saturated steam is, for example, introduced through a rotary union into the rotating roll. The steam is sprayed through several jets against the inside wall where it condenses. An especially good cooling or heating efficiency is accomplished with peripherally drilled rolls (see page 363), where each series of three connected cooling channels is arranged approximately 25 mm under the roll surface.

The condensate forming during the heating process flows out of the roll via the rotary union. A temperature control and indication is not customary. However, in order to regulate the temperature, the condensate has to be discharged through a condensate container in order to attain a high pressure inside the roll. When the rolls are heated with hot water, a steam-water mixer is used.

Cooling is of great importance because of the heat generated due to viscous flow which is generated continuously inside the elastomer in the roll nip. The cooling water also enters and leaves the inside of the roll through a rotary union.

Laboratory rolls are sometimes heated electrically where the control is accomplished using adjustable transformers.

The *drive* is generally by electric motor with a reduction *gear box* and a pair of gears to the stationary back roll. A gear is affixed to the other end of the back roll which mates with a gear on the front roll (the so-called coupling gears). The ratio of the teeth of these two gears determines the friction ratio.

The *friction* ratio is defined by the difference of the surface speed of the two rolls (the back roll is running faster). It promotes tearing, kneading and mixing of the rubber mass and the compounding ingredients in the roll nip. Variable friction ratios are also available. Two coupling gears are sometimes available for production mills, one for creating a friction ratio, one for synchronous operation.

All mills are equipped with *side guides* which prevent the rubber material to flow from the roll nip into the bearings.

The *roll nip* can be adjusted by moving the front roll (by hand or electrically operated screws, seldom hydraulically).

A *pressure plate* is placed in between the head of the adjusting screw and the mating part of the bearing in order to protect the rolls from breaking. This pressure plate is punched through when the nip pressure is too high and lets the front roll slide forward. For example, bearing pressures up to 1 MN are permissible for a two roll mill of 550 × 1500 mm.

As long as the material placed onto the rolls does not form a continuous band (does not band), part of it falls down into a *bottom pan* below the roll nip. This material is placed back onto the rolls by the operator with a shovel. Attempts have been made to automate the process as much as possible, since the mixing or pre-heating on the two roll mill is discontinuous and since the quality of the finished compound depends very much on the care of the operator; the stock blender is used for this purpose (see Fig. 5.3).

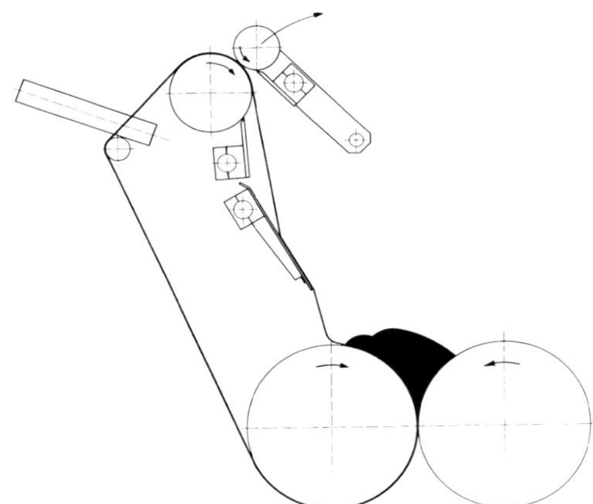

Figure 5.3 Stockblender (Principle: Berstorff)

The *stock blender* consists essentially of two deflection rolls of the same width as the mixing rolls which are arranged next to each other on the mill frames above the roll nip and two guide rolls which move backwards and forwards.

At the beginning of the mixing process, the compound is allowed to run on the front roll until it has banded sufficiently well. The band is then cut, pulled up and lead back to the roll nip over the deflection rolls of the stock blender. The driven deflection rolls assure the transport of the band. The band is deformed and uniformly and intensively mixed by the forward and backward movements of the guide rolls without any additional activity of the operator. After the predetermined mixing time, the band is cut off the stock blender and removed as usual in the form of slabs. The advantages of this process are in the rapid cooling of the band during processing and the better homogeneity and plasticity of the compound for a given mixing time.

A so-called *batch-off* device may be employed to automate the process further (see page 366). Its purpose is to dip the band coming from the mill into water or a non-stick solvent, to cool and then dry it, to cut it in conjunction with cutting equipment into plates of a certain size and then to either stack the plates or to deposit bands in wig-wag form.

In general the batch-off device is arranged so that the band is taken on a conveyor directly to the follow-up equipment.

Aside from mixing mills, which are available in production and laboratory sizes, the following machinery is also employed, depending on the processing requirements:

- washing mills
- fracture mills
- grinding mills
- refiners

Refiners are short, fast running two roll mills with a narrow gap and a friction ratio of, for example, 2.5:1 for the lump-free homogenizing of rubber solutions or regenerates.

Five different processing operations are carried out on mixing mills:

- mastication
- mixing
- cooling
- finishing of rubber compounds and
- pre-heating of already mixed compounds

5.1.1.2 Mastication on the Mixing Mill

[4.29–4.35]

NR has to be plasticized (broken down) by mastication before the mixing operation (see pages 25 and 217 ff). The mastication without peptizers is carried out at low roll temperatures with full cooling of the rolls. However, today the addition of mastication aids and higher temperatures are preferred. Since most SR types are already offered at mixing viscosities and since they are scarcely masticatable, a separate mastication step is superfluous.

For the *mastication without peptizer* the raw rubber is placed upon the rolls which are cooled and adjusted for a narrow gap, whereby the partially disintegrated material which passed through the rolls falls into the pan placed under the rolls. After the second pass most often a continuous band is formed which is pulled over the front roll back into the roll nip after the roll gap has been enlarged.

Because of the friction ratio of the rolls, the elastomer is torn and rotates around the front roll as a disjointed band. After repeated passages through the roll gap a masticated mass is formed which surrounds the roll as a continuous band. The final roll position is adjusted so that a consistently rotating bank of material forms in the roll nip. This bank aids the mastication process considerably. The degree of mastication attainable depends on roll temperature, the size of the gap and the number of passes. If a stock blender is available the banded material sheet is routed as a continuous ribbon over the deflection rolls through air for cooling reasons before it is returned to the roll gap. Blowing compressed air on the ribbon enhances the cooling effect.

When *mastication aids* are used, the rubber and the peptizers are placed on the rolls which are cooled to 40–60 °C and are adjusted for a narrow gap. Once the material has banded, the gap is opened to 8–10 mm. Now the band is cut regularly right and left while the rubber heats up to 120–130 °C. The dose of mastication aid determines the degree of mastication.

The *elastic component,* i.e., the tendency for elastic recovery of the elastomer (nerve), has to be reduced during mastication at least so much that the rubber is deformed easily during the mixing operation so that a continuous band is formed during mixing on the mill. The elimination of the nerve is of utmost importance for the following processing operations such as calendering or extrusion. A material with too much nerve would give rise to an excessive elastic recovery.

An excessive degradation (over-mastication) has to be avoided since the vulcanizate properties, especially the aging properties and mechanical properties, are adversely affected by it. Depending on the various applications of the elastomer, various degrees of mastication are desired. For example, for a sponge rubber compound a well masticated rubber is required so that the resistance to the gas formation due to the blowing agent is not hindered. On the other hand, the elastomer for the production of profiles which have to maintain their slope during vulcanization in the autoclave, should be masticated only so far that fillers can be incorporated and that it can be extruded smoothly. It is important that the degree of mastication for a given compound must be reproduced from batch to batch. Therefore to receive an uniform mastication effect, an uniform viscosity, molecular weight distribution and gel content of the rubber is of importance. In NR these conditions play a smaller roll. With application of constant viscosity NR-types (see page 18) mastication has lost some of its earlier importance.

5.1.1.3 Compound Preparation on the Mill

It is important during compound preparation on the mill to disperse as homogeneously as possible all compound ingredients which are present in powder form or in the liquid phase. The dispersion is based upon the kneading action in the roll nip and especially upon the friction ratio between the two rolls. Because of the higher circumferential speed of the back roll the material bank in the roll nip is kept in a rotating movement while the back roll as friction roll tears off pads continuously and redeposits them into the band rotating around the front roll.

The mixing process can be undertaken right after the mastication. If pre-masticated rubber is used, it has to be milled first on rolls with a temperature of 40–50 °C to a continuous band. Then the accelerators, aging and light protectors, the resins, bitumen, factices, colorants, fillers and plasticizers are added. The vulcanization chemicals are preferably added separately. The accelerators which are most often only used as fractions of a percent and have to be perfectly dispersed in the rubber are added in the beginning over the whole length of the roll; the sulfur, which can be dispersed relatively easily is added at the end of the mixing cycle. During the addition of the components the band must not be cut.

With increasing addition of the ingredients the roll gap is increased so that the rotating bank in the roll nip is maintained.

At the end the compound is cut off repeatedly in order to homogenize it, then rolled up, redeposited, cut on the left and right sides and folded over. If a stock blender is available, it is utilized to homogenize the compound. Finally, single bands are cut off the roll and perhaps cooled in baths, then dried and stored.

The rolls have to be cooled intensively because a large amount of dissipated heat warms up the material during the mixing cycle. In addition, a well defined sequence has to be followed while adding the compound ingredients, e.g., the following one: vulcanization accelerators, aging and light protectors, resins, bitumen, factice, regenerate, colorants, fillers, plasticizers, sulfur. The incorporation of the plasticizer before that of the fillers usually prolongs the mixing time because the band is disrupted by the addition of the plasticizer, the shear forces are reduced and thus the filler dispersion is impaired. Therefore the plasticizers are mostly added together with the fillers or shortly thereafter, especially if large quantities are used. This applies especially to plasticizers which are taken up by the elastomer only with difficulty. If only small amounts of plasticizer are used, one adds them quite often with the residue collected from the mill pan.

The complete mixing process lasts mostly 20–30 min. In exceptional cases it can extend to 60 min or even longer.

5.1.1.4 Pre-Heating on the Mixing Mill

Pre-heating or re-mastication is necessary when prepared compounds have to be stored for some time. At this time, that remainder of the vulcanization system is added which had been left out previously to prevent scorching during storage. The compound has to be pre-heated in any case for further processing on calenders or extruders so that the required viscosity is obtained. Frequently, several pre-heating mills are installed next to each other which are connected overhead via conveyors (see page 378).

5.1.2 Operation in the Internal Mixer

[5.27–5.50]

In large rubber factories, especially tire factories, the internal mixer has practically replaced the two roll mill for the preparation of compounds (see Fig. 5.4). For reasons of efficiency, also small rubber factories are forced to prepare compounds in internal mixers and only in exceptional cases, on the mixing mill.

The mixing process in the internal mixer is accomplished inside a closed chamber by rotating kneading rotors. In contrast to the mill, more uniform compounds and large batches can generally be prepared by the internal mixers. Freedom from dust, especially during the preparation of carbon black filled compounds, and reduction of the chance of accidents due to the completely enclosed mixing chamber are further advantages. Internal mixers are superior regarding capital outlay, energy and labor costs as well as floor space. The mixing in modern, optimized internal mixers is economically superior to the time consuming and labor intensive milling process, even though one can often achieve a better dispersion of the compound ingredients on the mill compared to the internal mixer.

5.1.2.1 Internal Mixers

[5.27–5.60]

One distinguishes between internal mixers with and without a ram. Internal mixers without a ram are not installed any more because high efficiency mixers with ram are superior, especially in automated mixing lines both from the point of view of economics and compound quality.

The internal mixer consists of an enclosed mixing chamber which is equipped with cooling ribs on its outside. In the two cylindrically formed (horizontally in the form

Figure 5.4 Internal Mixer GK 300N (Photo: Werner & Pfleiderer)

of a horizontal figure eight) parts of the chamber two rotors of pear shaped cross-section are turning at different speeds (friction ratio generally very high). Designs without friction ratio are also used.

The older internal mixers without ram had inclined rotors above each other whereby screw formed rotors interpenetrated each other.

The *rotors* of today's mixers are, in contrast, arranged horizontally. One distinguishes between tangential and intermeshing rotor arrangements (see Fig. 5.5). The rotor placement and geometry are of utmost importance for the mixing efficiency [5.38, 5.39, 5.42, 5.49].

The internal mixers generally exhibit the following design details:

The base plate or the *base frame* is a box of cast iron or of welded construction with a clearance in the center for the discharging of the compound.

The *mixing chamber* which is built according to a variety of designs has an abrasion resistant wall which is formed by arc welding a layer of material. The walls also contain cavities for cooling or, if required, heating. Modern mixing chambers of high efficiency mixers are often equipped with injection jets for the addition of plasticizers.

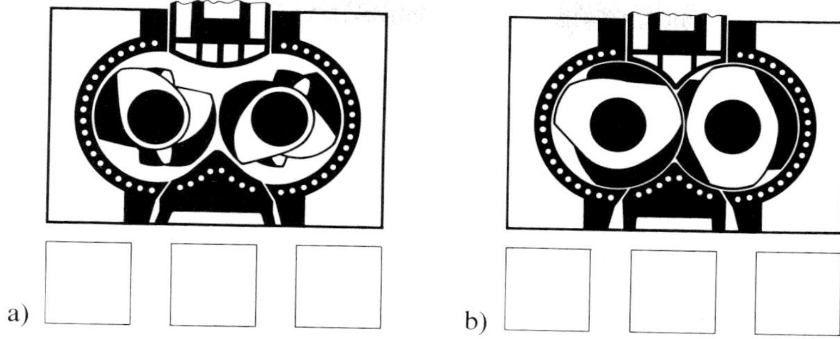

Figure 5.5 Mixing systems of internal mixers (Principle: Werner & Pfleiderer) a) tangential, b) intermeshing blade arrangement

The *filling chute,* mostly a steel sheet construction with a lid which is opened to add ingredients, is arranged over the whole width of the rotors. The chute is closed during the mastication or mixing cycle by a pneumatically operated ram from which the name "ram mixer" derives. The material is prevented from climbing up the chute by the ram. The ram mixer thus mixes very intensively and fast. Specific ram pressures of 2–12 bar are usual. Ram mixers with ram pressures of 6–12 bar are called high pressure mixers or high efficiency mixers.

The *kneading rotors* have to transmit high forces. They are manufactured from cast or alloyed steel and are also clad with a hard coating. The ratio of diameter to rotor length (d/l) varies between 1:1.4 and 1:1.7. The rotors also have bores or cavities for cooling or heating in the form of ring or spray cooling (see Fig. 5.6, see also page 357). They have either two or four blades arranged so that the compound is mixed intensively and is turned over from one side to the other.

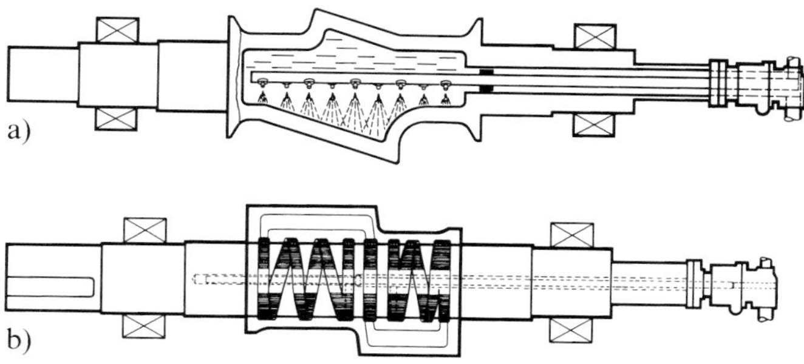

Figure 5.6 Ring cooling (a) and spray cooling (b) (Principle: Werner & Pfleiderer)

The *ram* is generally 2–4 mm smaller than the chute. Towards the filling lid it is tapered so that the material to be added can slide down. The ram can also often be cooled. It is connected via a rod to the piston inside the ram cylinder which moves the ram, mostly pneumatically by a compressor up to 12 bar. A sufficient *lubrication* of the piston rod guides is important since otherwise the danger of binding exists.

The *discharge opening* of internal mixer is closed by a rectangular, coolable or heatable hinged door, older designs have a sliding door which is operated hydraulically. A sliding door empties the chamber only slowly in smaller lumps while hinged doors permit the emptying in shortest time in the form of large lumps.

The size of an internal mixer is given in liters as the *total volume* which is determined by filling it with a liquid, the volume is then measured. The *nominal volume* (the free volume) is determined by the rotor geometry (mixing system) as well as the density and the viscosity of the rubber compound. The *fill volume* has to be smaller than the total volume so that the compound ingredients can also be mixed in the axial direction. The optimal fill volume for a new compound is generally determined by experiment. This volume is the one for which the physical vulcanizate properties exhibit the best values for otherwise constant conditions. The optimal free volume of a ram mixer is generally 70% of the total volume.

Laboratory mixers of 1 to 5 liters and production mixers of 15 to 650 liters total volume are available. After a long time in service the total volume of the mixer increases somewhat because of wear.

The *drive* of the mixer is by electric motor of relatively high power of 8 to 12 kw per liter mixing capacity, depending on the rotor speed, via gear reducers directly to the rotor shafts. While motors up to 800 kW are required for internal mixers of customary size, power of up to 2000 kW is required for 450 liter mixers [5.45].

Older models had variable ratio gear reducers with two to four reductions. A continuously controllable motor or two motors with different speed of rotation on one reducer are preferred today over a simple drive.

Modern internal mixers are equipped with *roller bearings*. In order to support the axial forces a double acting thrust bearing per rotor is installed. The roller bearing lubrication is primarily automatic using a grease lubrication system, in special cases by oil recirculating oil lubrication. The rotor shafts are sealed using a slide ring system, are lubricated by grease or plasticizer oil and have to seal the mixing chamber toward the outside. The seal system assures that no grease or oil enters the mixing chamber.

Seals of various design seal the rotor shafts with the mixing chamber walls. These seals are constantly supplied with oil or grease which mixes with the occuring dust and exudes as a paste.

Internal mixers have to be *cooled* very well in order to remove the heat generated during the mixing cycle. For this purpose the mixing chamber, the rotors and the discharge gate are, as already mentioned, equipped with bores and cavities for cooling water. Very modern mixers have cooling channels instead of drilled holes in the rotors and the housing.

The temperature control is via temperature probes in the mixer housing or discharge port area and are, as other control devices, connected to recorders in the control panel.

The *mixing times* are at multi-step operation in automatically controlled machines from 90 to 180 sec. That means not only that the compound ingredients have to be filled into the mixer in the shortest possible time but also that the compound has to be discharged very rapidly as well. A considerable improvement in modern machines is the hinged discharge port compared to the older sliding one whereby the discharge time has been reduced and also compound losses are avoided.

Modern internal mixers are often connected with automatic weighing equipment [5.19b, 5.20c, 5.20d, 5.25–5.25c]. Large volume ingredients, e.g., fillers and plasticizers and in few large factories also pelletized rubber is weighed, partially computer controlled, from silo installations and put into the mixer. Minor ingredients such as the chemicals are often added by hand but increasingly also in half-automatic small component weighing and transporting systems [5.19b, 5.20c, 5.25d].

5.1.2.2 Mastication in Internal Mixers

It can be accomplished only with mastication chemicals (see page 217 ff) because of the high temperatures. Because of the small quantities involved and because of the required good dispersion, formerly masterbatches of peptizers were frequently used. However, today mastication aids are offered in a form which permits their direct addition [2.121].

In order to limit heat generation during mastication as much as possible, the internal mixer is operated with full cooling. In addition, it is under-filled by about 10%. The operating conditions (speed, time) are chosen so that a rubber temperature of 180 °C is not exceeded. The entire process lasts approximately 3–5 minutes, depending on the desired degree of mastication, which depends in turn on the amount of mastication chemicals, the energy uptake (dependent on the filling level and the rotor speed), the rubber temperature and time.

The NR mixed in an internal mixer is cooled down to 80 °C on a follow-up machine and formed on a calender, extruder, strainer, or pelletizer, powdered or treated with anti-stick media.

The mixing process can follow immediately the mastication if internal mixers with synchronous inter-meshing rotors are used.

5.1.2.3 Compound Preparation in the Internal Mixer

[5.27–5.50]

The filling of a ram mixer is done through the chute with the ram in the raised position. The split or pelletized and weighed rubber is generally brought to the mixer on a conveyor which might be equipped with continuous weighing equipment or in special containers. After a short mastication, the fillers, plasticizers and chemicals are added where often fully or partially automatic equipment is used. Then the ram is lowered. A hard bottoming out of the ram can be heard after 15–25 seconds if the fill volume and the ram pressure are correct. The complete incorporation of the fillers into the rubber and a good dispersion can be detected by "sucking" noises. Then the temperature rises and the compound is discharged rapidly. Mixing times of only 2–3 minutes are no rarity today any more.

A variation of this procedure is the so-called upside-down method. Here the fillers are added first, then the rubber in order to obtain high shear forces [5.34].

Mostly a rubber compound is not mixed completely in an internal mixer. This would lead to scorch problems (see also page 240). Rather, frequently a masterbatch is prepared. The missing materials are then added in a second process (either after cooling in the mixer or on a pre-heating mill). There are also mixing systems that can mix a compound in a one-step procedure.

Efficiency increases with internal mixers, for example by further increases in the rotor speed (today often 40 rpm) or temperatures and thus decreases in mixing times are limited. A good dispersion [5.51–5.60] of the compounding ingredients has to be obtained since the degree of dispersion is rarely increased in following processing steps. This makes the addition of dispersing agents desirable [3.269, 3.860 a] (see page 300).

5.1.2.4 Homogenization in Follow-up Equipment

[5.50]

After discharge from the internal mixer, the compound is in the form of lumps and has to be homogenized, cooled and sheeted out on follow-up equipment.

Sheeting Mills. The sheeting mills are the oldest and most commonly used follow-up devices which, as already mentioned, are preferred for cooling down the compound, for sheet formation with perhaps simultaneous addition of minor ingredients and for cutting the sheets for cooling. In detail, these processes occur in the following way.

In those cases in which the internal mixer is placed above the mill, the hot compound falls in the form of irregular chunks either freely or over an inclined slide in the roll gap or is transported (if the machines are on the same level) via a conveyor. The rolls are cooled intensively. After banding, a stock blender is utilized (if available) in order to improve the heat transfer to the air and the homogeneity. Sometimes, compressed air exiting a jet is blown over the strip moving toward the stock blender roll. The band is contracted by the stock blender roll and returned to the roll nip in zig-zag form. After sufficient cooling and homogenizing, the band is cut off the rolls.

Batch-off Equipment. The band removal is accomplished in modern lay-outs by the batch-off equipment. In it, for example, a 600 mm wide strip is brought on a conveyor to the batch-off equipment where the strip runs through a water or anti-stick bath. After blow drying, the strip is then hung on the rods of a double chain conveyor or deposited in single sheets after cutting. The strips are cooled in a channel with blowers attached to the sides and then deposited at the end by a wig-wag (see Fig. 5.7).

Figure 5.7 Batch-off equipment (Landshuter Werkzeugbau; Photo: Dr. Küttner)

Forming Extruder. Forming extruders instead of sheeting mills are increasingly used as follow-up equipment of larger mixers or fully automated mixing systems with low personnel requirements. These can also, at least partially, be used as mixing extruders where part of the mixing effort is done in the post-mixing extruder. These forming extruders can be utilized as roller die, roller head or pelletizer units.

Figure 5.8 Sheet extrusion line (Photo: Troester)

The *roller die* extruders (sheet extrusion equipment, see Fig. 5.8) are specialty extruders which are used to manufacture an endless strip after the mixing operation. The compound pieces fall from the internal mixer through a big funnel onto the end of the screw where they are pressed into the screw flights by a hydraulically operated ram. The cylinder and the screw are cooled intensively. Either a wide strip is extruded through a sheet die or a thick walled tube which is then automatically slit and exits then as a wide strip which can then be carried for cooling to a batch-off machine just as is the case for the sheeting mill (see page 366).

In *roller head* systems [5.91, 5.92] the rubber compound is plasticated and homogenized well by the extruder and pressed into the roller head die. From here the compound goes over the whole width directly in the roll gap of a two-roll mill. In this way the roll bank normally found in calendering is avoided. Thus no air pockets are created which is the essential prerequisite for the production of bubble-free thicker strips. While sheets with a thickness of only about 3 mm can be prepared on classical calenders (see page 373) because of the danger of air pockets, thicker ones have to be prepared by stacking up thinner ones. The application possibilities of roller dies extend to thicknesses up to 20 mm and more. The application of roller head equipment thus produces endless compound sheets which are smooth and of high precision and are calibrated calendered sheets (see Fig. 5.9).

Another development is the *pelletizer,* also an extruder which extrudes the compound through a perforated hole plate to thumbthick strands which are cut into

Figure 5.9 WP-Roller-Head-Extruder (Photo: Werner & Pfleiderer)

pieces, the so-called pellets, by a rotating knife. This procedure is especially efficient since the pellets can be transported pneumatically, for example, to the final processing equipment and can be fed to it automatically.

5.1.3 Other Mastication and Mixing Methods

5.1.3.1 The Gordon Plasticator

The Gordon plasticator serves as a mastication machine and is according to its design a big extruder with a cylindrical barrel and a screw which is conical toward the die end. The rubber is pre-plasticated and warmed up in the cylindrical part. In the conical part of the screw increasingly strong shear gaps form between barrel wall and screw which have a masticating effect; their effect can be compared to the roll nip in the mixing mill. By shifting the screw, the shear gap can be adjusted to a desired thickness. This very robust machine can be charged with a whole bale of rubber (see also page 387, 392).

5.1.3.2 Powdered Rubber Processing

[5.61–5.75]

5.1.3.2.1 General Comments on Powdered Rubber Processing

Because of the high investment costs of today's mixing equipment and the high energy consumption during the mixing process from compacted rubber bales, attempts are being made to find a simplified process starting from powdered elastomers. Several elastomer types are already being offered in powder form.

The powdered rubber technology would probably gain importance first in the following product types:

- technical injection molded parts
- profiles and hoses
- cable sheaths
- calendered plates
- fabric coatings

For the continuous production processes for the articles named above the preparation of compounds from powdered rubber before the processing step (extrusion, calendering, vulcanization) consists of the following steps:

- weighing the powder coming from storage containers
- automatic weighing
- preparation of powdery mixture in a dry blender
- compacting the powdered mixture

A process system such as this could be operated with a minimum of personnel. However, for a variety of reasons the very efficient processing procedure described below could not find wide acceptance.

5.1.3.2.2 Compound Preparation in the Dry Blender

[5.70, 5.71]

The first step is the preparation of powder compounds in one step using a dry-blender.

A shaft to affix the mixing tool is protruding through the bottom of a vessel. The tool consists of a bottom scraper, a hammer knife in the center and a rotating suction knife above. The drive of the tool has a stepless control whereby the speeds of rotation decrease as the machines increase in size. The lid can be swung out of the way for manual filling. In case of continuous filling the lid is equipped to accept feeding pipes.

The degree of filling is an important operating variable. When the mixer is filled correctly, a continuous material flow develops. A sheet metal guide assures that the material stream is constantly directed toward the mixing tool. The optimum rotation speeds depend on the compound and are between 1000 and 1500 rpm. The mixing sequence when filling the mixer is arbitrary. It is important that the plasticizer is injected after the material stream is well developed. The total mixing cycle is between 3 and 4 minutes. During the mix preparation, essentially granulation and mixing work is done while the ingredients are simultaneously affixed to the rubber pellets. Considering the bulk specific volume, the fill mass equals 0.3 of the maximum usable volume.

In order to eliminate this first step, filler batches in powder form are available [5.68].

5.1.3.2.3 Compacting

[5.65, 5.66, 5.70, 5.71]

The second step during the powder processing is the compacting of the finished compound in powder form or of a powdery batch containing the chemicals. This step is required for most of the subsequent processing steps. One can also work without compacting in some processing technologies, e.g., the horizontal injection molding of simple compounds [5.71 a].

For the compacting of powdered completed compounds, two roll mills are applied [5.65, 5.66]. The mode of operation and compacting effectiveness is, however, not favorable because of the following reasons. The mill has to be adjusted to a very narrow gap to accept the powder since otherwise no compacting will take place at the beginning of the operation. In spite of this it cannot be avoided that the powder will fall through the roll gap. The narrow adjustment also results in a dramatic reduction of the mixing speed. If internal mixers are employed for compaction, two mill rolls to form banded sheets and follow-up batch-off equipment for cooling the bands are needed. Therefore the use of internal mixers should be excluded completely from the concept of powder technology and that of two roll mills as far as possible because of the creation of dust.

After examining several methods of compaction of the powdered mixture the conical screw as well as the twin screw compactor proved to be suitable approaches.

The *conical screw* offers the possibility to reach the required compression ratio. If one starts with a bulk specific weight of 0.4 g/cm^3 to produce a compacted compound of a specific weight of 1.2 g/cm^3, then the powdered compound has to be compressed by a ratio of 1:3. A compacting procedure using suitable mixing and screw designs seems reasonable. Because of the low energy and investment costs the additional possibility exists to combine such a machine with every extruder or every injection molding machine.

The material compacted in this fashion has not been plasticated yet which has to be accomplished, for example, during the extrusion process. A new trend is to simultaneously compact and plasticate the compound on *planetary roll extruders* whereby centrally located mixing units are desired.

5.1.3.2.4 Extrusion

[5.70, 5.71]

For the extrusion of such compacted compounds, extruders are required which provide a shear and de-gassing process (see page 391). Newly developed screws are especially suitable.

The extruded compound is finally vulcanized continuously. Thereby the possibility is given for the first time to produce rubber articles fully automatically and continuously. Of course, one can also process the compacted rubber compounds on calenders, conventional extruders and mills.

5.1.3.3 Other Continuous Mixing Methods

[5.76–5.87]

In the last years several continuous mixers were developed. However, the introduction of such mixing equipment is problematic as long as the raw material form of the elastomers does not correspond to this processing principle. Because of the numerous compound ingredients (20 or more) and their varying physical form, an economic and sufficiently accurate proportioning of the compounds in a continuous mixer is barely possible. In contrast, batches, for example, one consisting of elastomer and filler and another one containing the chemicals, can be well mixed when special mixing screws are used. Also, continuous mixing methods can be carried out with sufficient accuracy by methods where the chemicals are adhered to the surface of the elastomer-filler granules. Continuous mixing lines are applied up to now preferably only for partial operations, e.g., the mixing of some few main compound constituents, i.e., the preparation of batches or the pre-heating for the calender or the final mixing and extrusion of compacted powder compounds in one step.

EVK-Principle. One arrangement is the EVK-principle by Werner & Pfleiderer. The EVK machine is a continuous mixing extruder which distinguishes itself clearly from the usual forming extruders by its special mixing and kneading action. The process is based principally on the special design of the screw which rotates in a cylindrical, smooth barrel. The special feature of the screw geometry is the shear and flow path distribution elements which are distributed over the whole length of the screw and lead to an intensive shear activity (see Fig. 5.10).

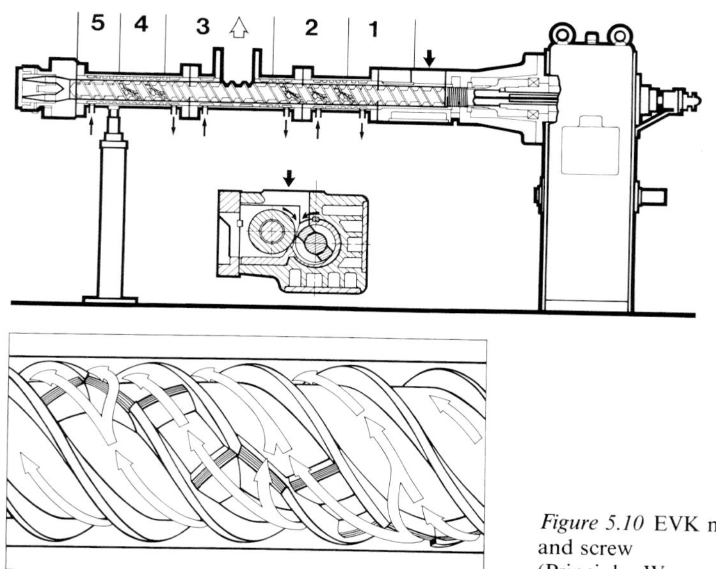

Figure 5.10 EVK mixing extruder and screw
(Principle: Werner & Pfleiderer)

Troester Powdered Rubber Mixing Screw. A special screw for the feeding of vacuum extruders with a pre-compacted powdered rubber compound in the form of strips has been developed by Troester. It is a combination of the familiar Troester mixing zones and shear sections. The first mixing zone is already placed into the feed zone of the screw. Different flow speeds and friction ratios are created by the different channel depths and compression ratios. The elastomer is well masticated at low temperature increases. The placement of the mixing section in the feed zone is made possible because this section can also transport the material. Along the screw, de-compression zones (transport zones) alternate with further short mixing zones. Thereby the orientation originating from the transport zones is repeatedly disturbed. The shearing section is arranged before the vacuum zone. It creates the large surface area necessary for de-gassing (see page 391).

After de-gassing the metering zone follows in which de-compression and mixing zones also alternate. The total screw length is 20 D for a screw diameter of 90 mm. The vacuum zone is placed after 8 D (see Fig. 5.11).

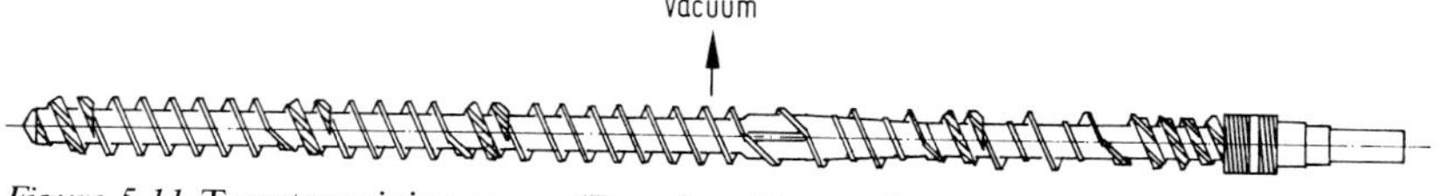

Figure 5.11 Troester mixing screw (Drawing: Troester)

Maillefer-Principle [5.80]. The Farrel-Bridge continuous mixer which has been used for a long time effectively in the plastics industry resembles in its operational principle an internal mixer (Fig. 5.12). A mixing chamber in the form of a cylinder is extended by two mixing rotors operating with a friction ratio toward the drive side. The rotors are formed as screws. The compound ingredients are fed into the feet port and then pressed by screws into the mixing rotor region, where the mixing process takes place. The residence time in the mixing chamber can be regulated by an adjustable out-put gate. The compound materials migrate through the mixing chamber and leave it continuously. The mixing quality and temperature are preferentially adjusted by the rotor speed and the size of the exit opening. The machine cannot be driven empty. Therefore the machine housing can be removed.

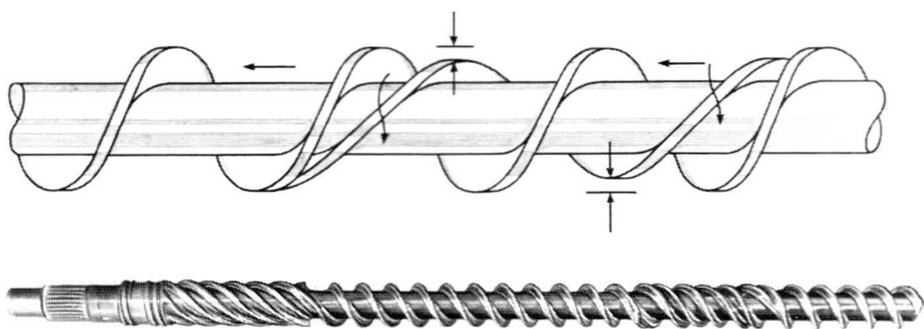

Figure 5.12 Maillefer mixing screw (Photo and principle: Berstorff)

Monomix [5.77]. Recently, a fast running complete mixer with the name Monomix has been offered for sale. The compound material is forced under the effect of shear to move from the side walls to the mixing chamber and perpendicular to it because of a specially counter-rotating thread-like extruder screw characteristic. At the same time it is pressed from between the space between rotor and cylinder by high hydraulic pressure via a ram. The mixing times are 3-4 minutes.

5.1.3.4 Preparation of Compounds from Liquid Polymers

[3.108-3.114a, 3.297, 5.74, 5.75, 5.87a-5.90]

In view of the fact that some of the elastomers are also offered in liquid form, attempts have been made to convert them to rubber compounds by processing methods analogous to polyurethane technology. Since sulfur vulcanization occurs for elastomers of low molecular weight very slowly if at all, one depends on reaction mechanisms via reactive side or end groups of the polymers. These so-called telechelics or reactive liquid polymers (RLPs) require an adequate crosslinking agent depending on their individual reactive groups, e.g. (-COOH, $-NH_2$, -OH, -BR).

An important problem during the preparation of rubber compounds from low viscosity polymers as starting products is the incorporation and even distribution of fillers and other ingredients, especially active carbon blacks. For this purpose intensive mixers have been developed, especially by RAPRA. Three roll devices have not proven successful. It could be shown that adequate reinforcement, as with carbon blacks, can be obtained when epoxides are used as cross-linking agents. Therefore the use of carbon black is not necessary in some cases [3.111].

The further processing of liquid rubber compounds depends upon their viscosity. High viscosity, carbon black containing compounds can be processed, for example, in injection molding machines, low viscosity ones in rotational casting machines. By the latter method, for example, Rugby balls can be manufactured cheaper compared to the former lay-up method.

5.2 Processing to Sheets and Rubberized Fabrics

5.2.1 Calendering

[5.45, 5.84, 5.91–5.97]

5.2.1.1 Rubber Calenders

Calenders are used to manufacture sheets of varying thickness to incorporate technical fabrics with rubber and to coat fabrics with thin rubber sheets.

The calender consists of two to four hard cast rolls of lengths up to 2300 mm, arranged in a frame. The rolls are hollow or drilled on their periphery and equipped for heating or cooling just as the mixing mills. The roll gap is adjustable.

During calendering of rubber compounds – in contrast to plastics – considerably lower temperatures have to be used because of the danger of scorching. Because of this reason and because the non-thermoplastic properties of the rubber compounds, very large forces have to be supported. This is the reason for the familiar massive construction of rubber calenders (see Fig. 5.13).

Figure 5.13 Rubber calender (Photo: Berstorff)

Calender Types. Depending on the number of rolls one distinguishes between two, three and four roll calenders.

Two Roll Calenders are suitable for rolling out rubber sheets and can be used, for example, for the production of shoe soles and floor coverings. However, these are seldom used anymore. An exception are roller head machines (see page 367), [5.91, 5.92], where a two roll calender flattens and calibrates a sheet originating from a sheeting extrusion die. Good quality, bubble free sheets up to 20 mm and more can be manufactured by this process. In two roll calenders one roll is stationary, the other can be adjusted.

A special design of the two roll calender is the so-called IT sheeting calender (see page 378).

Three roll calenders can be used for all calendering applications with the exception of two-sided coating. They are mostly utilized to form sheets of 1 mm down to 0.3 mm thickness as well as to coat fabrics. The center roll is, as a rule, stationary, the upper and lower ones are adjustable (see Fig. 5.14).

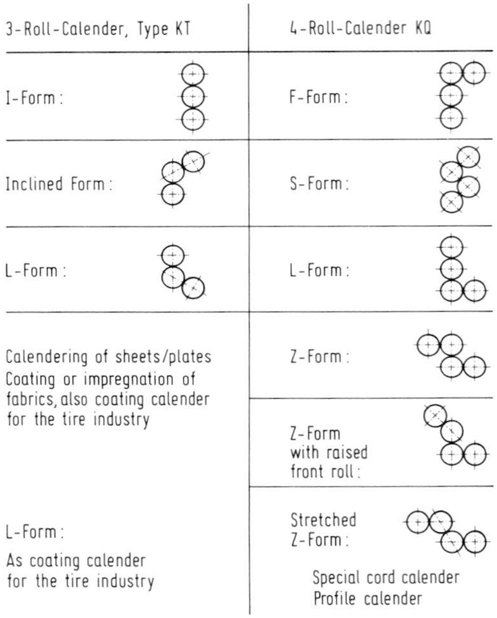

3-Roll-Calender, Type KT	4-Roll-Calender KQ
I-Form:	F-Form:
Inclined Form:	S-Form:
L-Form:	L-Form:
Calendering of sheets/plates Coating or impregnation of fabrics, also coating calender for the tire industry	Z-Form:
	Z-Form with raised front roll:
L-Form: As coating calender for the tire industry	Stretched Z-Form: Special cord calender Profile calender

Figure 5.14 Roll arrangement for three and four roll rubber calenders (Principle: Troester)

Three roll calenders in tandem arrangement. When no four roll calender is available, two three roll calenders of the same size and construction are arranged one after the other and thus form the tandem arrangement as is done in large factories to rubberize on both sides cord fabrics for the production of automobile tires. Aside from this application the possibility is given to use both single calenders to manufacture sheets or to coat fabrics.

Four roll calenders are most often used today in the rubber industry since they permit universal application due to their design.

Four roll calenders are seldom constructed in the so-called I formation (vertically stacked rolls) any more because of difficulties feeding them. Either the upper or the lower roll is located in front of the neighboring roll (F or L calender). In newest designs only the two middle rolls are above each other while the other two are horizontal with the two center ones, one in front, the other behind (Z-calender) (see Fig. 5.14).

The different calenders are used according to the following principles:

- *I-calender:* These calenders are of old construction which are rarely built any more. Otherwise they can be used as other four roll calenders, e.g. for the manufacture of sheets and the impregnation of fabric.
- *F-calenders* are called for when primarily calendered sheets from 0.1 mm to 1–2 mm in thickness shall be produced, more infrequently for coating fabrics. One nip is easily fed, ancillary equipment is easily attached and the forming nips are easily accessible. They are used especially frequently in the rubber industry.
- *L-calenders* are mostly small units, e.g. laboratory calenders. They are employed for the same tasks as F-calenders. The lower forward arranged roll makes operation easy.
- *Z-calenders* are employed when primarily fabrics or cord shall be coated on both sides. Both nips are easily fed and the bending of the rolls affects only one roll gap each. Even though the construction height of these calenders is greatly reduced, the base is considerably larger which is disadvantageous in many factories.
- *Z-calenders with raised forward roll* are applied as normal calenders. However, ancillary equipment can be attached more easily and the forming nips have better accessibility.
- *Three and four roll calenders* are also manufactured as profile calenders.

Construction of the Calenders. Rubber calenders are constructed in the following way:

The calendering tools per se, the rolls, are supported in strong *frames* with a box like cross section made from cast steel or cast Mechanite (cast iron with extremely fine and even graphite distribution of excellent quality). The frames have to take up the working pressures which can be very high.

The *rolls* are made primarily for production calenders from hard form castings. Laboratory calenders are nearly exclusively equipped with steel rolls. The advantage of steel rolls, i.e., lower bending because of the nearly double modulus of elasticity compared to cast iron is compensated for by the difficulties during calendering because of the higher degree of sticking of the compound to the rolls. The hard form cast rolls are therefore typical for rubber calenders.

Any *bending* of the calender roll in the last nip yields thickness variations of the sheet from the center to its edges. In order to reduce the thickness variations, the rolls are frequently not ground cylindrically but thicker in the center, i.e., they have a roll crown (see Fig. 5.15).

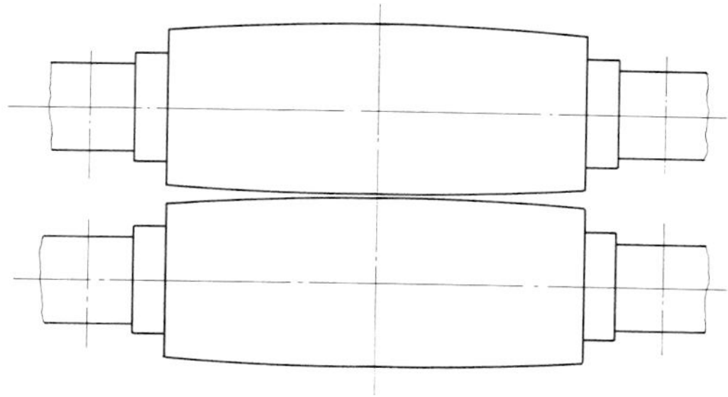

Figure 5.15 Roll crown, exaggerated representation (Principle: Troester)

One roll, usually the driven one, remains cylindrical while the other one receives a correspondingly higher crown.

The crown is practically invariable and can be only a compromise since the operating conditions such as the viscosity of the compound at calendering temperature, sheet thickness and circumferential speed are constantly changing, great difficulties are caused by the roll crown during impregnation and coating since the nip pressures are low for these operations and thus different layer thickness can result easily.

A variable *roll bending* compensation is accomplished by roll crossing. The roll is rotated around its central axis so that the resulting gap is narrow in the center, wider at the ends. Under production loads the crossed rolls bend and the roll gap becomes parallel, i.e., the sheet will be equally thick over its width. Roll crossing is, of course, only possible when the rolls are driven from a special gear box via universal joints.

Another method to avoid an increase in the roll gap due to a deflection of the rolls toward the center is called roll bending. This device is based upon two ancillary bearings which are attached to the extended roll journals to which a counter bending moment is applied (see Fig. 5.16).

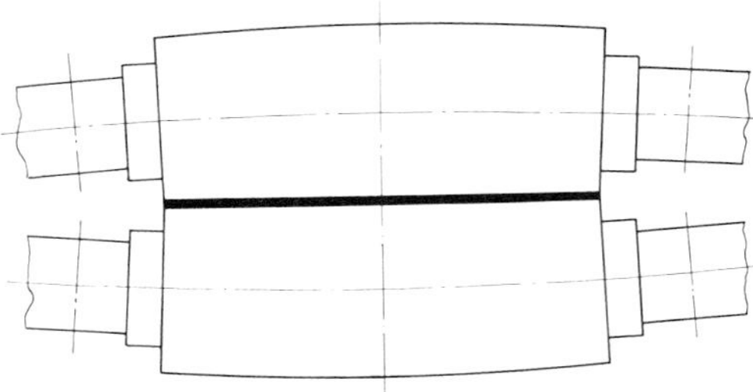

Figure 5.16 Roll-Bending (Principle: Troester)

The hydraulic cylinders are connected to two massive consoles which are attached to the *calender frame*. Since the rolls experience an additional load the counter bending moment is limited.

While the *surface* of the mixing mill rolls is left somewhat rough to increase the mixing effect, calender rolls are ground and polished.

The *bearings* are either antifriction or roller bearings.

Antifriction bearings are simple and cheap. The bearing surfaces are made from high quality bearing bronze. Their use causes high friction losses and the roll can "float", i.e., the journal can move backward and forward depending on the direction of the forces.

Roller bearings exhibit, in contrast, only low friction losses and high precision. That means a play free support (pre-tensioned bearings).

Antifriction and roller bearings are used in modern calenders. For example, the movable rolls of four roll calenders can be supported by antifriction bearings, the fixed rolls, in contrast, in roller bearings. Thereby the stationary roll is supported without play in the bearing while the movable ones have some play.

The *lubrication* of the bearings is recirculating high pressure oil. The return flow lines are equipped with monitors. Oil reservoirs on both sides of the bearings collect the exiting oil which is used to cool the bearings. In the beginning the oil is preheated (for example to 50 °C) and is cooled later on to carry away the bearing heat.

The *roll gap adjustment* of rubber calenders is accomplished by sliding conical bearing supports with the help of gear motors either in unison or separately. If each bearing of the same roll has its own gear motor, they are connected to each other through a synchronizing shaft. However, in order to permit corrections of the parallel position, they can also be adjusted separately. Hydraulic gap adjustments have not proven themselves and are not used anymore for rubber calenders.

Heating and *cooling* of the calender rolls, depending on the design of the rolls, is accomplished by steam-water mixtures up to 80 °C for simple hollow rolls or with hot water up to 150 °C and above from special hot water units for peripherally drilled holes which are constructed analogous to the rolls of mixing mills (see page 357, and Fig. 5.6, page 363). Predetermined water temperatures are always used since all kinds of temperature sensors either on the roll surface or in the roll wall have proven useless.

For steam-water mixing units a large amount of hot water of a given temperature is fed into the roll via a central spraying pipe. A certain roll temperature will develop because of the energy dissipation in the roll nip which is above that of the water.

For peripherially drilled holes one hot water unit is used per roll. The re-circulating hot water system yields good and even temperature control. Each three channels (approximately 50 mm under the roll surface) are connected in series so that no temperature gradient over the length of the roll occurs. The temperature sensors are placed in the return line.

The rolls of laboratory calenders are sometimes heated electrically in contrast to production calenders.

Older calenders, especially those with I form, are *driven* from a central drive. An adjustable DC motor drives via a gear reduction the second roll. This roll is mounted in a fixed position to the frame. Opposite to the drive end are the coupling gears with very long teeth on the extended roll journals. For equal number of gear teeth one obtains synchronous roll speeds.

For operation using a friction ratio, a second row of coupling gears is provided next to those yielding synchronous operation. The coupling gears are secured to the drive roll, those of the other rolls are movable, i.e., they rotate in antifriction bearings on the roll journals. In between the coupling gears, coupling rings with teeth on both of their sides are positioned. They can be moved in the direction of the roll axis but cannot turn with respect to the roll. The counter parts of the teeth in the coupling rings are matching teeth in the coupling gears. Once the teeth of a coupling ring mate with those of a gear, the gear is connected firmly with the journal and can thus drive the roll. The number of teeth of the coupling gears determines the friction ratio.

In modern calenders each roll is driven by a separate DC motor. The motors are flange mounted onto the special gear box housing with separate gear drives. The drive shafts are connected to the rolls via double universal joint shafts. This type of drive has considerable advantages over the coupling gear drive system.

Every calendered sheet has to be cut to an exact width after leaving the last roll nip. If cutting takes place directly on the calender roll, spring loaded circular *knives* are used. These are made from hard bronze. More modern is the cutting with a spring loaded circular steel knife on a softer steel roll which is driven at calendering speed. The edge trimmings cut off are fed back to the pre-heating mill.

The *sheet thickness* is measured, e.g. with a β ray resp. Laser measuring device before the sheet is taken away on the follow-up conveyor.

Behind the calender is the *take-up equipment* [5.96] with several cooling rolls, a take-off conveyor and a wind-up. A regulated drive assures that take-up equipment and calender operate at the same speed.

Cord Coat Calender. Calenders which are used exclusively to coat cord, form frequently with the equipment in front of the calender and behind it a unified production line. A calender line starts with the double un-winder followed by the splicing press to connect quickly (by vulcanization) the single sheets. A sheet storage follows which permits the splicing without interrupting the calendering process. Following the cord fabric passes through a heated calender for heating and passes then through the calender. Following the calender another storage is provided, then a cooling device and finally a wind-up utilizing interlayers of fabric (see Fig. 5.17).

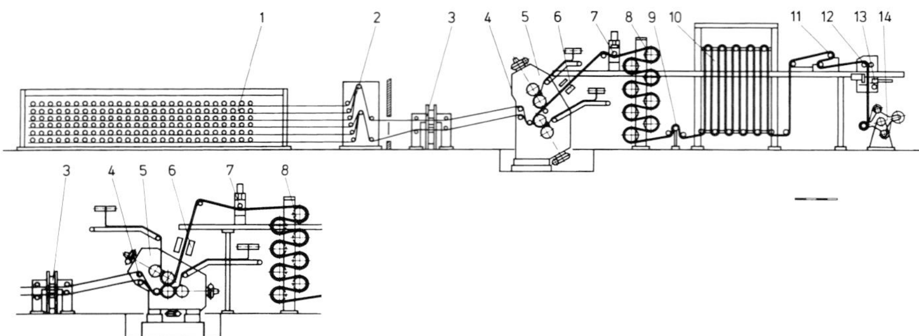

Figure 5.17 Double sided coating with a double two roll or a four roll calender (Principle: Berstorff)
1 Roll guide, 2 rolls for measuring tension, 3 two-stage press, 4 groove roll system, 5 double two roll or four roll calender, 6 thickness gauge, 7 tensioning and guide roll for measuring tension, 8 cooling line, 9 take-off, 10 take-off wig-wag, 11 take-off, 12 holding and stretching device, 13 cutting device, 14 double swing wind-up

IT-Sheet Calender. A special design is the so-called IT sheet calender. Calenders to manufacture IT-sheets have two rolls with different diameters and vertical or horizontal placement of the frame. The larger heated roll serves as the working roll, it is mounted in a fixed position and is connected through a pair of coupling gears to the pressure roll located above. The gap in between the rolls is thus adjustable.

5.2.1.2 Working on Calenders [5.95] and Influence of Compound Composition on Calenderability

Working on Rubber Calenders. *Feeding of the calender* is by sufficiently pre-heated compounds. For continuous production processes the compounds are frequently taken from the follow-up machines of the internal mixer (see page 366 ff) directly onto the pre-heat mill and are fed from there directly to the calender. In contrast, if the compounds have been stored, they are broken down first on the ground rolls of a break down mill and then on several parallel arranged normal mills with smooth rolls (up to six, the capacity of which determines the calender production rate) which are connected over-head by conveyors. Such a material transport from the first pre-heat roll to the last one is established which represents the cutting mill and feeds the material to the calender. The design of the pre-heat and cutting mills is according to standard practice for mixing mills (see page 357).

A strip of warm compound runs from the cutting mill to the feed nip of the calender. The strip was cut either by adjustable fixed or rotary knives. A conveyor moving from left to right deposits the strip in zig-zag form onto a feeding band. Thereby the nip receives material evenly over its width. In order to protect the expensive calender rolls, the pre-heated strip runs through a metal detector which shuts down the line instantaneously if a piece of metal is detected. Automatic feeding guarantees equal material quantities at constant material temperatures, a pre-requisite for consistent calendering results [5.94].

For *working* on the calender it is extremely important that the compounds are heated to the correct temperature and that the roll temperature is properly maintained. If the compounds or the rolls are too cold, cold spots on the surface of the calendered sheets form, so-called "crows feet". At excessively high temperatures, bubbles form. Only within a rather narrow temperature region which varies from elastomer to elastomer and compound to compound one obtains a flawlessly smooth sheet. Finding the correct temperature gradient of the calender rolls and the most favorable working speed depends on the skill of the calender operator.

Rubber-Impregnating Fabrics. A very important process on the calender is the rubber impregnation of fabrics since a large fraction of rubber articles consists of a combination of rubber and fabric. Very soft compounds are required for this procedure so that they can be worked intimately into the fabrics. They also have to adhere firmly to the calender roll surfaces. The compound is fed to the upper roll nip of a three roll calender and runs around the middle roll. This roll has a circumferential speed ca. 30–50% higher than that of the other outer rolls, i.e., it turns at a friction ratio of up to 1:1.5. The pre-dried fabric is introduced into the lower roll nip and assumes the speed of the lower roll. The soft and tacky rubber compound is now worked thoroughly into the fabric because of the higher relative velocity of the center roll. The nip between the center roll and the fabric is adjusted so that a small compound bank forms on the fabric. By double sided impregnation the first side always takes up more rubber compound. The deposited thickness depends on the type of fabric and compound and is difficult to determine.

Coating of Fabrics with Thin Layers. Frequently fabrics shall be coated with thin layers on both sides. The familiar example for this process is the coating of tire cord or conveyor belt fabrics. The fabric has to be made rubber compatible by a previous impregnation step (see page 443 ff) so that sufficient adhesion between fabric and rubber can be achieved. In the first nip the sheet is formed with a small friction ratio of 1:1.1 to 1:1.2, and in the second one it is pressed in contact with the fabric. Thus a three roll calender is sufficient for one-sided coating. A four roll calender can coat on both sides by forming a sheet in the third roll nip which is then combined with the fabric in the center nip.

Calendering speeds have increased more and more and are around 10–25 m/min. Coating of tire cord is done at velocities of approximately 70 m/min.

At these high velocities the feeding of the calender via rolls alone is not sufficient any more. Therefore, internal mixers are used for break-down and pre-heating or a roller head extruder (see page 367) [5.91, 5.92]. The cord layers which may be, for example, 400 m long are then connected to each other via vulcanization. A large number of storage rolls assures a continuous process. To obtain better adhesion, the fabric is most often fed to the calender pre-heated. A system of cooling rolls behind the calender assures that the rubber coated fabric is wound up cool to avoid scorching and sticking to the separating fabric layers [5.17].

Calendering of Profiles. For the calendering of flat profiles such as treads for bicycle tires, soles of rubber shoes etc., calenders are employed one roll of which is profiled. The rolls of modern profile calenders are equipped with interchangeable mold inserts the exchange of which does not take longer than four to five minutes.

Follow-up Equipment [5.96]. After leaving the calender the material is brought to take-up equipment or to an automatic vulcanization device. Before the take-off the calendered material is first in contact with several *cooling rolls*. At the end of the *take-off conveyor* the calendered sheet or the impregnated or coated fabric are wound up with interlayers of separating fabrics to avoid sticking of adjacent rubber layers. Recently the application of disposable films (e.g. 0.04 mm thick textured polyethylene) has been introduced successfully. The *wind-up* is usually via a friction coupling which is adjusted manually while the separating fabric (or film) is tensioned. Alternatively, the winding roll rests on the conveyor on which the calendered sheet arrives, thus yielding constant circumferential speeds of the wind-up roll.

Commonly sheets of 1 to 3 mm can be calendered. Thicker sheets are prepared by *doubling*. Doubling can be done easily on the wind-up. The bubble free sheets are pressed by soft rubber coated rolls onto the take-away conveyor, also called the doubling conveyor, around which they are wound. After a desired number of layers has been obtained the sheets are cut and pulled off the length of the conveyor belt. The doubling procedure is then immediately repeated.

Another method to manufacture bubble free calendered sheets of high precision and of larger thicknesses is the roller head system (see page 367) [5.91, 5.92].

Influence of the Rubber Compounds. Aside from the temperature, the compound composition is very important to obtain smooth calendered sheets. Compounds with too much nerve exhibit after calendering a strong elastic recovery, called memory effect. The shrinking is perpendicular to the calendering direction. This phenomenon is called *"calendering effect"*. Perfectly smooth and shiny calendered sheets can be obtained after sufficiently good mastication, by the use of inactive or semi-active fillers, plasticizers, process aids and especially factice, etc. The calender effect is also beneficially influenced by the application of loosely pre-crosslinked elastomer types.

Vulcanization. In most cases the calendered products are intermediary products and are taken unvulcanized, are then finished to products and finally vulcanized. In rare cases, when calendered sheets are used as end products, a continuous vulcanization process follows which will be discussed elsewhere (see page 386).

5.2.2 Impregnating, Skimming and Coating with Solutions of Rubber Compounds

5.2.2.1 Impregnating and Coating

In order to increase the adhesion of rubber compounds to cord or woven fabrics, i.e., to make them "rubber friendly", they are impregnated before the calendering process when no self-adhering rubber compounds are used. This is especially required for cord fabrics made from artificial silk (Rayon) or other synthetic fibers. Self adhering compounds can be calendered onto untreated fabrics in most cases (see page 313 ff).

Impregnation. For impregnation, either a water based latex or resin mixture or a benzene based isocyanate solution is used (see page 314). The impregnation is carried out by dipping the fabric into the solution and then subsequent drying using sometimes a stretching process.

The dip, stretching and drying devices are frequently arranged so that the impregnated fabric is brought directly to the calender.

The fabric is unwound on a double unwinding block which is equipped with an adjustable mechanical or electrical brake system. After forming a large fabric feed storage between a splicing press or a sewing machine and the dip trough it is running under tension through the dip solution. A teflon coated blade removes the excess solution which runs back into the trough. The fabric is now led via a vacuum chamber in the drier, a steam or pressurized water heated air chamber with air temperatures of 160–180 °C. The impregnating trough is formed, for example, V-shaped and manufactured from stainless steel. It is equipped with a fill level adjustment control with automatic dip solution replenishment. It is supplied with a lifting device which can control the dip depth of the fabric running over a roll. The residence time of the fabric in the dryer is, for example, 100–120 seconds. The required final moisture content should be 0.5–1.5%. Contact drying by steam heated roll is only seldom utilized today. Before wind-up or feeding into the calender a sufficiently dimensioned take-off supply device is provided.

The entire impregnation process is usually carried out with the fabric under tension. For this purpose the following units are used: the brakes of the wind-off block, the tension of the wind-up block as well as hydraulically or pneumatically controlled floating rolls and stretching devices for fabric covers with heatable or coolable hard chromed rolls which can be arranged either in a horizontal or vertical position. The fabric tension is measured with the help of pressure transducers attached to rolls which are supported by swing arms. If the stretching is carried out at high temperatures, for example, for synthetic fibers, temperatures of 180–225 °C are applied to stretch the material by 15–20%. The remaining strain after cooling is 10–30%.

After impregnation the fabric is passed through a heated calender, a cooling device, the winder with interlaced plastic sheets or the calender.

Solution Coating. The highly viscous solution of a rubber compound in organic solvents is applied to both sides of the fabric in a synchronously operating, horizontally arranged two roll mill, the so-called "solution" or "coating" calender. Only fabrics with high tensile strength can be used for this method. The height of the upper roll of the solution calender can be adjusted and thus controls the application thickness, the accuracy of which is, however, very limited. One can, however, obtain higher coating thicknesses during one pass than with the skimming machine (see page 382). Drying takes place in a drying tower. Formerly, coating and solution calenders with a follow-up drying drum were used.

5.2.2.2 Skimming with Solutions of Rubber Compounds

Rubberized fabrics for technical applications are prepared primarily on the calender, finer fabrics (e.g., coat fabrics, rubber cloths) are rubberized on the skimming machine. A paste like solution of a rubber compound in organic solvents (sometimes erroneously called a rubber solution) is skimmed onto a fabric. One can obtain very thin coating thicknesses and very good gas permeability for thicker coatings. For thicker coatings the skimming process has to be repeated several times.

Gasoline is most frequently used as a solvent. The solutions are prepared in solution mixers after the rubber has been swelled in a solvent for 8 to 24 hrs.

For skimming of latex see [1.16].

Skimming machine. The skimming machine consists of a leathery hard rubberized or chrome plated roll over which a so-called doctor blade is arranged with adjustable gap. The thickness of the blade depends on the fineness of the fabric to be coated and ranges from 2 to 6 mm.

Several blade arrangements and forms are in use. The most frequently used blade form is the knife blade. It is attached to a stiff beam which can be tilted around two pins which in turn are supported by two supports of the machine frame and are adjustable in height. The tilting and height adjustment of the knife beam is done manually.

These blade arrangements can be distinguished:
- Roll blade
- Band blade
- Membrane blade
- Air blade

For the *roll blade system* the doctor blade is arranged above a roll which is strong but hollow inside and can be cooled. It is driven via gears or a chain drives. The fabric is pulled in between the roll and the blade and the pasty solution is applied before the blade in between side guides which serve as side limits. The coating thickness applied to the fabric is determined by adjusting the doctor blade. This method affords the highest accuracy.

In the *band blade* the fabric is running over a rubber conveyor belt above which the height adjustable doctor blade is arranged. The fabric is pulled by the conveyor belt under the blade and the side guards which retain the solution.

The *membrane blade* is similar in construction. However, the doctor blade is located above an air filled membrane.

Finally, for an *air blade* the blade is resting freely on the fabric which is highly tensioned by two transport rolls which move the fabric under the blade. The height adjustment is adjusted by regulating the textile tension which is adjusted best by self regulating braking rolls. Side guards are not common for this type of machine and the solution runs off the sides.

Skimming. The fabric is unwound and runs through the gap in between the doctor blade and the roll and carries the dough-like solution in desired gap thickness along. This process is repeated after each drying step following the application zone by a drying roll, until the rubberization has reached the desired thickness.

5.2.2.3 Other Coating Methods

Deposition Method. Another coating method for fabrics is based upon the deposition by deposition rolls. One roll is dipping into the solution and transfers a solution film onto the fabric. Numerous possibilities to arrange the roll with respect to the fabric exist.

Dipping and Spray Method. Aside from the above mentioned methods, rubberized textiles can also be manufactured by the dipping or the spray method whereby - as the name implies - the textile is dipped into the solution in the one case and is sprayed by spray guns in the other.

Laminating. When another sheet is pressed onto the coated fabric, laminating is involved. One of the various variations of this process is the so-called lamination pressing. The rubberized fabric and the coating are led together to a joining roll and carried under it under tension. Then the fabric, joined under pressure, is dried on a drum and wound up.

5.2.2.4 Secondary Operations

Take Away Equipment. After every application the fabric is taken away over a steam heated table, a steam heated drying drum or through a heat chamber. Infra-red heating is also possible. The drying tract (4–12 m) and the process speed are dimensioned so that the fabric is dried in a single pass. The fabric is wound up onto a roll using an interspersing fabric layer.

Vulcanization. After the last frictioning or coating and after drying, vulcanization on rotation vulcanization machines or in hot air follows (see page 386).

Measuring the Coating Thickness. After drying the coating thickness is determined using a β- or Laser-radiation gage.

Solvent Recovery. For reasons of economy and ecology solvent recovery is important. Two processes are used. The *condensation method* is based upon pumping the solvent vapors through water cooled condensers where they are liquified. During the *adsorption method* the solvent vapors are blown through various adsorbers, e.g., A-carbon and are adsorbed. Recovery is by steam distillation.

While the recovery rate of the condensation method is only relatively low and its operation dangerous because of the possibility of explosions, one recovers approximately 90% of the solvent by the adsorption process. The danger of explosion is less also.

Explosion Prevention. It is important that the partially toxic and explosive solvent vapors are exhausted during drying. The explosion limit of a gasoline-air mixture is between 35 and 320 g gasoline per m^3 air. It has to be assured that the solvent concentration remains under the explosion limit. All motors, switches, lights, etc. have to be explosion proof, also all drives, rolls etc. have to be protected and well grounded. Finally, secure sprinkler systems are legally required. Spark generation by the rubberized fabric is, for example, eliminated because the static electricity is eliminated by ionozation methods.

5.2.2.5 Preparation of Rubber Compound Solutions

Unvulcanized rubber compound solutions in organic solvents are used for the impregnation (see page 379), dough, doctor blade, dipping (see page 443) or spray process, also for the immersion or adhesion process and partially for the manufacture of rubber threads.

Solvents. Organic solvents capable of dissolving the elastomers are used for the preparation of rubber compound solutions.

Gasoline, chlorinated hydrocarbons and ester are used. Which solvent is used depends on the elastomer type to be dissolved and the end use (see page 321).

For example, NR compounds are soluble in gasoline, NBR and CR compounds, in contrast, are not.

Especially good solvents for these are aromatic and chlorinated hydrocarbons as well as ester. Quite often solutions of rubber compounds are also prepared in a mixture of solvents, where solvating and only strongly swelling solvents can be combined, which, as combination, exhibit stronger solvating properties and permit the preparation of more concentrated solutions than the single solvents by themselves. This is, for example, especially important for the preparation of adhesives [5.97 a].

Solutions of NR compounds where the solvent should evaporate especially rapidly are frequently prepared from gasoline, especially light gasoline, while aromatic or chlorinated hydrocarbons are used for solutions with high adhesion because of their good swelling behavior.

Since organic solvents exhibit usually an unpleasant odor, are toxic or are sometimes prone to explode, the proper design of the machines used for the process and a ventilation of the working place has to be carefully considered.

The following equipment is used to prepare solutions of rubber compounds:

- double shafted solution kneaders
- rotational mixers
- planetary mixers
- stirrers

Solution Kneaders. The double shafted solution kneaders are machines (see Fig. 5.18) which are manufactured with working volumes of 1–4300 liters. They consist of a trough formed from two cavities in a frame with a lid and two kneading paddles. The machine is either equipped with a tilting device or a discharge screw.

Figure 5.18 Two-shafted solution mixer (Photo: Werner & Pfleiderer)

The trough is formed by two semi-circular tubs in which the *kneading paddles* turn with differential circumferential speed (friction ratio), the forward one 1.5 to 2 times faster than the one in the back. The kneading paddles are operating so that the preswollen elastomer is not only disintegrated but is also constantly wetted by the solvent until total solvation has taken place. The trough enclosure is double walled to permit heating or cooling. The trough is closed by a movable lid which is counter weighted for easy opening. For working with an open lid, a protective screen is provided also.

The kneading paddles are most often Z-shaped and have journals on both sides for bearing support and provision for a drive. The bearings are sealed.

To empty the trough, the tilting device is operated either by hand, electrically or hydraulically. For *trough* designs with discharge screws, the screw works during kneading in direction of the trough, for emptying the direction of rotation is reversed.

Change gears with two or three working speeds are used in the drives.

Rotational Mixers consist of drive box with gear box, working vessel with lid and rotor with feeding and emptying ladles. They operate with a vertically arranged shaft which carries a rotor with a system of feeding arms and discharge blades with a coolable or heatable working vessel, where usually three speeds of rotation can be chosen. In connection with stationary deflectors the rotating kneading elements fragment the rubber compound and assure good mixing with the solvent. By this arrangement the solvent flow is turbulent with a cross-over flow pattern which assures rapid solvation of the rubber compound. This mixer design yields lump-free solutions in relatively short mixing times.

Planetary Mixers consist of the machine frame with drive, the stirring and mixing unit and the mixing vessel. The machine head with the stirrers is raised hydraulically for the filling of the vessel and then lowered again. The stirrers rotate around a common axis and also around their own axis, thereby an especially effective mixing effect is achieved.

Stirrers are primarily used to homogenize or dilute prepared compound solutions as well as for the mixing of solutions which were stored for an extended period of time. They consist of the solution vessel and the stirring shaft protruding from above with stirring paddles of varying design.

Heating and Cooling. It is important for all solution equipment that it is, aside from heating, equipped with an effective cooling capability of the mixing container in order to remove the heat generated during the mixing process. During the solvating process considerable quantities of heat are generated which can lead to premature scorching when the compounds contain rapid accelerators. The lid of the mixer has to be carefully sealed so that the solvents cannot evaporate because of the mixing heat.

Compound Formulation. For the compound formulation of rubber compounds which are processed in solution applies even more what applies to calendering and extrusion compounds (see page 380, 393f), namely that the elastomers are carefully masticated before the solution preparation and that all raw materials have to be kept painstakingly free of contamination. The ability to go into solution and the adhesion themselves depend on the type of elastomer used and the degree of mastication. Masticated rubbers adhere better than unmasticated ones. For light, nearly colorless solutions pale crepe is usually used. Since SR types are generally more difficult to masticate, types with low viscosity are used to prepare solutions. The adhesive strength of SR types depends to a large degree on the molecular weight distribution. The wider it is, the stronger is the adhesion.

In order to avoid lumps, the solutions should be passed several times through a narrow gap refiner or they should be strained. All additives which strongly increase the viscosity of the rubber compound, as, for example, fillers have to be avoided. If required, inactive fillers can be used within limits. The Mooney viscosity of the rubber compound should be as low as possible, but always constant.

The higher the viscosity of a rubber compound to be solvated, the higher will also be the viscosity of the solution, i.e., the lower the solids content at constant solution viscosity. By adding a small amount of alcohol to gasoline (approximately 0.5%) one can obtain a marked lowering of the viscosity. For the preparation of solution with high solids content quite often factice containing compounds are used.

Most often ultra accelerators (e.g., dithiocarbamate) are used for the manufacture of articles made from solution. Premature gelation has to be expected when acceleration is too rapid. Therefore the vulcanization speed during the processing of solutions – in contrast to latex processing – has to be carefully adjusted to their shelf life. Sometimes the accelerator has to be left out of the compound (at least partially) and is added to the solution before it is processed. Dip solutions are usually renewed daily because of the danger of premature scorching. However, they have to be kept stationary for some time to eliminate bubbles where the viscosity has to be low enough so that bubbles can rise to the surface fast enough.

5.2.3 Vulcanization of Sheets and Rubberized Fabrics

5.2.3.1 Auto Vulcanization, Hot Air Vulcanization

In some cases calendered sheets, impregnations or coatings (e.g., corrosion protection of large vessels, roof coverings in some cases) are accelerated so rapidly that they vulcanize at ambient temperature (auto vulcanization). Most of the sheet products such as roof coverings, raincoat fabrics, etc. are vulcanized, however, pressureless in hot air (heating chambers at 60–70 °C) (see page 447f).

5.2.3.2 Rotational Vulcanization

In former times the calendering or coating process and the vulcanization, for example in heating chambers, were completely separate process steps. In contrast, thin calendered sheets and coated fabrics, belts, bands etc. can nowadays be continuously vulcanized immediately following the calendering or coating process in one step when rotational vulcanization machines are employed. Also, preformed articles, as for example V-belts, can be vulcanized on rotational vulcanization machines. The working speed of a rotational vulcanization machine is, however, most often slower than that of a calender. For this reason, it cannot be coupled directly to a calendering process.

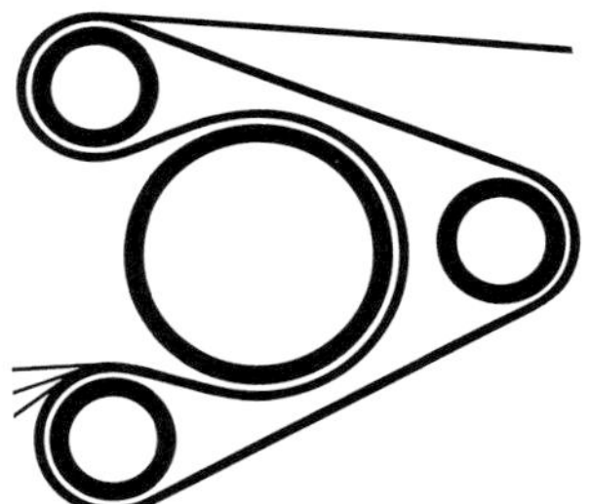

Figure 5.19 Rotational vulcanization (Drawing: Berstorff)

In rotational vulcanization machines on endlessly welded steel sheet band or steel fabric runs over about two thirds of the circumference of a slowly rotating, steam heated steel drum which may be covered by different profile inserts. The steel band runs around two deflection rolls and a hydraulically tensioned roll and is thus pressed against the drum (see Fig. 5.19). The sheet to be vulcanized runs between drum and steel band and is pressed firmly against the drum with a pressure of, for example, 6 bar. Thus the sheet is slowly moved along because of the rotation of the drum. Vulcanization occurs because of the temperature of the drum on that sheet side facing the drum made under the influence of the pressure created by the steel band. Welding of two sheets fed in the machine separately can be accomplished also. Since heat is supplied from one side only, the thickness of the sheet to be vulcanized should not exceed 5 mm. In spite of this, vulcanization over the thickness of the plate can be obtained even for slow curing compounds with a residence time of, for example, 10 min. or less at a drum temperature of up to 190 °C. Even though the specific pressure on such rotational vulcanization machines is less than during other vulcanization processes (e.g. in the press) this pressure is completely sufficient because of the pliability of the steel band to obtain a homogeneous structure and a smooth surface and to avoid porosity.

5.3 Manufacture of Extruded Products

Extruders serve to manufacture rods, hoses and profiles, treads for tires and sheets as well as to coat cables and wires. In addition, large machines are used as strainers to plasticate (Gordon plasticator) (see page 368). Thus, extrusion encompasses a rather large area of processing.

5.3.1 Rubber Extruders

[5.98-5.125a]

5.3.1.1 Screw Extruders

Rubber extruders work according to the principle of a meat grinder. The screw extruder consists of a heatable or coolable screw which rotates via a suitable drive inside a heatable or coolable cylinder which, in turn, carries the extrusion die which is forming the extrudate; the cylinder has a feed port through which the preheated compound is fed to the screw.

Design. The screw, in conjunction with the cylinder, has to transport the compound from the feed port to the die, to masticate and to heat it so that it reaches the viscosity required for the final forming process. It also has to build up the pressure necessary for the out-put through the die (see Fig. 5.20). Depending on its design it is fulfilling this role more or less well and thus determines the out-put of the machine.

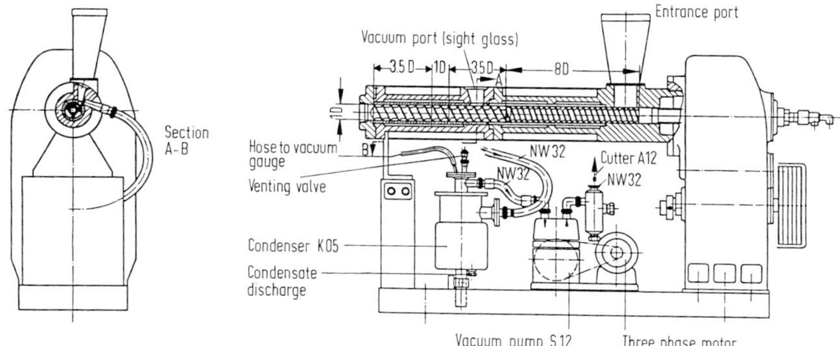

Figure 5.20 Extruder (Principle: Berstorff)

In addition, the temperature of screw and cylinder, the coefficient of friction between compound and screw and between compound and cylinder as well as the design of the extruder head, its surface quality and especially also the cross-sectional area of the die are important to obtain a high out-put efficiency.

A certain pressure is required for the through-put of the compound and thus the corresponding extrusion speed which is created by the *screw* and which has to be kept constant with a possibly small tolerance by balancing the head design and that of the die with the screw diameter. Important for fulfilling this requirement are:

- the *geometry*, i.e., pitch and depth of the screw flights
- the *cross-sectional area* of the die and
- the *speed of rotation* of the screw as well as
- the *clearance* between the screw flights and the cylinder wall which depends on the screw diameter.

The axial component of the force which acts upon the to be transported compound is the larger, the smaller the *screw pitch* is. Because of this reason a screw with possibly smallest pitch and high speed of rotation and thus large axial force is desirable. This goal is limited in practical applications because of the frictional heat fluctuation which can lead to premature vulcanization of the compound.

It is, therefore, desirable to build-up the pressure in the screw so that a significant pressure build-up occurs only toward the end of the screw in order to avoid premature vulcanization and a back flow of the compound toward the feed port. Such a pressure increase can be achieved by a corresponding screw geometry, for example a gradual decrease of the *channel area* of the screw toward the head, which in turn can be arrived at by an increasing screw core diameter (conical screw core) or by a decreasing pitch (decreasing pitch design) toward the end of the screw.

The profile of the screw core cross-section is most often formed so that the compound experiences a rotating relative motion as it is moving axially forward within the channel. Screws are seldom machined to have a *single channel*, more frequently they have *multiple channels*. In the feed zone of the extruder, pulsating of the compound can occur when screws with too few channels are used.

Important for good efficiency of the machine, for the pressure build-up and a good quality of the end product is to a certain extent also the *fill factor* of the screw in the feed zone which should be as high as possible. One can achieve this through deep cut screw channels, by using feed packets or feed curves in the cylinder at the entrance port or – perhaps most effectively – by the use of so-called feed rollers or a feed ram. The roll is arranged so at the entrance port, that it forms with the screw flights a nip – similar to a two roll mill – through which the material is pulled in and is pressed into the screw channels.

The screw should not be longer than is necessary for the heating and plastication as well as the build-up of the necessary extrusion pressure. An excessice *screw length* would only lead to disadvantages like unnecessary energy consumption, the danger of a too rapid increase of the friction heat and the loss in drive energy.

The screw length is defined as a multiple of the screw diameter. Often one chooses as the screw length, for example for the production of tire treads, a five fold of the screw outside diameter ($=5 \cdot D$). Treads can also be manufactured satisfactorily with screws of length $3 \cdot D$. Formerly one tried to use screws as short as possible in contrast to the processing of plastics where screw lengths up to $15 \cdot D$ and above are common. Extruders with screw lengths of 3 to $6 \cdot D$ have a mastication effect and have to be fed with preheated strips or slabs. Such machines are called *hot feed extruders*. Extruders with screw lengths of $12–24 \cdot D$ can be fed with cold strips or cold pellets. No preheating is necessary for these machines. They are called *cold feed extruders*. In these machines the pressure is built up in the last part of the screw as in the hot feed machine; the first section heats and plasticizes.

Until a few years ago the trend went certainly toward cold feed machines [5.98, 5.133]. However, in the tire industry one has learned the benefits of the hot feed machine for the processing of NR in conjunction with continuous processes, i.e., feeding of extruders directly from the mixing equipment whereby the nerve of the rubber – especially during processing of NR – is preserved and an elastic vulcanizate with low heat build-up can be obtained. Thus, for the manufacture of tire treads, frequently hot feed extruders are employed while cold feed extruders are preferentially used for the manufacturing of technical extrudates. Recently, however, the trend towards hot feed extruders in the tire industry is reversing again [5.131]. Here a new trend has started, namely the use of *cold feed machines with pin barrels,* i.e., with mixing cylinders (see Fig. 5.21) [5.114–5.120].

Figure 5.21 Pin barrel extruder QSM 200 K/18D, ESV 45K/12D for tire treads up to 800 mm width (Photo: Troester)

The great advantages offered by cold feed extruders were up to some years ago only taken advantage of for screw diameters of up to 150 mm. The processing advantage of cold feeding shrank with increasing extruder size because of which this principle was even inferior for a long time to the classical hot feeding system. Very important advances were made by development of ever newer screw geometries in sometimes very short order (see page 371 f), but all these systems were lacking the universal characteristics in order to apply them for a larger portion of the rubber compound varieties. This universal efficiency seems to be an unreachable goal for the conventional cold feed extruder.

This situation changed principally in 1978 with the introduction of the *pin barrel cold feed extruder QSM* [5.118]. This machine has proven itself so convincingly for the continuous processing of nonpreheated rubber compounds that one can consider the QSM extruder as the trend setting extruder principle for the area of technical rubber articles, conveyor belts and tire manufacturing. The QSM extruder is a single screw extruder with a cylinder equipped with up to 10 pin areas with 6–10 pins located in each pin area. These cylinder pins protrude into the specially designed screw, namely nearly down to the screw core. The first pin area is in the region of the pressure maximum, i.e., where the filling of the screw channels has been completed with certainty. The cylinder pins cause a portion of the laminar, rotating compound flow and thus force a material exchange between the different material layers. Thus, the formation of a cold compound core as is the case in a simple screw is prevented, and best mixing and extrusion results are obtained.

Aside from the length of a screw of, for example, $5 \cdot D$, the *channel depth* is chosen often as D/5 to D/6 and the pitch as D.

In order to obtain possibly small slippage between compound and *cylinder* one has to pay attention to the coefficient of friction between compound and cylinder wall on the one hand and compound and screw on the other. The latter shall be kept as small as possible by highly polishing the surface of the screw channel so that the compound will not rotate with the screw. The cylinder surface should not be polished as much.

Another means to obtain low coefficients of friction during cold feeding is to adjust the screw temperature to higher values (e.g., 80–100 °C) than the cylinder wall.

The *temperature* of the screw must, of course, not be so high that scorching will set in, at the same time the cylinder temperature most not be so low that the compound cannot reach processing temperatures.

Since the extrusion temperature depends upon the polymer and compound type, a heater in the cylinder zone of hot feed extruders was provided in former times, heated with saturated steam of up to 4 bar. The temperature was measured with simple bi-metallic or expansion thermometers while the regulation of steam heating and the water cooling was manual. The extrusion die is heated with a gas flame or it is equipped with an electrical heater band since this is very effective on the small available surface and is also easily installed. The screw is usually only equipped for water cooling which is also manually controlled. In more modern machines recirculating warm water systems and temperature controllers are used because of tightened specifications.

The *screw support* or that of the screw holder from which the screw can be easily removed during screw changes is built very solidly because of the considerable axial pressures.

In most cases a step-less regulation of the screw speed is superfluous. Therefore, an electric *motor* and *gear* reduction with 2, 3, or 4 fixed manually adjusted ratios is provided. The screw speed is usually not higher than 60 rpm and depends upon the screw diameter.

When preparing the *extrusion die* it has to be considered that the emerging profile does not have the same dimensions as the die opening but that a certain extrudate swell occurs (see page 393). The correct dimensioning of the extrusion die is done after experimental extrusion runs.

Because of the wide variety of extruded products, different designs of extrusion heads are required. The most frequently used ones are:

- *profile heads* for profiles
- *hose heads* with torpedo and spider to manufacture hollow products
- *sheeting heads* for sheets, belts, etc.
- *cross heads* for the coating of cables, wires, fabric hoses, etc. These are usually special order items.
- *shear heads* (see Fig. 5.22) in which the heat necessary for vulcanization is created by mechanical dissipation [5.112, 5.120a–5.128]. This occurs within the compound mass shortly before the forming process, i.e., there where it is needed to be effective.

Because the extrudate leaves the die of shear head extruders at vulcanization temperature other heating systems are not necessarily needed. Because of this the new shear head technology simplifies the processing operations.

For the production of hoses with fabric reinforcement, usually one extruder is placed before a braiding machine and one behind it.

Figure 5.22 Shear head extruder SK 90/70 (Photo: Troester)

The size of an extruder is determined by the screw diameter and length. Screw diameters of 30, 60, 90, 120, 150, 200, 250 and 300 to 600 mm are customary. While the machines with screw diameters up to 250 mm are available as cold feed extruders, those with screw diameters of 350–600 mm are only built as hot feed types.

Modern extruder lines are extensively controlled and automated through the use of microprocessors [5.120].

Vacuum Extruder. Gas (air) or moisture inside the compound can lead to porosity of the extrudate when vulcanization proceeds without pressure. De-gassing is therefore a prerequisite for continuous vulcanization processes which follow the extrusion process (see also page 371).

Since 1962 extruders with de-gassing ports (vacuum zones) for the de-gassing of rubber compounds for the directly following continuous vulcanization have been built [5.101].

Such extruders had already been used in the plastics industry. Their application to the rubber industry requires, however, a special effort since the de-gassing of the rubber compounds which are even highly viscous at high temperatures presents a bigger problem.

The vacuum extruder consists of the feed port, a feed zone which exhibits the shallow cut screw typical for cold feeding. A *ring (dam)* is arranged between the feeding and the vacuum zone.

In the *de-gassing zone* which is connected to a vacuum pump, the vacuum has to be between 3 and 5 Torr. This part of the screw is cut either deeper or wider or the space for the vacuum zone has been enlarged by widening the cylinder. The free space in this section of the screw has to be so big that it is only half filled with material during extrusion. This is assured by the ring or dam between the feed and vacuum zones which lets only a small "hose" of compound enter the vacuum zone of the machine and thus also seals the vacuum section. The "hose" is repeated by perpendicularly arranged grooves whereby the surface area is greatly enlarged. After passing the vacuum zone, the extrusion material enters the *extrusion section.*

Even the thin compound layer which enters the vacuum zone cannot be completely de-gassed because of the high viscosity of the rubber compounds in contrast to certain plastic compounds. A complete de-gassing is only accomplished inside the machine because the compound rolls off repeatedly from the screw flights of the vacuum zone whereby new surfaces are continuously formed from which air and moisture enclosures can be removed by the vacuum. Due to the relative motion of the compound it rolls off continuously while creating ever new surfaces [5.101].

It is therefore of crucial importance that the temperatures of the screw and the barrel in all parts of the machine, but especially in the vacuum zone, are measured accurately and be controlled. This machine thus requires a degree of temperature control and adjustment which is much higher than normally found in extruders. The accurate adjustment of the temperature at the dam is in addition crucial for the viscosity of the material entering the vacuum zone.

Other Extruders. Aside from the types discussed so far, others are also important:

- the *double or tripel extruder* where, e. g. the base compound and the tread compound for a protector are found and thus combined in one die to one extrudate.
- the *mill strainer,* a machine which produces pre-warmed strips, for example for the feeding of calenders and extruders.
- the *strainer,* which carries a screen before the material discharge port with which foreign particles are removed from the rubber compound.
- the *pelletizer* where the rubber compound is extruded in the form of strands which are chopped by rotating knives.
- the *Gordon plasticizer* where the front part of the screw is formed so that the rubber is forced to pass several screw flights whereby the rubber is continuously masticated (heavy mastication machine, see page 368).
- the *roller model* extruder and the *sheets extruder* which are used to extrude sheets or slabs after the compound preparation (see page 367) [5.109].
- the *roller head extruder* where an QSM cold feed extruder is feeding a two roll calender directly out of a sheeting die (see page 367) [5.92].
- the *rubber thread extruder* where a paste made from a rubber compound in an organic solvent is extruded through a multiple die into a leaching bath. They barely play a role today.

5.3.1.2 Ram Extruders

In ram extruders the compound is not transported by screws but, as the name implies, by a ram. The pre-heated compound is placed into the extrusion chamber in the form of rolled up slabs and pressed through the die by a piston which is, for example, driven hydraulically. One obtains hereby a more uniform pressure build-up in the die compared to screw extruders with short screws where the pressure build-up is accomplished by the rotational movement of the screw. For the same reason the dimensional stability of the extrudate when leaving the die is better than for screw extruders. The same compound can exhibit a very different extrudate swell when worked on either machine. Because of the possibility to manufacture especially accurate extrudates, the ram extruder is preferred for the preparation of pre-forms for the manufacture of molded articles. Because of the disadvantage of the inherently discontinuous operation mode this machine has not found universal application in spite of its undisputed advantages.

5.3.2 Working with Extruders and Influence of the Compound Ingredients on the Extrudability of Compounds

[5.126–5.143]

As can be seen from the statements above, the temperature profile of the rubber compound deriving from the extrusion process has to be carefully considered.

For machine types with long screws (cold feed machines) a preheating of the compound is not necessary. These can, for example, be fed continuously with pellets or cold strips.

For the frequently used hot feed extruders the same applies as to preheating of the compounds as for calendering (see page 378). For example, strips are cut with the help of circular knife pairs from preheat mills and then led to the feed port of the extruder by a conveyor. But a preheated compound strip may also be prepared by a mill strainer and fed into the extruder. The application of mill strainers has the advantage that the compound is freed of all possible foreign particles, filler agglomerations, rubber gel etc. by the screen.

During the extrusion process such contaminants are more detrimental to the production process than at other processing operations. A contamination which hangs up at the extrusion die can form grooves over the whole length of the extrudate which would yield the material as scrap.

Because of these reasons extrusion compounds have to be especially carefully prepared [5.129]. All fillers and chemicals should be sifted beforehand so that they are grit free. It should also be carefully observed if hardened filler or chemical agglomerations have formed during the compound preparation. The formation of good milled slabs without rubber knots (nonmasticated spots) before the start of mixing is a pre-requisite.

The lower the elastic component of a rubber compound, i.e., the better the rubber was masticated before the mixing process, the smoother it can be extruded. In the ideal case of only viscous behavior a rubber compound would exhibit exactly the same dimensions as the die opening after leaving the extrusion die. However, rubber compounds exhibit smaller or larger elastic components. Because of this "elastic memory effect" of the rubber compound it exhibits, similar to calendering, a more or less pronounced shrinkage which translates into an increase of the cross-section of the extrudate. This so-called extrudate swell is the larger, the higher the elastic component of the rubber compound is. This influences often the surface smoothness (see also page 368). Thus, the rheological properties of the rubber compound are of great importance for the extrusion results [5.134–5.143].

Because of these reasons, filler free or low filler content compounds cannot or only barely be extruded into smooth profiles. A careful *compound* design is thus important to produce exact extrudates a high production rates.

The prerequisite for good extrudability is a faultless premastication or correct choice of the viscosity of the *polymers*. Loosely pre-cross-linked polymers usually exhibit an especially good extrusion behavior. *Fillers* should preferably be inactive or half active ones. Many of the half active carbon blacks, but also, for example, chalk or kaolin yield very good extrusion results. Among the *plasticizers* especially the processing plasticizers, e.g., mineral oils, are effective, but also certain greases and *process aids*. For example, wool grease belongs to the most effective extrusion aids. Also *factices* belong to the most important aids during the design of extrusion compounds since they not only improve the smoothness of the extrudates and the extrusion speed but also, on the other hand, as "structure formers", the dimensional stability of the extruded profile or hose during vulcanization. Finally, also the *vulcanization system* has to be chosen carefully since no scorching must occur at the process temperatures in the extruder while a possibly fast vulcanization at only slightly increased temperatures shall occur after processing. Vulcanization accelerators with sufficiently delayed onset of vulcanization (high processing safety) but high vulcanization rates are applied.

After extrusion, the extrudate, when further processed as an intermediate, is powdered or pulled through an anti-stick bath and wound onto sheet metal drums, often with support devices to avoid deformations and vulcanized in a separate processing step, for example, in steam or under lead (see page 406, 448). Most often, however, continuous vulcanization processes are used in extrusion lines which will be discussed next.

5.3.3 Vulcanization of Extrudates

[5.144–5.189]

5.3.3.1 General Discussion of the Vulcanization of Extrudates

Formerly extrudates were nearly exclusively vulcanized in autoclaves, hot air or steam.

During the last years continuous vulcanization methods [5.144–5.156] for extruded products have been introduced because of economic reasons. While part of the extrudates are processed as intermediates (e.g. tire treads) before they are vulcanized and other parts are brought to a separate vulcanization process (for example because of small production volume), the vulcanization of a considerable part of extruded products occurs immediately after leaving the extrusion die in a continuous fashion. Also the continuous vulcanization in high pressure steam (steam pipe, CV curing, see page 406) is important in this connection; it is preferred for the manufacturing of cables. Finally, also the vulcanization under lead plays a certain role for extruded products.

5.3.3.2 Vulcanization in Autoclaves

For the vulcanization of extrudates in non-continuous processes the heating in hot air or steam autoclaves under pressure is still of the highest importance. This topic will be described elsewhere (see page 448). In order to avoid a deformation of the extrudates during vulcanization, the compounds are filled with sufficient amounts of factice or the profiles are placed into support structures or into talc powder. Thin walled hoses are placed onto mandrels in order to avoid deformations and are vulcanized under pressure in vulcanization autoclaves, of for example, 30 m length and more. It is also possible to heat the extrudates submerged in water within the autoclave (see page 450). Since these types of vulcanization are always discontinuous, they are only applied when continuous processes are excluded or are uneconomical.

5.3.3.3 Continuous Vulcanization in Liquid Baths (LCM Vulcanization)

[5.153, 5.155, 5.156, 5.164]

Years ago a continuous vulcanization method for extruded profiles, hoses, etc. has been developed by Du Pont. This method vulcanizes the rubber in a hot liquid. It is called LCM (**L**iquid **C**uring **M**ethod).

Principle. The principle is very simple: the profiles are brought into a long hot liquid bath immediately after leaving the extrusion die via a conveyor belt. They are kept submerged in the liquid and transported through the bath by a steel conveyor belt. The profiles arrive at the end of the bath fully vulcanized.

The main advantages of the liquid curing method compared to the vulcanization in vulcanization autoclaves is not only that continuous profile lengths can be manufactured but also that:

- scrap rate is lower
- profiles have better appearance
- the vulcanization times are shorter

The lowering of the scrap rate is based upon the fact that a considerably lower distortion and depreciation of the profiles occurs in the most frequently used heating media of the LCM method compared to vulcanization in the autoclave.

A further decrease of the scrap rate is achieved for this method because the dimensions and appearance of the finished vulcanizate can be controlled and supervised immediately after leaving the vulcanization bath while one can evaluate them for profiles vulcanized in autoclaves only batchwise and one has to scrap the whole batch if faults are detected. Furthermore, the efficiency of LCM equipment is higher; during vulcanization in autoclaves heat is always lost when they are opened.

Since powdering materials usually do not have to be used for liquid bath vulcanization and condensed water spots which occur for autoclave vulcanization do not form, one obtains by this method especially good looking extrudates.

Design. The following *equipment* is required for LCM vulcanization:

The salt bath is contained in a long *trough* through which a conveyor belt, for example from steel, runs. This trough is located close to the extruder. The extrudate is submerged into the salt bath and transported through it by the belt.

The required length of a fluid bath is not only dependent upon the vulcanization rate of the profiles and the temperature of the salt but also upon the output rate of the attached extruder, i.e., the maximum speed. This, in turn, is dependent upon the dimensions of the extruded profile. An excessive time beyond the optimal vulcanization time is dangerous for the profile in view of the high vulcanization temperatures. Therefore, a careful balance between extrusion speed and residence time in the heated bath has to be established.

The *heating media* for vulcanization baths are, for example, salt mixtures as well as (more seldom) metal mixtures, polyalkyl glycols, glycerine, silicone oil, etc. The method is also called "salt bath vulcanization" because of the most frequently used salt mixtures.

The eutectic, molten salt mixture most frequently used is composed of (by weight percent)

53% Potassium Nitrate
40% Sodium Nitrite
7% Sodium Nitrate

Advantages and Disadvantages of LCM. Using these baths yields very clean looking profiles. The disadvantage of the salt mixture is its high *density*. Since the density of the rubber compound is practically always considerably lower than that of the salt mixture, a more or less strong buoyancy develops which has to be overcome by the steel band pressing the profile into the bath. For very soft compounds of low density and complicated profile cross-section, deformations can result. In such cases, one can work better, for example, with polyalkyl glycol or silicone oil as the heating bath medium because of their lower density or fluid bed vulcanization (see page 398) or UHF vulcanization (see page 399) are preferred.

The *heat transfer* of the heating media is generally quite good and since high vulcanization temperatures can be chosen (normally 210–240 °C) one obtains rather short vulcanization times. Of the bath liquids the so-called salt mixture has especially good heat transfer characteristics. It is considerably higher than for other continuous vulcanization methods. However, the relatively slow heat transfer, especially for thick walled articles or for hollow profiles with partitions, is problematic because of the low heat conductivity of the rubber compounds. The larger the profile to be heated, the longer has to be the residence time of the profile in the liquid bath so that it will be heated through. However, this causes a corresponding severe over-vulcanization of the profile on the surface, especially at very high heating temperatures. Finally, with increasing thickness, an ever increasing gradient of the degree of vulcanization results, thus also an anisotropy of the vulcanizate properties from the outside to the center. Here the limits of the LCM method are met. For the so-called UHF method (see page 399 ff) a better heating effect over the area of the profile is obtained compared to LCM.

Another problem for the liquid bath vulcanization is avoiding porosity which can arise because of the very low vulcanization pressure (only the small liquid columns). Porosity arises from air and moisture contained within the rubber compound. It is obvious that this depends to a large degree on the kind and magnitude of the fillers and the other ingredients as well as on the processing technology. The higher the hardness of the unvulcanized compound, the lower is normally the porosity.

The major part of porosity is due to the water content of the compound. Using dessicants in the compound like, e.g. CaO, the porosity may be reduced significantly. However, the vulcanization behavior of the compound will be affected.

Elimination of the remaining porosity caused by entrapped air is more difficult. By using extruders with vacuum zones in the screw cylinder (page 391), the remaining air may be eliminated. Also by using small amounts (5–10 phr) of factice the degassing of compounds during mixing is much more efficient (page 306).

Porous profiles with fine pores can also be manufactured continuously in LCM equipment even though their production with the help of UHF equipment is advantageous. The blowing agents contained in the compound decompose at the bath temperatures and blow the profile to the desired height before the onset of vulcanization. A premature gas loss in the extruder should be avoided. The temperature history and choice of blowing agent have to be coordinated (see Section 4.6, page 307 ff, and 5.4.6.4, page 440).

The following is a list of the advantages of LCM vulcanization:

- no interruption of the continuous extrusion process
- no transporting semi-finished goods
- only small temperature losses
- no expensive support structure of the extrudates
- no anti-stick powder, therefore cleaner operation
- low reject rate
- labor saving
- better appearance of the profiles
- possibility to vulcanize peroxide containing compounds
- no oxidation of oxidation prone elastomers, e.g., NR

The following is a list of the disadvantages of the LCM vulcanization:

- necessity to use expensive extruders with de-gassing zones
- therefore often lower extrusion rate compared to conventional processes
- certain salt loss in the salt bath depending on the kind of profile, the extrusion speed and the temperature
- therefore maintenance of the bath necessary
- need to clean profiles
- higher danger level working with the salt bath
- occasional deformation of the profiles
- formation of nitrosoamines

Influence of Compound Composition. Naturally, the compound composition is of strong influence on the salt bath vulcanization. All elastomer types can be vulcanized without great difficulty pressureless as long as the compound has been designed properly. When determining the maximum bath *temperature* one has to consider the heat stability as well as other technological considerations.

From experience the following temperature limits can be given (see Table 5.1).

Table 5.1: Maximum Vulcanization Temperatures of Various Elastomer Types During LCM-Vulcanization

Elastomer Type	Max. Vulcanization Temperature	Comment
NR	up to 210 °C	above this: surface tackiness, reversion
SBR	up to 240 °C	perhaps even higher
OE-SBR	up to 240 °C	above this: formation of pores
NBR	up to 240 °C	perhaps even higher
CR	up to 240 °C	in exceptions higher
EAM	up to 220 °C	above this great pores because of peroxide decomposition

In the beginning relatively fast *accelerator* combinations were used as vulcanization accelerators for the salt bath vulcanization to accommodate the short vulcanization time available. Later experimentation showed that this is not necessary or even not practical since the differences in the heating time are relatively minor at the very high vulcanization temperatures, however, the possible danger of reversion is increased. The less rapidly accelerated compounds have, above all, the advantage of higher scorch safety in the extruder. Sulfenamides in combination with mercapto accelerators, thiurams and dithiocarbamates are employed. For CR, for example, ETU-thioureas are combined with polyethylene-polyamines. For large cross-sections the use of quick accelerated compounds is preferred because of the low heat conductivity of the rubber compounds and thus the danger of severe under-vulcanization at the center of the profiles. For elastomers which tend to revert, accelerators are preferred, because of the high vulcanization temperatures, which improve the reversion resistance.

For the vulcanization of compounds with *peroxides* the vulcanization temperature has to be limited to 200–220 °C since a sudden decomposition of the peroxide at higher temperatures can give rise to bubbles.

Among the *fillers*, furnace, acetylene and oil blacks do not create bubbles and thus do not create difficulties. In contrast, it is practically impossible to manufacture bubble free vulcanizates with channel blacks. Negative results are also obtained with Kaolin. By using CaO as well as factice the porosity can be minimized in most cases.

The effect of *plasticizers* of low volatility on porosity is relatively minor during salt bath vulcanization. Most plasticizers, especially the mineral oil plasticizers, do not cause porosity even at temperatures up to 240 °C. Use of emulsion plasticizers, in contrast, almost always results in bubble formation since the water content remaining in the compound cannot be eliminated even in the vacuum extruder.

5.3.3.4 Fluid Bed Vulcanization

[5.151, 5.166]

Aside from the salt bath vulcanization the so-called "fluid bed vulcanization" is used to a lesser extent for the continuous manufacturing of hoses and profiles.

The principle of the fluid bed vulcanization is very similar to that of the liquid bath vulcanization. Also the problems faced by the operator of these two methods are similar. The principle is the following:

Principle. When one lets a gas stream flow upwards through a layer of solid, small particles, the latter are lifted. At a sufficient flow velocity the layer expands and every particle is being suspended by the moving gas and exhibits through the "gas cushioning" a considerable fluidity. In this state the system is called a fluid bed because the small suspended particles behave in many respects like a Newtonian liquid: It fills the volume of the container, exhibits a horizontal surface, can be transported through a pipe and the hydrostatic pressure increases proportional to the depth. It is noteworthy that a profile to be vulcanized can be pulled through this apparent Newtonian liquid analogous to the salt bath. Also, the heat transfer characteristics correspond to those of a liquid and are approximately 50 times higher than those of air.

Such fluid beds may be utilized analogous to the liquid baths for the continuous vulcanization of extrudates. As particles, small glass spheres of 0.13 to 0.25 mm diameter are used, for example. They are called Ballotini. The layer has in loose packing about 40% empty space; when gas is flowing the layer expands by about 10% and the empty space increases to about 45%. This expansion is sufficient to reach a liquid like state and to transform the glass beads into a fluid bed. The extrudate to be vulcanized can be pulled through it.

Heat Transfer. The heat transfer of a fluid bed depends only to a small extent on the material from which the particles are made, it is determined much more by the degree of fluidization and the specific heat of the gas, e.g., steam. The degree of polymerization in turn, determines the "density" of the fluid bed which can be changed and can be adjusted to the density of the vulcanizate. The fluid bed vulcanization is therefore especially useful for the vulcanization of complicated profiles which are sensitive to deformations since the profile can "float" freely through the fluid bed.

The problems of porosity are the same as those for the liquid bath vulcanization and are corrected by the same means (see page 394ff).

Arrangement. Fluid beds are built in either horizontal or vertical positions.

5.3.3.5 Continuous Vulcanization in Hot Air after Pre-Heating in Ultra High Frequency AC Field (UHF Vulcanization)

[5.157–5.165, 5.167–5.174]

Hot Air Vulcanization. The vulcanization in hot air is one of the oldest vulcanization methods (see page 447). During continuous operation it is only applied without pressure. The extrudate is pulled through hot air tunnels which can be up to 150 m long; the total length is divided into different sections in order to permit good temperature control and to obtain good speed regulation.

In attempting to reduce the length of the hot air tunnels drastically for the newest continuous vulcanization, a pre-heating system for the extrudate in the form of an ultra high frequency "oven" was developed which is coupled with a relatively short hot air tunnel. This continuous vulcanization in hot air after pre-heating in an ultra-high frequency AC field is called – not quite correctly – "UHF" or "micro wave" vulcanization. After UHF preheating the hot air tunnels can be shortened to, for example, 22 m or shorter. They serve only to maintain the temperature reached in the UHF field (see Fig. 5.23 on next page, see also page 429).

Principle of the UHF Method. For UHF vulcanization an extruded profile is brought (after leaving the extruder die) through a micro wave oven with an ultra high frequency AC field where it is heated because of the dielectric losses.

The *dielectric heating* which permits a rapid and even heating of all electrically non-conducting substances is based upon the fact that the non-conducting substances brought into the electric high frequency field are subject to a polarization due to the ultra high frequency AC field whereby dielectric losses are created in these materials which in turn heat up the substances.

These dielectric losses are, according to *Debye,* depending on the frequency of the electric AC field, caused by the different types of polarization which consume part of the energy of the field in the form of internal friction which manifests itself in the generation of heat.

One has to distinguish between the following types of polarization:

- *electron polarization,* created by a shifting of electrons with respect to the positive nucleus
- *dipole polarization,* caused by the influence of the field of molecules with dipole character
- *Interface or ion polarization,* created by the accumulation of free ions at the interface between materials of different conductivity and dielectric constant.

For pure NR, SBR, EPDM, IIR etc. where one deals with nonpolar substances, ultra high frequency heating is only possible by *electron polarization* which requires an extraordinarily high frequency. A practical and economical use of pre-heating is therefore not feasible in this case. In contrast, for elastomer types with *dipole character,* i.e., polar substances such as NBR and CR, considerably lower energy levels are necessary and they are therefore much better suited for an even heating in a high frequency field.

In many cases a heating by interface or *ion polymerization* with relatively low frequencies compared to electron polarization is possible because one is dealing in practical use of high frequency heating with rubber compounds, i.e., with *heterogeneous material systems.* Hereby the explanation of the heating mechanism is still being debated. The "Wagner theory" of the dielectric gains importance which declares the electrical heterogeneity as the cause for the energy loss whereby the

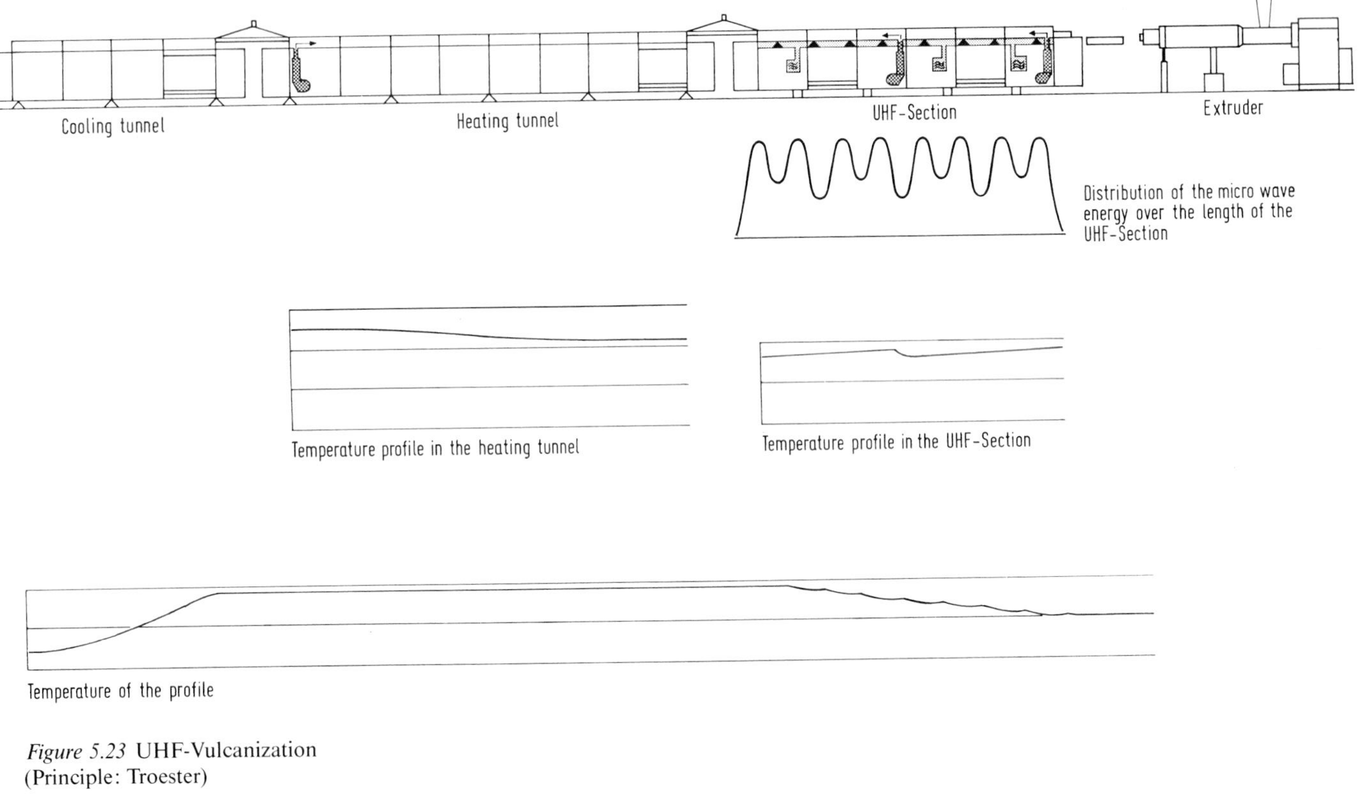

Figure 5.23 UHF-Vulcanization (Principle: Troester)

comparison is made that an electrically disturbed surrounding absorbs energy in a form similar to an optically disturbed system absorbing light energy.

The heatability of a compound is dependent heavily upon the *fillers* aside from the type of polymer.

With increasing amounts of carbon black the heating rate is raised. But also the reactivity of the black is of influence. At equal loadings more active carbon blacks give rise to a more rapid heating than less active ones. White fillers yield less rapidly heatable compounds than blacks. Among the white fillers, ZnO yields faster heatable compounds than, for example, $MgCO_3$ and this, in turn, is faster than chalk or silicas.

The energy dissipation in a lossy dielectric by a ultra high frequency field is described by the following equation:

$$P = K \cdot E_2 \cdot f \tan \delta \cdot \varepsilon'_r$$

where P is the loss energy in W/cm^3, K a constant ($0.556 \cdot 10^{-12}$), E the field strength in V/cm, f the frequency of the field, tan δ the loss factor and ε'_r the dielectric constant. The product of $\tan \delta \cdot \varepsilon'_r$ is the loss number ε'_r a characteristic material property. Tan δ as well as also ε'_r are frequency and temperature dependent. The loss number ε'_r is important for the heating effect of a compound in the UHF field, i.e., it affords an evaluation of the vulcanization rate in the UHF field. It can be measured in a small cylindrical cavity resonator with a measuring bridge to evaluate the usefulness of rubber compounds.

Design. For UHF pre-heating the following equipment is required. The generator is a transmitter according to the *oscillating magnetron* with a frequency of 2450 MHz which is the only one authorized for industrial purposes in Europe (in addition in USA and England 915 MHz are applicable).

For the wave guide principle the ultra high frequency waves created by the magnetron are fed into a wave guide whereby waves are created which propagate axially in the wave guide and which form standing waves by the calibration of the wave guide length with the help of reflectors. By special added devices like coaxial conductors or a ferrite cage, provisions are made that during the pre-heating of rubber-metal combinations, for example cables, a distortion of the fields is avoided and no stray radiation of electrical energy takes place.

For the *resonator chamber* principle the micro wave energy is developed in the pre-heating region as scattered radiation which is distributed by reflection from the chamber walls over the entire area of the chamber. In contrast to the standing wave generated in the wave guide, the diffused energy in the resonator chamber is, of course, weakened because of which the production capacity, based on the same cross-sectional area of the profiles is lower by 20–30%. However, with such equipment considerably bigger profiles can be manufactured whereby the production rate may be higher even though the production speed could be lower. But since the heat transfer is more uniform in the resonator chamber compared to the wave guide, the resonator chamber principle is especially suitable for the manufacture of bigger profiles as well as sponge rubber for which a uniform heat creation is a prerequisite for even cell formation. The micro wave guides or the resonator chambers are the pre-warm channel. Through this channel of, for example, 4 m length a silicone or teflon coated glass fiber belt transports the unvulcanized material. The equipment contains, furthermore, the controls to adjust the speed of the conveyer belt and to center the wave guide which permits to lead the profile, depending on its size and form, exactly through the middle of the electric field.

Furthermore, equipment is available to control how much energy of the wave is absorbed by the rubber compound [5.170]. *Photocells* at the end of the vulcanization tunnel turn off the magnetron when no material passes through the channel anymore.

In order to avoid heat losses of the profile by radiation, a hot air blower is provided which at the same time carries away the developing gasses and prevents any condensation.

The relatively short length of the *pre-heating chamber* requires the addition of an infrared or hot air tunnel in order to finish vulcanization at the required production speeds. This combination has the advantage of a uniform degree of vulcanization. Naturally, a complete vulcanization would also be possible with a correspondingly long UHF tunnel, but this would require uneconomical efforts.

In contrast to other vulcanization methods, the *heat* is not supplied in the UHF equipment from the outside but the heat is generated here simultaneously on the surface and the interior so that a pronounced gradient of the degree of vulcanization does not develop. A uniform degree of vulcanization results and an isotropic property field over the area of the vulcanizate. Therefore this method requires a considerably lower level of energy compared to the other methods [5.146] (yield of energy 50 to 60%) and provides, in addition, a substantial shortening of the vulcanization time for thicker articles for which, for example, the LCM vulcanization time is problematic.

It is necessary also for the UHF equipment, analogous to the LCM method, to use an extruder with a de-gassing zone in order to obtain bubble free vulcanizates.

Compound Design. It is necessary to consider the energy to be absorbed, i.e. the developing vulcanization temperature, during the design of compounds which are to be vulcanized with the help of UHF equipment.

The heating of a rubber compound occurs the faster, as already mentioned, the more polar the rubber is or the higher the proportion of *polar compound ingredients* like, for example, carbon black. Thus, it is not very difficult to design rapidly vulcanizing compounds based upon polar elastomers like NBR or CR which have a high polarity by themselves, or from non-polar polymers like SBR, NR, EPDM etc. as long as they are filled with suitable carbon blacks [5.167, 5.171]. Problems arise with non-polar elastomers filled with white fillers. In order to solve this problem, the following solutions are available [5.167]:

- the blending of non-polar with *polar elastomers*
- addition of *polar materials* to non-polar compounds.

Blends of non-polar NR or EPDM with polar *NBR* or *CR* lead, for example, to heating rates which make it possible at already small amounts of NBR or CR to develop compounds for continuous vulcanization. Also, an addition of PVC or PVDC has been applied.

When polar *plasticizers* are added, for example chloroparaffin, in the customary amounts, no sufficient heating rates can be obtained.

A certain increase in the heating rate is obtained by employing *factice*, especially castor oil factice [4.564]. Especially high increases are obtained when *filler activators* are used like diethylene-glycol and triethanol-amine. It is often already sufficient to add 10 weight percent of the filler loading of diethylene-glycol and triethanol-amine in equal amounts in order to obtain a marked increase of the heating rate [5.167]. Also, *process aids* like Zn soaps, fatty alcohol esters etc. are beneficial for the continuous UHF vulcanization. As a rule, no one single step is sufficient to reach satisfactory degrees of heating. Therefore, usually several of the above mentioned procedures are applied.

When silica fillers are used, always a part of the water is bound so intimately that it cannot be removed in the vacuum zone of the extruder. This water can be set free at the high vulcanization temperatures and can thus lead to porous vulcanizates unless CaO is added. However, when diethylene-glycol or triethanol-amine is used, the water normally bound to the active filler surface can be split off and be removed in the vacuum zone. One obtains, therefore, often bubble free vulcanizates even without the incorporation of CaO.

It should be, furthermore, considered that in some cases a change in the polarity of the compound is caused by the vulcanization which results in an increase in the energy absorption after the onset of and during the vulcanization.

5.3.3.6 Vulcanization by High Energy Radiation

[1.13, 1.14, 3.400, 5.175–5.188]

Since the formation of free radicals on the polymer chains can initiate their cross-linking, essentially all those methods are suitable for cross linking which form radicals on the polymer chain. A radical formation on the polymer chain by dehydration and thus a cross-linking of the polymers can also be accomplished by high energy radiation, which one can thus consider to be an indirect vulcanization "material". The energy of the radiation has to be, naturally, higher than the bond energy of the most labile carbon-hydrogen bond of the elastomer. This cross-linking principle which is used frequently in the plastics industry, e.g. for the cross-linking of PE and PVC, is also applicable to the crosslinking of elastomers.

Radiation Sources. *Cobalt sources* (Co^{60}), *Van-de-Graaff generators* and *resonance transformers,* cascaded and linear accelerators or *betatrons* are used.

The *Co^{60} source* is the weakest radiation source. It delivers radiation with an average total radiation energy of approx. 60–140 W, this means a particle energy of 1.17 to 1.33 MeV and a particle energy of 0.306 MeV. The radiation energy is naturally limited for the Co^{60} source since the half time of Co^{60} is about 5 years.

Electron accelerators work according to the following principle: Electrons are generated inside the generator which are rapidly accelerated in vacuo because they pass through a potential gradient. At the end of the radiation channel, the exit port through a Titanium widow, the electrons have a high energy.

The first publications in this field, up to the middle of the sixties, were of more academic interest, because up to that time only a few electron accelerators or other radiation sources existed in industry and delivered only little energy. But starting in the middle of the sixties the electron accelerators became more powerful and thus the radiation cross-linking more economical, because of which the industry became more interested in this technology [3.400].

The radiation energy of the linear accelerators has increased over the last 20 years by a factor of 40, from 5 to 200 kW. Therefore such equipment became economical.

It is estimated that approximately 30 to 40 generators with an average energy of about 35 to 40 kW each and a total energy of 1300 kW were installed in Western Europe in 1981. For 1987 it was estimated that the number of units with an energy of 55 kW each increased to 85 with a total installed energy of 4300 kW in Western European industry. These numbers reflect the extremely rapid transition from classical vulcanization to radiation crosslinking which occurs especially in the cable industry and other factories with large volumes of continuously produced thin walled articles. The electron beam can be controlled by a magnetic deflection system so that one obtains a radiation source in the form of either a spot or a wide band.

Compared to a Co^{60} source, the electron accelerator yields an accurately focusable and constant radiation. The Co^{60} source creates always a diffuse radiation, the energy of which decreases with time according to the half time of the material.

Cross-Linking. Cross-linking due to radiation is normally carried out within the electron accelerator so that the article to be cross-linked is carried through it by a conveyer belt. For example, when a 4 mm thick plate needs for full cross-linking an absorbed radiation energy of 8 Mrad* and the generator supplies one of only 4 Mrad during one pass, then one pass is required for each side of the plate.

During the irradiation with high energy irradiation (>0.5MeV), radicals are created in the polymer chain which cause C-C crosslinking. It can, therefore, be expected according to the discussions on page 227 that the elastomers will have good heat stability. Reduced to the simplest formula, the cross-linking reaction can be described by the following scheme:

$$2-\overset{|}{\underset{|}{C}}H + 2e \longrightarrow 2-\overset{|}{\underset{|}{C}}^{*} + H_2$$

$$2-\overset{|}{\underset{|}{C}}^{*} \longrightarrow -\overset{|}{\underset{|}{C}}-\overset{|}{\underset{|}{C}}-$$

In reality, the cross-linking reaction is more complicated.

The radiation energy and thus the type of electron accelerator to be used are dependent on the required penetration depth, the density of the irradiated material and the chosen irradiation system. If one measures the density (d) in g/cm^3 and the layer thickness (S) in mm, one can determine the radiation energy (E) necessary for optimal homogeneity approximately from:

$$E=\frac{S\cdot d}{3} \text{ (MeV) for one sided irradiation,}$$

or

$$E=\frac{S\cdot d}{8} \text{ (MeV) for two sided irradiation.}$$

For a correspondingly chosen radiation energy, the production speed (V) is then a function of the radiation current (i), the dose (D) and the type of irradiation or the irradiation device, characterized by a constant (k):

* For the definition of the irradiation of technical materials the unit "Roentgen" is not suitable (the radiation energy of a Roentgen tube which in 1 cm^3 of air at 760 Torr at 18^0 while utilizing all secondary electrons, causes such a conductivity that the electrostatic unit of $3.3359\ 10^{-10}$ is measured at saturation current). Rather, analogous units were created to define the absorbed radiation energy, referred to unit weight. When air under standard conditions is irradiated with a radiation dose of one Roentgen, an energy of 83.6 erg is absorbed by each 1 g of air. This energy is 1 rep (Roentgen equivalent physical) and correspondingly one speaks during the irradiation of materials by the same dose also of 1 rep = 83.6 erg/g. However, it has proven more advantageous when the absorbed energy is chosen in the decimal system per gram of material. Therefore, one speaks of one rad, when per g of material 100 erg are absorbed. One rad is therefore 100 erg/g, 1 Mrep = 10^6 rep, 1 Mrad = 10^6 rad. In SI units, the units Gy (Gray) or kJ/kg are used. The conversion is done by: 1 Mrad = 10^6 = 10^4 Gy = 10 kJ/kg = 10^8 erg/s.

$$V = k \cdot \frac{i}{D} \text{ or } D = k \cdot \frac{i}{V}.$$

If the ratio i/V is kept constant, a constant speed can be maintained.

When an electron beam of high energy enters a material which is to be vulcanized, it is slowed down. When it has lost so much energy, that its energy corresponds to that necessary for crosslinking, cross-linking takes place. This means, that the essential cross-linking reaction takes place on the inside of the material to be cross-linked, depending on the level of radiation. With increasing penetration depth more and more energy is lost until the radiation energy is not sufficient anymore to initiate cross-linking. The intensity of the effectiveness decrease depends primarily on the density of the compound to be cross-linked. The effectiveness loss increases with increasing density for a given constant thickness. Is the effectiveness depth of the radiation in a material of a density of $1\,g/cm^3 = n$ millimeter, it is in a material of a density of $1.2\ g/cm^3 = \frac{n}{1 \cdot 2}$ millimeter. Beyond that, also the maximum depth which can be reached by the radiation, i.e. the thickness multiplied by the density, for which the radiation has been reduced to 10% of its original intensity is of importance. Since a certain minimum amount of energy is required to form polymer radicals on the one hand, but the radiation energy decreases with the penetration depth of the radiation on the other hand, it is understandable that a uniform degree of cross-linking above a certain thickness of the article is difficult to reach even when the irradiation is carried out from both sides. These problems are of a processing technical nature. Several technical concepts were developed for the irradiation of cables in a high energy field [3.400].

The absorbed radiation energy is considerably lower than that required for the radio-active excitation of the metal atoms contained in the compound.

Compound Design. No vulcanization aids are required for the radiation curing of elastomers like, for example, peroxides. It is sufficient, therefore, to add the usual fillers, plasticizers, colorants etc. to the elastomers. Sometimes activators which are also used as co-activators for the peroxide cure are utilized, like, for example, EDMA or TPTA (see page 259) in order to lower the required amount of radiation energy [3.400] or to achieve at the same level of radiation energy an increased state of cure. For Q, where the radiation cross-linking seems to be especially promising, the average required radiation energy is, for example, approximately 10 Mrep. However, the required dose can vary considerably according to the type of Q used. Vinyl-free types require generally a somewhat higher dose than vinyl containing products. For methyl phenyl polysiloxanes the energy requirements rise sharply with increasing fraction of phenyl groups. For the same reason, therefore, phenyl-group containing polysiloxanes are also more stable against radiation than dimethyl polysiloxane.

Vulcanizate properties. At a given crosslink-density, thin-walled vulcanizates cross-linked either by radiation or by peroxide are barely distinguishable as far as their mechanical properties are concerned. For higher thicknesses – especially at a simultaneous high density of the compounds – it is difficult with the application of *older weak equipment* to obtain by high energy radiation a state of cure which is uniform over the thickness since the radiation, as already mentioned, is slowed down when entering the elastomer, i.e. it loses kinetic energy. Above certain wall thicknesses the inside of the part can be severely under- vulcanized or remain even completely uncured. The latter is the case when the radiation energy has become lower, because of absorption, than the energy of the most labile bonds. For *newer equip-*

ment with high energy output a considerably more uniform degree of cure can be obtained because of the greater penetration depth even for only one-sided irradiation.

During radiation cross-linking no decomposition products are created by any vulcanization chemicals which could influence the property spectrum or the physiology of the vulcanizates.

Economics. The use of high energy electron accelerators seems to have become more economical, for example, for the continuous vulcanization of sheets (according to statements by the manufacturers) than rotational vulcanization [5.185].

5.3.3.7 Continuous Vulcanization in the Steam Pipe

The vulcanization in high pressure steam can be applied only to those extruded products which have a metallic insert like metallic conductors in cables in order to be able to pull the extrudate. This type of vulcanization is thus nearly exclusively used in the cable industry.

At sufficiently high steam pressure very short vulcanization times can be obtained because of the high temperatures.

Principle. The principle of vulcanization in the steam pipe is the following: The cable, after having been coated with the jacket is brought into a double walled pipe which is directly attached to the cross head in which the steam pressure is normally 5 to 12 bar, sometimes 20 bar or even more. The steam pipe can be connected via a suitable seal to a cooling water pipe; or it may go over into a cooling trough.

The required length of the pipe depends, aside from the gauge of the cable and the thickness of its jacket, primarily on the steam pressure as well as on the choice of the vulcanization accelerators. One cannot vulcanize arbitrarily rapidly since one of the variables is the heat transfer of the rubber compound just as for the LCM vulcanization (see page 396).

The continuous vulcanization is only applicable up to certain jacket thicknesses of the cables because of the low heat conductivity of the rubber compounds. However, jackets of medium gauge can be manufactured in this manner just as cables of small diameter.

Arrangement. The steam pipes are normally arranged horizontally. But suspended and vertical designs are also known.

Vulcanization Accelerators. For the vulcanization in the steam pipe only vulcanization accelerators are suitable which have a sufficiently broad plateau and a high scorch safety but can still vulcanize within seconds. Such accelerators are, for example, the sulfenamides which permit a long flow period in the extruder but act rapidly in steam, especially in combination with thiuram or dithiocarbamates (see page 242, 246).

5.3.3.8 Vulcanization under Lead

[5.126, 5.189]

In order to avoid a deformation of extruded articles, for example cables or hoses of especially large cross-sectional area, during vulcanization they are sometimes enclosed in lead before vulcanization. The lead sheath also protects against other influences, for example the hydrolytic effects of the steam.

The lead sheaths can be applied discontinuously by hydraulic presses or continuously using extruders [5.189].

The extrudates enclosed in lead are normally vulcanized discontinuously, rolled up onto large drums which are placed into autoclaves. The vulcanization time is, of course, longer than for a simple steam vulcanization since the lead has to be heated.

After vulcanization the sheathed vulcanizate is cooled as fast as possible, for example by a water spray. After cooling, the lead is peeled off after the sheath has been slit, separated from the vulcanizate and returned to the lead press.

5.3.3.9 Silane Cross-Linking

[3.400, 5.147]

Recently, polymers are also silane modified which self-cross-link in a moist atmosphere. Such methods are employed, for example, in EPDM for the cable and construction industry. The addition of silane is said to improve the water resistance, for example, of cables [3.400].

5.4 Manufacture of Molded Articles

[5.190–5.294]

Molded articles to which typical mass produced articles belong, like stoppers, suction cups, bottle stops, soles, household goods as well as numerous articles, which find their use in industry, e.g. seals, gaskets, membranes, valve balls, bumpers of all types and finally tires which represent a combined article based on hand manufacturing and molding are very important for the rubber processing industry.

Molded rubber articles can be produced by four different manufacturing methods:

- Compression method (Press-Method)
- Transfer molding method
- Flashless method
- Injection molding method

whereby the flashless method is a special case of the transfer molding method. Vulcanization occurs during the molding process. However, many articles have to be assembled from several compounds before vulcanization. Confectioning is important for this process.

5.4.1 Confectioning

The confectioning methods for different rubber articles, for example automotive tires, hoses with reinforcement, conveyor belts, V-belts, roll covers, rubber shoes, balls and many others are remarkably varied. Attempts have been made to eliminate by the development of proper machinery the expensive manual labor for the construction of these articles. Many processes operate fully or partially automatic today. Others, in contrast, still require extensive manual labor. The manufacture of molded articles is therefore often still quite labor intensive.

Some methods of confectioning will be discussed in connection with the manufacture of the most important rubber articles. For each type of confectioning a certain tackiness of the calandered sheet or the rubber coated fabric is necessary. NR compounds normally have this required tackiness. It can be improved for SR by the addition of resins or plasticizers depending on the requirements. Since the surface may be soiled by the inter-layers, by oxidation or sulfur bloom, it has to be cleaned first by rubbing it with solvents like, for example, gasoline. For NBR which is insoluble in gasoline, other solvents like methyl ethylketone etc. have to be used, perhaps with additions of a resin. In some cases rubber solutions are used during confectioning (see also page 380).

5.4.2 Compression Method and Vulcanization Presses

[5.190–5.203]

5.4.2.1 General Comments about the Manufacture of Molded Articles

A large number of rubber articles like molded goods, conveyor belts, tires etc. is vulcanized in compression presses which are closed either hydraulically, by spindles or by toggles. A preform of the rubber compound is prepared by extrusion, stamping, cutting or building-up. It is then formed by pressing it into molds under the application of heat. Vulcanization occurs at the same time.

All articles manufactured by this method have the disadvantage that they have a more or less pronounced orientation because of which their dimensional stability in the direction of pressing is affected.

5.4.2.2 Presses

The compression presses are subdivided into hydraulic, toggle and specialty presses.

Hydraulic Presses. The hydraulic presses (see Fig. 5.24) consist of two or more press platens which are contained within a frame or a pair of uprights. They are *heated* by steam, hot water or electrically. The most common heating method is by steam. The

Figure 5.24 Hydraulically operated multi platen press (Photo: Berstorff)

presses are connected to hydraulic systems where the one, the low pressure system with 30–40 bar is used to close and open the press while the other one, the high pressure system, supplies 200 bar to the piston. Since the pistons are considerably smaller than the press platens, one can normally expect pressures of 35 to 100 bar. These high pressures are required to assure a uniform press deformation of the rubber compounds. Larger presses, as used, for example, for the manufacture of conveyer belts, have several cylinders.

The *control* of the presses is by programmed controllers which automate the closing, repeated "bumping" (short time opening to permit the air to escape) and opening of the press.

The preforms to be vulcanized are placed into closable *molds* which are pushed in between the press platens. After completion of vulcanization the molds are pulled out of the press and opened. The article is then taken from the mold.

Toggle Presses. Instead of hydraulic presses, electrically operated toggle presses are quite often used in order to simplify the molding process for high volume articles. Opening and closing is mechanical via toggles after which the presses are named. The molds are built into the press. Therefore they are quite often specialty presses. The inclined position of the heated platens in the open position facilitate a faster loading and emptying of the molds. The most important press of this type is the so-called single tire press.

In *comparison* to the other three molding methods the compression molding method has the advantage that the molds for the manufacture of the molded articles are simple to machine. Thereby the possibility is given to manufacture even relatively short production runs at comparatively low cost. In addition, the investment costs for the presses are low compared to the injection molding method.

Specialty Presses. Aside from the presses mentioned above which are preferred for the manufacture of small or medium sized molded articles, a multitude of specialty presses exists which are mostly designed for the production of certain articles. Among others the following ones shall be mentioned, which are partly of older construction.

- *Tire Presses* with a high degree of automation are mostly single or double presses in which two mold halves containing the tire shape are permanently attached to the upper and the lower part of the press (see page 435).
- *Jaw Presses* the working space of which is open to all three sides due to the use of a cantilever frame. They are preferred for the manufacture of sheets, floor coverings, mats etc.
- *Conveyor Belt Presses* which are equipped with clamping and stretching devices in order to permit the vulcanization of the pre-fabricated, still unvulcanized conveyer belt under tension. Vulcanization takes place in sections. An over-curing at the overlaps is avoided by cooling zones at the end of the press.
- *V-Belt Presses* with a design similar to that of the conveyer presses.
- *Shoe and Boot Presses* in which rubber shoes or boots which were confectioned on a form are molded and vulcanized.
- *Shoe Sole Presses* on which soles are molded onto leather, felt or plastic parts and then vulcanized.
- *Steam Vessel Presses* where the presses are enclosed in a steam vessel for the production of large parts like, for example, spring elements, for which the radiation of heat would be too high in a conventional press.
- *Autoclave Presses* which work according to the same principle as the steam vessel presses. However, they consist of a steam autoclave which is vertically sunk into

the ground and a press cylinder which presses in axial direction against the secured lid and then against several molds which are arranged above each other.

The presses for vulcanization are mostly heated with steam which is forced through channels of the press platens. Since saturated steam is used, the vulcanization temperatures are often given by the steam pressure. The statement, vulcanization at 4 bar pressure above normal means, for example, that the compound is heated to 150 °C. Electric presses are used when higher temperatures are required. Articles with a large area like conveyor belts, mats, gasket plates etc. are vulcanized directly in between the press platens or are sandwiched between polished steel sheets with steel bars added as spacers.

5.4.3 Transfer Molding and the Flashless Method

5.4.3.1 Transfer Molding

[5.203 a-5.205]

Transfer molding is a refinement of the compression molding method and is related to injection molding.

Principle. In its simplest form, transfer molding consists of the use of a mold having three parts, the upper and lower part of which are attached to the platens of a hydraulic press while the middle part is removable and can be pulled out on a pair of tracks for larger presses. The upper part of the mold is often designed as a piston in more modern machines, the middle part contains the cylinder which receives the compound to be molded (fill cavity) as well as the injection nozzle, while the lower part is the mold itself. These are specialty presses which are used for the mass production of one and the same article.

In the transfer molding process the rubber compound is "transferred" during the closure of the three part mold by the pressure created for this purpose from the "fill cavity" (upper mold part) through channels or flat grooves (in the mold's middle section) into the mold section proper (lower part). This is the reason for the name of this molding process.

The use of such molds consisting of three or more parts has, however, certain disadvantages, namely time losses when removing the mold and the fact that the often considerable pressures during the transfer phase can open the mold because the full closure pressure is not yet fully applied to the mold plates during this phase. This can give rise to stronger flash formation. Since the compound is brought into the cavity at somewhat lower temperatures than for the injection molding process, the heating times are, therefore, correspondingly longer.

Equipment. In order to overcome these problems, the *injection transfer molding process* was developed [5.203 a-5.205]. It has a tool in the piston cavity into which in the barely opened position a small volume of material is placed which is pressed into the mold cavity by the closure pressure exerted onto the mold (see Figs. 5.25 and 5.26). This method can be used today with any injection molding machine and requires only a corresponding tool design and machine operation. In addition, there is the so-called *injection forming method.* During this process the compound is placed in the place between the plates in a slightly opened position and the molded parts are formed by applying the closure pressure. This method can be applied with any injection molding machine.

Molds. The molds for this method are more expensive than two part vulcanization molds for normal hydraulic presses and vulcanized scrap is created only because a

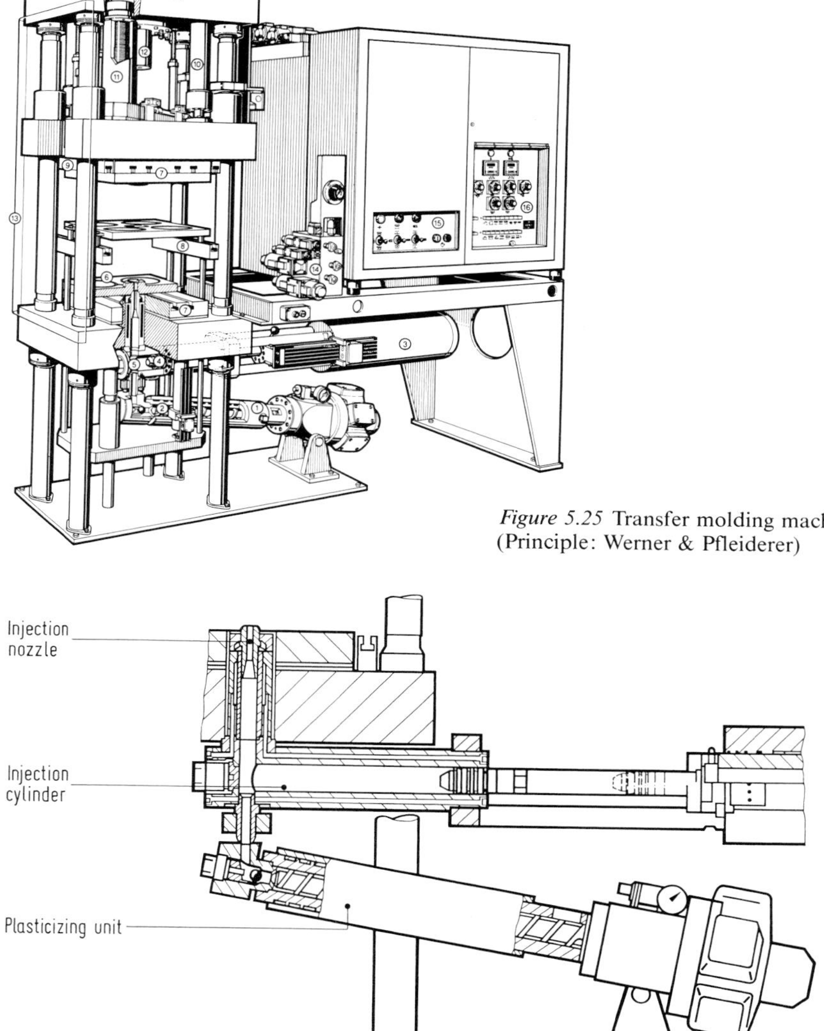

Figure 5.25 Transfer molding machine (Principle: Werner & Pfleiderer)

Figure 5.26 Transfer molding machine (Principle: Werner & Pfleiderer)

certain amount of compound in the fill chamber of the mold and the channels through which it has to run remains. In another modification of this method, the injection forming, the parts can be removed so that no secondary finishing operations are required.

The transfer molding method simplifies the loading of the molds which saves costs especially during the production of complicated parts. It also reduces the danger of air entrapment and permits the molding of dimensionally accurately and nearly flash free articles.

Problems. In contrast to the compression molding method, bumping is not required. The transfer molding method combines two problems: Flow forming and compression vulcanization. The problem of *flow forming* is very similar to that encountered during injection molding and will be discussed there (see page 414ff), while the *compression forming* will be discussed in Section 5.4.2 (see page 408ff). For this reason a discussion of specific vulcanization problems during transfer molding can be dispensed of here.

5.4.3.2 Flashless Method

[5.205]

Principle. The flashless method has been developed with the goal to manufacture articles without flash which do not need any secondary processes for deflashing and to manufacture parts which meet the ever increasing demands for close dimensional tolerances. It is a modified transfer molding process whereby in principle the flash is eliminated by the fact that all dimensions of the mold can adjust to the bending of the platens. The ring surfaces of the cavities are laid out so that a specific pressure of 200 bar is developed. In addition, the ring surfaces are manufactured with a certain surface roughness during grinding so that the air can escape through the ring surfaces but not the compound.

Equipment (see Figs. 5.27, 5.28). The *flashless method* can be used on presses which are also used for compression molding. The transfer cylinder is, in contrast to the transfer molding tool, integrated into the press mechanics. In an improved version of the flashless method, the *wasteless process*, the piston is cooled and the piston ring insulated from the hot mold. The *calibers* are inside the surface of the piston surface and are always hardened inserts. They are installed so, that they are movable within the plate, i.e. they are fastened with the help of retaining rings or Bellview washers. These inserts are protruding at least 0.05 mm from the plate surface. Thus, the cavities are only sealed via the ring surfaces and the mold plates do not experience any stress. The piston is manufactured from cast iron in order to avoid

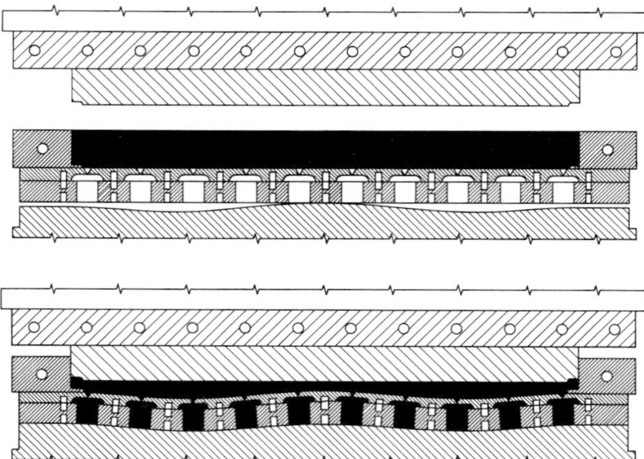

Figure 5.27 Flashless mold; above: before closure, below: after closure (from: *Juergeleit, M.F.:* Flashless Injection Molding. Rubber Age, 1962.)

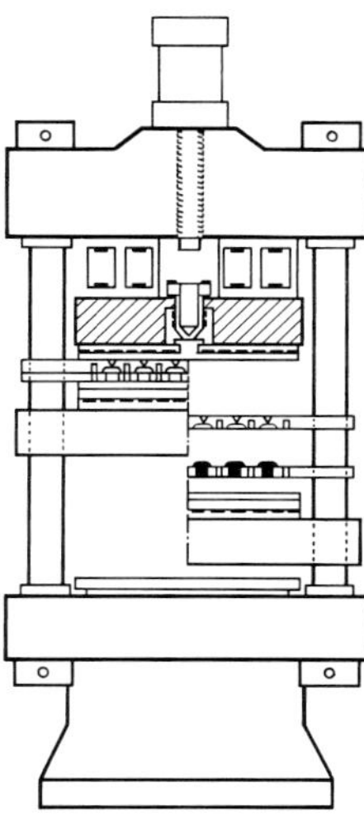

Figure 5.28 Vertical injection molding machine equipped with a central flashless injection mold
(from: *Juergeleit, M. F.:* Flashless Injection Molding. Rubber Age, 1962.)

binding with the piston ring. The play between piston and piston ring should be at least 0.1 mm per 100 mm diameter. The piston ring is supported by 4 telescoping brackets and is pulled from the piston in the opened position by four Belleview washers and is kept at a distance. During closure of the press the piston ring presses on the mold because of the Bellview washer packet and closes it during the closure mode. The piston ring has on the inside a steel lip which acts like a sealing gasket. Thus it is prevented that the compound escapes between the underside of the piston and the mold. During closure, the compound in between the piston and the mold acts like a hydraulic cushion. The *closure force* of the press and thus also the pressure on the mold as well as on the edge of the caliber can be determined by simple calculation. The single caliber which floats in the plate adjusts to the deflection of the platen (see Fig. 5.27). Because of the uniform closure pressure over the entire area of the platens and the absolute flatness of the edge surfaces of the calibers flashless parts are obtained.

Mode of Operation. Rubber compound sufficient for two or three cycles is placed into the piston ring. The layer should be at least 0.8 mm thick. During opening of the press the Belleview washers pull the piston ring away from the piston whereby the mold is laid open. The mold can now be pushed out. When the mold is flipped open, the part is separated from the runners. Now the mold itself is opened and the parts removed from the cavity or they are pushed out if a center plate is provided for. The mold parting lines are chosen so that the parts do not adhere randomly either to the upper or lower mold half. The parting line can be chosen arbitrarily since its effect on the part cannot be seen with the naked eye, as long as the mold is securely centered. After every part removal all mating surfaces of the mold as well as the calibers have to cleaned since rubber scrap, especially on the caliber surfaces can lead to tooling damage and flash. Special attention has to be paid to the mold filling phase. The air contained within the cavities has to be able to escape through the mating caliber surfaces. Injection times of 10 to 15 seconds are the rule. Closing the press too fast leads to air pockets. Because of the short flow paths and the generally small injection volumes the acceleration can be adjusted to be very fast so that relatively short cycle times may be obtained. The fashless method is especially suited to manufacture small parts.

The higher mold price is compensated for by the high production rate and especially by the the fact that no secondary processing steps are required. The tooling can also be expected to have a longer life time when it is treated with care.

5.4.4 Injection Molding

[5.206–5.253]

Aside from all other methods to manufacture molded articles, the injection molding method has proven itself especially and has become the mainstay of productive manufacturing.

Starting with presses which are suitable for compression molding, machines were first developed for the injection molding process to which a plasticating and injection unit is attached either above or below the mold carrier. Later, machines were added which injected in the parting line, so-called horizontal machines whose injection unit was equipped with a plastication device suitable for the processing of elastomers. Because of the required high pressures, the clamping plates and frames for the processing of rubbers are heavily built. Also the fact that the molds have to be heated up to 230 °C poses problems because of the coefficient of expansion and the insulation against the main body of the machine. Parts with little flash can be produced on injection molding machines [5.238].

Types of Machines [5.232–5.234, 5.235a–5.235e, 5.237–5.238, 5.240–5.242]. The following distinctions may be made between the machines used:

- The *toggle or hydraulic* mold closing mechanism
- The *plunger or screw* type plasticating and injection unit
- The *plunger screw* type plastication unit may be further subdivided into those with and those without reciprocating movement (see Fig. 5.29).

The fully *hydraulic mold* clamping mechanism is preferred today.

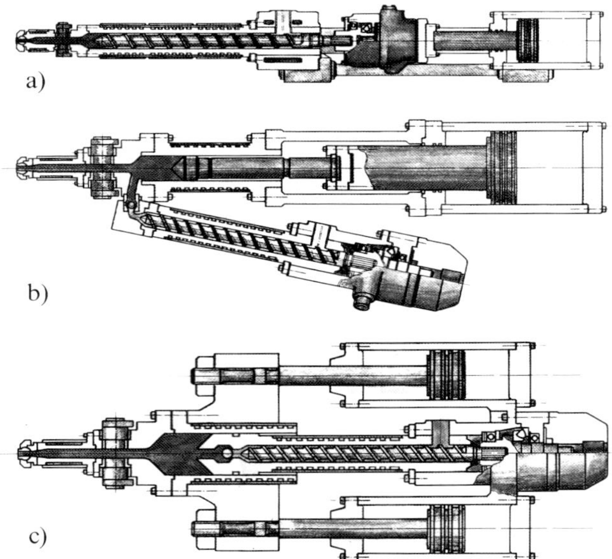

Figure 5.29 Different plastication units with different accuracy of the temperature control (Principle: Desma)
a screw type,
b plunger type,
c screw plunger type.

The *screw plasticating unit* pre-heats that amount of rubber compound which is necessary to fill the mold and brings it into a space in front of it while the screw is retracting (see Fig. 5.29, upper part). Thus, during plastication the length of the screw changes, i.e. the thermal history of the compound is not the same in all locations. Once a sufficient amount of material has been plasticated, a hydraulically operated plunger moves the screw forward and injects the plasticated material into the hot, closed mold.

In the *plunger screw injection* molding machine the rubber compound is first plasticated in a pre-chamber. The plasticated rubber compound is then injected into the mold by an axial movement of the plunger (see Fig. 5.29, middle part). As far as temperature variations are concerned, this principle is superior to screw injection molding and is better suited for the molding of larger volumes.

In the screw plunger unit the plunger is placed in front of the screw. The plasticated material reaches the plunger chamber by axial movement of the plunger and the screw. The plunger chamber is closed with respect to the screw area by a ball check valve once the material is injected into the mold. The uniformity of the plasticated and injected material is the best of the different principles (see Fig. 5.29, lower part).

In order to make the injection molding process even more economical, some further developments have taken place:

- The *slide table machine,* in which the lower part of the mold is moved sideways in order to achieve faster part removal. The double slide table machine is equipped with two lower parts of the mold which can be unloaded on the sides of the machine while the other lower mold half mates with the upper mold half and is filled.
- The *shuttle machine* contains two complete molds which are attached to a forming station. The molded parts are heated up to about two thirds in the central unit, by one third under reduced pressure in the final forming station. Molds of different geometries can also be used on these machines. Only the compound type has to be the same for both molds.
- A machine to *process two types of compounds* which are partially vulcanized either one after the other or in separate molds and are then bonded together in the parting plane with the help of a turning mechanism.
- Aside from these machines the *rotary mold table* has remained important for mass production. On a rotary table several different molds may be injected, however only with the same kind of compound. The individual degree of cross-linking has to be adjusted for each mold by the vulcanization temperature since the vulcanization time is, of course, the same for all the molds. The rotary table is very useful when a high degree of automation is desired.
- A special form of the injection molding machine is used to mold and *vulcanize corners* onto mitered profiles (see Fig. 5.30, next page).

Operation Principles. Three types of injection molding operations can be distinguished according to the differences in tool design and process sequence:

- the conventional,
- the injection forming and
- the transfer-injection molding methods.

These different systems are described in Fig. 5.31. The latter have already been dealt with in Chapter 5.4.3 (see page 410).

Figure 5.30 Injection molding machine for the molding of corner of window frames
(Landshuter Werkzeugbau, photo: Dr. Küttner)

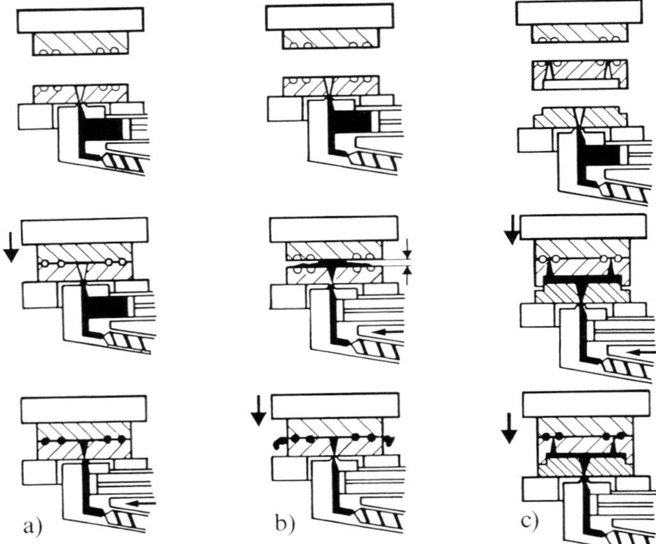

Figure 5.31 Rubber injection molding method with screw plastication
(Drawing: Werner & Pfleiderer)
a) conventional method, b) injection molding method, c) transfer molding method

Methods of Operation [5.229a–5.243]. While the efficiency of the compression molding method is limited because the rubber compound has to be heated in the closed mold before it is vulcanized which, of course, excludes a rapid molding cycle, the economic advantage of the injection molding method is based on the fact that the rubber compound is brought up to a temperature which is only a few degrees below that of the mold by the processes occurring in the plastication unit.

This would be the ideal situation to inject the rubber compound at such a temperature that the further heating which occurs as a result of vulcanization reactions will just raise near to that of the mold cavities. Under these conditions there is virtually no further heat flow, and the uniformity of the degree of crosslinking over the cross-section of the vulcanizate is optimized.

For example, it has been found in practice that if a rubber mix is to be vulcanized at 140 °C, injecting it at 130 °C gives uniform vulcanization throughout [5.213a] as the mix can be heated by the vulcanization reaction to 140 °C. If instead the mix is preheated to the cavity temperature, the heat produced during vulcanization causes over-cure in the interior of the molded article. In order to achieve correct dissipation (i.e. conversion of mechanical energy to heat), it was necessary to design systems with transfer channels such that no scorch occurs within them, but which nevertheless give sufficient heating-up. The ideal situation is that friction during transfer should raise the mix near to the cavity temperature. The adiabatic temperature rise $\Delta\vartheta_{\text{diss}}$ can be calculated from the dissipation equation [5.213d]:

$$\Delta\vartheta_{\text{diss}} = \frac{\Delta p}{\rho \cdot c_{\text{p}}},$$

(Δp: pressure difference in injection channel, ρ: density, c_{p}: specific heat). Inserting units into this equation gives:

$$\frac{\Delta p/\text{bar}}{\dfrac{\rho}{\text{g}\cdot\text{cm}^{-3}} \cdot \dfrac{c_{\text{p}}}{\text{J}\cdot\text{g}^{-1}\cdot\text{K}^{-1}}} \cdot 10^{-1}.$$

The actual temperature rise is less than this owing to heat conduction in the reverse direction during the residence time in the distribution system. This effect can be quite considerable.

Injection Systems [5.213b]. With regard to distribution channels generally, there is a distinction between *hot channel* and *cold channel designs*. Whereas in thermoplastics processing the desire to minimize losses at the inlets leads to hot channel systems, in the case of rubber compounds the cold channel technique gives the least inlet losses. For this reason developments in rubber injection molding are moving increasingly towards the cold channel technology, which will be dealth with in a separate section in view of its importance (see page 418). While it is claimed that in the USA about 90% of rubber processing installations are fitted with cold channel equipment, the proportion in the Federal Republic of Germany is estimated as at most 10%, although there is an increasing trend in this direction. Moreover, an increase in the use of cold channel technology is based on the grounds that rubber injection molding by this technique can be successfully automated.

In rubber injection molding, *hot channel systems* are those in which the flowing mix is not cooled, and can therefore heat up as a result of friction. Care is needed in designing their shapes and dimensions so that the temperature reached by frictional heating is below that which would cause scorch. The final part of the channel system consists of the distribution manifold and the channel terminations.

Filling of the cavities is fastest if channels with a relatively large diameter are used, whereas in order to obtain sufficient heating of the mix by dissipation one needs small diameters and a large pressure drop. However, the choice of diameter has a direct effect on the amount of wastage.

That part of the rubber mix which is left behind in the distribution manifold and channel terminations in a hot channel system becomes vulcanized along with the molding. This fraction increases with the number of cavities, and the material must usually be discarded when the mold is opened. In hot channel technology one is therefore concerned to minimize the volume of the manifold, often at the expense of increasing the injection time [5.213 b]. It is quite possible for the manifold volume to be greater than that of the molding. This seriously affects the economics of the process, particularly for expensive mixes. The manifold volume in hot channel rubber processing is about the same as in injection molding of thermoplastics (without hot channels). However, thermoplastics sprue wastes do not have to be discarded, but can be re-granulated and fed back into the process.

In contrast to hot channels, the channels in *cold channel blocks* [5.213 b] are maintained at such a temperature that there is no possibility of vulcanization being initiated. The material in the distribution channels remains plastic and can be injected into the cavities in the next cycle. The volume in the cold channel should be the same as in the mold; so that in theory the channel contents are replaced once in each cycle. In extreme cases the transit time of the rubber mix from its entry into the plasticizing barrel to its emergence from the nozzles can be as long as 20 to 40 cycles [5.213 c]. In this respect, cold channel systems in rubber injection molding are analogous to hot channel systems in the case of thermoplastics. The heating through dissipation is not allowed to take effect until the mix reaches the injection nozzles, with the result that the mix is injected into the cavities at a higher temperature than when it is in the distribution channels. After injection, scorch can only take place in the outermost part of the injection nozzle, as a result of heat conduction from the body of the tool to the nozzle. The extent of vulcanization depends on the contact time between nozzles and heat cavities. During demolding this precured compound is drawn out of the nozzle with the waste from the inlet port, and is discarded with it. However, the total wastage in the cold channel method is considerably less than with a hot channel system.

The design of a cold channel tool calls for a considerable amount of experience, and high precision is needed in its construction for reliable and economical operation. Although cold channel technology has been known for several years, design is still at an early stage of development. Novel principles had to be worked out so that a knowledge of rheological and thermodynamic processes, sometimes requiring further research for their clarification, could be incorporated into the design of tools. Various experimental prototypes and precisely controlled machines needed to be developed for this work [5.229 a, 5.229 b, 5.238, 5.240 a].

A cold channel tool consists basically of the cold channel block, with sometimes also a sub-channel block, nozzle holders, and the injection nozzles themselves. So far as is practicable, a cold channel block is constructed so that it can be used in different molding tools to produce a variety of moldings, or with automatic mold-changing systems to reduce the time for setting up and changing molds. This makes it possible to quickly recoup the cost of these tools, which are often very expensive.

Advantages and Disadvantages of Various Machine Types. The machine types discussed above have the following advantages and disadvantages regarding their mode of operation:

Because of short plastication times, extremely high nozzle exit temperatures can be achieved with *screw injection molding machines*. The temperature before the screw can, for example, be 120 °C, the nozzle exit temperature even above 140 °C. During the flow through the runners another increase in temperature of 30–40 °C can be expected, depending on the injection speed [5.209]. However, an increasing variation of the temperature along the injected part has to be expected. The essential advantage of screw plasticating machines is the very short cycle time which is very important for high volume production, for example using rotary table mold arrangements. These high production rates cannot be obtained with the other designs. A fast responding temperature regulation in the cylinder is required.

In both variations of the *plunger injection molding machine* the material can be heated up to approximately 90 degrees because of the longer residence time. The further temperature increase during the injection cycle is limited by the injection speed so that the achievable temperatures are normally lower than in the screw type machine. The plunger injection molding machines have the advantage, however, of higher temperature homogeneity, where the injection machine type without flow reversal has proven to be the one with the highest temperature homogeneity. In so-called one station machines, where maximal plastication speed is not important because of the expected vulcanization times, this type of machine has proven itself. In order to control this process satisfactorily, corresponding features of the control system are required:

- constant screw speed,
- control of the nozzle pressure and
- control of the injection speed.

With the help of *proportional hydraulic systems* very constant conditions can be achieved and the start-up period kept to a minimum. The shortest molding cycles can only be achieved when all parameters are kept constant. This is only possible when *process controls* are used [5.244–5.251] (see Fig. 5.32). A thorough understanding of all process dependent parameters is necessary for this. Recently, micro-pro-

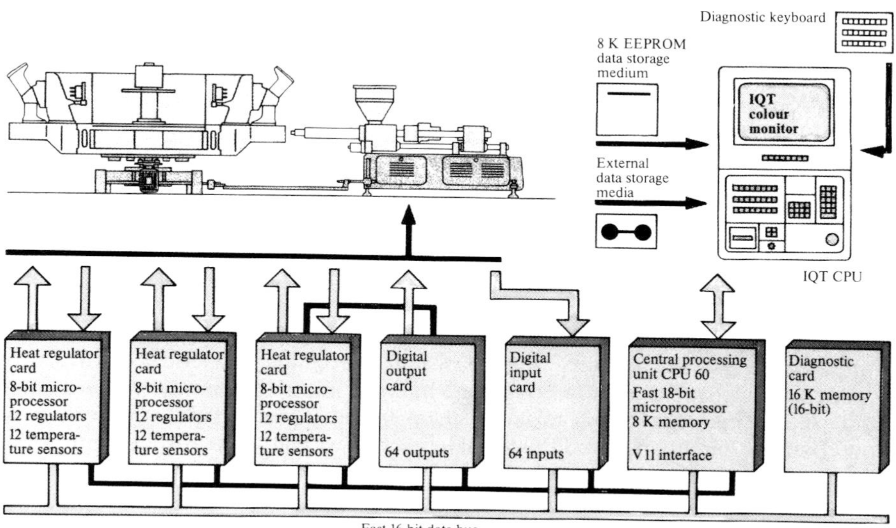

Figure 5.32 Fully process controlled injection molding unit (Principle: Desma)

cessors (e.g. a software called CADGUM [5.249a–5.251] as basis for the production of precision articles) have been developed which shall help the process optimization of elastomer injection molding tooling. It is thus, for example, possible to obtain the mold filling history, the pressure, velocity and shear rate distributions and also crosslink density gradients. It is also possible to obtain a simulation of the heating and the operating behavior of the mold and to analyze experimental data for material property predictions [5.251]. Regarding process control, e.g. the heating time regulation (see Section 5.4.5, page 424).

For the manufacture of nearly flash free articles by the injection molding method the post-injection or post-pressure phase has to be controlled carefully based upon an evaluation of the energy expended during injection [5.238]. In order to obtain this control, the process has to be operated with a so-called velocity profile.

Trend in Injection Molding Tools [5.213b]. Main objective of developments of rubber injection molding processes is the *automation of production processes*. Automation of the rubber injection molding process means providing technology which avoids the need for control by an operator, either constantly or at regular intervals in the cycle. There are certain essential conditions for achieving this. These concern the preparation of the rubber mix and of the blanks, and most importantly the mechanical operations and control devices.

The rubber processing industry is, for the most part, still a long way from this ideal vision of an automated future, although some installations with a high degree of automation already exist. A first requirement for automated processing is very precise control of all the process variables which affect the quality of the injected compound. Intensive work in this direction is being undertaken by the machinery and rubber processing industries.

One of the requirements for automated rubber injection molding is to produce a rubber compound of the highest possible *uniformity*, which is still quite a difficult task even today. In older type internal mixers which are still used in many installations, inhomogeneities in the chemical and rheological condition of the compound cannot be avoided, which presents a serious obstacle to completely automating the injection molding process [5.10a]. It is evident that for computer controlled flow and vulcanization processes one also needs automatically controlled methods for preparing the compound. In order to meet the requirements for *statistical process control* (SPC) [5.244–5.251], the process machinery industry has in recent years developed computer controlled internal mixers which make it possible to control energy input, material turnover etc. during the mixing process, thereby giving a highly uniform compound. The temperature of the compound is controlled by the rotor speed, the viscosity is measured by the rotor torque, and the vulcanization properties are partly controlled by the mixing time [5.246a–5.249d].

Other important factors depending on controlled mixing are those of optimum and uniform *rheological properties*, the *dispersion* of the components in the compounds, and *computer-aided* optimization of the *vulcanization process*, especially for thick-walled moldings. Compounds with consistent processing properties can now be produced using automatically controlled internal mixers. Whereas it was at one time usual to incorporate vulcanization chemicals by a second mixing process, it is now common to prepare two different batches, one with and one without vulcanization chemicals, and to blend these just before processing; this ensures the production of a series of compounds which are completely identical.

Another problem is that of automatically *feeding* the raw material into the injection molding machine. The strip feeding method in its original form is too imprecise and needs constant manual attention. Recently, however, the strip feeding method has been refined by using strips of precise dimensions wound on drums, allowing limited periods of automated production. Another method used is to feed the compound into the injection molding machine in granulated form. This may prove to be the best method for fully automated production in the future. There are difficulties, however, as the soft, gummy granules tend to stick together. This must be counteracted in the compounding process, and by using vibrating containers or channels.

A further requirement is that the operation of the injection molding machine must be precisely controllable, by signals derived from *sensors* based on the rheological and chemical processes occurring [5.249 a]. For example, when the measured viscosities, temperatures or flow rates deviate from the specified values, online control signals are immediately generated to alter the pressures, temperatures or displacements in such a way as to correct the deviations. Intensive work on these computer-aided control systems is being undertaken by the rubber industry, research institutes concerned with process engineering, firms of engineering consultants, and the processing machinery industry [5.213 b]. Whereas computer-aided investigations of the vulcanization process were formerly of doubtful value, the personal computers now available permit on-line monitoring of mold heating, even for thick-walled moldings, by matching the measurements to boundary conditions, and even heat flows which change with time can be handled [5.213 a]. These techniques, in conjunction with further studies, now allow long production runs using computer-aided vulcanization control.

Finally, steps must be taken to prevent foreseeable production faults, e.g. by minimizing or elimination flash and preventing tool fouling as far as possible. One must aim at stable operating conditions, so that the machine remains in a constant state. Flash can be greatly reduced or prevented by suitable process design measures [5.239] and by correct tool design. On the other hand, deposition of material in the molding tool is affected by details of the flow front pattern, filling of the cavities, venting, and the shape of the cavities [5.239], and also to a large extent by the preparation of the compound. Since it is not possible to completely eliminate fouling, work is in hand [5.239] to develop tools for different classes of moldings which can be dismantled for rapid automatic replacement of the affected parts. The downtime involved in changing tools can thus be kept to a minimum.

Injection molding of rubber is at a stage of rapid development, and is emerging from being a crude mechanical process to being a precise and refined production technique. Although rubber technology is much older than thermoplastics processing, injection molding in the latter case is considerably more advanced owing to the fact that the process is simpler. In the case of rubber, not only is the flow behaviour complicated by its being considerably dependent on compounding, but also the vulcanization process has a complex dependence on the temperature and other variables. To fully develop injection molding for rubber therefore demands a considerably greater understanding of the rheology, thermodynamics and vulcanization chemistry involved, and correspondingly sophisticated designs, e.g. with regard to precisely controlling the temperatures within the molding tool.

Considerable advances in the automation of rubber injection molding have been made in the last few years. In a number of areas of manufacturing, this has already led to automated plants being installed, e.g. in the production of shaft seal rings. Within a few years one can expect to see fully automated injection molding processes becoming a reality for rubber processing, allowing production of components with a high level of quality assurance. The remaining problems call for close collaboration between research groups, machine manufacturers, compounding suppliers and rubber manufacturers.

Compound Design. No systematic, only sporadic papers deal with the design of compounds which are used for injection molding [5.206, 5.208, 5.211, 5.212a, 5.213b, 5.213c, 5.218, 5.220]. In the past, a general survey was made difficult because the specific design, especially that of the screw, had to be taken into account for the compound design. This is especially true for the compound viscosity and the adjustment of the acceleration. For formerly used screws with compression and special shear elements, the temperature stability, especially for highly viscous compounds, could not be maintained. When such machines are still used today, one has to take care to obtain a low viscosity compound and a delayed onset of vulcanization. Often, degraded qualities of the compounds are unavoidable, especially the compression set and the stress relaxation behavior of the compounds.

For the modern plastication systems the adjustment of the viscosity is not of the same importance as it used to be. It affects essentially the optimization of the process cycle whereby the choice of the acceleration system is of great importance. After adjusting the acceleration system only slightly, today most of those rubber compounds which were developed for the compression molding process can also be processed on injection molding machines without problems.

Therefore it is often necessary to optimize the property profile of the compound for working with a particular tool. The tool design and the compounding thus need to be matched to each other. As there are no general rules for preparing the compound, this will be discussed by the author's experience.

The problem of air entrainment cannot be solved by tool design alone, but must also involve the preparation of the compound. For example, a familiar problem is that air bubbles often form when rings are made by injecting at one side, or when the tool has undercuts; these interfere with the flow of the compound and cause rejects. It is sometimes necessary, for example, to work with higher viscosities or with *pre-crosslinked types of rubber,* although less viscous types are preferred because of their better flow properties.

It is less commonly known, although reliably reported by the rubber industry, that mixes containing 5 to 10 phr of *factice* are already sufficiently vented during their preparation that practically no flow failures occur (see page 306). These small quantities of factice usually have no detrimental effect on the property profile of the vulcanizate, and usually even improve the surface appearance of the molding. The addition of a small quantity of factice can therefore often avoid the need to alter the tool. Adding small amounts of factice also improves the preparation process and the storage life of the compound, and reduces the stickiness of rubber compounds in granule form, making them easier to use in injection molding processes, according to reports from the rubber industry.

One can, of course, alter the injection and flow properties of rubber compounds by adjusting the viscosity, starting with the *viscosity of the raw rubber* used as the basis, continuing with the choice of *fillers and plasticizers,* and lastly by using *process aids.* The addition of stereo specific BR or IR also sometimes gives an increased flow behavior. It is not widely known, however, that small quantities (2 to 5 phr) of process aids affect the flow properties to a much greater extent than is possible by using a rubber basis of lower viscosity or by using similar quantities of plasticizers. Small amounts of suitably chosen process aids make it possible to considerably increase the quantities of compound injected, without altering the molding tool [3.267, 3.268]. A decrease of the mixing temperature reduces the danger of scorch. If the type of process aid and the dosage are correctly chosen, no confluence faults arise.

It is desirable that rubber compounds, for injection molding should remain mobile for as long a time as possible during the initial phase of vulcanization, so as to avoid scorch problems, even with hot channels. Subsequently they should undergo full vulcanization as rapidly as possible, with little or no tendency towards reversal, e.g. sulfenamide accelerators perhaps in combination with thiurames, dithiocarbamates or guanidines have sometimes been proven successfully in the presence of retarders. Many different *vulcanization systems* are available for this purpose, including accelerator mixtures specially developed for the vulcanization of EPDM, and optimized for the type of molding [3.400] (see page 97).

The time for the onset of vulcanization should be such that the compound remains freely flowing long enough to prevent scorch in the plasticizing barrel or in the injection channels. Even a small amount of prevulcanization will cause the compound to flow less freely. In determining the incubation time of the compound from data measured in the laboratory, the values must be corrected to take into account the thermal conductivity of the mix and the time for heating up in the rheometer.

To reduce mold fouling [5.239, 5.252a–5.253] one can, in addition to avoiding air bubbles, add a small quantity of *factice* so as to reduce those components of the compound which are volatile at the tool temperature. Such volatile components include particularly mineral oils and plasticizers, and the amounts of these should be limited to what is absolutely necessary. If this reduces the flowability, it is preferable to improve it by adding small quantities of process aids with low volatility, rather than large quantities of volatile plasticizers.

Chlorine-containing rubbers often cause mold corrosion, especially when ETU and lead oxide are used for crosslinking. In such cases one should definitely consider incorporating neutralizing agents into the compound. For example, a carefully measured amount of well distributed *MgO* can often greatly reduce tool contamination. Tool contamination cannot be entirely eliminated, but the measures described here can reduce it to a tolerable level.

Advantages and Disadvantages of Injection Molding Process. A comparison of injection molding with compression molding shows the following advantages and disadvantages:

Advantages are for example:

- *Reduced preparatory labor.* Generally, preforms have to be cut from milled slabs or extruded strips for the compression molding operation. The preparation of preforms is not necessary for injection molding. Continuous feeding with granules or, like for the extrusion process, with continuous strips is possible.
- *Transport and storing* of the cut preforms is eliminated. Scrap remains after cutting the preforms which is not discarded but put back onto the mill and then extruded again.
- *The placement of the preforms* into the opened mold is eliminated. It has to be considered in this connection that the incorrect placement of preforms increases the scrap rate.
- *Considerably shorter vulcanization times* because of homogeneously preheated material. Depending on the part geometry, the compound and the machine, the vulcanization times can be reduced by 70–90%.
- *More homogeneous degree of vulcanization* of the finished part at high injection

temperatures, therefore an isotropic property spectrum over the whole cross section for thick walled articles [5.213 a].

- *No bumping.* During compression molding the mold has to be opened (bumped) once or several times.
- *Faster removal* of the finished parts which can be accomplished without heavy manual labor which is required for the compression molding process, especially for flat molds in multi-layered presses. Difficulties with the ejection of vulcanized parts inhibits, however, the full automation of the full production process during injection molding of elastomeric parts compared to the production of plastic parts.
- *Elimination or simplification of de-flashing* of the finished parts. In contrast to compression molding, the use of mold release agents is often not required for injection molding.
- *Lowering of the waste and scrap rate* which is, on average, between 10 and 40% for compression molding to 5–10% for injection molding. In special cases these differences are even higher. The part to waste ratio can be effectively influenced by the use of cold runner systems. Currently, the use of cold runners with flashless parts is being discussed.
- *Possibility of automation.*

These advantages are contrasted by the *disadvantages* of *considerably higher investment costs* for molds and machinery. The injection molding process can also not handle as high volumes as compression molding.

Aside from the manufacture of molded articles of all kinds the injection molding method has also been introduced to manufacture shoe soles and heels as well as to the molding of soles onto shoe uppers.

While the shoe sole or the heel can be produced like a normal technical formed part separate from the shoe, special molds and mold carriers are necessary for the molding of soles and heels onto shoe uppers. Since the molding of soles onto shoe uppers is normally done on fast running rotary machines, screw injection systems are normally used because of the high production rates.

Economic Efficiency [5.230]. Which method is most economical for the production of molded parts, compression molding with low investment costs but with higher production costs or injection molding with higher investment costs, depends to a high degree on the number of parts to be produced. Below certain limits it is advantageous to work with low investments but higher wage costs while for high volume production runs higher investments may be preferable.

5.4.5 Vulcanization in Vulcanization Presses, Machine Control

[5.253 a–5.257]

The processing tolerances of rubber compounds in the plastication units and during injection have also an effect on the vulcanizate properties. The determining parameters are here primarily the temperature distribution in the unvulcanized compound as well as the temperature distribution over the mold surface. Variations of these parameters have a negative effect on the part quality due the formation of anisotropies in the cross link density. In order to avoid this, optimal control systems are utilized. Another possibility is based on using, for example, a proportional hydraulic system with an exact machine control system as well as a process sequence which fulfills these prerequisites. This is required in order to optimize the process cycle. The process control determines on the one hand the after- pressure by measuring the injection energy and the heating time by monitoring the heat energy which flowed into the part as well as the boundary conditions, as the heat loss by the temperature increase of the plastication unit, that of the part and heat radiation and the heat conductivity of the rubber compound etc.

Because of the great variety of manufacturing methods of molded parts, not all vulcanization and machine control problems can be discussed.

5.4.5.1 Influence of Temperature on the Rate of Vulcanization [5.253 a–5.257]

Rate of reaction. As all chemical reactions, the rate of vulcanization increases, of course, with increasing temperature. The *van't Hoff* rule can be applied approximately according to which the rate of reaction doubles roughly for each increase in temperature by 8–10 °C (corresponding to approximately 1 bar steam pressure increase). This rule can be expressed by the following equation:

$$RG_2 = RG_1 \cdot 2^{0,1 \Delta T}$$

where RG is the rate of reaction and T the temperature. According to this equation the rate of reaction would quadruple for a temperature increase of 20 °C and increase eightfold after an increase of 30 °C. The rate increase with temperature expected according to this rule (temperature coefficient = 2, i. e. doubling of the rate for every temperature increase of 10 °C) is, however, only partially true for the vulcanization of rubber since it depends strongly upon the composition of the rubber compound.

Some already commercially available control systems are based on the mathematical formulation of *van't Hoffs* principle.

Activation Energy. In reality, however, practically every compound has a slightly different temperature coefficient which ranges, for example, from 1.86 (for NR compounds with chalk as a filler) up to 2.50 (for example for hard rubber compounds). In addition, the temperature coefficient of the rate of vulcanization is itself dependent on the reaction temperature. Because of these reasons one has gone over to the use of the *Arrhenius* equation during the last years more and more for the determination of the temperature dependence of the vulcanization rate. In this equation the activation energy A has been introduced as a coefficient:

$$RG_2 = RG_1 \cdot e^{-\frac{A}{R}} \cdot \left(\frac{1}{T_2} - \frac{1}{T_1}\right)$$

or

$$\ln RG_2 = RG_1 - \frac{A}{R} \cdot \left(\frac{1}{T_2} - \frac{1}{T_1}\right)$$

In this equation R is the gas constant and T the absolute temperature. This equation is not more complicated than the *van't Hoff* equation. It is even simpler to evaluate it graphically.

The activation energy according to *Arrhenius* is an exact measure for the temperature dependence of the vulcanization rate which can also be used in every day life. One obtains it directly from the isochronal values (i. e. the times which lead to equal degrees of vulcanization) by plotting their values against the reciprocal absolute temperatures and a subsequent determination of the slope.

The equation for the activation energy has also limits since it is only valid, strictly speaking, for low and medium conversions. At high conversions the results become inaccurate; here corresponding corrections have to be applied. In spite of this, the application of the *Arrhenius* equation leads to more accurate control devices than the *van't Hoff* equation. Therefore, some commercial control systems are based on the *Arrhenius* equation.

Limits of the Vulcanization Temperature. Because of reasons of efficiency one is of course interested in choosing vulcanization temperatures as high as possible in order to achieve correspondingly short vulcanization times. However, these attempts are limited, especially for thick walled articles. An arbitrary shortening of the vulcanization time by raising the temperature is already not possible because the compounds have to flow in the molds until all cavities of the mold are filled before vulcanization is allowed to commence.

A limit to the vulcanization temperature which is equally important for all vulcanization systems is, for example, given by the *width of the plateau* of the vulcanizate (see Chapter 4.2.1.1, page 221 ff). While compounds with a wide plateau permit high vulcanization temperatures, those with only a narrow one can only be heated up to relative low temperatures.

Differences in the highest permissible temperatures also exist between NR and SR. The more *heat resistant* the SR is, the higher may be the temperature which can be applied during vulcanization without causing damage to the vulcanizates due to heat. NR must be heatet at relative low temperature, due to its reversion behaviour.

Since the *mechanical properties of the vulcanizate* depend on the vulcanization temperature [5.253 a, 5.256 a], it is possible in special cases, for example when certain specifications have to be met, that a temperature limit exists which must not be exceeded.

Another temperature limit results, for example, from the wall thickness of the articles (see page 427).

During vulcanization in the press when certain dimensions have to be maintained for a given mold, the vulcanization temperature may not be changed because one obtains only at one temperature accurate dimensions because of the thermal shrinkage.

During vulcanization in transfer and injection molding machines the temperature is limited as a rule by the resulting temperature in the plastication unit which must not be chosen too high because of the danger of scorching or degradation of accelerators or other rubber chemicals. The mold temperature must not be too high either, because an excessively high gradient of the crosslink density between outside and inside may result from short vulcanization times with negative effects on the compression set.

For molded parts with undercuts it has to be, furthermore, considered that the tear resistance decreases with increasing vulcanization temperature so that the removal of such parts which were vulcanized at too high temperatures can present problems. Such problems can also be solved by the choice of heat resistant elastomers or suitable vulcanization systems or low undercure.

The choice of fillers and process aids has to be also adjusted to the vulcanization temperature used, especially for the process aids, plasticizers and the water content of fillers etc. If the boiling point is exceeded, any vaporization would lead to porosity.

Thus the vulcanization temperature can be raised only in some cases to such a degree that one obtains very short vulcanization times. This is accomplished during continuous vulcanization (see page 394 ff).

5.4.5.2 Flow Period

Considering the high investment costs for a press department, the high mold costs and the relatively high steam and electricity consumption, a far reaching streamlining of the molding operation is necessary for the economical production of molded articles. An essential prerequisite for this is that the vulcanization times are kept as short as possible.

In order to obtain possibly short vulcanization times the flow period should be as short as possible. It depends mainly on the flow paths which have to be transversed, the compound viscosity, the number of required bumpings etc. The vulcanization following this period should be as short as possible as well. The ideal of the vulcanization curve for heating in the mold is therefore the one shown in Fig. 5.33. It is obvious that such a vulcanization curve can be only obtained as an approximation. One gets closest to this curve by using benzothiazole or dithiocarbamylsulfenamide accelerators (see page 239, 248). Depending on the choice of sulfenamide accelerator types and other compound ingredients one has the possibility to vary the flow period over a wide range and thus to accommodate the process requirements. After the onset of vulcanization the vulcanization rate is very fast for sulfenamide accelerators. They also yield a very broad vulcanization plateau.

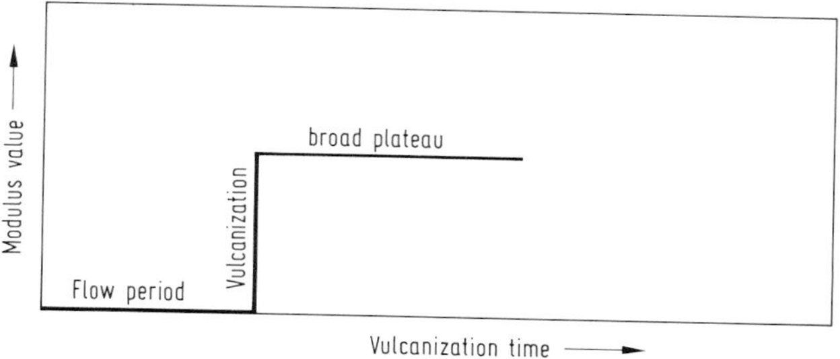

Figure 5.33 Idealized vulcanization curve for the manufacture of molded articles

The flow period can also be modified be the choice of the vulcanization temperature. At very high temperatures one can only obtain a sufficiently long flow period with the most scorch resistant accelerators or by the use of retarders without an increase in the scrap rate due to incorrect mold flow.

5.4.5.3 Vulcanization of Thick Walled Articles

[5.213 a, 5.256, 5.257]

Temperature. Most articles to be vulcanized are bigger than the samples used to determine the optimal vulcanization conditions in laboratories. One can, therefore, often only estimate the vulcanization times needed for production based upon the optimal vulcanization times determined on e.g. 6 mm thick laboratory samples. The determination of the correct vulcanization temperature and time is often difficult, especially for articles of large volume. Generally, the rule is applied that the optimal vulcanization time determined on, for example, 6 mm thick samples is increased for thick walled articles independent of the vulcanization temperature per millimeter of increased wall thickness by one minute in order to permit the vulcanization heat to penetrate to the inside of the part. The extension of the heating time which is called for due the low heat conductivity of the rubber causes, of course, a more or less severe over-vulcanization on the surface of the vulcanizate which is dependent on the level of the vulcanization temperature. In its center the part can still be under-vulcanized or may have reached just an optimal state of cure.

The conclusion may be derived from this, that a part has to be vulcanized at lower temperatures and longer times the larger it is, within certain limits. The determination of the optimal vulcanization temperature and time requires considerable experience and is only possible by trial and error and experimental vulcanization runs followed by testing of the part. The cure time for thick-walled rubber parts could only be calculated in recent times [5.213 a, 5.256, 5.257].

However, the vulcanization simulation with temperature programmed curometers which have been developed in the last years is gaining in importance. This method permits the determination of the optimal vulcanization conditions in the laboratory even for large articles (see page 473).

Importance of Uniform State of Cure. A part of large dimensions which is subject to dynamic loadings has to be vulcanized well and uniformly since an insufficient state of cure, even in a limited area in the inside, can lead during strong dynamic loading because of the internal friction to stronger heat generation and thus destruction from the inside [6.5 a]. For foundation shock mounts an undervulcanization can lead after longer times to a sagging of the supported part (building, machine, etc.). The state of cure can be tested by a determination of the spring rate without destroying the part [6.5 a].

Achieving a Uniform State of Cure. The size of the vulcanization mold, the type of press, the mold material etc. affect, of course, the state of cure considerably. For high vulcanization temperatures a more or less pronounced heat loss has an adverse effect on the vulcanization of the part. Very large parts are best vulcanized in an autoclave press where the directly heated press platens are surrounded by an autoclave in which additional heating by steam is provided for.

In order to obtain a uniform state of cure of thick walled articles, the following procedures may be applied:

- *Stepwise increase* of the press temperature
- Construction of the parts from *compounds with different degrees of acceleration*
- *Preheating* of the preform in an oven at intermediate temperatures
- *High frequency preheating*.

While the first three methods do not require any special technology, the high frequency pre-heating, which is very important today, is much more problematic. It will therefore be discussed later (see next page).

Vulcanization Accelerators. For the vulcanization of thick-walled articles, mostly slow accelerators or those with an especially wide plateau are employed because of the long heating times.

Back Rinding [1.12, 1.13]. During the vulcanization of articles of large volume, crevices and blemishes can easily form next to the parting lines of the molds which can reduce the commercial value of these parts. These phenomena, called back rinding, have not been explained yet fully. The most plausible explanation reduces this effect to an inner heating of the compound caused by the exothermic vulcanization reaction above the vulcanization temperature through which an excessive pressure is created inside the mold. This pressure can open the mold during vulcanization at the parting line and a certain amount of material, above the normally expected one, exits through the separated mold halves. Some of this material may have already been vulcanized. When the internal pressure subsides after a certain amount of time and the mold is closed again by the closing pressure, the exuded material is pinched in-between the mold halves and sticks to them. During fracture of this exuded material the above mentioned effects occur.

They can be eliminated by avoiding the inside pressure build-up, for example by reducing the volume of the preform and also especially by reducing the vulcanization temperature and perhaps an extension of the flow period or avoiding volatile compound ingredients. A partial success can also be achieved by a suitable pressure profile control.

Finally, a solution to back rinding can be achieved by a special design of the molds whereby the mold is not divided horizontally but parallel to the mold. In this way no material can enter the parting surfaces and thus the problem is avoided.

5.4.5.4 High Frequency Pre-heating

[1.12, 1.13]

Because of the long heating times caused by the low heat conductivity of the elastomer at relatively low vulcanization temperatures of large volume rubber articles, the capacity of presses and autoclaves is lowered to such a degree that the manufacture of rubber articles becomes relatively expensive.

By the application of dielectric preheating of the preforms before placing them into the molds, higher vulcanization temperatures can be chosen, the heating times can be reduced considerably and thus the manufacturing costs can be lowered.

Basic Principle. The basic principle of the preheating of molded articles corresponds to a large extent to the statements made on page 399 regarding the UHF-vulcanization.

The dielectric heating which permits the rapid and even heating of all electrically non-conducting substances is based upon the fact that a non-conducting substance placed in the electric high frequency field in between the two capacitor plates of a high frequency generator is subject to a polarization due to the influence of the high frequency alternating field whereby dielectric losses occur in these parts by which the material is heated.

Equipment. *High frequency generators* of different designs are commercially available. In order not to disturb normal broad-casting with such equipment certain frequency bands are assigned to the high frequency generators for heating.

Rubber preforms are preheated with a high frequency generator outside the mold without pressure up to approximately 110 to 115 °C and are then vulcanized inside a mold.

As *generator load* approximately 1 kg compound per delivered generator power has proven to be still permissible. For larger amounts of compound the pre-heating time gets to long.

In order to always obtain short pre-heating times it is furthermore required to adjust the *form of the electrodes* to the outer shape of the pre-forms (e.g. contact surfaces with parallel or oval cross section). Also, an efficient utilization of the contact area of the electrodes is important (e.g. several preforms simultaneously). Through these measures one does not only obtain a good utilization of the equipment but also minimizes field losses.

The *preforms* have to be cut smooth and have to be dry so that arcing due to tip discharges or scorching inside the unvulcanized articles is prevented.

Vulcanization by dielectric heating in the press is not carried out in the rubber industry; it is done inside vulcanization presses. This is because of several reasons such as:

- the question of the mold material which must not be conducting,
- insufficient heating rate of some compounds, especially those without carbon black,
- uneven heating due to increasing energy uptake because of dipole formation during binding of the sulfur,
- local overheating of the compound due to inhomogeneities.
- missing of pressure.

During pre-heating of rubber compounds most of the listed difficulties are not observed.

5.4.5.5 Shrinkage During the Manufacture of Molded Articles

[1.12, 1.13]

An article molded from a rubber compound always ends up smaller than the mold in which it was vulcanized. The difference between the dimensions of the finished article and the mold, measured at room temperature and expressed in per cent is called the degree of shrinkage. Since the dimensions of the molded articles have to be very often within very close tolerances, it is important to know the degree of shrinkage so that it can be considered during the design of the mold.

On the one hand the shrinkage of the rubber during cooling can be welcomed. Had the rubber and the mold the same coefficient of expansion, removal of the part from the mold would be problematic. On the other hand, this phenomenon complicates the manufacture of articles of correct dimensions.

Basic Principle. The degree of shrinkage is primarily determined by the difference of the coefficient of expansion between the vulcanizate and the mold material as well as by the vulcanization temperature. The rubber part which fills the mold completely during vulcanization contracts during cooling to room temperature more than the mold because it has a considerably higher coefficient of expansion than the latter. Thus the degree of shrinkage increases with increasing difference between the coefficients of expansion of the rubber compound and the mold material as well as between vulcanization and room temperature. The difficulties during the calculation of the degree of shrinkage arise mostly from the fact that the coefficient of expansion of vulcanizates changes with the compound composition. In spite of this, degrees of shrinkage can be determined approximately without preparing sample moldings.

Determination of the Degree of Shrinkage. Aside from the influence of the vulcanization temperature the *elastomer content* of the compound is of importance for the degree of expansion and thus the degree of shrinkage. The coefficient of expansion is the higher, the higher the elastomer content of the compound. The coefficients of expansion of the filler are of the same order of magnitude as those of the mold materials because of which essentially only the elastomer content of the compound is considered during the calculation.

Acetone soluble chemicals and additives like accelerators, antioxidants, plasticizers, resins, waxes and sulfur, also factices and reclaim have, in contrast, a coefficient of expansion similar to that of the rubber so that these substances are considered as part of the elastomer during the calculation of the degree of shrinkage.

The *vulcanization time* and the mastication time during the preparation of the compound are practically without influence on the degree of shrinkage.

Some of the *coefficients of expansion* important for the calculation of the degree of shrinkage are summarized in Table 5.2.

Table 5.2: Linear Expansion Coefficients of some Elastomer Types, Fillers and Mold Materials

	Linear Expansion Coefficient
Elastomers	
NR	216×10^{-6}
SBR	216×10^{-6}
NBR	196×10^{-6}
CR	200×10^{-6}
IIR	194×10^{-6}
Fillers	Order of Magnitude: 5 to 10×10^{-6}
Mold Materials	
Steel	11×10^{-6}
Light Metals	22×10^{-6}

The *calculation of the degree of shrinkage* from the composition of the compound can lead, according to the statements above, only to approximate values the accuracy of which is, however, sufficiently accurate. The following equation can be used for this purpose.

$$S \text{ (degree of shrinkage in \%)} = \Delta T \cdot \Delta A \cdot K \cdot \Delta F \cdot \Delta H$$

where

ΔT = the difference between vulcanization and room temperature

ΔA = the difference between the coefficient of expansion of the rubber and that of the mold material

K = vol.% rubber + acetone soluble chemicals

ΔF = difference between the coefficient of expansion of the fillers and that of the mold material

ΔH = difference between the coefficient of expansion of the acetone soluble chemicals and that of the elastomer.

If one neglects the insignificant factors F and H, one arrives at the following approximate equation:

$$S = \Delta T \cdot \Delta A \cdot K$$

This approximate equation is valid under the assumption that the filler has the same coefficient of expansion as the mold material and that the acetone soluble components have the same coefficient of expansion as the rubber and thus do not contribute to the degree of shrinkage.

It follows from this equation that the degree of shrinkage is the larger for a constant compound composition, the higher one chooses the vulcanization temperature. One also obtains another degree of shrinkage when one switches from one mold material to another one.

The temperature independent *contraction* which is a result of the chemical reaction between sulfur and the elastomer is of little consequence for the degree of shrinkage. Experiments have shown that it accounts for only 0.1% for sulfur dosages which are common for soft vulcanizates but is larger for hard rubber compounds. Since the contraction is further reduced in filled compounds because of the reduced elastomer content, it is within experimental error and can generally be neglected.

5.4.6 Some Special Molded Articles

5.4.6.1 Vehicle Tires

[5.258–5.275]

5.4.6.1.1 General Comments About Vehicle Tires

Vehicle tires are the most important rubber articles regarding volume and importance. They are also the most important design and spring element of the vehicle. More than half of the NR and SR produced in the world are consumed in the tire industry.

Nature of the Tire. The tire transmits the motor forces necessary for propulsion to the road. In conjunction with the suspension it smoothes the unevenness of the surface and thus provides for driving comfort. However, the spring element when driving is not the rubber but the air. The tire only serves as its container and keeps it under pressure. The tire has to have also spring properties, i.e. it should complement the spring action of the air. As a matter of fact, one can drive on air itself as the air cushion vehicles prove, however, with considerably higher effort to replace the escaping air volume.

Tire development had to keep up with the constantly changing vehicle constructions, increasing motor power, higher accelerations and driving speeds. Therefore, a big difference exists between the first invention of the tire by *Thomsen* in 1845 or the practical improvements in 1888 by the English veterinarian *John Dunlop*, via the first fully synthetic tire in 1912 to today's high speed tire. The load which an average passenger tire has to withstand is, for example, in excess of 4,700 N at an internal pressure of 1.9 bar and a maximum speed of 180 km/hr and more over a temperature range from Sibirian cold to +110 °C in continuous service. Truck tires are exposed to even higher loads of, for example, 30,000 N at an internal pressure of 7.5 bar and a speed of 100 km/hr over the same temperature range. These forces cannot be taken up by a rubber layer alone; the load carrier, the carcass, is here of special importance which is rendered air tight by encasement in rubber and is also protected from moisture, damage and abrasion.

Classification of Tires. The tire represents the rubber article with the most complicated construction. Construction and materials are the outcome of years of development and experience. Depending on the construction or the method of manufacture, the vehicle tires are, for example, classified into diagonal, radial, and bias belted tires. The tires can also be classified according to the kind of load carrying material into textile, steel cord or glass fiber designs (the latter, however, of only limited importance). Further classification principles are according to winter or summer tires, the maximum permissible speed, the height to width ratio etc.

According to the multitude of vehicles with different functions and requirements, a large variety of tire types are available which differ in size, profile shape, and inner construction. These varieties, speed limits, profile configurations, nomenclatures etc. cannot be considered within the following discussion.

Important differentiations of the tires, for example, according to the type of vehicle are cycle tires, those for passenger cars, light trucks, trucks, earth-moving equipment, agricultural vehicles, aircraft and military vehicles.

5.4.6.1.2 Diagonal Tires

Diagonal tires are constructed from a highly abrasion resistant tread compound, a side wall compound which can withstand especially severe dynamic loads, the carcass consisting of textile cord and the bead wire. The latter ensures a proper fit at the heel of the tire and a seal to the rim and serves as a lower fastening of the carcass.

Construction. For passenger cars the carcass consists of 2 to 4 layers and for truck tires depending upon size of 6 to 10 layers.

The cord layers are joined so that the fiber direction of two successive layers are crossing under an angle. The name of the tire derives from this diagonal arrangement of the cord fibers. The fiber angle applied during the construction of the tire has to correspond to the force diagram air pressure/centrifugal force or air pressure/tracking force. With increasing speed the fiber angle has to be increasingly pointed since the effect of the speed is according to the square.

Manufacture. The tire is built-up on so-called tire building machines from the single elements. These machines are semiautomatic. All pre-fabricated components are brought to the tire building machines via servicers. The manufacturing machine itself is a collapsible build-up drum. First the carcass layers are deposited upon the drum under the previously determined angle. The bead rings (wire rings) are enclosed in the free ends of the carcass and fixed. Finally, the tread with the side walls is applied. All single components are well rolled onto their substrate because the manufacture has to be done without any inclusion of bubbles and the layers have to adhere to each other well. Finally, the drum is collapsed so that the green tire can be removed. The tire coming off the tire building machine has a cylindrical form. The green preform obtains its proper tire shape during a separate forming process (earlier) or during the vulcanization in modern tire presses.

5.4.6.1.3 Radial Tires

The radial tire was introduced in 1948 by *Michelin* even though its development goes as far back as 1933. While it has been very successful in Europe, its market share is much less in the U.S.A. One of the reasons for the slow development was that the American driver demands great driving comfort because of the large distances. In the main, however, economic reasons are responsible which are based upon the necessary retooling of the whole tire building process.

Construction. Radial tires are constructed according to principles different from those for bias tires. Here, preferentially four layers of a textile belt or two steel belts, also with Nylon bandages (cord in circumferential direction), are positioned under the tread, or three to five layers as belts for truck tires. These belts, which end at the shoulder region of the tire, stabilize the tread. The single belt layers are also, according to a force vector diagram, arranged under an angle, which is considerably smaller than that for the diagonal tire. Since the tire is considerably stiffer because of the stabilization of the tread area, the side walls have to be constructed correspondingly softer. They consist for radial tires of a sidewall compound which is especially protected against flexing and aging. Under it 1 to 3 carcass layers which are wrapped around the bead wire are positioned. Their cord fibers are 80 to 90 degrees for several carcass layers and 90 degrees for single layer carcasses from bead to bead. The tread is applied to the belt. For radial tires the bead of the tire is especially subject to friction and has to be protected by hard, abrasion resistant rubber compounds or suitable textiles, for example mono-filament Nylon fabric.

Manufacturing. The carcass of the radial tire is also built on a cylindrical tire building machine according to the principles as for the diagonal tire. In contrast to the diagonal tire the green carcass has to be taken before its final production steps to another machine where it is brought into the form of a tire with the help of compressed air in a secondary building step. Only after this forming step the belts and the tread are applied. For both building steps a careful centering is required in order to obtain the required concentricity. In newly developed modern production facilities both of these steps can be accomplished on one machine (see Fig. 5.34).

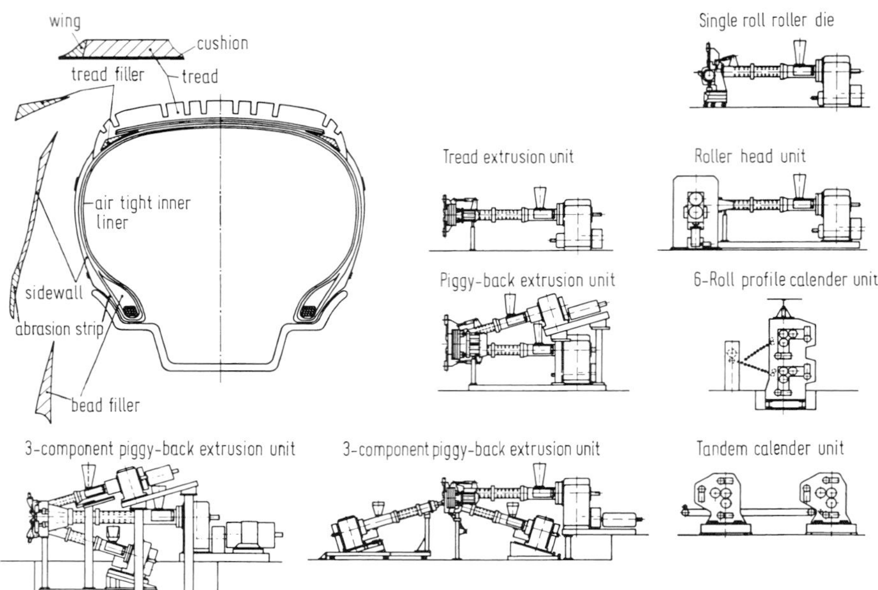

Figure 5.34 Manufacturing line for the one step manufacture of tire treads (Drawing: Troester)

Technology has modernized the manufacture of tires by the application of microprocessors and modern sensor technology. For example, the changes occurring during the processing of the components are measured during processing and the machine is controlled correspondingly. Thereby tires can be manufactured considerably faster, more uniformly and can be built up more accurately than by human control. Automated equipment can already be imagined which has a minimum of mechanical devices but a high degree of electronics, which needs man only for control, supervisory and corrective functions, but works otherwise fully automatically. These very compact machines meet at the same time the requirements for a fast change-over from one tire size or type to another, whereby the need for the constantly rising number of tire sizes will be met.

5.4.6.1.4 Bias Belted Tire and Other Designs

The so-called *bias belted tire* was developed in the U.S.A. as a compromise, a combination of diagonal and radial tire. It can be built in one step on conventional tire building machines.

Also, *other designs* are customary, e.g. two carcass layers of rayon at an angle of 34 to 60° and a steel belt with an angle 4 to 10° less than the carcass. The cylindrical green tire is then formed and vulcanized like a diagonal tire. Because of the difficulty to center exactly during the forming process, the uniformity of this type of tire is not satisfactory.

5.4.6.1.5 Tire Textiles

[4.598–4.611 a]

The cord material used to build a tire has undergone radical changes during the last few years. Cotton, used nearly exclusively up to World War II has been abandoned in favor of semi- or fully synthetic textile materials. First it was replaced by artificial silk (Rayon). Higher strength, higher dynamic damping and good flex resistance have some years ago favored the use of Polyamide cord (Nylon 66 and Nylon 6) for passenger and truck tires and also that of Polyester cord. These materials promote, however, flat-spotting of the tire, a flattening of the foot print during parking of the vehicle which only disappears after longer periods of driving. Also a stronger growth of diagonal tires can often be observed when fully synthetic tires are utilized. This effect can be minimized during tire design (stabilization of the side wall shoulder by the belt).

The introduction of the radial tire also changed the use of the tire fabrics. Rayon, Polyamide, glass, and steel cord materials are available as well as Aramides. Suitable cord materials are selected primarily according to their strength, elongation, fatigue strength and rubber adhesion. Also the change of the mechanical cord properties at vulcanization and running temperatures as well as the chemical influences due to substances from the rubber compound have to be considered. For the manufacture of belts, only cord materials with small elongation, high strength and high bending stiffness can be used. Steel meets these requirements best and is therefore a nearly ideal belt material [4.652–4.668]. Also Aramides which exhibit a low bending stiffness but high damping, which assures a high driving comfort, are very well suited. This applies because of the high strength also to cord materials for truck carcasses while rayon, Polyamide and Polyester are also considered for passenger carcasses. Because of price considerations, Rayon is today often preferred.

Co-Naphthenate (only for brass coated steel) or adhesion systems on the basis of resorcinol formaldehyde resins are applied to bond the rubber to the cords (see page 312ff) [4.577–4.628].

5.4.6.1.6 Tire Vulcanization

[5.261]

For the manufacture of tires special tire presses are employed with a high degree of automation, for either single or double molds. The tire molds are built into the presses. For diagonal tires, molds are used which are divided in the middle, for radial tires those with radial partitions.

The uniqueness compared to other molded articles is given by the fact that these molded articles of large volume are not only heated from the outside, i.e. from the mold surface, but also from the inside, i.e. from the curing bladder or the heating membrane.

Formerly, a green tire preform which had been built-up on the winding drum and had thus the form of a cylinder was always preformed separately while simultaneously a heating hose was inserted for the subsequent application of pressure and heat during vulcanization.

In modern heating devices a rubber cylinder is permanently installed as a heating bellows in the automated heating system which assumes the form of the green tire, presses it into the heated mold which carries the tire profile and heats it simultaneously from the inside.

A preforming of the green tire with simultaneous introduction of the heating hose and a removal of the heating hose after finishing vulcanization is thus superfluous. Only for radial tires preforming of the green tire during its construction is still required for the exact centering of the belt; a heating hose is not used during this process anymore either.

In tire presses of most modern construction, especially for the vulcanization of radial tires, instead of the heating bellows a permanently installed heating membrane is also used as substitute for the heating tube. This heating membrane is pumped up just like the heating bellows so that it presses the green tire into the profile area of the mold which is divided into sections in order to permit an easy introduction of the preformed tire and supplies heat from the inside. After completion of vulcanization the heating membrane is pulled into a cylinder located underneath the mold and thus the introduction of the new green tire preform is not hindered by a protruding heating bellows.

Heating of the presses is by steam of, for example, 180 °C, while the heating from the inside via the heating tubes, heating bellows or membranes is carried out preferentially with the help of hot water of approximately 200 °C. Special steam, water, and/or compressed air equipment is required for these tasks.

While in former times the vulcanization times for a four ply passenger tire were around 45–60 minutes they have been reduced in modern vulcanization presses at high vulcanization temperatures to 8–16 minutes or less.

The length of the heating time is, of course, also dependent on the thickness of the heating bellows or membrane. The thinner these are, the better is the heat transfer.

The course of the heating sequence is determined by process controllers. All partial sequences of the heating processes, turning on or off of steam or water, inside or outside, opening and closing of the molds, is automatically controlled.

Diagonal tires which contain fully synthetic fibers as reinforcements tend to shrink during cooling after vulcanization. In order to avoid this, they are mounted after vulcanization and are inflated until they have cooled down. Through this so-called post inflation process it is avoided that the tire shrinks. In modern tire vulcanization plants a device is provided which accomplishes this process automatically.

The compound properties of the tire have to be adjusted to the heat flux, that from the outside as well as that from the inside, because the heat flux in the thinner side wall is different from that in the tread or the shoulder of the tire. The adjustment of the vulcanization systems, i.e. the sulfur and accelerator dosage have thus to correspond to the amount of heat effective in the different layers of the tire. By using novel temperature programmable vulcameters, also called vulcanization simulators, the correct adjustment of the vulcanization rates of the compounds and of the optimal vulcanization rates has been simplified (see page 473).

Furthermore, the correct pressure during vulcanization is especially important. Only when it reaches a certain value (around 20 bar) can it be avoided that air or air bubbles are enclosed between the outer shell and the carcass. Such inclusions do not only give rise to unsightly but also mechanically inferior “weak spots” on the outside skin of the tire. If they are inside the tire, they can be the cause of separation during service. It is also important that the pressure is high enough right at the beginning of heating because the rubber layers can be deformed plastically and undergo flow only at the beginning of vulcanization in order to compensate for irregularities. The pressure can, by the way, be applied only from the inside since the two mold shells support each other completely on the outside.

5.4.6.1.7 Tire Comparison, Trends in Tire Development

Diagonal tires are simple to manufacture since they are built in one step.

The driving comfort they provide is relatively high. The tires exhibit, however, a strong deformation during service. They form a rolling bulge the size of which increases with speed. Because of this, a strong heat build-up results, also a permanent vertical movement of the foot print (foot print ellipse) on the road surface which is related to an unfavorable abrasion behavior (especially during cornering) and bad road holding as well as bad braking behavior, rolling resistance, higher fuel consumption and less longevity.

Radial tires exhibit, because of the stabilization of the tread surface, a smaller deformation of the tire and a firmer elliptical footprint on the road which gives rise to lower vertical movements on the road. Therefore, the application of block tread designs with more traction edges is possible for radial tires in contrast to diagonal ones. Thereby one obtains a larger traction and an approximately 15% larger foot print ellipse which adapts to the ground in kidney shaped fashion. For radial tires one obtains, in contrast to diagonal ones, a more even load distribution, lower specific surface stress, shorter braking distances, lower abrasion, lower heat build-up, and lower rolling resistance. This results in higher longevity, higher safety and lower fuel consumption. The complicated, i.e. two step construction, which is necessary because of the exact centering of the belt is not advantageous. Better possibilities are seen here because of the new development of a one step building machine.

Intensive research regarding the improvement of the rolling resistance has been carried out recently [3.92, 3.145, 5.262–5.275].

Bias belted tires are with their abrasion and rolling resistance in between diagonal and radial tires.

Textile belts exhibit, compared to steel belts, the following disadvantages: Worse self centering (more difficult steering), considerably higher abrasion, thus lower life expectancy, higher tire growth, thus lower uniformity. Aramids in conjunction with steel cord increase the driving comfort.

Safety Tires. Constructions with inside hard rims have not proven successful in the area of safety tires. One has gone over to tires with a strongly reduced height to width ratio of 0.6 (normally 0.8) with reinforced shoulders and to special rim constructions where the tire bead cannot come off the rim even when the air is lost. After pressure loss a lubricated fluid is released which prevents a destruction. One can drive with such tires in the case of a defect at reduced speed to a service facility. However, such tires are expensive (four safety tires are more expensive than five conventional ones). Newer developments use also support inner tubes.

Winter Tires. The problem of tire tread adhesion to ice became of great interest after spikes were forbidden in Germany and some other countries. Good ice adhesion could be obtained in the beginning by the incorporation of silicas and silanes instead of carbon black. Such tires exhibited, however, unfavorable adhesion on wet roads, had high abrasion and were not suitable for sustained service. Tires with silicate and carbon black [1:1] as fillers were also offered as ice tires and offered a compromise regarding ice adhesion and wet skid resistance. Newer developments are based upon an increase of the BR component in a blend with SBR resp. the partial use of 1.2 BR [3.95] while retaining carbon black as the sole filler and a tread design with many traction edges. Winter tires are equipped with special traction promoting profiles and deep grooves.

Cast Tires. The age old dream of tire technology to be able to manufacture a carcass free tire from liquid starting materials has not died. Again and again manufacturing methods using polyurethane raw materials are being introduced and are being abandoned. Recently, this topic has gained interest again. A carcass-less tire, especially one without a belt, has to exhibit necessarily a larger growth rate than a tire with load carrying fibers which increases necessarily with heat build-up, resp. with increasing speed and will thus also have more unfavorable tracking forces. It also exhibits generally unfavorable adhesion values and durability and cannot be considered at this time as an adequate substitute for radial tires. Cast tires with belts, however, seem to show long range promise as a partial replacement, for example, for slow and for military vehicles [3.762].

Trends in Passenger Tires. The steel belted radial tire has proven completely successful, at least in Western Europe. The height to width ratio is steadily declining. Attempts are being made to reduce the number of compounds and design components per tire, in order to permit even more advanced automation of the tire manufacturing process. In addition, attempts are made to develop tires for universal applications *(all season tires)* as well as tires with low rolling resistance (fuel savings).

Trends for Truck Tires. Here the steel belted radial tire dominates the market. Here too, the trend is toward smaller height/width ratios, coupled with a changeover to tubeless tires, which are mounted on one part rims. By eliminating the inner tube and the bead etc. the wheel weight and thus the un-sprung mass is reduced. Another trend is to replace twin tires by one *"super single tire"*.

5.4.6.2 Belts

[5.276–5.284]

Conveyor Belts [5.276–5.279]. For the manufacture of conveyor belts, drive belts and the like frictioned heavy fabrics of 750–1100 g/m^2 or those with a skimmed thin layer of a rubber compound are used including the two rubber plates serving as cover layers and the two edge protectors. Here, the application of proper rubber to fabric bonding agents is important. Vulcanization takes place in conveyer belt presses or, for thin belts, in rotational vulcanization presses. The belts are stretched in the vulcanization presses up to a remaining elongation of 10 to 16%. Steel cord inlays are used for heavy duty belts [4.652–4.668, 5.276, 5.278].

Drive Belts. Drive belts are most often manufactured from less heavy fabric and without cover layer. They are either cut to the required width from a wide band or are manufactured by folding of the single layers to the desired width.

V-Belts [5.279a–5.284]. *Wrapped V-belts* are manufactured so that a specified number of rubberized cord layers are laid up in an endless construction. Onto these a V-belt cushion is placed and finally the whole structure is wrapped with frictioned bias fabric.

Recently, so-called *open flank V-belts* without bias fabric [5.280–5.284] have been manufactured. They have taken over the market in European mass produced vehicles nearly 100%. Of the European open flank V-belts nearly 90% are manufactured in the toothed version, a percentage considerably higher than that in the U.S.A.

The open flank, toothed V-belt consists of a textile reinforced CR belt upper structure, the tension cord assembly, a CR inlay and the belt upper structure. The application of polyester cords specially developed for the tension member permit the belt to transmit high forces at low elongations. The tension member has been especially prepared for good adhesion to the interlayer compound which can withstand high dynamic loads. The lower structure consists of a CR compound with fibers which are oriented perpendicular to the belt direction. Thereby the V-belt becomes very flexible in the direction of rotation and very stiff against twisting in the other direction. Furthermore, the flanks become very abrasion resistant. A good adhesive strength between all construction components is a prerequisite for good dynamic performance.

Open flanked V-belts of newer construction last in comparison to wrapped ones 2 to 3 times longer. They exhibit a considerably lower hysteresis and thus a correspondingly lower heat build-up; they assume the shape of the pulley better because of the lower bending stiffness, even for small pulley diameters and thus they are much less prone to slip. Recently self-stretched V-belts (based on prestretched yarns, which tend to relax at increasing temperature) have been invested with drastically increased longevity.

The so-called *multi-ribbed V-belts* can be looked upon as a further development of the open flanked belts. They are also called *"serpentine belts"* or S-belts. Because of the large number of auxiliary units on the engines of American cars they have found a larger market share there than in Europe. However, the market in Europe will grow rapidly in the near future.

V-belts and multi-ribbed belts are generally vulcanized in special presses or on modern automated vulcanization machines. Regarding the use of textiles and their pre-treatment see [4.577-4.668].

5.4.6.3 Rubber-Metal Elements

[2.136-2.140, 4.629-4.651, 5.285-5.287]

For the manufacture of rubber to metal elements the rubber compound has to be vulcanized to the metal.

Vulcanization of rubber compounds to metal parts was, in former times, only possible by brass coating or via a hard rubber layer. Today, mostly chemical adhesives are employed (see page 315f). The metal surface, cleaned by degreasing and sand blasting, is coated with a thin layer of the adhesive and the preform, possibly well formed, is vulcanized onto it after thorough drying. Vulcanization occurs after careful placement into molds, mostly by the compression molding method, but increasingly also by injection molding. During the construction of the molds one has to avoid that the parting line is in the same plane with the adhesive layer since otherwise the rubber can shift during molding. For the same reason the injection molding method has been used very rarely in the past for the manufacture of rubber-metal elements with the help of chemical adhesives.

Since *vibration isolators* have to be in many cases very soft and have to have high tensile properties, these parts are often manufactured from NR [2.136-2.140]. In addition, according to the required oil resistance or damping properties, also NBR, IIR or other rubbers are used.

Recently, especially in Europe, so-called *hydro-bearings* have been penetrating the market, in which, for example, a spring element made from a low damping compound is combined with a hydraulically damping element in one bearing. Thereby good noise isolation and good vibration behavior, for example of a motor exited from the road surface, is achieved. A very big section of rubber-metal parts consists of so-called *Simmer rings* which are mainly made of NBR, ACM, FKM etc. by transfer or injection molding processes.

5.4.6.4 Cellular Elastomers

[4.571-4.576, 5.288]

Cellular elastomers are formed when a rubber contains a blowing agent (see page 307 ff) which sets free a gas at the beginning or during the vulcanization whereby pores are formed which expand the compound. Depending on the pore structure one distinguishes between open and closed cells. *Moss rubber* (closed cell sponge rubber with small, closed and open cells) has small pores which are not all closed and depending on the manufacturing technique a firm vulcanized skin. While moss and *sponge rubber* (with open cells) are produced by blowing in the press or after extrusion during continuous pressureless vulcanization in the salt bath or the UHF line, micro cellular rubber is manufactured in the press according to the expansion method.

Blowing Method. The *principle* of the blowing method is comparable to baking a cake with baking powder: the blowing agent decomposes and expands the compound (volume increase up to 1000%). Then vulcanization sets in.

In order to obtain a large volume increase the compound must have good flow properties. This is achieved by good mastication or uniform strong degradation of the elastomer, the utilization of inactive fillers, factice and process aids. It is also important that blowing temperature and scorch temperature are well adjusted to each other. When using benzolsulfohydrazide with its gas formation temperature of 80 °C this is normally no problem. To obtain good flow behavior and blowing of the compound, accelerators with delayed onset of vulcanization are used. The blowing agent has to be decomposed to a large extent before vulcanization starts. Otherwise the article is blown irregularly or incompletely. For large articles which can be warmed only with difficulty, it is required to apply stepwise heating, i.e. one heats up to a temperature first at which the blowing agent will decompose but vulcanization will not begin yet. It must also be considered that vulcanization must not start too late since otherwise the already blown compound can collapse again.

For the *manufacture of sponge rubber* one fills the mold which has been covered with fabric and paper to approximately 10-50% with the rubber compound which contains so much blowing agent that the mold will be filled after expansion. One thus obtains a product with partially open and partially closed cells. The mold frame is covered at its upper part also with paper or fabric so that the air which is displaced during the blowing phase can escape and thus air enclosures are avoided.

The vulcanization of sponge rubber sheets for *sanitary sponges* is generally carried out in vulcanization presses at low pressure, but can also be carried out in hot air. After cooling of the vulcanized sponge rubber sheet the sponge collapses since the gas contained within the still closed cells contracts upon cooling. It is thus required to break open the thin cell walls which is accomplished by repeated milling at zero friction ratio and to peel off the surface skin. By these measures the sponge regains its original volume and becomes able to absorb water.

During the manufacture of *soles for house shoes* a sponge rubber compound is placed in a sole mold of a shoe press where it expands just as during the manufacture of sponge rubber and vulcanizes whereby the compound joins at the same time the fabric upper part. In contrast to the manufacture of expanded soles the house shoe sole production is entirely a blowing process.

The *moss rubber manufacture* takes place after extrusion during a continuous vulcanization process. A constant pore geometry can only be obtained when all process parameters are kept constant.

Expansion Method. The *principle* of the expansion method can be compared to the bubble formation in a bottle of mineral water after it has been opened. The gas is dissolved in the rubber under high pressure and expands the compound after depressurization because of gas formation.

For the *manufacture of porous soles* and *microcellular rubber* (with closed cells) the rubber compound containing a blowing agent with higher gas formation temperature (see page 310) is placed in a mold with tapered corners. In contrast to the blowing method where the inserted material fills the mold only partially, one fills here so much compound that the volume of the mold is completely occupied (+3% excess to seal the mold). During the vulcanization process the blowing agent decomposes; the developing nitrogen is partially dissolved under the high press pressure in the elastomer. The press is opened before vulcanization is completed. The partially vulcanized compound expands because of the high internal gas pressure, one obtains a material with closed cells which can then be completely vulcanized in a second, somewhat larger mold or in open air. The volume enlargement which can be attained depends, aside from the formed gas volume, i.e. the kind and the proportion of the blowing agent used, on the degree of vulcanization at which the press is opened. The longer one vulcanizes at the first step, the lower will be the volume expansion of the expanding plates; for shorter heating times higher volume increases will be obtained. That means, that the vulcanizate will become softer and lighter at shorter times of the first vulcanization step. Therefore, a well balanced ratio of the blowing agent dose and the duration of vulcanization is important for the adjustment of the density and the pore structure. Also the press pressure during vulcanization is of essential importance in order to avoid gas losses. During vulcanization of these porous plates a considerable pressure is created by the gasses which can in some cases open the press platens during vulcanization.

In contrast to the sponge rubber compounds which have to expand heavily in the mold and should be therefore soft and pliable, the use of heavily plasticated rubber is not necessary for the production of expanded cellular rubber. Quite the opposite, it is even desirable that the unvulcanized compounds are as stiff and hard as possible, in order to assure good mold definition. A certain amount of styrene containing resins are used in such compounds.

After removal of pre-vulcanized plates from the molds they tend to shrink considerably. After longer storage the shrinkage progresses. In order to compensate for the shrinkage and to obtain a full vulcanization, it is necessary to post-vulcanize them.

The *post-curing* during which the plates are fully vulcanized can be done using hot air, or in a not fully closed vulcanization press with a correspondingly larger cavity, so that no pressure is applied anymore to the vulcanizate.

A cellular rubber of especially fine pores *(microcellular)* is obtained using the so-called *Pfleumer method*. In this method the rubber compound is vulcanized under high pressure in a nitrogen atmosphere. A certain amount of nitrogen dissolves in the rubber compound. After reaching a certain degree of vulcanization the pressure is released and the dissolved nitrogen expands the elastomer. Since the cellular rubber produced by this method has only closed cells it is, for example, well suited for diving suits.

5.4.6.5 Hollow Articles

[5.289]

Hollow articles are sometimes assembled from calandered sheets and then vulcanized in free steam or in molds, sometimes with the application of a blowing agent. Articles include: Toy balls, toy figures, heating bottles, soccer ball bladders and inflatable cushions. The calandered sheets are, after being cut to size, formed by a machine into a hollow closed article. A blowing agent, for example a mixture of ammonium chloride, sodium nitrite and water or an organic blowing agent, is placed inside of the hollow article, for example balls, so that the rubber compound is pressed with sufficient pressure against the mold and also has a sufficient internal gas pressure.

5.4.7 Mold Cleaning

[1.12, 1.13, 5.252]

After longer use of the molds organic components of the rubber compounds or the mold release deposit in the form of crusts or carbonizations [5.239]. Even when silicone based mold releases are used dirt deposits form after a longer time.

For cleaning, the molds are first placed in a cleaning bath in order to soften the dirt layer. Formerly a chromic acid bath, heated to 60 °C, was frequently used. But since such baths attack the mold material, today basic baths are used. Among these is, for example, sodium hydroxide in which the molds are kept for one hour. A short cleaning with water follows.

After this pre-cleaning the mold is cleaned in a water jet device. This apparatus contains small glass beads suspended in water which are pressed upon the mold through a jet under a pressure of up to 8 bar. For sensitive molds a pressure of only, for example, 3 bar is applied. After this cleaning procedure the tool is rinsed under a high pressure jet stream. The mold is then dried and often protected on its surface with a layer of silicone. In order to prevent rust formation it is advantageous to add some oil to the equipment; also other rust preventatives can be used. An immediate use of the mold after cleaning is advantageous in order to avoid rust formation. Otherwise the mold should be dipped into a corrosion protector.

5.4.8 Deflashing

[5.290–5.294]

Unless molded articles have been manufactured without flash by the flachless method [5.205] or according to certain injection molding methods [5.238] either without or with very little flash, they carry more or less pronounced amounts of flash which have to be removed. This process was in former times carried out by hand and was very expensive; the deflashing costs can exceed the manufacturing costs, especially for smaller articles. Therefore, several processes and machines to automate this production step have been developed. For rubber articles which are not too elastic, the so-called cryogenic deflashing is considered the most modern method whereby the whole article is exposed to low temperatures (for example liquid nitrogen [5.290, 5.291, 5.294] or carbon dioxide) so long until the flash is already frozen but the thicker sections are not frozen yet. Through mechanical handling the flash is then broken off.

With other machines flash can be fully or semi-automatically removed by either cutting or stamping.

Deflashed molded articles are often ground at the parting lines if that has not already been done in the rubber cutting machines.

For smaller molded articles the flash is often removed by socalled drum treatments. The articles are turned together with rough stones in a drum or spun in water.

5.5 Manufacture of Hand Made Articles and Dipped Articles from Solutions of Rubber Compounds

5.5.1 Hand Made Articles

(see also Section 5.4.1, page 407)

Aside from the already described articles, so-called hand made articles are manufactured in the rubber industry. Among these hand made items are, for example, bellows, flat and profile rings, stamped articles and frames of all kinds. These are those rubber articles which, because of their form or their size, cannot be manufactured because of financial reasons in a mold but are confectioned on templates or are hand assembled and then vulcanized. A template is here a body made from steel sheet onto which one places the rubber compound and from which one removes it after vulcanization.

It is important for the manufacture of hand made articles that the compound has a sufficient tackiness and green strength during vulcanization since they are vulcanized unsupported. During heat-up, until the onset of vulcanization which brings about a dimensional stability, the confectioned article must not deform.

The tackiness required for this process and the green strength are obtained by a well balanced ratio of resins (tackifiers), plasticizers, factices, and fillers.

In order to enhance the dimensional stability during vulcanization some confectioned free hand articles, for example hoses and roll covers, are wrapped in wet fabric wraps before they are vulcanized. Among the articles which were hand confectioned in former times but are manufactured today automatically are, for example, roll covers and pressure hoses as well as belts, straps and tires (where, however, the latter are vulcanized as molded articles after they have been hand confectioned).

5.5.2 Manufacture of Dipped Articles from Solutions of Rubber Compounds

By dipping forms into solutions of rubber compounds (frequently simply called solutions, see page 383 ff, production of parts out of latex is not dealed with here) and subsequent processing, rubber articles are manufactured which are called dipped articles. Since they do not have any parting seem in contrast to molded or hand confectioned articles, they are also often called seamless articles. Among these are pacifiers, prophylactics, surgical gloves, industrial and electricians gloves, finger covers, balloons and specialty rubber articles for medical applications.

Form Material. Forms made from glass, glazed porcelain or light metals have proven to be good form materials. In special cases, forms made from plastic are also used. The surface of the forms has to be smooth and must not have any porous sections since these parts would give rise to bubbles. The forms must neither have sharp edges nor must they have a pointed shape since the solution would run off at these locations after dipping and thin walls would result.

The forms are placed in steel frames or cassettes. The dipping machines are operated with these.

Dipping Machinery. The dipping apparatus consists of a sheet metal housing the upper cylindrical part of which contains a two or four part cross, depending on the size of the form, to accept the form frame. The solution resides in a solution container, a movable dipping cart which can be raised or lowered for dipping or withdrawal. This is done using hydraulic equipment.

Dipping. During *dipping* the dipping cart is raised so that the form tips, and for gloves also the roots of the fingers, enter the solution very slowly. Then the dipping speed may be raised until the marks affixed to the forms have been reached, however, the speed has to be adjusted so that the more or less viscous solution does not double up and thus entrap air which would lead to undesirable rings of bubbles. The adjustment of the correct dipping speed requires a well developed finger tip feeling.

During *withdrawal* the dipping cart can be lowered at constant very slow speed so that the excess rubber solution can run off.

After the solution stream has been broken, the dipping cart is lowered completely and covered with a lid. Now the *rotary cross* is rotated at a speed of approximately nine RPM. Because of the special gearing the direction of rotation is changed every minute and thus a one directional running of the solution during the drying cycle is prevented. After sufficient *drying* of the film, the opposite frame is dipped in a similar fashion.

A uniformly thin film has to be formed on the forms since solution films which are too thick will shift during the rotation and dry only very slowly and the remaining solvent will give rise to bubbles during vulcanization.

The *number of dipping and drying cycles* is repeated until the desired wall thickness has been reached. Very thin articles such as prophylactics are dipped once or twice, surgical gloves and finger covers two to four times and thick walled articles often 20 to 30 times. The rubber film of the first dip is mostly thinner than the films of subsequent dippings since the solution adheres less firmly to the smooth form than to the already present rubber film. Each subsequent dipping can only be done after complete drying of the former one where one has to consider that the solvent does not only diffuse toward the outside but also into the lower layers. Therefore drying is slower and slower the thicker the film gets.

Edging. After drying the dipped articles are edged, i.e. the lower edge of the film of the rubber compound is rolled up to the edge marking on the form. The bulge thus formed makes the removal of the dipped articles from the form after vulcanization easier and prevents tearing during part removal and also during use.

Vulcanization. Vulcanization takes place in hot air ovens.

The manufacture of dipped articles from solutions is a very laborious process. Therefore the manufacture of dipped articles from latex has gained great importance. This method will not be dealt with in this book [1.16]. Articles such as prophylactics, balloons and surgeon's gloves are manufactured today mostly from latex. For the manufacture of electrician's gloves for which mineral fillers, emulsifiers and other contaminants contained in latex would reduce the electric insulation properties are still today produced from solution.

5.5.3 Free Heating (Vulcanization in Free Steam, Hot Air, or Water)

[1.12, 1.13, 5.295–5.303]

5.5.3.1 General Comments About Free Heating

[1.12, 1.13]

Most of the confectioned articles and also some molded ones are not vulcanized in molds but freely. Free vulcanization is after vulcanization in the mold the most frequently applied vulcanization method. In a lot of cases it occurs in the autoclave under pressure. Since the heat is transferred to the part in this heating method slower than in molds, deformations can occur here quite easily. In order to avoid these, the products are, for example, packed into talc powder, are supported by support bars or templates or are, for example for hoses, roll covers etc., wrapped with wet cotton strips which shrink when they dry and thus press the article firmly against its form. Articles which cannot tolerate the influence of steam or where a smooth surface finish is important, for example lacquered boots, rubberized fabric, dipped articles etc. are vulcanized in hot air.

During free heating, in contrast to the vulcanization in the press, the onset of vulcanization has to be quite rapid so that the tendency for deformation during heating is counteracted by a rapid hardening. However, the rate of vulcanization must not be so fast that the danger of scorching may be present at the given processing temperatures. Thus, generally only a very narrow window between good deformability at the processing temperatures on the one hand and the deformation stability at the vulcanization temperatures on the other exists. This requires a careful compound design especially as far as the vulcanization chemicals are concerned, particularly for hot air vulcanization.

Differences During Hot Air and Steam Vulcanization. The compound design has also to be designed according to what heat transfer medium is to be used. During utilization of hot air or steam the following principal differences exist:

Air has in contrast to steam only a very low *heat capacity*. Also, the *heat transfer coefficient* is much lower. Because of this, the heat transfer of hot air to rubber articles to be vulcanized is only low and the heating times for air heating have to be for the same compound approximately double as long as for those in saturated steam for which in turn an extension of the heating times or a temperature increase (mostly 1 bar) are required compared to heating in the mold.

The accelerators in compounds which shall be vulcanized in hot air have to be active already at relatively low temperatures because of the low *rate of vulcanization* so that the surface of the articles obtains a certain green strength and thus does not deform. Their vulcanization rate has to be greater compared to compounds which are vulcanized in steam.

Due to the oxygen content of the hot air, rubber compounds can be endangered at high temperatures because of *oxidation*. The upper limit for hot air vulcanization e.g. for NR is considered to be approximately 150 °C. At higher temperatures the oxidative attack progresses faster than the vulcanization. Therefore, NR compounds do not fully vulcanize in too hot air. Because of the oxygen content the oxidation of compounds which are vulcanized in hot air is also dependent on the vulcanization pressure (air pressure). Because of the absence of air during the steam vulcanization, higher temperatures, for example up to 200 °C and above may be permitted. SR types, which are stable against oxidizing, can be heated of course up to higher temperatures.

Another limit of the applicability of hot air arises when *peroxides* are used as vulcanization chemicals. Since the peroxide reacts partially with the oxygen in the air, and is thus not available to crosslink the elastomers, problems with the vulcanization can occur (surface tackiness, see page 258).

While vulcanizates heated in hot air often impress because of their beautiful and *smooth surface,* steam vulcanizates often exhibit condensed water spots and have a considerably rougher surface.

Hot air is not dependent on the application of *pressure* while the saturated steam temperature always correlates with a corresponding pressure. During the beginning of the heating period with steam, first only low and then rising pressures are obtained. In articles of large dimensions porosity can develop for low heating rates in steam because the pressure does not build up fast enough.

While directly heated steam generators are sufficient for steam heating, indirectly heated vessels are required for heating with air or air/steam mixtures.

In order to combine the relative advantages of hot air and steam heating, one sometimes combines the two methods.

Vulcanization Phases During Free Heating. The heating time can be divided into four sections, the rise time, the pre-heating time, the vulcanization time and the release time. During the free heating these sections play a special role (compare also page 222).

The *rise time* occurs necessarily for every vulcanization because cold parts are placed into the vessel and it takes a certain time to overcome the resulting temperature loss.

For parts of large dimensions a preheating time or a considerably extended rise time may be required (see Fig. 5.35.).

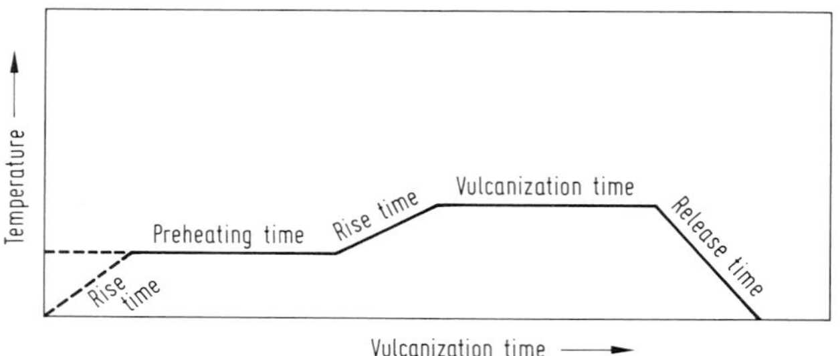

Figure 5.35 Vulcanization temperature as a function of the vulcanization time, especially for articles of large dimensions (rise time, pre-heat time, vulcanization time, release time)

The *pre-heating* time is the time during which the temperature remains constant or slowly rising below the onset of vulcanization in order to heat the parts to be vulcanized evenly. Its role is the more important the thicker walled the article is.

The *vulcanization time* begins when the desired temperature has been reached and ends with the releasing or cooling of the vulcanization medium. For vessels which are filled heterogeneously or for heterogeneously composed articles the vulcanization time is determined by the vulcanization rate of that compound which vulcanizes most slowly.

It is customary to empty the vessel after vulcanization has been completed as fast as possible *(cooling time)* in order to avoid a strong decrease of the vessel temperature and thus to avoid the necessity for subsequent heating up. The time necessary for this is determined by the pipes installed for this purpose. Only for parts based on Q a very rapid cooling can be disadvantageous. Because of the high gas permeability of this material more or less pronounced amounts of the vulcanization medium, e.g. steam, can dissolve in Q during vulcanization as a function of the pressure which leads to bubble formation during a spontaneous pressure reduction. Therefore a slow pressure adjustment has to be provided for, for example 1 atmosphere in 5 to 10 minutes, when Q is heated freely.

5.5.3.2 Vulcanization in Hot Air

[1.12, 1.13]

Vulcanization Equipment. Hot air vulcanization can take place in vessels under pressure or in pressure less ovens. Hot air vessels are pressurerized autoclaves, generally in a vertical arrangement with interior or external hot air circulation devices. Those with *internal air circulation devices* contain a heating element which heats the air in the vessel and a blower inside the vessel is circulating the air constantly and thus assures an even temperature distribution.

For vessels with *exterior air circulation* the heat exchanger is located outside. The air is blown into the one side of the vessel and simultaneously exhausted on the other side. After half the vulcanization time has passed the flow direction is reversed. Because of this reversal the articles are especially uniformly vulcanized.

When the cold vulcanization material is introduced into the vessel a noticeable cooling takes place. Now the correct vulcanization temperature has to be reached as soon as possible. This can be accomplished with the help of sufficiently dimensioned heating elements and corresponding heat circulation. It is, of course, important that the temperature is equal in all locations of the vessel; therefore the heating elements and the circulation speed of the hot air have to correspond to each other. Automatic valve adjustment assures the maintenance of the desired vulcanization temperature, which is controlled by recorders. The number of necessary temperature sensors depends on the size of the vessel. With the quality level of today's commercially available vessels close temperature limits can be maintained (approximately 2 °C accuracy), which is required for good uniformity of the rubber products.

Aside from ovens which can be, for example, up to 150 m long, also hot air shafts and tunnels are being used.

Vulcanization. Vulcanization in hot air can be carried out pressureless or under pressure.

The *pressureless heating* is especially applicable for very thin walled articles. Since pressureless vulcanization leads to bubble formation in thick walled articles these should always be vulcanized under pressure.

As already mentioned, for the *hot air vulcanization in the autoclave* any desired pressure can be used in contrast to saturated steam vulcanization. For example, it is possible to work with high vessel pressure and low temperature or conversely with low vulcanization temperature and high pressure, which is not possible in saturated steam. Porosity and insufficient layer adhesion in confectioned articles caused by vulcanization can be avoided. It has been observed that in praxis a vessel pressure of maximal 8 to 10 bar is sufficient to obtain pore free vulcanizates even for thick walled articles.

However, the hot air vulcanization cannot be applied in all cases because of the reasons mentioned on page 445.

At sufficiently high temperatures or correspondingly long times the hot air vulcanization can also be carried out continuously. The maximum applicable temperature depends primarily on the oxidation behavior of the elastomer and the contact time with the hot air. Thin walled profiles with fast acting accelerators can be vulcanized even from oxidation-prone elastomers without special aids in hot air. For profiles of higher cross-sectional area and/or lower vulcanization rate, a pre-heating of the profiles, for example in an ultra high frequency field (UHF vulcanization, see page 399ff) before entry into the hot air tunnel may be called for. Especially heat and oxidation resistant elastomers can tolerate very high air temperatures and may be vulcanized continuously in hot air without UHF pre-heating.

When hot air vulcanization occurs without pressure it can only be applied for thin walled articles such as profiles, rubberized fabrics, dipped articles etc. or for thicker ones when their compound has been de-gassed.

The increasing application of newer polymers, as for example, ACM, EAM, FKM, Q etc. requires often after the vulcanization in the press a *post-vulcanization* (also called tempering) in hot air. For these elastomers vulcanization in the press is only the forming pre-vulcanization; the final cross-linking occurs as a pressure-less post-vulcanization in special heated ovens. Tempering occurs usually at considerably higher temperatures than the pre-vulcanization. In the case of Q a constant fresh air supply is required in addition to high temperatures and long post-heating times. During this post-heating the decomposition products of the rubber chemicals often escape which can affect the aging properties adversely.

5.5.3.3 Vulcanization in Steam

Vulcanization Equipment. For the vulcanization in free steam one uses horizontal vessels or autoclaves.

The *horizontal vulcanization vessel* has rails inside on which a cart can be driven into it. The lid is equipped with a snap closure mechanism. Lip seals are used to seal the vessel. The lid can be swung around a pair of hinges or a swing arm.

The *autoclave* stands upright with the lid on the top. In most cases it is sunk into the floor and can thus be loaded easily from the shop floor. Modern units also have a snap action closure mechanism. A lip seal provides for a seal in-between the lid and the vessel also. The lid is lifted with the help of a swing arm and then moves aside.

Pressure vessels of this type can also be used for the vulcanization in water as long this is carried out under pressure.

Vulcanization. The vulcanization in steam assures good *heat transfer* and thus shorter heating times compared to the hot air vulcanization can be expected.

When saturated steam is used, for which it is well known that the water and the steam are in equilibrium, the heat transfer is very even because of the relatively high free heat of condensation. However, in large vessels the air present can under certain circumstances be trapped in some locations so that air sacks with bad heat transfer are formed. In order to avoid this and the resulting partial under-vulcanization, the vessel should be blown out with steam before vulcanization.

When the steam supply connections are sufficiently large, the desired steam pressure and thus the required vulcanization temperature are reached very fast and they can be controlled easily and efficiently. A big disadvantage of the vulcanization with saturated steam is the dependence of the steam pressure on the temperature and conversely that of the temperature on pressure.

Another disadvantage of the vulcanization with saturated steam is the large amount of *condensed water* which forms when the vessel is cold at the beginning of the shift which prevents an even heating of the equipment. The formation of this condensed water can be minimized or even eliminated by pre-heating of the vessel or by installing additional heating elements inside the vessel.

Because of the formation of this condensed water, so-called *water spots* can be formed on the surface of the vulcanizates which do not only have the disadvantage of being unsightly but can also lead to local under-vulcanization. This undesired phenomenon can be partially avoided by treating the unvulcanized rubber with a wetting agent.

The vulcanization in steam permits considerably higher *temperatures* than the air vulcanization since steam, in contrast to air, acts like an inert gas. Thus one can employ during the vulcanization in steam without difficulties vulcanization temperatures up to 200 °C and above.

Since the tendency to revert increases rapidly with increasing temperature for elastomers which are prone to revert, the steam pressure should be limited for such materials to ten bar above ambient. Some SR types, in contrast, can tolerate considerably higher steam pressures.

If, however, the articles are manufactured from elastomer types which are subject to hydrolysis, as for example AU, the application of steam should be avoided. Also for Q not too high temperatures should be applied. At high steam temperatures one can reach vulcanization times of less than a minute. These short vulcanization times are utilized, for example, for the continuous vulcanization *(CV-vulcanization)* during the manufacture of cables in the steam pipe (see page 406).

The application of *superheated steam* which can be generated from saturated steam in directly or indirectly heated vessels with additional heating elements would eliminate some of the disadvantages of the saturated steam vulcanization. One could, for example, during the application of superheated steam choose a pressure setting which is independent of the temperature. Also, the formation of condensed water is considerably less than when saturated steam is used. These important advantages are contrasted by the disadvantage that the temperature of the superheated steam can be controlled with the required accuracy of +/− 2° only with much more effort compared to the vulcanization with saturated steam. Finally, the superheated steam is at higher temperatures such an aggressive gas, that it corrodes the vessel walls easily and rapidly unless these are constructed from a correspondingly resistant and thus expensive material. For these reasons, supersaturated steam is used for vulcanization only seldom in contrast to the regeneration methods which utilize superheated steam quite often.

5.5.3.4 Vulcanization in Air/Steam Mixtures

The air/steam vulcanization for which the same technical installations are required as for the hot air or steam vulcanization (see page 446) combines the advantages of the hot air and the steam vulcanization. Aside from the intense heat conductivity which is due to the steam, the applied air pressure permits a rapid adjustment of the pressure at the beginning of the vulcanization, independent of the temperature, whereby undesirable bubble formation can be prevented.

5.5.3.5 Vulcanization in Water

The advantage of the water vulcanization is given by the fact that the heat transfer is even better than for steam and that the danger of distortion of the profile is much less because of the higher density of the water in comparison to steam or air and because of the higher vulcanization rate. Furthermore, the water vulcanization can also be carried out on very large parts, for example large containers, for which the normal vulcanization equipment would be too small.

The vulcanization in water can be carried out in different ways:

- *Under atmospheric pressure*
- *Under pressure*, where the articles to be vulcanized are placed in water filled vessels or are placed into the vessels in waterfilled troughs.

Vulcanization Under Atmospheric Pressure. Under atmospheric pressure the maximum temperature is 100 °C. When, for example, sodium chloride is added, the boiling point can be raised somewhat. At these low vulcanization temperatures the vulcanization rate is very low in spite of the excellent heat transfer and one needs long heating times.

For vessels of large dimensions which cannot withstand any great pressure, the vessels are filled with water of room temperature and then heated up by steam which is blown into the water. It is necessary, however, that the vulcanization takes place under a column of water of at least 2 m hight. In order to obtain this pressure, a stand pipe of the corresponding length is attached to the vessel. Then the vessel is filled with water. The heating time of the water with the help of steam pipes is relatively long. It should not exceed 5 to 6 hours. The vulcanization time requires often another 48 hours or even more until the article - most often hard rubber liners - has been fully vulcanized and has reached the required hardness. For soft rubber liners this method is more and more replaced by the use of self-vulcanizing compounds.

Vulcanization Under Pressure. During the vulcanization in water, the material to be vulcanized is, for example, placed into a water filled vessel. For this purpose an autoclave has to be available which can be filled with water so far, that the articles to be vulcanized are covered completely with water. Parts protruding from the water would later have another degree of vulcanization. Then the steam is introduced.

The maximum water temperature is determined by the applied steam pressure.

After vulcanization the water has to be cooled to 100 °C before the autoclave is opened since it would otherwise start to boil and would spill over.

As can be seen, a rather complex piece of equipment is required for the vulcanization in water. Therefore, this method is only seldom used, for example for the vulcanization of soft printer's rolls. The energy consumption is very high, especially when the equipment is not used continuously and the heating times are considerably longer than for the other methods described above because the water takes up a large amount of energy. In order to obtain an even temperature over the whole heating medium, it has to be assured that the water can flow freely in between the parts to be vulcanized.

5.5.3.6 Rubber Liners and Hard Rubber Articles

Liners made from rubber compounds are used to a large extent to protect surfaces against corrosion, especially pipes, reaction vessels, storage tanks etc. The largest amount of lining is done with hard rubber because of its especially good chemical resistance.

Hard rubber is formed by vulcanizing diene based elastomers with more than 25% sulfur. A NR vulcanizate with 15 phr sulfur is leather like; it becomes increasingly harder and brittle with increasing sulfur loading. With 50 phr all double bonds are saturated in NR.

The vulcanization time of hard rubber is considerably longer than that of soft rubber. Since the vulcanization is very exothermic after the onset of the reaction it has to be carried out at low temperatures. For thick-walled hard rubber articles a low vulcanization temperature has to be chosen and thus a correspondingly long heating time. At excessively high vulcanization temperatures the reaction can become so rapid that it can lead to an explosion which can be so violent that presses and autoclaves can be destroyed.

For surface protection the metal surfaces have to be first freed of rust or contaminations by blasting with quartz or steel sand and, for example, coated with a solution of a hard rubber compound. After complete evaporation of the solvent a 3 to 5 mm thick coating made from a hard rubber compound, only in rare cases from a soft rubber, is applied. The lining is as a rule applied by hand operation which requires great experience and dexterity.

Vulcanization is then carried out in big autoclaves, for hard rubber linings mostly in hot air since this way a lacquer like, dense and damage resistant surface is formed. In special cases vulcanization is carried out in steam where an especially good warm-up of the metal and the rubber compound occurs but where the danger of porosity is given. Especially big containers can also be vulcanized in hot water (see page 450) with hard rubber compounds which have an especially fast acceleration system. This method is especially applied when the vessels to be lined do not fit into any autoclave and have to be vulcanized on location.

Today, self vulcanizing soft rubber compounds are increasingly employed, with the proper use of the correct adhesive also to concrete, wood or plastics.

After the corrosion protection lining has been applied, it has to be tested for tightness. This is done with arc inductors which are operated at high voltage.

Hard rubber is also suitable for the manufacture of molded articles, even though these have been replaced by plastics to a large extent. Typical articles are, for example, flow meters and other parts in chemical engineering, smoking pipe tips, high quality combs, and articles for the electronics industry with outstanding electrical insulation properties, as well as sheets, rods etc. as materials which can be machined.

5.5.3.7 Roll Covers

For the production of rubberized rolls, first the usually metallic core has to be prepared. This is done by sand blasting or machining on the lathe and the subsequent application of an adhesive layer. After careful drying the rubber layer is applied bubble free. This can be done by hand, is done today, however, by machine by rolling the material onto the surface with the help of special machines. Also, the application of a strip from an extruder onto the rotating roll is carried out. The roll core is covered thicker than its desired dimensions require. Often the rubber layer is then wrapped with wet fabric strips and vulcanized in steam. For very large rolls a considerable time is required before the roll core has been heated. Therefore their vulcanization may take 12 hours or more in steam. Low temperatures are used. After vulcanization, the wrap is removed and the roll ground to its desired diameter on a grinding machine.

5.5.3 Cold Vulcanization

[5.263]

The cold vulcanization can be either carried out according to the *Peachy reaction* under the effect of sulfur dioxide and then hydrogen sulfide or the effect of *sulfur dichloride*. While the first method is without any technical importance, the sulfur chloride vulcanization, which was very popular during the twenties, is still employed to a limited degree, as, for example, for the manufacture of shower caps.

Self Vulcanization [5.304]. When very fast acting accelerators are employed (ultra accelerators), the sulfur vulcanization can proceed under certain circumstances at room temperature (self vulcanization) which is of great importance for the lining of large vessels with soft rubber compounds.

Sulfur Dichloride Vulcanization. The sulfur dichloride vulcanization (see also page 233) can only be used for thin walled articles. The following methods are used:

- Application of sulfur dichloride *vapor*
- Application of sulfur dichloride *solution*.

In order to treat rubber articles with sulfur dichloride *vapor*, the finished plates or the dipped articles manufactured from solution, which do not contain any vulcanization chemicals, are placed into ovens the air within which is as dry as possible, where they come into contact with sulfur chloride which evaporates from a lead or iron pan.

The application of sulfur dichloride *solutions* in benzene with the help of rollers was applied to the manufacture of rubberized fabrics. A contact of the rubber compound with the sulfur dichloride is sufficient in both cases in order to bring about vulcanization at room temperature. The formerly frequently used carbon disulfide with the help of which a cold vulcanization was also possible for dipped articles by subsequent dipping into sulfur dichloride solution is not used anymore.

For *neutralization* of any traces of hydrochloric acid which might have formed during the sulfur dichloride vulcanization through reactions between the sulfur dichloride and the elastomer, MgO is added to the compounds before the vulcanization. Furthermore, a good, non-staining phenolic antioxidant is generally required. Other substances are not required for the design of the compound.

Often the finished articles are dipped into ammonia, in order to neutralize the formed acidic compounds. Even though, the *aging characteristics* and thus the durability of sulfur dichloride vulcanizates is worse than those of hot vulcanized articles, while the surfaces of the finished products have an especially brilliant appearance.

5.6 Finishing Rubber Articles

[5.292]

After vulcanization many rubber articles are not ready to be sold yet. Many gaskets are, for example, stamped from flat sheets or are cut using templates or have to be varnished. Finishing of molded articles has already been discussed in Section 4.9.2.3, page 319f.

5.7 References on Processing of Rubber*

(see also [1.1-1.4, 1.6, 1.9, 1.18, 1.21, 1.28, 2.8, 3.4, 3.15-3.17, 3.19])

5.7.1 General References

[5.1] *Alfrey, T.:* Mechanical Behaviour of High Polymers. Interscience Publ., New York, 1948.

[5.2] *Barron, H.:* Modern Synthetic Rubber. Chapman u. Hall, London, 1949.

[5.3] *Berry, J. P.:* Advances in Rubber Processing. IRC 85, Oct. 15-18, 1985, Kyoto, Proc. p. 9.

[5.3 a] *Berry, J. P.:* The Future of Rubber Processing. KGK *39* (1986), pp. 199-201.

[5.3 b] *Bristow, J. D.:* Computer Integrated Manufacturing. Rubbercon '87, June 1-5, 1982, Harrogate, GB, Proceed, B 51 and P 14.

[5.4] *Brown, J.:* Developments in Rubber and Plastics Machinery. Rubber and Plastics Age *37* (1956), p. 400.

[5.5] *Byam, J. D., Souffie, R. D., Ziegel, K. D.:* Energiebedarf bei der Verarbeitung von Polymerwerkstoffen. GAK *34* (1981), p. 724.

[5.6] *Eichstädt, H.:* Maschinenkunde. Fachbuchverlag GmbH, Leipzig, 1953.

[5.7] *Heiss, K. G., Allalouf, J. W., Schmid, H. M.:* Technological Benefits Through Process Control. Paper 11, 128th ACS-Conf., Rubber-Div., Oct. 1-4, 1985, Cleveland, OH.

[5.8] *Lang, M.:* Die mischungstechnischen Grundlagen der Kunststoff- und Gummi-Industrie. C. Marhold-Verlag, Halle, 1950.

[5.9] *Lehnen, J. F.:* Maschinenanlagen und Verfahrenstechnik in der Gummi-Industrie. Verlag Berliner Union, Stuttgart, 1968.

[5.9 a] *Luscalu, R.:* Einsatz von Leitrechnersystemen in Verarbeitungsanlagen der Kautschukindustrie. KGK *39* (1986), pp. 823-829.

[5.10] *Menges, G., Wortberg, J., Geisbüsch, P., Targiel, G.:* Fließverhalten von Kautschukmischungen. KGK *34* (1981), p. 631.

[5.10 a] *Menges, G., Sommer, F.:* Elastomerverarbeitung heute und morgen. KGK *40* (1987), pp. 1160-1167.

[5.10 b] *Mierisch, V.:* Rechnergestützte Gummiwarenfertigung. KGK *39* (1986), pp. 323-326.

[5.11] *Moakes, R. C. W., Wake, W. C.:* Rubber Technology. Butterworth Scient. Publ., London, 1951.

[5.12] *Pearson, J. R. A.:* Mechanisms of Polymer Processing. Applied Polymer Science Publ. 1985.

[5.13] *Rapetak, W. A.:* Moderne Entwicklungen bei der Verarbeitung von synthetischem Kautschuk. KGK *35* (1982), p. 671.

[5.14] *Seyderhelm, W., Frey, J.:* Zusammenfassende Darstellung der Rohlingsvorbereitung in der Gummi-Industrie. KGK *28* (1975), p. 335.

[5.15] *Springer, A.:* Werkstoffkunde. Fachbuchverlag GmbH, Leipzig, 1952.

[5.15 a] *Werner, H.:* High Tech in der Kautschukindustrie. KGK *41* (1988), pp. 29-32.

[5.16] ...: Machinery and Equipments for Rubber and Plastics, Vol I. India Rubber World, New York, 1952.

[5.17] ...: Maschinen und Apparate in der Gummi-Industrie. Dtsch. Verlag für die Grundstoff-Industrie, Leipzig, 1970.

---

* Literature on Processing of Latex see [1.16]

5.7.2 References on Mixing Methods

Automation and Process Control

[5.18] *Acquarulo, jr., L.A., Notte, A.J.:* Automatic Control of Polymer Mixing. Paper 61, 123rd ACS-Conf., Rubber-Div., May 10-12, 1983, Toronto, Ont.

[5.19] *Brautley, jr., H.L.:* Control of Laboratory Mixing by Use of a Power Integrator, Paper 21, 121st ACS-Conf., Rubber-Div., May 4-7, 1982, Philadelphia, PA.

[5.19a] *Brichta, A.M.:* Energy Balance on and Control of a Rubber Mill. KGK *39* (1986), pp.319-322.

[5.19b] *Ebers, F.:* EDV-unterstützte Herstellung von Kautschukmischungen. DKG-Tagung Bezirksgruppe Hamburg/Schleswig-Holstein, West Germany, Feb.18, 1988.

[5.19c] *Giffin, H.D.:* Mixer Control - the Basis for a more Efficient Rubber Industry. Rubbercon '87, June 1-5, 1987, Harrogate, GB, Proceed, B 6.

[5.20] *Göhler, G.:* Steuerung von Großmischsälen mit Prozeßrechnern und Mikroprozessoren. KGK *34* (1981), p.206.

[5.20a] *Göhler, D.:* Bümatik-Mikroprozessorsystem für eine Kautschukmischlinie. KGK *41* (1988), pp.160-163.

[5.20b] *Hardy, H.:* The Use of Fluidizing Techniques in Handling Carbon Blacks and Process Materials. Rubbercon '87, June 1-5, 1987, Harrogate, GB, Proceed. B 5.

[5.20c] *Hartmann, O.:* Manuelle und automatische Kleinkomponentenverwiegesysteme in On- und Off-Line-System. dkt '88, July 4-7, 1988, Nürnberg, West Germany.

[5.20d] *Hoppe, H.:* New Technical Tendencies in the Handling of Bulk Materials. Rubbercon '87, June 1-5, 1987, Harrogate, GB, Proceed. B2.

[5.21] *Mueller, D., Sorgatz, V.:* Automation of Rubber Mixing Plants. Paper 25, 124th ACS-Conf., Rubber-Div., Oct.25-28, 1983, Houston, TX.

[5.22] *Scheffler, U.:* The Step by Step Introduction of Automation in the Rubber Industry. Rubber World, July 1985.

[5.23] *Schmid, H.-M., Kraus, R.:* Qualitätskonstanz bei der Mischungsaufbereitung und -Verarbeitung durch Rechnereinsatz. ikt 85, June 24-27, 1985, Stuttgart. KGK *39* (1986), pp.127-130.

[5.23a] *Schmid, H.-M.:* Der wirtschaftliche Mischbetrieb des Gummiformteilherstellers. KGK *40* (1987), pp.820-825.

[5.23b] *Simmonis, R.:* A Comparison of Different Techniques for the Pneumatic conveying of Carbon Black and the Impact of Fluidization of these Techniques. Rubbercon '87, June 1-5, 1987, Harrogate, GB, Proceed. B 3.

[5.24] *Sorgatz, V.:* Experience Gained with Fully Automatic Mixing Rooms. IRC 85, Oct.15-18, 1985, Kyoto, Proc. p.733.

[5.25] *Sorgatz, V.:* Automatisierungstendenzen im Mischsaal. ikt 85, June 24-27, 1985, Stuttgart, West Germany. KGK *39* (1986), pp.410-412.

[5.25a] *Sorgatz, V.:* Experience gained with Fully-Automatic Mixing Rooms. Rubbercon '87, June 1-5, 1987, Harrogate, GB, Proceed. B 1.

[5.25b] *Sorgatz, V.:* A Newly Patented Conveying System for Carbon Black. Rubbercon '87, June 1-5, 1987, Harrogate, GB, Proceed. B 4.

[5.25c] *Sorgatz, V.:* Kleinkomponentenverwiegung. dkt '88, July 4-7, 1988, Nürnberg, FRG.

[5.26] *Weinberg, H.:* Microcomputergesteuerte Wägetechnik in der Kautschuk-Industrie. KGK *34* (1981), p.735.

Mixing in Internal Mixers

[5.27] *Asai, T.:* High Speed Mixing for High Productivity/Quality. IRC 85, Oct.15-18, 1985, Kyoto, Proc. p.715.

[5.28] *Ellwood, H.:* Neue Entwicklungen in der Konstruktion, der Fertigung und in der Anwendung von Banbury-Innenmischern. ikt 85, June 24-27, 1985, Stuttgart, West Germany.

[5.28a] *Ellwood, H.:* New Developments in the Design, Manufacture and Operation of Banbury Mixers. KGK *39* (1986), pp.711-714.

[5.29] *Funt, J.M.:* Rubber Mixing. RCT *53* (1980), p.772, Review.

[5.30] *Freakley, P.K., Patel, S.R.:* Internal Mixing: A Practical Investigation of the Flow and Temperature Profiles During the Mixing Cycle. RCT *58* (1985), p.751.

[5.30a] *Freakly, P.K.:* Batch Condition Control for Internal Mixers. Rubbercon '82, June 1-5, 1987, Harrogate, GB, Proceed. A 39.

[5.31] *Griffin, H.:* Rubber Processing Machines in the Eighties, Influence of Energy Cost. SRC 83, May 19-20, 1983, Bergen, Norway.

[5.32] *Hold, P.:* Das Mischen von Polymeren - Ein Überblick. KGK *34* (1981), p.1027.

[5.33] *Johnson, P.S.:* A Review of Developments in the Application of Science to Rubber Mixing and Milling. Paper at the 120th. ACS-Conf., Rubber-Div., Oct. 13-16, 1981, Cleveland, OH.

[5.34] *Lehnen, J.P.:* Kunstst. u. Gummi *3* (1964), pp.65, 132, 183, 257.

[5.35] *Macleod, D.W.:* Der Einsatz von temperiertem Wasser in modernen Banbury-Mischern. GAK *34* (1981), p.286.

[5.36] *Matthews, G.:* Polymer Mixing Technology. Applied Polymer Science Publ., London, 1982.

[5.37] *Manas-Zloczower, I., Tadmor, Z.:* Scale-Up of Internal Mixers. RCT *57* (1984), p.48.

[5.38] *Melotto, M.A.:* The History and State of Art in the Internal Mixer. Paper at the 120th ACS-Conf., Rubber-Div., Oct. 13-16, 1981, Cleveland, OH.

[5.39] *Melotto, M.A.* Rotor Design and Mixing Efficiency. Paper 8, 128th ACS-Conf., Rubber-Div., 1985, Cleveland, OH.

[5.39a] *Melotto, M.A.:* Control and Design Innovations in Modern Banbury Mixers. Rubbercan '87, June 1-5, 1987, Harrogate, GB, Proceed. A 7 E.

[5.39b] *Menges, G., Grajewski, F.:* Process Analysis of a Laboratory Internal Mixer. KGK *40* (1987), pp.467-471.

[5.40] *Min, K., White, J.L.:* Transverse and Longitudinal Flow Visualization of Gum Elastomers and Their Mixing with Carbon Black in an Internal Mixer. Paper 7, 128th ACS-Conf., Rubber-Div., Oct. 1-4, 1985, Cleveland, OH.

[5.41] *Ponce, M.A., Ramirez, R.R.:* Mixing Process of Natural and Synthetic Polyisoprene Rubbers. RCT *54* (1981), p.211.

[5.42] *Schiesser, W.:* Recent Improvements in Efficient Mixing Technology. Paper at the 120th ACS-Conf., Rubber-Div., Oct. 13-16, 1981, Cleveland, OH.

[5.43] *Schmid, H.-M.:* Quality and Productivity Improvements Using the Newly Developed Intermeshing Rotor Mixing System. Paper 39, 124th ACS-Conf., Rubber-Div., Oct. 25-28, 1983, Houston, TX.

[5.44] *Schmid, H.-M.:* Optimieren des Mischprozesses im Innenmischer. KGK *35* (1982), p.674.

[5.44a] *Schmid, H.-M.:* Produktivitätssteigerungen durch Modernisierung tangierender Kneterlinien. dkt '88, July 4-7, 1988, Nürnberg, West Germany.

[5.45] *Werner, H.:* Stand der Verfahrenstechnik und der Maschinenentwicklung bei Misch- und Kalanderanlagen für die Reifenproduktion. KGK *16* (1963), p.42.

[5.46] *Whitaker, P.:* Parameters Affecting Mixing in an Internal Mixer. KGK *34* (1981), p.295.

[5.47] *White, J.L.:* Basic Studies of Rubber Processing: Flow in an Internal Mixer and in Extruders. IRC 85, Oct. 15-18, 1985, Kyoto, Proc. p.62.

[5.48] *Wiedmann, W.M., Schmid, H.-M.:* Optimization of Rubber Mixing in Internal Mixers. RCT *55* (1982), p.363.

[5.49] *Wiedmann, W.M., Schmid, H.-M.:* Optimierung tangierender und ineinandergreifender Rotorgeometrien von Gummiknetern. KGK *34* (1981), p.479.

[5.50] ...: Der Mischbetrieb in der Gummi-Industrie. VDI-Verlag, Düsseldorf, 1984.

Mixing Energy and Influence of Mixing on Dispersion and Other Properties

[5.51] *Basir, K.M., Freakley, P.K.:* The Influence of Internal Mixer Variables an the Flow Properties of a PVC/Nitrile Rubber Compound. KGK *35* (1982), p.202.

[5.52] *Cotten, G.R.:* Mixing of Carbon Black with Rubber, I. Measurement of Dispersion Rate by Changes in Mixing Tarque. RCT *57* (1984), p.118; II. Mechanisms of Carbon Black Incorporation, RCT *58* (1985), p.774.

[5.53] *Lee, M. C. H.:* Effects of Degree of Mixing on the Properties of Filled Elastomers, Paper 62, 123rd ACS-Conf., Rubber-Div., May 10–12, 1983, Toronto, Ont.

[5.54] *Manas-Zloczower, I., Nir, A., Tadmor, Z.:* Dispersive Mixing in Internal Mixers. RCT *55* (1982), p. 1250.

[5.55] *Manas-Zloczower, I., Nir, A., Tadmor, Z.:* Dispersive Mixing in Rubber and Plastics. RCT *57* (1984), p. 583.

[5.56] *Melotto, M.:* Importance of Dispersive Mixing in Preventing Premature Rubber Failure. Paper at the 125th ACS-Conf., Rubber-Div., May 8–11, 1984, Indianapolis, IN.

[5.57] *Nakajima, N.:* Energy Measures of Efficient Mixing. RCT *55* (1982), p. 931.

[5.58] *Nakajima, N., Harrell, E. R., Seil, D. A.:* Energy Balance and Heat Transfer in Mixing of Elastomer Compounds with the Internal Mixer. RCT *55* (1982), p. 456.

[5.59] *Nakajima, N.:* Deformation Behavior and Energy Mixing of Elastomer Carbon Black in Internal Mixtures. Paper at the 128th. ACS-Conf., Rubber-Div., Oct. 1–4, 1985, Cleveland, OH.

[5.59 a] *Den Otter, J. L.:* The Efficiency of Mixing with Carbon Black. IRC '86, June 2–6, 1986, Göteborg, Sweden.

[5.60] *Topcik, B.:* The Energy and Material Advances of Predispersed Additives. Paper at the 120th ACS-Conf., Rubber-Div., Oct. 13–16, 1981, Cleveland, OH.

Mixing and Processing of Powdered Elastomers

[5.61] *Accella, A., Vergnaud, J. M.:* Calculation of the Temperature and Extent of Reaction during the Vulcanization of Powdered Rubber. RCT *56* (1983), p. 689.

[5.62] *Antal, I.:* Powder Rubber. Int. Polym. Sci. Technol. *4* No. 12 (1977), p. T7

[5.63] *Bleyie, P. L.:* Einfluß von Morphologie und Korngröße von Kautschuken in Pulverform bei Verarbeitung auf dem Walzwerk. KGK *27* (1974), p. 336; RCT *48* (1975), p. 254.

[5.64] *Burford, R. P., Pittolo, M.:* Characterization and Performance of Powdered Rubber. RCT *55* (1982), p. 1233.

[5.65] *Evans, C. W.:* Energy Considerations Using Particulate Polymers. Paper 18, 117th ACS-Conf., Rubber-Div., May 20–23, 1980, Las Vegas, NV.

[5.66] *Evans, C. W.:* Powdered and Particulate Rubber Technology. Applied Polymer Science Publ., London, 1978.

[5.67] *Hindmarch, R. S., Gale, G. M., Berry, J. P.:* The Application of the Cavity Transfer Mixer to Rubber Processing. IRC 85, Oct. 15–18, 1985, Kyoto, Proc. p. 721.

[5.68] *Huhn, G.:* Pulverbatchtechnologie, ein Weg in Grenzgebiete der Elastomertechnik. ikt 85, June 24–27, 1985, Stuttgart, West Germany. KGK *39* (1986), pp. 625–628.

[5.69] *Inoue, K., Kuriyama, A., Hayashi, T.:* Development of a New Continuous Mixer and its Application to Mixing Powdered Rubber. IRC 85, Oct. 1985, Kyoto, Proc. p. 727.

[5.70] *Lehnen, J. P.:* Neue Fertigungsverfahren zur Herstellung technischer Gummiwaren aus Pulverkautschuk. KGK *31* (1978), p. 25.

[5.71] *Lehnen, J. P.:* Pulverkautschukverarbeitung, eine Technik für die Zukunft. GAK *35* (1982), p. 253.

[5.71 a] *Menges, G., Weyer, G., Spenser, G.:* Ein Konzept zur Direktverarbeitung von Pulverkautschuk auf Spritzgießmaschinen. KGK *41* (1988), pp. 3–70.

[5.72] *Millauer, C.:* Compounding and Processing of Powder and Crumby Rubber. IRC 79, Oct. 3–6, 1979, Venice, Italy, Prov. p. 710.

[5.73] *Nakajima, N., Kumler, P. R., Harrell, E. R.:* Effect of Pressure and Shear on Compaction of Powdered Rubber with Carbon Black. RCT *58* (1985), p. 392.

[5.74] *Schroeder, H. E.:* Prospects for Powdered, Liquid and Thermoplastic Elastomers. KGK *35* (1982), p. 661.

[5.75] . . .: Powdered and Liquid Elastomers. Bibliography 8, Publ. by RCT.

Continuous Mixing

[5.76] *Allison, K.:* Rubber World *154* No. 5 (1966), p. 67.

[5.77] *Bökmann, A.:* KGK *32* (1979), p. 330.

[5.78] *Capelle, G.:* Mischen im Extruder. KGK *33* (1980), p. 191.

[5.79] *Capelle, G.:* Kontinuierliche Herstellung von Kautschukmischungen. SRC 83, May 19–20, 1983, Bergen, Norway.
[5.80] *Ellwood, H.:* Integrated Mixing and Extrusion for Particulated Rubber Compound. IRC 79, Oct. 3–6, 1979, Venice, Italy, Proc. p. 738.
[5.80a] *Ellwood, H.:* Continuous Processing of Rubber Compounds. dkt '88, July 4–7, 1988, Nürnberg, West Germany.
[5.81] *Evans, C. W.:* Rubber Age *101* No. 9 (1969), p. 61.
[5.82] *Gohlisch, H. J.:* Transfermix für die Verarbeitung von Kautschukmischungen. SGF Arsmöte, June 1, 72. Tylösand, Schweden.
[5.83] *Lehnen, J. P.:* Kunstst. u. Gummi *6* (1967), p. 267.
[5.84] *Luers, W.:* GAK *26* (1975), pp. 248, 410.
[5.85] *Morrell, S. H.:* Progr. Rubb. Technol. *41* (1978), p. 9.
[5.86] *Ponshall, C. P., Saulino, A.:* Rubber World *156* No. 2 (1967), p. 78.
[5.87] ...: Use of Continuous Mixers in the Ruber Industry, Bibliography 4, Publ. by RCT.

Mixing and Processing of Liquid Elastomers (see also [3.108–3.114, 3.297, 5.74, 5.75]).

[5.87a] *Borchert, A. E.:* Poly BD Resin: Structure, properties and Application. Paper 25, 118th ACS-Conf., Rubber-Div., Oct. 7–10, 1980, Detroit, MI.
[5.88] *Coleman, J. F.:* Castable Reinforced Elastomers. Paper 57, 107th ACS-Conf., Rubber-Div., May 1975.
[5.88a] *Dontsov, A. A., Khabarova, E. V., Fedjukin, D. I.:* Crosslinking of Compositions based on Liquid Rubbers with Functional End-Groups and Vulcanized Rubber Properties. Paper 5, 124th ACS-Conf., Rubber-Div., Oct. 25–28, 1983, Houston, TX.
[5.88b] *Musch, R.:* Polyisobutylene Telechelics for Rubber Networks. RTC *58* (1985), p. 45.
[5.88c] *Riew, C. K.:* Amine Terminated Reactive Liquid Polymers-Modification of Thermoset Resins. RCT *54* (1981), p. 374.
[5.89] *Szczesio, M., Slusarski, L.:* Vulkanisation und Eigenschaften flüssiger Kautschuke. KGK *34* (1981), p. 921.
[5.90] *Zajicek, M., Zahradnickova, K.:* Liquid Rubbers. Int. Polym. Sci. Technol. *4* No. 12 (1977), p. T16.

5.7.3 References on Calenders and Calendering (see also [5.45])

[5.91] *Anders, D.:* Roller-Head-Anlagen, Neue Entwicklungen und Einsatzgebiete. KGK *34* (1981), p. 371.
[5.92] *Capelle, G.:* Roller-Head-Anlagen für Innerlining. KGK *37* (1984), p. 212.
[5.93] *Decker, H.:* Kalander für Gummi und thermoplastische Massen. PTH 1943.
[5.94] *Gronstedt, M.:* Automatisierung von Kalanderlinien in der Gummi-Industrie. ikt 85, June 24–27, 1985, Stuttgart, West Germany. KGK *39* (1986), pp. 413–419.
[5.95] *Kopsch, H.:* Kalandertechnik. Carl Hauser Verlag, München, Wien, 1978.
[5.96] *Seidler, E.:* Nachfolgeanlagen für Extruder, Kalander und ähnliche Maschinen. ikt 85, June 24–27, 1985, Stuttgart, West Germany.
[5.97] *Willshaw, H.:* Calenders for Rubber Processing. Ed.: IRI, Lakeman, London, 1956.
[5.97a] ...: Perbunan C für Klebstoffe. Techn. Bulletin of Bayer, May 1, 1963, Tables 2–9, pp. 10–19.

5.7.4 References on Extruders and Extrusion

5.7.4.1 References on Extruders (see also [5.31, 5.76–5.87, 5.91, 5.92])

General References

[5.98] *Baumgarten, W.:* Kaltbeschickung von Kautschuk-Schneckenpressen. KGK *18* (1965), p. 670; Rubber Plast. Age *43* (1962), p. 349.
[5.99] *Capelle, G.:* Maschinen-Neuentwicklungen für die Gummi-Profilherstellung. KGK *34* (1981), p. 744.

[5.100] *Capelle, G., Hunziker, P.F.:* Newest Development and Technology in Multiplex Extrusion Systems. Paper 95, 128th ACS-Conf., Rubber-Div., Oct. 1–4, 1985, Cleveland, OH.
[5.100a] *Capelle, G.:* Kautschukverarbeitungsextruder für die Beschickung von Kalandern. dkt '88, July 4–7, 1988, Nürnberg, West Germany.
[5.100b] *Capelle, G.:* Extrusion und Vulkanisation von Profilen. KGK *40* (1987), pp. 1058–1066.
[5.101] *Fellenberg, K.:* Spezialaufbereitungen von Kautschuk-Mischungen im Vakuum. KGK *18* (1965), p. 665.
[5.101a] *Grajewski, F., Limper, A., Schwenzer, C.:* Von der Mischung zum Profil-Hilfsmittel für die Extrusion von Elastomerprofilen. KGK *39* (1986), pp 1198–1214.
[5.102] *Green, W.:* Energiebedarf verschiedener Extruderkonzepte für die Kautschukindustrie. KGK *35* (1982), p. 405.
[5.102a] *Green, W.K., Anisic, L., Grotkasten, K., Rickmann, G.:* Die Verkettung der Verfahrensschritte Extrusion – Konfektion – Vulkanisation. dkt '88, July 4–7, 1988, Nürnberg, West Germany.
[5.103] *Haney, J.C., Kemper, D.C.:* An Overview of Modern Extrusion Technology. Paper 10, 128th ACS-Conf., Rubber-Div., Oct. 1–4, 1985, Cleveland, OH.
[5.104] *Harms, E.G.:* Kautschuk-Extruder, Aufbau und Einsatz aus verfahrenstechnischer Sicht. Verlag Krauskopf, Mainz, 1974.
[5.105] *Iddon, M.I.:* Extrusion Today and Tomorrow with High Performance Screws. SRC 83, May 19–20, 1983, Bergen, Norway.
[5.105a] *Iddon, M.I.:* The Evolution of Computahose. Rubbercon '87, June 1–5, 1987, Harrogate, GB, Proceed. B7.
[5.106] *Johnson, P.S.:* Development in Extrusion Science and Technology. RCT *56 (1983)*, p. 575, Review.
[5.107] *Kaufmann, A.A., Pinsolle, F.:* Entwicklungen bei Einschneckenextrudern. GAK *35* (1982), p. 328.
[5.108] *Lehnen, J.P.:* Gegenwärtiger Stand und Zukunftsperspektiven der Kautschukextrusion bedeutender Gummiartikel. KGK *37* (1984), p. 25.
[5.108a] *Limper, A.:* Extrudieren von Elastomeren. GAK *39* (1986), pp. 448–453.
[5.108b] *Limper, A.:* Investigations on the Feed Zone of Rubber Extruders. IRC '86, June 2–6, 1986, Göteborg, Sweden.
[5.109] *May, W.:* Das Einwalzenkopf-System EWK, eine neue Technologie in der Kautschukverarbeitung. KGK *37* (1984), p. 505.
[5.109a] *May, W., Ramm, H.F.:* Produktionsführung und Überwachung von Profilanlagen durch Mikroprozessorführung. dkt '88, July 4–7, 1988, Nürnberg, West Germany.
[5.109b] *Menges, G., Grajewski, F., Limper, A., Greve, A.:* Mischteile für Kautschukextruder. KGK *40* (1987), pp. 214–218.
[5.110] *Meyer, P.:* The Development Potential of Different Concepts of Rubber Extruders. Paper 27, 124th ACS-Conf., Rubber-Div., Oct. 25–28, 1983, Houston, TX.
[5.110a] *Morgan, J., Stoten, D.P., Turner, D.M.:* A Teaching Company Programme to Establish Control of a Tread Extrusion Line. Rubbercon '87, June 1–5, 1987, Harrogate, GB, Proceed. A41.
[5.110b] *Rautenbach, R., Triebeneck, K.:* Untersuchungen zum Einzugsbereich von kaltgefütterten Einschnecken-Kautschukextrudern. KGK *39* (1986), pp. 210–215.
[5.111] *Rüger, W.:* Anlage zur Herstellung von Zwei- oder Drei-Komponenten-Laufstreifen und -Seitenstreifen für die Reifenindustrie. KGK *36* (1983), p. 27.
[5.111a] *Schiesser, W.:* Der Kaltfütter-Blend-Extruder. GAK *39* (1986), pp. 676–678.
[5.111b] *Seidler, E.:* Nachfolgeeinrichtungen für Extrusionsanlagen zur Profilherstellung. KGK *39* (1986), pp. 420–423.
[5.112] *Targiel, G.:* Der Extruder, eine Alternative zum Kalander? VDI-Tagung: Extrudieren von Elastomeren, April 23–24, 1986, Göttingen, West Germany.
[5.113] *Tokita, N.:* Laboratory Simulation of a Factory Extrusion Process by the Die Swell Tester. RCT *54* (1981), p. 453.

References on Pin Barrel Extruders (QSM)

[5.114] *Capelle, G.:* Der Stiftzylinderextruder in Versuch und Praxis. GAK *36* (1983), p. 286.

[5.115] *Capelle, G., Hunziker, P. F.:* Performance of Cold Feed Pin Barrel Extruders. Paper 29, 124th ACS-Conf., Rubber-Div., Oct. 25–28, 1983, Houston, TX.

[5.115 a] *Capelle, G.:* Neuentwicklungen bei Stiftzylinder-Extrudern und Multiplex-Extrusionssystemen. KGK *39* (1986), pp. 202–209.

[5.116] *Green, W.:* Einfluß der Betriebsbedingungen der Kautschuk-Extrusion auf die Gestaltung von Stiftextrudern und Extrusionsköpfen. ikt 85, June 1985, Stuttgart, FRG.

[5.117] *Gohlisch, H. J.:* Der QSM-Extruder – Eine Entwicklung für die kautschukverarbeitende Industrie. GAK *32* (1979), p. 744.

[5.118] *Harms, E. G.:* Modellübertragung für Kautschukextruder, Anwendungsfall: Stiftextruder. KGK *36* (1983), p. 470.

[5.118 a] *Laake, H.-J.:* Druck- und Temperaturentwicklung in einem kautschukverarbeitenden Stiftextruder. dkt ' 88, July 4–7, 1988, Nürnberg, West Germany.

[5.119] *Limper, A.:* Optimieren und Auslegen von Kautschukextrudern unter besonderer Berücksichtigung des Stiftextruders. VDI-Tagung: Extrudieren von Elastomeren, April 23–24, 1986, Göttingen, West Germany.

[5.120] *May, W.:* New Philosophics in Design of Pin-Type Extruders and Extrusion Heads. Paper at the 128th ACS-Conf., Rubber-Div., Oct. 1–4, 1985, Cleveland, OH.

References on Shear Head Extruders (see also [5.112, 5.120, 5.128]).

[5.120 a] *Jepsen, C., Räbiger, N.:* Scherkopftechnologie – eine Verfahrensstudie. KGK *40* (1987), pp. 1177–1190.

[5.121] *May, W.:* Stand der Scherkopftechnologie, VDI-Tagung: Extrudieren von Elastomeren. April 23–24, 1986, Göttingen, West Germany.

[5.122] *Niehus, G.:* Shear Head Technology – A Presentation of the Possibilities and Applications of a New Krupp Development for Better Extrusion. Paper 28, 124th ACS-Conf., Rubber-Div., Oct. 25–28, 1983, Houston, TX.

[5.122 a] *Schröber, W.-D., Bernard, W.:* Praxisorientierte Berechnung von Druckverläufen in Extruderspritzköpfen der Kabel- und Leitungsindustrie. KGK *41* (1988), pp. 169–172.

References on Control and Automation of Extruder Lines (see also [5.96])

[5.123] *Brunner, M.:* Möglichkeiten der In-Line-Produktionsüberwachung unter Verwendung von Lasern. VDI-Tagung: Extrudieren von Elastomeren. April 23–24, 1986, Göttingen, West Germany.

[5.124] *Gronstedt, M.:* Automatisieren einer Extrusionslinie für technische Produkte. VDI-Tagung: Extrudieren von Elastomeren, April 23–24, 1986, Göttingen, FRG.

[5.124 a] *May, W.:* Processor – Controlled Extrusion Lines for the Production of Multicomponent Profiles in Type Manufacturing. IRC '86, June 2–6, 1986, Göteborg, Sweden.

[5.125] *Ramm, H. F.:* Anwendung von Mikroprozessoren bei der Extrusion. ikt 85, June 24–27, 1985, Stuttgart, West Germany.

[5.125 a] *Ramm, A. F.:* Anwendung von Mikroprozessoren in Extrusionsanlagen. KGK *39* (1986), pp. 719–722.

5.7.4.2 References on Extrusion

[5.126] *Evans, C. W.:* Hose Technology. Elsevier Applied Science Publ., London, 1979.

[5.127] *Kristukat, P.:* Beschicken von Extrudern mit Granulat oder Streifen. VDI-Tagung: Extrudieren von Elastomeren, April 23–24, 1986, Göttingen, West Germany.

[5.128] *Kroksness, F., Anisic, L.:* Kontinuierliche Herstellung von armierten Kautschukprodukten mit Hilfe der Scherkopftechnologie. KGK *37* (1984), p. 780.

[5.129] *Rohde, E.:* Anforderungen an das Verarbeitungsverhalten und Aufbauprinzipien von Extrusionsmischungen. VDI-Tagung: Extrudieren von Elastomeren, April 23–24, 1986, Göttingen, West Germany.

[5.130] *Schiesser, W.H.:* Die Extrusion von Profilen und Schläuchen aus Siliconkautschuk. GAK *35* (1982), p.122.
[5.131] *Schiesser, W.H.:* Kaltfutterextruder können Reifenproduktion verbessern. GAK *35* (1982), p.334.
[5.132] *Stevens, M.S.:* Extruder Principles and Operation. Elsevier Applied Science Publ., London, 1985.
[5.133] ...: Zur Verarbeitung kaltgefütterter Kautschukmischungen auf Extrudern. KGK *32* (1979), p.987.

References on Flow Behavior of Rubber Compounds in Extruders, Design of Extruders

[5.134] *Kaunabiram, R.:* Application of Flow Behavior to Design of Rubber Extrusion Dies. Paper 15, 128th ACS-Conf., Rubber-Div., Oct. 1-4, 1985, Cleveland, OH.
[5.135] *Ma, C.-Y., White, J.L., Weissert, F.C., Isayer, A.I., Nakajima, N., Min, K.:* Flow Pattern in Elastomers and their Carbon Black Compounds During Extrusions Through Dies. RCT *58* (1985), p.815.
[5.136] *Menges, G., Targiel, G.:* Prozeßanalyse an einem Kautschukextruder. KGK *35* (1982), p.733.
[5.137] *Menges, G., Limper, A., Neumann, W.:* Rheologische Funktionen für das Auslegen von Kautschuk-Extrusionswerkzeugen. KGK *36* (1983), p.684.
[5.138] *Menges, G., Limper, A.:* Die Modelltheorie, ein wertvolles Hilfsmittel für den Kautschukextruder. GAK *37* (1984), p.430.
[5.139] *Menges, G., Limper, A., Grajewski, F.:* Ein Prozeßmodell für die Extrusion von Kautschuken. KGK *37* (1984), p.314.
[5.140] *Menges, G., Limper, A., Weyer, G., Benfer, W.:* Extrusion. KGK *38* (1985), p.382.
[5.141] *Michaelis, W.:* Auslegen von Kautschuk-Extrudierwerkzeugen. VDI-Tagung: Extrudieren von Elastomeren, April 23-24, 1986, Göttingen, West Germany.
[5.142] *Targiel, G., Menges, G.:* Thermodynamic Analysis of a Rubber Extruder. Paper 47, 120th ACS-Conf., Rubber-Div., Oct. 13-16, 1981, Cleveland, OH.
[5.143] *White, J.L., Ma, C.-Y., Brzokowski, R., Weissert, F.C., Isayer, A.I., Nakajima, N., Min, K.:* Flow Visualization and Flow Mechanisms of Rubber Compounds in the Screw and Die Regions of an Extruder. Paper 6, 128th ACS-Conf., Rubber-Div., Oct. 1-4, 1985, Cleveland, OH.

5.7.4.3 References on Continuous Vulcanization

General References

[5.144] *Crowther, B.G., Morrell, S.H.:* Continous Production. Progr. Rubb. Technol. *36* (1972), p.37.
[5.145] *Evans, C.W.:* Continuous Vulcanization in Europe-Present and Future. Rubber Age *103* (1971), p.53.
[5.146] *Gardiner, R.A.:* Energy Considerations in the Continuous Vulcanization of Elastomer Products. Paper 20, 117th ACS-Conf., Rubber-Div., May 1980, Las Vegas, NV.
[5.147] *Hindmarch, R.S., Gale, G.M.:* Silane Moisture-Curing - A One-shot Extrusion Process Using the Cavity Transfer Mixer. Paper 25, 128th ACS-Conf., Rubber-Div., Oct. 1-4, 1985, Cleveland, OH.
[5.148] *Krieger, B.:* New Curing Technology Enhances Rubber Processing. Paper 24, 128th ACS-Conf., Rubber-Div., Oct. 1-4, 1985, Cleveland, OH.
[5.149] *Luers, W.:* Die kontinuierliche Vulkanisation von Gummierzeugnissen. GAK *26* (1973), pp.628, 732, 858, 870.
[5.150] *Manus, J.Mc:* Continuous Vulcanization - A Practical Appricrisal of Existing Methods. J. IRI *5* (1971), p.109; GAK *25* (1972), pp.798, 835.
[5.151] *Manus, J.Mc:* Continuous Atmospheric Cure of Dual Durometer in Fluidized Bed. Paper 21, 124th ACS-Conf., Rubber-Div., Oct. 25-28, 1983, Houston, TX.
[5.152] *Mai, W.:* Neueste Entwicklungen auf dem Gebiet der kontinuierlichen Vulkanisationssysteme für die Erzeugung von Kautschukprofilen und -Schläuchen. SRC 83, May 19-20, 1983, Bergen, Norway.

[5.153] *Sommer, F.:* Verfahren der kontinuierlichen Vulkanisation. VDI-Tagung: Extrudieren von Elastomeren, April 23–24, 1986, Göttingen, West Germany.
[5.154] *Stockman, C.H.:* Some Factors Affecting Quality in Continuous Manufacture of Rubber Hose. Paper 36, 128th ACS-Conf., Rubber-Div., Oct. 1985, Cleveland, OH.
[5.155] ...: Continuous Curing. Rubber J. *153* (1971), p. 38.
[5.156] ...: Use of Continuous Vulcanization Processes for Rubber. Bibliography 7 of ACS, Publ. by RCT.

References on UHF Vulcanization and Pre-Heating

[5.157] *Anders, D.:* Continuous Cure by Microwave. Rubber J. *152* No. 3 (1970), p. 19.
[5.158] *Boonstra, B.B.:* Dielectric Heating. Rubber Age *103* No. 4 (1970), p. 49.
[5.158 a] *Camnerin, C., Terselius, B.:* Assessment of Microware Heating Characteristics of Rubber Compounds Using Wave-Guide Technique. KGK *40* (1987), pp. 210–213.
[5.159] *Chabinsky, I.J.:* Practical Applications of Microwave Energy in the Rubber Industry. Paper 22, 122nd ACS-Conf., Rubber-Div., Oct. 5–7, 1982, Chicago, MI.
[5.160] *Chabinsky, I.J.:* Tire Curing – Microwave Preheating and Alternations – Past, Present and Future. Paper 53, 128th ACS-Conf., Rubber-Div., Oct. 1–4, 1985, Cleveland, OH.
[5.161] *Fecht:* Ein Jahrzehnt UHF-Anlagen in der Gummiprofilherstellung. GAK *32* (1979), p. 622.
[5.162] *Frère, G.:* Entwicklungstendenzen bei der kontinuierlichen UHF-Vulkanisation. VDI-Tagung: Extrudieren von Elastomeren, April 23–24, 1986, Göttingen, West Germany.
[5.162 a] *Frère, G.:* Modernes Produktionssystem zur Vulkanisation von Profilen. dkt '88, July 4–7, 1988, Nürnberg, West Germany.
[5.163] *Gaccione, V., Braus, H., Mitchell, J.M.:* Economic Considerations of Microwave Continuous Vulcanization. Paper 23, 124th ACS-Conf., Rubber-Div., Oct. 25–28, 1983, Houston, TX.
[5.164] *Gohlisch, H.J.:* Salt Bath and UHF-Methods. Rubber Age *103* (1971), p. 49.
[5.165] *Gregor, H.:* Neuentwicklungen an Mikrowellenanlagen für die kontinuierliche Vulkanisation extrudierter Kautschukprofile. GAK *32* (1979), p. 732.
[5.166] *Hughes, B.G., et al.:* Rubber World *147* No. 1 (1962), p. 82; *149* No. 2 (1963), p. 57.
[5.167] *Ippen, G.:* Additives Enables Microwave Curing of Most Elastomers. Chem. Engng. News *48* No. 46 (1970), p. 34 SGF-Arsmöte 72, June 1, 1972, Tylösand, Sweden.
[5.168] *Krieger, B., Allen, R.D.:* Improvement in Microwave-Technology. Paper 7, 118th ACS-Conf., Rubber-Div., Oct. 7–10, 1980, Detroit, MI.
[5.169] *Luxpaert, P.:* A New Look at Continuous Vulcanization: The Ultra Hertz-System (UHS). Paper 24, 124th ACS-Conf., Rubber-Div., Oct. 25–28, 1983, Houston, TX.
[5.170] *Naumann, G.:* Prüfung der dielektrischen Erwärmbarkeit von Gummi- und Kunststoffprodukten. GAK *35* (1982), p. 513.
[5.171] *Oettner, C.:* Behaviour of Elastomers and their Blends in a High Frequency Field. Rev. Gén. Caoutch. *46* (1969), p. 973.
[5.172] *Probst, N., Iker, J.:* Characterization of the Effects of Carbon Black on Dielectric Parameters and their Microwave Curing of Rubbers. KGK *37* (1984), p. 385.
[5.173] *Tedesko, P., Hausman, J.M., Duch, J.:* The Uses of Microwave Energy in Curing of Rubber Products. Paper 114, 128th ACS-Conf., Rubber-Div., Oct. 1–4, 1985, Cleveland, OH.
[5.174] *Wang, C.S.:* Processing Parameters of Continuous Microwave Heating of Ethylene Propylene Terpolymer. Paper 13, 124th ACS-Conf., Rubber-Div., Oct. 25–28, 1983, Houston, TX.

References on Radiation Crosslinking (see also [1.13, 1.14])

[5.175] *Becker, R.C.:* Electron Beam Irradiation. Paper 54, 128th ACS-Conf., Rubber-Div., Oct. 1–4, 1985, Cleveland, OH.
[5.176] *Böhm, G.G.A., Taffkrem, J.O.:* Radiation Chemistry of Elastomers and its Industrial Applications. RCT *55* (1982), p. 575, Review.

[5.177] *Grossmann, R. F.:* Compounding for Radiation Crosslinking. Radiation Phys. Chem. *9* (1977), p. 659.

[5.178] *Heinze, H. D., Hofmann, E. G.:* Anwendung, Wirtschaftlichkeit und Auswahl von Strahlenquellen. Kerntechnik *3* (1961), p. 475.

[5.179] *Hofmann, E. G.:* Industrielle Bestrahlungstechnik – Stand und Zukunft einer neuen Technologie. Bulletin of AEG-Telefunken, Wedel, West Germany, 1980.

[5.180] *Hollain, G. de:* Aspects of Long Term Properties of Irradiated Polymers. J. Ind. Irr. Technol. *1* No. 1 (1983), p. 89.

[5.181] *Mohammed, S. A. H., Timar, J., Walker, J.:* Green Strength Development by Electron Beam Irradiation of Halobutyl Rubber. Paper 47, 121st ACS-Conf., Rubber-Div., May 4–7, 1982, Philadelphia, PA.

[5.182] *Mohammed, S. A. H., Walker, J.:* Applications of Electron Beam Radiation Technology in Tire Manufacturing. Paper 55, 128th ACS-Conf., Rubber-Div., Oct. 1–4, 1985, Cleveland, OH.

[5.183] *Morgenstern, K. H.:* Radiation Vulcanization. Rubber Age *103* (1971), p. 49.

[5.184] *Morgenstern, K. H.:* Radiations Time Has Arrived for Plastics and Rubber. Paper FC-540, Int. Konf. Strahlenvernetzung, SME, 1974, Atlanta, GA.

[5.185] *Morgenstern, K. H., Becker, R. G.:* The Technology and Economics of Radiation Cure. Paper at the 107th ACS-Conf., Rubber-Div., May 6–9, 1975, Cleveland, OH.

[5.186] *Sonnenberg, A. M.:* Electron Beam Vulcanization of Elastomers. KGK *37* (1984), p. 864.

[5.187] *Spenadel, L.:* Radiation Crosslinking of Polymer Blends. Paper 2, 2nd Conf. on Radiation Processing. 1978, Miami, FL.

[5.188] *Wiegand, G.:* Strahlenvernetzung von Kabelmänteln und -Isolierungen. GAK *38* (1985), p. 364.

References on Lead Sheath Vulcanization (see also [5.126])

[5.189] *Czerny, H., Fahrner, F.:* Schläuche mit Einlage. KGK *16* (1963), p. 93.

5.7.5 References on Manufacturing of Molded Articles

5.7.5.1 References on Compression Molding

[5.190] *Devek, H., Menges, G.:* Verfahrenskontrolle und Automatisierung beim Warmpreßverfahren. GAK *35* (1982), p. 335.

[5.191] *Eule, W.:* Vor- und Nachteile bei der Anwendung des Kompressions-, Transfer- und Injections-Molding-Verfahrens. KGK *31* (1978), p. 637.

[5.192] *Franke, A., Hafner, K., Kern, W. F.:* Untersuchungen zur Aktivierungsenergie der Vulkanisation. KGK *13* (1960), p. WT 392.

[5.193] *Isayer, A. I., Azari, A. D.:* Viscoelastic Effect in Compression Molding of Elastomers; Shear-Free Squeesing Flow. Paper 14, 128th ACS-Conf., Rubber-Div., Oct. 1–4, 1985, Cleveland, OH.

[5.194] *Isayer, A. I., Kochar, L. S.:* Properties of Rubber Vulcanized under High Pressure. Paper 35, 128th ACS-Conf., Rubber-Div., Oct. 1–4, 1985, Cleveland, OH.

[5.195] *Jarkowski, B., Kubis, J.:* The Determination of the Kinetic Constants of the Crosslinking Reaction of Rubber Compounds. KGK *38* (1985), p. 515.

[5.196] *Hempel, J.:* GAK *28* (1975), p. 792.

[5.197] *Ludwig, H.-J.:* Werkzeugwerkstoffe, ihre Oberflächenbehandlung, Verschmutzung und Reinigung. GAK *35* (1982), p. 72.

[5.198] *Matula, S.:* Hydraulic Curing Press with Sequence Timer. Paper 56, 128th ACS-Conf., Rubber-Div., Oct. 1–4, 1985, Cleveland, OH.

[5.199] *Morrisson, B.:* Rev. Gén. Caoutch. *49* (1972), p. 155.

[5.200] *Raies, N. D.:* Important Factors in Mould Design for Compression, Transfer and Injection Molding of Rubbers. Paper 22, 128th ACS-Conf., Rubber-Div., Oct. 1–4, 1985, Cleveland, OH.

[5.201] *Sommer, J. G.:* Moulding of Rubber for High-Performance Applications. RCT *58* (1985), p. 662, Review.
[5.202] *Weir, Th.:* Rubber World *169* No. 2 (1973/1974), p. 52.
[5.203] *Werner, J.:* Rubber Age *103* No. 7 (1971), p. 48.

5.7.5.2 References on Transfer Molding (see also [5.191, 5.192, 5.195, 5.197, 5.200, 5.201])

[5.203 a] *Cottancin, G.: Spritztransfer für die Fertigung von Gummi/Metall-Artikeln. GAK 40* (1987), pp. 124–134.
[5.204] *Jentzsch, J., Michael H., Rümmler, L., Babatz, D.:* Compression Injection Moulding of Rubber Mixtures. Plaste u. Kautsch. *41* (1984), p. 24.
[5.205] *Müller, R. H.:* Das Transfer-Verfahren unter besonderer Berücksichtigung der Formenkonstruktion für die Herstellung austriebfreier Gummi-Formteile. KGK *20* (1967), p. 83.

5.7.5.3 References on Injection Molding (see also [5.191, 5.192, 5.195, 5.197, 5.200, 5.201])

General References on Injection Molding and Compound Design (see also [3.267, 4.548].

[5.206] *Boozer, C. E., Hoek, E. G., Karg, R. F.:* Elastomers for Injection Molding. Paper 57, 120th ACS-Conf., Rubber-Div., Oct. 13–16, 1981, Cleveland, OH.
[5.207] *Byan, M. E., Chang, T. S., Abdalla, S. Z.:* Understanding Injection Mold Filling. Soc. Plast. Engng., Techn. Paper, Reg. Techn. Conf., 1980, p. 116.
[5.208] *Camnerin, C.:* Injection Molding of Rubber-Optimizing of Cure Characteristics. IRC 85, Oct. 15–18, 1985, Kyoto, Proc. p. 745.
[5.209] *Christy, R.:* Generation and Dissipation of Pressure in Rubber Extruders. Paper at the 128th ACS-Conf., Rubber-Div., Oct. 1–4, 1985, Cleveland, OH.
[5.210] *Domininghaus, H.:* Spritzgießen von Elastomeren. Plastverarb. *30* (1979), p. 390.
[5.211] *Eckert, R. C., Du Puis, I. C., Leibu, H. J.:* Injection Molding of CSM Synthetic Rubber. Paper 58, 120th ACS-Conf., Rubber-Div., Oct. 13–16, 1981, Cleveland, OH.
[5.212] *Fink, C.:* Spritzgießen von Kautschuken. Kunstst. *75* (1985), p. 654.
[5.212 a] *Gastrow, H.:* Der Spritzgießwerkzeugbau, Chapter 6: Werkzeuge für Duroplaste und Elastomere. Carl Hanser Verlag, München Wien, 1982, pp. 205–215.
[5.213] *Grossmann, R. F.:* Direct Injection Molding of Rubber Dry-Blends. Paper 93, 127th ACS-Conf., Rubber-Div., April 23–26, 1985, Los Angeles, CA.
[5.213 a] *Härtel, V., Steinmetz, C.:* Instationäre Wärmeströmung in dickwandigen Gummiartikeln. KGK *38* (1985), pp. 34–38.
[5.213 b] *Hofmann, W.:* Werkzeuge für das Kautschuk-Spritzgießen – Lagebericht. Kunststoffe *77* (1987), pp. 1211–1226.
[5.213 c] *Holm, D.:* Aufbau von Werkzeugen für Spritzgießmaschinen. In: Spritzgießen von Elastomeren. VDI-Verlag, Düsseldorf, 1978, pp. 63–79.
[5.213 d] *Janke, W.:* Rechnergestütztes Spritzgießen von Elastomeren. Diss. RWTH Aachen, 1985.
[5.214] *Jentzsch, J.:* Verarbeitung von Elastomeren, Spritzgießen von Mischungen. Plaste u. Kautsch. *24* (1977), p. 49.
[5.215] *Johnson, P. S.:* Processibility Testing and Injection Molding. Paper 56, 120th ACS-Conf., Rubber-Div., Oct. 13–16, 1981, Cleveland, OH.
[5.216] *Karg, R. F., Boozer, C. E., Renefield, R. E.:* Injection Moulding of Elastomers. Rubber World 1985, p. 14.
[5.217] *Larsen, L. C., Klingensmith, W. H., Danilowicz, P. A.:* Processing Agents to Improve Injection Molding. Paper 60, 120th ACS-Conf., Rubber-Div., Oct. 13–16, 1981, Cleveland, OH.
[5.218] *Paris, W. W., Dillhoefer, J. R., Woods, W. C.:* Acceterator Systems for Injection Molding. RCT *55* (1982), p. 494.
[5.219] *Penn, W. S.:* Injection Moulding of Elastomers. Elsevier Applied Science Publ., London, 1969.
[5.220] *van Pul, J.:* Some Reflections on the Preparations of Rubber Compounds for Injection Moulding. SGF-Publ. 1981, pp. 3–32.

[5.221] *Schlueter, R. H.:* Injection Molding - Review of Nontraditional Techniques and Concepts. Paper 63, 120th ACS-Conference, Rubber-Div., Oct. 13-16, 1981, Cleveland, OH.

[5.222] *Townson, G.:* Injection Molding - Where it is Going. Paper 62, 120th ACS-Conf., Rubber-Div., Oct. 13-16, 1981, Cleveland, OH.

[5.223] *Wheelan, A., Craft, J. L.:* Developments in Injection Moulding. Elsevier Applied Science Publ., London, 1.) 1978. 2.) 1981.

[5.224] *Wheelan, A.:* Injection Moulding Materials. Elsevier Applied Science Publ., London, 1982.

[5.225] *Wolff, T. J.:* Rubber Injection Molding: Equipment, Tooling, Process, Soc. Plast. Engng., Techn. Paper *25* (1979), p. 103.

[5.226] . . .: Spritzgießen von Elastomeren. Ed.: WdK, Frankfurt/M., Grünes Buch No. 25, 1966. No. 26, 1967.

[5.227] . . .: Injection Moulding: Procedure, Procederal Parameters, Process. Carl Hanser-Verlag, München, Wien, 1979.

[5.228] . . .: Spritzgießen von Elastomeren. VDI-Verlag, Düsseldorf, 1978.

[5.229] . . .: Spritzgießen technischer Gummi-Formteile. VDI-Verlag, Düsseldorf, 1981.

Process Technology of Injection Molding

[5.229 a] *Barth, P., Benfer, W., Fischbach, G., Schneider, W., Weyer, G.:* Moderne Fertigung von Gummiformteilen. GAK *39* (1986), pp. 540-547; 669-675; *40* (1987), pp. 14-19.

[5.229 b] *Barth, P., Benfer, W.:* Automation in Rubber Injection Moulding. Rubbercon '87, June 1-5, 1987, Harrogate, GB, Proceed. B 10.

[5.229 c] *Bowers, S., Dickin, P., Simpson, R.:* A New Computer-Aided Design System for Rubber Moulders. KGK *40* (1987), pp. 953-957.

[5.230] *Byam, J. D., Colbert, G. P.:* Applying Science to Processing - Profitability Injection Molding. Paper 21, 112th ACS-Conf., Rubber-Div., Oct. 1977, Chicago, MI.

[5.231] *Byam, J. D., Colbert, G. P., Hagman, J. F.:* Designing Injection Molding Systems for Ethylene/Acrylic Elastomer. Elastomerics *113* (1981), pp. 26, 40.

[5.231 a] *Carcano, L.:* Automatic Moulding on a Four-Nozzle Machine. IRC '86, June 2-6, 1986, Göteborg, Sweden.

[5.232] *Coscia, G. A., Ceppino, L.:* Injection Molding Machines for Rubber Utilizing Technologies for Maximum Saving in Raw Materials. KGK *34* (1981), p. 565.

[5.233] *Graf, H. J.:* High Quality Articles in Injection Molding by Means of Process and Mold Design. Paper 12, 128th ACS-Conf., Rubber-Div., Oct. 1985, Cleveland, OH.

[5.234] *Graf, H. J.:* Verfahren und Werkzeugauslegung beim Spritzgießen dünnwandiger Elastomerartikel. Kunststoffberater *7/8* (1985), p. 31.

[5.235] *Graf, H. J., Gierschewski, F.:* Qualitätserzeugung von Artikeln im IM-Verfahren mittels genauer Prozeßführung und Werkzeugauslegung. ikt 85, June 24-27, 1985, Stuttgart, West Germany.

[5.235 a] *Graf, H.-J., Gierschewski, F.:* Ein modernes Spritzgießverfahren zur vollautomatischen Herstellung unvernetzter Formteile. KGK *39* (1986), pp. 1094-1097.

[5.235 b] *Graf, H.-J., Gierschewski, F.:* Qualitätserzeugung von Artikeln im IM-Verfahren mittels genauer Prozeßführung und Werkzeugauslegung. KGK *39* (1986), pp. 524-527.

[5.235 c] *Graf, H.-J., Richter, K. P.:* Injection Moulding Process for Plastic-Rubber Applications. Rubbercon '87, June 1-5, 1987, Harrogate, GB, Proceed. B 12.

[5.235 d] *Graf, H.-J.:* Austriebsfreie Herstellung von Gummiformteilen auf Spritzgießmaschinen. KGK *40* (1987), pp. 829-836.

[5.235 e] *Graf, H.-J.:* Präzisionsspritzgießen durch Zuhaltung, Werkzeugkonzeption und Prozeßregelung. dkt '88, July 4-7, 1988, Nürnberg, West Germany.

[5.236] *Hull, J. L., Exner, W. E.:* Latest Advances in Injection Molding of Rubber. Paper 13, 128th ACS-Conf., Rubber-Div., Oct. 1-4, 1985, Cleveland, OH.

[5.237] *Kliever, L. B., Cornell, W. H., Simpson, B. D.:* Effects of Machine and Compounding Variables on the Injection Molding Characteristics of Styrene Butadiene Thermoplastic Elastomers. Paper 61, 120th ACS-Conf., Rubber-Div., Oct. 13-16, 1981, Cleveland OH.

[5.237 a] *Lampl, A.:* Automatisches Spritzgießen von Elastomerformteilen. KGK *41* (1988), pp. 71–75.

[5.237 b] *Lampl, A.:* Wirtschaftliche Fertigung kleiner Serien von Gummiformteilen im Spritzgießverfahren. dkt '88, July 4–7, 1988, Nürnberg, West Germany.

[5.238] *Menges, G., Buschhaus, F.:* Spritzgießen von Elastomeren, Herstellung gratfreier Formartikel aus Elastomeren. KGK *32* (1979), p. 869.

[5.239] *Menges, G., Benfer, W.:* Formenverschmutzung beim Spritzgießen von Elastomeren. GAK *36* (1983), p. 161.

[5.240] *Menges, G., Haack, W., Benfer, W.:* Der Weg vom Kautschuk zum Elastomerprodukt, Rechenmodelle ergänzen die Erfahrung. GAK *38* (1985), pp. 13, 53, 100, 222.

[5.240 a] *Menges, G., Barth, P.:* Automatisierung beim Kautschukspritzgießen. KGK *39* (1986), pp. 43–46.

[5.240 b] *Menges, G., Sercer, M., Wölfel, U.:* Zur Angußauslegung bei Elastomerformteilen. KGK *40* (1987), pp. 139–142.

[5.240 c] *Sercer, M., Wölfel, U., Menges, G.:* Effects of Runner System Design on the Quality of Rubber Injection Molding. Rubbercon '87, June 1–5, 1987, Harrogate, GB, Proceed. B 11.

[5.240 d] *Wilgenbus, A.-S.:* Korrektur der Heizzeiten beim Spritzgießen von Gummi. dkt '88, July 4–7, 1988, Nürnberg, West Germany.

[5.241] *Wheelans, A.:* Mix and Machine Adjustment for Safe and Injection Moulding with Good Vulcanisate Properties. NR-Technol *11* (1980), pp. 11.

[5.242] *Wheelans, A.:* Injection Moulding Machines. Elsevier Applied Science Publ., London, 1984.

[5.243] *Yang, P. H., Chang, W. V., Salovey, R.:* Flow Induced Anisotropy in Injection Moulded O-Rings. 6th Ann. Pac. Tec., Aug. 1981, Los Angeles, CA, Proc. p. 51.

Process Control

[5.244] *Fink, L.:* Automatiken beim Spritzgießen von Elastomeren. KGK *35* (1982), p. 847.

[5.245] *Gissing, K., Lampl, A.:* Monitoring the Injection Moulding Process by Measuring Viscosity-Dependent Characteristic Data. Plastverarb. *34* (1983), p. 427.

[5.246] *Isaac, J. L., Farrer, R. S. H.:* REP Injection Moulding, Press and Microprocessor. IRC 81 (Rubbercon), June 8–12, 1981, Harrogate, GB.

[5.246 a] *Krehwinkel, Th., Schneider, Ch.:* Verarbeitungsfenster für den Elastomerspritzgießprozeß. KGK *41* (1988), pp. 164–168.

[5.246 b] *Lampl, A.:* Moderne Steuerungskonzepte beim Spritzgießen von Gummiformteilen. KGK *39* (1986), pp. 723–728.

[5.247] *Menges, G., Buschhaus, F.:* Prozeßsteuerung beim Spritzgießen von Elastomeren. KGK *35* (1982), p. 202.

[5.248] *Menges, G., Matzke, A., Janke, W.:* Geregelte Spritzgießmaschine fertigt Teile gratfrei in vollautomatischem Prozeß. MM *89* (1983), p. 51.

[5.249] *Menges, G.:* Umfassende Automatisierung beim Kautschukspritzguß. ikt 85, June 24–27, 1985, Stuttgart, West Germany.

[5.249 a] *Menges, G., Benfer, W., Groth, S.:* Cadgum – ein Programm zur Auslegung von Spritzgießwerkzeugen für Elastomere. KGK *40* (1987), pp. 337–342.

[5.249 b] *Pettit, D., Page, J.:* Computer Monitoring of Shoe-Sole Vulcanizing Machines. Rubbercon '87, June 1–5, 1987, Harrogate, GB, Proceed. B 50.

[5.249 c] *Schmid, H.-M.:* Moderne Spritzgießmaschinensteuerung ermöglicht Produktionsanalyse und -überwachung im Hinblick auf statistische Prozeßkontrolle (SPC). dkt '88, July 4–7, 1988, Nürnberg, West Germany.

[5.249 d] *Schneider, Ch.:* Verarbeitungsfenster – eine Methode zur Bewertung der Prozeßsicherheit beim Elastomerspritzgießen. dkt '88, July 4–7, 1988, Nürnberg, FRG.

[5.250] *Souschagrin, B.:* Process Control of Injection Moulding. Polym. Engng. Sci. *23* (1983), p. 431.

[5.251] . . .: Forschung für die Praxis am IKV. GAK *39* (1986), p. 60.

Mold Fouling and Shrinkage (also see [5.239])

[5.252 a] *Braun, D., Bezdadea, E.:* Analyse von Ablagerungen bei der Verarbeitung von Elastomeren. KGK *39* (1986), pp. 191-195.

[5.252] *MacLean, A.:* Mould Fouling - A Literature Survey. RAPRA-Members J. *2* (1974), p. 296.

[5.253] *Beatty, J. R.:* Einfluß des Rezepturaufbaues auf den Schrumpf von Gummiformartikeln. GAK *32* (1979), p. 688.

5.7.5.4 References on Vulcanization Cycles

[5.253 a] *Eckel, R., Grewe, E.:* Einsatz eines Tischrechners zur Überwachung und Steuerung des Ausheizgrades von Gummi-Metall-Achsfedern. dkt '83, June 13-16, 1983, Wiesbaden, West Germany.

[5.253 b] *Eckelmann, W., Reichenbach, D., Sempf, H.:* Über die Abhängigkeit der Eigenschaften von Vulkanisaten von Vulkanisationszeit und -temperatur. KGK *22* (1969), pp. 5-13.

[5.254] *Hands, D., Horsfall, F.:* An Accurate Method of Calculating Cure Cycles for Rubber Products. Paper 18, 120th ACS-Conf., Rubber-Div., Oct. 13-16, 1981, Cleveland, OH.

[5.255] *Hands, D., Horsfall, F.:* Calculation Cure Cycles. Paper 11, 124th ACS-Conf., Rubber-Div., Oct. 25-28, 1983, Houston, TX.

[5.256] *Härtel, V.:* Berechnung der Vulkanisationszeiten dickwandiger Artikel mit einem programmierbaren Kleinrechner. Paper of DKG-Section Rheinland-Westfalen, Münster, West Germany, 1981.

[5.256 a] *Reichenbach, D., Eckelmann, W.:* Zur Konzentrationsabhängigkeit der Vulkanisation. KGK *24* (1971), pp. 443-450.

[5.257] *Vandoren, P.:* Minicomputer Technique for Simulation of Heat Transfer and Cure Level in Thick Rubber Elements. KGK *37* (1984), p. 398.

5.7.5.5 References on Some Special Molded Articles

General References on Vehicle Tires

[5.258] *Asaka, T.:* Structural Mechanics of Radial Tires. RCT *54* (1981), p. 461, Review.

[5.259] *Beatty, J. R., Miksch, B. J.:* Some Effects of Tire Inflation Pressure or Radial Tire Performance. Paper 45, 117th ACS-Conf., Rubber-Div., May 20-23, 1980, Las Vegas, NV.

[5.259 a] *Chabinsky, I. J.:* Tire Curing - Microwave Preheating and Alternatives - Past, Present and Future. KGK *39* (1986), pp. 424-426.

[5.260] *Klingmann, H.:* The Development of the Radial Tire. Paper 14, 121th ACS-Conf., Rubber-Div., May 4-7, 1982, Philadelphia, PA.

[5.260 a] *Rabitsch, E.:* Herstellung und Verarbeitung von Mischungen für Autoreifen. KGK *31* (1978), p. 149.

[5.260 b] *Saito, Y.:* New Polymer Development for Low Rolling Resistane Types. KGK *39* (1986), pp. 30-32.

[5.260 c] *Schiesser, W.:* Wirtschaftliche Reifenfertigung. GAK *39* (1986), pp. 66-71.

[5.261] *Walker, L. A., Helt, W.:* High Temperature Curing of Radial Passenger Tires. Paper 44, 128th ACS-Conf., Rubber-Div., Oct. 1-4, 1985, Cleveland, OH.

References on Tire Rolling Resistance

[5.262] *Browne, A. I., Potts, G. R.:* Tire Power Loss: The Effect of Heat Conduction Through the Wheel. Paper 41, 122nd ACS-Conf., Rubber-Div., Oct. 5-7, 1982, Chicago, MI.

[5.263] *Clark, S. K.:* A Brief History of Rolling Resistance. Paper 29, 122nd ACS-Conf., Rubber-Div., Oct. 5-7, 1982, Chicago, MI.

[5.264] *Keefe, R. I., Koralek, A. S.:* Precision Measurement of Tire Rolling Resistance. Paper 33, 122nd ACS-Conf., Rubber-Dir., Oct. 5-7, 1982, Chicago, MI.

[5.265] *Knight, R. E.:* Trend in Rolling Resistance and Fuel Economy for Advanced Design Radial Highway Truck Tires. Paper 64, 122nd ACS-Conf., Rubber-Div., Oct. 5-7, 1982, Chicago, MI.

[5.266] *Luchini, J. R.:* Rolling Resistance Test Methods. Paper 32, 122nd ACS-Conf., Rubber-Div., Oct. 5-7, 1982, Chicago, MI.

[5.267] *Luchini, J. R., Simonelli, L. A.:* Rolling Resistance Tests on Smooth and Textured Surface Test Wheels. Paper 39, 122nd ACS-Conf., Rubber-Div., Oct. 5-7, 1982, Chicago, MI.

[5.268] *Prevorsek, D. C., Kwon, Y. D.:* Feasibility of A Low Rolling Resistance Tire Suitable High Speed Driving. Paper 44, 117th ACS-Conf., Rubber-Div., May 20-23, 1980, Las Vegas, NV.

[5.269] *Saito, Y.:* New Polymer Development for Rolling Resistance Tires. ikt 85, June 24-27, 1985, Stuttgart, West Germany.

[5.270] *Shackleton, J. S., Chang, L. Y.:* An Overview on Rolling Resistance. Paper 30, 122nd ACS-Conf., Rubber-Div., Oct. 5-7, 1982, Chicago, MI.

[5.271] *Schuring, D. J.:* The Rolling Loss of Pneumatic Tires. RCT *53* (1980), p. 600, Review.

[5.272] *Takao, H., Imai, A.:* Polymer Design for Lower Rolling Resistance Rubbers. IRC 85, Oct. 15-18, 1985, Kyoto, Proc. p. 465.

[5.273] *Warholic, T. C.:* Rolling Resistance Performance of Passenger Tires During Warm-Up (Speed, Load, and Inflation Pressure Effects). Paper 40, 122nd ACS-Conf., Rubber-Div., Oct. 5-7, 1982, Chicago, MI.

[5.274] *Wilder, C. R., Haws, J. R., Middlebrook, T. C.:* Rolling Loss of Tires Using Tread Polymers of Variable Characteristics with Compounding Variations. Paper 81, 124th ACS-Conf., Rubber-Div., Oct. 25-28, 1983, Houston, TX.

[5.275] *Williams, F. R., Dudek, T. J.:* Load-Deflection and its Relationship to Tire Rolling Resistance. Paper 38, 122nd ACS-Conf., Rubber-Div., Oct. 5-7, 1982, Chicago, MI.

References on Conveyor Belts

[5.276] *Gozdiff, M.:* Factors Relating to Vulcanized Splice Reliability for Steel Cable Reinforced Conveyor Belting. Paper 54, 125th ACS-Conf., Rubber-Div., May 8-11, 1984, Indianapolis, IN.

[5.277] *Silvey, D. H.:* Fire Retardant Belting. Paper 53, 125th ACS-Conf., Rubber-Div., May 8-11, 1984, Indianapolis, IN.

[5.278] *Thompson, S. B.:* Kevlar Aramid Fiber Reinforced Conveyor Belting. Paper 52, 125th ACS-Conf., Rubber-Div., May 8-11, 1984, Indianapolis, IN.

[5.279] *Vieweg, H.:* Form- und Spritzartikel. KGK *16* (1963), p. 33.

References on V-Belts (see also [3.317, 3.318])

[5.279 a] *Borchardt, H.:* Herstellung von flankenoffenen Keilriemen. GAK *40* (1987), pp. 543-544.

[5.280] *Hendersen, J. L., Campbell, J. H.:* Polyester Elastomer in Power Transmission Belts - A Review of Current Technology. Paper 81, 125th ACS-Conf., Rubber-Div., May 8-11, 1984, Indianapolis, IN.

[5.281] *Olivier jr., D.:* Power Transmission Belting Industry: Trends - Past, Present, Future. Paper 70, 125th ACS-Conf., Rubber-Div., May 8-11, 1984, Indianapolis, IN.

[5.282] *Schroer, Th. E.:* Improving the Heat Resistance of Power Transmission Belt Formulations. Paper 84, 125th ACS-Conf., Rubber-Div., May 8-11, 1984, Indianapolis, IN.

[5.283] *Stanhope, H. V.:* V-Belt Reinforcement - Polyester the Most Popular Fiber. Paper 82, 125th ACS-Conf., Rubber-Div., May 8-11, 1984, Indianapolis, IN.

[5.284] *Yarnell, L.:* Power Transmission Belting Design and Materials - A Historical Overview. Paper 69, 125th ACS-Conf., Rubber-Div., May 8-11, 1984, Indianapolis, IN.

References on Rubber-Metal Elements [see also [2.136-2.140, 4.629-4.651])

[5.285] *Kurr, K., Ticks, G. H.:* Das dynamisch gedämpfte Gummilager. GAK *38* (1985), p. 148.

[5.286] *Madigowski, W. M.:* Dynamic Properties: Measurement, Modeling and Design Implications. Paper 36, 127th ACS-Conf., Rubber-Div., April 23-26, 1985, Los Angeles, CA.
[5.286 a] *Pohlmann, K.:* Die Luftfederung im Fahrzeugbau. KGK *39* (1986), pp. 220-227.
[5.287] *Rogers, L.:* Vibration Damping Augmentation in Aerospace Structures. Paper 35, 127th ACS-Conf., Rubber-Div., April 23-26, 1985, Los Angeles, CA.

References on Sponge and Cellular Rubber (see also [1.14, 4.571-4.576])

[5.288] *Wilson, A. D.:* Progr. Rubber Technol. *35* (1971), p. 97.

References on Hollow Articles

[5.289] *Reiner, S.:* Die Herstellung von Gummihohlkörpern. Kautschuk u. Gummi *3* (1950), p. 18.

5.7.5.6 References on Deflashing of Molded Articles

[5.290] *Clames, J., Geyer, V.:* Neues Verfahren zur Tieftemperatur-Strahlentgratung mit Flüssig-Stickstoff. GAK *34* (1981), p. 314.
[5.291] *Geyer, V.:* Neue Verfahren zur Tieftemperatur-Strahlentgratung mit Flüssig-Stickstoff. KGK *36* (1983), p. 577.
[5.292] *Hofherr, S.:* Rubber World *172* No. 5 (1975), p. 35.
[5.293] *Köhler, A., Grund, P.:* Maschinelle Entgratung von Gummi- und Kunststoff-Formteilen. KGK *37* (1984), p. 965.
[5.294] *Rebhan, D.:* Verfahren und Anlagen zur Gummi-Kaltentgratung mit Stickstoff. KGK *39* (1986), pp. 47-50.

5.7.6 References on Free Heating (see also [5.144-5.188])

References on Equipment Protection with Hard Rubber Products

[5.295] *Boehnert, J.:* Anwendung von Kautschuk im Oberflächenschutz. KGK *33* (1980), p. 95; see also: *Klement, G.:* KGK *20* (1967), pp. 462-465.
[5.296] *Davies, R. L.:* Ann Rep. Progr. Rubber Technol *17* (1953), p. 143; *22* (1958), p. 97; *23* (1959), p. 115; *24* (1960), p. 129.
[5.297] *Gall, W. M.:* RCT *23* (1950), p. 266.
[5.298] *Ippen, J.:* Oberflächenschutz mit Weichgummi. SGF-Arsmöte, June 5, 1969, Tammerfors, No. 35.
[5.299] *Neumann, L.:* Gummiauskleidungen. In: *Boström, S. (Ed.).:* Kautschuk-Handbuch. Verlag Berliner Union, Stuttgart, Vol. 4, 1961, p. 132.
[5.300] *Peters, H.:* Ind. Engng. Chem. *44* (1952), p. 2344; *49* (1957), p. 1604; *51* (1959), p. 1176.
[5.301] *Sonnemann, E.:* Auskleiden von Apparaten mit Polymerwerkstoffen. GAK *32* (1979), p. 672.
[5.302] *Vennels, W. G.:* The Manufacture and Properties of Ebonite. NR-Technol. 1971 No. 1, p. 2.
[5.303] *Vennels, W. G.:* Rubber and Ebonite Linings for Chemical and Industrial Equipment. NR-Technol. 1971, No. 12, p. 2.

References on Low Temperature Vulcanization (see also [5.175-5.188])

[5.304] ...: Low Temperature Vulcanization of Rubber. Bibliography, Publ.: ACS, Rubber-Div. Library, Akron, OH, 1935-1960.

6 Elastomer Testing and Analysis

6.1 General Comments about Elastomer Testing

[6.1–6.18]

Rubber, because of its variety of properties, has a special position among organic materials: High deformability paired with a high elasticity, are the outstanding properties. Added is an extraordinarily high variability which is caused on the one hand by the many possibilities of compound formulation and on the other hand because of different responses at different excitations. Simple material property constants are thus not sufficient to describe the spectrum of properties. Instead, it is necessary to determine the relation between excitation and response.

The requirements of a product characterization on the one hand and of a behavior description during the raw material processing and the rubber application on the other hand have led to a multitude of testing methods. Among these the formerly often applied single data point tests can lead to wrong conclusions; because of this, one has to vary the temperature and deformation rate for the so-called static tests and the temperature, frequency and amplitude for the dynamic tests.

Physical-technological testing extends to the raw rubber, the intermediate product (the unvulcanized compound) and the vulcanizate.

6.2 Mechanical-Technological Testing

[6.1–6.18]

6.2.1 Testing the Unvulcanized Material

6.2.1.1 Plasticity and Viscosity Test Methods

[6.19–6.79]

Uncrosslinked high molecular materials such as elastomers can be deformed irreversibly. They behave as non-Newtonian liquids of high viscosity. Their flow behavior thus includes in contrast to Newtonian liquids sometimes considerable components of elastic behavior (visco-elastic behavior). The elastic behavior is the most important characteristic in vulcanizates and the plastic contributions are generally undesirable even though unavoidable. However, for the processing of elastomers and elastomer compounds the flow behavior is the most important property for the forming process, and the unavoidable elastic contributions are most often disturbing.

This ambivalent behavior of uncrosslinked elastomers and elastomer compounds (their visco-elasticity) has led to a great number of different testing methods. The majority of these methods is used to determine the viscous component of the flow properties [6.19–6.28], which are of special importance to predict the processibility. Some other methods are used to determine the plastic and elastic behavior through the change of crosslinking of a sample after heating or deformation as well as the dependence of the viscosity of crosslinked compounds on the residence time, especially at higher temperatures through degradation (chemo-rheology).

The viscosities of elastomers and rubber compounds can change in an unpredictable fashion with increasing shear rate [4.547]. Therefore, test results from one point measurements at only one shear rate cannot be correlated to the behavior during processing operations; they are not sufficient. Because of these reasons numerous test procedures and testing devices for the determination of the processing behavior of elastomers and rubber compounds have been developed which yield most often only very limited information. Aside from the laboratory processing equipment (e.g. two roll mill, kneader, extruder, calender, vulcanization press, injection molding machine), as well as miniature versions as simulation equipment for such machines, the following measuring methods are available (the listing and the literature cited are not necessarily complete):

Compression viscometers [6.29–6.33]. The oldest viscometers for rubber technology were squeeze film devices or compression viscometers, e.g. the Defo-device [6.29], the machines by *Hoekstra* [6.30], by *Scott* [6.31], by *van Rossem* [6.32], by *Williams* [6.32, 6.33] and by *Wallace*.

Defo Device [6.29]. The Defo device, formerly the most frequently used one in Germany is of the squeeze film type. The cylindrical rubber sample, 10 mm high and 10 mm in diameter, is preheated for 20 min. at 80 °C and is then loaded in between two parallel plates of 10 mm diameter. The deformation weight which can compress the sample from 10 mm height to 4 mm within 30 sec is called the Defo hardness. 30 sec after removal of the load the height of the stress-free sample is determined to within 1/10th of a mm and the elastic recovery in height is defined as the Defo elasticity. Recently the Defo method has attracted new attention [6.29 a, 6.29 b].

Rotating Disc Viscometer [6.33 a–6.39]. The perhaps most frequently used apparatus to determine viscosity is the rotating disc viscometer according to *Mooney* [6.37].

The most important part of the apparatus is a disc shaped rotor which rotates within a flat cylindrical cavity at 2 rpm. The surfaces of the rotor are grooved to avoid slippage between the rotor and the rubber surrounding it. The upper and lower part of the chamber are bounded by electrically heated plates. The chamber is grooved on its faces and circumference just as the rotor. The rotor is rotated by a synchronous motor and the rubber sample is thus sheared. The resistance of the rubber to this rotation causes a thrust upon a floating, horizontal worm which presses against a cantilever spring and thus bends it. The amount of bending is determined by a dial gage or other device after predetermined time intervals and corresponds to a torque. It is defined as the Mooney viscosity. In modern devices the values are monitored and plotted on a recorder.

The determination of the Mooney viscosity and also of the Defo values permit the differentiation between different batches of the same compound and also predictions about the processibility of rubber compounds at low shear rates. However, because of the low shear rates and the absence of slippage at the wall during the determination of the Mooney viscosity, it is not possible to predict correctly behavior during processes occuring at high shear rates or with slippage at the wall (e.g. extrusion, injection molding etc.) [4.547].

During the measurement of the torque as a function of time at elevated temperatures in rotating disc viscometers according to Mooney of rubber compounds which are compounded to undergo vulcanization, information regarding the scorch and vulcanization behavior of the compounds can be obtained [6.38, 6.39]. However, more refined measurement techniques have been introduced (see Section 6.2.1.3).

Sphere Indentation Viscometer [6.40]. Especially for the determination of compounds of low viscosity or of rubber solutions, the sphere indentation viscometer, e.g. the *Hoeppler* Konsistometer [6.40] is applied. Such viscometers are, however, seldom used for rubber technology applications.

Cup Viscometers, Simple Cappilary Viscometers [6.42–6.46]. Already very early, cup and simple capillary viscometers were introduced into the field of rubber technology, e.g. those operated with gas pressure. Examples are those by *Behre* [6.41, 6.43], *Dillon* [6.42], *Karrer* [6.44] and *Marzetti* [6.45]. These can be considered to be prerunners of the high pressure capillary viscometers described in the next section.

High Pressure Capillary Rheometer (Processibility Tester) [6.47–6.66]. Modern high pressure capillary rheometers are certainly high precision instruments with considerable software yielding very reproducible results. The formerly used constant gas pressure has been often replaced by a constant velocity drive.

The dependence of the viscosity of elastomers or rubber compounds upon shear rate or output rate can be measured with the help of high pressure capillary viscometers over a very wide range.

For non-Newtonian liquids such as high polymers, the change of viscosity η with shear rate $\dot{\gamma}$ is not linear but exponential according to a flow index n, in the equation

$$\eta = \left[\frac{\tau}{\dot{\gamma}}\right]^n$$

(where τ is the shear modulus [4.547] depending on the structure dependence of the flow behavior of the particular polymer). Double logarithmic plots of the data yield quite often straight lines of slope n. Since n can be different from compound to compound, differently inclined – not parallel – straight lines will result when different rubber compounds are measured. Because of this fact [3.267] it is important for the evaluation of different rubber compounds to know the dependence of their viscosities upon shear rate. For this purpose, viscosity measurements in the high pressure capillary rheometer (analogous to measurements of the extrusion volume) as a function of shear rate (or the pressure), that means complete viscosity function, are especially useful for the evaluation of the behavior of rubber compounds during most of the processing operations. Therefore, these devices are also called processibility testers and they are superior to single point measurements. Because of the visco-elastic behavior of rubber compounds in the high pressure capillary rheometer, a direct determination of the viscosity is not possible. First, apparent viscosities are obtained from which the real viscosities have to be determined using complicated correction calculations, e.g. according to *Rabinowitsch* [4.547]. However, in modern high pressure rheometers the real viscosity values can be determined directly and reliably by built-in computers and then plotted graphically.

An extrapolation of the viscosities of rubber solutions to those of the rubber itself or compounds prepared by them (as was done frequently in the past) is less reliable and can lead to wrong conclusions.

Rheovulcameter [6.67, 6.68]. Rheovulcameters are capillary viscometers which include a mold of complicated shape. The rubber compounds, containing the cure system, are injected through the capillary at processing temperatures into the mold, where they vulcanize. Such devices, which resemble miniature injection molding machines, permit the measurement of the flow behavior and the superimposed curing mechanism of compounds. Therefore, such devices may gain wider acceptance in the future. Rheovulcameters can also be used as simple rheometers, when not accelerated compounds are extruded in open air.

Mixing and Extrusion Simulation Devices [6.69–6.79]. It has been proven advantageous to study rubber compounds in miniature mixers or extruders which are well instrumented before the compounds are studied in laboratory processing machines.

From these experimental results conclusions can often be drawn as to the processing behavior on large machines. Scale-up is often difficult.

6.2.1.2 Solubility and Tack of Elastomers and Elastomer Compounds

Solubility. An important test which has to be performed with the gum rubber as well as the unvulcanized compound is the determination of the solubility in organic solvents. General guidelines as to the choice of solvents cannot be given. First, low volume fraction solutions, e.g. 1% solution in cyclohexane should be formed in order to clearly see possible signs of birefringence. Aside from the purely visual investigation of the solution, the determination of its viscosity is also important. For this purpose, cup viscometers or those with a falling rigid body such as the *Höppler* viscometer [6.40] are suitable. More viscous solutions and rubber cements – as found frequently in real life – are preferably measured using consistometers or Rotation Viscometers [6.80].

Tack. Many rubber articles are manufactured by adhering many layers to each other. The determination of the tack of rubber compounds is a rather complicated story which has not been solved satisfactorily up to date. One has to distinguish between "Tack" (the ability of different layers to adhere to each other) and "Tackiness", the adhesion force itself. Several devices have been developed which compress several layers of compound under standardized conditions and then measure the force necessary to seperate them [6.81–6.84].

6.2.1.3 Vulcanization Behavior

[6.85–6.99]

The vulcanization behavior is important to the processor, he has to know after what time and at what temperature scorching occurs and vulcanization is complete. A rubber compound is transformed from a viscous to an elastic material during vulcanization so that the mechanical properties such as hardness, tensile strength, damping, modulus etc. change. Monitoring one or several of these data permits the evaluation of the degree of vulcanization as a function of time and temperature. This procedure is the most reliable but the most complicated one as well.

Step-Wise Heating. The formerly most often applied system of step-wise heating permits only a discontinuous measurement of the mechnical properties. Samples were vulcanized to different times and their mechanical properties measured afterwards. This complicated method is because of the relatively wide error range of the single measurements rather inaccurate. This method is also rather cumbersome for compounds with fast cure systems because the data points cannot be obtained in sufficiently short order. The point of optimal degree of vulcanization cannot be found sufficiently accurate even if modern computation methods are employed [6.85].

Defo-Scorch Determination [6.29–6.29b]. The behavior at the beginning of the cure cycle can be studied by controlling the change of the viscous state as a function of time as well as temperature. Formerly, the Defo device (see page 470) was used for this purpose. This procedure is exact but requires more time than can be justified for everyday quality control procedures.

Mooney Scorch [6.38–6.39]. The scorch behavior can be determined considerably faster and using less sample volume with the Mooney viscometer (see page 470). The procedure is modified by not using the big rotor (38.1 mm diameter) but the small one (30.5 mm diameter). A temperature of for example 125 °C or 150 °C is used. The so-called scorch time is defined as that time when the Mooney value rose above the minimum value by, for example, 5 points [6.86].

T-50-Test. The vulcanization optimum, especially of SR, was frequently determined in the US using the so-called T-50-Test [6.87]. Rectangular test samples from different levels of vulcanization are stretched to 100 or 200% strain and then cooled to −70 °C in the stretched condition. Then the clamps on one side are removed and the samples allowed to warm up slowly. That temperature for which the sample contracted to 50% of its original deformation is the test value. For optimum state of cure one obtains the lowest temperature. The same test also yields information regarding the low temperature behavior.

Vulcameter [6.88–6.99]. A complete vulcanization curve can be drawn automatically using the most modern method to determine the degree of vulcanization [6.95]. One obtains information about flow time, scorch, optimal vulcanization time, reversion and after-vulcanization. This device has originally been manufacutred and sold by Agfa-Physik under the name "Agfa-Vulcameter". Meanwhile, several technical variations exist world-wide especially torsional-based methods, e.g. Monsanto, Göttfert etc., some of which show considerable technical improvements [6.92].

The change in shear modulus is measured on a sample. One part of the heated sample holder is therefore moved by an excenter backwards and forwards whereby a shear force is forced upon the sample. The other part is connected to a force transducer which measures the transmitted force and records it. In modern devices the initial sine wave impulses are processed so that only the change in amplitude is recorded which is proportional to the change of the shear modulus as a function of vulcanization time.

The type of shear deformation, linear shear and torsional shear, distinguish between the two most important types of instruments. Recently also rotorless instruments came on the market [6.90].

Newer devices also plot, aside from the vulcanization curve, the differential curve, i.e. the first derivative of the vulcanization curve [6.88–6.91] which yields information about the rate of vulcanization (Göttfert-Elastograph). Also, temperature programmed vulcameters for the simulation of vulcanization processes have been developed [6.93] (see page 428, 436).

Because of different ranges of the dynamometer, hard as well as soft rubber types can be measured with equal accuracy. The Vulcameter yields in shortest time definitive and accurate information about the times and temperatures which have to be used during the vulcanization process. Progress has been made in the last years to continue the development of the Vulcameters [6.88, 6.90, 6.94, 6.96, 6.98].

6.2.2 Mechanical-Technological Testing of Vulcanizates

The properties of rubber vulcanizates are at the temperatures of technical importance half way between the liquid and the solid state. When studying the deformation mechanisms, one has to consider the laws of liquid and solid materials.

Like in all material testing, one distinguishes between static and dynamic testing. The static test yields essentially a measure for one time deformation and defines values which characterize up to what limits a material can be loaded. The dynamic test determines essentially, how often a material can be loaded under defined conditions and defines values which characterize the durability, i.e. generally the lifetime. The test specimen for most mechanical-technological tests are produced by stamping, cutting or grinding from larger samples. This step has to be executed carefully since the reproducibility of the test will otherwise be affected [6.100].

Since many rubber products are exposed during use to a wide variety to temperatures and their properties are dependent strongly upon temperature, the properties have to be determined over a wide temperature range in order to determine in which range the material may be applied.

6.2.2.1 The Tensile Test

[6.101]

The tensile test in a tensile testing machine is the most frequently used type of test. One obtains a characteristic sigmoidal stress-strain curve when rubber is extended up to large deformations. *Hooke's* law cannot be applied in this case.

In order to obtain a material property, nonetheless, it is customary to define the stress which is required to obtain a certain deformation (the stress value). Frequently, one chooses the stress value σ_{300} at 300% deformation.

The stress-strain curve can be quantitatively described for smaller strains as:

$$\sigma = \frac{E_o}{3}(\lambda - \lambda^{-2})$$

This equation also applies to the compression behavior. It can be transformed into the form:

$$\sigma = \left[E_o(\lambda - 1)\right] \left[\frac{1}{\lambda}\right] \left[(\lambda/3)(1 + \frac{1}{\lambda} + \frac{1}{\lambda^2})\right]$$

where $\lambda = L/L_o$.

E_o is *Young's* modulus (analogous to the shear modulus G) for small strains of the stress-strain curve, L_o the undeformed sample length, $\lambda - 1$ a measure for the deformation. The first factor represents Hooke's law, the second one corrects for the cross sectional area change of the sample (σ is the stress of the undeformed area of the sample), and the third factor represents a correction which follows from the statistical theory of rubber elasticity.

For the determination of the tensile strength and the elongation at break in the testing machine, one uses also frequently (aside from rectangular samples) ring samples which simplify a direct recording of stress-strain diagram. The tensile strength found using rectangular samples is frequently higher than that found for ring samples. The tensile strength is dependent upon the number of flaws found in a sample which is proportional to the volume of the sample. The volume of rings is larger than that of rectangular samples.

Aside from the tensile strength the force is measured which corresponds to a certain strain and is calculated to correspond to the original cross-sectional area. This stress value is a measure for the stiffness of a rubber sample and one of the most important measures for the evaluation of vulcanizates because it is independent of the changes upon which the tensile strength depends. The stress value is sometimes called a modulus, analogous to Young's modulus for metals and other hard materials. The use of the word modulus is incorrect, however, since the stress value is always taken in an area where Hooke's law does not apply any more [6.102].

A rubber specimen subjected to a tensile test shows after deformation a remaining deformation (the permanent set) because rubber in the vulcanized state is not an ideally elastic material. The permanent set can be determined in two ways. Either the deformation after the application of a certain load over a certain time span is measured or that force (tension) which leads to a certain value of the deformation [6.103] (see section on permanent set, page 476).

For compression or shear loading the shear stress-strain curve shows a larger linear range than the stress-strain curve in tension [6.104]. Because of the incompressability of elastomers, Young's modulus is as a rule three times the shear modulus. The form factor, i.e. the height/width ratio is important for the tensile and shear modulus determination. For tension a small from factor is advantageous, for shear a large one.

Both the tensile and compression stress-strain curves are strongly dependent upon the test temperature, the deformation rate and any previously applied loading [6.105]. Modern tensile testing is highly automated today [6.106].

6.2.2.2 Tear Strength

[6.107–6.111]

Tear strength is defined as the resistance which a rubber sample, modified by cutting or slitting, offers to the propagation of the tear. For this type of test the multitude of test specimen configurations exceeds that for the tensile test [6.107, 6.109]. Recently, the term tear energy has gained importance, a quantity which is largely independent of sample geometry [6.108, 6.110].

The needle pull strength is closely related to the tear strength. This test is derived from those in the leather industry and permits a qualitative evaluation of soles and heels made from rubber when these are sewed or nailed to a substrate [6.111].

6.2.2.3 Hardness Measurement

[6.112–6.120]

The hardness measurement correlates to a certain degree with the determination of the stress value because a well defined penetrating body is pressed into the material under well defined conditions. The penetration depth is dependent upon the form of the penetrating body, the type of loading and Young's modulus [6.118].

First, only a spherical penetrator was internationally standardized for the measurement of the so-called ISO hardness. However, the widely used hardness measurement according to Shore has become an international standard also.

Shore-Hardness. A very simple and inexpensive test device which does not have to be installed stationary but can be carried around very easily can be used to measure the Shore hardness [6.12, 6.119]. The hardness of soft rubber is mostly measured with the Shore A instrument, harder ones with the Shore D instrument.

The penetrator is in the form of a truncated cone. By deflecting a leaf spring with calibrated stiffness characteristics the resistance against penetration of the cone can be measured. At a theoretical Shore A hardness of 100, an extremely hard material, the spring is compressed to its maximum position without any penetration of the truncated cone. The spring loading is then 822 g. At a theoretical Shore A hardness of 0, an extremely soft material, the spring is not deflected, i.e. the loading of the spring is then 56 g. Intermediate values are obtained by linear interpolation. The hardness is related to Young's modulus [6.120].

In order to obtain reproducable hardness values, the following device constants have been defined:

- characteristic of the spring
- the cone angle of the penetrator and the diameters on both planes of the truncated cone
- the free length of the protruding pin.

Pusey-Jones Hardness. The hardness according to *Pusey-Jones* has to be distinguished from the shore hardness [6.114], and is used to measure very high hardnesses.

ISO Hardness. ISO hardness is measured over the range from 30 to 85 IHRD (International Hardness Degree). The range results from the difference in penetration depth of a steel ball of 2.5 mm diameter under a pre-load of 0.3 N and a main load of 5.4 N, i.e. a total force of 5.1 N [6.112].

Brinell Hardness. The hardness of hard rubber or hard plastics is measured as Brinell hardness. It is based upon the same principle as the ISO hardness with the difference that the ball is loaded with a higher force.

6.2.2.4 Permanent Set, Relaxation and Creep Test

Permanent Set [3.218–3.224, 6.121, 6.122]. Permanent set is a measure for the viscous behavior of the elastomers. The compression set is given at constant deformation by the relation:

$$R = \frac{h_o - h_2}{h_o - h_1}$$

where h_o is the initial height of the sample before deformation, h_1, the height during deformation and h_2 the height after a certain amount of time after deformation. For the experiment one chooses most often a deformation of, for example, 50%. The samples are compressed in between two parallel plates, the amount of deformation is determined by spacers and the compression set is measured after a relaxation time of, for example, half an hour. Frequently, the samples are stored in the compressed state at an elevated temperature in order to simulate the requirements of gasket materials where changes due to aging effects play a role [6.121–6.123].

Relaxation and Creep Tests [3.216, 6.124–6.126]. During relaxation and creep tests the time dependent change of the stress or the deformation are monitored directly. During the relaxation test the strain is kept constant and the change in stress is monitored, during the creep test the stress is kept constant and the time dependent strain is measured. The stronger the viscous component, the more pronounced are relaxation or creep.

For gasket materials, only the magnitude of the permanent set has been specified up to now. However, it becomes more and more obvious that it is better to measure the relaxation itself. It is of interest for the functioning of a gasket how the sealing stress has decreased after a certain time, i.e. how strongly does the stress fall with time.

6.2.2.5 Rebound Elasticity

[6.127–6.130]

During the determination of the rebound elasticity the energy is measured which is lost during a short, impact like deformation. A pendulum hammer (according to *A. Schob*) falls from a certain height on the specimen and the height to which the pendulum hammer returns after the impact is measured [6.127].

Here, the term "elasticity" is used as a quantitative measure which indicates to which extent the material behaves elastically and to what extent in a viscous fashion. The elasticity is the higher, the less deformation energy is dissipated into heat. According to a proposal by *A. Schob* the following ratio is called the "elastic efficiency":

$$\frac{\text{Rebound Height}}{\text{Fall Height}} = \frac{\text{Recovered Energy}}{\text{Dissipated Energy}}$$

As a matter of fact, this test method belongs to the dynamic tests since it consists of a vibration caused by an impact. The test values range for different rubber vulcanizates from 10 to 75% or even higher.

Highly elastic compounds exhibit values from between 60 to 75% or higher, intermediate elasticity values from 40 to 60% and low ones from 10 to 40%. The rebound elasticity is strongly dependent upon temperature.

6.2.2.6 Visco-Elastic Behavior, Dynamic Damping

[6.14, 6.131–6.145]

General. The deformation behavior can be reduced to the following two basic functions: Resistance to deformation and deformation energy loss. For the following statements it is assumed that the deformation mechanism is in the linear region between stress and strain.

For the description of the stress-strain behavior one distinguishes between two ideal cases: the solid material behaves according to *Hooke's* law

$$\sigma = E \cdot \gamma$$

in which σ is the stress (often also the shear modulus τ is used), E is the *Young's* modulus and γ is the deformation. The viscous liquid follows *Newton's* law

$$\sigma = \eta \cdot \dot{\gamma}$$

In the first case the stress σ in the material is proportional to the deformation γ, in the second one to the rate of deformation, the shear rate $\dot{\gamma}$. The constant of proportionality is for a Hookean solid Young's modulus E and for the Newtonian Liquid the viscosity η [4.547].

Elastomers do not behave like any of these ideal cases but occupy an intermediate position. According to the way an elastomer is loaded during a test procedure, one can extract from the test a material constant corresponding to a modulus or a viscosity. From dynamic experiments one obtains simultaneously both values. This fact can be illustrated using the following models:

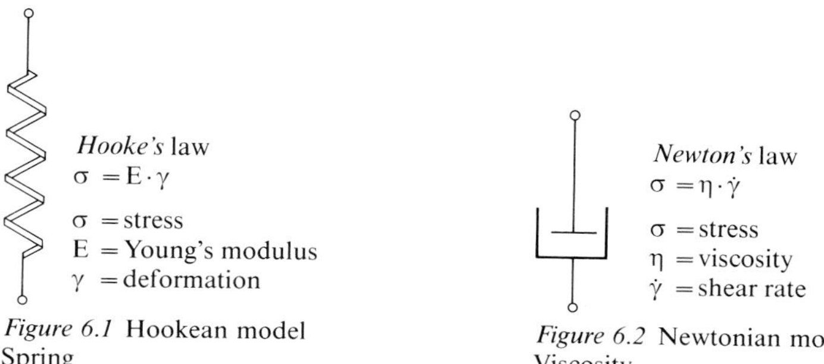

Figure 6.1 Hookean model Spring

Figure 6.2 Newtonian model Viscosity

When spring and dashpot are combined, the two primary models result:

$\sigma = E \cdot \gamma + \eta \cdot \dot{\gamma}$

viscoelastic deformation

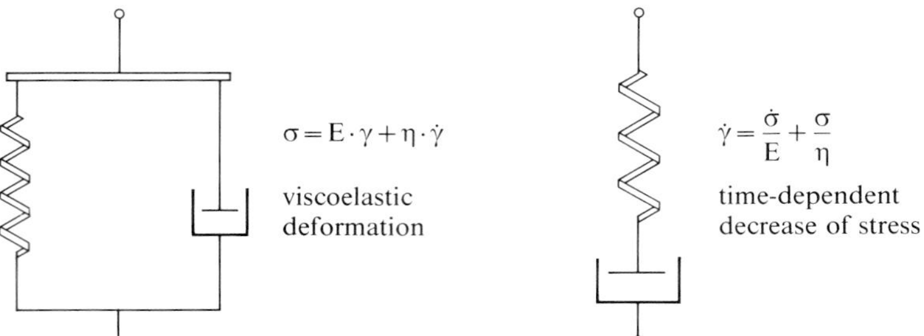

Figure 6.3 *Voigt* model

$\dot{\gamma} = \frac{\dot{\sigma}}{E} + \frac{\sigma}{\eta}$

time-dependent decrease of stress

Figure 6.4 *Maxwell* model

The time dependent deformation can be seen in Fig. 6.5. A relation between the elastic and the viscous components during dynamic loading can be represented by

$$E^* = E_1 + iE_2$$

E^* is called the complex modulus and it can be shown that the real component is identical to *Young's* modulus and that the imaginary component characterizes the viscosity.

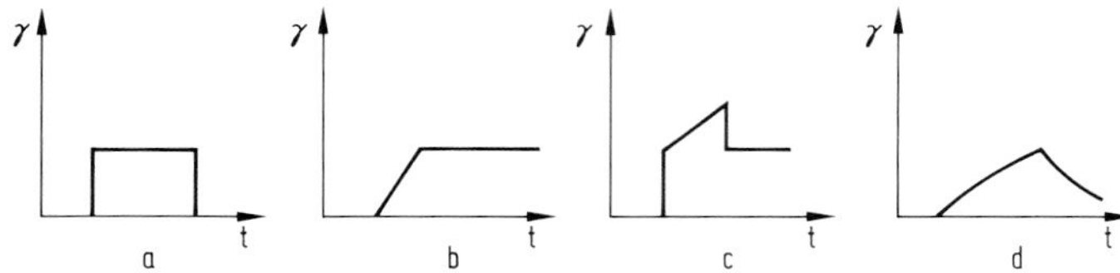

Figure 6.5 Time dependent deformation of various models;
a) Hookean model
b) viscous Newtonian flow
c) *Maxwell* model
d) *Voigt* model

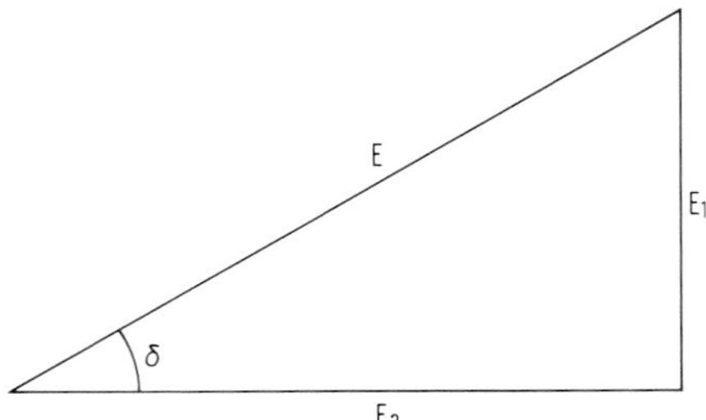

Figure 6.6 Representation of the mechanical loss angle

Frequently, the real and imaginary components are not reported but the ratio of E_2 and E_1 as the tangent of the mechanical loss angle (see Fig. 6.6):

$$\tan\delta = E_2/E_1$$

This quantity is analogous to the dielectric loss angle.

From the equations above, the following quantities may be derived:

$\sigma E^* \cdot \gamma = (E_1 + iE_2)\ \varepsilon$ (analogous to *Hooke's* law)

$\sigma = E\gamma + \eta\dot{\gamma}$ (Voigt model)

$$\sigma = E\gamma + \eta\ \frac{d\gamma}{dt}$$

For periodic deformations one obtains:

$$\gamma = \gamma_o \cdot e^{i\omega\eta}$$

thus

$$\frac{d\gamma}{dt} = i\omega\ \underbrace{E \cdot e^{i\omega\eta}}_{=\gamma}$$

$$\sigma = E\gamma + i\omega\eta\gamma = (E + i\omega\eta)\gamma$$

Accordingly: $E_1 = E$

$E_2 = \omega\eta$ the imaginary part corresponds to the product of circular frequency and viscosity.

$\tan\delta = \frac{\omega\eta}{E}$ can be determined from damping measurements

Measurement Techniques [6.131–6.134]. The visco-elastic properties are best obtained utilizing dynamic loading through free or forced vibrations. In the generally used *apparatus* developed by *H. Roelig* the specimen is deformed periodically by an excenter. The resulting sinusoidal force is measured by a dynamometer. The deformation diagram is obtained through optical representation. Since a phase shift between stress and strain exists, one obtains an ellipse the slope of which represents the complex modulus E* [6.134]. One obtains the mechanical loss factor from the area of the ellipse. The area of the ellipse represents the energy dissipated into heat per deformation cycle. When this heat energy is related to the expended energy per cycle, one obtains the percent damping [6.131].

The loss angle δ can also be derived from the stress-time and strain-time diagrams. Because of the phase shift between stress and strain the strain reaches its maximum somewhat later than the stress or the zero crossing of the strain is delayed with respect to the zero crossing of the stress. This time delay has the value δ/ω with $\omega = 2\pi f$ with ω the circular frequency and f the frequency of the vibration.

The specimen receives energy from outside at all times while in the dynamic tester. Therefore, one speaks of forced vibrations in contrast to free vibrations where energy is supplied to the vibration system only once at the beginning of the test. A decrease in the vibration amplitude can then be observed.

The *Torsion Pendulum* [6.133, 6.142, 6.143] operates according to the principle of free vibrations, where an oblong sample is deformed in torsion. The vibrations are registered by an optical system. The dynamic torsion modulus is obtained from the frequency. The logarithmic decrement Δ is obtained from the difference in two successive vibration amplitudes. For small amounts of damping one obtains $\Delta = \pi \tan\delta$.

The visco-elastic material properties are dependent on the form factor and the amplitude, aside from the frequency and the temperature. The amplitude dependence is the stronger, the more filler the specimen contains.

Considerable advances in the area of visco-elastic measurements have been made lately [6.137, 6.140, 6.143].

Heat Build-Up. Since the dissipated mechanical energy during the dynamic loading is transformed into heat because of molecular friction, the viscous component may be measured directly by monitoring the increase in heat of the sample (heat build-up). The experiment should be conducted using forced vibrations of large amplitude. Flexometers are used for these experiments. Through the use of flexometers (e.g. the *Goodrich Flexometer* [6.146, 6.147] or the *St. Joe-Flexometer*) one can also determine the structural integrity after long periods. The limiting loading is determined at which the sample will blow out.

Ball Fatigue according to *Martens*. This test belongs to the same categorie. Rubber balls are being rotated under an ever increasing load. At excessive loads the rubber in between the balls is degraded due to high temperatures.

6.2.2.7 Abrasion and Aging

Performing tests described so far do not require any great length of time. Insight into the performance of a product is gained in real life only after years. During short time testing the specimen is over-loaded and this over load is expected to predict the natural destructive effects which cause failure in practical applications. This is often very difficult and has never been accomplished satisfactorily. In spite of this, such methods have to be applied, for example, for the testing of abrasion and aging behavior.

Abrasion Resistance [6.148–6.153]. In order to test the abrasion resistance the rubber specimen is pressed against a rotating drum coated with an abrasive. The loss (in mm^3) obtained under standardized conditions is a measure for the abrasion resistance, which is often compared to that obtained for a standard rubber specimen. The result is not always in agreement with in-service conditions. Methods of this kind are, however, most often useful enough for development work.

Ageing Tests [4.257a–4.275, 6.125, 6.154–6.158]. The accelerated nature of tests is most important for ageing testing. Different manifestations of ageing are known for the different classes of elastomers (see also page 168). For example, NR tends to get soft and tacky with age while many SR types, e.g. SBR or NBR harden with age and become brittle. In both cases the tensile strength and elongation at break decrease. Therefore, one determines the decrease of these two properties with storage time at elevated temperatures for an accelerated ageing test [6.154]. Preferred are three methods:

- For the *Geer* aging test the specimen are aged in a space with circulating hot air.
- For the *Bierer-Davis* test the rubber samples are exposed to an oxygen atmosphere, e.g. under 20 bar oxygen pressure at 70 °C. The duration of the ageing is 24 h or a multiple thereof [6.155].
- The fastest ageing test is carried out using compressed *air* with 5 bar pressure above ambient at, for example, 125 °C [6.156].

Samples for un-like elastomers stored simultaneously in the same space can influence each other by migration of certain components (e.g. antioxydants, plasticizers, etc.). Therefore, devices have been developed wherein each specimen is placed separately in its own compartment [6.157].

After storage under the described conditions, the most important vulcanizate properties are determined and then compared to those of an unaged sample.

The ageing process can also be determined with the help of relaxation measurements [6.125, 6.158]. The sample is subject to a constant strain or stress in oxygen and the change in length is measured. The observed stress relaxation or tensile creep are caused by chemical reactions which destroy primary valence bonds or cause chain scission (chemorelaxation).

6.2.2.8 Crack Formation

Ozone Crack Formation [4.276–4.281, 6.159–6.164a]. Among the ageing phenomena is the formation of cracks on the surface of extended samples which are caused by ozone in the presence of oxygen. This phenomenon is investigated in ozone radiation chambers, weatherometers or through weathering outside. The latter two methods include other influences such as water and ozone.

Ozone leads in NR and most of the SR to a rapid destruction. For testing, the stretched rubber sample is exposed to a stream of air containing ozone. The ozone resistance is determined from the crack formation as a function of the strain [6.159–6.164a]. Below a certain strain (e.g. 20%) no crack formation is visible.

Light Crack Formation. When white or light colored vulcanizates without the application of a tensile strain are exposed under atmospheric conditions to sunlight, fine cracks in random orientation form on the surface. These are called elephant skin. After longer exposure a hard, inelastic skin forms which chan be leached out by water so that only the filler (chalk) remains. An accepted procedure to quantify this phenomenon called "frosting" does not yet exist.

Dynamic Crack Formation [4.281, 4.333–4.335, 6.162, 6.165–6.173]. During dynamic loading of vulcanizates, e.g., by compression, tension, bending, shear etc., cracks are formed on the surface (fatigue cracks) as well. During testing the number of load cycles necessary to form the first surface cracks is determined.

For such tests, for example, the *De Mattia fatigue tester* is applied [6.165, 6.166]. The crack formation and their growth are measured [6.166]. During *flex cracking* the sample is tested undamaged, one observes the formation of the first cracks and their growth. In the second case the test specimen is cut in a prescribed way before the test commences and one observed the *cut growth rate* during the test.

The onset of crack formation is strongly dependent on environmental conditions, e.g. temperature, humidity, ozone level, intensity and wave length of incident light as well as on the mechanical loading, its amplitude and frequency [6.171]. During fatigue testing cracks are only formed above a critical strain level [4.280]. The crack formation is especially pronounced under the influence of ozone.

While the *de Mattia* machine applies a constant strain, the *Du Pont fatigue tester* [6.168] utilizes constant load amplitudes.

A belt consists of several test specimen, which as members of a chain are connected to each other. The specimen themselves have different wide and deep ridges so that different modes of compression and tension result when they rotate over a set of small rollers. The evaluation is by counting the number of cracks.

Correlation to fatigue in tires is relatively good.

6.2.2.9 Low Temperature Behavior

[6.131–6.134, 6.174–6.177a]

Principle. The low temperature behavior of elastomers is characterized by three temperature regions [6.174] (see page 170):

Region 1 corresponds to the frozen-in state (glassy state) with high modulus and low damping.

Region 2 is the critical transition region from the glassy to the highly elastic state. A leathery consistence prevails. The modulus drops sharply and damping shows a maximum.

Region 3 is characterized by the rubber-elastic state with low modulus and low damping.

Correspondingly one can distinguish between the following characteristic temperatures:

- *The low temperature brittleness temperature* (T_S) is the temperature at which a glassy material will just not break, splinter or be damaged otherwise any more, when subject to an impact test. The low temperature brittleness point characterizes the behavior under impact conditions.
- At the low temperature characteristic value (T_G), the glass transition temperature, the damping exhibits a maximum and the modulus an inflection point.
- The *limiting temperature* (T_R) is that temperature at which the modulus changes by a factor compared to that at room temperature (e.g. by a factor of 2,5 or 10).

With thermoplastic elastomers of higher temperatures the modulus curve drops down to zero T_M, the *softening point* (see page 146, Fig. 3.7).

All three definitions are necessary and important for characterizing the behavior of polymers at low temperatures. The test results depend to a large extent on the testing conditions. For example, the low temperature resistance value shifts to higher temperatures as the frequency is increased (see page 170). Correspondingly, the most suitable testing method has to be chosen from the multitude of test procedures available which makes the performance requirements in service best [6.174].

Measurement Methods. The determination of the *brittleness point* (T_S) can be carried out by two methods: by impact loading [6.175] or by a bending test. In both cases the tests are repeated at decreasing temperatures until the sample breaks.

The low *temperature characteristic value* (T_G) can be determined by three methods: by the determination of the damping maximum as a function of temperature [6.131], by the torsion pendulum test according to *Clash* and *Berg* [6.133] as well as by the determination of the impact elasticity minimum at decreasing temperatures [6.127], which leads to relatively similar figures. The glass transition temperature can characterize in some respect the status of the elastomer [6.177 a].

Determination of the *limiting temperature* (T_R) can – as also like T_G – be accomplished through the visco-elastic properties as a function of temperature [6.131] as well as the torsion pendulum test and in addition also by the change of the permanent set with decreasing temperature [6.121, 6.122] or the determination of the Shore hardness with decreasing temperature [6.122] as well as the determination of the change of *Young's* modulus [6.134].

6.2.2.10 Swelling Behavior and Permeability

Swelling Behavior. Solvents especially, but also an assortment of gases and vapors, affect rubber and swell it and tend to degrade it. Most often the volume before and after the effect of the swelling chemicals is determined [6.178]. The mechanical properties such as tensile strength, elongation at break and hardness should be measured before and after the exposure as well, since they are affected by swelling.

Permeability. This is important for many rubber articles (e.g. for inner tubes, inner liners of tires, fuel permeation through fuel hoses). The pressure or volume differences on both sides of a membrane are measured [3.201 a, 6.179, 6.180].

6.2.2.11 Heat Conductivity

The heat conductivity of elastomers is determined by the heat energy which passes through a cross-section perpendicular to the flux direction per unit time. It is proportional to the temperature gradient. The determination with calorimeters requires experience. The heat conductivity constant R is most often quoted in Kcal/mh °C [6.181, 6.182].

6.3 Electrical Testing

[6.183–6.193 a]

6.3.1 General Comments about Electrical Testing

[6.188]

NR and a variety of SR types (e.g. SBR, IR, BR, EPDM, IIR, MQ) exhibit a very low electrical conductivity and are, therefore, suitable as electrical insulating materials. However, some other types like CR and NBR contain electrically polarizable groups or dipoles and are, therefore, less suitable as electrical insulators.

The range of electrical conductivity of all elastomers can be affected over a wide range by the composition of the compound or by the addition of insulating (e.g. light) fillers or of conducting substances (especially carbon blacks (see page 283 [4.416–4.424]), anti-static plasticizers (see page 301 [1.14]) etc.).

Also, rubber articles with such a high electric conductivity can be produced, that any static electricity build-up can be prevented. This is important, for example, for gas pump hoses, conveyor belts as well as for tires.

6.3.2 Measurement of the Surface and Transmitted Resistance

[6.183]

Different methods may be employed to measure the resistivity:

- Discharge of a capacitor
- Direct current and voltage measurement
- Bridge arrangement
- Electrometer tube
- Potential measurement system

6.3.2.1 Discharge of a Capacitor

In this first method the dielectric of a capacitor consists of the material to be measured. The capacitor is charged up to a certain potential and the ensuing voltage drop vs. time is measured using an electrostatic volt meter. For well insulating materials the time of measurement is too long. Thus other measurement techniques have to be employed.

6.3.2.2 Current and Voltage Measurements

During the determination of the resistance by direct current and voltage measurements, a negative voltage is applied to one electrode, the other one is connected to ground. The connection leading from the negative electrode to the galvanometer has to be shielded well so that no noise currents can reach the galvanometer.

6.3.2.3 Application of Bridges

One can also connect the sample to one leg of a *Wheatstone* Bridge and can utilize for the determination of the balancing current a vacuum tube voltmeter instead of a galvanometer. Difficulties arise with proper grounding for DC measurements. Therefore, this method is seldom used.

6.3.2.4 Application of an Electrometer Tube

The most important and currently most frequently used method is the measurement using an Electrometer Tube. For this measurement, the shielded electrode is applied directly to the grid. The insulation current flowing through the sample generates a voltage drop which determines the grid voltage and thus the anode current. One encounters difficulties for conducting rubber, when rectangular samples and electrodes are used. It has proven advantageous to utilize samples having the forms of strips in order to obtain a homogeneous current path.

6.3.2.5 Potential Determination

Where the potential determination method is employed, the current is applied at two electrodes at the end of the sample. A potential differential is determined using two additional electrodes, the so-called potential electrodes. The resistance is calculated from the voltage across the potential electrodes. This method is especially suited for conductive elastomers.

The measurement of the resistance can be determined rather exactly if one operates - aside from the most appropriate electrode form - with a shielding ring. The exact determination of the surface resistance is compared to this rather problematic, since one cannot prevent the current paths to enter the inside of the probe. In addition, surface structures such as cracks, bloomed materials, moisture etc. influence the surface resistance to a large extent.

6.3.3 Determination of Dielectric Constant and Dielectric Loss Factor

[6.184]

For the determination of the dielectric constant and of the dielectric loss factor one can use a bridge arranged parallel or in series of several elements of a loss free capacitance of known quantity and the sample as an Ohm's resistance in the form of a *Schering* bridge and one measures in an AC frequency range from 15 to 500 KHz. Generally, plate capacitors are used the capacitance of which can be calculated in vacuo. In the bridge arrangement the resistance is varied until the bridge is balanced, i.e. until the null meter does not measure any current any more, then the dielectric constant and the dielectric loss factor can be calculated from the parameters of the experiment.

At higher frequencies one cannot employ the *Schering* bridge any more. For the region between 50 and 100000 Hz the Bridge according to *Giebe-Zickner* is recommended.

At even higher frequencies (0,1-100 MHz) a bridge arrangement is not used any more but one employs resonant circuits which are tuned by variable capacitances and inductors to the measurement frequency.

6.3.4 Measurement of the Dielectric Strength, Leakage Current Resistance and Arcing Resistance

Analogous to the bulk resistance and the surface resistance one has to distinguish between dielectric strength and leakage current.

Dielectric Strength. To measure the dielectric strength, a voltage is raised until dielectric failure. This maximum voltage depends strongly on the experimental set-up. Because of this, measured values are often not very specific and scatter considerably. Therefore, exact experimental procedures have to be followed carefully. The form of the electrodes [6.189, 6.193], the dielectric with which the experiment is carried out in order to avoid arcing (e.g. silicone oil), the time of electric charging and other variables have to be carefully considered.

Leakage Current. Analogous considerations also apply to the leakage current determination. The surface configuration is very unimportant during this test. Two charged electrodes are applied to the surface of the sample. A solution, for example NaCl, is allowed to drop between the two electrodes. The number of drops is counted until a short circuit because of leakage current has formed. This can lead to a visible destruction of the surface and the formation of the short circuit current path.

Arcing Resistance. In order to measure the arcing resistance, two carbon electrodes are brought in contact with the insulating material. Then the electrodes, after applying a voltage to them, are moved apart and the effect of the arc on the rubber surface is observed. For rubber, with the exception of silicone rubber, this test leads to carbonization and an increase of the electric conductivity. For silicone rubber an insulating silica layer forms.

6.3.5 Measurement of Electro-Static Charging

[6.190–6.192]

The insulating material rubber can be charged electro-statically when mechanically deformed, especially when friction is involved. The amount of charge does not only depend on the composition of the rubber and thus its conductivity but is affected strongly by the total system. It is determined by the opposite material against which rubber is rubbed, by the arrangement of the rubber as a conductor, by the humidity and the dust content of the surrounding air etc.

The measurement of the static charges may be carried out, for example, by positioning an insulated plate over the charged area. Through induction, charges in the plate are separated. The like charges are then removed from the back side of the plate by short time grounding. Now the plate is placed into a *Faraday* cage and the charge is measured with an electrometer.

Such measurements are only suitable for the laboratory. For in-service measurements field strength measurements by electronic means have been introduced. A charge is transferred via an electric field to a plate which is grounded via a resistor. The charge flows through a second, grounded plate after the field has been shielded. This causes a voltage drop across the resistor.

6.4 Analysis of Rubber

[6.194–6.304]

The analysis methods, especially the modern ones, for the identification of the elastomer type and the most important compounding ingredients like stabilizers, antioxidants accelerators, and plasticizers will be discussed.

6.4.1 Identification of Compounding Chemicals and Additives

6.4.1.1 Methods of Separation

The German industry standards (DIN) as well as the American (ASTM) and British ones (BS) describe a procedure for the full analysis of rubber vulcanizates. However, the combination of the single procedures is completely left to the analytical chemist.

6.4.1.1.1 Sampling

[6.205]

During sampling, special attention has to be paid to the question if a representative average of the sample is really obtained. *Unvulcanized* rubber or compound samples are thinly sheeted out on a mill or are cut into small pieces. *Vulcanizates* have to be sufficiently cut into small pieces so that they can be sifted completely through a mesh of $40/cm^2$. Rubber adhesives and solutions have to be dried first in vacuo and then the dried rubber is treated like an unvulcanized one.

6.4.1.1.2 Determination of the Moisture Content

[6.206]

One or two grams of the chopped up sample are dried on a watch glass in an evacuated desicator over concentrated sulfuric acid until constant weight has been obtained. All results deriving from the following rules have to be referred to the dried material. Because of reasons of practicality it is advantageous to determine the moisture loss after drying for 2 hrs at 105 °C in a drying oven. In this way, however, all components volatile at this temperature will be removed. Therefore, this method is only applicable in the absence of any other volatile components.

6.4.1.1.3 Determination of Water Soluble Components

[6.207]

By treatment in warm water the following components are dissolved partially from the compound or vulcanizate: Proteins, starch, glue, glycerine, several accelerators and sugar.

6.4.1.1.4 Determination of the Components Soluble in Acetone

[6.208]

The following components can be extracted from rubber and vulcanizates: rubber resins, free sulfur, waxes, added resins, plasticizers, paraffines, cellulose ester and ether, accelerators and protection chemicals, lanolin, some other organic materials. Partially extracted are oils, their oxydation products as well as heavily masticated rubber.

6.4.1.1.5 Determination of the Components Soluble in Chloroform

Only vulcanizates are extracted with chloroform. Chloroform dissolves from the vulcanizates already extracted with acetone bituminous substances, which are recognizable by their dark color as well as by their fluorescence. Because vulcanized rubber always contains only small amounts of components extractable by chloroform, a chloroform extract up to 4% referred to the dried amount of rubber is considered normal. If this value is higher and if the color of the extract does not point towards the presence of bitumous substances, considerable amounts of rubber have gone into solution. This indicates that the sample has not been fully vulcanized, contains regrind or has been depolymerized considerably before vulcanization. The chloroform extract is applicable to NR, BR, SBR, and CR and well vulcanized NBR. This method cannot be applied to IIR, TM or strongly undervulcanized vulcanizates as well as those containing reclaim. The duration of the extraction has to be, for example, eight hours for carbon black containing vulcanizates.

6.4.1.1.6 Determination of the Components Soluble in Alcoholic KOH

Alcoholic KOH dissolves the factice from vulcanizates already extracted with acetone or chloroform. In addition, proteins and components of hardened phenolic resins are dissolved. When the nitrogen content of the extract is multiplied by 6.25 one obtains an approximate value for the protein content. One can determine the amount of factice approximately by subtracting the amount of sulfur and protein from the total extract.

6.4.1.2 Paper and Thin Layer Chromatography Separation

[6.210–6.215]

It is assumed that the paper and thin layer chromatography [6.211–6.214] is known. The thin layer chromatography [6.210] has replaced the paper chromatography [6.213, 6.214] to a large extent.

The procedure for the thin layer chromatographical separation is carried out according to [6.210]. It is especially suited for the detection of stabilizers and antioxidants [6.215] as well as plasticizers [6.222, 6.225]. While monomeric plasticizers migrate, polymeric ones normally remain in place.

6.4.1.3 Identification of the Extractable Components

[6.215–6.228]

The extracted compound components, especially the stabilizers, antioxidants, accelerators and plasticizers [6.208, 6.216] can be identified and determined according to the following reaction, spectographic and chromatographical methods, after the solvents have been evaporated.

6.4.1.3.1 Identification of Stabilizers and Antioxidants

[6.200, 6.215–6.220, 6.224, 6.226]

The identification of stabilizers and antioxidants is preferrably carried out using thin layer chromatography [6.215]. Identification methods are described in [6.216–6.220, 6.224, 6.226]. See [6.200, 6.226] regarding quantitative methods.

6.4.1.3.2 Identification of Accelerators and their Decomposition Components

[6.200, 6.213, 6.214, 6.216, 6.217, 6.221, 6.227, 6.228]

The identification of accelerators can be done by paper or thin layer chromatography. The procedures have been described in detail [6.210, 6.213, 6.214, 6.227]. The quantitative determination can follow [6.228]. See also [6.258, 6.269, 6.288, 6.292].

6.4.1.3.3 Identification of Plasticizers

[6.200, 6.222, 6.225]

The analysis of plasticizers is carried out using chemical analysis, thin layer chromatography, gas chromatography and IR spectroscopy. These methods are described in detail in a book by Wandel, Tengler and Ostromow [6.225] and are also applied to plasticizer mixtures.

6.4.1.4 Ash Content

6.4.1.4.1 Identification of Fillers

[6.229–6.234]

The ashing [6.229] does not yield quantitative results, quite often, because of the decomposition of fillers. In spite of this, this method is applied quite often because it is rather simple and can yield, if caution is exercised, quantitative results when the ashing temperature is not higher than 500–600 °C.

Methods are widely used for the determination of carbon black content for which the rubber component and other compound ingredients are degraded by not HNO_3 [6.233]. CR, IIR and TM-compounds cannot be analyzed by this method since they are resistant to HNO_3. If the samples, for example of SBR, are not completely soluble in HNO_3, the procedure has to be modified somewhat. A disadvantage of the procedure is the difficult filtration of the carbon black which passes even asbestos filters easily because of its fine particle size.

In another procedure [6.232] the elastomer is oxydized to well soluble low molecular weight decomposition products by tert. butylhydroperoxyde in the presence of osmium tetroxyde. Acid soluble fillers are washed out of the residue with diluted nitric acid and water. This procedure yields good results for vulcanizates of NR and SR with higher fraction of conjugated divinyl bonds.

A complete identification procedure is contained in [6.234]. By applying thermogravimetric analysis, the carbon black content can be determined quantitatively [6.230]. It has been possible only recently to determine carbon blacks by the gas chromatography method [6.254a].

6.4.1.4.2 Determination of Sulfur

[6.235–6.238]

The determination of the total amount of sulfur proceeds according to *Wurzschmitt* or to *Schoeninger* [6.235, 6.236].

By the addition of Perchloric acid or Bromine to smoking HNO_3 and heating in a sand bed, vulcanizates of NR, SBR, IIR, and NBR are cracked in a relatively short time. In boiling HNO_3 sulfates are formed in the presence of ZnO. For the determination of the total amount of sulfur the aqueous deposition of the HNO_3 yield is combined with $BaCl_2$, the sulfates are percipitated by $BaCl_2$, then isolated and

$BaSO_4$ is weighed. Analysis results of equal reliability result from the oxydation with Na-peroxyde in a reactor. The method has to be modified for CR and NBR, sometimes also for SBR; for IIR values are frequently found which are too low by 0.1–0.5%. In CSM, TM and CR-thiuram types the sulfur bound to the polymer is included in the measurement results. However, the method is applicable to soft rubber, not to hard rubber.

The corresponding BS-method favors the determination according to *Carius*. The formed sulfur dioxyde is oxydized by Hydrogen peroxide to sulfuric acid which titrates either directly with NaOH or precipitates with 4-Amino-4′-chloro-diphenyl as with the combustion of NBR and CR. The residue is titrated with NaOH utilizing an indicator mixture of phenol red and bromothymol blue.

For the determination of the *free sulfur* [6.237] the thinly sheeted out vulcanizate is treated in a boiling sodium sulfite solution. The free sulfur is thus converted to thiosulfate and gets determined by an iodine analysis.

Also, a new potentiometric method according to *Gross* [6.238] may be applied.

For determination of the acetone soluble sulfur the sample is extracted at least for 16 hrs in acetone. The dried acetone extract is then oxydized with either Br/HNO_3 or with Na-Peroxide. The resulting sulfate can be determined gravimetrically or by photometrical tritration with barium chloride to the maximum point of turbidity.

6.4.2 Identification of the Elastomer Type by Infra-Red Spectroscopy

[6.239–6.247]

6.4.2.1 Identification of Raw Rubbers and Rubbers in Unvulcanized Compounds

The first publication which describes the use of infra-red spectroscopy for the identification of elastomer types is the one by *Barnes, Williams, Davis* and *Giesecke* [6.239]. These authors separated the polymer from the various fillers, especially carbon black, by solvation in high boiling solvents and by removing the fillers by filtration. Important is, that the sample be first extracted by acetone or petrolether in order to remove disturbing components such as plasticizers, paraffines, antioxydants etc. Also, the extraction of any uncrosslinked portions is important [6.245]. From the solution the polymer is precipitated with alcohol, dried and then identified as a film or in solution in a solvent suitable for IR spectroscopy. This preparation is tedious and time consuming but leads to an unambiguous identification of the polymers because a spectrum extending over the whole range is measured and then compared to a suitable set of comparison spectra.

6.4.2.2 Identification of the Elastomer Type in Vulcanizates by Dissolving

Dinsmore and *Smith* [6.241] are utilizing also a high boiling solvent (e.g. o-Dichlorobenzene) in order to dissolve the elastomer, however, they removed the fillers by centrifuging or filtration, then concentrate the solutions to a highly viscous consistency and spread them between two metal bars as distance marks on a sodium chloride or potassium bromide crystal. The spectrum is taken after the solvent has completely evaporated. Since the sample does not completely dissolve in o-dichlorobenzene, those fractions which have been dissolved are assumed to be representative of the total.

6.4.2.3 Identification of the Elastomer Type in Vulcanizates by Infra-Red Spectroscopy of Pyrolysis Products

A faster method to determine the type of elastomer in vulcanizates is based on infra-red spectroscopy of thermal decomposition products. This method was introduced by *Harms* [6.243] and *Kruse* and *Wallace* [6.244] for the identification of insoluble plastics and resins.This technique was then used by *Tyron, Korowitz* and *Mondel* [6.246] for the quantitative analysis of mixtures of NR and SBR. National Bureau of Standards and an ISO standard [6.247] describe the analysis of vulcanizates by pyrolysis infra-red spectroscopy in detail.

The pyrolysis technique and its theoretical basis to identify thermal degradation products has been described in detail by *Fiorenza* and *Bonani* [6.242]. This paper is one of the most important ones in this field.

Most authors describe the pyrolysis in air which results in a very dirty, heavily oxidized pyrolisate, the composition of which changes only insufficiently for many aliphatic hydrocarbon polymerizates. Because of this, ISO recommends a pyrolysis under Nitrogen. Bentley [6.240] recommends to carry out the decomposition in vacuo. This method was extended by other authors to a wide variety of currently used elastomer types and includes absorption data for some typical compounds and for a variety of polymers and co-polymers which can be used as compound components. Infra-red spectroscopy is a technique which is normally insensitive to the presence of minor components, especially when the major components exhibit strong absorption. It must be pointed out, therefore, that a concentration of 10% of the minor component could be present (depending on its absorption behavior) and that it might still not be detected. Examples of such compounds are the following:

- up to 10% CSM in EPDM
- up to 15% CR in SBR-NBR mixtures

It is obvious, therefore, that for the analyses of compounds containing many components as many characteristic absorption bands as possible should be measured. In addition, the chemist should study the rest of the spectrum as carefully as possible and should be familiar with spectra of the standard chemicals. In this connection a quick Beilstein test is useful, when the presence of small amounts of chlorine containing polymers, e.g. CSM or CR shall be determined.

6.4.3 Pyrolysis in Connection with a Paper Chromatographical Identification of Elastomer Types

[6.248]

Feuerberg [6.248] has described a pyrolysis method carried out in an electrically heated pyrolysis pipe under exactly defined conditions as well as the preparation and the identification via paper chromatography of Hg adducts of volatile rubber pyrolisates. The pyrolizates taken from the cold trap are mixed with solid Hg acetate and then shaken for at least two hours. The adducts obtained are chromatographed on round paper filters of 30 mm diameter. One to three ml solution are applied.

The carrier is $CH_3OH/CH_3COC_2H_5/1.5nNH_4OH/1.5n(NH_4)_2\ CO_3$ (6:6:1:1 v) and the elution time is 4–6 hrs. A 1% solution of diphenylcarbazone in methanol is used to color the surfaces. Blue or violet ring segments, characteristic for the elastomer, are formed.

6.4.4 Identification of Elastomer Types with the Help of Pyrolysis-Gas Chromatography

[6.249–6.255]

Pyrolysis-gas chromatography is understood to be a combination of pyrolytic decomposition and gas chromatographic separation of the decomposition products. The chromatograms can be assigned like finger prints to certain polymerizates. The method of pyrolysis-gas chromatography complements the pyrolysis-IR-spectroscopy method and is indispensible when IR spectroscopy is yielding insufficiently sensitive data for minor components. The high frequency pyrolysator (*Curie*-point-pyrolysis) is most often used as the pyrolysis system. The sample is placed in the form of a thin thread into the inside of a ferro-magnetric spiral. The most suitable pyrolysis temperature is 700 °C. The gas-chromatic separation is best obtained with a 30 m long thin layer capillary column using Polysev as the polar separation fluid.

A quantitative examination of the composition of elastomer mixtures with the help of pyrolysis-gas chromatography is only then possible, when a representative sample is available and the starting components are known. There is no theoretical possibility to detect the BR fraction in an unknown vulcanizate made from a BR-SBR blend with the exception that the two starting components are known (e.g. a mixture of cis-1,4 BR and SBR with a styrene content of approximately 23.5% by weight is present).

For an extensive literature survey about pyrolysis-gas chromatography see [6.200, 6.249]. Recently, further advances have been made [6.250–6.254].

6.4.5 High Pressure Liquid Chromatography and Automation of the Analysis Methods

[6.256]

The high pressure liquid chromatography (HPLC) has been known for some time and gains increasing importance for the analysis of organic compounds. It has been used, recently, for analyzing elastomers.

The sorption media are smaller than 1 μm, so that they have very large surface areas (200–500 m^2/g). The eluents are pressed through columns under high pressure (up to 300 bar). The fluids used are for example di-iso propyl ether, methanol, dioxan, tetrahydrofuran, water etc. HPLC is, because of the high number of analytical columns, very accurate.

This technique promises to simplify the analysis and has the possibility of automation. Because of the short time required it may be possible in the future to carry out the quality control of compounds and elastomers by chemical means. It serves to identify quantitatively and qualitatively additives, where the identification can be carried out, for example, with IR spectroscopy. Once the most important substances with their known spectra have been stored in a data base, the substances to be tested can be compared because of their functional groups by computer with the stored spectra and thus be identified with high probability. The identification can also be done by mass spectroscopy.

6.4.6 Thermal Analysis of Elastomers

[6.256 a–6.268]

These methods are less suitable to determine the composition of elastomers and compounds but serves to detect structural phenomena.

The principal of *differential thermal analysis (DTA)* [6.256 a–6.262] is that an enthalpy difference exists at constant temperature between the test specimen and a reference sample which shows up as a temperature difference between the two samples. This enthalpy difference gives, for example, information about crystallization, structural changes, glass transition temperature etc.

This test has the disadvantage that the heat changes cannot be determined qualitatively.

Another method, *differential scanning calorimetry (DSC),* overcomes this disadvantage [6.262 a–6.265]. There, instead of the temperatures, the difference of the heat flux between the specimen and the standard is measured. Therefore, the DSC method determines primarily changes in the enthalpy.

The *thermo-gravimetric analysis (TGA)* [6.266–6.268] is based upon the continuous monitoring of the weight (density) of the sample as a function of a programmed temperature change. This method serves primarily to detect decomposition reactions, impurities etc.

Thermo-mechanical analysis (TMA) is a one-dimensional dilatometric measurement for which the length of the sample is measured as a function of temperature. This method is suitable for expansion and penetration measurements. It is, furthermore, useful for the determination of the linear coefficient of expansion, the glass transition temperature, anisotropic effects etc.

With the help of newly developed *thermo analysis (TA) detectors* rather accurate determinations of Nitrosoamines are possible.

6.4.7 Mass Spectroscopy

[6.268 a–6.276]

With the increasing use of mass spectrometers and the required auxiliary equipment including even Laser-Pyrolysis devices [6.275] in laboratories, considerable progress has been made in the last few years to understand vulcanization [6.269] and degradation [6.270] mechanisms as well as to obtain a direct analysis of vulcanizates [6.271–6.276]. Herewith the old dream of rubber technologists to determine directly the composition of vulcanizate samples seems to be coming true [6.275, 6.276].

6.4.8 Nuclear Magnetic Resonance

[6.277–6.292]

The same applies to the method of nuclear magnetic resonance (NMR) which according to recent suggestions can be employed to analyze compounds and elastomers [6.280, 6.282, 6.284–6.286, 6.290–6.292]. In addition, the NMR method is utilized to elucidate polymer structures [6.281, 6.289] and networks [6.277. 6.288, 6.292] and to analyze degradation mechanisms [6.283] as well as the deformation behavior of elastomers [6.278, 6.279].

6.4.9 Analysis of Polymer Structure

[1.7, 6.268 a, 6.281, 6.289, 6.293-6.302]

and Network Structure

[1.13, 4.65, 6.303, 6.304]

The extraordinarily numerous new literature on structural investigations of polymers is described in [6.293, 6.294].

The methods applied are: High Pressure Liquid Chromatography [6.256], Gel-Permeation-Chromatography (GPC) [6.295-6.297], Thermal Analysis [6.256 a-6.268], Mass Spectroscopy [6.268 a-6.276], Torsional Braid Analysis (TBA), Electron Spectroscopy for Chemical Analysis (ESCA), Nuclear Magnetic Resonance [6.277-6.292], Electron Spin Resonance, Raman Spectroscopy, Neutron Scattering [6.298, 6.299], X-Ray Diffraction [6.300], Electron Microscopy, i.e. the direct visualization of single molecules [6.301, 6.302], acustics, turbodimetric titration and molecular weight determination by viscosity measurements [1.7].

The former chemical determination of sulfur cross-link structures is extremely time consuming. Recently, newer methods to determine cross-link structures have been proposed [6.277, 6.288, 6.292, 6.303, 6.304].

6.5 References on Elastomer Testing and Analysis

6.5.1 References on Elastomer Testing

6.5.1.1 General References (see also [1.2, 1.21, 3.15, 3.17])

[6.1] *Bach, H.:* Reproduzierbarkeit der Kautschukprüfung. KGK *40* (1987), pp. 651-655.

[6.1 a] *Brown, R. P.:* Physical Testing of Rubber. Elsevier Applied Science Publ., London, 1979.

[6.2] *Clamroth, R.:* Prüfung und Beurteilung der Elastomere. In: *Gohl, W. (Ed.):* Elastomere - Dicht- und Konstruktionswerkstoffe. Expert-Verlag Grafenau, 2nd Ed., 1980, p. 49.

[6.3] *Ecker, R.:* Mechanisch-Technologische Prüfung von Kautschuk und Gummi. In: *Boström, S. (Ed.):* Kautschuk-Handbuch. Verlag Berliner Union, Stuttgart, Vol. 5, 1962, p. 82.

[6.4] *Ecker, R.:* Entwicklungstendenzen in der Prüftechnik für Elastomere. KGK *16* (1963), p. 73.

[6.5] *Frank, K.:* Prüfungsbuch für Kautschuk und Kunststoffe. Verlag Berliner Union, Stuttgart, 1955.

[6.5 a] *Härtel, V., Schreiber, F., Theisen, D.:* Kurz- und Langzeitprüfungen von Werkstoffen bzw. Bauteilen und die Korrelation zur Praxis. KGK *40* (1987), pp. 656-664.

[6.5 b] *Härtel, V.:* Physikalische Methoden zur zerstörungsfreien Prüfung von Elastomeren. KGK *40* (1987), pp. 837-843.

[6.5 c] *Helmers, H.:* Fähigkeit der Gummi-Prüfverfahren. dkt '88, July 4-7, 1988, Nürnberg, West Germany.

[6.6] *Hofmann, W.:* Kautschuk, Prüfung, Ullmanns Encyclopädie der technischen Chemie. Verlag Chemie, Weinheim, 4th Ed., Vol. 13, p. 701.

[6.7] *Kainradl, P.:* Kritische Überlegungen zu den physikalischen und technologischen Prüfmethoden in der Gummiindustrie. KGK *31* (1978), p. 341.

[6.8] *Krevelen, D. W., Hoftyzer, P. J.:* Properties of Polymers. Elsevier Scientific Publishing Company, Amsterdam, 1976.

[6.9] *Lautenschlaeger, K. F.:* Das Konzept der Maximumeigenschaften elastischer Materialien. GAK *35* (1982), pp. 498, 680.

[6.10] *Menges, G., Grajewski, F., Limper, A., Weyer, G.:* Bessere Qualitätskontrolle durch Nachrüstung von Standardprüfgeräten. GAK *39* (1986), p. 126.

[6.11] *Payne, A.R.:* Physics and Physical Testing of Polymers. In: Progr. in High Polymers. Heywood Books, London, Vol. 2, 1968, pp. 1–93.

[6.12] *Schmitz, J.M.:* Testing of Polymers. Interscience Publ., New York, Vol. 3, 1965–1967.

[6.13] *Scott, J.R.:* Physical Testing of Rubbers. MacLaren, London; Palmerton, New York, 1965.

[6.14] *Timm, Th.:* Elemente des viskoelastischen Eigenschaftsbildes von Kautschuk und Elastomeren im Hinblick auf einige Probleme der Praxis. KGK *31* (1978), p. 901.

[6.15] *Varga, O.H.:* Stress-Strain Behavior of Elastic Materials. Interscience Publ., New York, 1966.

[6.16] *Welding, G.N.:* Rubber and Plastics Testing. Chapman and Hall, London, 1963.

[6.17] ...: Annual Book of ASTM Standards; Part 37: Rubber, Natural and Synthetic – General Test Methods; Carbon Black. Part 38: Rubber-Products, Industrial – Specifications and Related Test Methods; Gaskets; Tires, Ed.: American Society for Testing and Materials.

[6.18] ...: DIN-Taschenbuch 47, Kautschuk und Elastomere 1, Prüfnormen für physikalische Prüfverfahren, Beuth-Verlag GmbH, Berlin, Köln.

6.5.1.2 References on Determination of Viscosity

General References on Elastomer Viscosity

[6.19] *Borzenski, F.J.:* Application of Viscoelastic Theory to Rubber Processing. KGK *35* (1982), p. 937.

[6.20] *Ferry, J.D.:* Viscoelastic Properties of Elastomers. Paper 66, 125th ACS-Conf., Rubber-Div., May 8–11, 1984, Indianapolis, IN.

[6.21] *Gleißle, W.:* Rheology, VIII. Int. Tagung der Rheologie, Naples, 1980, Plenum Press, New York, Proc. Vol. 2, 1980.

[6.22] *Gleißle, W.:* Flow Behavior of Concentrated Suspensions at High Shear Stresses and Shear Rates. 8th Int. Technology-Conference on Slurry Transportations, March 15–18, 1983, San Francisco, CA, Proc. p. 103.

[6.23] *Koopmann, R.:* Verfahren zur genauen und umfassenden rheologischen Materialcharakterisierung. KGK *36*, (1983), p. 108.

[6.23 a] *Koopmann, R.:* The Rheology of Rubber Polymers and Mixes. IRC '86, June 2–6, 1986, Göteborg, Sweden.

[6.24] *Kramer, H., Koopmann, R.:* Der Einfluß thermisch-mechanischer Beanspruchung auf die Polymer-Viskosität bei der Probenvorbehandlung. KGK *37* (1984), p. 874.

[6.25] *Han, C.D.:* Rheology in Polymer Processing. Academic Press, New York, 1976.

[6.26] *Lenk, R.S.:* Polymer Rheology. Elsevier Applied Science Publ., London, 1978.

[6.27] *Soong, D.S.:* Time Dependent Nonlinear Viscoelastic Behavior of Polymer Fluids – A Review of Current Understanding. RCT *54* (1981), p. 641.

[6.27 a] *Poltersdorf, B., Schwambach, D.:* Rheologisches Verhalten gefüllter Kautschukmischungen. 1. Stationäres Fließverhalten. KGK *40* (1987), pp. 454–457; 2. Dynamisches Fließverhalten. KGK *41* (1988), pp. 40–43.

[6.28] ...: Praktische Rheologie der Kunststoffe. VDI-Ingenieurwissen, VDI-Verlag, Düsseldorf, 1978.

Squeeze Film Viscometer

[6.29] *Baader, Th.:* Die Kontrolle der Verarbeitbarkeit von Kautschukmischungen – Probleme und Ziele. Kautschuk *14* (1938), p. 223; DIN 53 514 (withdrawn).

[6.29 a] *Koopmann, R., Schnetger, J.:* Eine schnelle rheologische Mischungskontrolle. KGK *39* (1986), pp. 131–141.

[6.29 b] *Schramm, G.:* Rubber Testing with Defo-Elastometers. KGK *40* (1987), pp. 756–764.

[6.30] *Hoekstra, J.:* Gummi-Ztg. *53* (1939), p. 35.

[6.31] *Kluckow, P.:* Die Praxis des Gummichemikers. Verlag Berliner Union, Stuttgart, 1962, p. 247.

[6.32] *Lassagne, P.:* Vergleich von Plastizitäten nach Williams, van Rossem und Mooney. Rev. Gén. Caoutch. *25* (1948), p. 92.

[6.33] *Williams, I.:* The Plasticity of Rubber and its Measurement. Ind. Engng. Chem. *16* (1924), p. 362; ASTM-D 926-79.

Rotary Viscometer (Rheometer) (see also [6.32] and Section *Vulcameter*, page 497)

[6.33 a] *Gilbert, A.:* New Flexibility in Mooney Viscosity and Oscillating Disc Rheometry Testing From a Single New Combined Function Instrument. Rubbercon '87, June 1-5, 1987, Harrogate, GB, Proceed. B 62.

[6.34] *Koopmann, R.:* Improvements for Mooney-Viscosity. KGK *38* (1985), p. 28.

[6.35] *Kramer, H.:* Zum Problem der Vergleichbarkeit der Mooney-Viskositätsmessung zwischen verschiedenen Prüfstellen. KGK *33* (1980), p. 20.

[6.36] *Nakajima, N., Harrell, E. R.:* Calculation of Complex Viscosity from Slow Speed Transient Torque Measurement with Mooney-Rheometer. RCT *55* (1982), p. 1426.

[6.37] *Mooney, M.:* A Shearing Disc Plastometer for Unvulcanized Rubber. Ind. Engng. Chem. Anal. Ed. *6* (1934), p. 147; RCT *35* (5/1962), p. XXVII; ASTM-D 1646-74; DIN 53523 (Part 1-3).

[6.38] *Shearer, R., Juve, A. E., Musch, J. H.:* Measurement of the Scorch and Cure Rate of Vulcanizable Mixtures Using the Mooney Plastometer. Indian Rubber World *117* (2/1947), p. 216.

[6.39] *Weaver, J. V.:* Determination of Scorch by Means of Mooney-Viscosimeter. RCT *14* (1941), p. 458.

Ball Indentation Viscometer

[6.40] *Höppler, F.:* Rheologische und elastometrische Messungen an Kautschukprodukten. Kautschuk *17* (1941), p. 17.

Cup Viscometer

[6.41] *Behre, J.:* Über Plastizitätsmessungen in der Gummiindustrie. Kautschuk *8* (1932), p. 167; *15* (1939), p. 112.

[6.42] *Dillon, J. H. et al.:* Physics *4* (1933), p. 225; *7* (1936), p. 73; Rubber Age *41* (1937), p. 306.

[6.43] *Greiner, W., Behre, J.:* Schwankungen von Plastizität. Kautschuk *2* (1926), p. 207.

[6.44] *Karrer, E.:* Z. techn. Physik *8* (1930), p. 326.

[6.45] *Marzetti, I.:* Giorn. Chim. Ind. Appl. *5* (1923), p. 342; *6* (1924), pp. 277, 567; Indian Rubber World *68* (1923), p. 776.

[6.46] *Mooney, M.:* Fundamental Analysis of Non-Newton Viscosity, Analysis of Capillary and Cuette Viscosimetry. J. Rheol. *2* (1931), p. 210; Physics *7* (1936), p. 413.

High Pressure Capillary Rheometer (Processibility Tester)
(see also [3.238, 3.267, 4.547, 4.548, 6.21-6.23, 6.25-6.28]

[6.47] *Amsden, C. S.:* Prediction of Rubber Processing Behavior with the Monsanto Processibility Tester. Paper at the DGF-Tagung, Vinter Moda, Denmark, 1980.

[6.47 a] *Bohlin, L.:* Rheometry and the Processing of Rubbers. IRC '86, June 2-6, 1986, Göteborg, Sweden.

[6.48] *Colbert, G. P.:* Capillary Rheometry and its Application. Paper at the ACS-Conf., Southern Rubber Group, Nov. 10, 1978, Tampa, FL.

[6.49] *Colbert, G. P., Ziegel, K. D.:* Estimation of End Corrections in Capillary Flow. Paper 366, Golden Jubilee Conf., Soc. of Rheology, Oct. 28-Nov. 1, 1979, Boston, MA.

[6.50] *Froelich, D., Muller, R., Zang, Y. H.:* New Extensional Rheometer for Elongational Viscosity and Birefringence Measurement. Paper 66, 128th ACS-Conf., Rubber-Div., Oct. 1-4, 1985, Cleveland, OH.

[6.51] *Gleißle, W.:* Hochdruck-Kapillar-Rheometer. Technical Bulletin of Göttfert Werkstoff-Prüfmaschinen GmbH, Buchen, West Germany.

[6.52] *Göttfert, O.:* Das Rheometer, ein Prüfgerät zur schnellen Aussage über das Verarbeitungsverhalten von Kautschukmischungen. KGK *35* (1982), p. 849.

[6.53] *Graessley, W. W., Ver Strate, G.:* A Simple Method for Estimating Viscosity of Highly Viscous Liquids. RCT *53* (1980), p. 842.

[6.54] *Johnson, P. S.:* General Overview of Processibility Testing. KGK *33* (1980), p. 725.

[6.55] *Leblanc, J. L., Pintens, E. A.:* An Automated Capillary Rheometer to solve Practical Processibility Problems. KGK *34* (1981), p. 34.

[6.56] *Menges, G., Wortberg, J., Geisbüsch, P., Targiel, G.:* Fließverhalten von Kautschukmischungen. KGK *34* (1981), p. 631.

[6.57] *Nakajima, N., Collins, E. A.:* Capillary Rheometry of Carbon Filled Butadiene-Acrylonitrile Copolymers. RCT *48* (1975), p. 615.

[6.58] *Norman, R. H.:* Some Considerations in the Use of RAPRA/Monsanto Stress Relaxation Processibility Tester. RAPRA-Member Report, No. 53 (1980).

[6.59] *Norman, R. H., Johnson, P. S.:* Processibility Testing. RCT *54* (1981), p. 493 (Review).

[6.60] *Patel, A. C.:* Capillary Rheology of Carbon Black. 2. Separation of Viscoelastic and Elastic Forces. KGK *37* (1984), p. 303.

[6.61] *Röthemeyer, F.:* Entwicklung und Erprobung eines Meßgerätes zur Erfassung des Verarbeitungsverhaltens von Kautschukmischungen. KGK *33* (1980), p. 1011.

[6.62] *Sezua, J. A., Di Mauro, P. J.:* Processibility Testing of Injection Molding Rubber Compounds. RCT *57* (1984), p. 826.

[6.63] *Starita, J. M.:* Rheological Characterization and Relation to Processing. Paper 12, 121st ACS-Conf., Rubber-Div., May 4–7, 1982, Philadelphia, PA.

[6.64] *Toussaint, H. E., Unger, W. N., Schäfer, H. O.:* Untersuchung der Verarbeitungseigenschaften rußgefüllter Kautschukmischungen mit einem Stempelextruder. KGK *25* (1972), p. 155.

[6.64 a] *Trim, R. S.:* Predicting Processibility – The TMS Rheometer and its Application, Rubbercon '87, June 1–5, 1987, Harrogate, GB, Proceed. B 58.

[6.65] *Wazer, J. R., Lyons, J. W., Kim, K. Y., Collwell, R. E.:* Viscosity and Flow Measurement. Interscience Publ., New York, 1963.

[6.66] *Wesche, H.:* Untersuchungen des Fließverhaltens von Kautschukmischungen mit einem Kapillarviskosimeter. KGK *31* (1978), p. 495.

Rheovulcameter

[6.67] *Göttfert, O.:* Das Rheovulkameter, ein Prüfgerät zur schnellen Aussage über das Verarbeitungsverhalten von Kautschukmischungen. KGK *35* (1982), p. 849.

[6.68] *Kilthau, C., Wiegand, N.:* Vergleich einiger Prüfverfahren zur Überprüfung des Verarbeitungsverhaltens von Kautschukmischungen. Paper at the 2nd Göttinger Tagung des Fachbeirates Elastomerverarbeitung, 1981; VDI-K: Spritzgießen technischer Formartikel. VDI-Verlag, Düsseldorf, 1981; Paper see KGK *34* (1981), p. 1040.

Mixing and Extrusion Simulation Devices

[6.69] *Barth, H.:* Meßextruderergebnisse. Kunstst. *69* (1979), p. 370.

[6.70] *Brabender:* Indian Rubber World *117* (1947), pp. 62, 74.

[6.70 a] *Brichzin, D. H., Krambeer, M.:* Verschiedene Laboratoriumsmethoden zur Untersuchung der Verarbeitbarkeit von Kautschukmischungen. KGK *40* (1987), pp. 1152–1155.

[6.71] *van Buskirk, O. P.:* Characterizing Rubber Compounds with Brabender Plasticorder Miniature Internal Mixer. Paper at the 119th ACS-Conf., Rubber-Div., June 2–5, 1981, Minneapolis, MN.

[6.72] *Cotton, G.:* Significance of Extensional Flow in Processing of Rubbers – Typical Extensional Viscosity Results, Significance of Extensional Flow in Extrusion; Significance of Extensional Flow in Mixing. RCT *54* (1981), p. 61.

[6.73] *Leblanc, J. L.:* Factors Affecting the Extrudate Swell and Melt Fracture Phenomena of Rubber Compounds. RCT *54* (1981), p. 905.

[6.74] *Schöne, L.:* Einsatz von Computern bzw. Microprocessoren in Prüfverfahren in der Gummiindustrie. Technical Bulletin of Brabender OHG, Duisburg, West Germany.

[6.74a] *Schöne, L.:* Testing the Die Swell Behaviour of Polymers During Extrusion. Rubbercon '87, June 1–5, 1987, Harrogate, GB, Proceed. B 57.

[6.74b] *Schöne, L.:* Erfahrungen mit Meßextrudern und kleinen Meß- und Laborknetern der Brabender-Versuchseinrichtung. KGK *40* (1987), pp. 1156–1159.

[6.75] *Sezua, J.A., Di Mauro, P.J.:* Synthetic Polymer Processibility Testing – Mixing. Paper 63, 123rd ACS-Conf., Rubber-Div., May 10–12, 1983, Toronto, Ont.

[6.76] *Sezua, J.A.:* Processibility Testing of Extrusion Compounds. Paper 22, 125th ACS-Conf., Rubber-Div., May 8–11, 1984, Denver, CO.

[6.77] ...: Extrusiometer. Technical Bulletin of Göttfert Werkstoff-Prüfmaschinen GmbH, Buchen, West Germany.

[6.78] ...: Plastograph, Plasti-Corder. Technical Bulletin of Brabender OHG, Duisburg, West Germany.

[6.79] ...: Torsiograph, der messende Laborkneter. Technical Bulletin of Göttfert Werkstoff-Prüfmaschinen GmbH, Buchen, West Germany.

6.5.1.3 References on Solubility and Tack Measurements

Solubility Measurements (see also [6.40])

[6.80] *Heinz, W.:* Kolloid-Z. *145* (1956), p. 119.

Tack Measurements (see also [4.525a–4.537])

[6.81] *Umminger, O.:* Kunstst. *42* (1952), p. 169.

[6.82] *Bauer, R.F.:* J. Polym. Sci. Polym. Phys. Ed. *10* (1972), p. 541.

[6.83] *Ghag, A.S.:* Rubber News *11* (1972), p. 18.

[6.84] ...: Bull. Lab. Rech. Caoutch. 90, pp. 1–34.

6.5.1.4 References on Vulcanization Reactions

Step-Wise Heating

[6.85] *Mengjiao, W., Dingxi, C.:* Studies on State of Cure by Programmed Temperature Press, IRC 81, June 8–12, 1981, Harrogate, GB.

Mooney Scorch (see also [6.38, 6.39])

[6.86] DIN 53523, Part 4, 1976; ASTM-D 1646–74

T-50-Test

[6.87] ASTM-D 599-40 T

Vulcameter

[6.88] *Göttfert, O.:* Der Elastograph – ein rotorloses Rotationsschub-Vulkameter. KGK *29* (1976), pp. 261, 341.

[6.89] *Hands, D., Horsfall, F.:* A New Method for Simulating Industrial Cure Process. KGK *33* (1980), p. 440.

[6.90] *Hands, D., Norman, R.H., Stevens, P.:* A New Curemeter. KGK *39* (1986), pp. 330–333.

[6.91] *Härtel, V.:* Differentielle Vulkametrie – Eine Methode zur direkten Registrierung von logarithmischen Vernetzungsisothermen und Reaktionsgeschwindigkeiten mit dem Vulkameter. KGK *31* (1978), p. 415.

[6.92] *Hofmann, W.:* Ein Überblick über 15 Jahre Vulkametrie. GAK *27* (1974), p. 265; Europ. Rubber J. *1* (1974), p. 42.

[6.93] *Ippen, J., Jahn, H.-J., Kramer, H.:* Vulkanisationsoptimierung voluminöser Gummikörper. KGK *28* (1975), pp. 647, 720; Bayer-Mitt. f. d. Gummi-Industrie *49* (1976), p. 21, Technical Bulletin of Bayer AG, Leverkusen, West Germany.

[6.93 a] *Markert, J.:* Beurteilung der Verarbeitungscharakteristik neuer Kautschuke mit Hilfe eines Rotations-Rheometer-Tests. dkt '88, July 4–7, 1988, Nürnberg, West Germany.

[6.94] *Pawlowski, H. A., Perry, A. L.:* A New Automatic Curemeter. PRI 84, March 12–16, 1984, Birmingham, GB.

[6.95] *Peter, J., Heidemann, W.:* Eine neue Methode zur Bestimmung der optimalen Vulkanisation von Kautschuk-Mischungen. KGK *10* (1957), p. 168; *11* (1958), p. 159.

[6.95 a] *Persson, S.:* Dielectric Vulcametry. IRC '86, June 2–6, 1986, Göteborg, Sweden.

[6.96] *Roslaniec, Z., Raczkiewicz, M.:* Eine neue Interpretation von Vulkameterkurven. KGK *33* (1980), p. 536.

[6.97] *Streit, G., Hofmann, G.:* Rechnerunterstützte Auswertung von Vulkameterkurven. KGK *35* (1982), p. 497.

[6.98] *Tosaki, C., Ito, K., Ninonuya, K.:* A New Oscillating-Die Type Curemeter for Quality Control-Testing. Paper 38, 126th ACS-Conf., Rubber-Div., Oct. 23–26, 1984, Denver, CO.

[6.99] DIN 53 529, Bl. 1 und 2.

6.5.1.5 References on Mechanical-Technological Testing of Vulcanizates

Sample Preparation

[6.100] ISO DP 2214; ASTM-D 3182-72; ASTM-D 3188-73; DIN 53 502.

Tensile Testing

[6.101] ISO R 37-1968; ASTM-D 412-75; DIN 53 504.

[6.102] *Müller, F. H.:* Betrachtungen zum Zug-Dehnungsverhalten von Kautschuk. Kautschuk u. Gummi *9* (1956), p. WT 197.

[6.103] ISO 2285-1975; ASTM-D 1456-61; DIN 53 518.

[6.104] *Kainradl, P., Händler, F.:* Verformungseigenschaften von Vulkanisaten. Kautschuk u. Gummi *10* (1957), p. WT 278.

[6.105] *Tobisch, K.:* Contribution to the Mathematical Description of Stress-Strain Behavior of Elastomers. RCT *53* (1980), p. 836.

[6.106] *Clamroth, R., Picht, W.:* Elektronische Datenverarbeitung und Automation bei Materialprüfung von Elastomeren. KGK *32* (1979), p. 660.

[6.106 a] *Akhtar, S., De, S. K.:* SEM Studies on Morphology and Tensile and Tear Failure of Blends of Polyethylene and trans-Polyoctenamer. KGK *39* (1986), pp. 327–329.

Tear Testing (Tear Resistance, Tear energy)

[6.107] ASTM-D 624-73; ASTM 3629-78; DIN 53 507; DIN 53 515.

[6.108] *Clamroth, R., Eisele, U.:* KGK *28* (1975), p. 433.

[6.109] *Clamroth, R., Kempermann, Th.:* Vergleich verschiedener Prüfmethoden zur Bestimmung des Weiterreißwiderstandes. KGK *38* (1985), p. 206.

[6.110] *Eisele, U.:* Zum Einfluß der Mikrostruktur auf die Weiterreißenergie von Synthesekautschuken. GAK *31* (1978), p. 724.

[6.110 a] *Kilian, H. G.:* Möglichkeiten und Grenzen der Charakterisierung von Elastomeren im Zugversuch. GAK *39* (1986), pp. 548–552.

[6.111] DIN 53 506.

Hardness Measurement

[6.112] ISO 868; ASTM-D 2240-75; BS 903: 19: 1950; DIN 53 505 (Shore).

[6.113] ASTM-D 1415-68 (International).

[6.114] ASTM-D 531-78 (Pusey-Jones).

[6.115] DIN 53 456 (Brinell).

[6.116] *Petzold, H.:* Die Härteprüfung als Methode zur Bestimmung des Elastizitätsmoduls. GAK *32* (1979), p. 824.

[6.117] *Pöllert, P.:* Härteprüfung an Kunststoffen und Gummi. KGK *32* (1979), p. 877.

[6.118] *Soden, A. L.:* A Practical Manual for Rubber Hardness Testing. MacLaren, London, 1951.
[6.119] *Späth, A.:* Gummi, Asbest *8* (1955), p. 418.
[6.120] *Tobisch, K.:* Über den Zusammenhang zwischen Härte und Elastizitätsmodul bei Elastomeren. KGK *34* (1981), p. 105.

Permanent Set (see also [3.218–3.224]), **Relaxation and Flow Test** (see also [3.216])

[6.121] ISO R 815-1969; ASTM-D 395-78; DIN 53 517 (Druckverformungsrest).
[6.122] ASTM-D 1229-62 (Druckverformungsrest bei tiefen Temperaturen).
[6.123] DIN 53 518 (Zugverformungsrest).
[6.124] DIN 53 537; ASTM-D 1390-76 (Spannungsrelaxation).
[6.125] *Clamroth, R., Ruetz, L.:* Spannungsrelaxation als Schnellmethode zur Vorhersage des Langzeitverhaltens. KGK *34* (1981), p. 836; RCT *56* (1983), p. 31.
[6.126] *Tobisch, K., Hoffmann, H.:* Langzeitverhalten der Spannungsrelaxation von Elastomeren für Dichtringe in Trink- und Abwasserrohren. KGK *37* (1984), p. 945.

Rebound Elasticity

[6.127] ISO/DIS 4662; ASTM-D 1054-66; DIN 53 512.
[6.128] ASTM-D 2634-74 (Vertical).
[6.129] *Meyer, R.:* Ein Rückprall-Elastizitätsprüfgerät mit erweiterten Möglichkeiten. KGK *34* (1981), p. 750.
[6.130] *Rehage, R.:* Gummielastizität und Struktur von Elastomeren. KGK *34* (1981), p. 715.

Visco-Elastic Behavior and Damping

[6.131] ISO/DIS 4664-1975; ASTM-D 945-77; DIN 53 513 (Dämpfung, Druck).
[6.132] ASTM-D 412-75 (Dämpfung, Zug).
[6.133] ASTM-D 945-77; DIN 53 520 (Dämpfung, freie Schwingung).
[6.134] ASTM-D 797-64 (Youngs-Modulus).
[6.134 a] *Bandel, P., Giuliani, G. P., Volpi, A.:* Analysis of the Dynamic Properties of Rubber Compounds and Cord-Rubber Composites in Tyres. KGK *39* (1986), pp. 793–799.
[6.134 b] *Devis, B., De Meersman, C., Peters, J.:* Digital Analysis of Relaxation Properties for Quality and Process Control in the Rubber Industry. Rubbercon '87, June 1–5, 1987, Harrogate, GB, Proceed. B 54.
[6.134 c] *Driscoll, S.:* Dynamic Rheological Measurements for On-Line and Off-Line Quality Control of Rubber. Rubbercon '87, June 1–5, 1987, Harrogate, GB, Proceed. B 55.
[6.135] *Flocke, H. A.:* Ein Beitrag zur Messung der viskoelastischen Eigenschaften von Vulkanisaten unter technischen Schwingungsbeanspruchungen. KGK *34* (1981), p. 376.
[6.136] *Funt, J. M.:* Dynamic Properties and Compound Performance. Paper 80, 128th ACS-Conf., Rubber-Div., Oct. 1–4, 1985, Cleveland, OH.
[6.136 a] *Funt, J. M.:* Dynamic Testing and Reinforcement of Rubber. IRC '86, June 2–6, 1986, Göteborg, Sweden.
[6.137] *Gaddum, F.:* Messung der diskontinuierlichen Spannungsrelaxation und des viskoelastischen Verhaltens von Kautschuken und Kunststoffen. KGK *32* (1979), p. 774.
[6.138] *Gall, P. S., Elia, A.:* Dynamic Mechanical Analysis for Characterization of Elastomer Viscoelastic Properties. Paper 48, 124th ACS-Conf., Rubber-Div., Oct. 25–28, 1983, Houston, TX.
[6.139] *Howgate, P. G.:* The Effect of Shape upon the Damping Characteristics of Elastomers. Paper 35, 118th ACS-Conf., Rubber-Div., Oct. 7–10, 1980, Detroit, MI.
[6.140] *de Meersman, C., Vandoren, P.:* Eine neue Methode zur Bestimmung des Verhaltens von Gummi unter dynamischer Beanspruchung. GAK *34* (1981), p. 280.
[6.141] *de Meersman, C., Peters, H. C. J., Devis, B.:* The Dynaliser: A New Dimension in Materials Testing Technology. Paper 87, 128th ACS-Conf., Rubber-Div., Oct. 1–4, 1985, Cleveland, OH.
[6.141 a] *Ng, T. S.:* Das Fließ- und Relaxationsverhalten unvulkanisierter Kautschukmischungen. KGK *39* (1986), pp. 1175–1181.

[6.142] *Seeger, M.:* The Measurement of Dynamic Properties of Elastomers with A Computerized Torsional Pendulum. IRC 79, Oct. 3-6, 1979, Venice, Italy, Proc. p. 442.
[6.143] *Seeger, W.:* Vollautomatische Torsionsschwingungsprüfung an polymeren Werkstoffen mit digitalem Rechner. KGK *33* (1980), p. 545.
[6.144] *Sullivan, J. L., Morman, K. N., Pett, R. A.:* A Non-Linear Viscoelastic Characterization of A Natural Rubber Gum Vulcanizate. RCT *53* (1980), p. 805.
[6.145] *Volpi, A., Guiliani, G.:* Development in Dynamic Testing Procedures: Effect of Prestrain Specimen Shape and Nonsinusoidal Waveform. Paper 79, 128th ACS-Conf., Rubber-Div., Oct. 1-4, 1985, Cleveland, OH.

Heat Build-Up

[6.146] ASTM-D 623-67.
[6.147] *Kempermann, Th.:* Der Heat Build-Up in Abhängigkeit vom Vernetzungssystem. GAK *31* (1978), p. 941; *32* (1979), p. 96.

Abrasion and Wear

[6.148] DIN 53516 (DIN-Abrieb).
[6.149] ASTM-D 2228-69 (Pico-Abrieb).
[6.150] ASTM-D 1630-61 (NBS-Abrieb).
[6.151] *Fischer, W., Mattil, K.:* Neues Verfahren zur Bestimmung des Verschleißes und des Gleitverhaltens von Sohlenwerkstoffen. GAK *32* (1979), p. 370.
[6.152] *Hoffmann, H., Tobisch, K.:* Das Langzeitverhalten von Vergleichselastomerplatten für die Abriebprüfung nach DIN 53516. KGK *33* (1980), p. 101.
[6.153] *Schallamach, A.:* Gummiabrieb. GAK *31* (1978), p. 502.

Ageing Tests (see also [4.258-4.275, 6.125])

[6.154] ISO 188-1976; ASTM-D 573-78; DIN 53508 (Wärmelagerung).
[6.155] ASTM-D 572-73 (Bierer-Davis).
[6.156] ASTM-D 454-53 (Druckluft).
[6.157] ASTM-D 865-52 (Zellenalterung).
[6.158] *Ecker, R.:* Archiv f. techn. Messen V 8276-7 (Febr.1960).

Ozone Cracking, Light Cracking (see also [4.276-4.281, 4.320-4.335])

[6.159] ISO/TC 45/WG 7; ISO 1431-1972; ASTM-D 1149-78a; ASTM-D 1171-68; DIN 53509 Part 1 and 2 (statische Messung).
[6.160] ASTM-D 3395-75 (dynamische Messung).
[6.161] ASTM-D 518-61 (unter Dehnung).
[6.162] *Ban, L. L., Doyle, M. G., Smith, G. R.:* Morphological Characterization of Fatigue and Ozone Crack-Growth in Sidewall Compounds. Paper 67, 128th ACS-Conf., Rubber-Div., Oct. 1-4, 1985, Cleveland, OH.
[6.163] *Kempermann, Th.:* GAK *26* (1973), p. 90.
[6.164] *Seeberger, D. B.:* Praxis der Ozonalterung. GAK *40* (1987), pp. 64-71.
[6.164a] *Wündrich, K.:* Ein einfaches direktes chemisches Verfahren zur Bestimmung der Ozonkonzentration in Prüfkammern. KGK *39* (1986), pp. 633-636.

Dynamic Crack Formation (see also [4.281, 4.333-4.335, 6.162])

[6.165] ISO 132-1975; ASTM-D 430-73; DIN 53522, Part 1 and 2 (Biegeermüdung, De Mattia).
[6.166] ISO 133-1975; ASTM-D 813-59; ASTM-D 1052-55; DIN 53522, Part 3 (Rißwachstum).
[6.167] *Bandel, P., Morando, A.:* Machine for Testing the Fatigue Resistance of Rubber Compounds. KGK *37* (1984), p. 34.
[6.168] *Buist, J. M.:* Fundamentals of Rubber Technology. ICI (Ed.), 1947, p. 162.

[6.169] *Knauss, W. G.:* The Effect of Viscoelastic Material Behavior on Fatigue Crack Propagation. Paper 65, 125th ACS-Conf., Rubber-Div., May 8-11, 1984, Indianapolis, IN.
[6.169 a] *Lionnet, R., Orband, A.:* A New Environmental Fatigue Tester. Rubbercon '87, June 1-5, 1987, Harrogate, GB, Proceed. A 51.
[6.170] *Minguez, M. G., Royo, J.:* Evaluation of Results of Tension Fatigue Resistance Tests on Vulcanized Rubber. In: *Brown, R. (Ed.):* Polymer Testing. Elsevier Applied Science Publ., London, Vol. 1, 1980, p. 287.
[6.170 a] *Muhr, A. H., Thomas, A. G.:* A New Machine for Determining Crack Growth Characteristics. Rubbercon '87, June 1-5, 1987, Harrogate, GB, Proceed. B 46.
[6.171] *Newton, R. G.:* Trans. IRI *15* (1939), p. 172.
[6.172] *Peters, H. C. J., de Meersman, C., Broeckx, J., Devis, B.:* A Rubber Fatigue Testing Device with Constant Energy Input Pro Cycle. Paper 88, 128th ACS-Conf., Rubber-Div., Oct. 1-4, 1985, Cleveland, OH.
[6.172 a] *Peters, J., Broeckx, J., de Meersman, C.:* A Fatigue Testing Device with Constant Elastic Energy-Input per Cycle. KGK *39* (1986), pp. 939-949.
[6.173] ...: Fatigue of Elastomers. Bibliography 10, Ed. RCT.

Low Temperature Behavior (see also [6.131-6.134])

[6.174] ASTM-D 1053-73; ASTM-D 3388-75; DIN 53545 (T_G, Glasübergang).
[6.175] ASTM-D 2137-75 (T_S, Brittleness Point).
[6.176] ASTM-D 1329-78 (T_R-Test).
[6.177 a] *Engelmann, E.:* Kälteverhalten von Vulkanisaten - Einfluß des Mischungsaufbaues und Wahl der Prüfmethode. KGK *25* (1972), p. 538; Bayer-Mitt. f. d. Gummi-Industrie *47* (1973), p. 31.
[6.177 a] *Nordsiek, K. H.:* Die Tg-Beziehung - Grundlagen und Bedeutung für die Kautschuktechnologie. KGK *39* (1986), pp. 599-611.

Swelling and Permeation (see also [3.190-3.211])

[6.178] ISO 1817; ASTM-D 471-79; BS 903: 27: 1950; DIN 53521 (Quellverhalten).
[6.179] ASTM-D 814-76; DIN 53536 (Gasdurchlässigkeit).
[6.180] *Schuck, H.:* Die Gaspermeabilität von Hochpolymeren, insbesondere Kautschuken. KGK *33* (1980), p. 705.

Heat Conductivity

[6.181] DIN 52612.
[6.182] *Fischer, F.:* Wärmeleitfähigkeit von amorphen und teilkristallinen Kunststoffen. GAK *32* (1979), p. 922.

6.5.1.6 References on Electrical Testing of Vulcanizates

[6.183] ASTM-D 991-75; DIN 53482; 53596 (Widerstand).
[6.184] DIN 53483 (dielektrischer Verlustfaktor).
[6.185] DIN 53481 (Durchschlagfestigkeit).
[6.186] DIN 53480 (Kriechstromfestigkeit).
[6.187] DIN 53484 (Lichtbogenfestigkeit).
[6.188] *Clamroth, R.:* Elektrische Prüfungen an Kautschuk und Gummi. In: *Boström, S. (Ed.):* Kautschuk-Handbuch. Verlag Berliner Union, Stuttgart, Vol. 5, 1962, p. 226.
[6.189] *Rogowski, W.:* Arch. Elektrotechn. *12* (1923), p. 1; *16* (1926), p. 76.
[6.190] *Schön, G., Vieth, G.:* Kautschuk u. Gummi *9* (1956), p. WT 159.
[6.191] *Schwenkhagen, H. F.:* Elektrizitätswirtsch. *42* (1943), p. 120.
[6.192] *Umminger, O.:* Kautschuk u. Gummi *11* (1958), p. WT 297.
[6.193] *Wagner, K. W.:* Arch. Elektrotechn. *39* (1948), p. 215.
[6.193 a] *Wardell, G. E.:* Measurement of Electrical Conductance for On-Line Rubber Process Control. Rubbercon '87, June 1-5, 1987, Harrogate, GB, Proceed. P 4.

6.5.2 References on Elastomer Analysis

6.5.2.1 References on Quantitative Chemical Rubber Analysis

General References (see also [6.258, 6.264, 6.266–6.276, 6.280, 6.282–6.286, 6.290–6.292])

[6.194] *Bark, L. S.:* Analysis of Polymer Systems. Elsevier Applied Science Publ., London, 1982.

[6.194a] *Chu, C. Y.:* Modern Analytical Methods Used in the Rubber Industry – The Polysar Experience. KGK *41* (1988), pp. 33–39.

[6.195] *Frey, H. E.:* Methoden zur chemischen Analyse von Gummi-Mischungen. Springer-Verlag, Berlin, Göttingen, Heidelberg, 1953.

[6.196] *Höss, E., Auler, A., Kirchhof, F.:* Chemische Prüfung von Kautschuk und Gummi. In: *Boström, S. (Ed.):* Kautschuk-Handbuch. Verlag Berliner Union, Stuttgart, Vol. 5, 1962, p. 1.

[6.197] *Kluckow, P.:* Die Praxis des Gummichemikers. Verlag Berliner Union, Stuttgart, 1954.

[6.197a] *Mersch, F., Zimmer, R.:* Analysis of Rubber Vulcanizates by Advanced Chemical Techniques. KGK *39* (1986), pp. 427–432.

[6.198] *Ostromow, H.:* Analyse von Kautschuken und Elastomeren. Springer-Verlag, Berlin, Heidelberg, New York, 1981.

[6.199] *Ostromow, H., Hofmann, W.:* Chemische Analyse von Kautschukmischungen und -vulkanisaten. Bayer-Mitteilungen f. d. Gummi-Industrie, 1. *34* (1964), p. 68; 2. *35* (1965), p. 78; 3. *36* (1965), p. 87; 4. *37* (1966), p. 72; 5. *38* (1966), p. 43; 6. *39* (1967), p. 65; 7. *40* (1967), p. 60; 8. *41* (1968), p. 53; 9. *42* (1968), p. 31; 10. *45* (1972), p. 27, Technical Bulletin of Bayer AG, West Germany.

[6.200] *Ostromow, H., Hofmann, W.:* Untersuchung von Bedarfsgegenständen aus Gummi. Beilage zur 18. bzw. 41. Mitteilung des Bundesgesundheitsamtes über die "Untersuchung von Kunststoffen, soweit sie als Bedarfsgegenstände im Sinne des Lebensmittelgesetzes verwendet werden," Bundesgesundh. bl. *14* (1971), No. 8, p. 104; MPV-Berichte *2* (1978), Dietrich Reimer-Verlag, Berlin. (ED.: BGA, Berlin).

[6.201] *Scheele, W., Gensch, Ch.:* Über die qualitative Bestimmung von Kautschukhilfsstoffen. Kautschuk u. Gummi *6* (1953), p. WT 147; *7* (1954), p. WT 122; *8* (1955), p. WT 55.

[6.201a] *Schnecko, H., Angerer, G.:* Developments and Problems in Rubber and Tyre Materials Analysis. Rubbercon '87, June 1–5, 1987, Harrogate, GB, Proceed. A 44.

[6.201b] *Schnecko, H., Angerer, G.:* Entwicklungstendenzen und Probleme bei der Analyse von Kautschuk und Gummi. KGK *41* (1988), pp. 149–153.

[6.201c] *Schuster, R. H.:* Genauigkeit von analytischen Prüfverfahren unter Qualitätsgesichtspunkten. KGK *40* (1987), pp. 642–650.

[6.202] *Wake, W. C.:* Die Analyse von Kautschuk und kautschukartigen Polymeren. Verlag Berliner Union, Stuttgart, 1960.

[6.203] *Wake, W. C., Tidd, B. K.:* Analysis of Rubber and Rubber-like Polymers. Elsevier Applied Science Publ., London, 3rd Ed. 1983.

[6.204] ...: DIN-Taschenbuch 131, Kautschuk und Elastomere 2, Prüfnormen für chemische Prüfverfahren, Bodenbeläge, Latex, Ruße und Schaumstoffe.

Separation Procedure

[6.205] DIN 53 525, Part 1–4, für Kautschuk; DIN 53 551, für chemische Prüfungen (Probenahme).

[6.206] DIN 53 554 (Bestimmung der Feuchtigkeit).

[6.207] DIN 53 536 (Bestimmung wasserlöslicher Anteile).

[6.208] DIN 53 553 (Bestimmung acton- und chloroformlöslicher Anteile).

[6.209] DIN 53 588 (Bestimmung Methanol/Kaliumhydroxid-löslicher Anteile).

Paper and Thin Layer Chromatographical Separation (see also [6.240a])

[6.210] DIN 53622, Part 1 (Dünnschichtchromatographie).

[6.211] *Bellamy, J.L., et al.:* Chromatographic Analysis of Rubber Compounding Ingredients and their Identification in Vulcanizates. Trans. IRI *22* (1955), p.308.

[6.212] *Gedeon, B.J., Chu, T., Copeland, S.:* The Identification of Rubber Compounding Ingredients Using Thin Layer Chromatography. RCT *56* (1983), p.1080.

[6.213] *Weber, K.:* Der papierchromatographische Nachweis einiger Vulkanisationsbeschleuniger. Plaste und Kautschuk *1* (1954), p.35.

[6.214] *Zijp, J.W.H.:* Papierchromatographische Identifizierung von Vulkanisationsbeschleunigern. Kautschuk u. Gummi *8* (1953), p. WT 160.

Determination of the Extractable Components

[6.215] ASTM-D 3156-73; DIN 53622, Part 2 (dünnschichtchromatographische Bestimmung von Stabilisatoren und Alterungsschutzmitteln).

[6.216] *Baker, K.M., et al.:* Analyse von Beschleunigern und Alterungsschutzmitteln in Kautschuken – ein Überblick über vorhandene Bestimmungsmethoden. KGK *33* (1980), p.175.

[6.217] *Brook, M.C., Louth, G.D.:* Identification of Accelerators and Antioxidants in Compounded Rubber Products. Analyt. Chem. *27* (1955), p.1575.

[6.218] *Burchfield, H.P.:* Qualitative Spot-Tests for Rubber Polymers. Ind. Engng. Chem. Anal. Ed. *17* (1945), p.806.

[6.219] *Burchfield, H.P., Judy, N.:* Color Reactions of Amine Antioxidants. Ind. Engng. Chem. Anal. Ed. *19* (1947), p.786.

[6.220] *Hilverley, R.A., et al.:* Detection of Some Antioxidants in Vulcanized Rubber Stocks. Analyt. Chem. *27* (1955), p.100.

[6.221] *Kress, K.E., Steevens Mees, F.G.:* Identification of Curing Agents in Rubber Products. Analyt. Chem. *27* (1955), p.528.

[6.222] *Lattimer, R.P., Hooser, R., Diem, H.E., Rhee, C.K.:* Analytical Characterization of Tackifying Resins. Paper 53, 120th ACS-Conf., Rubber-Div., Oct. 13–16, 1981, Cleveland, OH.

[6.223] *Lussier, F.E.:* Analysis of Compounded Elastomers – A Practical Approach. Paper 27, 123rd ACS-Conf., Rubber-Div., May 10–12, 1983, Toronto, Ont.

[6.224] *Potter, N.M., Mehta, R.K.S., Wyzgoski, M.G.:* Determination of p-Phenylenediamine Antiozonants in Neoprene. Paper 23, 123rd ACS-Conf., Rubber-Div., May 10–12, 1983, Toronto, Ont.

[6.225] *Wandel, M., Tengler, H., Ostromow, H.:* Die Analyse von Weichmachern. Springer-Verlag, Berlin, Heidelberg, New York, 1976.

[6.226] Ref. [6.199], No. 6 and 7.

[6.227] Ref. [6.199], No. 8.

[6.228] Ref. [6.199], No. 9.

Determination of Fillers

[6.229] DIN 53568 (Veraschung).

[6.230] *Charlesby, E.L., Dunn, D.G.:* The Application of Thermogravimetry (TG) to the Characterization and Quantitative Determination of Carbon Blacks. RCT *55* (1982), p.382.

[6.231] *Hummel, K., Groyer, S., Lechner, H.:* Bestimmung von Füllstoffen in vernetztem Polybutadien durch Abbau mittels Olefin-Metathese. KGK *35* (1982), p.731.

[6.232] *Kolthoff, J.N., Gutmacher, R.G.:* J. Analyt. Chem. *22* (1950), p.1002.

[6.233] Ref. [6.195], p.56.

[6.234] Ref. [6.199], No. 3.

Determination of Sulfur

[6.235] DIN 53561 (Schwefelaufschluß, Methode A nach Wurzschmitt, Methode B nach Schöninger).

[6.236] Ref. [6.199] No. 5.
[6.237] Ref. [6.199] No. 4.
[6.238] *Zimmer, H., Kretzschmar, H. J., Strauß, K., Gross, D.:* Potentiometrische Bestimmung des Gesamtschwefelgehaltes in Elastomeren. KGK *33* (1980), p. 175.

6.5.2.2 References on Spectroscopic and Chromatographic Methods

Infra Red Spectroscopy (see also [6.269, 6.282, 6.284])

[6.239] *Barnes, R. L., et al.:* Ind. Engng. Chem. Anal. Ed. *16* (1944), p. 9.
[6.240] *Bentley, Freeman, F.:* Techn. Rep. 54-268, Washington, DC, 1956.
[6.240 a] *Brück, D.:* Analysis of Unvulcanized Rubber by Physical Methods, with Particular Reference to IR-Spectroscopy and Thin-Layer Chromatography. KGK *39* (1986), pp. 1165-1174.
[6.241] *Dinsmore, H. L., Smith, D. C.:* Analyt. Chem. *20* (1948), p. 11.
[6.242] *Fiorenza, A., Bonani, G.:* RCT *36* (1963), p. 1129.
[6.243] *Harms, O. L.:* Analyt. Chem. *25* (1953), p. 1140.
[6.244] *Kruse, P. F., Wallace, W. B.:* Analyt. Chem. *25* (1953), p. 1156.
[6.245] *Schnetger, J., et al.:* Einsatz von Extraktionsverfahren und Infrarotspektroskopie zur quantitativen Bestimmung von unvernetzten EPDM-Anteilen in Vulkanisat-Verschnittmischungen. KGK *33* (1980), p. 175.
[6.246] *Tyron, M., et al.:* Rubber World *134* (1956), p. 421; Kautschuk u. Gummi *10* (1957), p. WT 167.
[6.247] ISO/TC-45-2852 (Identifikation von Kautschuk durch IR-Spektroskopie).

Pyrolysis in Connection with Paper or Thin Layer Chromatography
(see also [6.210-6.214])

[6.248] *Feuerberg, H.:* Die Identifizierung und die quantitative Bestimmung von Elastomeren in Gummi. Kautschuk u. Gummi *14* (1961), p. WT 33.

Pyrolysis-Gaschromatography (see also [6.200, 6.271])

[6.249] *Canji, E.:* Angew. Makromol. Chem. *29/30* (1973), p. 491; *33* (1973), p. 143; *35* (1974), p. 27; *36* (1974), pp. 67, 75.
[6.250] *Kretzschmar, H.-J., Kelm, J., Tengicki, H., Gross, D.:* Die Analyse von EPDM/SBR-Vulkanisaten durch Pyrolyse-Gaschromatographie und Spektrometrie. KGK *34* (1981), p. 846.
[6.251] *Krishen, A.:* Chromatographic Techniques for Analysis of Rubber - Concepts and Applications. Paper 45, 120th ACS-Conf., Rubber-Div., Oct. 13-16, 1981, Cleveland, OH.
[6.252] *Levy, E. J.:* Application of A Pyrolysis Tuning Procedure in the Development of Pyrolysis-Gas Chromatography Pattern. Paper 43, 120th ACS-Conf., Rubber-Div., Oct. 13-16, 1981, Cleveland, OH.
[6.253] *Schrafft, R.:* Automatische Curie-Punkt-Pyrolysegaschromatographie mit Kapillarsäulen zur Analyse von Kautschukmischungen. KGK *36* (1983), p. 851.
[6.254] *Sullivan, A. B., Kuhls, G. H.:* The Evolution of Chromatographic Techniques for Determining the Fate of Curatives During Sulfur Vulcanization. Paper 9, 123rd ACS-Conf., Rubber-Div., May 10-12, 1983, Toronto, Ont.
[6.254 a] *Schuster, R. H.:* Identifizierung von Rußen durch Gaschromatographie. dkt '88, July 4-7, 1988, Nürnberg, West Germany.
[6.255] ASTM-D 3452-78.

High Efficiency Liquid Chromatography

[6.256] *Gross, D., Strauß, K.:* Hochleistungsflüssigkeitschromatographie in der Gummi-Analyse. KGK *32* (1979), p. 18.

6.5.2.3 References on Thermo Analysis

General References

[6.256 a] *Baker, K., Leckenby, J.:* Improved Quality Assurance of Elastomers with Thermal Analysis. KGK *40* (1987), pp. 223–227.

[6.257] *Brazier, D. W.:* Applications of Thermal Analytical Procedures in the Study of Elastomers and Elastomer Systems. RCT *53* (1980), p. 437, Review.

[6.258] *Brazier, D. W., Nickel, G. H., Szentgyorgyi, Z.:* Enthalpic Analysis of Vulcanization by Calorimetry, Thiuram Monosulfide/Sulfur Vulcanization of NR, BR and SBR. RCT *53* (1980), p. 160.

[6.259] *Cassel, B.:* Thermal Analytical Test Methods for Characterizing Elastomers. Paper 26, 117th ACS-Conf., Rubber-Div., May 20–23, 1980, Las Vegas, NV.

[6.259 a] *Leckenby, J. N.:* Improved Quality Assurance of Elastomers with Thermal Analysis. Rubbercon '87, June 1–5, 1987, Harrogate, GB, Proceed. B 52.

[6.260] *Maurer, J. J., Brazier, D. W.:* Applications of Thermal Analysis in the Rubber Industry. KGK *36* (1983), p. 37.

[6.261] *Sircar, A. K.:* Basic Concepts of Thermal Analysis in Elastomer Systems. Paper 54, 120th ACS-Conf., Rubber-Div., Oct. 13–16, 1981, Cleveland, OH.

[6.262] . . .: Rubber-Review. J. Analyt. Chem. *51* (1979), p. 303.

Differential Scanning Calorimetry (DSC)

[6.262 a] *Bolder, G., Meier, M.:* Schnelle Ermittlung des Vernetzungsgrades von vernetztem Polyethylen mittels DSC-Analyse. KGK *39* (1986), pp. 715–718.

[6.263] *Gonzalez, V.:* Thermooxidation of Elastomers by Differential Scanning Calorimetry. RCT *54* (1981), p. 134.

[6.264] *Rugo, R., et al.:* The Use of Differential Scanning Calorimetry for Analysis of Rubber Compounds and Components. KGK *39* (1986), pp. 216–219.

[6.265] *Sircar, A. K., Wells, J. L.:* Thermal Conductivity of Elastomer Vulcanizates by Differential Scanning Calorimetry. RCT *55* (1982), p. 191.

Thermal Gravimetry (TGA) (see also [6.230])

[6.266] *Jaroszynska, D. J., Kleps, T.:* Uses of Thermogravimetric Methods for Quantitative Analysis for Rubber Vulcanizates. Int. Polym. Sci. Technol. *5* (1978), p. T 125.

[6.267] *Schwarz, N. V.:* Analysis of Rubber Compounds by Means of Thermogravimetry and Pyrolysis. Paper 10, 123rd ACS-Conf., Rubber-Div., May 10–12, 1983, Toronto, Ont.; GAK *37* (1984), p. 274.

[6.268] *Sickfeld, J., Neubert, D., Gross, D.:* Thermogravimetrie von Vulkanisaten. KGK *36* (1983), p. 760.

6.5.2.4 References on Mass Spectroscopy

[6.268 a] *Czybulka, G.:* Anwendung der Pyrolyse-Feldionenmassenspektrometrie zur Untersuchung des thermischen Abbaus von Copolymeren und Polymermischungen sowie zur Identifizierung von Zusatzstoffen. Dissertat. 1985, Köln.

[6.269] *Devlin, E. F., Folk, T. L.:* A Mass Spectrometric Fourier Transform Infrared Study of the Vulcanization of Chlorosulfonated Polyethylene. RCT *57* (1984), p. 1098.

[6.269 a] *Düssel, H.-J., Czybulka, G., Holl, G.:* Anwendung der linear temperaturprogrammierten Pyrolyse-Massenspektrometrie zur Untersuchung des thermischen Abbauverhaltens von Elastomeren. KGK *40* (1987), pp. 439–446.

[6.270] *Hummel, D. O., Düssel, H.-J.:* Anwendung der linear temperaturprogrammierten Pyrolyse-Massenspektrometrie zur Untersuchung des thermischen Abbauverhaltens von Elastomeren. ikt 85, June 24–27, 1985, Stuttgart, West Germany.

[6.271] *Khalil, J. H., Koski, U.:* Gas Chromatography/Mass Spectrometry and Assessories: Characterization of Residual Trace Organics in Finished Rubber Articles. Paper 51, 120th ACS-Conf., Rubber-Div., Oct. 13–16, 1981, Cleveland, OH.

[6.272] *Lattimer, R.P., Welch, K.R.:* Direct Analysis of Polymer Chemical Mixtures by Field Desorption Mass Spectroscopy. RCT *53* (1980), p.151.

[6.273] *Lattimer, R.P., Harris, R.E., Ross, D.B., Diem, H.E.:* Identification of Rubber Additives by Field Desorption and Fast Atom Bombardment Mass Spectroscopy. RCT *57* (1984), p.10-13.

[6.274] *Lattimer, R.P., Harris, R.E., Schulten, H.-R.:* Application of Mass Spectrometry to Synthetic Polymers. RCT *58* (1985), p.577, Review.

[6.275] *Lattimer, R.P., Pausch, J.B., McCleunen, W.H., Richards, J.M., Menzelaar, M.L.C.:* Direct Characterization of Solid Rubber Samples by Laser Pyrolysis Mass Spectrometry (Py-Ms). Paper 81, 128th ACS-Conf., Rubber-Div., Oct.1-4, Cleveland, OH.

[6.276] *Pausch, J.B., Lattimer, R.P., Menzelaar, H.L.C.:* A New Look at Direct Compound Analysis. RCT *56* (1983), p.1031.

6.5.2.5 References on Nuclear Magnetic Resonance

[6.277] *Ackerman, J.L.:* Study of Networks Using ^{29}S: NMR-Spectroscopy. Gordon Res. Conf. of Elastomers, July 16-20, 1984, New London, NH.

[6.278] *Gronski, W.:* Deuterium Magnetic Studies on Strained Elastomers. Gordon Res. Conf. of Elastomers, July 16-20, 1984, New London, NH.

[6.279] *Gronski, W.:* Analytik des Deformationsverhaltens ein- und mehrphasiger Kautschuke mittels Deuterium NMR. ikt 85, June 24-27, 1985, Stuttgart, West Germany.

[6.280] *Gross, D., Kelm, J.:* Hochauflösende Festkörper-^{13}C-NMR-Spektrometrie an Elastomeren. [1]KGK *38* (1985), p.1089; [2]*40* (1987), pp.13-16.

[6.281] *Harwood, H.J.:* Characterization of the Structure of Diene Polymers by NMR. RCT *55* (1982), p.769, Review.

[6.282] *Hirst, R.C.:* Application of Computerized Nuclear Magnetic Resonance and Infrared Spectroscopic Techniques for Rubber Analysis. RCT *55* (1982), p.913.

[6.283] *Kelm, J., Gross, D.:* Fluorine NMR - A Method for Investigating the Decomposition of Vulcanization Accelerators. RCT *58* (1985), p.37.

[6.283 a] *Kelm, J., Gross, D.:* Hochauflösende NMR-Spektroskopie an Blends. dkt '88, July 4-7, 1988, Nürnberg, West Germany.

[6.284] *Johnson-Plauman, M.E., Plauman, H.P., Keeler, S.E.:* Analysis of EPDM Elastomers by Infrared and NMR-Methods. Paper 84, 128th ACS-Conf., Rubber-Div., Oct.1-4, 1985, Cleveland, OH.

[6.285] *Komorowski, R.A.:* Magic Angel Spinning ^{13}C-NMR Studies of Filled Vulcanizate. RCT *56* (1983), p.959.

[6.286] *Komorowski, R.A., Shockcov, J.P., Gregg, E.C., Savoca, J.L.:* Characterization of Elastomers, Rubber Chemicals and Related Materials by Modern NMR-Techniques. Paper 83, 128th ACS-Conf., Rubber-Div., Oct.1-4, 1985, Cleveland, OH.

[6.287] *van Meerwall, E.D.:* Pulsed and Steady Field Gradient NMR Diffusion Measurement in Polymers. RCT *58* (1985), p.527, Review.

[6.288] *Patterson, D.J., Koenig, J.L., Shelton, J.R.:* Vulcanization Studies of Elastomers Using Solid-State Carbon ^{13}C-NMR. RCT *56* (1983), p.971.

[6.288 a] *Perera, M.C.S.:* NMR Study of Chemical Modification of Natural Rubber. IRC '86, June 2-6, 1986, Göteborg, Sweden.

[6.288 b] *Spiess, H.W.:* Festkörper-NMR zur Untersuchung von Elastomeren. dkt '88, July 4-7, 1988, Nürnberg, West Germany.

[6.289] *Visintainer, J.:* Quantitative Determination of Polymer Microstructure Using ^{13}C-NMR. Gordon Res. Conf. of Elastomers, July 18-22, 1983, New London, NH.

[6.290] *Wardell, G.E., McBrierty, V.J., Marsland, V.:* NMR of Reinforced Elastomers, The Influence of Carbon Black Dispersion. RCT *55* (1982), p.1095.

[6.291] *Werstler, D.D.:* Analysis of Cured, Filled Elastomeric Compounds by ^{13}C-NMR. Paper 12, 117th ACS-Conf., Rubber-Div., May 20-23, 1980, Las Vegas, NV.

[6.292] *Zaper, A.M., Koenig, J.L.:* Application of Solid State ^{13}C-NMR Spectroscopy to Sulfur Vulcanized Natural Rubber. Paper 82, 128th ACS-Conf., Rubber-Div., Oct.1-4, 1985, Cleveland, OH.

6.5.2.6 References on Modern Polymer Structure Analysis
(see also [1.7, 6.277, 6.281, 6.289])

General References

[6.293] *Dawkins, J. V.:* Developments in Polymer Characterization. Elsevier Applied Science Publ., London, 1. (1978); 2. (1980); 3. (1982); 4. (1983).

[6.294] . . .: Analysis of High Polymers. Analytical Chem. *51* (1979), p. 287 Review.

Gel Permeation Chromatography (GPC)

[6.295] *Jordan, R.:* Characterization of Branching in Elastomers Via GPC and Low Angel Laser Light Scattering. Gordon Res. Conf. of Elastomers. July 14–18, 1980, New London, NH.

[6.296] *Ouano, G.:* Characterization of the Molecular Weight Distribution of Linear and Randomly Branched Polymers by GPC and On-Line Low Angel Laser Light Scattering Photometry/Viscosity Techniques. Gordon Res. Conf. of Elastomers, July 14–18, 1980, New London, NH.

[6.297] *Ouano, A. C.:* Recent Advances in Gel Permeation Chromatography. RCT *54* (1981), p. 535, Review.

Neutron-Scattering Investigations

[6.298] *Picot, C., Bastide, J.:* Polymer Network Structure by Small Angel Neutron-Scattering, A Neutron Scattering Study of Networks Containing Labelled Paths. Gordon Res. Conf. of Elastomers, July 15–19, 1985, New London, NH.

[6.299] *Richards, R. W.:* Neutron Scattering from Polystyrene and Black-Copolymer-Networks. Gordon Res. Conf. of Elastomers, July 16–20, 1984, New London, NH.

X-Ray Investigations

[6.300] *Laning, S. H.:* An X-Ray Analytical Procedure for the Study and Reconstruction of Rubber Compounds. Paper 25, 123rd ACS-Conf., Rubber-Div., May 10–12, 1983, Toronto, Ont.

Electron Microscopy

[6.301] *Shiibashi, T., Hirose, K., Tagata, N.:* Direct Observation of Individual Polymer Molecules by Electron Microscopy. IRC 85, Oct. 15–18, 1985, Kyoto, Proc. p. 147.

[6.302] *Takata, N.:* The Direct Observation of Individual Polymer Molecules by Electron Microscopy. Gordon Res. Conf. of Elastomers, July 16–20, 1984, New London, NH.

6.5.2.7 References on Modern Elucidation of Cross-Link Structure
(see also [1.7, 4.65, 6.277, 6.298, 6.299])

[6.303] *Barr-Howell, B. D., Peppas, N. A.:* Analysis of the Structure of Highly Crosslinked Polymeric Networks. Paper 33, 125th ACS-Conf., Rubber-Div., May 8–11, 1984, Indianapolis, IN.

[6.304] *Hallensleben, M. L.:* Chemische Analytik von Polymernetzwerken, Chemischer Abbau. ikt 85, June 24–27, 1985, Stuttgart, West Germany.

7 Trade Names and Manufacturers of Synthetic Rubbers and Rubber Chemicals

The following list gives the trade names of the most important synthetic rubbers and rubber chemicals in alphabetical order. The names of the manufacturing companies (in italics) are given for quick reference. This selection does not claim to be complete, since changes due to company acquisitions, or production changes may have occurred since the list was compiled. Also no guarantee can be given that the listed trade names are up to date. In some cases only a selection of products has been included.

In this section, only abbreviations of the manufacturers' names are given. The list in Section 7.5 gives most of the company names in full.

In most cases the trade names encompass a number of different products: for example numerous special types in the field of synthetic rubbers (SBR, NBR, etc.), and various handling forms, such as powders, dust-free powders, granulates (oil or polymer bound), dry liquids, different concentrations, etc. The limited space of this book prohibits the listing of the thousands of substances in the authors' files, which would fill a book of their own. Since such books already exist, reference has been made to them [7.1-7.3]. The following list is intended only as a broad survey. Not dealt with in the list are, for example, fillers, pigments, powders, resins, reclaims, mould release agents, solvents, etc. Additional information can be found in the suppliers list on page 556 [7.4].

7.1 Synthetic Rubber

7.1.1 Butadiene Rubber (BR), Density 0,90-0,93

Afdene BR, *Carbochem*
Ameripol CB*, *BFGoodrich*
Asadene, *Asahi Chemicals*
Budene, *Goodyear*
Buna CB, *Bunawerke Hüls*
Cariflex BR, *Shell*
Cis, *VEB Buna*
Cisdene*, *American Chemicals*
Diene*, *Firestone*
Duragene*, *General Tire*
Europrene Cis, *Enichem*
Finaprene, *Petrochim*
Intene, Intolene*, *Enichem*
JSR-BR, *Japan Synthetic Rubber*
Nipol, *Nippon Zeon*
Phillips Cis 4*, *Phillips Petroleum*
Solprene, *Calatrava*
Synpol EBR, *Synpol* (low cis content)
Taktene*, *Polysar*

7.1.2 Styrene Butadiene Rubber (SBR), Density 0,91-0,96

Afsol, *Carbochem*
Ameripol, *BFGoodrich*
Austrapol, *Australian Synthetic Rubber*
Buna EM, *Bunawerke Hüls*
Buna SB, *VEB Buna*
Cariflex S, *Shell*
Carom, *Danubiana*
Copo, *Copolymer*
Diapol, *Mitsubishi*
Duranit, *CWH* (Styrene resin)
Europrene, Unidene, *Enichem*
Finaprene, *Petrochim*
Humex, *Hules Mexicanos*
Intol, Intex, *Enichem*
KER, *Ciech-Stomil*
Kratex, *Chemopetrol*
Krylene, Krynol*, *Polysar*
Nipol, *Nippon Zeon*
Polysar S, *Polysar*
Sirel*, *SIR*
Sumitomo SBR, *Sumitomo*
Tufprene, *Asahi Chemicals*

* earlier used trade names

7.1.3 Butadiene Acrylonitrile Rubber (NBR), Density 0,96-1,01

Breon N*, *BP*
Buna N, *VEB Buna*
Butacril*, *Ugine Kuhlmann*
Chemigum, *Goodyear*
Elaprim*, *Enichem*
Europrene N, *Enichem*
Hycar, *BFGoodrich*
Humex N, *Hules Mexicanos*
JSR-N, *Japan Synthetic Rubber*
Krynac, *Polysar*
Nipol N, *Nippon Zeon*
NYsyn, *Copolymer*
Paracril, *Uniroyal*
Perbunan N, *Bayer*
SIR*, *Sir*

7.1.4 Nitrile Rubber PVC Blends (NBR/PVC), Density 1,05-1,40

Breon*, *BP*
Elaprim S Mix*, *Enichem*
Europrene S Ozo, *Enichem*
Hycar, *BFGoodrich*
Nipol DN, P, *Nippon Zeon*
Paracril Ozo, *Uniroyal*
Perbunan NVC, *Bayer*
Rhenoblend, *Rhein-Cemie*

7.1.5 Hydrogenated Nitrile Rubber (H-NBR), Density 0,95-1,00

Therban, *Bayer*
Tornac, *Polysar*
Zetpol, *Nippon Zeon*

7.1.6 Chloroprene Rubber (CR), Density 1,23

Baypren, *Bayer*
Butaclor, *Distugil*
Denka, *Denka Kagaku*
Neoprene, *Du Pont*
Perbunan C*, *Bayer*
Skyprene, *Toya Soda*
Nairit, *USSR*

7.1.7 Isoprene Rubber (IR), Density 0,90-0,94

Afprene IR, *Carbochem*
Cariflex IR, *Shell*
Carom, *Danubiana*
Europrene SOL, *Enichem*
Natsyn, *Goodyear*
Nipol, *Nippon Zeon*

7.1.8 Isobutylene Isoprene Rubber (IIR), Density 0,92

Esso Butyl, *Exxon*
Polysar Butyl, *Polysar*
Soca Butyl, *Japan Butyl*

7.1.9 Halogenated IIR (BIIR, CIIR), Density 0,92-0,93

Esso Bromobutyl, *Exxon*
Esso Chlorobutyl, *Exxon*
Polysar Bromobutyl, *Polysar*
Polysar Chlorobutyl, *Polysar*

7.1.10 Ethylene Propylene Rubber (EPM, EPDM), Density 0,86-0,89

Buna AP, *Bunawerke Hüls*
Dutral, *Montedison*
Epcar*, *BFGoodrich*
EPsyn, *Copolymer*
Eptotal*, *Socabu*
Esprene, *Sumitomo*
Intolan, *Enichem*
Keltan, *DSM*
Nordel, *Du Pont*
Polysar EPDM, *Polysar*
Royalene, *Uniroyal*
Trilene, *Uniroyal* (liquid)
Vistalon, *Exxon*

---

* earlier used trade names

7.1.11 Ethylene Vinylacetate Rubber (EVM), Density 0,98–1,00

Escorene Ultra, *Exxon*
Evatane, *Atochem*
Levaprene, *Bayer*
Orevac, *Atochem*
Ultrathene, *USI Chemicals*
Vynathene, *USI Chemicals*

7.1.12 Chlorinated Polyethylene (CM), Density 1,10–1,25

Bayer CM, *Bayer*
Elaslen, *Showa Denko*
Daisolac, *Osaka Soda*
Dow CPE*, *Dow Corning*
Hostapren*, *Hoechst*
Kelrinal, *DSM*

7.1.13 Chlorosulphonated Polyethylene (CSM), Density 1,10–1,16

CS PE, *China*
Hypalon, *Du Pont*

7.1.14 Acrylic Rubber (ACM), Density 1,03–1,10

Acralen, *Bayer* (Latex)
Cyanacryl, *Cyanamid*
Elaprim AR*, *Montedison*
Europrene AR, *Enichem*
Hycar*, *BFGoodrich*
HyTemp, *BFGoodrich*
Nipol AR, *Nippon Zeon*

7.1.15 Ethylene Acrylate Rubber (EAM), Density 1,10

Vamac, *Du Pont*

7.1.16 Epichlorohydrin Rubber (CO, ECO, ETER), Density 1,24–1,38

Epichlormer, *Osaka Soda*
Herclor*, *Hercules*
Hydrin, *BFGoodrich*

7.1.17 Propylene Oxide Rubber (PO, GPO), Density 1,01

Parel, *BFGoodrich*

7.1.18 Fluorinated Rubber (FPM resp. FKM, TFE/PP resp. TFE-FMVE copolymer), Density 1,50; 1,70–1,90

Aflas, Asahi Glass, *3M* (TFE/PP)
Dai-El, *Daikin*
Dai-El Perfluor, *Daikin* (TFE-FMVE)
Fluorel, *3M*
Kalrez, *Du Pont* (TFE-FMVE)
Technoflon, *Montedison*
Viton, *Du Pont*

7.1.19 Polyfluorphosphazene Rubber (PNF), Density 1,75–1,85

Eypel-F, *Ethyl*

7.1.20 Polynorbornene Rubber (PNR), Density 0,96

Norsorex, *CdF-Chimie*

7.1.21 Polyoctenamer Rubber (TOR), Density 0,91

Vestenamer, *CWH*

* earlier used trade names

7.1.22 Polysiloxane Rubber (Q), Density 1,06-1,38

Blensil, *General Electric*
Elastosil, *Wacker*
FRV, *General Electric*
J-Sil, *J-Sil*
NRC, *Nordmann, Rassmann*
Por-A-Mold, *Compounding Ingredients*
RP, *Rhône Poulenc*
RTV, *General Electric*
SE, *General Electric*
Silastic, *Dow Corning*
Silopren, *Bayer*
Tufsel, *General Electric*

7.1.23 Fluorosilicone Rubber (FVQ), Density 1,10-1,45

FSE, *General Electric*
R, *Wacker*
RP, *Rhône Poulenc*
Silastic, *Dow Corning*
VP-R, *Wacker*

7.1.24 Polyurethane Rubber (AU, EU), Density 1,03-1,24

Adiprene, *Du Pont*
Castomer, *Baxenden*
Daltoped, *ICI*
Davathane, *Krahn*
Diorez, Diprane, *Briggs + Townsend*
Elastothane, *Thiokol*
Igulan, *Hokuchin Kagaku*
Millathane, *Notedome*
Monothane, *Compounding Ingredients*
Urepan, *Bayer*
Vibrathane, *Uniroyal*
Texin, *Bayer*

7.1.25 Polysulfide Rubber (TM), Density 1,28-1,34

Thiokol, *Thiokol*
Thiokol LP, *Thiokol*

7.2 Thermoplastic Elastomers (TPE)

7.2.1 Styrene Butadiene (Isoprene) Types (TPE-S, SBS, SIS), Density 0,94-0,95

Cariflex TR, *Shell*
Europrene SOL T, *Enichem*
Finaprene, *Petrochim*
Humex, *Hules Mexicanos*
Solprene, *Calatrava*

7.2.2 Styrene Olefine Types (TPE-S, SEBS/O), Density 0,90-1,17

Elexar, *Shell*
Kraton, *Shell*

7.2.3 Thermoplastic Olefin Elastomers (TPE-O, EPDM/PP), Density 0,87-1,06

Dutral TP, *Montedison*
Keltan TP, *DSM*
Kelprox, *DSM*
Levaflex, *Bayer*
Moplen, *Montedison*
Santoprene, *Monsanto*
Trefsin, *Exxon*
Vistaflex, *Exxon*

7.2.4 Thermoplastic Nitrile Elastomer (TPE-NBR), Density 1,05-1,09

Geolast, *Monsanto*

7.2.5 Thermoplastic Chloroolefin Elastomer (Alcryn), Density 1,21-1,25

Alcryn, *Du Pont*

7.2.6 Thermoplastic Polyurethane Elastomers (TPE-U), Density 1,10-1,25

Desmopan, *Bayer*
Elastollan, *BASF*
Estane, *BFGoodrich*
PU Elastomer, *Briggs + Townsend*

7.2.7 Thermoplastic Copolyester (TPE-E), Density 1,17-1,25

Arnitel, *Akzo*
Hytrel, *Du Pont*

7.2.8 Thermoplastic Copolyether (TPE-A), Density 1,01

Pebax, *Atochem*

7.3 Vulcanizing Chemicals

7.3.1 Curatives (Crosslinking Agents)

7.3.1.1 Sulfur

Crystex, *Kalichemie*
Deosulf, *DOG*
Kenmix Sulfur, *Kenrich*
Mahlschwefel, *Kalichemie*
Manox, *Manchem*
MC Sulfur, *Anchor*
Perkacit S, *Akzo*
Polydispersion ASD, *Croxton & Garry*
Rhenogran S, *Rhein-Chemie*
Struktol SU, *Schill & Seilacher*
Velvet, *Lehmann & Voss*

7.3.1.2 Sulfur Donors

Dithiodimorpholine (DTDM), WTR No. 67

Deovulc M, *DOG*
Ekaland MDS, *Landaise*
Naugex SD-I, *Uniroyal*
Perkacit DTDM, *Akzo*
Polydispersion P, *Croxton & Garry*
Rhenocure M, *Rhein-Chemie*
Robac DTDM, *Robinson*
Sulfasan R, *Monsanto*
Vanax A, *Vanderbilt*

Dipentamethylenethiuramtetra (hexa) sulfide (DPTT), WTR No. 68

DPTTS, *Lehmann & Voss*
Ekaland TSPM, *Landaise*
Perkacit DPTT, *Akzo*
Polydispersion DPTT, *Croxton & Garry*
Rhenogran DPTT, *Rhein-Chemie*
Robac P 25, *Robinson*
Soxinol TRA, *Sumitomo*
Sulfads, *Vanderbilt*
Tetrone A, *Du Pont*

Tetramethylthiuram disulfide (TMTD)
see Section 7.3.2.6

2-Morpholinodithiodibenzothiazole (MBSS), WTR No. 23e

Morphax, *Goodyear*
Vulcuren 2, *Bayer*

Caprolactamdisulfide (CLD)

Rhenocure S, *Rhein-Chemie*

7.3.1.3 Peroxides

(each trade name represents a different type)

DiCup, *Hercules*
Interox, *Peroxid-Chemie*
Lucidol, Luperco, *Luperox*
Luperox, *Luperox*
Trigonox, *Akzo*
Perkadox, *Akzo*
Peroximon, *Montedison*
Polydispersion E, *Croxton & Garry*
Varox, *Vanderbilt*
Vul-Cup, *Hercules*
Silopren-Vernetzer, *Bayer*

7.3.1.4 Crosslinking Agents for Butyl Rubber

Quinonedioxime (CDO)

BQD, *Krahn*
GMF, *Uniroyal*
Kenmix GMF, *Kenrich*

Dibenzoquinonedioxime (Dibenzo CDO)

DBQD, *Krahn*
Dibenzo GMF, *Kenrich*
Dibenzo GMF, *Uniroyal*
Rhenocure BQ, *Rhein-Chemie*

Resins

SP-Types, *Schenectady*

7.3.1.5 Crosslinking Agents for Fluoro Rubber

Curing Agents P_2, P_3, V_1, *Daikin*
Diak 1, 2, 3, 7, *Du Pont*
Fluorel FC, *3 M*
Technoflon M, *Montedison*
Technosin, *Montedison*
Viton Curatives, *Du Pont*

7.3.1.6 Crosslinking Agents for Polyurethane Rubber

Caytur, *Du Pont*
Curalon, *Uniroyal*
Desmodur, *Bayer*
Desmorapid, *Bayer*
Moka, *Du Pont*
Vernetzer 30/10, *Bayer*
Vibracure, *Uniroyal*

7.3.1.7 Other Crosslinking Agents

Luazo ABA, AP, *Luperox*
Novor (Urethane Crosslinker), *Durham*
Polydispersion G, T, *Croxton & Garry*
Tonox, *Uniroyal*
Vulklor, *Uniroyal*

7.3.2 Vulcanization Accelerators

7.3.2.1 Benzothiazole Accelerators

2-Mercaptobenzothiazole (MBT), WTR No. 23

Accelerator M, *Ciech-Stomil*
Ancap, Anchor MBT, *Anchor, Nordmann, Rassmann*
Captax, *Vanderbilt*
Ekaland MBT, *Landaise*
Good Rite MBT, *BFGoodrich*
MBT, *Krahn*
MBT, *Lehmann & Voss*
MBT, *Uniroyal*
Pennac MBT, *Pennwalt*
Perkacit MBT, *Akzo*
Polydispersion TMD, *Croxton & Garry*
Rhenogran MBT, *Rhein-Chemie*
Rotax, *Vanderbilt*
Soxinol M, *Sumitomo*
Thiotax, *Monsanto*
Vulkacit Merkapto, *Bayer*
Wobecit Merkapto, *VEB Bitterfeld*

Sodium 2-Mercaptobenzothiazole (SMBT, NaMBT), WTR No. 23a

Ekaland NaMBT, *Landaise*
Na-MBT, *Bayer*
Sodium MBT, *Monsanto*
Sodium MBT, *Uniroyal*
Soxinol M-Na-G, *Sumitomo*

Zinc-2-Mercaptobenzothiazole (ZMBT), WTR No. 23b

Anchor ZMBT, *Anchor, Nordmann, Rassmann*
Bantex, *Monsanto*
Ekaland ZMBT, *Landaise*
Merazin MBTZ, *Bozzetto*
Naftocit ZMBT, *Metallgesellschaft*
Naugatex OXAF, *Uniroyal*
Perkacit ZMBT, *Akzo*
Pennac ZMBT, *Pennwalt*
Polydispersion E, *Croxton & Garry*
Rhenogran ZMBT, *Rhein-Chemie*
Soxinol MZ, *Sumitomo*
Vanax ZMBT, *Vanderbilt*
Vulkacit ZM, *Bayer*
Zefax, *Vanderbilt*
Zink Ancap, *Anchor*
ZMBT, *Krahn*

Dibenzothiazoledisulfide (MBTS), WTR No. 23d

Anchor MBTS, *Anchor, Nordmann, Rassmann*
Ekaland MBTS, *Landaise*
Good Rite MBTS, *BFGoodrich*
MBTS, *Ciech-Stomil*
MBTS, *Krahn*
MBTS, *Lehmann & Voss*
MBTS, *Uniroyal*
Merasulf MBTS, *Bozzetto*
Pennac MBTS, *Pennwalt*
Perkacit MBTS, *Akzo*
Polydispersion EAD, *Croxton & Garry*
Rhenogran MBTS, *Rhein-Chemie*
Soxinol DM, *Sumitomo*
Thiofide, *Monsanto*
Vanax MBTS, *Vanderbilt*
Vulkacit DM, *Bayer*
Wobecit DM, *VEB Bitterfeld*

Dinitrophenyl-thio-benzothiazole

Ureka Base, *Monsanto*

Thiazole Blends

Anchor TGB, 1, 2, 3, *Anchor, Nordmann, Rassmann*
F, *Krahn* (MBTS/DPG)
Mix EXP, *Bozzetto*
Radax, *Monsanto* (TMTD/MBT)
Ureka White F, *Monsanto* (MBTS/DPG)
Vulkacit F, *Bayer* (MBTS/DPG)
Vulkacit MT, *Bayer* (MBT/TMTD)

7.3.2.2 Benzothiazole Sulfenamides

Cyclohexylbenzothiazole-2-sulfenamide (CBS), WTR No. 19

Anchor CBS, *Anchor, Nordmann, Rassmann*
CBS, *Ciech-Stomil*
CBS, *Krahn*
CBS, *Lehmann & Voss*
Delac S, *Uniroyal*
Durax, *Vanderbilt*
Ekaland CBS, *Landaise*
Furbac, *Anchor*
Good Rite CBTS, *BFGoodrich*
Meramid C, *Bozzetto*
Pennac CBS, *Pennwalt*
Perkacit CBS, *Akzo*
Polydispersion EC (CBS), *Croxton & Garry*
Rhenogran CBS, *Rhein-Chemie*
Santocure, *Monsanto*
Soxinol CZ, *Sumitomo*
Sulfenax CB, *Petrimex*
Vanax CBS, *Vanderbilt*
Vulkacit CZ, *Bayer*

Dicyclohexylbenzothiazole-2-sulfenamide (DCBS), WTR No. 20

DCBS, *Krahn*
Ekaland DCBS, *Landaise*
Meramid DCN, *Bozzetto*
Perkacit DCBS, *Akzo*
Vulkacit DZ, *Bayer*

Benzothiazol-2-sulfen-morpholid (MBS), WTR No. 22

Amax, *Vanderbilt*
Delac MOR, *Uniroyal*
Ekaland NOBS, *Landaise*
Good Rite OBTS, *BFGoodrich*
MBS, *Krahn*
Meramid M, *Bozzetto*
NOBS Spezial, *Anchor*
Pennac MBS, *Pennwalt*
Perkacit MBS, *Akzo*
Santocure MOR, *Monsanto*
Soxinol MBS, *Sumitomo*
Vanax OBTS, *Vanderbilt*
Vulkacit MOZ, *Bayer*

2-Morpholinodithiobenzothiazole (MBSS)
see Section 7.3.1.2

Tert. Butylbenzothiazole-2-sulfenamide (TBBS), WTR No. 21

Conac NS, *Du Pont*
Delac NS, *Uniroyal*
Good Rite BBTS, *BFGoodrich*
Santocure NS, *Monsanto*
Pennac TBBS, *Pennwalt*
Vanax NS, *Vanderbilt*
Vulkacit NZ, *Bayer*

Diisopropylbenzothiazole-2-sulfenamide (DIBS), WTR No. 18a

Ekaland DIBS, *Landaise*
Pennac DIBS, *Pennwalt*
Santocure IPS, *Monsanto*
Vulkacit LZ, *Bayer*

Tert. Amylbenzothiazole-2-sulfenamide (5 BS)

Vulkacit AMZ, *Bayer*

7.3.2.3 Zinc Dithiocarbamates

Zinc dimethyldithiocarbamates (ZDMC), WTR No. 36

Anchor ZDMC, *Anchor, Nordmann, Rassmann*
Ekaland ZDMC, *Landaise*
JO 4012, *Bozzetto*
Methasan, *Monsanto*
Methazate, *Uniroyal*
Methyl Zimate, *Vanderbilt*
Naftocit Di 4, *Metallgesellschaft*
Pennac ZDMC, *Pennwalt*
Perkacit ZDMC, *Akzo*
Polydispersion E (MC), *Croxton & Garry*
Rhenogran ZDMC, *Rhein-Chemie*
Robac ZMD, *Robinson*
Soxinol PC, *Sumitomo*
Vulkacit L, *Bayer*
ZDMC, *Krahn*
ZDMC, *Lehmann & Voss*

Zinc diethyldithiocarbamate (ZDEC), WTR No. 38

Anchor ZDEC, *Anchor, Nordmann, Rassmann*
Ekaland ZDEC, *Landaise*
Ethasan, *Monsanto*
Ethazate, *Uniroyal*
Ethyl Zimate, *Vanderbilt*
JOS 4026, *Bozzetto*
Naftocit Di 7, *Metallgesellschaft*
Pennac ZDEC, *Pennwalt*
Perkacit ZDEC, *Akzo*
Polydispersion E (EZ), *Croxton & Garry*
Rhenogran ZDEC, *Rhein-Chemie*
Robac ZDC, *Robinson*
Soxinol EZ, *Sumitomo*
Vulkafor ZDEC, *Vulnax*
Vulkacit LDA, *Bayer*
ZDEC, *Krahn*
ZDEC, *Lehmann & Voss*

Zink dibutyldithiocarbamate (ZDBC), WTR No. 40

Ancazate BW, Anchor ZDBC, *Anchor, Nordmann, Rassmann*
Butasan, *Monsanto*
Butazate, *Uniroyal*
Butyl Zimate, *Vanderbilt*
Ekaland ZDBC, *Landaise*
JO 4013, *Bozzetto*
Naftocit Di 13, *Metallgesellschaft*
Pennac ZDBC, *Pennwalt*
Perkacit ZDBC, *Akzo*
Polydispersion T (BZ), *Croxton & Garry*
Rhenogran ZDBC, *Rhein-Chemie*
Robac ZBUD, *Robinson*
Vulkacit LDB, *Bayer*
ZDBC, *Krahn*
ZDBC, *Lehmann & Voss*

Zinc ethylphenyldithiocarbamate (ZEPC), WTR No. 44

Ekaland ZEPC, *Landaise*
Rhenogran ZEPC, *Rhein-Chemie*
Soxinol PX, *Sumitomo*
Vulkacit P extra N, *Bayer*
Wobecit, P extra N, *VEB Bitterfeld*
ZEPC, *Lehmann & Voss*

Zinc benzylethyldithiocarbamate (ZBEC), WTR No. 44a

Arazate, *Uniroyal*
Perkacit ZBEC, *Akzo*
Robac ZBED, *Robinson*

Zinc pentamethylenedithiocarbamate (Z5MC), WTR No. 45

Robac ZPC, *Robinson*
Vulkacit ZP, *Bayer*
Z5MC, *Krahn*

7.3.2.4 Other Dithiocarbamates

Ammonium (piperidyl) dithiocarbamate (PPC), WTR No. 35

Accelerator 552, *Du Pont*
Robac PPC, *Robinson*
Vanax 552, *Vanderbilt*
Vulkacit P, *Bayer*

Bismuth dimethyldithiocarbamate (BiDMC)

Bismate, *Vanderbilt*
Perkacit BCMC, *Akzo*

Cadmium dithiocarbamates

Amyl Cadmate, *Vanderbilt*
Cadmate, *Vanderbilt* (CdDEC)

Copper dithiocarbamates (CuDMC, CuDBC), WTR No. 386

CuDMC, *Lehmann & Voss*
Cumate, *Vanderbilt* (CuDMC)
Cupsac, *Cyanamid*
Ekaland CDBC, *Landaise*
Pennac CDMC, *Pennwalt*
Robac CuDD, *Robinson* (CuDMC)

Lead dithiocarbamates (PbDMC, PbDEC, PbD5C)

Amyl Ledate, *Vanderbilt*
Ethyl Ledate, *Vanderbilt*
Ledate, *Vanderbilt* (PbDMC)

Nickel dithiocarbamates, WTR No. 40a

Antigene NC, *Sumitomo*
Ekaland NDBC, *Landaise*
Methylniclate, *Vanderbilt*
NBC, *Du Pont*
Pennac NDBC, *Pennwalt*
Perkacit NDBC, *Akzo*
Robac NiBUD, *Robinson*
Robac NiDD, *Robinson*

Selenium dithiocarbamates (SeDMC, SeDEC)

Ethyl Selenac, *Vanderbilt*
Selenac, *Vanderbilt* (SeDMC)

Sodium dithiocarbamates ((SDMC) (NaDMC), SDEC (NaDEC), SDBC (NaDBC)), WTR No. 37, 39

Na DBC, *Lehmann & Voss*
Pennac SDBC, *Pennwalt*
Perkacit SDBC, *Akzo*
Perkacit SDEC, *Akzo*
Perkacit SDMC, *Akzo*
Robac SBUD, *Robinson*
Robac SED, *Robinson*
Robac SMD, *Robinson*
Soxinol ESL, *Sumitomo*
Vulkacit WL, *Bayer*

Tellurium dithiocarbamates (TeDEC, TED5C), WTR No. 38a

Ekaland TeDEC, *Landaise*
Ethyl Tellurac, *Vanderbilt*
Pennac TDEC, *Pennwalt*
Perkacit TDEC, *Akzo*
Rhenogran TDEC, *Rhein-Chemie*
TDEC, *Lehmann & Voss*
TeDEC, *Krahn*
Tellurac, *Vanderbilt*
Soxinol TE, *Sumitomo*

Activated dithiocarbamate combinations

Anchor DBD, *Anchor*
Butyl Eight, *Vanderbilt*

7.3.2.5 Xanthogenates (ZIX)

CPB, *Uniroyal*
Propyl Zithate, *Vanderbilt*

7.3.2.6 Thiurames

Tetramethyl thiuram disulfide (TMTD), WTR No. 46

Ancacide ME, Anchor TMTD, *Anchor, Nordmann, Rassmann*
Ekaland TMTD, *Landaise*
Good Rite TMTD, *BFGoodrich*
Methyl Tuads, *Vanderbilt*
Methyl Tuex, *Uniroyal*
Naftocit Thiuram 16, *Metallgesellschaft*
Pennac TMTD, *Pennwalt*
Perkacit TMTD, *Akzo*
Polydispersion T (MT), *Croxton & Garry*
Robac TMT, *Robinson*
Soxinol TT, *Sumitomo*
Thiuram M, *Du Pont*
Thiurad, *Monsanto*
TMTD, *Lehmann & Voss*
TMTD, *Krahn*
TMTS, *Bozzetto*
Tuex, *Uniroyal*
Vulkacit Thiuram, *Bayer*
Wobecit Thiuram, *VEB Bitterfeld*

Tetramethyl thiuram monosulfide (TMTM), WTR No. 47

Anchor TMTM, *Anchor, Nordmann, Rassmann*
Ekaland TMTM, *Landaise*
Monex, *Uniroyal*
Mono Thiurad, *Monsanto*
Perkacit TMTM, *Akzo*
Polydispersion E (MX), *Croxton & Garry*
Rhenogran TMTM, *Rhein-Chemie*
Robac TMS, *Robinson*
Soxinol TS, *Sumitomo*
Thionex, *Du Pont*
TMTM, *Bozzetto*
TMTM, *Krahn*
TMTM, *Lehmann & Voss*
Tuads, *Vanderbilt*
Vulkacit Thiuram MS, *Bayer*

Tetraethyl thiuram disulfide (TETD), WTR No. 48

Ekaland TETD, *Landaise*
Ethyl Thiurads, *Vanderbilt*
Ethyl Tuex, *Uniroyal*
Pennac TETD, *Pennwalt*
Perkacit TETD, *Akzo*
Rhenogran TETD, *Rhein-Chemie*
Robac TET, *Robinson*
Soxinol TET, *Sumitomo*
TETD, *Krahn*
TETD, *Monsanto*
TETS, *Bozzetto*
Thiuram E, *Du Pont*

Other Thiurames

Butyl Tuads, *Vanderbilt* (TBTD)
D 5 MTD, *Krahn*
D 5 MTT, *Krahn*
Methyl Ethyl Tuads, *Vanderbilt* (METD)
PETS, *Bozzetto*
TBTD, *Krahn*
TBTS, *Bozzetto*
Vulkacit J, *Bayer* (PMTD)

Dipentamethylenethiurame tetra (hexa) sulfide (DPTT)
see Section 7.3.1.2

7.3.2.7 Dithiocarbamylsulfenamide (OTOS)

Cure Rite 18, *BFGoodrich*
Cure Rite XL, *BFGoodrich*

7.3.2.8 Guanidines

Diphenylguanidine (DPG), WTR No. 27

Ekaland DPG, *Landaise*
DPG, *Krahn*
DPG, *Monsanto*
DPG, *Lehmann & Voss*
Naftocit DPG, *Metallgesellschaft*
Perkacit DPG, *Akzo*
Polydispersion E (DPG), *Croxton & Garry*
Rhenogran DPG, *Rhein-Chemie*
Robac DPG, *Robinson*
Soxinol D, *Sumitomo*
Vanax DPG, *Vanderbilt*
Vulkacit D, *Bayer*

Di-o-Tolylguanidine (DOTG), WTR No. 28

Ekaland DOTG, *Landaise*
DOTG, *Krahn*
Perkacit DOTG, *Akzo*
Rhenogran DOTG, *Rhein-Chemie*
Robac DOTG, *Robinson*
Soxinol DT, *Sumitomo*
Vanax DOTG, *Vanderbilt*
Vulkacit DOTG, *Bayer*

o-Tolylbiguanide (OTBG)

Rhenogran OTBG, *Rhein-Chemie*
Vulkacit 1000, *Bayer*
Wobecit 1000, *VEB Bitterfeld*

7.3.2.9 Thiourea Accelerators

Ethylenethiourea (ETU), WTR No. 31

Ekaland ETU, *Landaise*
END, GND, *Wyrough & Loser*
ETU, *Krahn*
NA 22, *Du Pont*
Naftocit Mi 12, *Metallgesellschaft*
Perkacit ETU, *Akzo*
Polydispersion END, *Croxton & Garry*
Rhenogran ETU, *Rhein-Chemie*
Robac 22, *Robinson*
Thiate N, *Vanderbilt*
Vulkacit NPV, *Bayer*
Warecure C, *Ware Chemical*

Diethylthiourea (DETU), WTR No. 30 a

Ekaland DETU, *Landaise*
DETU, *Krahn*
DETU, *Lehmann & Voss*
Perkacit DETU, *Akzo*
Polydispersion S (DETU), *Croxton & Garry*
Rhenogran DETU, *Rhein-Chemie*
Robac DETU, *Robinson*
Thiate H, *Vanderbilt*

Dibutylthiourea (DBTU)

DBTU, *Krahn*
Rhenogran DBTU, *Rhein-Chemie*
Robac DBTU, *Robinson*
Thiate U, *Vanderbilt*

Diphenylthiourea (DPTU), WTR No. 30

A 1, *Monsanto*
Ekaland DPTU, *Landaise*
DPTU, *Krahn*
NA 101, *Du Pont*
Rhenocure CA, *Rhein-Chemie*
Rhenogran DPTU, *Rhein-Chemie*

Trimethylthiourea (TPTU)

Thiate EF 2, *Vanderbilt*

Tributylthiourea (TBTU)

Santowhite TBTU, *Monsanto*
TRIB TU, *Krahn*

7.3.2.10 Amine Accelerators

Hexamethylenediamine (HEXA, HMT), WTR No. 34

Ekaland HMT, *Landaise*
Hexa, *Krahn*
Hexa K, *Degussa*
Hexamethylenetetramine, *Union Carbide*
Polydispersion TDH, SDH, *Croxton & Garry*
Rhenogran Hexa, *Rhein-Chemie*
Vulkacit H 30, *Bayer*

Aldehydamines (WTR No. 32)

Accelerator 808, 833, *Du Pont*
Beutene, *Uniroyal*
Ekagom VS, *Ugine Kuhlmann*
Heptene Base, *Uniroyal*
Meramid D/E, *Bozzetto*
Ridakto, *Kenrich*
Vanax 808, 833, *Du Pont*
Vulkacit CT-N, 576, *Bayer*

Secondary Amines

DBA, *Uniroyal*
Rhenofit B, 2642, 3555, *Rhein-Chemie*
Silacto, *Kenrich*
TA 11, *Du Pont*
Trimene Base, *Uniroyal*
Vulkacit HX, *Bayer*
Vulkacit TR, *Bayer*

7.3.2.11 Dithiophosphate Accelerators

Rhenocure AT, CUT, TP, ZAT, *Rhein-Chemie*
Vocol, *Monsanto*

7.3.2.12 Special Accelerators for CR (CM, ECO)

Meramid CR, *Bozzetto* (Activated amine phenol complex)
Permalux, *Du Pont* (DOTG salt of dicatechole borate)
Rhenocure TDD, *Rhein-Chemie* (Thiadiazole derivative)
Rhenofit KE, *Rhein-Chemie*
Robac 44, *Robinson*
R 240, *Lehmann & Voss* (Oxadiazine thioketone derivative)
Vanax CPA, *Vanderbilt* (Isophthalate derivative)
Vanax NP, *Vanderbilt* (Thiadiazine derivative)
Vanax PML, *Vanderbilt* (DOTG salt of dicatechole borate)
Vulkacit CRV, *Bayer* (3-Methylthiazoline-thione-2)

7.3.2.13 Special Accelerators for EPDM (Accelerator Blends)

Deovulc BG 1/87, *DOG*
Deovulc BL 6, *DOG*
Deovulc EG L, *DOG*
Deovulc EG 3, 40, 28, *DOG*
JO 4061, *Bozzetto*
Perkacit 2, *Akzo*
Rhenocure CMT, CMU, EPC, *Rhein-Chemie*
Vulkafor 2, *Vulnax*
Vulkaperl 2, *Vulnax*

7.3.3 Organic Crosslinking Activators

7.3.3.1 Methacrylates

Ethylene glycol dimethacrylate (EDMA)

Actigran, *Kettlitz*
Ancomer ATM3, *Ancomer*
EDMA DL, *Lehmann & Voss*
Perkalink 401, *Akzo*
Rhenofit EDMA, *Rhein-Chemie*
SR-206 Drimix, *Kenrich*
Sartomer 206, *Ancomer*

Butylene glycol dimethacrylate (BDMA)

Ancomer ATM 9, *Ancomer*
Chemlink 20, *Ware Chemicals*
Rhenofit BDMA, *Rhein-Chemie*
SR-297 Drimix, *Kenrich*

Trimethylolpropane trimethacrylate (TMPT)

Ancomer ATM 11, *Ancomer*
Chemlink 30, *Ware Chemicals*
Perkalink 400, *Akzo*
Rhenofit TRIM, *Rhein-Chemie*
SR-350 Drimix, *Kenrich*
TMPT DL, *Lehmann & Voss*

Other Methacrylates

Ancomer ATM 16, *Ancomer*
Ancomer ATM 17, *Ancomer*
Saret 500 Drimix, *Kenrich*
Saret 515 Drimix, *Kenrich*

7.3.3.2 Triallyl Derivatives

Triallyl cyanurate (TAC)

Activator OC, *Degussa*
Alcatal, *Safic*
Perkalink 300, *Akzo*
Primix TAC, *Kenrich*
Rhenofit TAC, *Rhein-Chemie*
TAC DL, *Lehmann & Voss*
TAC GR, *Kettlitz*

Triallyl Isocyanurate (TAIC)

Diak 7, *Du Pont*
Perkalink 301, *Akzo*
Kasai TAIC, *Nippon Kasai*
TAIC DL, *Lehmann & Voss*
TAIC DL, *Nordmann, Rassmann*

Triallyl phosphate (TAP)

Activator OP, *Degussa*

Triallyl trimellitate (TATM)

Chemlink TATM, *Lehmann & Voss*
Chemlink TATM, *Ware Chemicals*
TATM DL, *Lehmann & Voss*

7.3.3.3 m-Phenylenedimaleimide

HVA 2, *Du Pont*

7.3.3.4 Silanes

Vinyl triethoxy silane

Dynasylan VTEO, *Dynamit Nobel*
Silane A-151, *Union Carbide*

Vinyl-tris-(β-methoxyethoxy) silane

Dynasylan VTMOEO, *Dynamit Nobel*
Ucarsil DSC 12, *Union Carbide*
Silanogran V, *Kettlitz*

γ-Methacryloylpropyltrimethoxysilane

Dynasylan MEMO, *Dynamit Nobel*
Silane A-174, *Union Carbide*

γ-Mercaptopropyltrimethoxysilane

Dynasylan MTMO, *Dynamit Nobel*
Ucarsil DSC 189, *Union Carbide*
Silanogran M, *Kettlitz*

Other Silanes

Polylanes, *Safic*
Reinforcing Agent Si 230, Si 69, X 50, *Degussa*
Silane A-171, 186, 187, 188, 1120, 1130, *Union Carbide*
Silane Y 9194, *Union Carbide*
Silanogran Si 69 GR, *Kettlitz*
Silcat 17, *Union Carbide*

7.3.3.5 Titanates

Kenreact, *Kenrich*

7.3.3.6 Other Organic Activators

Actiol, *Kettlitz*
Rhenofit 1987, 2009, 2642, 3555, *Rhein-Chemie*
Struktol FA 541, IB 531, *Schill & Seilacher*
Vanax PY, *Vanderbilt*

7.3.4 Inorganic Curing Activators

Zinc oxide (ZnO)

Hansa Ultra, *Lehmann & Voss*
Luvozink, *Lehmann & Voss*
Polydispersion EZD, SZD, *Croxton & Garry*
Rhenogran ZnO, *Rhein-Chemie*
Rhenosol ZnO, *Rhein-Chemie*
Struktol Neopast, *Schill & Seilacher*
Struktol Perlzink, *Schill & Seilacher*
Struktol WB 700, *Schill & Seilacher*
Zinkoxid, *Harwick*
Zinkoxid, *Malmsten & Bergvall*
Zinkoxid, *Zinkwit*
Zinkoxid aktiv, *Bayer*
Zinkoxid transparent, *Bayer*

Magnesium oxide (MgO)

Deomag Granulat 70, *DOG*
Elastomag 170, *Morton*
Garospers, *Croxton & Garry*
Luvomag, *Lehmann & Voss*
Maglite, *Merck*
Magnox, *Basic*
Magnolux, *Lehmann & Voss*
Plastomag, *Morton*
Polymag, *Safic*
Rhenomag, *Rhein-Chemie*
Scorchguard, *Malmsten & Bergvall*
Struktol WB 900, 902, *Schill & Seilacher*

Zinc oxide/Magnesium oxide Combinations

Deomag Zn, *DOG*
Liquisperse MBZ, *Basic*
Magnozink, *Lehmann & Voss*
Polymix ZM, *Safic*
Struktol WB 890, *Schill & Seilacher*
Struktol Zimag 29/43, *Schill & Seilacher*

Lead oxides, phosphites and phthalates

Eagle Picher 98, *Eagle Picher*
Dyphos, *NL Industries*
Dythal, *NL Industries*
Kenlastik K, *Kenrich*
Kenmix Litharge, *Kenrich*
Kenmix Red Lead, *Kenrich*
Lead phosphite, *ALM*
Lead phthalate, *ALM*
Mix Pb, *Bozzetto*
Polydispersion KLD, PLD, TLD, *Croxton & Garry*
Polydispersion ERD, PRD, *Croxton & Garry*
Polydispersion K (DYT), *Croxton & Garry*
Polyminimum G, *Safic*
Polytharge G, *Safic*
Rhenogran Pb 0, Pb_3O_4, *Rhein-Chemie*

Other Inorganic Activators

Rhenofit CF, *Rhein-Chemie* (CaOH)
Rhenogran Ba CO_3, *Rhein-Chemie*

7.3.5 Retarders

7.3.5.1 Cyclohexylthiophthalimide (CTP)

Polydispersion S, *Croxton & Garry*
Rhenogran CTP, *Rhein-Chemie*
Robac CTP, *Robinson*
Santogard PVI, *Monsanto*

7.3.5.2 Phthalic Anhydride (PTA), WTR No. 56

Phthalic anhydride, *Atochem*
Phthalic anhydride, *VEB Buna*
Retarder PD, *Anchor*
Vulkalent B, *Bayer*

7.3.5.3 Salicylic Acid (SA), WTR No. 55

Retarder TSA, *Monsanto*
Salicylic acid, *Rhône Poulenc*

7.3.5.4 Other Retarders

Benzoic acid K, *Bayer*
Retarder J, *Uniroyal*
Tonox, *Uniroyal*
Vulkalent A, *Bayer*
Vulkalent E, *Bayer*

7.3.6 Antioxidants

7.3.6.1 p-Phenylenediamines

N-Isopropyl-N′-phenyl-p-phenylenediamine (IPPD), WTR No. 1

Antigene 3C, *Sumitomo*
Eastozone 34, *Eastman Chemicals*
Flexzone 3C, *Uniroyal*
Permanax IPPD, *Akzo*
Rhenogran IPPD, *Rhein-Chemie*
Santoflex IP, *Monsanto*
Vulkanox 4010 NA, *Bayer*

N-1.3-dimethylbutyl-N′-phenyl-p-phenylenediamine (6PPD), WTR No. 2

Antigene 6C, *Sumitomo*
Antozite 67, *Vanderbilt*
Anto 3E, *Pennwalt*
Flexzone 7-F, *Uniroyal*
Permanax 6PPD, *Akzo*
UOP 588, *Universal Oil*
Santoflex 13, *Monsanto*
Vulkanox 4020, *Bayer*
Wingstay 300, *Goodyear*

N-1.4-dimethylphenyl-N′-phenyl-p-phenylenediamine (7PPD), WTR No. 2a

Flexzone 11L, *Uniroyal*
Santoflex 14, *Monsanto*
Vulkanox 4050, *Bayer*

N,N′-Bis (1,4-dimethylphenyl)-p-phenylenediamine (77PD), WTR No. 3

Antocite MPD, *Vanderbilt*
Anto 3G, *Pennwalt*
Eastozone 33, *Eastman Chemicals*
Flexzone 4C, *Uniroyal*
Santoflex 77, *Monsanto*
UOP 788, *Universal Oil*
Vulkanox 4030, *Bayer*

N, N′-Diphenyl-p-phenylenediamine (DPPD), WTR No. 4

Agerite DPPD, *Vanderbilt*
Altofane DIP, *Ugine Kuhlmann*
Antioxidant DPPD, *Anchor*
Ekaland DPPD, *Landaise*
DPPD, *Monsanto*
Good Rite AO 3152, *BFGoodrich*
J-Z-F, *Uniroyal*
Permanax DPPD, *Akzo*

N,N′-Di-β-naphthyl-p-phenylenediamine (DNPD)

Age Rite White, *BFGoodrich*
Antioxidant 123, *Anchor*

N,N′-Diaryl-p-phenylenediamines, WTR No. 4a

Naugard 496, *Uniroyal*
Vulkanox 3100, *Bayer*
Wingstay 100, *Goodyear*

Other p-Phenylenediamines

Antocite 1, 2, *Vanderbilt*
Flexzone 6H, 10L, 12L, 18L, *Uniroyal*

7.3.6.2 Phenylamine Antioxidants

Octylated diphenyl amine (ODPA), WTR No. 8

Agerite Gel, Hipar, *Vanderbilt*
Agerite Stalite, S, *Vanderbilt*
Anox NS, *Bozzetto*
Cyanox 8, *Anchor*
Flectol ODP, *Monsanto*
Good Rite AO 3190 × 29, *BFGoodrich*
Lowinox ODE, *Lowi*
Oxtamine, *Uniroyal*
Pennox ODP, *Pennwalt*
Permanax OD, *Akzo*
Vanox 12, *Vanderbilt*
Vulkanox OCD, *Bayer*

Styrenated diphenylamine (SDPA), WTR No. 86

Lowinox SDA, *Lowi*
Vulkanox DDA, *Bayer*
Wingstay 29, *Goodyear*

Acetone-diphenylamine condensation products (ADPA), WTR No. 9

Agerite Superflex, *Vanderbilt*
Aminox, *Uniroyal*
BLE, *Uniroyal*
Flexamin, *Uniroyal*
Good Rite AO 3146, *BFGoodrich*
Naugard 492, *Uniroyal*
Permanax BBL, BLN, *Akzo*

Phenyl-α-(β)-naphthylamines (PAN, PBN), WTR No. 10

Agerite Powder, *Vanderbilt* (PBN)
Altofane A, *Ugine Kuhlmann* (PAN)
Altofane MC, *Ugine Kuhlmann* (PBN)
Anchor PBN, *Anchor*
Antigene PA, *Sumitomo* (PAN)
Antioxidant 116 X, *Du Pont* (PBN)
Naugard PAN, *Uniroyal*
PAN, *Union Carbide*
PBN, *BASF*
PBN, *VEB Bitterfeld*
Rhenogran PBN, *Rhein-Chemie*
Vulkanox PAN, PBN, *Bayer*

Other Phenylamines

Agerite Nepa, *Vanderbilt*
Altofane PCL, *Ugine Kuhlmann*
Aranox, *Uniroyal*
Betanox, *Uniroyal*
Naugard 445, *Uniroyal*
Permanax HD, *Akzo*
Polylite, *Uniroyal*
Vanax 200, *Vanderbilt*
Wingstay 29, *Goodyear*

7.3.6.3 Dihydroquinoline Derivatives

6-Ethoxi-2,2,4,-trimethyl-1,2-dihydroquinoline (ETMQ), WTR No. 6

Antigene AW, *Sumitomo*
Anox W, *Bozzetto*
Santoflex AW, *Monsanto*
Vulkanox EC, *Bayer*

2,2,4-Trimethyl-1,2-dihydroquinoline (polymerized) (TMQ), WTR No. 7

Agerite Resin D, MA, *Vanderbilt*
Anox HB, *Bozzetto*
Antigene RD, *Sumitomo*
Cyanox 12, *Anchor*
Flectol H, *Monsanto*
Good Rite AO 3140, *BFGoodrich*
Lowinox ACP, *Lowi*
Naftonox TMQ, *Metallgesellschaft*
Naugard Q, *Uniroyal*
Pennox HR, *Pennwalt*
Permanax TQ, *Akzo*
Ralox TMQ, *Raschig*
Vulkanox HS, *Bayer*

Other Quinone Derivates

Lowinox AH 25, *Lowi*
Santoflex DD, *Monsanto*
Santovar A, *Monsanto*

7.3.6.4 Bisphenol Derivatives

2,2′-Methylene-bis-(4-methyl-6-tert. butyl-phenol) (BPH), WTR No. 14

Antigene MDP, *Sumitomo*
Antioxidant 2246, *Anchor*
Bisoxol D, *CDF-Chimie*
Cyanox 2246, *Cyanamid*
Lowinox 22 M 46, *Lowi*
Naftonox 2246, *Metallgesellschaft*
Naruxol 15, *Uniroyal*
Ralox 2246, *Raschig*
Santowhite PC, *Monsanto*
Vanox 2246, *Vanderbilt*
Vulkacit BKF, *Bayer*

2,2′-Methylene-bis-(4-methyl-6-cyclohexyl-phenol) (CPH)

Permanax WSL, *Akzo*
Vulkanox ZKS, *Bayer*

2,2′-Isobutylidene-bis-(4,6-dimethylphenol) (IBPH)

Rhenogran NKF, *Rhein-Chemie*
Vulkanox NKF, *Bayer*

Other Bisphenols

Agerite Geltrol, Superlite, *Vanderbilt*
Anox G1, *Bozzetto*
Antigene WS-R, BBM, *Sumitomo*
Antioxidant 425, *Cyanamid*
Antioxidant 431, *Uniroyal*
Antioxidant 555, *Pitt Consol*
Bisoxol SM, *CDF-Chimie*
Good Rite AO 3112, 3113 X 1, *BFGoodrich*
Lowinox 002, 44 B 25, 22 CP 46, 22, B 46, 44 M 26, 44 M 36, P 22 B, *Lowi*
Naruxol, *Uniroyal*
Permanax CNS, WSP, *Akzo*
Santowhite Crystals, Powder, *Monsanto*
Santowhite 54, *Monsanto*
Vanox 1290, 1360, *Vanderbilt*
Vulkanox CS, *Bayer*
Vulkanox SKF, *Bayer*
Wingstay 2, *Goodyear*
Zalba, Zalba Special, *Du Pont*

7.3.6.5 Monophenol Derivatives

Butylated hydroxytoluene (BHT), WTR No. 15

Antigene BHT, *Sumitomo*
Antioxidant BHT, *Lehmann & Voss*
Bisoxol 220, *CDF-Chimie*
Ionol, *Shell*
Lowinox BHT, *Lowi*
Naftonox BHT, *Metallgesellschaft*
Naugard BHT, *Uniroyal*
Ralox BHT, *Raschig*
Vulkanox KB, *Bayer*

Styrenated phenols (SPH), WTR No. 16

Agerite Spar, *Vanderbilt*
Anox G2, *Bozzetto*
Antigene S, *Sumitomo*
Antioxidant SP, *Anchor*
Arconex SP, *Uniroyal*
Montaclere, *Monsanto*
Naugard SP, Uniroyal
Stabilite SP, *Reichold*
Vanox 102, *Vanderbilt*
Vulkanox SP, *Bayer*
Wingstay S, *Goodyear*

Alkylated and alkylated-styrenated phenols (APH, SAPH), WTR No. 16a, 16b

Agerite GT, *Vanderbilt*
Antioxidant N3, *BASF*
Anox T, *Bozzetto*
Bisoxol 24, MGB, *CDF-Chimie*
Cyanox LF, *Cyanamid*
Lowinox P 24 A, P 24 S, 001, *Lowi*
Naugard 431, 451, K, T, *Uniroyal*
Permanax WSO, *Akzo*
Stabilizer K1, K2, *BASF*
Stabilizer T, *Bayer*
Vanox 100, 1320, *Vanderbilt*
Vulkanox DS, KSM, TSP, *Bayer*
Wingstay C, T, *Goodyear*

Other Phenol Derivatives

Antioxidant MP, *Anchor*
Anox PP 18, *Bozzetto*
Irganox 565, 1010, 1076, 1081, 1313, *Ciba-Geigy*
Irgastab 2002, *Ciba-Geigy*
Naftonox ZMP, *Metallgesellschaft*
Ralox types, *Raschig*
Santowhite 54, MK, L, *Monsanto*
Vanax 13, *Vanderbilt*
Wingstay L, *Goodyear*

7.3.6.6 Imidazole Derivatives

2-Mercaptobenzimidazole (MBI), WTR No. 12

Altofane MTB, *Ugine Kuhlmann*
Antigene MB, *Sumitomo*
Naugard MB, *Uniroyal*
Rhenogran MBI, *Rhein-Chemie*
Vanox MTI, *Vanderbilt*
Vulkanox MB, *Bayer*

Methyl-mercaptobenzimidazole (MMBI), WTR No. 12a

Ekaland DES, *Landaise*
Rhenogran MMBI, *Rhein-Chemie*
Vulkanox MB2, *Bayer*

Zinc-2-mercaptobenzimidazole (ZMBI)

Altofane MTBZ, *Ugine Kuhlmann*
Vulkanox ZMB, *Bayer*

Zinc-methyl-mercaptobenzimidazole (ZMMBI), WTR No. 12b

Ekaland DESZ, *Landaise*
Vanox ZMTI, *Vanderbilt*
Vulkanox ZMB2, *Bayer*

7.3.6.7 Trisnonyl Phenyl Phosphite, WTR No. 17

Antigene TNP, *Sumitomo*
Good Rite AO 3113, *BFGoodrich*
Irgafos TNPP, *Ciba-Geigy*
Lowinox TNPP, *Lowi*
Polygard, *Uniroyal*
Stavinor TNP, *Atochem*
Vanox 13, *Vanderbilt*

7.3.6.8 Other Antioxidants

Chimassorb, *Ciba-Geigy*
Irganox PS 800, *Ciba-Geigy*
Lowinox BHA, DLTDP, *Lowi*
Tinuvin 622, 765, 770, *Ciba-Geigy*
Vanox AT, *Vanderbilt*

7.3.6.9 Copper Inhibitors

Rheomet types, *Ciba-Geigy*

7.3.6.10 UV-Absorbers

Tinuvin types, *Ciba-Geigy*

7.3.6.11 Antiozonant Waxes

Antilux types, *Rhein-Chemie*
Carnauba wax *Schütz*
Ceresin types, *Schütz*
Controzon types, *DOG*
Heliozone, *Du Pont*
Lunacera, *Fuller*
Mikrowachs, *BP*
Montanwachs, *Braunkohlen-Werk*
Negozone, *Campbell*
Okerin, *Astor*
Ozokerit, *Kemi*
Paraffin, *Malmsten & Bergvall*
Protektor G, *Fuller*
Sunolite, *Witco*
Sunproof, *Uniroyal*
Vanwax, *Vanderbilt*

7.3.7 Peptizing Agents

7.3.7.1 Pentachlorthiophenol Derivatives (PCTP, ZPCTP)

Cicclizzante 3C, *Bozzetto*
Endor, *Du Pont*
Renacit 4, 5, 7, 9, *Bayer*

7.3.7.2 Zinc Salts of Fatty Acids (Special Types)

Aktiplast F, *Rhein-Chemie*
Dispergum 24, *DOG*
Noctizer, *Ouchi Shinko*
Pepton, *Cyanamid*
Struktol A 86, *Schill & Seilacher*

7.3.7.3 Other Peptizing Agents

Aktiplast 6N, *Rhein-Chemie*
Pepton 22, 44, *Anchor*
Renacit 8, *Bayer*
Struktol A 82, *Schill & Seilacher*

7.3.8 Ester and Ether Plasticizers (Selection)

7.3.8.1 Adipates

Dioctyladipate (DOA)

Adimoll DO, *Bayer*
Bisoflex DOA, *BP*
Rheomol DOA, *Ciba-Geigy*
Witamol 320, *Dynamit Nobel*

Didecyladipate (DDA, DIDA)

Bisoflex DIDA, *BP*
Rheomol DIDA, *Ciba-Geigy*

Other Adipates

Adimoll BO, DB, DN, *Bayer*
Mediaplast NB 4, *Kettlitz*
Ultramoll I, II, III, *Bayer*
Witamol 615, *Dynamit Nobel*

7.3.8.2 Phthalates

Dibutylphthalates (DBP, DIBP)

Bisoflex DBP, DIBP, *BP*
DBP, *Atochem*
DIBP, *Atochem*
Palatinol C, *BASF*
Unimoll DB, *Bayer*

Dioctylphthalates (DOP, DIOP)

Bisoflex DOP, DIOP, *BP*
DOP, *Atochem*
Jayflex DOP, *Exxon*
Palatinol AH, *BASF*
Witamol 100, *Dynamit Nobel*

Dinonylphthalates (DNP, DINP)

Bisoflex DNP, *BP*
Jayflex DINP, *Exxon*
Vestinol 9, *CWH*

Tridecylphthalates (DTDP)

Bisoflex DTDP, *BP*
Rheomol DTDP, *Ciba-Geigy*
Witamol 190, *Dynamit Nobel*

Other Phthalates

Bisoflex DAP, DCHP, DEP, DMP, D 79 A, L 79 P, L 911 P, *BP*
Rheomol P, *Ciba-Geigy*
Unimoll BB, DM, 66, *Bayer*
Witamol 110, 600, *Dynamit Nobel*

7.3.8.3 Dioctylsebacat (DOS)

DOS, *Ashland*
Rheomol DOS, *Ciba-Geigy*
Witamol 500, *Dynamit Nobel*

7.3.8.4 Dioctylacelates (DOZ, DIOZ)

DOZ, *Ashland*
Triplast 3010, *Unichema*
Rheomol D 10Z, *Ciba-Geigy*

7.3.8.5 Trimellitates

Rheomol LTM, ATM, *Ciba-Geigy*
Trimellitate types, *Amoco*

7.3.8.6 Glycol Esters

Priplast 1562, *Unichema*
Vulkanol 88, 90, *Bayer*
Witamol 60, *Dynamit Nobel*

7.3.8.7 Phosphate Esters

Celluflex, *Celanese*
Disflamoll DPK, DPO, TOF, TCA, DKP, TP, *Bayer*
Rheofas 50, 65, 95, *Ciba-Geigy*
Rheomol 249, TPP, *Ciba-Geigy*

7.3.8.8 Other Ester Plasticizers

Koremoll CE 5185, *BASF*
Melonil, *Henkel*
Mesamoll, *Bayer*
Priplast 3117, 3131, *Unichema*
Rheoplex 245, 430, 903, 1102, GF, GL, MD, *Ciba-Geigy*
Struktol KW 400, KW 500, WB 300, *Schill & Seilacher*
Vulkanol 81, SF, *Bayer*

7.3.8.9 Ether Plasticizers

Antistaticum RC 100, *Rhein-Chemie*
Antistatic plasticizer WM, *Bayer*
Deotack 70 DL, *DOG*
FH DL, *Lehmann & Voss*
Rheomol BCF, *Ciba-Geigy*
Plasticizer TP-90 B, *Thiokol* (BCF)
Plasticizer TP-95, *Thiokol*
Vulkanol FH, *Bayer* (XF-Resin)
Vulkanol 85, BA, OT, *Bayer*

7.3.8.10 Epoxidized Plasticizers

AC-Types, *Ashland*
Ecepox-Types, *Atochem*
Rheoplast 38, 39, 392, *Ciba-Geigy*
Priplast 1431, *Unichema*

7.3.8.11 Chloro Paraffines

Arubren CP, *Bayer*
Cereclor, *ICI*
Chloroparaffin, *Hüls*
Chlorowax, *Diamond*
Witaclor 171, *Dynamit Nobel*

7.3.9 Mineral Oils (Selection)

7.3.9.1 Paraffinic Oils

Enerpar, *BP*
Flexon, *Exxon*
Ingraplast, *Fuchs*
Naftolen, *Metallgesellschaft*
Nyflex, *Nynas*
Sunpar, *Sun Oil*
Tudalen, *Dahleke*

7.3.9.2 Naphthenic Oils

Circolight Process Oil, *Sun Oil*
Circosol 4240, *Sun Oil*
Enerthene, *BP*
Flexon, *Exxon*
Ingraplast, *Fuchs*
Naftolen, *Metallgesellschaft*
Nytene, *Nynas*
Tudalen, *Dahleke*

7.3.9.3 Aromatic, High Aromatic Oils

Dutrex 6, *Shell*
Enerflex, *BP*
Ingralen, *Fuchs*
Kenplast G, *Kenrich*
Kenflex N, *Kenrich*
Naftolen, *Metallgesellschaft*
Sundex, *Sun Oil*
Tudalen, *Dahleke*

7.3.10 Other Plasticizers (Selection)

Bondogen, *Vanderbilt*
Citroflex, *Sun Oil*
Hercoflex 600, *Hercules*
Indopol, *Amoco*
K-Stay, *Vanderbilt*
Monoplex types, *Rohm & Haas*
Leegen, *Vanderbilt*
Mediaplast types, *Kettlitz*
Nesires types, *NCC*
Paraplex types, *Krahn*
Perkaflex 12, *Akzo*
Picco Plasticizers, *Hercules*
Plastogen, *Vanderbilt*
Rheogen, *Vanderbilt*
Santicizer types, *Monsanto*

7.3.11 Fatty Acids and Derivatives (Selection), Process Aids

7.3.11.1 Stearic Acids

Pristerene, *Unichema*
Safacit G, R, *Jahres*
Stearic Acid, *Foster*
Stearic Acid, *Henkel*
Stearin, *Malmsten & Bergvall*

7.3.11.2 Other Fatty Acids

Dehydol types, *Henkel*
Edenor types, *Henkel*
Oil Fatty Acid, *Foster*
Oleïn, *Malmsten & Bergvall*
Oleinic Acid, *Foster*
Priolene, *Unichema*
Talg Fatty Acid, *Foster*

7.3.11.3 Fatty Acid Salts

Aktiplast T, *Rhein-Chemie*
Aflux R, *Rhein-Chemie*
Deoflow S, *DOG*
Dispergum L, T, N, C, *DOG*
PAT-Additives, *Würtz*
Polyplastol 6, *Bozzetto*
Struktol types, *Schill & Seilacher*
Zinc Stearate, *Malmsten & Bergvall*
Zinc Starate, *Witco*

7.3.11.4 Pentaerythritol Tetrastearate (PET)

Aflux 54, *Rhein-Chemie*
Deoflow 821, *DOG*
Struktol WB 222, *Schill & Seilacher*

7.3.11.5 Fatty Alcohol Esters

Aflux 42, 44, *Rhein-Chemie*
Dehypon conc, *Henkel*
Deoflow A, D, *DOG*

7.3.11.6 Other Surface Active Substances

Activin types, *Kettlitz*
Deogum 80, *DOG*
Deosol H, VE, *DOG*
Dispersing Agent DS, FL, KB, OX, *Kettlitz*
Haftolat, *Kettlitz*
Interlube A, P, *Anchor*
Ken Reacts Drimix KR, *Kenrich*
Mediaplast WH, *Kettlitz*
PAT Additives, *Würtz*
Struktol types, *Schill & Seilacher*
TE-types, *Technical Processing*
Vanfre types, *Vanderbilt*

7.3.12 Other Process Aids

7.3.12.1 Process Aids for FPM (FKM)

Deoflow 821, *DOG*
Dynamar PPA 790, 791, *3 M*
Fluorel FC 2171, *3 M*
VPA 1, 2, *Du Pont*

7.3.12.2 Process Aids for CSM

AC Polyethylene, *Allied*
Carbowax, *Union Carbide*
Deoflow 821, *DOG*
Lunacerin, *Fuller*

7.3.12.3 Process Aids for EAM

Armeen 18D (n-Octadecylamin), *Armak*
Gafac RL 210 (Polyoxyethylene octadecylether phosphate), *GAE*
Vanvre (Fatty alcohol phosphate), *Vanderbilt*

7.3.12.4 Factices

Factice types, *DOG*
Factice types, *Anchor*
Factice types, *Rhein-Chemie*
Factice types, *Lefrant*

7.3.13 Blowing Agents

7.3.13.1 Azodicarbonamide (ADC), WTR No. 61

Azobul, *Atochem*
Celogen AZ, *Uniroyal*
Ekaland AZO, *Landaise*
Genitron AC, CR, *FBC*
Polydispersion types, *Croxton & Garry*
Porofor ADC/R, ADC/K, *Bayer*

7.3.13.2 Sulfohydrazide Derivatives (BSH, DBSH, TSH, etc.)

Celogen OT, *Uniroyal*
Genitron BSH, *FBC*
Luvopor OB, *Lehmann & Voss*
Luvopor TSH, *Lehmann & Voss*
Porofor BSH, *Bayer*
Porofor B 13, *Bayer*

7.3.13.3 Dinitroso Pentamethylene Tetramine (DNPT), WTR No. 62

Ekaland Dipentax 75, *Landaise*
Polydispersion S (DNPT), *Croxton & Garry*
Porofor DNO, *Bayer*

7.3.13.4 Other Blowing Agents

BIK, *Uniroyal*
Poreks, *Ciech-Stomil*
Rhenopor 1843, *Rhein-Chemie*

7.3.13.5 Activators for Blowing Agents

Attivante 4030, *Bozzetto*
Rhenofit 1100, 1600, *Rhein-Chemie*

7.3.14 Bonding Agents

7.3.14.1 Rubber to Metal Bonding Agents

Braze, *Vanderbilt*
Chemlock, *Hughson*
Chemosil, *Henkel*
DC-Primers 1200, 2260, *Dow Corning* (Q)
Duralink, *Monsanto*
Dynamar 5150, 5155, *3 M* (FKM)
Megum, *Metallgesellschaft*
Parlock, *PAr*
Rhodorsil, *Rhône Poulenc* (Q)
Thixon, *Thixon*
Ty Ply, *Anchor*
VP 3243, *Wacker*

7.3.14.2 Rubber to Textile Bonding Agents

Bonding Agent M3, R 4, *Uniroyal*
Bunatex VP, *Bunawerke Hüls*
Cohedur, *Bayer*
Desmodur R, RF, *Bayer*
Intene 181, *Enichem*
Megum, *Metallgesellschaft*
Parflock, *PAr*
Pliocord, *Goodyear*
Pliogrip, *Goodyear*
Rhenogran Resorcin, *Rhein-Chemie*
Thixon, *Thixon*

7.4 Trade Names of Synthetic Rubbers, Thermoplastic Elastomers and Vulcanizing Chemicals

Name	Company (Addresses see Section 7.5)	Type of Rubber rsp. Chemical see Section:
A 1	Monsanto	7.3.2.9
AC-Types	Ashland	7.3.8.10
Accelerator 552	Du Pont	7.3.2.4
Accelerator 808, 833	Du Pont	7.3.2.10
Accelerator M	Ciech-Stomil	7.3.2.1
AC Polyethylene	Allied	7.3.12.2
Acralen (Latex)	Bayer	7.1.14
Actigran	Kettlitz	7.3.3.1
Actiol	Kettlitz	7.3.3.6
Activator OC	Degussa	7.3.3.2
Activator OP	Degussa	7.3.3.2
Activin types	Kettlitz	7.3.11.6
Adimoll BO, DB, DN	Bayer	7.3.8.1
Adimoll DO	Bayer	7.3.8.1
Adiprene	Du Pont	7.1.24
Afdene BR	Carbochem	7.1.1
Aflas, Asahi Glass (TFE/PP)	3M	7.1.18
Aflux R	Rhein-Chemie	7.3.11.3
Aflux 42, 44	Rhein-Chemie	7.3.11.5
Aflux 54	Rhein-Chemie	7.3.11.4
Afprene IR	Carbochem	7.1.7
Afsol	Carbochem	7.1.2
Age Rite White	BFGoodrich	7.3.6.1
Agerite DPPD	Vanderbilt	7.3.6.1
Agerite Gel, Hipar	Vanderbilt	7.3.6.2
Agerite Geltrol, Superlite	Vanderbilt	7.3.6.4
Agerite GT	Vanderbilt	7.3.6.5
Agerite Nepa	Vanderbilt	7.3.6.2
Agerite Powder (PBN)	Vanderbilt	7.3.6.2
Agerite Resin D, MA	Vanderbilt	7.3.6.3
Agerite Spar	Vanderbilt	7.3.6.5
Agerite Stalite, S	Vanderbilt	7.3.6.2
Agerite Superflex	Vanderbilt	7.3.6.2
Aktiplast F	Rhein-Chemie	7.3.7.2
Aktiplast T	Rhein-Chemie	7.3.11.3
Aktiplast 6N	Rhein-Chemie	7.3.7.3
Alcatal	Safic	7.3.3.2
Alcryn	Du Pont	7.2.5
Altofane A (PAN)	Ugine Kuhlmann	7.3.6.2
Altofane DIP	Ugine Kuhlmann	7.3.6.1
Altofane MC (PBN)	Ugine Kuhlmann	7.3.6.2
Altofane MTB	Ugine Kuhlmann	7.3.6.6
Altofane MTBZ	Ugine Kuhlmann	7.3.6.6
Altofane PCL	Ugine Kuhlmann	7.3.6.2
Amax	Vanderbilt	7.3.2.2
Ameripol	BFGoodrich	7.1.2
Ameripol CB*	BFGoodrich	7.1.1

* earlier used trade name

Name	Company (Addresses see Section 7.5)	Type of Rubber rsp. Chemical see Section:
Aminox	Uniroyal	7.3.6.2
Amyl Cadmate	Vanderbilt	7.3.2.4
Amyl Ledate	Vanderbilt	7.3.2.4
Ancacide ME, Anchor TMTD	Anchor, Nordmann, Rassmann	7.3.2.6
Ancap, Anchor MBT	Anchor, Nordmann, Rassmann	7.3.2.1
Ancazate BW, Anchor ZDBC	Anchor, Nordmann, Rassmann	7.3.2.3
Anchor CBS	Anchor, Nordmann, Rassmann	7.3.2.2
Anchor DBD	Anchor	7.3.2.4
Anchor MBTS	Anchor, Nordmann, Rassmann	7.3.2.1
Anchor PBN	Anchor	7.3.6.2
Anchor TGB, 1, 2, 3	Anchor, Nordmann, Rassmann	7.3.2.1
Anchor TMTM	Anchor, Nordmann, Rassmann	7.3.2.6
Anchor ZDEC	Anchor, Nordmann, Rassmann	7.3.2.3
Anchor ZDMC	Anchor, Nordmann, Rassmann	7.3.2.3
Anchor ZMBT	Anchor, Nordmann, Rassmann	7.3.2.1
Ancomer ATM3	Ancomer	7.3.3.1
Ancomer ATM 9	Ancomer	7.3.3.1
Ancomer ATM 11	Ancomer	7.3.3.1
Ancomer ATM 16	Ancomer	7.3.3.1
Ancomer ATM 17	Ancomer	7.3.3.1
Anox G1	Bozzetto	7.3.6.4
Anox G2	Bozzetto	7.3.6.5
Anox HB	Bozzetto	7.3.6.3
Anox NS	Bozzetto	7.3.6.2
Anox PP 18	Bozzetto	7.3.6.5
Anox T	Bozzetto	7.3.6.5
Anox W	Bozzetto	7.3.6.3
Antigene AW	Sumitomo	7.3.6.3
Antigene BHT	Sumitomo	7.3.6.5
Antigene MB	Sumitomo	7.3.6.6
Antigene MDP	Sumitomo	7.3.6.4
Antigene NC	Sumitomo	7.3.2.4
Antigene PA (PAN)	Sumitomo	7.3.6.2
Antigene RD	Sumitomo	7.3.6.3
Antigene S	Sumitomo	7.3.6.5
Antigene TNP	Sumitomo	7.3.6.7
Antigene WX-R, BBM	Sumitomo	7.3.6.4
Antigene 3C	Sumitomo	7.3.6.1
Antigene 6C	Sumitomo	7.3.6.1
Antilux types	Rhein-Chemie	7.3.6.11
Antioxidant BHT	Lehmann & Voss	7.3.6.5
Antioxidant DPPD	Anchor	7.3.6.1
Antioxidant MP	Anchor	7.3.6.5

Name	Company (Addresses see Section 7.5)	Type of Rubber rsp. Chemical see Section:
Antioxidant N3	BASF	7.3.6.5
Antioxidant SP	Anchor	7.3.6.5
Antioxidant 116 X (PBN)	Du Pont	7.3.6.2
Antioxidant 123	Anchor	7.3.6.1
Antioxidant 425	Cyanamid	7.3.6.4
Antioxidant 431	Uniroyal	7.3.6.4
Antioxidant 555	Pitt Consol	7.3.6.4
Antioxidant 2246	Anchor	7.3.6.4
Antistatic plasticizer WM	Bayer	7.3.8.9
Antistaticum RC 100	Rhein-Chemie	7.3.8.9
Anto 3E	Pennwalt	7.3.6.1
Anto 3G	Pennwalt	7.3.6.1
Antocite MPD	Vanderbilt	7.3.6.1
Antocite 1, 2	Vanderbilt	7.3.6.1
Antozite 67	Vanderbilt	7.3.6.1
Aranox	Uniroyal	7.3.6.2
Arazate	Uniroyal	7.3.2.3
Arconex SP	Uniroyal	7.3.6.5
Armeen 18 D	Armak	7.3.12.3
Arnitel	Akzo	7.2.7
Arubren CP	Bayer	7.3.8.11
Asadene	Asahi Chemicals	7.1.1
Attivante 4030	Bozzetto	7.3.13.5
Austrapol	Australian Synthetic Rubber	7.1.2
Azobul	Atochem	7.3.13.1
Bantex	Monsanto	7.3.2.1
Bayer CM	Bayer	7.1.12
Baypren	Bayer	7.1.6
Benzoic acid K	Bayer	7.3.5.4
Betanox	Uniroyal	7.3.6.2
Beutene	Uniroyal	7.3.2.10
BIK	Uniroyal	7.3.13.4
Bismate	Vanderbilt	7.3.2.4
Bisoflex DAP, DCHP, DEP, DMP, D 79 A, L 79 P, L 911 P	BP	7.3.8.2
Bisoflex DBP, DIBP	BP	7.3.8.2
Bisoflex DIDA	BP	7.3.8.1
Bisoflex DNP	BP	7.3.8.2
Bisoflex DOA	BP	7.3.8.1
Bisoflex DOP, DIOP	BP	7.3.8.2
Bisoflex DTDP	BP	7.3.8.2
Bisoxol D	CDF-Chimie	7.3.6.4
Bisoxol SM	CDF-Chimie	7.3.6.4
Bisoxol 24, MGB	CDF-Chimie	7.3.6.5
Bisoxol 220	CDF-Chimie	7.3.6.5
BLE	Uniroyal	7.3.6.2
Blensil	General Electric	7.1.22
Bonding Agent M3, R 4	Uniroyal	7.3.14.2
Bondogen	Vanderbilt	7.3.10
BQD	Krahn	7.3.1.4
Braze	Vanderbilt	7.3.14.1

Name	Company (Addresses see Section 7.5)	Type of Rubber rsp. Chemical see Section:
Breon*	BP	7.1.4
Breon N*	BP	7.1.3
Budene	Goodyear	7.1.1
Buna AP	Bunawerke Hüls	7.1.10
Buna CB	Bunawerke Hüls	7.1.1
Buna EM	Bunawerke Hüls	7.1.2
Buna N	VEB Buna	7.1.3
Buna SB	VEB Buna	7.1.2
Bunatex VP	Bunawerke Hüls	7.3.14.2
Butaclor	Distugil	7.1.6
Butacril*	Ugine Kuhlmann	7.1.3
Butasan	Monsanto	7.3.2.3
Butazate	Uniroyal	7.3.2.3
Butyl Eight	Vanderbilt	7.3.2.4
Butyl Tuads (TBTD)	Vanderbilt	7.3.2.6
Butyl Zimate	Vanderbilt	7.3.2.3
Cadmate (CdDEC)	Vanderbilt	7.3.2.4
Captax	Vanderbilt	7.3.2.1
Carbowax	Union Carbide	7.3.12.2
Cariflex BR	Shell	7.1.1
Cariflex IR	Shell	7.1.7
Cariflex S	Shell	7.1.2
Cariflex TR	Shell	7.2.1
Carnauba wax	Schütz	7.3.6.11
Carom	Danubiana	7.1.2
Carom	Danubiana	7.1.7
Castomer	Baxenden	7.1.24
Caytur	Du Pont	7.3.1.6
CBS	Ciech-Stomil	7.3.2.2
CBS	Krahn	7.3.2.2
CBS	Lehmann & Voss	7.3.2.2
Celluflex	Celanese	7.3.8.7
Celogen AZ	Uniroyal	7.3.13.1
Celogen OT	Uniroyal	7.3.13.2
Cereclor	ICI	7.3.8.11
Ceresin types	Schütz	7.3.6.11
Chemigum	Goodyear	7.1.3
Chemlink TATM	Lehmann & Voss	7.3.3.2
Chemlink TATM	Ware Chemicals	7.3.3.2
Chemlink 20	Ware Chemicals	7.3.3.1
Chemlink 30	Ware Chemicals	7.3.3.1
Chemlock	Hughson	7.3.14.1
Chemosil	Henkel	7.3.14.1
Chimassorb	Ciba-Geigy	7.3.6.8
Chlorowax	Diamond	7.3.8.11
Cicclizzante 3C	Bozzetto	7.3.7.1
Circolight Process Oil	Sun Oil	7.3.9.2
Circosol 4240	Sun Oil	7.3.9.2
Cis	VEB Buna	7.1.1
Cisdene*	American Chemicals	7.1.1

* earlier used trade name

Name	Company (Addresses see Section 7.5)	Type of Rubber rsp. Chemical see Section:
Citroflex	Sun Oil	7.3.10
Cohedur	Bayer	7.3.14.2
Conac NS	Du Pont	7.3.2.2
Controzon types	DOG	7.3.6.11
Copo	Copolymer	7.1.2
CPB	Uniroyal	7.3.2.5
Crystex	Kalichemie	7.3.1.1
CS PE	China	7.1.13
CuDMC	Lehmann & Voss	7.3.2.4
Cumate (CuDMC)	Vanderbilt	7.3.2.4
Cupsac	Cyanamid	7.3.2.4
Curalon	Uniroyal	7.3.1.6
Cure Rite 18	BFGoodrich	7.3.2.7
Curing Agents P_2, P_3, V_1	Daikin	7.3.1.5
Cyanacryl	Cyanamid	7.1.14
Cyanox LF	Cyanamid	7.3.6.5
Cyanox 8	Anchor	7.3.6.2
Cyanox 12	Anchor	7.3.6.3
Cyanox 2246	Cyanamid	7.3.6.4
D 5 MTD	Krahn	7.3.2.6
D 5 MTT	Krahn	7.3.2.6
Dai-El	Daikin	7.1.18
Dai-El Perfluor	Daikin	7.1.18
Daisolac	Osaka Soda	7.1.12
Daltoped	ICI	7.1.24
Davathane	Krahn	7.1.24
DBA	Uniroyal	7.3.2.10
DBP	Atochem	7.3.8.2
DBQD	Krahn	7.3.1.4
DBTU	Krahn	7.3.2.9
DC-Primers 1200, 2260 (Q)	Dow Corning	7.3.14.1
DCBS	Krahn	7.3.2.2
Dehydol types	Henkel	7.3.11.2
Dehypon conc	Henkel	7.3.11.5
Delac MOR	Uniroyal	7.3.2.2
Delac NS	Uniroyal	7.3.2.2
Delac S	Uniroyal	7.3.2.2
Denka	Denka Kagaku	7.1.6
Deoflow A, D	DOG	7.3.11.5
Deoflow S	DOG	7.3.11.3
Deoflow 821	DOG	7.3.11.4, 7.3.12.1, 7.3.12.2
Deogum 80	DOG	7.3.11.6
Deomag Granulat 70	DOG	7.3.4
Deomag Zn	DOG	7.3.4
Deosol H, VE	DOG	7.3.11.6
Deosulf	DOG	7.3.1.1
Deotack 70 DL	DOG	7.3.8.9
Deovulc BG 1/87	DOG	7.3.2.13
Deovulc BL 6	DOG	7.3.2.13
Deovulc EG L	DOG	7.3.2.13
Deovulc EG 3, 40, 28	DOG	7.3.2.13

Name	Company (Addresses see Section 7.5)	Type of Rubber rsp. Chemical see Section:
Deovulc M	DOG	7.3.1.2
Desmodur	Bayer	7.3.1.6
Desmodur R, RF	Bayer	7.3.14.2
Desmopan	Bayer	7.2.6
Desmorapid	Bayer	7.3.1.6
DETU	Krahn	7.3.2.9
DETU	Lehmann & Voss	7.3.2.9
Diak 1, 2, 3, 7	Du Pont	7.3.1.5
Diak 7	Du Pont	7.3.3.2
Diapol	Mitsubishi	7.1.2
Dibenzo GMF	Kenrich	7.3.1.4
Dibenzo GMF	Uniroyal	7.3.1.4
DIBP	Atochem	7.3.8.2
DiCup	Hercules	7.3.1.3
Diene*	Firestone	7.1.1
Diorez, Diprane	Briggs & Townsend	7.1.24
Disflamoll types	Bayer	7.3.8.7
Dispergum L, T, N, C	DOG	7.3.11.3
Dispergum 24	DOG	7.3.7.1
Dispersing Agent types	Kettlitz	7.3.11.6
DOP	Atochem	7.3.8.2
DOS	Ashland	7.3.8.3
DOTG	Krahn	7.3.2.8
Dow CPE*	Dow Corning	7.1.12
DOZ	Ashland	7.3.8.4
DPG	Krahn	7.3.2.8
DPG	Lehmann & Voss	7.3.2.8
DPG	Monsanto	7.3.2.8
DPPD	Monsanto	7.3.6.1
DPTTS	Lehmann & Voss	7.3.1.2
DPTU	Krahn	7.3.2.9
Duragene*	General Tire	7.1.1
Duralink	Monsanto	7.3.14.1
Duranit (Styrene resin)	CWH	7.1.2
Durax	Vanderbilt	7.3.2.2
Dutral	Montedison	7.1.10
Dutral TP	Montedison	7.2.3
Dutrex 6	Shell	7.3.9.3
Dynamar PPA 790, 791	3 M	7.3.12.1
Dynamar 5150, 5155 (FKM)	3 M	7.3.14.1
Dynasylan MEMO	Dynamit Nobel	7.3.3.4
Dynasylan MTMO	Dynamit Nobel	7.3.3.4
Dynasylan VTEO	Dynamit Nobel	7.3.3.4
Dynasylan VTMOEO	Dynamit Nobel	7.3.3.4
Dyphos	NL Industries	7.3.4
Dythal	NL Industries	7.3.4
Eagle Picher 98	Eagle Picher	7.3.4
Eastozone 33	Eastman Chemicals	7.3.6.1
Eastozone 34	Eastman Chemicals	7.3.6.1
Ecepox-Types	Atochem	7.3.8.10

* earlier used trade name

Name	Company (Addresses see Section 7.5)	Type of Rubber rsp. Chemical see Section:
Edenor types	Henkel	7.3.11.2
EDMA DL	Lehmann & Voss	7.3.3.1
Ekagom VS	Ugine Kuhlmann	7.3.2.10
Ekaland AZO	Landaise	7.3.13.1
Ekaland CBS	Landaise	7.3.2.2
Ekaland CDBC	Landaise	7.3.2.4
Ekaland DCBS	Landaise	7.3.2.2
Ekaland DES	Landaise	7.3.6.6
Ekaland DESZ	Landaise	7.3.6.6
Ekaland DETU	Landaise	7.3.2.9
Ekaland DIBS	Landaise	7.3.2.2
Ekaland Dipentax 75	Landaise	7.3.13.3
Ekaland DOTG	Landaise	7.3.2.8
Ekaland DPG	Landaise	7.3.2.8
Ekaland DPPD	Landaise	7.3.6.1
Ekaland DPTU	Landaise	7.3.2.9
Ekaland ETU	Landaise	7.3.2.9
Ekaland HMT	Landaise	7.3.2.10
Ekaland MBT	Landaise	7.3.2.1
Ekaland MBTS	Landaise	7.3.2.1
Ekaland MDS	Landaise	7.3.1.2
Ekaland NaMBT	Landaise	7.3.2.1
Ekaland NDBC	Landaise	7.3.2.4
Ekaland NOBS	Landaise	7.3.2.2
Ekaland TeDEC	Landaise	7.3.2.4
Ekaland TETD	Landaise	7.3.2.6
Ekaland TMTD	Landaise	7.3.2.6
Ekaland TMTM	Landaise	7.3.2.6
Ekaland TSPM	Landaise	7.3.1.2
Ekaland ZDBC	Landaise	7.3.2.3
Ekaland ZDEC	Landaise	7.3.2.3
Ekaland ZDMC	Landaise	7.3.2.3
Ekaland ZEPC	Landaise	7.3.2.3
Ekaland ZMBT	Landaise	7.3.2.1
Elaprim*	Enichem	7.1.3
Elaprim AR*	Montedison	7.1.14
Elaprim S Mix*	Enichem	7.1.4
Elaslen	Showa Denko	7.1.12
Elastollan	BASF	7.2.6
Elastomag 170	Morton	7.3.4
Elastosil	Wacker	7.1.22
Elastothane	Thiokol	7.1.24
Elexar	Shell	7.2.2
END, GND	Wyrough & Loser	7.3.2.9
Endor	Du Pont	7.3.7.1
Enerflex	BP	7.3.9.3
Enerpar	BP	7.3.9.1
Enerthene	BP	7.3.9.2
Epcar*	BFGoodrich	7.1.10
Epichlormer	Osaka Soda	7.1.16
EPsyn	Copolymer	7.1.10

* earlier used trade name

Name	Company (Addresses see Section 7.5)	Type of Rubber rsp. Chemical see Section:
Eptotal*	Socabu	7.1.10
Escorene Ultra	Exxon	7.1.11
Esprene	Sumitomo	7.1.10
Esso Bromobutyl	Exxon	7.1.9
Esso Butyl	Exxon	7.1.8
Esso Chlorobutyl	Exxon	7.1.9
Estane	BFGoodrich	7.2.6
Ethasan	Monsanto	7.3.2.3
Ethazate	Uniroyal	7.3.2.3
Ethyl Ledate	Vanderbilt	7.3.2.4
Ethyl Selenac	Vanderbilt	7.3.2.4
Ethyl Tellurac	Vanderbilt	7.3.2.4
Ethyl Thiurads	Vanderbilt	7.3.2.6
Ethyl Tuex	Uniroyal	7.3.2.6
Ethyl Zimate	Vanderbilt	7.3.2.3
ETU	Krahn	7.3.2.9
Europrene AR	Enichem	7.1.14
Europrene Cis	Enichem	7.1.1
Europrene N	Enichem	7.1.3
Europrene S Ozo	Enichem	7.1.4
Europrene SOL	Enichem	7.1.7
Europrene SOL T	Enichem	7.2.1
Europrene, Unidene	Enichem	7.1.2
Evatane	Atochem	7.1.11
Eypel-F	Ethyl	7.1.19
F (MBTS/DPG)	Krahn	7.3.2.1
Factice types	Anchor	7.3.12.4
Factice types	DOG	7.3.12.4
Factice types	Lefrant	7.3.12.4
Factice types	Rhein-Chemie	7.3.12.4
FH DL	Lehmann & Voss	7.3.8.9
Finaprene	Petrochim	7.1.1
Finaprene	Petrochim	7.1.2
Finaprene	Petrochim	7.2.1
Flectol H	Monsanto	7.3.6.3
Flectol ODP	Monsanto	7.3.6.2
Flexamin	Uniroyal	7.3.6.2
Flexon	Exxon	7.3.9.1
Flexon	Exxon	7.3.9.2
Flexzone 11L	Uniroyal	7.3.6.1
Flexzone 3C	Uniroyal	7.3.6.1
Flexzone 4C	Uniroyal	7.3.6.1
Flexzone 6H, 10L, 12L, 18L	Uniroyal	7.3.6.1
Flexzone 7-F	Uniroyal	7.3.6.1
Fluorel	3M	7.1.18
Fluorel FC	3 M	7.3.1.5
Fluorel FC 2171	3 M	7.3.12.1
FRV	General Electric	7.1.22
FSE	General Electric	7.1.23
Furbac	Anchor	7.3.2.2

* earlier used trade name

Name	Company (Addresses see Section 7.5)	Type of Rubber rsp. Chemical see Section:
Gafac RL 210	GAF	7.3.12.3
Garospers	Croxton & Garry	7.3.4
Genitron AC, CR	FBC	7.3.13.1
Genitron BSH	FBC	7.3.13.2
Geolast	Monsanto	7.2.4
GMF	Uniroyal	7.3.1.4
Good Rite AO 3112, 3113 X 1	BFGoodrich	7.3.6.4
Good Rite AO 3113	BFGoodrich	7.3.6.7
Good Rite AO 3140	BFGoodrich	7.3.6.3
Good Rite AO 3146	BFGoodrich	7.3.6.2
Good Rite AO 3152	BFGoodrich	7.3.6.1
Good Rite AO 3190 × 29	BFGoodrich	7.3.6.2
Good Rite BBTS	BFGoodrich	7.3.2.2
Good Rite CBTS	BFGoodrich	7.3.2.2
Good Rite MBT	BFGoodrich	7.3.2.1
Good Rite MBTS	BFGoodrich	7.3.2.1
Good Rite OBTS	BFGoodrich	7.3.2.2
Good Rite TMTD	BFGoodrich	7.3.2.6
Haftolat	Kettlitz	7.3.11.6
Hansa Ultra	Lehmann & Voss	7.3.4
Heliozone	Du Pont	7.3.6.11
Heptene Base	Uniroyal	7.3.2.10
Herclor*	Hercules	7.1.16
Hercoflex 600	Hercules	7.3.10
Hexa	Krahn	7.3.2.10
Hexa K	Degussa	7.3.2.10
Hexamethylenetetramine,	Union Carbide	7.3.2.10
Hostapren*	Hoechst	7.1.12
Humex	Hules Mexicanos	7.1.2
Humex	Hules Mexicanos	7.2.1
Humex N	Hules Mexicanos	7.1.3
HVA 2	Du Pont	7.3.3.3
Hycar	BFGoodrich	7.1.3
Hycar	BFGoodrich	7.1.4
Hycar*	BFGoodrich	7.1.14
Hydrin	BFGoodrich	7.1.16
Hypalon	Du Pont	7.1.13
HyTemp	BFGoodrich	7.1.14
Hytrel	Du Pont	7.2.7
Igulan	Hokuchin Kagaku	7.1.24
Indopol	Amoco	7.3.10
Ingralen	Fuchs	7.3.9.3
Ingraplast	Fuchs	7.3.9.1
Ingraplast	Fuchs	7.3.9.2
Intene, Intolene*	Enichem	7.1.1
Intene 181	Enichem	7.3.14.2
Interlube A, P	Anchor	7.3.11.6
Interox	Peroxid-Chemie	7.3.1.3
Intol, Intex	Enichem	7.1.2

* earlier used trade name

Name	Company (Addresses see Section 7.5)	Type of Rubber rsp. Chemical see Section:
Intolan	Enichem	7.1.10
Ionol	Shell	7.3.6.5
Irgafos TNPP	Ciba-Geigy	7.3.6.7
Irganox PS 800	Ciba-Geigy	7.3.6.8
Irganox 565, 1010, 1076, 1081, 1313	Ciba-Geigy	7.3.6.5
Irgastab 2002	Ciba-Geigy	7.3.6.5
J-Sil	J-Sil	7.1.22
J-Z-F	Uniroyal	7.3.6.1
Jayflex DINP	Exxon	7.3.8.2
Jayflex DOP	Exxon	7.3.8.2
JO 4012	Bozzetto	7.3.2.3
JO 4013	Bozzetto	7.3.2.3
JO 4061	Bozzetto	7.3.2.13
JOS 4026	Bozzetto	7.3.2.3
JSR-BR	Japan Synthetic Rubber	7.1.1
JSR-N	Japan Synthetic Rubber	7.1.3
K-Stay	Vanderbilt	7.3.10
Kalrez	Du Pont	7.1.18
Kasai TAIC	Nippon Kasai	7.3.3.2
Kelprox	DSM	7.2.3
Kelrinal	DSM	7.1.12
Keltan	DSM	7.1.10
Keltan TP	DSM	7.2.3
Ken Reacts Drimix KR	Kenrich	7.3.11.6
Kenflex N	Kenrich	7.3.9.3
Kenlastik K	Kenrich	7.3.4
Kenmix GMF	Kenrich	7.3.1.4
Kenmix Litharge	Kenrich	7.3.4
Kenmix Red Lead	Kenrich	7.3.4
Kenmix Sulfur	Kenrich	7.3.1.1
Kenplast G	Kenrich	7.3.9.3
Kenreact	Kenrich	7.3.3.5
KER	Ciech-Stomil	7.1.2
Koremoll CE 5185	BASF	7.3.8.8
Kratex	Chemopetrol	7.1.2
Kraton	Shell	7.2.2
Krylene, Krynol*	Polysar	7.1.2
Krynac	Polysar	7.1.3
Lead phosphite	ALM	7.3.4
Lead phthalate	ALM	7.3.4
Ledate (PbDMC)	Vanderbilt	7.3.2.4
Leegen	Vanderbilt	7.3.10
Levaflex	Bayer	7.2.3
Levaprene	Bayer	7.1.11
Liquisperse MBZ	Basic	7.3.4
Lowinox ACP	Lowi	7.3.6.3
Lowinox AH 25	Lowi	7.3.6.3

* earlier used trade name

Name	Company (Addresses see Section 7.5)	Type of Rubber rsp. Chemical see Section:
Lowinox BHA, DLTDP	Lowi	7.3.6.8
Lowinox BHT	Lowi	7.3.6.5
Lowinox ODE	Lowi	7.3.6.2
Lowinox P 24 A, P 24 S, 001	Lowi	7.3.6.5
Lowinox SDA	Lowi	7.3.6.2
Lowinox TNPP	Lowi	7.3.6.7
Lowinox 002, 44 B 25, 22 CP 46, 22, B 46, 44 M 26, 44 M 36, P 22 B	Lowi	7.3.6.4
Lowinox 22 M 46	Lowi	7.3.6.4
Luazo ABA, AP	Luperox	7.3.1.7
Lucidol, Luperco	Luperox	7.3.1.3
Lunacera	Fuller	7.3.6.11
Lunacerin	Fuller	7.3.12.2
Luperox	Luperox	7.3.1.3
Luvomag	Lehmann & Voss	7.3.4
Luvopor OB	Lehmann & Voss	7.3.13.2
Luvopor TSH	Lehmann & Voss	7.3.13.2
Luvozink	Lehmann & Voss	7.3.4
Maglite	Merck	7.3.4
Magnolux	Lehmann & Voss	7.3.4
Magnox	Basic	7.3.4
Magnozink	Lehmann & Voss	7.3.4
Mahlschwefel	Kalichemie	7.3.1.1
Manox	Manchem	7.3.1.1
MBS	Krahn	7.3.2.2
MBT	Krahn	7.3.2.1
MBT	Lehmann & Voss	7.3.2.1
MBT	Uniroyal	7.3.2.1
MBTS	Ciech-Stomil	7.3.2.1
MBTS	Krahn	7.3.2.1
MBTS	Lehmann & Voss	7.3.2.1
MBTS	Uniroyal	7.3.2.1
MC Sulfur	Anchor	7.3.1.1
Mediaplast NB 4	Kettlitz	7.3.8.1
Mediaplast types	Kettlitz	7.3.10
Mediaplast WH	Kettlitz	7.3.11.6
Megum	Metallgesellschaft	7.3.14.1
Megum	Metallgesellschaft	7.3.14.2
Melonil	Henkel	7.3.8.8
Meramid C	Bozzetto	7.3.2.2
Meramid CR (Activated amine phenol complex)	Bozzetto	7.3.2.12
Meramid DCN	Bozzetto	7.3.2.2
Meramid D/E	Bozzetto	7.3.2.10
Meramid M	Bozzetto	7.3.2.2
Merasulf MBTS	Bozzetto	7.3.2.1
Merazin MBTZ	Bozzetto	7.3.2.1
Mesamoll	Bayer	7.3.8.8
Methasan	Monsanto	7.3.2.3
Methazate	Uniroyal	7.3.2.3
Methyl Ethyl Tuads (METD)	Vanderbilt	7.3.2.6

Name	Company (Addresses see Section 7.5)	Type of Rubber rsp. Chemical see Section:
Methyl Tuads	Vanderbilt	7.3.2.6
Methyl Tuex	Uniroyal	7.3.2.6
Methyl Zimate	Vanderbilt	7.3.2.3
Methylniclate	Vanderbilt	7.3.2.4
Mikrowachs	BP	7.3.6.11
Millathane	Notedome	7.1.24
Mix EXP	Bozzetto	7.3.2.1
Mix Pb	Bozzetto	7.3.4
Moka	Du Pont	7.3.1.6
Monex	Uniroyal	7.3.2.6
Mono Thiurad	Monsanto	7.3.2.6
Monoplex types	Rohm & Haas	7.3.10
Monothane	Compounding Ingredients	7.1.24
Montaclere	Monsanto	7.3.6.5
Montanwachs	Braunkohlen-Werk	7.3.6.11
Moplen	Montedison	7.2.3
Morphax	Goodyear	7.3.1.2
Na DBC	Lehmann & Voss	7.3.2.4
Na-MBT	Bayer	7.3.2.1
NA 22	Du Pont	7.3.2.9
NA 101	Du Pont	7.3.2.9
Naftocit Di 4	Metallgesellschaft	7.3.2.3
Naftocit Di 7	Metallgesellschaft	7.3.2.3
Naftocit Di 13	Metallgesellschaft	7.3.2.3
Naftocit DPG	Metallgesellschaft	7.3.2.8
Naftocit Mi 12	Metallgesellschaft	7.3.2.9
Naftocit Thiuram 16	Metallgesellschaft	7.3.2.6
Naftocit ZMBT	Metallgesellschaft	7.3.2.1
Naftolen	Metallgesellschaft	7.3.9.1
Naftolen	Metallgesellschaft	7.3.9.2
Naftolen	Metallgesellschaft	7.3.9.3
Naftonox BHT	Metallgesellschaft	7.3.6.5
Naftonox TMQ	Metallgesellschaft	7.3.6.3
Naftonox ZMP	Metallgesellschaft	7.3.6.5
Naftonox 2246	Metallgesellschaft	7.3.6.4
Nairit	USSR	7.1.6
Naruxol	Uniroyal	7.3.6.4
Naruxol 15	Uniroyal	7.3.6.4
Natsyn	Goodyear	7.1.7
Naugard BHT	Uniroyal	7.3.6.5
Naugard MB	Uniroyal	7.3.6.6
Naugard PAN	Uniroyal	7.3.6.2
Naugard Q	Uniroyal	7.3.6.3
Naugard SP	Uniroyal	7.3.6.5
Naugard 431, 451, K, T	Uniroyal	7.3.6.5
Naugard 445	Uniroyal	7.3.6.2
Naugard 492	Uniroyal	7.3.6.2
Naugard 496	Uniroyal	7.3.6.1
Naugatex OXAF	Uniroyal	7.3.2.1
Naugex SD-I	Uniroyal	7.3.1.2
NBC	Du Pont	7.3.2.4
Negozone	Campbell	7.3.6.11

Name	Company (Addresses see Section 7.5)	Type of Rubber rsp. Chemical see Section:
Neoprene	Du Pont	7.1.6
Nesires types	NCC	7.3.10
Nipol	Nippon Zeon	7.1.1
Nipol	Nippon Zeon	7.1.2
Nipol	Nippon Zeon	7.1.7
Nipol AR	Nippon Zeon	7.1.14
Nipol DN, P	Nippon Zeon	7.1.4
Nipol N	Nippon Zeon	7.1.3
NOBS Spezial	Anchor	7.3.2.2
Noctizer	Ouchi Shinko	7.3.7.2
Nordel	Du Pont	7.1.10
Norsorex	CdF-Chimie	7.1.20
Novor	Durham	7.3.1.6
NRC	Nordmann, Rassmann	7.1.22
Nyflex	Nynas	7.3.9.1
NYsyn	Copolymer	7.1.3
Nytene	Nynas	7.3.9.2
Octamine	Uniroyal	7.3.6.2
Oil Fatty Acid	Foster	7.3.11.2
Okerin	Astor	7.3.6.11
Olein	Malmsten & Bergvall	7.3.11.2
Oleinic Acid	Foster	7.3.11.2
Orevac	Atochem	7.1.11
Ozokerit	Kemi	7.3.6.11
Palatinol AH	BASF	7.3.8.2
Palatinol C	BASF	7.3.8.2
PAN	Union Carbide	7.3.6.2
Paracril	Uniroyal	7.1.3
Paracril Ozo	Uniroyal	7.1.4
Paraffin	Malmsten & Bergvall	7.3.6.11
Paraplex types	Krahn	7.3.10
Parel	BFGoodrich	7.1.17
Parflock	PAr	7.3.14.2
Parlock	PAr	7.3.14.1
PAT Additives	Würtz	7.3.11.6
PAT-Additives	Würtz	7.3.11.3
PBN	BASF	7.3.6.2
PBN	VEB Bitterfeld	7.3.6.2
Pebax	Atochem	7.2.8
Pennac CBS	Pennwalt	7.3.2.2
Pennac CDMC	Pennwalt	7.3.2.4
Pennac DIBS	Pennwalt	7.3.2.2
Pennac MBS	Pennwalt	7.3.2.2
Pennac MBT	Pennwalt	7.3.2.1
Pennac MBTS	Pennwalt	7.3.2.1
Pennac NDBC	Pennwalt	7.3.2.4
Pennac SDBC	Pennwalt	7.3.2.4
Pennac TBBS	Pennwalt	7.3.2.2
Pennac TDEC	Pennwalt	7.3.2.4
Pennac TETD	Pennwalt	7.3.2.6
Pennac TMTD	Pennwalt	7.3.2.6

Name	Company (Addresses see Section 7.5)	Type of Rubber rsp. Chemical see Section:
Pennac ZDBC	Pennwalt	7.3.2.3
Pennac ZDEC	Pennwalt	7.3.2.3
Pennac ZDMC	Pennwalt	7.3.2.3
Pennac ZMBT	Pennwalt	7.3.2.1
Pennox HR	Pennwalt	7.3.6.3
Pennox ODP	Pennwalt	7.3.6.2
Pepton	Cyanamid	7.3.7.2
Pepton 22, 44	Anchor	7.3.7.3
Perbunan C*	Bayer	7.1.6
Perbunan N	Bayer	7.1.3
Perbunan NVC	Bayer	7.1.4
Perkacit BCMC	Akzo	7.3.2.4
Perkacit CBS	Akzo	7.3.2.2
Perkacit DCBS	Akzo	7.3.2.2
Perkacit DETU	Akzo	7.3.2.9
Perkacit DOTG	Akzo	7.3.2.8
Perkacit DPG	Akzo	7.3.2.8
Perkacit DPTT	Akzo	7.3.1.2
Perkacit DTDM	Akzo	7.3.1.2
Perkacit ETU	Akzo	7.3.2.9
Perkacit MBS	Akzo	7.3.2.2
Perkacit MBT	Akzo	7.3.2.1
Perkacit MBTS	Akzo	7.3.2.1
Perkacit NDBC	Akzo	7.3.2.4
Perkacit S	Akzo	7.3.1.1
Perkacit SDBC	Akzo	7.3.2.4
Perkacit SDEC	Akzo	7.3.2.4
Perkacit SDMC	Akzo	7.3.2.4
Perkacit TDEC	Akzo	7.3.2.4
Perkacit TETD	Akzo	7.3.2.6
Perkacit TMTD	Akzo	7.3.2.6
Perkacit TMTM	Akzo	7.3.2.6
Perkacit ZBEC	Akzo	7.3.2.3
Perkacit ZDBC	Akzo	7.3.2.3
Perkacit ZDEC	Akzo	7.3.2.3
Perkacit ZDMC	Akzo	7.3.2.3
Perkacit ZMBT	Akzo	7.3.2.1
Perkacit 2	Akzo	7.3.2.13
Perkadox	Akzo	7.3.1.3
Perkaflex 12	Akzo	7.3.10
Perkalink 300	Akzo	7.3.3.2
Perkalink 301	Akzo	7.3.3.2
Perkalink 400	Akzo	7.3.3.1
Perkalink 401	Akzo	7.3.3.1
Permalux (DOTG salt of dicatechole borate)	Du Pont	7.3.2.12
Permanax 6PPD	Akzo	7.3.6.1
Permanax BBL, BLN	Akzo	7.3.6.2
Permanax CNS, WSP	Akzo	7.3.6.4
Permanax DPPD	Akzo	7.3.6.1
Permanax HD	Akzo	7.3.6.2

* earlier used trade name

Name	Company (Addresses see Section 7.5)	Type of Rubber rsp. Chemical see Section:
Permanax IPPD	Akzo	7.3.6.1
Permanax OD	Akzo	7.3.6.2
Permanax TQ	Akzo	7.3.6.3
Permanax WSL	Akzo	7.3.6.4
Permanax WSO	Akzo	7.3.6.5
Peroximon	Montedison	7.3.1.3
PETS	Bozzetto	7.3.2.6
Phillips Cis 4*	Phillips Petroleum	7.1.1
Phthalic anhydride	Atochem	7.3.5.2
Phthalic anhydride	VEB Buna	7.3.5.2
Picco Plasticizers	Hercules	7.3.10
Plasticizer TP-90 B (BCF)	Thiokol	7.3.8.9
Plasticizer TP-95	Thiokol	7.3.8.9
Plastogen	Vanderbilt	7.3.10
Plastomag	Morton	7.3.4
Pliocord	Goodyear	7.3.14.2
Pliogrip	Goodyear	7.3.14.2
Polydispersion ASD	Croxton & Garry	7.3.1.1
Polydispersion DPTT	Croxton & Garry	7.3.1.2
Polydispersion E	Croxton & Garry	7.3.2.1
Polydispersion E	Croxton & Garry	7.3.1.3
Polydispersion E (DPG)	Croxton & Garry	7.3.2.8
Polydispersion E (EZ)	Croxton & Garry	7.3.2.3
Polydispersion E (MC)	Croxton & Garry	7.3.2.3
Polydispersion E (MX)	Croxton & Garry	7.3.2.6
Polydispersion EAD	Croxton & Garry	7.3.2.1
Polydispersion EC (CBS)	Croxton & Garry	7.3.2.2
Polydispersion END	Croxton & Garry	7.3.2.9
Polydispersion ERD, PRD	Croxton & Garry	7.3.4
Polydispersion EZD, SZD	Croxton & Garry	7.3.4
Polydispersion G, T	Croxton & Garry	7.3.1.7
Polydispersion K (DYT)	Croxton & Garry	7.3.4
Polydispersion KLD, PLD, TLD	Croxton & Garry	7.3.4
Polydispersion P	Croxton & Garry	7.3.1.2
Polydispersion S	Croxton & Garry	7.3.5.1
Polydispersion S (DNPT)	Croxton & Garry	7.3.13.3
Polydispersion T (BZ)	Croxton & Garry	7.3.2.3
Polydispersion T (MT)	Croxton & Garry	7.3.2.6
Polydispersion TDH, SDH	Croxton & Garry	7.3.2.10
Polydispersion TMD	Croxton & Garry	7.3.2.1
Polydispersion types	Croxton & Garry	7.3.13.1
Polydispersion S (DETU)	Croxton & Garry	7.3.2.9
Polygard	Uniroyal	7.3.6.7
Polylanes	Safic	7.3.3.4
Polylite	Uniroyal	7.3.6.2
Polymag	Safic	7.3.4
Polyminimum G	Safic	7.3.4
Polymix ZM	Safic	7.3.4
Polyplastol 6	Bozzetto	7.3.11.3
Polysar Bromobutyl	Polysar	7.1.9

* earlier used trade name

Name	Company (Addresses see Section 7.5)	Type of Rubber rsp. Chemical see Section:
Polysar Butyl	Polysar	7.1.8
Polysar Chlorobutyl	Polysar	7.1.9
Polysar EPDM	Polysar	7.1.10
Polysar S	Polysar	7.1.2
Polytharge G	Safic	7.3.4
Por-A-Mold	Compounding Ingredients	7.1.22
Poreks	Ciech-Stomil	7.3.13.4
Porofor ADC/R, ADC/K	Bayer	7.3.13.1
Porofor B 13	Bayer	7.3.13.2
Porofor BSH	Bayer	7.3.13.2
Porofor DNO	Bayer	7.3.13.3
Primix TAC	Kenrich	7.3.3.2
Priolene	Unichema	7.3.11.2
Priplast 1431	Unichema	7.3.8.10
Priplast 1562	Unichema	7.3.8.6
Priplast 3117, 3131	Unichema	7.3.8.8
Pristerene	Unichema	7.3.11.1
Propyl Zithate	Vanderbilt	7.3.2.5
Protektor G	Fuller	7.3.6.11
PU Elastomer	Briggs & Townsend	7.2.6
R	Wacker	7.1.23
R 240 (Oxadiazine thio-ketone derivative)	Lehmann & Voss	7.3.2.12
Radax (TMTD/MBT)	Monsanto	7.3.2.1
Ralox BHT	Raschig	7.3.6.5
Ralox TMQ	Raschig	7.3.6.3
Ralox types	Raschig	7.3.6.5
Ralox 2246	Raschig	7.3.6.4
Reinforcing Agent Si 230, Si 69, X 50	Degussa	7.3.3.4
Renacit 4, 5, 7, 9	Bayer	7.3.7.1
Renacit 8	Bayer	7.3.7.3
Retarder J	Uniroyal	7.3.5.4
Retarder PD	Anchor	7.3.5.2
Retarder TSA	Monsanto	7.3.5.3
Rhenoblend	Rhein-Chemie	7.1.4
Rhenocure AT, CUT, TP, ZAT	Rhein-Chemie	7.3.2.11
Rhenocure BQ	Rhein-Chemie	7.3.1.4
Rhenocure CA	Rhein-Chemie	7.3.2.9
Rhenocure CMT, CMU, EPC	Rhein-Chemie	7.3.2.13
Rhenocure M	Rhein-Chemie	7.3.1.2
Rhenocure S	Rhein-Chemie	7.3.1.2
Rhenocure TDD (Thiadiazole derivative)	Rhein-Chemie	7.3.2.12
Rhenofit B, 2642, 3555	Rhein-Chemie	7.3.2.10
Rhenofit BDMA	Rhein-Chemie	7.3.3.1
Rhenofit CF (CaOH)	Rhein-Chemie	7.3.4
Rhenofit EDMA	Rhein-Chemie	7.3.3.1
Rhenofit KE	Rhein-Chemie	7.3.2.12
Rhenofit TAC	Rhein-Chemie	7.3.3.2
Rhenofit TRIM	Rhein-Chemie	7.3.3.1

Name	Company (Addresses see Section 7.5)	Type of Rubber rsp. Chemical see Section:
Rhenofit 1100, 1600	Rhein-Chemie	7.3.13.5
Rhenofit 1987, 2009, 2642, 3555	Rhein-Chemie	7.3.3.6
Rhenogran Ba CO_3	Rhein-Chemie	7.3.4
Rhenogran CBS	Rhein-Chemie	7.3.2.2
Rhenogran CTP	Rhein-Chemie	7.3.5.1
Rhenogran DBTU	Rhein-Chemie	7.3.2.9
Rhenogran DETU	Rhein-Chemie	7.3.2.9
Rhenogran DOTG	Rhein-Chemie	7.3.2.8
Rhenogran DPG	Rhein-Chemie	7.3.2.8
Rhenogran DPTT	Rhein-Chemie	7.3.1.2
Rhenogran DPTU	Rhein-Chemie	7.3.2.9
Rhenogran ETU	Rhein-Chemie	7.3.2.9
Rhenogran Hexa	Rhein-Chemie	7.3.2.10
Rhenogran IPPD	Rhein-Chemie	7.3.6.1
Rhenogran MBI	Rhein-Chemie	7.3.6.6
Rhenogran MBT	Rhein-Chemie	7.3.2.1
Rhenogran MBTS	Rhein-Chemie	7.3.2.1
Rhenogran MMBI	Rhein-Chemie	7.3.6.6
Rhenogran NKF	Rhein-Chemie	7.3.6.4
Rhenogran OTBG	Rhein-Chemie	7.3.2.8
Rhenogran Pb 0, Pb_3O_4	Rhein-Chemie	7.3.4
Rhenogran PBN	Rhein-Chemie	7.3.6.2
Rhenogran Resorcin	Rhein-Chemie	7.3.14.2
Rhenogran TDEC	Rhein-Chemie	7.3.2.4
Rhenogran TETD	Rhein-Chemie	7.3.2.6
Rhenogran TMTM	Rhein-Chemie	7.3.2.6
Rhenogran ZDBC	Rhein-Chemie	7.3.2.3
Rhenogran ZDEC	Rhein-Chemie	7.3.2.3
Rhenogran ZDMC	Rhein-Chemie	7.3.2.3
Rhenogran ZEPC	Rhein-Chemie	7.3.2.3
Rhenogran ZMBT	Rhein-Chemie	7.3.2.1
Rhenogran ZnO	Rhein-Chemie	7.3.4
Rhenograns	Rhein-Chemie	7.3.1.1
Rhenomag	Rhein-Chemie	7.3.4
Rhenopor 1843	Rhein-Chemie	7.3.13.4
Rhenosol ZnO	Rhein-Chemie	7.3.4
Rheofas 50, 65, 95	Ciba-Geigy	7.3.8.7
Rheogen	Vanderbilt	7.3.10
Rheomet types	Ciba-Geigy	7.3.6.9
Rheomol BCF	Ciba-Geigy	7.3.8.9
Rheomol D 10Z	Ciba-Geigy	7.3.8.4
Rheomol DIDA	Ciba-Geigy	7.3.8.1
Rheomol DOA	Ciba-Geigy	7.3.8.1
Rheomol DOS	Ciba-Geigy	7.3.8.3
Rheomol DTDP	Ciba-Geigy	7.3.8.2
Rheomol LTM, ATM	Ciba-Geigy	7.3.8.5
Rheomol P	Ciba-Geigy	7.3.8.2
Rheomol 249, TPP	Ciba-Geigy	7.3.8.7
Rheoplast 38, 39, 392	Ciba-Geigy	7.3.8.10
Rheoplex 245, 430, 903, 1102, GF, GL, MD	Ciba-Geigy	7.3.8.8
Rhodorsil (Q)	Rhône Poulenc	7.3.14.1

Name	Company (Addresses see Section 7.5)	Type of Rubber rsp. Chemical see Section:
Ridakto	Kenrich	7.3.2.10
Robac CTP	Robinson	7.3.5.1
Robac CuDD (CuDMC)	Robinson	7.3.2.4
Robac DBTU	Robinson	7.3.2.9
Robac DETU	Robinson	7.3.2.9
Robac DOTG	Robinson	7.3.2.8
Robac DPG	Robinson	7.3.2.8
Robac DTDM	Robinson	7.3.1.2
Robac NiBUD	Robinson	7.3.2.4
Robac NiDD	Robinson	7.3.2.4
Robac P 25	Robinson	7.3.1.2
Robac PPC	Robinson	7.3.2.4
Robac SBUD	Robinson	7.3.2.4
Robac SED	Robinson	7.3.2.4
Robac SMD	Robinson	7.3.2.4
Robac TET	Robinson	7.3.2.6
Robac TMS	Robinson	7.3.2.6
Robac TMT	Robinson	7.3.2.6
Robac ZBED	Robinson	7.3.2.3
Robac ZBUD	Robinson	7.3.2.3
Robac ZDC	Robinson	7.3.2.3
Robac ZMD	Robinson	7.3.2.3
Robac ZPC	Robinson	7.3.2.3
Robac 22	Robinson	7.3.2.9
Robac 44	Robinson	7.3.2.12
Rotax	Vanderbilt	7.3.2.1
Royalene	Uniroyal	7.1.10
RP	Rhône Poulenc	7.1.22
RP	Rhône Poulenc	7.1.23
RTV	General Electric	7.1.22
Safacid G, R	Jahres	7.3.11.1
Salicylic acid	Rhône Poulenc	7.3.5.3
Santicizer types	Monsanto	7.3.10
Santocure	Monsanto	7.3.2.2
Santocure IPS	Monsanto	7.3.2.2
Santocure MOR	Monsanto	7.3.2.2
Santocure NS	Monsanto	7.3.2.2
Santoflex AW	Monsanto	7.3.6.3
Santoflex DD	Monsanto	7.3.6.3
Santoflex IP	Monsanto	7.3.6.1
Santoflex 13	Monsanto	7.3.6.1
Santoflex 14	Monsanto	7.3.6.1
Santoflex 77	Monsanto	7.3.6.1
Santogard PVI	Monsanto	7.3.5.1
Santoprene	Monsanto	7.2.3
Santovar A	Monsanto	7.3.6.3
Santowhite Crystals, Powder	Monsanto	7.3.6.4
Santowhite PC	Monsanto	7.3.6.4
Santowhite TBTU	Monsanto	7.3.2.9
Santowhite 54	Monsanto	7.3.6.4
Santowhite 54, MK, L	Monsanto	7.3.6.5
Saret 500 Drimix	Kenrich	7.3.3.1

Name	Company (Addresses see Section 7.5)	Type of Rubber rsp. Chemical see Section:
Saret 515 Drimix	Kenrich	7.3.3.1
Sartomer 206	Ancomer	7.3.3.1
Scorchguard	Malmsten & Bergvall	7.3.4
SE	General Electric	7.1.22
Selenac (SeDMC)	Vanderbilt	7.3.2.4
Silacto	Kenrich	7.3.2.10
Silane A-151	Union Carbide	7.3.3.4
Silane A-171, 186, 187, 188, 1120, 1130	Union Carbide	7.3.3.4
Silane A-174	Union Carbide	7.3.3.4
Silane Y 9194	Union Carbide	7.3.3.4
Silanogran M	Kettlitz	7.3.3.4
Silanogran Si 69 GR	Kettlitz	7.3.3.4
Silanogran V	Kettlitz	7.3.3.4
Silastic	Dow Corning	7.1.22
Silastic	Dow Corning	7.1.23
Silcat 17	Union Carbide	7.3.3.4
Silopren	Bayer	7.1.22
Silopren-Vernetzer	Bayer	7.3.1.3
SIR*	Sir	7.1.3
Sirel*	SIR	7.1.2
Skyprene	Toya Soda	7.1.6
Soca Butyl	Japan Butyl	7.1.8
Sodium MBT	Monsanto	7.3.2.1
Sodium MBT	Uniroyal	7.3.2.1
Solprene	Calatrava	7.1.1
Solprene	Calatrava	7.2.1
Soxinol CZ	Sumitomo	7.3.2.2
Soxinol D	Sumitomo	7.3.2.8
Soxinol DM	Sumitomo	7.3.2.1
Soxinol DT	Sumitomo	7.3.2.8
Soxinol ESL	Sumitomo	7.3.2.4
Soxinol EZ	Sumitomo	7.3.2.3
Soxinol M	Sumitomo	7.3.2.1
Soxinol M-Na-G	Sumitomo	7.3.2.1
Soxinol MBS	Sumitomo	7.3.2.2
Soxinol MZ	Sumitomo	7.3.2.1
Soxinol PC	Sumitomo	7.3.2.3
Soxinol PX	Sumitomo	7.3.2.3
Soxinol TE	Sumitomo	7.3.2.4
Soxinol TET	Sumitomo	7.3.2.6
Soxinol TRA	Sumitomo	7.3.1.2
Soxinol TS	Sumitomo	7.3.2.6
Soxinol TT	Sumitomo	7.3.2.6
SP-Types	Schenectady	7.3.1.4
SR-206 Drimix	Kenrich	7.3.3.1
SR-297 Drimix	Kenrich	7.3.3.1
SR-350 Drimix	Kenrich	7.3.3.1
Stabilite SP	Reichold	7.3.6.5
Stabilizer K1, K2	BASF	7.3.6.5
Stabilizer T	Bayer	7.3.6.5

* earlier used trade name

Name	Company (Addresses see Section 7.5)	Type of Rubber rsp. Chemical see Section:
Stavinor TNP	Atochem	7.3.6.7
Stearic Acid	Foster	7.3.11.1
Stearic Acid	Henkel	7.3.11.1
Stearin	Malmsten & Bergvall	7.3.11.1
Struktol A 82	Schill & Seilacher	7.3.7.3
Struktol A 86	Schill & Seilacher	7.3.7.2
Struktol FA 541, IB 531	Schill & Seilacher	7.3.3.6
Struktol KW 400, KW 500, WB 300	Schill & Seilacher	7.3.8.8
Struktol Neopast	Schill & Seilacher	7.3.4
Struktol Perlzink	Schill & Seilacher	7.3.4
Struktol SU	Schill & Seilacher	7.3.1.1
Struktol types	Schill & Seilacher	7.3.11.3
Struktol types	Schill & Seilacher	7.3.11.6
Struktol WB 222	Schill & Seilacher	7.3.11.4
Struktol WB 700	Schill & Seilacher	7.3.4
Struktol WB 890	Schill & Seilacher	7.3.4
Struktol WB 900, 902	Schill & Seilacher	7.3.4
Struktol Zimag 29/43	Schill & Seilacher	7.3.4
Sulfads	Vanderbilt	7.3.1.2
Sulfasan R	Monsanto	7.3.1.2
Sulfenax CB	Petrimex	7.3.2.2
Sumitomo SBR	Sumitomo	7.1.2
Sundex	Sun Oil	7.3.9.3
Sunolite	Witco	7.3.6.11
Sunpar	Sun Oil	7.3.9.1
Sunproof	Uniroyal	7.3.6.11
Synpol EBR (low cis content)	Synpol	7.1.1
TA 11	Du Pont	7.3.2.10
TAC DL	Lehmann & Voss	7.3.3.2
TAC GR	Kettlitz	7.3.3.2
TAIC DL	Lehmann & Voss	7.3.3.2
TAIC DL	Nordmann, Rassmann	7.3.3.2
Taktene*	Polysar	7.1.1
Talg Fatty Acid	Foster	7.3.11.2
TATM DL	Lehmann & Voss	7.3.3.2
TBTD	Krahn	7.3.2.6
TBTS	Bozzetto	7.3.2.6
TDEC	Lehmann & Voss	7.3.2.4
TE-types	Technical Processing	7.3.11.6
Technoflon	Montedison	7.1.18
Technoflon M	Montedison	7.3.1.5
Technosin	Montedison	7.3.1.5
TeDEC	Krahn	7.3.2.4
Tellurac	Vanderbilt	7.3.2.4
TETD	Krahn	7.3.2.6
TETD	Monsanto	7.3.2.6
Tetrone A	Du Pont	7.3.1.2
TETS	Bozzetto	7.3.2.6
Texin	Bayer	7.1.24

* earlier used trade name

Name	Company (Addresses see Section 7.5)	Type of Rubber rsp. Chemical see Section:
Therban	Bayer	7.1.5
Thiate EF 2	Vanderbilt	7.3.2.9
Thiate N	Vanderbilt	7.3.2.9
Thiate H	Vanderbilt	7.3.2.9
Thiate U	Vanderbilt	7.3.2.9
Thiofide	Monsanto	7.3.2.1
Thiokol	Thiokol	7.1.25
Thiokol LP	Thiokol	7.1.25
Thionex	Du Pont	7.3.2.6
Thiotax	Monsanto	7.3.2.1
Thiurad	Monsanto	7.3.2.6
Thiuram E	Du Pont	7.3.2.6
Thiuram M	Du Pont	7.3.2.6
Thixon	Thixon	7.3.14.1
Thixon	Thixon	7.3.14.2
Tinuvin types	Ciba-Geigy	7.3.6.10
Tinuvin 622, 765, 770	Ciba-Geigy	7.3.6.8
TMPT DL	Lehmann & Voss	7.3.3.1
TMTD	Krahn	7.3.2.6
TMTD	Lehmann & Voss	7.3.2.6
TMTM	Bozzetto	7.3.2.6
TMTM	Krahn	7.3.2.6
TMTM	Lehmann & Voss	7.3.2.6
TMTS	Bozzetto	7.3.2.6
Tonox	Uniroyal	7.3.1.7
Tonox	Uniroyal	7.3.5.4
Tornac	Polysar	7.1.5
Trefsin	Exxon	7.2.3
TRIB TU	Krahn	7.3.2.9
Trigonox	Akzo	7.3.1.3
Trilene (liquid)	Uniroyal	7.1.10
Trimellitate types	Amoco	7.3.8.5
Trimene Base	Uniroyal	7.3.2.10
Triplast 3010	Unichema	7.3.8.4
Tuads	Vanderbilt	7.3.2.6
Tudalen	Dahleke	7.3.9
Tuex	Uniroyal	7.3.2.6
Tufprene	Asahi Chemicals	7.1.2
Tufsel	General Electric	7.1.22
Ty Ply	Anchor	7.3.14.1
Ucarsil DSC 12	Union Carbide	7.3.3.4
Ucarsil DSC 189	Union Carbide	7.3.3.4
Ultramoll I, II, III	Bayer	7.3.8.1
Ultrathene	USI Chemicals	7.1.11
Unimoll BB, DM, 66	Bayer	7.3.8.2
Unimoll DB	Bayer	7.3.8.2
UOP 588	Universal Oil	7.3.6.1
UOP 788	Universal Oil	7.3.6.1
Ureka Base	Monsanto	7.3.2.1
Ureka White F (MBTS/DPG)	Monsanto	7.3.2.1
Urepan	Bayer	7.1.24

Name	Company (Addresses see Section 7.5)	Type of Rubber rsp. Chemical see Section:
Vamac	Du Pont	7.1.15
Vanax A	Vanderbilt	7.3.1.2
Vanax CBS	Vanderbilt	7.3.2.2
Vanax CPA (Isophthalate derivative)	Vanderbilt	7.3.2.12
Vanax DOTG	Vanderbilt	7.3.2.8
Vanax DPG	Vanderbilt	7.3.2.8
Vanax MBTS	Vanderbilt	7.3.2.1
Vanax NP (Thiadiazine derivative)	Vanderbilt	7.3.2.12
Vanax NS	Vanderbilt	7.3.2.2
Vanax OBTS	Vanderbilt	7.3.2.2
Vanax PML (DOTG salt of dicatechole borate)	Vanderbilt	7.3.2.12
Vanax PY	Vanderbilt	7.3.3.6
Vanax ZMBT	Vanderbilt	7.3.2.1
Vanax 13	Vanderbilt	7.3.6.5
Vanax 200	Vanderbilt	7.3.6.2
Vanax 552	Vanderbilt	7.3.2.4
Vanax 808, 833	Du Pont	7.3.2.10
Vanfre types	Vanderbilt	7.3.11.6
Vanox AT	Vanderbilt	7.3.6.8
Vanox MTI	Vanderbilt	7.3.6.6
Vanox ZMTI	Vanderbilt	7.3.6.6
Vanox 12	Vanderbilt	7.3.6.2
Vanox 13	Vanderbilt	7.3.6.7
Vanox 100, 1320	Vanderbilt	7.3.6.5
Vanox 102	Vanderbilt	7.3.6.5
Vanox 1290, 1360	Vanderbilt	7.3.6.4
Vanox 2246	Vanderbilt	7.3.6.4
Vanvre UN	Vanderbilt	7.3.12.3
Vanwax	Vanderbilt	7.3.6.11
Varox	Vanderbilt	7.3.1.3
Velvet	Lehmann & Voss	7.3.1.1
Vernetzer 30/10	Bayer	7.3.1.6
Vestenamer	CWH	7.1.21
Vestinol 9	CWH	7.3.8.2
Vibracure	Uniroyal	7.3.1.6
Vibrathane	Uniroyal	7.1.24
Vistaflex	Exxon	7.2.3
Vistalon	Exxon	7.1.10
Viton	Du Pont	7.1.18
Viton Curatives	Du Pont	7.3.1.5
Vocol	Monsanto	7.3.2.11
VP-R	Wacker	7.1.23
VP 3243	Wacker	7.3.14.1
VPA 1, 2	Du Pont	7.3.12.1
Vul-Cup	Hercules	7.3.1.3
Vulcuren 2	Bayer	7.3.1.2
Vulkacit AMZ	Bayer	7.3.2.2
Vulkacit BKF	Bayer	7.3.6.4
Vulkacit CRV	Bayer	7.3.2.12
Vulkacit CT-N, 576	Bayer	7.3.2.10

Name	Company (Addresses see Section 7.5)	Type of Rubber rsp. Chemical see Section:
Vulkacit CZ	Bayer	7.3.2.2
Vulkacit D	Bayer	7.3.2.8
Vulkacit DM	Bayer	7.3.2.1
Vulkacit DOTG	Bayer	7.3.2.8
Vulkacit DZ	Bayer	7.3.2.2
Vulkacit F (MBTS/DPG)	Bayer	7.3.2.1
Vulkacit H 30	Bayer	7.3.2.10
Vulkacit HX	Bayer	7.3.2.10
Vulkacit J (PMTD)	Bayer	7.3.2.6
Vulkacit L	Bayer	7.3.2.3
Vulkacit LDA	Bayer	7.3.2.3
Vulkacit LDB	Bayer	7.3.2.3
Vulkacit LZ	Bayer	7.3.2.2
Vulkacit Merkapto	Bayer	7.3.2.1
Vulkacit MOZ	Bayer	7.3.2.2
Vulkacit MT (MBT/TMTD)	Bayer	7.3.2.1
Vulkacit NPV	Bayer	7.3.2.9
Vulkacit NZ	Bayer	7.3.2.2
Vulkacit P	Bayer	7.3.2.4
Vulkacit P extra N	Bayer	7.3.2.3
Vulkacit Thiuram	Bayer	7.3.2.6
Vulkacit Thiuram MS	Bayer	7.3.2.6
Vulkacit TR	Bayer	7.3.2.10
Vulkacit WL	Bayer	7.3.2.4
Vulkacit ZM	Bayer	7.3.2.1
Vulkacit ZP	Bayer	7.3.2.3
Vulkacit 1000	Bayer	7.3.2.8
Vulkafor ZDEC	Vulnax	7.3.2.3
Vulkafor 2	Vulnax	7.3.2.13
Vulkalent A	Bayer	7.3.5.4
Vulkalent B	Bayer	7.3.5.2
Vulkalent E	Bayer	7.3.5.4
Vulkanol FH (XF-Resin)	Bayer	7.3.8.9
Vulkanol 81, SF	Bayer	7.3.8.8
Vulkanol 85, BA, OT	Bayer	7.3.8.9
Vulkanol 88, 90	Bayer	7.3.8.6
Vulkanox CS	Bayer	7.3.6.4
Vulkanox DDA	Bayer	7.3.6.2
Vulkanox DS, KSM, TSP	Bayer	7.3.6.5
Vulkanox EC	Bayer	7.3.6.3
Vulkanox HS	Bayer	7.3.6.3
Vulkanox KB	Bayer	7.3.6.5
Vulkanox MB	Bayer	7.3.6.6
Vulkanox MB2	Bayer	7.3.6.6
Vulkanox NKF	Bayer	7.3.6.4
Vulkanox OCD	Bayer	7.3.6.2
Vulkanox PAN, PBN	Bayer	7.3.6.2
Vulkanox SKF	Bayer	7.3.6.4
Vulkanox SP	Bayer	7.3.6.5
Vulkanox ZKS	Bayer	7.3.6.4
Vulkanox ZMB	Bayer	7.3.6.6
Vulkanox ZMB2	Bayer	7.3.6.6
Vulkanox 3100	Bayer	7.3.6.1

Name	Company (Addresses see Section 7.5)	Type of Rubber rsp. Chemical see Section:
Vulkanox 4010 NA	Bayer	7.3.6.1
Vulkanox 4020	Bayer	7.3.6.1
Vulkanox 4030	Bayer	7.3.6.1
Vulkanox 4050	Bayer	7.3.6.1
Vulkaperl 2	Vulnax	7.3.2.13
Vulklor	Uniroyal	7.3.1.7
Vynathene	USI Chemicals	7.1.11
Warecure C	Ware Chemical	7.3.2.9
Wingstay C, T	Goodyear	7.3.6.5
Wingstay L	Goodyear	7.3.6.5
Wingstay S	Goodyear	7.3.6.5
Wingstay 2	Goodyear	7.3.6.4
Wingstay 29	Goodyear	7.3.6.2
Wingstay 29	Goodyear	7.3.6.2
Wingstay 100	Goodyear	7.3.6.1
Wingstay 300	Goodyear	7.3.6.1
Witaclor 171	Dynamit Nobel	7.3.8.11
Witamol 60	Dynamit Nobel	7.3.8.6
Witamol 100	Dynamit Nobel	7.3.8.2
Witamol 110, 600	Dynamit Nobel	7.3.8.2
Witamol 190	Dynamit Nobel	7.3.8.2
Witamol 320	Dynamit Nobel	7.3.8.1
Witamol 500	Dynamit Nobel	7.3.8.3
Witamol 615	Dynamit Nobel	7.3.8.1
Wobecit DM	VEB Bitterfeld	7.3.2.1
Wobecit Merkapto	VEB Bitterfeld	7.3.2.1
Wobecit, P extra N	VEB Bitterfeld	7.3.2.3
Wobecit Thiuram	VEB Bitterfeld	7.3.2.6
Wobecit 1000	VEB Bitterfeld	7.3.2.8
Z5MC	Krahn	7.3.2.3
Zalba, Zalba Special	Du Pont	7.3.6.4
ZDBC	Krahn	7.3.2.3
ZDBC	Lehmann & Voss	7.3.2.3
ZDEC	Krahn	7.3.2.3
ZDEC	Lehmann & Voss	7.3.2.3
ZDMC	Krahn	7.3.2.3
ZDMC	Lehmann & Voss	7.3.2.3
Zefax	Vanderbilt	7.3.2.1
ZEPC	Lehmann & Voss	7.3.2.3
Zetpol	Nippon Zeon	7.1.5
Zinc Starate	Witco	7.3.11.3
Zinc Stearate	Malmsten & Bergvall	7.3.11.3
Zink Ancap	Anchor	7.3.2.1
Zinkoxid	Harwick	7.3.4
Zinkoxid	Malmsten & Bergvall	7.3.4
Zinkoxid	Zinkwit	7.3.4
Zinkoxid aktiv	Bayer	7.3.4
Zinkoxid transparent	Bayer	7.3.4

7.5 Manufacturers, Suppliers

Akzo	**Akzo Chemie bv** Stationsstraat 48 NL-3800 AE Amersfoort Netherlands
ALM	**Associated Lead Manufacturers Ltd** Crescent House Newcastle upon Tyne NE99 IGE Great Britain
Allied	**Allied Chemical Corp.** P.O. Box 1087 R Morristown, NJ 07 960 USA
Amoco	**Amoco Chemicals Europe SA** 15, rue Rothschild CH-1211 Geneva 21 Switzerland
Anchor	**Anchor Chemical (UK) Ltd** Clayton Manchester M11 4 SR Great Britain
Ancomer	**Ancomer Ltd** Clayton Manchester M11 4 SR Great Britain
Ancomer	**Ancomer Ltd** Clayton Manchester M11 4 SR Great Britain
Armak	**Armak Company** P.O. Box 1805 Chicago, IL 60690 USA
Asahi Chemicals	**Asahi Chemicals Industry Co. Ltd** 1-2, Yurakucho 1-chome Chiyoda-ku Tokyo 100 Japan
Asahi Glass	**Asahi Glass Co. Ltd** 1-2, Marunouchi 2-chome Chiyoda-ku Tokyo 100 Japan
Ashland	**Ashland Chemical (France) SA** 104, Bureaux de la Colline F-92213 Saint Cloud France
Astor	**Astor Chemical Ltd** Tavistock Road West Drayton, Middlesex UB7 7RA Great Britain
Atochem	**Atochem** La Défense 10, Cedex 24 F-92091 Paris-La Défense France
Australian Synthetic Rubber	**Australian Synthetic Rubber Co. Ltd.** P.O. Box 33 Altona Vic 3018 Australia
BASF	**BASF Aktiengesellschaft** Carl-Bosch-Strasse D-6700 Ludwigshafen/Rh. West Germany
Basic	**Basic Chemicals** 845 Hanna Building Cleveland, OH 44115 USA
Baxenden	**Baxenden Chemical Co. Ltd** Accrington, Lancashire BB5 2SL Great Britain
Bayer	**Bayer AG** D-5090 Leverkusen-Bayerwerk West Germany
Bunawerke Hüls	see CWH
Bozzetto	**Bozzetto Industrie Chimiche SpA** Via Mazzini 11 I-24066 Pedrengo (Bergamo) Italy
BP	**BP Chemicals Ltd** Belgrave House 76 Buckingham Palace Road London SW1 W OS4 Great Britain
Braunkohlenwerk	**VEB Braunkohlenwerk** "Gustav Sobottka" DDR-4256 Röblingen/See GDR

Briggs + Townsend	**B + T Polymers Ltd** Station Road Birch Vale, New Mills Stockport Cheshire SK12 5BP Great Britain
Calatrava	**Calatrava** Empresa Para la Industria Petroquimica, SA Apart. de Correos No 388 Santander Spain
Campbell	**Dussek Campbell Ltd** Thames Road Crayford, Kent DA1 4QJ Great Britain
CdF-Chimie	**CdF Chimie S. A.** Tour Aurore 18, Place des Reflets, Cedex 5 F-92080 Paris-La Défense 2 France
Celanese	**Celanese Corp. of America** Chem. Div. New York 16, NY USA
Ciba Geigy	**Ciba-Geigy AG** Klybeckstrasse 141 Postfach CH-4002 Basel Switzerland
Ciech-Stomil	**Ciech-Stomil** ul. 22-Go Lipca 74 P.O. Box 118 90-646 Lodz Poland
Compounding Ingredients	**Compounding Ingredients Ltd** Byrom House, Quay Street Manchester M3 3HS Great Britain
Copolymer	**Copolymer Rubber and Chemical Corp.** P.O. Box 2591 Baton Rouge, LA 70821 USA
Croxton + Garry	**Croxton + Garry Ltd** Curtis Road Dorking, Surrey RH4 1XA Great Britain
CWH	**Hüls AG** Paul-Baumann-Strasse 1 D-4370 Marl West Germany
Cyanamid	**Cyanamid B. V.** P.O. Box 1523 NL-3000 BM Rotterdam Netherlands
Dahleke	**Klaus Dahleke** Heiholtkamp 11 D-2000 Hamburg 60 West Germany
Daikin	**Daikin Kogyo Co. Ltd, Japan** c/o Nordmann, Rassmann GmbH & Co Kajen 2 D-2000 Hamburg 11 West Germany
Danubiana	**I. S. C. E. Danubiana** 202A Splaiul Independentei Bucharest Romania
Degussa	**Degussa AG** P.O. Box 110533 D-6000 Frankfurt 11 West Germany
Denka Kagaku	**Denka Kagaku Kogyo K. K.** Sanshin Bldg. 4-1 Yuraku-cho Chiyoda-ku, Tokyo 100 Japan
Diamond	**Diamond Alkali Co.** Cleveland 14 Ohio USA
Distugil	**Distugil, Rhône Poulenc Polymères** 45-47, Rue de Villiers 92527 Neuilly sur Seine - Cedex France
DOG	**DOG Deutsche Oelfabrik** Ellerholzdamm 50 P.O. Box 111929 D-2000 Hamburg 11 West Germany
Dow Corning	**Dow Corning Ltd** Arco House, Castle Street Reading, Berkshire RG1 7DZ Great Britain
DSM	**DSM Sales Office Rubber** P.O. Box 43 NL-6130 AA Sittard Netherlands

Du Pont	**Du Pont de Nemours & Co (Inc.)** Elastomer Division Wilmington, DE 19898 USA
Durham	**Durham Chemicals Ltd** Birtley, Chester-le-Street County Durham DH3 1QX Great Britain
Dynamit Nobel	**Dynamit Nobel AG** P.O. Box 1209 D-5210 Troisdorf-Oberlar West Germany
Eagle Picher	**Eagle Picher Industries, Inc.** 580 Walnut St., P.O. Box 779 Cincinnati, OH 45201 USA
Eastman Chemical	**Eastmann Chemical Products, Inc.** Kingsport, TN USA
Enichem	**ENI Chemical SA** Seestrasse 42 CH-8802 Kilchberg, Zürich Switzerland
Ethyl	**Ethyl Corporation** 451 Florida Boulevard Baton Rouge, LA 70801 USA
Exxon	**Exxonchem Europe Inc.** 2, Nieuwe Nijverheidslaan B-1920 Machelen Belgium
FBC	**FBC Ltd** Hauxton Cambridge CB2 5HU Great Britain
Firestone	**Firestone Tire & Rubber Co.** 1200 Firestone Parkway Akron, OH 44317 USA
Foster	**Foster & Co (Stearines) Ltd** Aire Place Mills, Kirkstall Road Leeds LS3 1JL Great Britain
Fuchs	**Fuchs Mineraloelwerke GmbH** Friesenheimer Strasse 15 P.O. Box 740 D-6800 Mannheim West Germany
Fuller	**LW-Fuller GmbH** An der roten Bleiche 2-3 P.O. Box 2050 D-2120 Lüneburg West Germany
GAF	**GAF Corp.** 1180 Av. of America New York, NY 10036 USA
General Electric	**International General Electric AB** Silicone Products Division P.O. Box 80 S-60102 Norrköping Sweden
General Tire	**General Tire & Rubber Co.** One General Street Akron, OH 44329 USA
BF Goodrich	**BFGoodrich Chemical Sales Co.** Veurse Achterweg 6 NL-2260 AB Leidschendam Netherlands
Goodyear	**Goodyear France** Avenue des Tropiques ZA de Courtabœuf F-91941 Les Ulis Cedex France
Harwick	**Harwick Chemical Corp.** 60 S. Seiberling Street Akron, OH 44305 USA
Henkel	**Henkel KGaA** Henkelstrasse 67 P.O. Box 1100 D-4000 Düsseldorf West-Germany
Hercules	**Hercules Incorporated** Wilmington, DE 19899 USA

Hughson	**Hughson Chemicals** Lord Corporation 2000 West Grand View Boulev. P.O. Box 1099 Erie, PA 16512 USA
Hules Mexicanos	see Lehmann & Voss
ICI	**ICI Polyurethanes** Hexagon House Blackley P.O. Box 42 Manchester M9 3DA Great Britain
Jahres	**Jahres Fabrikker A.S.** P.O. Box 2051 3201 N-3200 Sandefjord Norway
Japan Butyl	see Japan Synthetic Rubber
Japan Synthetic Rubber	**Japan Synthetic Rubber Co., Ltd** 2-11-24 Tsukiji, ehno-ku Tokyo Japan
J-sil	**J-sil Silicones Ltd** Holloway Drive Worsley Manchester M28 4LA Great Britain
Kali-Chemie	**Kali-Chemie AG** Hans-Böckler-Allee 20 P.O. Box 220 D-3000 Hannover 1 West Germany
Kemi	**Kemi-Interessen AG** Vintergatan 1 P.O. Box 6018 S-17206 Sundbybers 6 Sweden
Kenrich	**Kenrich Petrochemicals, INC** P.O. Box 32 Bayonne, NJ 07002-0032 USA
Kettlitz	Kettlitz-Chemie GmbH & Co. KG P.O. Box 80 D-8859 Rennertshofen West Germany
Krahn	**Krahn Chemie GmbH** Grimm 10 D-2000 Hamburg 11 West Germany
Landaise	**Manufacture Landaise des Produits Chimiques** F-40 30 70 Rion des Landes France
Lefrant	**Lefrant-Rubco SA** P.O. Box 25, Muille F-80400 Ham France
Lehmann & Voss	**Lehmann & Voss & Co.** Alsterufer 19 D-2000 Hamburg 36 West Germany
Lowi	**Chemische Werke Lowi GmbH** Teplitzer Strasse, P.O. Box 1660 D-8264 Waldkraiburg West Germany
Luperox	**Luperox GmbH** Denzinger Strasse 7–9 P.O. Box 1354 D-8870 Günzburg West Germany
3 M	**3M Center** St. Paul, MN 55144 USA
Malmsten & Bergvall	**Malmsten & Bergvall AG** Bromstegenvägen 172 P.O. Box 58 S-16391 Spånga Schweden
Manchem	**Manchem Ltd** Ashton New Road, Clayton Manchester M11 4AT Great Britain
Merck	**Merck & Co. Inc** P.O. Box 2000 Rahway, NJ 07065 USA
Metall-gesellschaft	**Chemetall GmbH** Reuterweg 14 P.O. Box 3724 D-6000 Frankfurt/Main 1 West Germany
Mitsubishi	**Mitsubishi Ind., Ltd** 2-5-2, Marunoudi, Chiyoda-ku Tokyo Japan

Monsanto	**Monsanto Ltd** Edison Road Swindon, Wilts Great Britain
Montedison	**Montedison S. P. A.** Largo G Donegani 1-2 I-20121 Milano Italy
Morton	**Morton Chemical Co.** 110, North Wacker Drive Chicago, IL 60606 USA
NCC	**Neuville Cindu Chemie BV** P.O. Box 9 NL-1420 AA Uithoorn Netherlands
Nippon Kasai	**Nippon Kasei Chemical Co., Ltd** 4-1, Yuraku-cho, 1-chome Chiyoda-ku Tokyo 100 Japan
Nippon Zeon	**Nippon Zeon Co. Ltd** Furukawa Sögö Bldg 6-1 Marunouchi 2-Chome Chiyoda-ku, Tokyo Japan
NL-Industries	**National Lead Industries, Inc.** P.O. Box 700 Hightown, NJ 08520 USA
Nordmann, Rassmann	**Nordmann, Rassmann & Co.** Kajen 2 D-2000 Hamburg 11 West Germany
Notedome	**Notedome Ltd** 108a Station Street East Coventry Great Britain
Nynäs	**AB Nynäs Petroleum** Norrlandsgatan 20 P.O. Box 7856 S-10399 Stockholm Sweden
Osaka Soda	**Osaka Soda** 1-53 Edobori Nishi-Ku Osaka-Shi Japan
Oushi Shinko	**Ouchi Shinko Chemical Ind.** Tokyo Japan
PAr	**PAr Oberflächenchemie GmbH** Ottostrasse 28 D-5142 Hückelhoven-Baal West Germany
Pennwalt	**Pennwalt Corp BV** Westblaak 96 NL-3012 KM Rotterdam Netherlands
Peroxid-Chemie	**Peroxid-Chemie GmbH** D-8023 Höllriegelskreuth West Germany
Petrimex	**Petrimex Ltd** Dr. Vl. Clementisa 10 CS-82602 Bratislava Czechoslovakia
Petrochim	**Petrochim N. V.** Scheldelaan 10 B-2030 Antwerpen Belgium
Phillips Petroleum	**N. V. Phillips, Petroleum S. A.** Hilton Tower, Bld. de Waterloo 39 B-1000 Brussels Belgium
Pitt Consol	**Pitt Consol Chemicals Co.** Newark, NJ USA
Polysar	**Polysar International S. A.** Route de Beaumont 10 CH-1701 Fribourg Switzerland
Raschig	**Raschig AG** Mundenheimer Strasse 100 P.O. Box 211128 D-6700 Ludwigshafen West Germany
Reichold	**Reichold Chemicals Inc.** White Plains, NY 10603 USA
Rhein-Chemie	**Rhein-Chemie Rheinau GmbH** Mülheimer Strasse 24-28 P.O. Box 810409 D-6800 Mannheim 81 West Germany

Rhône-Poulenc	**Rhône-Poulenc Chimie de Base** 25, Quai Paul Doumer F-92408 Courbevoie Cedex France
Robinson	**Robinson Brothers Ltd** Phoenix Street, West Bromwich West Midlands B70 0AH Great Britain
Rohm & Haas	**Rohm & Haas Co.** Philadelphia, PA USA
Safic	**Safic-Alcan & Co.** 3, Rue Bellini F-92806 Puteaux Cedex France
Schenectady	**Schenectady-Midland Ltd** Four Ashes, Wolverhampton WV10 7BT Great Britain
Schill & Seilacher	**Schill & Seilacher GmbH & Co.** Moorfleeterstrasse 28 D-2000 Hamburg 74 West Germany
Schütz	**Georg Schütz GmbH** Erste Süddeutsche Ceresinfabrik P.O. Box 80 West Germany
Shell	**Shell Chemical Int. Trading Comp.** Shell Center London SE1 7PG Great Britain
Showa Denko	**Showa Denko K.K.** 13-9, Shiba daimon I Chome Minatoku Tokyo 105 Japan
SIR	**S.I.R. Consorzio Industriale S.p.A.** Via Grazioli 33 I-20161 Milan Italy
Sumitomo	**Sumitomo Chemical Co., Ltd** 15,5-chome, Kitihama Higashi-ku Osaka 500 Japan
Sun Oil	**Sun Oil Co.** 1608 Walnut Street Philadelphia, PA 19103 USA
Synpol	**Synpol Inc.** Two Greenwich Plaza Greenwich, CT 06830 USA
Technical Processing	**Technical Processing Inc.** 106 Railroad Avenue Paterson, NJ 07509 USA
Thiokol	**Thiokol Corp** Trenton, NJ 08650 USA
Thixon	**Compounding Ingredients Ltd** Byrom House, Quay Street Manchester M3 3HS Great Britain
Toyo Soda	**Toyo Soda Manufacturing CO Ltd** 1-Chome, Akasaka, Minato-Ku Tokyo, Japan
Ugine Kuhlmann	**Ugine Kuhlmann** 25, bd d'Admiral Bruix F-75016 Paris France
Unichema	**Unichema International** P.O. Box 1280 D-2420 Emmerich West Germany
Union Carbide	**Union Carbide Europe SA** 11, Avenue Choiseul CH-1290 Versoix/Geneva Switzerland
Uniroyal	**Uniroyal Chemical** Div. of Uniroyal Ltd Trafford Park Manchester M17 1DT Great Britain
USI Chemicals	**USI Chemicals Co.** 99 Park Avenue New York NY 10016 USA

Vanderbilt	**Vanderbilt Export Corp.** 30 Winfield Street Norwalk, CT 06855 USA	Ware Chemical	**The Ware Chemical Corp.** Bridgeport, CT USA
VEB Bitterfeld	**VEB Chemiekombinat Bitterfeld** Zörbiger Strasse DDR-4400 Bitterfeld GDR	Witco	**Witco Chemical Corp.** 213 West Bowery Street Akron 8, OH 44305 USA
VEB Buna	**VEB Chemische Werke Buna** DDR-4212 Schkopau/Merseburg GDR	Wyrough & Loser	**Wyrough & Loser, Inc.** P.O. Box 5047 1008 Whitehead Road Ext. Trenton, NJ 08638 USA
Vulnax	**Vulnax International Ltd** 321 Bureaux de la Colline 92213 Saint-Cloud Cedex France	Würtz	**E. & P. Würtz GmbH & Co. KG** Industriegebiet D-6530 Bingen/Rh. 17 West Germany
Wacker	**Wacker-Chemie GmbH** Prinzregentenstrasse 22 D-8000 Munich 22 West Germany	Zinkwit	**Zinkwit Nederland B.V.** Maastricht Netherlands

7.6 References

[7.1] Häggström B., Kinnhagen S., Laurell L. G.: Gummiteknisk Handbok, The Scandinavian Rubber Handbook, Publ. Sveriges Gummitekniska Förening, SGF Publ. 62. 7th Ed.

[7.2] Rubber Blue Book, Publ.: Rubber World.

[7.3] Rubber Red Book, 1983, 35th Edition. Communication Channels, Inc., Publ.: Elastomerics.

[7.4] The author intends to publish an extended and updated trade name index shortly. This index will be available on disks which can be used with commercially available data bank software systems on PCs. They can be updated individually by the user. Further informations can be obtained from the author:
Dr. Werner Hofmann, Kappelerstr. 5, D-4000 Düsseldorf 13, West Germany.

8 Appendix

8.1 Physical Units, Selection

8.1.1 U.S. Customary/SI Conversions

SI Units	×	Multiply by	→	U.S. Units	×	Multiply by	→	SI Units
kJ/kg·K	×	0.2388	→	BTU/lb °F	×	4.187	→	kJ/kg·K
W/K·m	×	6.944	→	BTU in/hr ft^2 °F	×	0.144	→	W/K·m
J/m	×	0.01873	→	ft lb/in	×	53.40	→	J/m
kJ/m^2	×	0.4755	→	ft lb/in^2	×	2.103	→	kJ/m^2
m/m K	×	0.5556	→	in/in °F	×	1.8	→	m/m K
MN/m^2 or MPa	×	144.9	→	lb/in^2	×	0.0069	→	MN/m^2 or MPa
Mg/m^3	×	62.50	→	lb/ft^3	×	0.0160	→	Mg/m^3
MV/m	×	25.38	→	V/mil	×	0.0394	→	MV/m

Temperature Conversion: °C → °F = (°C × 1.8) + 32 °F → °C = (°F − 32) ÷ 1.8
°C → K = °C + 273.15

W/K·m = W/mK $MN/m^2 = N/mm^2$ $MG/m^3 = g/cm^3$

8.1.2 Conversion of Typical Units in Use in the USA into Metric Units

Description	US Unit	Metric Unit	Factor
Adhesion metal	lb/in	kN/m	0.1751
Adhesion textile	psi	MPa	0.006895
Angle tear resistance	ppi (lb/in)	kN/m	0.1751
Dynam. modulus	psi	MPa	0.006895
Impact strength	ft·lb/in	J/cm	0.534
Pressure	psi	MPa	0.006895
Tear strength	lb/in	kN/m	0.1751
Tensile modulus	psi	MPa	0.006895
Tensile strength	psi	MPa	0.006895
Torque	in·lb	N·m	0.1130

8.2 List of Abbreviations*

ABR	Acrylic ester - butadiene rubber
ACM	Acrylate rubber
ACS	American Chemical Society
ADC	Azodicarbonamide
ADPA	Acetone-diphenylamine condensation product
AFMU	Nitroso rubber
ANM	Acrylic ester acrylonitrile copolymers
APH	Alkylated phenols
ASR	Alkylene sulfide rubber
ASTM	American Standard Methods
AU	Urethane rubber, based on polyesters
AU-I	Isocyanate corsslinkable AU
AU-P	Peroxidic crosslinkable AU
BAA	Butyraniline aldehyde
BD	Benzofurane derivative
BDMA	Butylene glycol dimethacrylate
BES	Benzoic acid
BET	Specific surface of fillers
BHT	2,6-di-Tert.butyl-p-cresole
BiDMC	Bismuth-dimethyl dithiocarbamate
BIIR	Bromobutyl rubber
BPH	2,2'-Methylene-bis-(4-methyl-6-tert.butylphenol)
BR	Butadiene rubber
BS	British Standard Methods
BSH	Benzene sulfo hydrazide
CBS	N-Cyclohexyl-2-benzothiazole sulfenamide
CdDMC	Cadmium-dimethyl dithiocarbamate
Cd5MC	Cadmium-pentamethylene dithiocarbamate
CDO	p-Benzoquinone dioxime
CEA	Cyclohexyl ethyl amine
CF	Conductive furnace black
CFM	Polychloro trifluoro ethylene (former FPM type)
CIIR	Chlorobutyl rubber
cis	molecular structure
CLD	Caprolactam disulfide
CM	chlorinated Polyethylene
CO	Epichlorohydrin homopolymer
Co-BR	BR, based on cobalt catalyst
CPH	2,2'-Methylene-bis-(4-methyl-6-cyclohexylphenol)
CR	Chloroprene rubber
CSM	chlorosulfonated polyethylene

---

* The following abbreviations are not considered:

- chemical symbols
- terms of equations
- physical units
- indices of trade names
- abbreviations of companies

CSR	Standardized NR from China
CTAB	Bromine hexadecyl trimethylammonia filler adsorption
CTFE	Chloro trifluoro ethylene
CTP	Cyclohexyl thio phthalimide
CuDBC	Copper dibutyl dithiocarbamate
CuDIP	Copper diisopropyl dithiophosphate
CuDMC	Copper dimethyl dithiocarbamate
CV	Constant viscosity
CV	continuous vulcanization
DAP	Diallyl phthalate
DBA	Dibutyl amine
DBP	Dibutyl phthalate
DBSH	Benzene-1,3-disulphohydrazide
DBTU	Dibutyl thiourea
DCBS	N,N-Dicyclohexyl-2-benzothiazole sulfenamide
DCP	Dicyclopentadiene
DDA	Didecyl adipate
DETU	N,N′-Dietyl thiourea
DIBP	Diisobutyl phthalate
DIDA	Diisodecyl adipate
DIN	German Industry Standards
DINP	Diisononyl phthalate
DIOP	Diisooctyl phthalate
DIOZ	Diisooctyl azelate
DIPS	N,N-Diisopropyl-2-benzothiazole sulfenamide
dkt	German Rubber Conference
DNP	Dinonyl phthalate
DNPD	N,N′-Di-β-naphthyl-p-phenylene diamine
DNPT	N,N′-Dinitroso pentamethylene tetramine
DOA	Dioctyl adipate
DOP	Dioctyl phthalate
DOPD	N,N′-bis-(1-ethyl-3-methylpentyl)-p-phenylene diamine
DOS	Dioctyl sebacate
DOTG	Di-o-tolylguanidine
DOZ	Dioctyl azelate
DPG	Diphenyl guanidine
DP-NR	Deproteinated NR
DPPD	N,N′-Diphenyl-p-phenylene diamine
DSC	Differential scanning calorimetry
DTA	Differential thermal analysis
DTDM	Dithio dimorpholine
DTDP	Tridecyl phthalate
DTOS	N-Dimethyl dithiocarbamyl-N′-oxidiethylene sulfenamide
DTPD	N,N′-Ditolyl-p-phenylene diamine
DPTT	Dipenta methylene thiuram-tetra(hexa)sulfide
DPTU	N,N′-Diphenyl thiourea
D5MTD	Dipentamethylene thiuram disulfide
EAM	Ethylene-acrylate copolymers
E-BR	Emulsion BR
ECO	Epichlorohydrin-ethylene oxide copolymer

EDMA Ethylene glycol dimethacrylate
EE Enolether
EMP Elastomer modified Plastomer
ENB Ethylidene norbornene
ENM Former proposed code for H-NBR
ENR Epoxidized NR
EPDM Ethylene propylene diene terpolymer (DCP, ENB or HX)
EPM Ethylene propylene copolymer
E-SR Emulsion synthetic rubber
E-SBR Emulsion SBR
ETER Epichlorohydrine-ethylene oxide-allylglycidyl ether terpolymer
ETMQ 6-Ethoxy-2,2,4-trimethyl-1,2-dihydroquinoline
ETU N,N′-Ethylene thiourea
EU Urethane rubber, based on polyethers
EV Efficient vulcanization (sulfurfree or sulfurpoor sulfurdonor containing systems)
EVK Extruder mixing screw system
EVM Ethylene vinylacetate copolymer (previous code EVA and EVAC)

FEF Fast extrusion furnace black
FF Fine furnace black
FKM Fluoro rubbers (see also FPM)
FMVE Perfluoro methyl vinyl ether
FMQ Fluoro silicone rubber
Fp Melting point
FPM Fluoro rubbers (previous code FKM)
FVMQ Fluoro vinyl group containing silicone rubber

GAK Journal Gummi, Asbest, Kunststoffe (rsp. Gummi, Fasern, Kunststoffe)
GP General purpose
GPF General purpose furnace black
GPO Propylenoxide-allylglycidyl ether copolymer
GR-S former code for SBR

HAF High abrasion furnace black
HEXA Hexamethylene tetramine (proposed code also HMT)
HFP Hexafluoro propylene
HFPE 1-Hydro penta fluoro propylene
HMF High modulus furnace black
HMT see HEXA
H-NBR Hydrogenated NBR (former proposed code ENM, other proposed code HSN)
HPLC High pressure liquid chromatography
HS High structure black
HSN Highly saturated nitrile rubber (see H-NRB)
HX Trans-1,4-hexadiene

IBPH 2,2′-Isobutylidene-bis-(4-methyl-6-tert.butylphenol)
ICR Initial concentrated rubber
IHRD International hardness degree
IIR Isobutylene Isoprene Copolymer (butyl rubber)
ikt International rubber conferences (in Germany)

IPPD	N-Isopropyl-N′-phenyl-p-phenylene diamine
IR	Infrared spectroscopy
IR	Isoprene rubber (synthetic)
IRC	International rubber conferences
ISAF	Intermediate super abrasion furnace black
ISO	International Standards Organization
IT	Asbestos rubber scalings
KGK	Journal Kautschuk u. Gummi, Kunststoffe
L-BR	Solution BR
LCM	Liquid curing method (salt bath vulcanization)
Li-BR	Butadiene rubber, based on lithium catalyst
LN-NR	Low nitrogen NR
LS	Low structure black
L-SBR	Solution SBR
LSR	Liquid silicone rubber
LV	Low viscosity
MBI	2-Mercapto benzimidazole
MBS	2-Benzothiazole-N-sulfene morpholide
MBSS	2-Morpholinodithio benzothiazole
MBT	2-Mercaptobenzothiazole
MBTS	Dibenzothiazole disulfide
MDI	4,4′-Diphenyl methanediisocyanate
Me	Metal ion
METD	Dimethyldiethyl thiuram disulfide
ML	Unit of mooney viscosity
MMBI	Methyl-2-mercapto benzimidazole
MOCA	Methylene bis chloro aniline
MOD	NR types with specified degree of curing
MPTD	Dimethyl diphenyl thiuram disulfide
MQ	Methyl silicone rubber
MT	Medium thermal black
NaDEC	Sodium diethyl dithiocarbamate
NaDMC	Sodium dimethyl dithiocarbamate
NaIX	Sodium isopropyl xanthogenate
NaMBT	Sodium-2-mercaptobenzothiazole
NBR	Butadiene acrylonitrile copolymer, nitrile rubber
NCR	Chloroprene acrylonitrile copolymer
Nd-BR	Butadiene rubber, based on neodymium catalyst
NDPA	N-Nitroso diphenylamine
Ni-BR	Butadiene rubber, based on nickel catalyst
NiDBC	Nickel dibutyl dithiocarbamate
NiDMC	Nickel dimethyl dithiocarbamate
NIR	Isoprene acrylonitrile copolymer
NMR	Nuclear magnetic resonance analysis
NR	Natural rubber
NS	Non staining
NV	Non discolouring
ODPA	Octylated diphenyl amine
OE-BR	Oil extended BR

OE-EPDM	Oil extended EPDM
OE-E-SBR	Oil extended emulsion SBR
OE-L-SBR	Oil extended solution SBR
OE-NR	Oil extended NR
OE-SBR	Oil extended SBR
OT	Polyglycolether
OTBG	o-Tolylbiguanide
OTOS	N-Oxidiethylene dithiocarbamyl-N′-oxydiethylene sulfenamide
OTTBS	N-Oxidiethylene dithiocarbamyl-N′-tert.butyl sulfenamide
PA	Polyamide
PA-NR	Superior processing NR with more than 50% crosslinked blends
PAN	Phenyl-α-naphthyl amine
PbDEC	Lead diethyl dithiocarbamate
PbDMC	Lead dimethyl dithiocarbamate
PbD5C	Lead pentamethylene dithiocarbamate
PBN	Phenyl-β-naphthyl amine
PBR	Pyridine butadiene rubber
PCD	Polycarbodiimide
PCTP	Pentachlorothiophenol
PE	Polyethylene
PEP	Polyethylene polyamine
PET	Pentaerythrilol tetrastearate
phr	Parts per hundred rubber (weight parts per hundred weight parts rubber)
PMQ	Silicone rubber with phenyl groups
PNF	Poly fluoro alkoxy phosphazenes
PO	Propylene oxide homopolymer
PP	Polypropylene
PPC	Piperidine pentamethylene dithiocarbamate
PRI	Plasticity retention index of NR
PS	Polystyrene
PSBR	Pyridine styrene butadiene rubber
PTA	Phthalic ankydride
PUR	Generic code for urethane elastomers
PVC	Polyvinyl chloride
PVDC	Polyvinylidene chloride
PVMQ	Silicone rubber with phenyl and vinyl groups
Q	Generic code for silicone rubber
QSM	Pin barrel cold feed extruder
RCT	Journal of Rubber Chemistry and Technology
RFK	Resorcinol formaldehyde bonding system
RFL	Resorcinol formaldehyde latex preparation
RI	Refraction intercept of mineral oils
RIM	Reaction injection moulding
RLP	Reactive liquid polymer
RTL	Reaction liquid types
SAF	Super abrasion furnace black
SAN	Styrene acrylonitrile copolymer
SAPH	Styrenated and alkylated phenol

SBR	Styrene butadiene rubber
SBS	Styrene butadiene styrene block copolymer
SCR	Styrene chloroprene copolymer
SCS	Salicylic acid
SDEC	see NaDEC
SDMC	see NaDMC
SDPA	Styrenated diphenyl amine
SEBS	Styrene ethyl butylene styrene block copolymer
SeDMC	Selenium dimethyl dithiocarbamate
Semi-EV	Semiefficient vulcanization (sulfur poor vulcanization)
SEP	Polysiloxan treated (grafted) EPDM
SIR	Standardized NR from Indonesia
SIR	Styrene isoprene rubber
SIS	Styrene isoprene styrene block copolymer
SIX	see NaIX
SLR	Standardized NR from Sri Lanka
SMBT	see NaMBT
SMR	Standardized NR from Malaysia
SPC	Statistical process control
SPH	Styrenated phenol
SP-NR	Superior processing NR (see also PA-NR)
SR	Generic code for synthetic rubber
SRC	Scandinavian rubber conference
SRF	Semi reinforcing furnace black
ST	Polythioglycol ether (see ASR)
STR	Standardized NR from Thailand
TA	Thermal analysis
TAC	Triallyl cyanurate
TAIC	Triallyl isocyanurate
TAM	Triallyl trimellithate
TAP	Triallyl phosphate
TBA	Torsional braid analysis
TBBS	N-tert. Butyl-2-benzothiazole sulfenamide
TBTD	Tetrabutyl thiuram disulfide
TBTU	Tributyl thiourea
TC	Technical classified NR
TCT	Tricrotonylidene tetramine
TDI	Toluylene diisocyanate
TeDEC	Tellurium diethyl dithiocarbamate
TeDMC	Tellurium dimethyl dithiocarbamate
TeD5C	Tellurium pentamethylene dithiocarbamate
TETA	Tetraethanol propane
TETD	Tetra ethyl thiuram disulfide
TFE	Tetra fluoro ethylene
T_G	Glass transition temperature
TGA	Thermo gravimetric analysis
Ti-BR	Butadiene rubber, based on titanium catalyst
TIOM	Triisooctyl trimellithate
Tm	Melting range of crystalline phases in TPE
TM	Thioplasts

TMA	Thermo mechanical analysis
TMPT	Trimethylolpropane trimethacrylate
TMQ	2,2,4-Trimethyl-1,2-dihydroquinoline, polymerized
TMTD	Tetramethylthiuram disulfide
TMTM	Tetramethylthiuram monosulfide
TNPP	Trisnonylphenyl phosphite
TOR	Transpolyoctenamer
TPA	Transpolypentenamer
TPE	Generic code for thermoplastic elastomers
TPE-A	Thermoplastic polyetheramide elastomers
TPE-E	Thermoplastic polyetherester elastomers
TPE-FKM	Thermoplastic fluoro elastomers
TPE-NBR	Thermoplastic elastomer, based on NBR
TPE-NR	Thermoplastic elastomer, based on NR
TPE-O	(see TPO)
TPE-S	see SBS, SIS
TPE-U	(see TPU)
TPO	Thermoplastic polyolefins, also TPE-O
TPR	Non acceptable generic name for TPE
TPTA	Trimethylolpropane trimethacrylate
TPU	Thermoplastic polyurethanes, also TPE-U
T_R	Low temperatur flexibility
T_S	Brittleness temperature
TSH	p-Toluylene sulfohydrazide
TSR	Technical classified NR
trans	molecular structure
T50	Low temperature test
U	Revolutions
UHF	Ultra high frequence (continuous vulcanization)
Upm	Revolutions per minute
UV	Ultra violet
V	Discolouring
VDK	Viscosity-density constant
VF_2	Vinylidene fluoride
VMQ	Silicone rubber with vinyl groups
VP	Vinyl Pyridine copolymer
WTR	Working group toxicology of rubber auxiliaries
X-CR	Chloroprene rubber with reactive groups
XIIR	Halogenated IIR
X-LPE	Cross linkable polyethylene
X-NBR	Nitrile rubber with reactive groups
Y-IR	Thermoplastic IR
Y-NBR	Thermoplastic NBR
Y-SBR	Thermoplastic SBR
ZBEC	Zinc dibenzyl dithiocarbamate
ZBX	Zinc butyl xanthogenate
ZDBC	Zinc dibutyl dithiocarbamate
ZDBP	Zinc dibutyl dithiophosphate

ZDEC	Zinc diethyl dithiocarbamate
ZDMC	Zinc dimethyl dithiocarbamate
ZEPC	Zinc ethyl phenyl dithiocarbamate
ZIX	Zinc isopropyl xanthogenate
ZMBI	Zinc-2-mercaptobenzimidazole
ZMBT	Zinc-2-mercaptobenzothiazole
ZMMBI	Methyl-zinc-2-mercaptobenzimidazole
ZnPCTP	Zinc pentachlorthiophenol
Z5MC	Zinc pentamethylene dithiocarbamate
5BS	tert. Amyl-2-benzothiazole sulfenamide
6PPD	N-(1,3-Dimethylbutyl)-N′-phenyl-p-phenylene diamine
7PPD	N-(1,4-Dimethyl pentyl)-N′-phenyl-p-phenylene diamine
77PD	N,N′-Bis-(1,4-dimethyl pentyl)-p-phenylene diamine

Index

Buyer's Guide

Materials, equipment and services for the rubber and ancillary industries. All entries are paid for by and are the responsibility of the companies concerned. No claim is made as to the completeness of the index and the editor assumes no responsibility for omissions or errors.

Further details from

Hanser Publishers, P. O. B. 860420, D-8000 München 86, FRG.

Group 1: Natural Rubber

Chem. modified natural rubber

Nordmann, Rassmann GmbH & Co.
Kajen 2, D-2000 Hamburg 11
T. 040/3687-0, Tx. 211122+212087
Fax. /3687-249, T#: nordrascos

Modified hevea natural rubber

Nordmann, Rassmann GmbH & Co.
Kajen 2, D-2000 Hamburg 11
T. 040/3687-0, Tx. 211122+212087
Fax. /3687-249, T#: nordrascos

Non-classified or -specified natural rubber

Nordmann, Rassmann GmbH & Co.
Kajen 2, D-2000 Hamburg 11
T. 040/3687-0, Tx. 211122+212087
Fax. /3687-249, T#: nordrascos

Techn. classified or specified natural rubber

Nordmann, Rassmann GmbH & Co.
Kajen 2, D-2000 Hamburg 11
T. 040/3687-0, Tx. 211122+212087
Fax. /3687-249, T#: nordrascos

Group 2: Synthetic Rubber

Elastomer

Acrylonitrile-butadiene elastomer, NBR

BFGoodrich Chemical
Görlitzer Str. 1, D-4040 Neuss
Tel. 02101/18050

Acrylate-based elastomer, ACM, EAM

BFGoodrich Chemical
Görlitzer Str. 1, D-4040 Neuss
Tel. 02101/18050

Epichlorohydrin elastomer, CIO, ECO, ETER

BFGoodrich Chemical
Görlitzer Str. 1, D-4040 Neuss
Tel. 02101/18050

Ethylene-propylene Elastomer, EPM, EPDM

DSM Marketing and Sales
Elastomers, P.O. Box 43
6130 AA Sittard, Netherlands
Tel. 04490/69222, Telex 36775
Fax. 04490/63853
Keltan

Bunawerke Hüls GmbH
D-4370 Marl
Sales and technical service
are handled by Hüls AG

Fluoroelastomer, FKM (FPM)

3M Deutschland GmbH
Abt. Chemische Produkte
PF 100422, D-4040 Neuss 1
Tel. 02101/142511, Tx. 8517511
AFLAS™ TFE-Elastomers
Fluorine containing elastomers
FLUOREL™ Fluorelastomers

Silicone elastomer

Wacker-Chemie GmbH
D-8000 München 22
Tel. (089) 2109-0
Tlx. 529121 56
Elastosil®, Wacker HDK®

Elastomer (cont.)

Styrene-butadiene elastomer, SBR

BFGoodrich Chemical
Görlitzer Str. 1, D-4040 Neuss
Tel. 02101/18050

Bunawerke Hüls GmbH
D-4370 Marl
Sales and technical service
are handled by Hüls AG

Thermoplastic elastomer

DSM Marketing and Sales
Elastomers, P.O. Box 43
6130 AA Sittard, Netherlands
Tel. 04490/69222, Telex 36775
Fax. 04490/63853
Kelprox

Hüls AG
Referat 1122/RTH
D-4370 Marl

Monsanto Europe S.A., Brussels
Avenue de Tervuren 270
Tel. 32-2-761.41.11, Tx. 62927
Santoprene®

Other elastomer types

Bunawerke Hüls GmbH
D-4370 Marl
Sales and technical service
are handled by Hüls AG

Polyethylene, chlorinated, CM

DSM Marketing and Sales
Elastomers, P.O. Box 43
6130 AA Sittard, Netherlands
Tel. 04490/69222, Telex 36775
Fax. 04490/63853
Kelrinal

Polyurethane (PUR)

BFGoodrich Chemical
Görlitzer Str. 1, D-4040 Neuss
Tel. 02101/18050

Resina Chemie B.V.
NL-9607 PS Foxhol (GN)
05980-17911/77246/05980-90437

Rubber compounds

Flevo Rubber B.V., Rubber compounds, P.O. Box 257
NL-8200 AG Lelystad - Holland
Tel. (0)3200-27290
Tlx. 47410
Fax. (0)3200-49170

Trans-Polyoctenamer

Hüls AG
Referat 1122/RTH
D-4370 Marl

Group 3: Rubber Chemicals (Additives)

Accelerators

Amine accelerators

Rhein-Chemie Rheinau GmbH
Mülheimer Str. 24-28
D-6800 Mannheim 81
Tel. 0621/89070. Tx. 462282

Benzimidazoles

Rhein-Chemie Rheinau GmbH
Mülheimer Str. 24-28
D-6800 Mannheim 81
Tel. 0621/89070, Tx. 462282

Benzothiazole accelerators

Monsanto Europe S.A., Brussels
Avenue de Tervuren 270
Tel. 32-2-761.41.11, Tx. 62927
Thiofide®, Bantex, Thiotax

Rhein-Chemie Rheinau GmbH
Mülheimer Str. 24-28
D-6800 Mannheim 81
Tel. 0621/89070, Tx. 462282

Dithiocarbamate accelerators

Monsanto Europe S.A., Brussels
Avenue de Tervuren 270
Tel. 32-2-761.41.11, Tx. 62927
Ethasan, Methasan

Rhein-Chemie Rheinau GmbH
Mühlheimer Str. 24-28
D-6800 Mannheim 81
Tel. 0621/89070, Tx. 462282

Dithiophosphate and xanthate accelerators

Monsanto Europe S.A., Brussels
Avenue de Tervuren 270
Tel. 32-2-761.41.11, Tx. 62927
Vocol®/ Vocol S

Accelerators (cont.)

Rhein-Chemie Rheinau GmbH
Mühlheimer Str. 24-28
D-6800 Mannheim 81
Tel. 0621/89070, Tx. 462282

Elastomer bound accelerators

Croxton & Garry Ltd.
Curtis Road
GB Dorking, Surrey RH4 1XA
Tel. 0306/886688, Tx. 859567
Fax. 0306/887780
MULTISPERSE

Guanidine and thiourea accelerators

Rhein-Chemie Rheinau GmbH
Mülheimer Str. 24-28
D-6800 Mannheim 81
Tel. 0621/89070, Tx. 462282

Special accelerators for EPDM

D.O.G. Deutsche Oelfabrik
P.B. 111929, D-2000 Hamburg
Tx. 212164, Fax. 040/3192748

Sulfenamide accelerators

BFGoodrich Chemical
Görlitzer Str. 1, D-4040 Neuss
Tel. 02101/18050

Monsanto Europe S.A., Brussels
Avenue de Tervuren 270
Tel. 32-2-761.41.11, Tx. 62927
Santocure®

Rhein-Chemie Rheinau GmbH
Mülheimer Str. 24-28
D-6800 Mannheim 81
Tel. 0621/89070, Tx. 462282

Accelerators (cont.)

Thiurame accelerators

Monsanto Europe S.A., Brussels
Avenue de Tervuren 270
Tel. 32-2-761.41.11, Tx. 62927
Thiurad, Monothiurad

Rhein-Chemie Rheinau GmbH
Mülheimer Str. 24–28
D-6800 Mannheim 81
Tel. 0621/89070, Tx. 462282

Other accelerators

Schering Ind. Chemicals
Mount Pleasant House
Huntingdon Rd., Cambridge, UK
Tel. 0223/323222, Tx. 81654
GENIPLEX

Activators

Aromatic polyethers

Rhein-Chemie Rheinau GmbH
Mülheimer Str. 24–28
D-6800 Mannheim 81
Tel. 0621/89070, Tx. 462282

Metal oxides

D.O.G. Deutsche Oelfabrik
P.B. 111929, D-2000 Hamburg
Tx. 212164, Fax. 040/3192748

Metall- und Farbwerke GmbH
Grillo-Werke AG, Werk Goslar
Postfach 2120
D-3380 Goslar
Tel. 05321/681-0, Tx. 953883
Telefax 05321/681-29

Rhein-Chemie Rheinau GmbH
Mülheimer Str. 24–28
D-6800 Mannheim 81
Tel. 0621/89070, Tx. 462282

Silanes

Hoffmann Mineral
Franz Hoffmann & Söhne KG
Münchener Str. 75,
D-8858 Neuburg (Donau)
Tx. 55214-18
T. 08431/53-1, Fax. 08431/53-330

Hüls AG
Referat 1122/RTH
D-4370 Marl

Activators (cont.)

Kettlitz-Chemie GmbH & Co. KG
POB 80, D-8859 Rennertshofen
Tlx. 55237, Fax. 08434/719

Vulcanization and filler activators

Kettlitz-Chemie GmbH & Co. KG
POB 80, D-8859 Rennertshofen
Tlx. 55237, Fax. 08434/719

Rhein-Chemie Rheinau GmbH
Mülheimer Str. 24–28
D-6800 Mannheim 81
Tel. 0621/89070, Tx. 462282

Other activators

Monsanto Europe S.A., Brussels
Avenue de Tervuren 270
Tel. 32-2-761.41.11, Tx. 62927
DPG, Diphenylguanidine

Adhesives and binders

Degussa AG
GB Industrie- und Feinchemikalien
PF 110533, D-6000 Frankfurt 11
T. 069/218-01, Tx. 412220 dg d
Fax. /218-3218, TTX 69917=DGHV
Hexa K, Hexa K oil-treated

Adhesives

Rubber fabric adhesives, chemically activated

Cyanamid B.V.
POB 1523, NL-3000 BM Rotterdam
Tel. 010/4116340
CYREZ®

Rubber-metal adhesives, chemically activated

3M Deutschland GmbH
Abt. Chemische Produkte
PF 100422, D-4040 Neuss 1
Tel. 02101/142511, Tx. 8517511
DYNAMAR™ Metal Bonding Agent

Cyanamid B.V.
POB 1523, NL-3000 BM Rotterdam
Tel. 010/4116340
CYREZ®

Adhesives (cont.)

Henkel KGaA
D-4000 Düsseldorf 1
T. 0211/797-4475, Fax. /797-8596
CHEMOSIL®, SIPIOL®

Monsanto Europe S.A., Brussels
Avenue de Tervuren 270
Tel. 32-2-761.41.11, Tx 62927
Duralink® HTS

Antioxidants

Bisphenoles

CIBA-GEIGY AG (AD 5.4)
4002 Basel/Switzerland
Tel. 061/6971111, Tx. 962355
Fax. 061/6964354
®IRGANOX

Chemische Werke Lowi GmbH & Co
Teplitzer Str., D-8264 Waldkraiburg
Tel. 08638/4011, Tlx. 17863884
LOWINOX®

Copper inhibitors

CIBA-GEIGY AG (AD 5.4)
4002 Basel/Switzerland
Tel. 061/6971111, Tx. 962355
Fax. 061/6964354
®IRGANOX

Dihydrochinolines

Monsanto Europe S.A., Brussels
Avenue de Tervuren 270
Tel. 32-2-761.41.11, Tx. 62927
Santowhite®, Flectol®

Diphenylamines

CIBA-GEIGY AG (AD 5.4)
4002 Basel/Switzerland
Tel. 061/6971111, Tx. 962355
Fax. 061/6964354
®IRGANOX

Monophenols

CIBA-GEIGY AG (AD 5.4)
4002 Basel/Switzerland
Tel. 061/6971111, Tx. 962355
Fax. 061/6964354
®IRGANOX

Chemische Werke Lowi GmbH & Co
Teplitzer Str., D-8264 Waldkraiburg
Tel. 08638/4011, Tlx. 17863884
LOWINOX®

Antioxidants (cont.)

Non-staining antioxidants

CIBA-GEIGY AG (AD 5.4)
4002 Basel/Switzerland
Tel. 061/6971111, Tx. 962355
Fax. 061/6964354
®IRGANOX

Rhein-Chemie Rheingau GmbH
Mülheimer Str. 24–28
D-6800 Mannheim 81
Tel. 0621/89070, Tx. 462282

p-Phenylenediamines

Monsanto Europe S.A., Brussels
Avenue de Tervuren 270
Tel. 32-2-761.41.11, Tx. 62927
Santoflex®

Staining antioxidants

Rhein-Chemie Rheinau GmbH
Mülheimer Str. 24–28
D-6800 Mannheim 81
Tel. 0621/89070, Tx. 462282

Tetrabisphenol derivatives

CIBA-GEIGY AG (AD 5.4)
4002 Basel/Switzerland
Tel. 061/6971111, Tx. 962355
Fax. 061/6964354
®IRGANOX

Trisphenol derivatives

CIBA-GEIGY AG (AD 5.4)
4002 Basel/Switzerland
Tel. 061/6971111, Tx. 962355
Fax. 061/6964354
®IRGANOX

Other antioxidants

CIBA-GEIGY AG (AD 5.4)
4002 Basel/Switzerland
Tel. 061/6971111, Tx. 962355
Fax. 061/6964354
®IRGANOX

Blowing agents

Luperox GmbH
PF 1354, D-8870 Günzburg
Tel. 08221/980, Tx. 531121

Rhein-Chemie Rheinau GmbH
Mülheimer Str. 24–28
D-6800 Mannheim 81
Tel. 0621/89070, Tx. 462282

Blowing agents (cont.)

Schering Ind. Chemicals
Mount Pleasant House
Huntingdon Rd., Cambridge, UK
Tel. 0223/323222, Tx. 81654
GENITRON

Elastomer bound blowing agents

Croxton & Garry Ltd.
Curtis Road
GB Dorking, Surrey RH4 1XA
Tel. 0306/886688, Tx. 859567
Fax. 0306/887780
MULTISPERSE

Cleaning agents, mold-

Julius Hoesch GmbH & Co.KG
PF 100855, D-5160 Düren
Tel. 02421/83081, FS 833896
HOESCHALIN 821

Schill & Seilacher (GmbH & Co.)
Moorfleeter Straße 28
D-2000 Hamburg 74, Tx. 212932
T. 040/733501-0, Fax. /7335094
"Struktol®"

Co-agents and co-reactants

Kettlitz-Chemie GmbH & Co.KG
POB 80, D-8859 Rennertshofen
Tlx. 55237, Fax. 08434/719

Cross-linking agents, organic peroxides

Luperox GmbH
PF 1354, D-8870 Günzburg
Tel. 08221/980, Tx. 531121

Peroxid-Chemie GmbH
D-8023 Höllriegelskreuth
T. 089/7279-0. Tx. 523482 pcm

Cross-linking aids

Degussa AG
GB Industrie- u. Feinchemikalien
PF 110533, D-6000 Frankfurt 11
T. 069/218-01, Tx. 412220 dg d
Fax. /218-3218, TTX 69917=DGHV
Triallylcyanurate, Butandiodimethacrylate (BDMA), Ethyleneglycoldimethacrylate (EDMA), Triethyleneglycoldimethacrylate (TEDMA), Trimethylolpropantrimethacrylate (TRIM)

Cross-linking aids (cont.)

Kettlitz-Chemie GmbH & Co.KG
POB 80, D-8859 Rennertshofen
Tlx. 55237, Fax. 08434/719

Cross-linking resins/tackifying resins

Rhein-Chemie Rheinau GmbH
Mülheimer Str. 24–28
D-6800 Mannheim 81
Tel. 0621/89070, Tx. 462282

Curing additives

amohr – Technische Textilien
PF 200336, D-5600 Wuppertal 2
Tx. 8591839, Tel. 0202/81046

Dessicants

D.O.G. Deutsche Oelfabrik
P.B. 111929, D-2000 Hamburg
Tx. 212164, Fax. 040/3192748

Dusting agents

Scheruhn Talkum-Bergbau
POB 1329, D-8670 Hof/Saale
Tx. 643746 fuh, Fax. 09281/66951

Fillers

Aluminium oxyhydrate

Nordmann, Rassmann GmbH & Co.
Kajen 2, D-2000 Hamburg 11
T. 040/3687-0, Tx. 211122+212087
Fax. /3687-249, T#: nordrascos

Bariumsulfate

Scheruhn Talkum-Bergbau
POB 1329, D-8670 Hof/Saale
Tx. 643746 fuh, Fax. 09281/66951

Calcites

Ulmer Füllstoff Vertrieb GmbH
PF 1126, D-7906 Blaustein
Tel. 07304/8171, Tx. 712464

Fibers

Monsanto Europe S.A., Brussels
Avenue de Tervuren 270
Tel. 32-2-761.41.11, Tx. 62927
Santoweb®

Schwarzwälder Textil-Werke
Heinrich Kautzmann GmbH
D-7623 Schenkenzell
Tel. 07836/2031, Tlx. 7525612

Fillers (cont.)

Silica

Degussa AG
GB Anorganische Chemieprodukte
PF 110533, D-6000 Frankfurt 11
T. 069/218-01, Tx. 412220 dg d
Fax. /218-3218, TTX 69917=DGHV

Silicates

Degussa AG
GB Anorganische Chemieprodukte
PF 11 0533, D-6000 Frankfurt 11
T. 069/218-01, Tx. 412220 dg d
Fax. /218-3218, TTX 69917=DGHV

Hoffmann Mineral
Franz Hoffmann & Söhne KG
Münchener Str. 75
D-8858 Neuburg (Donau)
Tx. 55214-18
T. 08431/53-1, Fax. 08431/53-330

Talcum

Scheruhn Talkum-Bergbau
POB 1329, D-8670 Hof/Saale
Tx. 643746 fuh, Fax. 09281/66951

Other fillers

Wacker-Chemie GmbH
D-8000 München 22
Tel. (089) 2109-0
Tlx. 529121 56
Elastosil®, Wacker HDK®

Finishing agents

Hans W. Barbe
Chemische Erzeugnisse GmbH
PF 130223, D-6200 Wiesbaden 13
T. 06121/22081, Tx. 4186470 hwb d
PROMOL

Flame retardants

Martinswerk GmbH
PF 1209, D-5010 Bergheim
Tel. 02271/9020, Tx. 888712
MARTINAL®

Nordmann, Rassmann GmbH & Co.
Kajen 2, D-2000 Hamburg 11
T. 040/3687-0, Tx. 211122+212087
Fax. /3687-249, T#: nordrascos

Flame retardants (cont.)

NORIMPEX A/S, Hochdahlerstr. 16
D-4000 Düsseldorf
T. 0211/292505, Tx. 8587955
Zincborate

Rhein-Chemie Rheinau GmbH
Mülheimer Str. 24-28
D-6800 Mannheim 81
Tel. 0621/89070, Tx. 462282

VAW Vereinigte
Alumium-Werke AG
Sparte Spezialoxide
P.O.B. 1860, D-8460 Schwandorf
Tel. 09431/53-461, Tx. 65341
"Apyral®"

Homogenizers

D.O.G. Deutsche Oelfabrik
P.B. 111929, D-2000 Hamburg
Tx. 212164, Fax. 040/3192748

Initiators, hydroperoxide

Hüls AG
Referat 1122/RTH
D-4370 Marl

Lead compounds, elastomer bound

Croxton & Garry Ltd.
Curtis Road
GB Dorking, Surrey RH4 1XA
Tel. 0306/886688, Tx. 859567
Fax. 0306/887780
MULTISPERSE

Lubricating agents

Hans W. Barbe
Chemische Erzeugnisse GmbH
PF 130223, D-6200 Wiesbaden 13
T. 06121/22081, Tx. 4186470 hwb d
PROMOL

Mastication and regeneration aids

D.O.G. Deutsche Oelfabrik
P.B. 111929, D-2000 Hamburg
Tx. 212164, Fax. 040/3192748

Rhein-Chemie Rheinau GmbH
Mülheimer Str. 24-28
D-6800 Mannheim 81
Tel. 0621/89070, Tx. 462282

Mastication and regeneration aids (cont.)

Schill & Seilacher (GmbH & Co.)
Moorfleeter Straße 28
D-2000 Hamburg 74, Tx. 212932
T. 040/735501-0, Fax. /7335094
"Struktol®"

Mold lubricants

Hans W. Barbe
Chemische Erzeugnisse GmbH
PF 130223, D-6200 Wiesbaden 13
T. 06121/22081, Tx. 4186470 hwb d
PROMOL

Phenolic resins/compounds

N.V. Occidental Chemical SA
Av. Louise 222, B-1050 Brussels
Tel. (2)6485635, Tx. 23046
DUREZ® Tackifier and reinforcing resins

Pigments

colored

Julius Hoesch GmbH & Co. KG
PF 100855, D-5160 Düren
Tel. 02421/83081, FS 833896

white

Julius Hoesch GmbH & Co. KG
PF 100855, D-5160 Düren
Tel. 02421/83081, FS 833896

Plasticizers

Adipates

Kettlitz-Chemie GmbH & Co.KG
POB 80, D-8859 Rennertshofen
Tlx. 55237, Fax. 08434/719

Unichema Chemie GmbH
P.O. Box 1280, D-4240 Emmerich
Tel. 02822/720, Tlx. 8125113
Fax. 02822/72276
PRIPLAST

Ester plasticizers

Schill & Seilacher (GmbH & Co.)
Moorfleeter Straße 28
D-2000 Hamburg 74, Tx. 212932
T. 040/733501-0, Fax. /7335094
"Struktol®"

Unichema Chemie GmbH
P.O. Box 1280, D-4240 Emmerich
Tel. 02822/720, Tlx. 8125113
Fax. 02822/72276
PRIPLAST

Plasticizers (cont.)

Naphthenic mineral oil plasticizer

Nynäs GmbH
Immermannstraße 65a
D-4000 Düsseldorf 1, BRD
Tel. 0211/352046, Tx. 8586352
Nynäs Naphthenics AB
P.O. Box 7856
S-10399 Stockholm, Schweden
Tel. 08/7885030, Tx. 19042
NYFLEX 800, 810, 820, 220
NYTEX 800, 810, 820, 830
NYTENE 800, 810, 830, 230

Phosphates

Monsanto Europe S.A., Brussels
Avenue de Tervuren 270
Tel. 32-2-761.41.11, Tx. 62927
Santicizer®

Phthalates

Julius Hoesch GmbH & Co. KG
PF 100855, D-5160 Düren
Tel. 02421/83081, FS 833896

Monsanto Europe S.A., Brussels
Avenue de Tervuren 270
Tel. 32-2-761.41.11, Tx. 62927
Santicizer®

Other plasticizers

Kettlitz-Chemie GmbH & Co. KG
POB 80, D-8859 Rennertshofen
Tlx. 55237, Fax. 08434/719

Monsanto Europe S.A., Brussels
Avenue de Tervuren 270
Tel. 32-2-761.41.11, Tx. 62927
Santicizer®

Unichema Chemie GmbH
P.O. Box 1280, D-4240 Emmerich
Tel. 02822/720, Tlx. 8125113
Fax. 02822/72276
PRIPLAST

Processing aids

Fatty acids

Rhein-Chemie Rheinau GmbH
Mülheimer Str. 24–28
D-6800 Mannheim 81
Tel. 0621/89070, Tx. 462282

Processing aids (cont.)

Unichema Chemie GmbH
P.O. Box 1280, D-4240 Emmerich
Tel. 02822/720, Tlx. 8125113
Fax. 02822/72276
UNIWAX/PRISORINE

Factice

D.O.G. Deutsche Oelfabrik
P.B. 111929, D-2000 Hamburg
Tx. 212164, Fax. 040/3192748

Rhein-Chemie Rheinau GmbH
Mülheimer Str. 24–28
D-6800 Mannheim 81
Tel. 0621/89070, Tx. 462282

Fat and oil derivatives

Rhein-Chemie Rheinau GmbH
Mülheimer Str. 24–28
D-6800 Mannheim 81
Tel. 0621/89070, Tx. 462282

Fatty acid and fatty alcohol derivatives

D.O.G. Deutsche Oelfabrik
P.B. 111929, D-2000 Hamburg
Tx. 212164, Fax. 040/3192748

Kettlitz-Chemie GmbH & Co. KG
POB 80, D-8859 Rennertshofen
Tlx. 55237, Fax. 08434/719

Rhein-Chemie Rheinau GmbH
Mülheimer Str. 24–28
D-6800 Mannheim 81
Tel. 0621/89070, Tx. 462282

Schill & Seilacher (GmbH & Co.)
Moorfleeter Straße 28
D-2000 Hamburg 74, Tx. 212932
T. 040/733501-0, Fax. /7335094
"Struktol®"

Fatty acid derivatives

Rhein-Chemie Rheinau GmbH
Mülheimer Str. 24–28
D-6800 Mannheim 81
Tel. 0621/89070, Tx. 462282

Unichema Chemie GmbH
P.O. Box 1280, D-4240 Emmerich
Tel. 02822/720, Tlx. 8125113
Fax. 02822/72276
UNIWAX/PRISORINE

Processing aids (cont.)

Processing aids for special elastomers

D.O.G. Deutsche Oelfabrik
P.B. 111929, D-2000 Hamburg
Tx. 212164, Fax. 040/3192748

3M Deutschland GmbH
Abt. Chemische Produkte
PF 100422, D-4040 Neuss 1
Tel. 02101/142511, Tx. 8517511
DYNAMAR™ Polymer Processing Aids

Kettlitz-Chemie GmbH & Co. KG
POB 80, D-8859 Rennertshofen
Tlx. 55237, Fax. 08434/719

Rhein-Chemie Rheinau GmbH
Mülheimer Str. 24–28
D-6800 Mannheim 81
Tel. 0621/89070, Tx. 462282

Schill & Seilacher (GmbH & Co.)
Moorfleeter Straße 28
D-2000 Hamburg 74, Tx. 212932
T. 040/733501-0, Fax. /7335094
"Struktol®"

Protectants

Hindered amine light stabilizers

CIBA-GEIGY AG (AD 5.4)
4002 Basel/Switzerland
Tel. 061/6971111, Tx. 962355
Fax. 061/6964354
®TINUVIN, ®CHIMASSORB

Physically active ozone protectants

D.O.G. Deutsche Oelfabrik
P.B. 111929, D-2000 Hamburg
Tx. 212164, Fax. 040/3192748

ELASTOCHEM
W. Koblizek & Co.
Schäferweg 5
D-3000 Hannover 51
Tel. 0511/5499410, Tx. 9230559
OSM ozone protecting waxes

Rhein-Chemie Rheinau GmbH
Mülheimer Str. 24–28
D-6800 Mannheim 81
Tel. 0621/89070, Tx. 462282

Protectants (cont.)

UV-absorbers

CIBA-GEIGY AG (AD 5.4)
4002 Basel/Switzerland
Tel. 061/6971111, Tx. 962355
Fax. 061/6964354
®TINUVIN, ®CHIMASSORB

Reinforcements

amohr - Technische Textilien
PF 200336, D-5600 Wuppertal 2
Tx. 8591839, Tel. 0202/81046
Ribbons and fabrics in cotton and synthetics

Rubber blacks

Degussa AG
GB Anorganische Chemieprodukte
PF 110533, D-6000 Frankfurt 11
T. 069/218-01, Tx. 412220 dg d
Fax. /218-3218, TTX 69917 = DGHV

Rubber curing salt

Goerig GmbH
POB 101362, D-6800 Mannheim 1
T. 0621/313051, Tx. 176211811
Fax. 0621/311320
"Effge®"

Separation agents

Hans W. Barbe
Chemische Erzeugnisse GmbH
PF 130223, D-6200 Wiesbaden 13
T. 06121/22081, Tx. 4186470 hwb d
PROMOL

Henkel KGaA
D-4000 Düsseldorf 1
T. 0211/797-4475, Fax. /797-8596
CHEMOSIL®, SIPIOL®

Kettlitz-Chemie GmbH & Co. KG
POB 80, D-8859 Rennertshofen
Tlx. 55237, Fax. 08434/719

Rhein-Chemie Rheinau GmbH
Mülheimer Str. 24-28
D-6800 Mannheim 81
Tel. 0621/89070, Tx. 462282

Scheruhn Talkum-Bergbau
POB 1329, D-8670 Hof/Saale
Tx. 643746 fuh, Fax. 09281/66951

Separation agents (cont.)

Schill & Seilacher (GmbH & Co.)
Moorfleeter Straße 28
D-2000 Hamburg 74, Tx. 212932
T. 040/733501-0, Fax. /7335094
"Struktol®"

Silanes

Degussa AG
GB Anorganische Chemieprodukte
PF 110533, D-6000 Frankfurt 11
T. 069/218-01, Tx. 412220 dg d
Fax. /218-3218, TTX 69917 = DGHV

Stabilisators

Monsanto Europe S. A., Brussels
Avenue de Tervuren 270
Tel. 32-2-761.41.11, Tx. 62927
Duralink® HTS as post-vulcanization stabilizer

Sulfur

Elastomer bound sulfur

Croxton & Garry Ltd.
Curtis Road
GB Dorking, Surrey RH4 1XA
Tel. 0306/886688, Tx. 859567
Fax. 0306/887780
MULTISPERSE

Insoluble sulfur

Kali-Chemie Stauffer GmbH
PF 220, D-3000 Hannover 1
Tel. 0511/857-1, Tx. 922755
CRYSTEX (R)

Rhein-Chemie Rheinau GmbH
Mülheimer Str. 24-28
D-6800 Mannheim 81
Tel. 0621/89070, Tx. 462282

Rubber maker's sulfur

Kali-Chemie AG
PF 220, D-3000 Hannover 1
Tel. 0511/857-1, Tx. 922755

Soluble sulfur

Koch & Reis N. V.
Ijzerlaan 3, B-2008 Antwerpen
Tel. 3/233.12.05, Tx. 32876
K R Ventilated sulfur
K R Triturated sulfur
Oil treated sulfur
Free flowing sulfur

Sulfur (cont.)

Rhein-Chemie Rheinau GmbH
Mülheimer Str. 24–28
D-6800 Mannheim 81
Tel. 0621/89070, Tx. 462282

Sulfur donors

Caprolactam disulfide

Rhein-Chemie Rheinau GmbH
Mülheimer Str. 24–28
D-6800 Mannheim 81
Tel. 0621/89070, Tx. 462282

Dithiomorpholine

D.O.G. Deutsche Oelfabrik
P.B. 111929, D-2000 Hamburg
Tx. 212164, Fax. 040/3192748

Monsanto Europe S.A., Brussels
Avenue de Tervuren 270
Tel. 32-2-761.41.11, Tx. 62927
Sulfasan®

Rhein-Chemie Rheinau GmbH
Mülheimer Str. 24–28
D-6800 Mannheim 81
Tel. 0621/89070, Tx. 462282

Tackifiers

D.O.G. Deutsche Oelfabrik
P.B. 111929, D-2000 Hamburg
Tx. 212164, Fax. 040/3192748

Titanium dioxide extenders, silicates

Degussa AG
GB Anorganische Chemieprodukte
PF 110533, D-6000 Frankfurt 11
T. 069/218-01, Tx. 412220 dg d
Fax. /218-3218, TTX 69917=DGHV

Vulcanization agents

Rhein-Chemie Rheinau GmbH
Mülheimer Str. 24–28
D-6800 Mannheim 81
Tel. 0621/89070, Tx. 462282

Vulcanization retarders

Monsanto Europe S.A., Brussels
Avenue de Tervuren 270
Tel. 32-2-761.41.11, Tx. 62927
Santogard® PVI, Retarder TSA

Rhein-Chemie Rheinau GmbH
Mülheimer Str. 24–28
D-6800 Mannheim 81
Tel. 06 21/89070, Tx. 462282

Water absorbers

Kettlitz-Chemie GmbH & Co. KG
POB 80, D-8859 Rennertshofen
Tlx. 55237, Fax. 08434/719

Waxes

H.B. Fuller GmbH
An der Roten Bleiche 2/3
D-2120 Lüneburg, Tx. 2182102
T. 04131/705-107, Fax. /705-227
PROTEKTOR® G

Rhein-Chemie Rheinau GmbH
Mülheimer Str. 24–28
D-6800 Mannheim 81
Tel. 0621/89070, Tx. 462282

Zinc oxides, elastomer bound

Croxton & Garry Ltd.
Curtis Road
GB Dorking, Surrey RH4 1XA
Tel. 0306/886688, Tx. 859567
Fax. 0306/887780
MULTISPERSE

Group 4: Machines for Elastomer Processing

CAD-standard elements software

HASCO, D-5880 Lüdenscheid
Tel. 02351/4320, Tx. 826842

Calender, rubber-

Berstorff Maschinenbau GmbH
D-3000 Hannover 1, Box 629
Tel. 0511/5702-0, Tx. 921348

Extruders

Lead extruders

H. Folke Sandelin AB
Box 6064, S-12606 Hägersten
Tel. 468880240, Tx. 17461
Fax. 468883230
Hanson-Robertson continous
lead extruders lead-melting
systems, -stripping machines

Extruders (cont.)

Rubber extruders

Berstorff Maschinenbau GmbH
D-3000 Hannover 1, Box 629
Tel. 0511/5702-0, Tx. 921348

C.A. Picard GmbH & Co.KG
Postfach 140440
D-5630 Remscheid 1
Tel. 02191/893-0, Tx. 8513906
Fax. 02191/893-111
Screws, Liners, Cage equipment for mechanical rubber dewatering presses

Extrusion plants

Berstorff Maschinenbau GmbH
D-3000 Hannover 1, Box 629
Tel. 0511/5702-0, Tx. 921348

Granulating equipment for rubber

Condux
Maschinenbau GmbH & Co. KG
Rodenbacher Chaussee 2
D-6450 Hanau 11
Tel.: 06181/506-0
Tx.: 4184158 cdx
Fax.: 06181/571270

Herbold GmbH, D-6922 Meckesheim
Tel. 06226/53-0, Tlx. 466524

Granulators

Herbold GmbH, D-6922 Meckesheim
Tel. 06226/53-0, Tlx. 466524

Injection molding equipment

Dieffenbacher
P.O. Box 1120, D-7519 Eppingen
T. 07262/650, Tx. 782317 jds d
Fax. 07262/65297
Hydr. Presses & Press Plants

Klöckner Ferromatik Desma
Desma, Werk Achim
Desmastr. 3-5, 2807 Achim, FRG
Tel. 04202/50-0, Fax. /50-210
Telex 249431 desm d

Krauss-Maffei
Kunststofftechnik GmbH
Krauss-Maffei-Str. 2
D-8000 München 50
Tel. 089/88990, Tx. 523163-41
Fax. 089/8899-3219

Injection molding equipment (cont.)

J. Wickert & Söhne KG
Injection Molding, Machines
PF 2267, D-6740 Landau/Pf.-West-Germany
T. 06341/30021-23, Fax. /32644

Injection molding equipment for rubber

REP-Deutschland GmbH
Poststr. 10, D-6948 Wald-Michelbach
Tel. 06207/5084
Ttx. 620794, Fax. 06207/6632

Injection molding machines

J. Wickert & Söhne KG
Injection Molding, Machines
PF 2267, D-6740 Landau/Pf.-West-Germany
T. 06341/30021-23, Fax. /32644

Manufacture of V-belts, machines

Berstorff Maschinenbau GmbH
D-3000 Hannover 1, Box 629
Tel. 0511/5702-0, Tx. 921348

Mechanical handling equipment

Gebrüder Bühler AG
Förder- u. Verfahrenstechnik
CH-9240 Uzwil/Schweiz
Tel. 073/501111, Tx. 883131

Mixer, solvation-

Hermann Linden Masch.-Fabr.
GmbH + Co. KG, Hauptstr. 123
D-5277 Marienheide
Tel. 02264/7002, Tx. 884112

Mixing equipment

Gebrüder Bühler AG
Förder- u. Verfahrenstechnik
CH-9240 Uzwil/Schweiz
Tel. 073/501111, Tx. 883131

Mold cleaning, machines

Graf - Reinigungsanlagen
Max-Eyth-Str. 23
D-7012 Fellbach-Oeffingen BRD
T. 0711/515085, Tlx. 7254728
Fax. /515087 - mold cleaning and polishing machines

Mold manufacture

HASCO, D-5880 Lüdenscheid
Tel. 02351/4320, Tx. 826842

Klöckner Ferromatik Desma
Desma, Werk Achim
Desmastr. 3-5, 2807 Achim, FRG
Tel. 04202/50-0, Fax. /50-210
Telex 249431 desm d

Molded parts production, machines

Dieffenbacher
P.O. Box 1120, D-7519 Eppingen
T. 07262/650, Tx. 782317 jds d
Fax. 07262/65297
Hydr. Presses & Press Plants

Klöckner Ferromatik Desma
Desma, Werk Achim
Desmastr. 3-5, 2807 Achim, FRG
Tel. 04202/50-0, Fax. /50-210
Telex 249431 desm d

Krauss-Maffei
Kunststofftechnik GmbH
Krauss-Maffei-Str. 2
D-8000 München 50
Tel. 089/88990, Tx. 523163-41
Fax. 089/8899-3219

J. Wickert & Söhne KG
Injection Molding, Machines
PF 2267, D-6740 Landau/Pf.-
West-Germany
T. 06341/30021-23, Fax /32644

Presses

Injection presses

Dieffenbacher
P.O. Box 1120, D-7519 Eppingen
T. 07262/650, Tx. 782317 jds d
Fax. 07262/65297
Hydr. Presses & Press Plants

Mold presses

Dieffenbacher
P.O. Box 1120, D-7519 Eppingen
T. 07262/650, Tx. 782317 jds d
Fax. 07262/65297
Hydr. Presses & Press Plants

Krauss-Maffei
Kunststofftechnik GmbH
Krauss-Maffei-Str. 2
D-8000 München 50
Tel. 089/88990, Tx. 523163-41
Fax. 089/8899-3219

Presses (cont.)

J. Wickert & Söhne KG
Injection Molding, Machines
PF 2267, D-6740 Landau/Pf.-
West-Germany
T. 06341/30021-23, Fax /32644

Transfer presses

Dieffenbacher
P.O. Box 1120, D-7519 Eppingen
T. 07262/650, Tx. 782317 jds d
Fax. 07262/65297
Hydr. Presses & Press Plants

J. Wickert & Söhne KG
Injection Molding, Machines
PF 2267, D-6740 Landau/Pf.-
West-Germany
T. 06341/30021-23, Fax /32644

Rolls

Rolls for calandering lines

Leonhard Breitenbach GmbH
Trupbach, Walzenweg 60,
D-5900 Siegen, T. 0271/37337
Ttx. 17-271308, Fax. /371906

Rolls for heating, cooling and other treatment

Drink & Schlössers - Rolls
D-4150 Krefeld - 29 Hüls
T. 02151/73876, Tx. 853435

Rolls for rubber roll mills

Leonhard Breitenbach GmbH
Trupbach, Walzenweg 60,
D-5900 Siegen, T. 0271/37337
Ttx. 17-271308, Fax. /371906

Standard elements for molds and tools

HASCO, D-5880 Lüdenscheid
Tel. 02351/4320, Tx. 826842

Static eliminators

Haug GmbH & Co. KG
PF 200333, D-7022 L.-Echterdingen
Tel. 0711/7979021-23
Telex 7-255242

Strainers

Berstorff Maschinenbau GmbH
D-3000 Hannover 1, Box 629
Tel. 0511/5702-0, Tx. 921348

UHF equipment

Berstorff Maschinenbau GmbH
D-3000 Hannover 1, Box 629
Tel. 0511/5702-0, Tx. 921348

Vulcanization equipment

Berstorff Maschinenbau GmbH
D-3000 Hannover 1, Box 629
Tel. 0511/5702-0, Tx. 921348

Vulcanization equipment (cont.)

Hoesch Maschinenfabrik
Deutschland AG
PF 101662, D-4600 Dortmund 1
T. 0231/8491-0, Fax. /8491-333
Tx. 822269 mfd d

Group 6: Measuring Instruments

Extrusiometers

Brabender oHG®
Kulturstr. 51–55, D-4100 Duisburg 1
Tel. 0203/738010
Fax. 0203/7380149, Tx. 855603

Flexometers

DOLI Elektronik GmbH
Herzogspitalstr. 10
D-8000 München 2
Tel. 089/2604013, Tx. 5212278
Fax. 089/2608968
Fullautomatic Flexometer,
Automation with PC's, CAQ

Measuring instruments

Density measuring instruments

Brabender oHG®
Kulturstr. 51–55, D-4100 Duisburg 1
Tel. 0203/738010
Fax. 0203/7380149, Tx. 855603

Monsanto Europe S.A., Brussels
Avenue de Tervuren 270
Tel. 32-2-761.41.11, Tx. 62927
Densitron

Hardness measuring instruments

Monsanto Europe S.A., Brussels
Avenue de Tervuren 270
Tel. 32-2-761.41.11, Tx. 62927
Duratron®

Mooney measuring instruments

Monsanto Europe S.A., Brussels
Avenue de Tervuren 270
Tel. 32-2-761.41.11, Tx. 62927
MV 2000/-E

Measuring instruments (cont.)

Viscosity measuring instruments

Brabender oHG®
Kulturstr. 51–55, D-4100 Duisburg 1
Tel. 0203/738010
Fax. 0203/7380149, Tx. 855603

Monsanto Europe S.A., Brussels
Avenue de Tervuren 270
Tel. 32-2-761.41.11, Tx. 62927
Mooney MV 2000/-E

Vulcanization measuring instruments

Brabender oHG®
Kulturstr. 51–55, D-4100 Duisburg 1
Tel. 0203/738010
Fax. 0203/7380149, Tx. 855603

Monsanto Europe S.A., Brussels
Avenue de Tervuren 270
Tel. 32-2-761.41.11, Tx. 62927
Monsanto® Rheometer

Tensile testers

DOLI Elektronik GmbH
Herzogspitalstr. 10
D-8000 München 2
Tel. 089/2604013, Tx. 5212278
Fax. 089/2608968
Fullautomatic Flexometer,
Automation with PC's, CAQ

Monsanto Europe S.A., Brussels
Avenue de Tervuren 270
Tel. 32-2-761.41.11, Tx. 62927
Tensometer 10

Testing equipment

Testing equipment for chemical properties

Rudolf Wechsler
CH-4127 Birsfelden-Basel

Testing equipment for corrosion/ erosion properties

Rudolf Wechsler
CH-4127 Birsfelden-Basel

Testing equipment for mechanical properties

Monsanto Europe S. A., Brussels
Avenue de Tervuren 270
Tel. 32-2-761.41.11, Tx. 62927
Tensometer 10

Testing equipment (cont.)

Carl Schenck AG
Landwehrstr. 55, D-6100 Darmstadt
Tx. 4196940 cs d
Tel. 06151/32-0, Fax. /893686
Hydropuls® High-Frequency
Elastomer Testing Machine

Torsion pendulum

Brabender oHG®
Kulturstr. 51-55, D-4100 Duisburg 1
Tel. 0203/738010
Fax. 0203/7380149, Tx. 855603

Vulcameter, rheo-

Monsanto Europe S. A., Brussels
Avenue de Tervuren 270
Tel. 32-2-761.41.11, Tx. 62927
Monsanto® Rheometer

Group 7: Laboratory Equipment

Laboratory mixer

Brabender oHG®
Kulturstr. 51-55, D-4100 Duisburg 1
Tel. 0203/738010
Fax. 0203/7380149, Tx. 855603

Herman Linden Masch.-Fabr.
GmbH + Co. KG, Hauptstr. 123
D-5277 Marienheide
Tel. 02264/7002, Tx. 884112

Laboratory presses

J. Wickert & Söhne KG
Injection Molding, Machines
PF 2267, D-6740 Landau/Pf.-
West-Germany
T. 06341/30021-23, Fax /32644

Planetary dissolver

Hermann Linden Masch.-Fabr.
GmbH + Co. KG, Hauptstr. 123
D-5277 Marienheide
Tel. 02264/7002, Tx. 884112

Planetary mixer

Hermann Linden Masch.-Fabr.
GmbH + Co. KG, Hautpstr. 123
D-5277 Marienheide
Tel. 02264/7002, Tx. 884112

Product information of the rubber industry

A 2–A 18

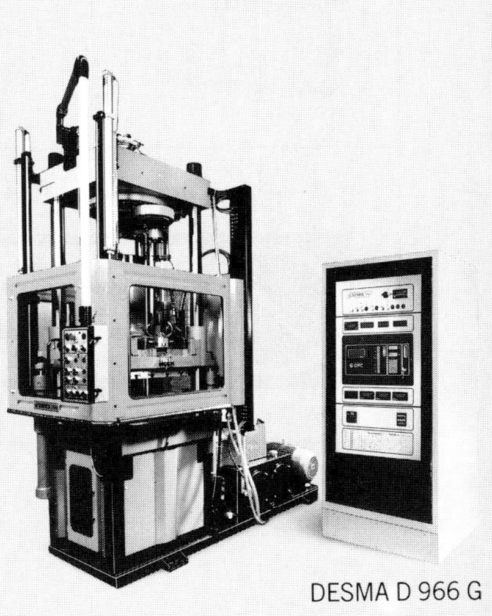

Effective.

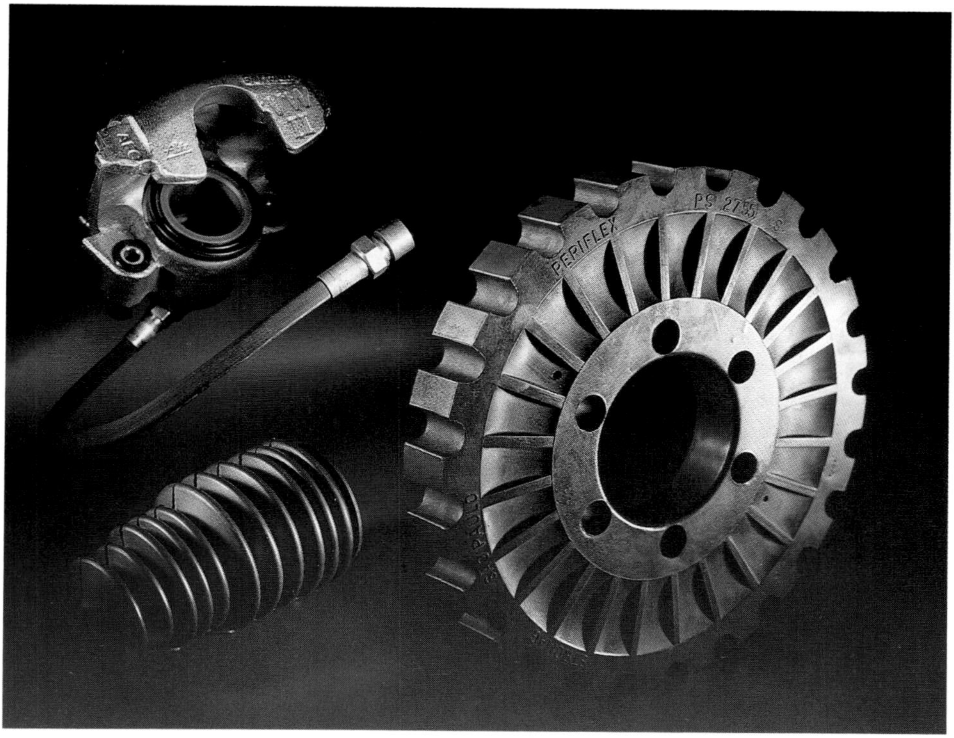

Rubber blend component VESTENAMER®
VESTENAMER, trans-polyoctenamer, is a rubber blend component with an impressive range of action: Improves the processability of rubber blends and optimises the physical properties of vulcanisates. **Extrusion:** VESTENAMER improves the flow properties of the blend and the dimensional stability of the product. **Rolls manufacture:** VESTENAMER increases vulcanisate adhesion and improves inter-layer bonding.

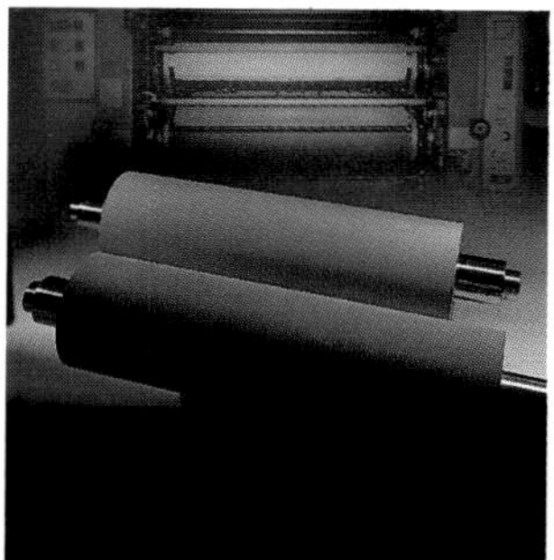

Injection moulding: VESTENAMER ensures highest precision and good flow conditions even with large-volume components. **Hose manufacture:** VESTENAMER effects increased collapse resistance of the uncured hose. VESTENAMER – the high-value material for rubber processing. We shall be pleased to send you our collection of formulations classified according to processing method.

Hüls AG, Referat 1122,
D-4370 Marl

4/7648 E

hüls

80 million years are behind it.
Sillitin and Aktisil
We do not want to claim such a long development time for rain apparel. This past is concealed in the raw material that is just as important for rainwear as it is for 589 other products – now and in the future: Neuburg Silica, which received its inimitable structure as a result of having been deposited on limestone beds dating from the late Jurassic period. A natural combination of corpuscular quartz and laminar kaolinite makes this filling material unique, because it can be mixed quickly and forms no filler agglomerates. It assures good extrusion and calandering behaviour as well as low permanent deformation of rubber articles. Neuburg silica in the form of Sillitin and improved Aktisil provides the appropriate solution for every filler problem. And if you should encounter new problems, we will simply find new formulas.
Ask for our brochure. Postcard suffices.
Neuburg Silica.
A raw material shows its character.
HOFFMANN
MINERAL
D-8858 Neuburg/Donau, Postfach 14 69
Telefon: + 84 31-5 30
Telex: 5 5 214-18
Telefax: + 84 31-5 33 30

HANSER
PUBLISHERS

WP programme for the rubber industry

Rubber compounding technology

Plants, machines and devices for continuous and discontinuous mixing and kneading processes for compounding raw materials.
Pelletizers,
Roll mills,
Rubber bale cutters.

Rubber processing technology

Precision blank extruders,
Strip cutters,
Rubber injection presses with plasticizing screw,
Special oil- and water-hydraulic presses.

Rubber compounding

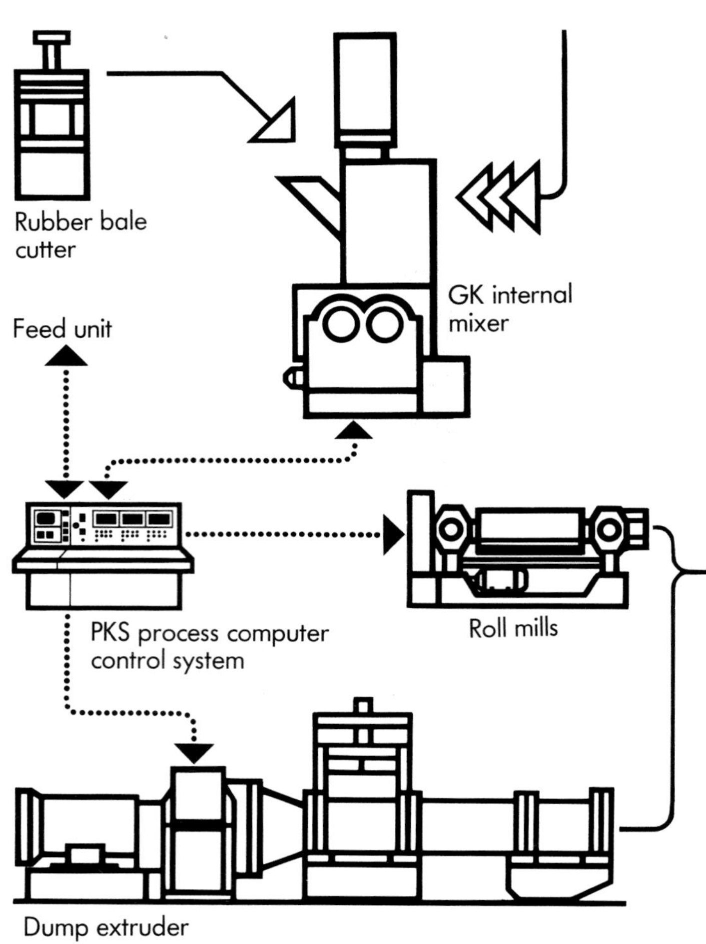

04 - 1705/8 SCH